Birkhäuser

Trends in Mathematics

Research Perspectives

Research Perspectives collects core ideas and developments discussed at conferences and workshops in mathematics, as well as their increasingly important applications to other fields. This subseries' rapid publication of extended abstracts, open problems and results of discussions ensures that readers are at the forefront of current research developments.

Paula Cerejeiras • Michael Reissig •
Irene Sabadini • Joachim Toft
Editors

Current Trends in Analysis, its Applications and Computation

Proceedings of the 12th ISAAC Congress, Aveiro, Portugal, 2019

Birkhäuser

Editors
Paula Cerejeiras
Department of Mathematics
University of Aveiro
Aveiro, Portugal

Michael Reissig
Institut of Applied Analysis
TU Bergakademie Freiberg
Freiberg, Germany

Irene Sabadini
Dipartimento di Matematica
Politecnico di Milano
Milano, Italy

Joachim Toft
Department of Mathematics
Linnaeus University
Växjö, Sweden

ISSN 2297-0215 ISSN 2297-024X (electronic)
Trends in Mathematics
ISSN 2509-7407 ISSN 2509-7415 (electronic)
Research Perspectives
ISBN 978-3-030-87501-5 ISBN 978-3-030-87502-2 (eBook)
https://doi.org/10.1007/978-3-030-87502-2

This book is published under the imprint Birkhäuser, www.birkhauser-science.com, by the registered company Springer Nature Switzerland AG
The registered company address is: Gewerbestrasse 11, 6330 Cham, Switzerland

Preface

This volume contains the contributions of the participants of the 12th ISAAC Congress which was held at the University of Aveiro, Portugal, from 29 July to 3August 2019. Although the International Society for Analysis, its Applications and Computation (ISAAC) supports several conferences, workshops, and other events of scientific nature, the ISAAC Congress is the principal bi-annual event of the society. The 12th ISAAC Congress continues this successful series of meetings initiated in Delaware, USA (1997), which have been held regularly every 2 years since, while rotating among different countries in Europe, America, and Asia.

With a long tradition in joining researchers form different areas of analysis and its applications, the 12th edition had 543 registered participants from 64 different countries who attended 433 talks spanned by 23 sessions, including sessions of the special interest groups of the society. As a novelty and to encourage collaboration between different special interest groups, additional joint sessions were organised around topics of common interest. Last but not least, there were also nine plenary talks given by

- Afonso Bandeira—Courant Institute, NYU, USA;
- Alexander Grigoryan—Universität Bielefeld, Bielefeld, Germany;
- Karlheinz Groechenig—Universität Wien, Wien, Austria;
- Håkan Hedenmalm—The Royal Institute of Technology, Sweden;
- André Neves—University of Chicago, USA;
- Roman Novikov—Centre de Mathématiques Appliquées, École Polytechnique, France;
- Tohru Ozawa—Waseda University, Tokyo, Japan;
- Samuli Siltanen—University of Helsinki, Finland;
- Durvudkhan Suragan—Nazarbayev University, Kazakhstan;

which gave a broad overview on recent developments in their respective areas. This greatly contributed to the high scientific level of the congress, combined with the warm and relaxed atmosphere which prevailed during those 5 days. It is noteworthy to notice that many young mathematicians used the opportunity to join ISAAC during this time.

As per tradition, an award was given to young scientists for their particular achievements in analysis, its applications and computation. For this edition, there was a particularly strong group of outstanding candidates. In consequence, two ISAAC Awards for young scientists were given to:

Afonso Bandeira (Courant Institute, NYU, USA). Although young, his contributions range from classic analysis to applications in inverse problems, random matrices, optimisation, and statistical physics. Among his major contributions are applications of methods from analysis to the theory of data science. He has publications in top journals in both applied and pure mathematics, statistics, information theory, and data science (e.g. Communications on Pure and Applied Mathematics, Applied and Computational Harmonic Analysis, and Annals of Probability). He was awarded a Sloan Fellowship in 2018 and is currently a full professor in the Department of Mathematics at ETH Zurich.

Fábio Pusateri (University of Toronto, Ontario, Canada), for his contributions to the study of dispersive and wave equations, fluid dynamics, and harmonic analysis. He made major contributions to the study of the water wave problem. In particular, he proved the existence of global smooth, nontrivial solutions of gravity water waves systems in 2D, and the existence of global solutions to initial value problems for water waves in 3D and the long-time regularity for the gravity-capillary water-wave model in 3D. He has published in first-class journals such as Inventiones Mathematicae, Advances in Mathematics, Memoirs of American Mathematical Society, Acta Mathematica, and Annales de l' Institute Henri Poincaré.

While the plenary lectures given at the Congress will appear in the independent volume: P. Cerejeiras, M. Reissig (Eds.) *Mathematical Analysis and Applications - Plenary Lectures, ISAAC 2019, Aveiro, Portugal*, Springer Proceedings in Mathematics and Statistics, Springer, the present volume contains selected contributions of talks delivered at the Congress. As in previous editions, some of the sessions or interest groups decided to publish their own volumes of proceedings, independently, and, therefore, these contributions do not appear in the present collection.

The following sessions contributed to the present volume:

- S1. *Applications of Dynamical Systems Theory in Biology*, organized by Torsten Lindström
- S2. *Complex Analysis and Partial Differential Equations*, organised by Okay Celebi, Sergei Rogosin
- S3. *Complex Geometry*, organised by Alexander Schmitt
- S4. *Complex Variables and Potential Theory*, organised by Tahir Aliyev Azeroglu, Massimo Lanza de Cristoforis, Anatoly Golberg, Sergiy Plaksa
- S5. *Constructive Methods in the Theory of Composite and Porous Media*, organised by Vladimir Mityushev
- S7. *Function Spaces and Applications*, organised by Alexandre Almeida, António Caetano, Stefan Samko
- S9. *Generalized Functions and Applications*, organised by Michael Kunzinger, Michael Oberguggenberger, Stevan Pilipović

- S10. *Geometric & Regularity Properties of Solutions to Elliptic and Parabolic PDEs*, organised by Pierre Bousquet, Lorenzo Brasco, Rolando Magnanini
- S11. *Geometries Defined by Differential Forms*, organised by Mahir Bilen Can, Sergey Grigorian, Sema Salur
- S14. *Partial Differential Equations on Curved Spacetimes*, organised by Anahit Galstyan, Makoto Nakamura, Karen Yagdjian
- S15. *Partial Differential Equations with Nonstandard Growth*, organised by Hermenegildo Borges de Oliveira, Sergey Shmarev
- S17. *Quaternionic and Clifford Analysis*, organised by Swanhild Bernstein, Uwe Kähler, Irene Sabadini, Franciscus Sommen
- S18. *Recent Progress in Evolution Equations*, organised by Marcello D'Abbicco, Marcelo Rempel Ebert
- S22. *Wavelet theory and its Related Topics*, organised by Keiko Fujita, Akira Morimoto

Finally, we wish to thank all organisers of all sessions of the Congress for their work. Their efforts—such as inviting and selecting speakers, arranging their sessions, and providing chairpersons—greatly contributed to the relaxed, friendly, and scientific atmosphere which prevailed during the Congress. The session organisers were also responsible for collecting and organising the refereeing process of the contributions of their session to this volume.

University of Aveiro Paula Cerejeiras
Aveiro, Portugal
TU Bergakademie Freiberg Michael Reissig
Freiberg, Germany
Politecnico di Milano Irene Sabadini
Milano, Italy
Linnaeus University Joachim Toft
Växjö, Sweden
July 2021

Contents

Part I
Applications of Dynamical Systems Theory in Biology

Bifurcation Analysis of the Topp Model

Valery A. Gaiko, Alef E. Sterk, and Henk W. Broer

Abstract In this paper, we study the 3-dimensional Topp model for the dynamics of diabetes. We show that for suitable parameter values an equilibrium of this model bifurcates through a Hopf-saddle-node bifurcation. Numerical analysis suggests that near this point Shilnikov homoclinic orbits exist. In addition, chaotic attractors arise through period doubling cascades of limit cycles.

Keywords Dynamics of diabetes · Topp model · Reduced planar quartic Topp system · Singular point · Limit cycle · Hopf-saddle-node bifurcation · Period doubling bifurcation · Shilnikov homoclinic orbit · Chaos

Mathematics Subject Classification (2010) Primary 34C23, 37G15, 37G35; Secondary 34C05, 34C07

1 Introduction

In this paper, we study a Topp model for the dynamics of diabetes:

$$\begin{aligned} \dot{G} &= a - (b + cI)G, \\ \dot{I} &= \frac{\beta G^2}{1 + G^2} - \alpha I, \\ \dot{\beta} &= (-l + mG - nG^2)\beta, \end{aligned} \tag{1}$$

V. A. Gaiko (✉)
National Academy of Sciences of Belarus, United Institute of Informatics Problems, Minsk, Belarus

A. E. Sterk · H. W. Broer
University of Groningen, Bernoulli Institute for Mathematics, Computer Science and Artificial Intelligence, Groningen, The Netherlands
e-mail: a.e.sterk@rug.nl; h.w.broer@rug.nl

P. Cerejeiras et al. (eds.), *Current Trends in Analysis, its Applications and Computation*, Research Perspectives,
https://doi.org/10.1007/978-3-030-87502-2_1

where G, I and β are glucose, insulin and β-cells variables; a, b, c, l, m, n and α are parameters [25].

On the short timescale, β is approximately constant and, relabelling the variables, the fast dynamics can be described by a planar system

$$\dot{x} = a - (b + c\,y)x,$$
$$\dot{y} = \frac{\beta x^2}{1 + x^2} - \alpha\,y. \tag{2}$$

By rescaling time, this can be written in the form of a quartic system:

$$\dot{x} = (1 + x^2)(a - (b + c\,y)x) \equiv P,$$
$$\dot{y} = \beta x^2 - \alpha\,y(1 + x^2) \equiv Q. \tag{3}$$

Together with (3), we also consider an auxiliary system (see [1, 10])

$$\dot{x} = P - \gamma Q, \qquad \dot{y} = Q + \gamma P, \tag{4}$$

applying to these systems new bifurcation methods and geometric approaches developed in [3, 5, 7, 10–20] and carrying out the qualitative analysis of (3).

In particular, using system (4) and applying results of [23], we have proved the following theorem [19].

Theorem 1 *The reduced Topp system* (3) *can have at most two limit cycles.*

In Sect. 2, we perform a numerical study of the Topp model (1).

2 Analysis of the 3-Dimensional Topp Model

In this section, we study numerically the dynamics of the 3-dimensional Topp model (1). Our particular interest is to identify the bifurcations leading to chaotic dynamics. We fix the following parameter values:

$$b = 1, \quad c = 1, \quad m = 2, \quad n = 1.$$

The remaining parameters α, a, and l will be used for bifurcation analysis.

We start by studying equilibrium solutions and their stability. The Topp system (1) has at most three equilibria which are given by

$$E_1 = (a, 0, 0),$$
$$E_{2,\pm} = \left(\xi_\pm, \frac{a - \xi_\pm}{\xi_\pm}, \frac{\alpha(a - \xi_\pm)(1 + \xi_\pm^2)}{\xi_\pm^3}\right),$$

where $\xi_\pm = 1 \pm \sqrt{1-l}$. Note that $E_{2,-}$ and $E_{2,+}$ coalesce in a saddle-node bifurcation which occurs for $l = 1$.

Now assume that $l = 1$. In this case it follows that

$$E_{2,+} = E_{2,-} = (1, a-1, 2\alpha(a-1)).$$

A straightforward calculation shows that the characteristic polynomial of the Jacobian matrix of (1) evaluated at $E_{2,\pm}$ is given by $p(\lambda) = -\lambda(\lambda^2 - T\lambda + D)$, where $T = \alpha + a$ and $D = \alpha(2a-1)$. Note that $\lambda = 0$ is a zero of $p(\lambda)$; indeed, this is the eigenvalue associated with the saddle-node bifurcation. For $0 < a < \frac{1}{2}$ and $\alpha = -a$ it follows that $T = 0$ and $D > 0$, which implies that $p(\lambda)$ also has two imaginary zeros $\lambda = \pm i\sqrt{-a(2a-1)}$. In conclusion, in the 3-dimensional (α, a, l)-parameter space there is a plane of saddle-node bifurcations given by $l = 1$ and a line segment of Hopf-saddle-node bifurcations given by $(-\alpha, \alpha, 1)$ where $-\frac{1}{2} < \alpha < 0$.

The possible unfoldings of the Hopf-saddle-node (HSN) bifurcation are presented in [22]. The HSN bifurcation is a codimension-2 bifurcation which forms an organizing centre in the 2-dimensional (a, l)-parameter plane. From the HSN point typically other bifurcation curves emanate, such as Hopf-Neĭmark-Sacker bifurcations which lead to quasi-periodic attractors. In addition, Shilnikov homoclinic bifurcations can occur subordinate to a HSN bifurcation [4]. In certain cases, Shilnikov homoclinic are associated with the existence of chaotic dynamics and strange attractors. The HSN bifurcation and related Shilnikov bifurcations occur in many atmospheric models [4, 6, 9, 24, 26].

We take cross sections in the parameter space by fixing α and study bifurcations and routes to chaos in the (a, l)-plane. The Lyapunov diagram in Fig. 1 shows a classification of the dynamical behaviour of the Topp model in different regions of the (a, l)-parameter plane where $\alpha = -0.2$ is kept fixed. The diagram suggests that periodic attractors and chaotic attractors with a positive Lyapunov exponent occur

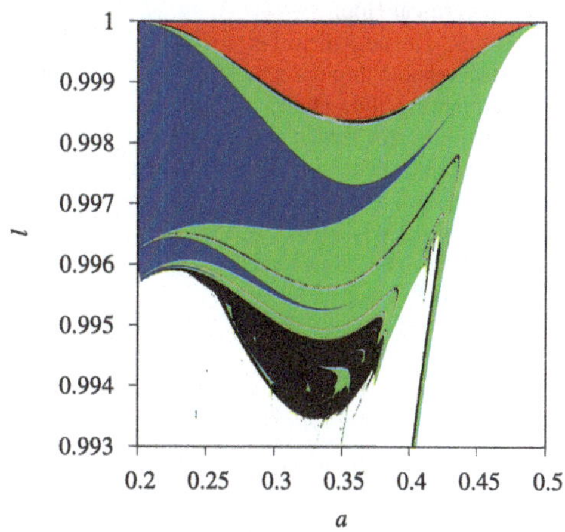

Fig. 1 Lyapunov diagram of attractors for the Topp model as a function of the parameters a and l, where $\alpha = -0.2$ is kept fixed. See Table 1 for the color coding. The Hopf-saddle-node bifurcation is located at the point $(a, l) = (0.2, 1)$

Table 1 Color coding for the Lyapunov diagram presented in Fig. 1

Color	Lyapunov exponents	Attractor type
Red	$0 > \lambda_1 \geq \lambda_2 \geq \lambda_3$	Stable equilibrium
Green	$0 = \lambda_1 > \lambda_2 > \lambda_3$	Periodic attractor (node)
Blue	$0 = \lambda_1 > \lambda_2 = \lambda_3$	Periodic attractor (focus)
Grey	$0 = \lambda_1 = \lambda_2 > \lambda_3$	2-torus attractor
Black	$\lambda_1 > 0 > \lambda_2 \geq \lambda_3$	Chaotic attractor
White		No attractor detected

for regions in the parameter plane with positive Lebesgue measure. For other values of $-\frac{1}{2} < \alpha < 0$ the Lyapunov diagrams look qualitatively similar (not shown) (Table 1).

Now we fix the parameters $\alpha = -0.2$ and $a = 0.33$ and perform a more detailed bifurcation analysis by varying the parameter l. For $l = 0.9999$ the equilibrium $E_{2,-}$ is stable. Continuation with decreasing l shows that $E_{2,-}$ becomes unstable through a supercritical Hopf bifurcation which occurs for $l \approx 0.99852$. Next, we continue the periodic orbit born at the Hopf bifurcation. For $l \approx 0.995641$ the periodic orbit becomes unstable through a period doubling bifurcation. Presumably this is the first period doubling of an infinite cascade.

Continuation of the periodic orbit beyond the first period doubling bifurcation reveals the following phenomenon. The unstable periodic orbit bifurcates further through a rapid succession of saddle-node bifurcations. Presumably, infinitely many saddle-node bifurcations occur. The newly born periodic orbits themselves may bifurcate through period doubling bifurcations. Figure 2 shows a bifurcation diagram in which the period of the orbit is plotted as a function of the continuation parameter l. Clearly, the diagram suggests that the periods of the periodic orbits born through the saddle-node bifurcations tend to infinity.

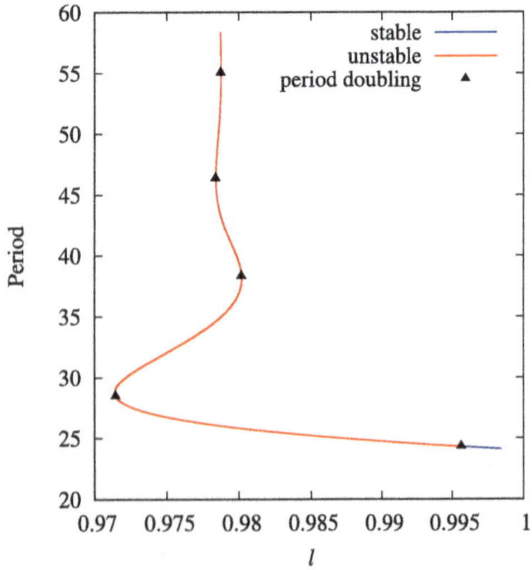

Fig. 2 Bifurcation diagram of a stable periodic orbit born at a supercritical Hopf bifurcation. The periodic orbit becomes unstable through a period doubling bifurcation and then bifurcates further through a rapid succession of saddle-node bifurcations. The periods of the newly born (unstable) periodic orbits tend to infinity as $l \to l_\infty \approx 0.978$

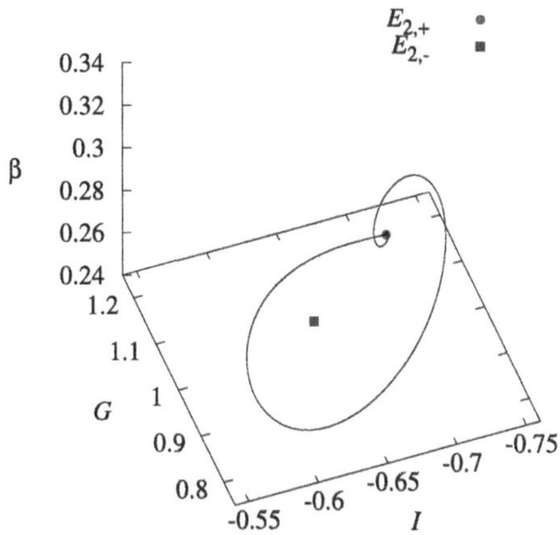

Fig. 3 A periodic orbit of large period which is close to a Shilnikov homoclinic orbit formed by the intersection of 1-dimensional unstable manifold and the 2-dimensional stable manifold of the equilibrium $E_{2,+}$

The phenomenon depicted in Fig. 2 can be explained as follows. During the continuation the periodic orbits born through the saddle-node bifurcations become arbitrarily close to an equilibrium. Hence, this bifurcation sequence leads to a homoclinic orbit. Figure 3 shows a periodic orbit which has a striking resemblance to a Shilnikov homoclinic orbit which is formed by an intersection of the 1-dimensional unstable manifold and the 2-dimensional stable manifold of the equilibrium $E_{2,+}$. Indeed, it is expected that these Shilnikov homoclinic orbits occur along a curve in the (a, l)-plane which emanates from the HSN bifurcation point [22]. Likewise, there may also be curve emanating from the HNS point along which there are Shilnikov homoclinic orbits which are formed by the 1-dimensional stable manifold and 2-dimensional unstable manifold of the equilibrium $E_{2,-}$. The numerical computation of these curves and performing a more detailed bifurcation analysis will be pursued in forthcoming work by the authors.

Finally, we explore the chaotic regime for $0.994775 < l < 0.993466$. From the flow of the Topp model we numerically compute a Poincaré map by computing the intersections of the integral curves with the plane $G = 0.9$. Figure 4 shows a bifurcation diagram of the Poincaré map. The period doubling bifurcations of periodic attractors are clearly visible. After what is presumably an infinite cascade of period doublings we find chaotic attractors. Figure 5 shows a chaotic attractor for the parameter values $(\alpha, a, l) = (-0.2, 0.35, 0.994)$.

The attractor in Fig. 5 seems to have the geometric structure of a "fattened curve". In fact, we conjecture that the attractor is Hénon-like, which means that the attractor is the closure of the 1-dimensional unstable manifold of a fixed point. For the classical Hénon map the existence of such attractors has been proven by [2]. Hénon-like attractors appear in many applications which range from climate models [6, 8] to control systems [21]. Their occurrence in the Topp model will be investigated in more detail by the authors in forthcoming work.

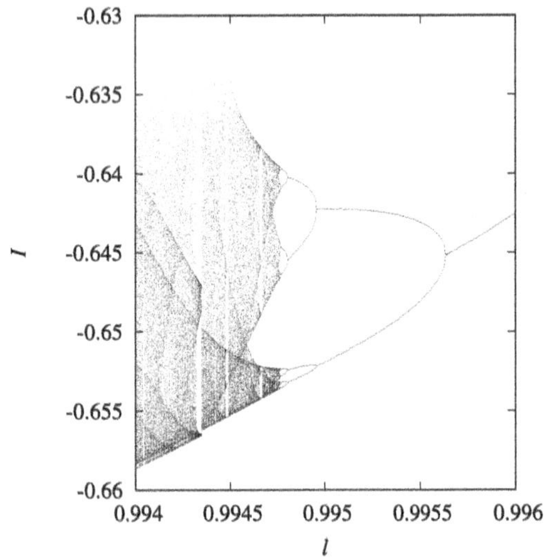

Fig. 4 Bifurcation diagram of attractors for the Poincaré map derived from the Topp model

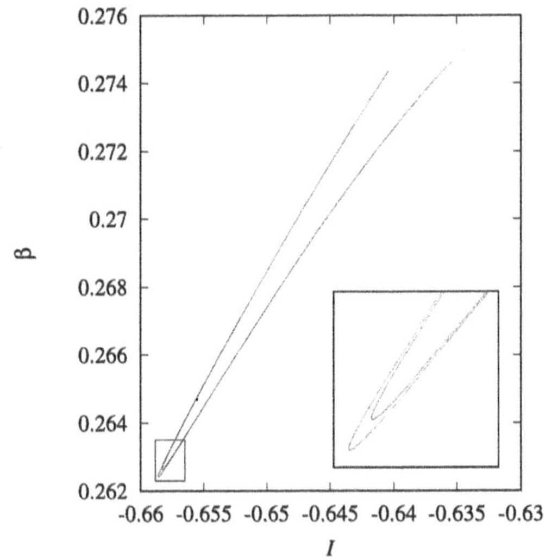

Fig. 5 Chaotic attractor of the Poincaré map of the Topp model for the parameters $(\alpha, a, l) = (-0.2, 0.35, 0.994)$. The inset shows a magnification of the attractor enclosed by the box

Acknowledgments The authors thank Prof. Robert MacKay (University of Warwick, UK) who initiated this research. The first author was supported by the London Mathematical Society (LMS) and the Netherlands Organization for Scientific Research (NWO).

References

1. N.N. Bautin, E.A. Leontovich, *Methods and Ways of the Qualitative Analysis of Dynamical Systems in a Plane*, Nauka, Moscow (1990) (in Russian)
2. M. Benedicks, L. Carleson, The dynamics of the Hénon map. Ann. Math. **133**, 73–169 (1991)

3. H.W. Broer, V.A. Gaiko, Global qualitative analysis of a quartic ecological model. Nonlinear Anal. **72**, 628–634 (2010)
4. H.W. Broer, G. Vegter, Subordinate Shilnikov bifurcations near some singularities of vector fields having low codimension. Ergod. Th. Dynam. Sys. **4**, 509–525 (1984)
5. H.W. Broer, R. Vitolo, Dynamical systems modeling of low-frequency variability in low-order atmospheric models. Discrete Contin. Dyn. Syst. Ser. B **10**, 401–409 (2008)
6. H.W. Broer, C. Simó, R. Vitolo, Bifurcations and strange attractors in the Lorenz-84 climate model with seasonal forcing. Nonlinearity **15**, 1205–1267 (2002)
7. H.W. Broer, V. Naudot, R. Roussarie, K. Saleh, Dynamics of a predator-prey model with non-monotonic response function. Discrete Contin. Dyn. Syst. Ser. A **18**, 221–251 (2007)
8. H.W. Broer, H.A. Dijkstra, C. Simó, A.E. Sterk, R. Vitolo, The dynamics of a low-order model for the Atlantic Multidecadal Oscillation. Discrete Contin. Dyn. Syst. Ser. B **16**, 73–107 (2011); Ergodic Th. Dyn. Sys. **4**, 509–525.
9. D.T. Crommelin, J. D. Opsteegh, F. Verhulst, A mechanism for atmospheric regime behaviour. J. Atmos. Sci. **61**, 1406–1419 (2004)
10. V.A. Gaiko, *Global Bifurcation Theory and Hilbert's Sixteenth Problem* (Kluwer, Boston, 2003)
11. V.A. Gaiko, Multiple limit cycle bifurcations of the FitzHugh–Nagumo neuronal model. Nonlinear Anal. **74**, 7532–7542 (2011)
12. V.A. Gaiko, On limit cycles surrounding a singular point. Differ. Equ. Dyn. Syst. **20**, 329–337 (2012)
13. V.A. Gaiko, The applied geometry of a general Liénard polynomial system. Appl. Math. Lett. **25**, 2327–2331 (2012)
14. V.A. Gaiko, Limit cycle bifurcations of a general Liénard system with polynomial restoring and damping functions. Int. J. Dyn. Syst. Differ. Equ. **4**, 242–254 (2012)
15. V.A. Gaiko, Limit cycle bifurcations of a special Liénard polynomial system. Adv. Dyn. Syst. Appl. **9**, 109–123 (2014)
16. V.A. Gaiko, Maximum number and distribution of limit cycles in the general Liénard polynomial system. Adv. Dyn. Syst. Appl. **10**, 177–188 (2015)
17. V.A. Gaiko, Global qualitative analysis of a Holling-type system. Int. J. Dyn. Syst. Differ. Equ. **6**, 161–172 (2016)
18. V.A. Gaiko, Global bifurcation analysis of the Kukles cubic system. Int. J. Dyn. Syst. Differ. Equ. **8**, 326–336 (2018)
19. V.A. Gaiko, Limit cycles of a Topp system, in *9th International Conference on Physics and Control*, Innopolis, Russia, September 8–11 (2019), pp. 96–99
20. V.A. Gaiko, C. Vuik, Global dynamics in the Leslie–Gower model with the Allee effect. Int. J. Bifurcation Chaos **28**, 31850151 (2018)
21. H. Ghane, A.E. Sterk, H. Waalkens, Chaotic dynamics from a pseudo-linear system. IMA J. Math. Control Inform. **2**, 1–18 (2019)
22. Yu.A. Kuznetsov, *Elements of Applied Bifurcation Theory* (Springer. New York, 2004)
23. L.M. Perko, Multiple limit cycle bifurcation surfaces and global families of multiple limit cycles. J. Diff. Equ. **122**, 89–113 (1995)
24. A.E. Sterk, R. Vitolo, H.W. Broer, C. Simó, H.A. Dijkstra, New nonlinear mechanisms of midlatitude atmospheric low-frequency variability. Physica D **239**, 702–718 (2010)
25. B. Topp, K. Promislow, G. Devries, R.M. Miuraa, D.T. Finegood, A model of β-cell mass, insulin, and glucose kinetics: pathways to diabetes. J. Theor. Biol. **206**, 605–619 (2000)
26. L. van Veen, Baroclinic flow and the Lorenz-84 model. Int. J. Bifurcation Chaos **13**, 2117–2139 (2003)

Optimal Harvesting in Age- and Size-Structured Population Models

N. Hritonenko, M. C. A. Leite, and Y. Yatsenko

Abstract This paper briefly surveys age- and size- structured linear and nonlinear population models and reveals their practical meaning. It focuses on optimal control problems with two types of harvesting and their impact on the sustainable harvesting.

Keywords Age- and size-structured population models · Harvesting rate and effort · Two-dimensional optimal controls · PDE · State constraints

Mathematics Subject Classification (2010) Primary: 92D25, 37N25; Secundary: 49K20, 49J22

1 Introduction

Various mathematical models have been suggested to capture population dynamics. PDE models are commonly used when age or size of individuals is relevant. They have been constantly extended to address practical needs, restrictions, and quotas. Sustainable harvesting is an important task in agriculture, aquaculture, forestry, fishery, and other applications. The paper aims to show relations between age- and size- structured models (Sect. 2).

N. Hritonenko (✉)
Prairie View A&M University, Prairie View, TX, USA
e-mail: nahritonenko@pvamu.edu

M. C. A. Leite
University of South Florida St. Petersburg, St. Petersburg, FL, USA

Y. Yatsenko
Houston Baptist University, Houston, TX, USA

P. Cerejeiras et al. (eds.), *Current Trends in Analysis, its Applications and Computation*, Research Perspectives,
https://doi.org/10.1007/978-3-030-87502-2_2

Harvesting in populations often depends on age or size of individuals. Such harvesting models have been offered and analyzed since 1980s, see surveys in [1, 13, 19], and references therein. Two types of harvesting control, harvesting rate and effort, and their practical relevance are analyzed in Sect. 3.

Finding effective harvesting regime in age- or size-structured population involves the optimal control of PDEs with two-dimensional controls and state constraints [1–3, 5, 9–17, 20, 23]. Analyzing optimal harvesting in linear and nonlinear population models, we explore the bang-bang form of the optimal harvesting and present applied interpretation of model assumptions and outcomes in Sect. 4.

2 Age- and Size-Dependent Population Models

The mathematical description of age- and size-structured populations has a long history, though the choice of the best model is still not always clear. Moreover, the relation between age and size in most biological populations is rather weak.

Let us briefly discuss connections and differences between two well-known linear PDE models.

The Lotka-McKendrick model describes the dynamics of an age-structured population under abundant resources [2, 10, 11, 13, 18, 19]

$$\frac{\partial x(t,a)}{\partial t} + \frac{\partial x(t,a)}{\partial a} = -\mu(t,a)x(t,a), \tag{2.1}$$

$$x(t,0) = p(t), \qquad x(0,a) = x_0(a), \qquad a \in [0,A), \quad t \in [0,\infty). \tag{2.2}$$

The function $x(t,a)$ represents the population density of individuals of age a at time t in the sense that the population size is given by $N(t) = \int_0^A x(t,a)da$ at time t, where A is the population maximum age. Then $\dfrac{\partial x(t,a)}{\partial a}$ is an "aging" term, $x_0(a)$ represents the population density at the initial moment, and $\mu(t,a)$ is the age-specific mortality rate. The influx of new individuals is determined by a fertility equation if the natural reproduction is allowed or by the density $p(t)$ of introduced new species in a fully managed population, in which all young individuals (trees, fish, etc.) are introduced from outside.

The related size-structured population model [7, 15] can be represented by

$$\frac{\partial x(t,l)}{\partial t} + \frac{\partial (g(t,l)x(t,l))}{\partial l} = -\mu(t,l)x(t,l), \tag{2.3}$$

$$g(t,l_0)x(t,l_0) = p(t),\ x(0,l) = x_0(l),\ l \in [l_0, l_{max}],\ t \in [0,\infty), \tag{2.4}$$

where $x(t,l)$, $\mu(t,l)$, $p(t)$, and $x_0(l)$ have the same meaning as in the model (2.1)–(2.2) but with respect to the size l that can be viewed as an individual's length, diameter, weight, volume, or other physiological quantity. The total population size at any time t in (2.3)–(2.4) is $N(t) = \int_{l_0}^{l_{max}} x(t,l)dl$ and $\dfrac{\partial(g(t,l)x(t,l))}{\partial l}$ is a "growth" term. The growth rate $g(t,l)$ can be interpreted by $\int_{l_1}^{l_2} \frac{1}{g(t,l)} dl$ as the time during which an individual grows from size l_1 to l_2, with $l_0 \le l_1 \le l_2 \le l_{max}$.

Equations (2.1)–(2.2) and (2.3)–(2.4) differ only by the presence of the growth rate $g(t,l)$. To justify g in the initial condition (2.4), let us consider a relation among the functions $x(t,l)$, $p(t)$, and $g(t,l)$ in a neighborhood of the initial time 0. The size of new individuals changes as $\Delta l \approx g(t,l)\Delta t$ during a small time period Δt, while the age changes as $\Delta a = \Delta t$. Then, $\Delta x \approx x(t,l_0)\Delta l \approx x(t,l_0)g(t,l_0)\Delta t$. Furthermore, new individuals of size l_0 brought into the population during Δt change their density by $\Delta x \approx p(t)\Delta t$ during Δt. Equating the last two statements we obtain (2.4). Similar reasoning explains the appearance of g in the second term of (2.3).

We consider examples of optimal control in age- and size-structured population models with relevant applied interpretation in the following sections.

3 Harvesting Control

Rational harvesting in populations is one of the most common applied problems in forestry, fishery, and agriculture [1–7, 9, 13, 16, 17, 19–22]. Depending on practical needs, harvesting objectives vary significantly and may include recommending the maximum sustainable yield or profit, minimizing environmental damage, preventing population extinction, and other requirements. Dependence of a harvesting control h on time, age, or size, or other factors greatly affects the investigation techniques. Let us consider two types of a two-dimensional harvesting control $h(t,a)$, the harvesting effort and rate.

Harvesting in fishery is often directly proportional to the density or abundance of a fish population. It can be measured by the overall effort of catching fish and the size of a stock.

Then the harvesting control $h(t,a) = u(t,a)x(t,a)$, where $u(t,a)$ is the *harvesting effort* that measures the expense made at time t to harvest fish of age a from all the available fish, and (2.1) can be extended to

$$\frac{\partial x(t,a)}{\partial t} + \frac{\partial x(t,a)}{\partial a} = -\mu(a)x(t,a) - u(t,a)x(t,a). \tag{3.1}$$

The model (3.1), (2.2) reflects the *catch-per-unit-effort* hypothesis and assumes the cost of harvesting to be proportional to the effort. In the case of a sole owner of a fish resource or in forestry, a certain portion of a population is harvested at a constant cost. Then the harvesting control $h(t,a) = u(t,a)$, where $u(t,a)$ is the *harvesting rate* or *intensity* at age a and time t, and the evolution equation (2.1) of a stock becomes

$$\frac{\partial x(t,a)}{\partial t}+\frac{\partial x(t,a)}{\partial a}=-\mu(a)x(t,a)-u(t,a). \tag{3.2}$$

The choice of a harvesting control is determined by the nature and objectives of a practical problem. The harvesting rate is commonly used in economics [11, 16], farming (agriculture, aquaculture), while the harvesting effort is preferred in modeling of wild populations, mostly forestry [6, 13, 16, 17, 23] and fishing. The Gordon-Schaefer, Beverton-Holt, and other bioeconomic models of open-access commercial fishing use the effort as an endogenous control.

4 Optimization in Age-, Size-Structured Models

Harvesting problems have been intensively investigated in [1–3, 8–17], and others. The corresponding optimal control problems (OCPs) are constructed to find the most effective harvesting policy. Formulation of a detailed objective function with meaningful applied interpretation often poses more challenges than the choice of a model. Investigation steps in OCPs usually include extremum conditions, the existence and uniqueness of solutions, qualitative and quantitative analyses of optimal trajectories, bang-bang solutions, and sustainable development of a system. Their outcomes are important in development of rational harvesting strategy. We will focus on OCPs with two-dimensional controls implemented in the Lotka-McKendrick model and its nonlinear modifications and analyze different bang-bang solutions depending on what harvesting control, rate or effort, has been chosen.

4.1 Optimization in Age-Structured Models

Let us consider the OCP of maximizing the harvesting profit in (3.1), (2.2).

$$\max_{u,p,x} I=\max_{u,p,x}\int_0^T e^{-rt}\left[\int_0^A c(t,a)u(t,a)x(t,a)da-b(t)p(t)\right]dt, \tag{4.1}$$

$$0\le u(t,a)\le u_{max},\qquad 0\le p(t)\le p_{max},\quad x(t,a)\ge 0, \tag{4.2}$$

where, $u\in L^{\infty}([0,T)\times[0,A)), p\in L^{\infty}[0,T)$, and $x_0\in L^{\infty}[0,A), x\in C([0,T),L^{\infty}[0,A))$. In this OCP $c(t,a)$ is the unit price of the harvesting output, $b(t)$ is the price of introduced individuals, and e^{-rt} is the discounting factor. The functional (4.1) describes the present value of the profit as the difference of harvesting revenue and operating costs to bring new individuals.

An important feature of harvesting optimization models is the occurrence of bang-bang solutions. A bang-bang optimal control is a solution u that switches between the boundaries 0 and u_{max} of the constraint-inequality $0 \le u \le u_{max}$. The strong bang-bang principle defines conditions under which the optimal control takes only boundary values, while the weak bang-bang principle allows u to take interior values.

The *strong bang-bang principle* [10] for the OCP (4.1), (3.1), (4.2), (2.2), shows that the optimal harvesting effort $u(t, a)$ is of the form

$$u^{\star}(t,a) = \begin{cases} 0, & 0 \le a < a^{\star}(t), \\ u_{max}, & a^{\star}(t) \le a \le A, \end{cases} \tag{4.3}$$

if

$$\frac{\partial c}{\partial a} > 0. \tag{4.4}$$

The *strong bang-bang principle* also occurs in the OCP (4.2), (4.1), (2.2) in the following nonlinear Gurtin-McCamy model

$$\frac{\partial x(t,a)}{\partial t} + \frac{\partial x(t,a)}{\partial a} = -\mu(E(t),a)x(t,a) - u(t,a)x(t,a), \tag{4.5}$$

where $E(t)$ reflects the environmental impact and intensity of the intra-specific competition. In this model the key assumptions are

$$\frac{\partial c}{\partial a} > 0, \quad \frac{\partial \mu}{\partial E} \ge 0, \quad \frac{\partial \mu}{\partial a} \ge 0. \tag{4.6}$$

The dependence of the mortality rate $\mu(E, a)$ on $E(t)$ reflects the intra-species competition. This is a common example of the *non-local nonlinearity* that represents non-local effects in the population. Although conditions (4.6) are stronger than condition (4.4), they are still realistic and reflect the increase of profit with individual's age and increase of the mortality caused by both, age and intra-species competition.

The age-structured population models with controlled harvesting rate are more natural in real applications, especially in forestry, though they lead to additional mathematical challenges caused by the active state constraint $x \ge 0$. The optimal control in the Lotka-McKendrik model (3.2) and (2.2) with two-dimensional optimal harvesting rate $u(t, a)$ and the objective

$$\max_{u,p,x} I = \max_{u,p,x} \int_0^T e^{-rt} \left[\int_0^A c(t,a)u(t,a)da - b(t)p(t) \right] dt \tag{4.7}$$

takes the *bang-bang form*

$$u^\star(t,a) = \begin{cases} 0, & 0 \le a < a^\star(t), \\ u_{max}, & a^\star(t) \le a \le a_e, (t), \\ 0, & a_e(t) < a \le A, \end{cases} \tag{4.8}$$

under restrictions

$$u_{max} >> 1 \quad \text{and} \quad \int_0^A c(t,a) e^{-\int_0^a \mu(t,\xi)d\xi} da > b(t), \quad t \in [0,\infty), \tag{4.9}$$

where $0 \le a^\star(t) < a_e(t) < A$, and the endogenous age $a_e(t)$ is determined from the condition $x(t, a_e(t)) = 0$ for each t.

The bang-bang (4.8) in the model with harvesting intensity is qualitatively different from the corresponding one (4.3) in the models with controlled effort. It states that the optimal strategy is to harvest individuals older than $a^\star(t)$ but younger than $a_e(t)$, that is, before they reach the maximum age A. This policy is more realistic than (4.3). For instance, in forestry, old trees, that do not have any market value, provide nutrition for younger trees when they die.

4.2 Optimization in Size-Structured Models

The existence of bang-bang solutions has been proven for some OCPs in size-structured models. Let us consider the following OCP in a Gurtin-McCamy model of forest management with a two-dimensional harvesting effort $u(t,l)$ that aims to find the functions $x(t,l)$, $u(t,l)$, $E(t)$, $p(t)$ for $t \in [0,\infty)$, $l \in [l_0, l_m]$, that maximize

$$\max_{u,p,x,E} J = \int_0^\infty e^{-rt} \left\{ \int_{l_0}^{l_m} c(t,l)u(t,l)x(t,l)dl - k(t)p(t) \right\} dt, \tag{4.10}$$

subject to the following constraints:

$$\frac{\partial x(t,l)}{\partial t} + \frac{\partial (g(E(t),l)x(t,l))}{\partial l} = -\mu(E(t),l)x(t,l)) - u(t,l)x(t,l), \tag{4.11}$$

$$E(t) = \chi \int_{l_0}^{l_m} l^2 x(t,l)dl, \quad g(E(t),l_0), x(t,l_0) = p(t), \tag{4.12}$$

$$0 \le u(t,l) \le u_{max}, \quad x(0,l) = x_0(l), \quad l \in [l_0, l_m], \quad t \in [0,\infty). \tag{4.13}$$

A forest is a renewable resource, which provides timber, offers recreation facilities, mitigates climate change, and improves air quality. Human intervention, natural disturbances, and climate change may cause irreversible and unfavorable changes in the forest dynamics. Therefore, modeling in forestry is vital to understand the dynamics of its development and predict negative consequences of climate change and human interaction.

The size of trees better suits biological and economic needs than their age. In applications to forestry, parameters and state variables of the model (4.10)–(4.13) are taken as follows: l is the tree diameter at breast height, $x(t, l)$ is the density of size-structured trees at time t, $E(t)$ reflects the environmental impact and intra-specific competition, $p(t)$ is the flux of young trees planted, and χ is a parameter related to a specific type of tree. The objective is to maximize the profit from logging. The OCP (4.10)–(4.13) introduced in [14] is a simplified version of a larger problem that maximizes profit from timber production and carbon sequestration studied in [9, 15–17]. The problem is of great importance in countries where forestry farms are compensated for keeping trees that sequester carbon from the atmosphere and store it in their trunks. Calibration of obtained theoretical outcomes from data on a Pinus Sylvestris forest in Catalonia, Spain, show good applicability of the model. It is shown in [9] and references therein that effects of climate change on a Pinus forest development can be captured by the growth rate

$$g(E, l) = (l_m - l)\hat{g}(E), \quad \hat{g}(E) = (\beta_0 - \beta_1)E, \quad \beta_0 > 0, \tag{4.14}$$

where β_0 and β_1 are parameters related to different climate change scenarios.

Assuming the growth rate (4.14) and

$$\frac{\partial c(t, l)}{\partial l} > 0, \qquad \frac{\partial \mu(E, l)}{\partial E} \geq 0, \qquad \frac{\partial \mu(E, l)}{\partial l} \geq 0, \tag{4.15}$$

the optimal control $u^\star(t, l)$ in OCP (4.10)–(4.13) has the following form

$$u^\star(t, l) = \begin{cases} 0, & l_0 \leq l < l^\star(t), \\ u_{max}, & l^\star(t) \leq l \leq l_m, \end{cases} \tag{4.16}$$

with, at most, one switching size $l^\star(t)$, $l_0 \leq l^\star(t) \leq l_m$, $t \in [0, \infty)$.

Conditions (4.15) are similar to conditions (4.6) and have a similar applied interpretation. The bang-bang solution (4.16) is similar to (4.3) and justifies advantages of selective harvesting over clear cutting for sustainable forest development, which is in agreement with reality. Even countries that have a mandatory clear-cutting regime in forestry tend to change the law.

The optimal control for the linear size-structured model with controlled harvesting rate and natural reproduction was proven to be bang-bang in [11]. The nonlinear case remains an open question.

Acknowledgments The authors are thankful to Chair Dr. Torsten Lindström and participants of the ISAAC session "Applications of dynamical systems theory in biology" for their useful comments and to anonymous reviewers for their valuable comments that essentially improve the paper. N. Hritonenko acknowledges support of PVAMU FIE and FEP programs. N. Hritonenko and M. Leite acknowledge support from NIMBioS, NSF Award # DBI – 1300426.

References

1. S. Aniţa, Optimal harvesting for a nonlinear ag-dependent population dynamics. J. Math. Analy. Appl. **226**(1), 6–22 (1998)
2. S. Aniţa, *Analysis and Control of Age-Dependent Population Dynamics* (Springer, Berlin, 2000)
3. V. Barbu, M. Iannelli, Optimal control of population dynamics. J. Optim. Theory Appl. **102**(1), 1–14 (1999)
4. G.B. Bonan, Forests and climate change: forcings, feedbacks, and the climate benefits of forests. Science **320**(5882), 1444–1449 (2008)
5. M. Brokate, Pontryagin's principle for control problems in age-dependent population dynamics. J. Math. Biol. **23**(1), 75–101 (1985)
6. A.M. de Roos, Numerical methods for structured population models: the escalator boxcar train. Numer. Meth. Part. D. E. **4**(3), 173–195 (1988)
7. B. Ebenman, L. Persson (eds.), *Size-Structured Populations* (Springer, Berlin, 1988)
8. R.U. Goetz, N. Hritonenko, Y. Yatsenko, The optimal economic lifetime of vintage capital in the presence of operating costs, technological progress, and learning. J. Econ. Dyn. Control **32**(9), 3032–3053 (2008)
9. R.U. Goetz, N. Hritonenko, R. Mur, A. Xabadia, Y. Yatsenko, Forest management for timber and carbon sequestration in the presence of climate change: the case of Pinus Sylvestris. Ecolog. Econ. **88**, 86–96 (2013)
10. N. Hritonenko, Y. Yatsenko, The structure of optimal time- and age-dependent harvesting in the Lotka–McKendrik population model. Math. Biosci. **208**(1), 48–62 (2007)
11. N. Hritonenko, Y. Yatsenko, Maintenance of age-structured populations: optimal control, state constraints and bang–bang regime. J. Biol. Syst. **17**(04), 793–816 (2009)
12. N. Hritonenko, Y. Yatsenko, Bang-bang, impulse, and sustainable harvesting in age-structured populations. J. Biolog. Syst. **20**(02), 133–153 (2012)
13. N. Hritonenko, Y. Yatsenko, *Mathematical Modeling in Economics, Ecology and the Environment* (Springer, Berlin, 2013)
14. N. Hritonenko, Y. Yatsenko, R.U. Goetz, A. Xabadia, Maximum principle for a size-structured model of forest and carbon sequestration management. Appl. Math. Lett. **21**(10), 1090–1094 (2008)
15. N. Hritonenko, Y. Yatsenko, R.U. Goetz, A. Xabadia, A bang–bang regime in optimal harvesting of size-structured populations. Nonlinear Analys. Theory, Methods Appl. **71**(12), e2331–e2336 (2009)
16. N. Hritonenko, Y. Yatsenko, R.U. Goetz, A. Xabadia, Sustainable dynamics of size-structured forest under climate change. Appl. Math. Lett. **25**(10), 1439–1443 (2012)
17. N. Hritonenko, Y. Yatsenko, R.U. Goetz, A. Xabadia, Optimal harvesting in forestry: steady-state analysis and climate change impact. J. Biol. Dyn. **7**(1), 41–58 (2013)
18. N. Hritonenko, N. Kato, Y. Yatsenko, Optimal control of investments in old and new capital under improving technology. J. Optim. Theory Appl. **172**(1), 247–266 (2017)
19. M. Iannelli, F.A. Milner, *The Basic Approach to Age-Structured Population Dynamics: Models, Methods and Numerics* (Springer, Berlin, 2017). OCLC: 1005114704
20. N. Kato, Optimal harvesting for nonlinear size-structured population dynamics. J. Math. Analy. Appll. **342**(2), 1388–1398 (2008)
21. G.R. Munro, The optimal management of transboundary renewable resources. Canadian J. Econ. **12**(3), 355 (1979)
22. J.H.M. Thornley, M.G.R. Cannell, Managing forests for wood yield and carbon storage: a theoretical study. Tree Physiol. **20**(7), 477–484 (2000)
23. G.F. Webb, Population models structured by age, size, and spatial position, in *Structured Population Models in Biology and Epidemiology*, ed. by J.M. Morel, F. Takens, B. Teissier, P. Magal, S. Ruan, vol. 1936 (Springer, Berlin, 2008), pp. 1–49

Part II
Complex Analysis and Partial Differential Equations

Dirichlet Problem for Inhomogeneous Biharmonic Equation in Clifford Analysis

Ümit Aksoy and A. Okay Çelebi

Abstract An integral representation formula in terms of the bi-Laplacian operator is obtained and Dirichlet problem for the bi-Poisson equation is solved in Clifford analysis.

Keywords Dirichlet problem · Clifford analysis · Integral representations

Mathematics Subject Classification (2010) Primary 31B10; Secondary 30G35

1 Introduction

Many problems in science and engineering are modeled via boundary value problems for partial differential equations. Methods of complex function theory, Cauchy-Pompeiu type integral representation formulas in particular, serve as an efficient tool in solving such problems in the plane. Clifford analysis is developed as a generalization of complex function theory to the higher dimensions and many properties of the complex function theory are maintained, see [2–4, 6–11, 13, 15–18].

In this article, we study the integral representation formulas related to Laplacian and bi-Laplacian operators to investigate the solution of the Dirichlet problems for inhomogeneous harmonic and biharmonic equations which are connected to the elasticity and shell theory models. In [3, 4], some representations in terms of powers of Dirac operator and Laplacian are introduced with integral operators providing particular solutions of higher-order Poisson and Dirac equations. The

Ü. Aksoy (✉)
Department of Mathematics, Atilim University, Ankara, Turkey
e-mail: umit.aksoy@atilim.edu.tr

A. O. Çelebi
Department of Mathematics, Yeditepe University, Istanbul, Turkey
e-mail: acelebi@yeditepe.edu.tr

P. Cerejeiras et al. (eds.), *Current Trends in Analysis, its Applications and Computation*, Research Perspectives,
https://doi.org/10.1007/978-3-030-87502-2_3

Dirichlet problem for the Poisson equation was studied and higher order boundary value problems were investigated for Clifford-valued functions in [12, 14]. In [1], a complex Clifford algebra form of an integral representation in terms of the Laplacian for universal Clifford algebra valued functions studied in [20], is presented and Dirichlet problem for Poisson equation is solved. Our first aim is to rewrite that formula to apply the iteration procedure in a natural way leading to a higher order kernel function and higher order Cauchy-Pompeiu type integral representation. In [1], a second-order Green type function is introduced with some of its properties and Dirichlet problem is solved for biharmonic equation. Another aim of the current paper is to present a representation formula with regard to the inhomogeneous Dirichlet problem for the bi-Poisson equation and to find the unique solution of the corresponding Dirichlet problem.

The preliminary results on basic concepts and integral representations in Clifford analysis are given in the next section. In Sect. 3, Dirichlet problems for inhomogeneous harmonic and biharmonic equations are studied.

2 Integral Representations in Complex Clifford Algebra $\mathbb{C}_m$

In this section, firstly some preliminary concepts and results on functions in Clifford analysis are given. Then, Green-Gauss theorem, Cauchy-Pompeiu type representation and integral representation related to Laplace operator are presented. For a detailed information on Clifford algebra and Clifford analysis, see [9, 11, 13].

Any $x = (x_1, \dots, x_m) \in \mathbb{R}^m$ is represented as $x = \sum_{k=1}^{m} x_k e_k$ with $\{e_k : 1 \leq k \leq m\}$ being an orthonormal basis for $m \geq 2$. The multiplication

$$e_j e_k + e_k e_j = -2\delta_{jk} \quad 1 \leq j, k \leq m \tag{2.1}$$

and the unit element $e_0 = 1$ lead to a 2^m-dimensional real linear, associative and non-commutative (universal) Clifford algebra $\mathbb{R}_{0,m}$.

For $a_A \in \mathbb{R}$, any $a \in \mathbb{R}_{0,m}$ may be written as $a = \sum_A a_A e_A$ where

$$e_A = e_0 = 1 \text{ if } A = \emptyset$$

$$e_A = e_{\alpha_1} e_{\alpha_2} \dots e_{\alpha_k} \text{ if } A = \{\alpha_1, \alpha_2, \dots \alpha_k\} \subseteq \{1, 2, \dots, m\}$$

with $1 \leq \alpha_1 < \alpha_2 < \cdots < \alpha_k \leq m$. Any $x = (x_0, x_1, \dots, x_m) \in \mathbb{R}^{m+1}$ is written as

$$x = x_0 e_0 + \sum_{k=1}^{m} x_k e_k.$$

If $a_A \in \mathbb{C}$, the corresponding algebra is denoted as $\mathbb{C}_m$ and called as complex Clifford algebra. The complex conjugate of $a = \sum_A a_A e_A$, $a_A \in \mathbb{C}$ is given by $\bar{a} = \sum_A \bar{a}_A \bar{e}_A$ where

$$\overline{e_0} = e_0 = 1, \ \overline{e_k} = -e_k, \quad 1 \leq k \leq m,$$

$\overline{e_A} = \overline{e}_{\alpha_k}\overline{e}_{\alpha_{k-1}} \ldots \overline{e}_{\alpha_1}$ and $\overline{e_A e_B} = \overline{e_B e_A}$. $|a| := \left(\sum_A |a_A|^2\right)^{1/2}$ is a norm for $a = \sum_A a_A e_A \in \mathbb{C}_m$ with $|a|_0 = 2^{m/2}|a|$ is an algebra norm.

For a complex valued function $f_A(z)$ defined in a domain D of $\mathbb{R}^{m+1}$, a $\mathbb{C}_m$-valued function f has the form $f(z) = \sum_A f_A(z)e_A$.

The Dirac operator ∂, its complex conjugate operator $\overline{\partial}$ and the Laplace operator Δ are given by

$$\partial := \sum_{k=1}^{m} e_k \partial_{x_k}\ , \quad \overline{\partial} := \sum_{k=1}^{m} \overline{e_k}\partial_{x_k} = \partial_{x_1} - \sum_{k=2}^{m} e_k \partial_{x_k}\ , \Delta = \partial\overline{\partial} = \overline{\partial}\partial = \sum_{k=1}^{m} \frac{\partial^2}{\partial x_k^2}\ .$$

For $f = \sum_A f_A e_A$,

$$\partial f = \sum_{k=1}^{m}\sum_{A} e_k e_A \frac{\partial f_A}{\partial x_k}, \quad f\partial = \sum_{k=1}^{m}\sum_{A} e_A e_k \frac{\partial f_A}{\partial x_k},$$

$$\overline{\partial} f = \sum_{k=1}^{m}\sum_{A} \overline{e_k} e_A \frac{\partial f_A}{\partial x_k}, \quad f\overline{\partial} = \sum_{k=1}^{m}\sum_{A} e_A \overline{e_k} \frac{\partial f_A}{\partial x_k}$$

hold. It can be seen that

$$\partial z = z\partial = 2 - m\ , \quad \partial \bar{z} = \bar{z}\partial = m,$$

$$\partial |z|^\alpha = |z|^\alpha \partial = \alpha |z|^{\alpha-2} z\ , \quad \overline{\partial} |z|^\alpha = |z|^\alpha \overline{\partial} = \alpha |z|^{\alpha-2} \bar{z},$$

$$\partial(\bar{z}^k + z^k) = (\bar{z}^k + z^k)\partial = 2k\bar{z}^{k-1},$$

$$\partial\left(\frac{\bar{z}}{|z|^m}\right) = \frac{\bar{z}}{|z|^m}\partial = 0$$

hold, see [3, 9, 11] for other properties.

The following Clifford algebra version of the Gauss theorem is the basis of the representation formulas for $\mathbb{C}_m$-valued functions, see [3, 4, 6].

Theorem 2.1 (Gauss Theorem) *Let* $f, g \in C^1(D; \mathbb{C}_m) \cap C(\overline{D}; \mathbb{C}_m)$. *Then*

$$\int_D ((f\partial)g + f(\partial g))dv = \int_{\partial D} f d\vec{\sigma} g$$

$$\int_D ((f\overline{\partial})g + f(\overline{\partial} g))dv = \int_{\partial D} f d\bar{\vec{\sigma}} g$$

where dv *denotes the volume element of* D, $d\sigma$ *the area element of* ∂D, $n = (n_1, n_2, \ldots, n_m)$ *the outward directed normal vector on* ∂D, $\vec{n} = \sum_{\mu=1}^{m} n_\mu e_\mu$ *the corresponding element in* $\mathbb{C}_m$, $d\vec{\sigma} = d\sigma\vec{n}$ *the directed area element on* ∂D *and* $d\bar{\vec{\sigma}} = d\sigma\bar{\vec{n}}$ *its complex conjugate.*

In the following subsection, we present the integral representation formula including the Dirac operator.

2.1 *Dirac Equation*

The Gauss theorem directly implies the Cauchy-Pompeiu representation formulae which are related to the Dirac operator, see [5, 9].

Theorem 2.2 (Cauchy-Pompeiu Type Representations) *Any* $w \in C^1(D; \mathbb{C}_m) \cap C(\overline{D}; \mathbb{C}_m)$ *can be represented as*

$$w(z) = \frac{1}{\omega_m} \int_{\partial D} \frac{\overline{\zeta - z}}{|\zeta - z|^m} d\vec{\sigma}(\zeta) w(\zeta) - \frac{1}{\omega_m} \int_D \frac{\overline{\zeta - z}}{|\zeta - z|^m} \partial w(\zeta) dv(\zeta) \tag{2.2}$$

or

$$w(z) = \frac{1}{\omega_m} \int_{\partial D} \frac{\zeta - z}{|\zeta - z|^m} d\bar{\vec{\sigma}}(\zeta) w(\zeta) - \frac{1}{\omega_m} \int_D \frac{\zeta - z}{|\zeta - z|^m} \overline{\partial} w(\zeta) dv(\zeta). \tag{2.3}$$

Here ω_m *denotes the area of the unit sphere in* $\mathbb{R}^m$.

The volume integral appearing in the formula (2.2) provides a particular solution to the inhomogeneous Dirac equation $\partial w = f$ in D which leads to the Teodorescu transform over D

$$Tf(z) = -\frac{1}{\omega_m} \int_D \frac{\overline{\zeta - z}}{|\zeta - z|^m} f(\zeta) dv(\zeta). \tag{2.4}$$

Complex plane analogue of this operator is the well-known Pompeiu operator which was extensively investigated by Vekua [19] in the theory of generalized analytic functions.

In the next subsections, we give the representation formulae in terms of Laplacian and bi-Laplacian operators.

2.2 Poisson Equation

A representation formula related to the Laplacian Δ is obtained by iterating the formulas (2.2) and (2.3) properly, see [2].

Theorem 2.3 *Let D be a bounded and smooth domain and $w \in C^2(D; \mathbb{C}_m) \cap C^1(\overline{D}; \mathbb{C}_m)$. Then for $z \in D$,*

$$w(z) = \frac{1}{\omega_m} \int\limits_{\partial D} \frac{\overline{\zeta - z}}{|\zeta - z|^m} d\vec{\sigma}(\zeta) w(\zeta) - \frac{1}{\omega_m} \int\limits_{\partial D} \frac{|\zeta - z|^{2-m}}{2-m} \overline{d\vec{\sigma}(\zeta)} \partial w(\zeta)$$
$$+ \frac{1}{\omega_m} \int\limits_{D} \frac{|\zeta - z|^{2-m}}{2-m} \Delta w(\zeta) dv(\zeta) \,. \tag{2.5}$$

A higher-order representation formula involving kth powers of Laplacian Δ^k in various dimension cases is given in [3, 5] via iterating the formula (2.5) inductively. With regard to the Dirichlet problem for the Poisson equation, the representation formula (2.5) is not suitable. A variant of the formula (2.5) is needed.

In the unit ball $\mathbb{B}_m$ of $\mathbb{R}^m$ for $m \geq 3$, an integral representation formula in terms of Laplacian is given by Zhang [20] for functions with values in universal Clifford algebra using a Green-type kernel function. A complex Clifford algebra analogue of a Green-type function and a representation formula in terms of the Laplacian for $\mathbb{C}_m$-valued functions in $\mathbb{B}_m$ studied in [1] are revisited below.

Definition 2.4 The function

$$G_1(z, \zeta) = \frac{1}{|\zeta - z|^{m-2}} - \frac{1}{\left| z|\zeta| - \frac{\zeta}{|\zeta|} \right|^{m-2}}, \quad z, \zeta \in \mathbb{B}_m, z \neq \zeta \tag{2.6}$$

is called the Green-type function for the unit ball $\mathbb{B}_m$.

Remark 2.5 By direct calculation, it can be observed that $G_1(z, \zeta)$ is a fundamental solution to the Laplace operator satisfying the following properties:

- $\Delta G_1(z, \zeta) = 0$, $z \in \mathbb{B}_m \backslash \{\zeta\}$,
- $G_1(z, \zeta) = G_1(\zeta, z)$ for $z \neq \zeta$, $z, \zeta \in \mathbb{B}_m$,

- $G_1(z,\zeta) = 0$ for $\zeta \in \partial\mathbb{B}_m$, $z \in \mathbb{B}_m$,
- $\partial G_1 = G_1\partial$ and $\overline{\partial} G_1 = G_1\overline{\partial}$,

where $\partial\mathbb{B}_m$ is the unit sphere.

The following theorem is the revised version of the one given in [1].

Theorem 2.6 *Any* $w \in C^2(\mathbb{B}_m;\mathbb{C}_m) \cap C^1(\overline{\mathbb{B}}_m;\mathbb{C}_m)$, $z \in \mathbb{B}_m$ *can be represented as*

$$w(z) = \frac{1}{2\omega_m}\int_{\partial\mathbb{B}_m} \partial_\nu G_1(z,\zeta)w(\zeta)d\sigma(\zeta) + \frac{1}{(2-m)\omega_m}\int_{\mathbb{B}_m} G_1(z,\zeta)\Delta w(\zeta)dv(\zeta) \tag{2.7}$$

where $\partial_\nu G_1 = (\partial G_1)\bar{\zeta} + (\overline{\partial} G_1)\zeta = 2\dfrac{1-|z|^2}{|\zeta - z|^m}$ *for* $|z| < 1, \zeta \in \partial\mathbb{B}_m$.

For the proof see [1].

2.3 Second-Order Poisson Equation

In [1], a second order Green function is defined by

$$G_2(z,\zeta) = \int_{\mathbb{B}_m} G_1(z,\tilde{\zeta})G_1(\tilde{\zeta},\zeta)dv(\tilde{\zeta}).$$

$G_2(z,\zeta)$ has the following properties:

- $\Delta^2 G_2(z,\zeta) = 0$ in $\mathbb{B}_m \backslash \{\zeta\}$ for any $\zeta \in \mathbb{B}_m$,
- $G_2(z,\zeta) = G_2(\zeta,z)$,
- $G_2(z,\zeta) = 0$ and $\Delta G_2(z,\zeta) = 0$ for $\zeta \in \partial\mathbb{B}_m$, $z \in \mathbb{B}_m$,
- $\partial G_2 = G_2\partial$ and $\overline{\partial} G_2 = G_2\overline{\partial}$,

where G_1 is given by (2.6).

An integral representation formula in terms of bi-Laplacian Δ^2 can be obtained by iteration the formula (2.7) in a proper way:

Theorem 2.7 *Any* $w \in C^4(\mathbb{B}_m;\mathbb{C}_m) \cap C^2(\overline{\mathbb{B}}_m;\mathbb{C}_m)$, $z \in \mathbb{B}_m$ *can be represented as*

$$\begin{aligned} w(z) = {} & \frac{1}{2(2-m)\omega_m}\int_{\partial\mathbb{B}_m} [(2-m)\partial_\nu G_1(z,\zeta)w(\zeta) + \partial_\nu G_2(z,\zeta)\Delta w(\zeta)]d\sigma(\zeta) \\ & + \frac{1}{(2-m)\omega_m}\int_{\mathbb{B}_m} G_2(z,\zeta)\Delta^2 w(\zeta)dv(\zeta). \end{aligned} \tag{2.8}$$

Proof Observe that

$$\int_{\mathbb{B}_m} G_2(z,\zeta)\Delta^2 w(\zeta)dv(\zeta)$$

$$= \frac{1}{2}\int_{\mathbb{B}_m} [(G_2\partial)\partial\overline{\partial}^2 w(\zeta) + G_2\Delta^2 w(\zeta) + (G_2\overline{\partial})\partial^2\overline{\partial} w(\zeta)]dv(\zeta)$$

$$+\frac{1}{2}\int_{\mathbb{B}_m} [G_2\Delta^2 w(\zeta) - (\overline{\partial}G_2\partial)\Delta w(\zeta) - \overline{\partial}G_2(\partial\Delta w(\zeta))]dv(\zeta)$$

$$-\frac{1}{2}\int_{\mathbb{B}_m} [(\partial G_2\overline{\partial})\Delta w(\zeta) + \partial G_2(\overline{\partial}\Delta w(\zeta)) - 2\Delta G_2\Delta w(\zeta)]dv(\zeta).$$

Gauss theorem implies

$$\int_{\mathbb{B}_m} G_2(z,\zeta)\Delta^2 w(\zeta)dv(\zeta) = \frac{1}{2}\left(\int_{\partial\mathbb{B}_m} G_2 d\vec{\sigma}\,\partial\overline{\partial}^2 w(\zeta) + \int_{\partial\mathbb{B}_m} G_2 d\vec{\overline{\sigma}}\,\partial^2\overline{\partial} w(\zeta)\right)$$

$$-\frac{1}{2}\left(\int_{\partial\mathbb{B}_m} \partial G_2 d\vec{\overline{\sigma}}\,\Delta w(\zeta) - \int_{\partial\mathbb{B}_m} \overline{\partial}G_2 d\vec{\sigma}\,\Delta w(\zeta)\right) + \int_{\mathbb{B}_m} G_1(z,\zeta)\Delta w(\zeta)dv(\zeta)$$

$$= \frac{1}{2}\int_{\partial\mathbb{B}_m} \partial_\nu G_2(z,\zeta)\Delta w(\zeta)d\sigma(\zeta) + \int_{\mathbb{B}_m} G_1(z,\zeta)\Delta w(\zeta)dv(\zeta).$$

Using Theorem 2.6 gives the required result. □

3 Dirichlet Problem

In this section, firstly we present the solution of the Dirichlet problem for Poisson equation given in [1] and then introduce the solution of the Dirichlet problem for inhomogeneous biharmonic equation. For the solution of the bi-Poisson equation under homogeneous Dirichlet conditions, we refer the reader to [1].

The representation (2.7) is used to solve the following Dirichlet boundary value problem for the Poisson equation [1]. This problem is solved for functions in universal Clifford algebra in [20].

Theorem 3.1 *Dirichlet problem for the Poisson equation*

$$\Delta w = f \ \ in \ \ \mathbb{B}_m \quad w = g \ \ on\ \partial\mathbb{B}_m$$

for $f \in L^1(\mathbb{B}_m; \mathbb{C}_m)$ *and* $g \in C(\partial\mathbb{B}_m; \mathbb{C}_m)$ *is uniquely solvable. The solution is given by*

$$w(z) = \frac{1}{2\omega_m} \int_{\partial\mathbb{B}_m} \partial_\nu G_1(z,\zeta) g(\zeta) d\sigma(\zeta) + \frac{1}{(2-m)\omega_m} \int_{\mathbb{B}_m} G_1(z,\zeta) f(\zeta) dv(\zeta). \tag{3.1}$$

In the case of inhomogeneous biharmonic equation, Dirichlet problem is solved by employing the representation (2.8).

Theorem 3.2 *Dirichlet problem for the bi-Poisson equation*

$$\Delta^2 w = f \ \ in \ \ \mathbb{B}_m \quad w = g_1(z),\ \Delta w = g_2(z) \ on\ \partial\mathbb{B}_m$$

for $f \in L^1(\mathbb{B}_m; \mathbb{C}_m)$, $g_1, g_2 \in C(\partial\mathbb{B}_m; \mathbb{C}_m)$, *is uniquely solvable. The solution is given by*

$$w(z) = \frac{1}{2\omega_m} \int_{\partial\mathbb{B}_m} \partial_\nu G_1(z,\zeta) g_1(\zeta) d\sigma(\zeta + \frac{1}{2(2-m)\omega_m} \int_{\partial\mathbb{B}_m} \partial_\nu G_2(z,\zeta) g_2(\zeta) d\sigma(\zeta)$$

$$+ \frac{1}{(2-m)\omega_m} \int_{\mathbb{B}_m} G_2(z,\zeta) f(\zeta) dv(\zeta)\,. \tag{3.2}$$

Proof Trivially, $w(z)$ is a solution of $\Delta^2 w = f$. It can be observed that

$$\lim_{z \to z_0} \frac{1}{\omega_m} \int_{\partial\mathbb{B}_m} \partial_\nu G_1(z,\zeta) g(\zeta) d\sigma(\zeta) = g(z_0)$$

holds for $z_0 \in \partial\mathbb{B}_m$, $g \in C(\partial\mathbb{B}_m; \mathbb{C}_m)$. Since

$$\Delta w(z) = \frac{1}{\omega_m} \int_{\partial\mathbb{B}_m} \partial_\nu G_1(z,\zeta) g_2(\zeta) d\sigma(\zeta) + \frac{1}{(2-m)\omega_m} \int_{\mathbb{B}_m} G_1(z,\zeta) f(\zeta) dv(\zeta)$$

we have

$$\lim_{z \to z_0} w(z) = g_1(z_0), \ \lim_{z \to z_0} \Delta w(z) = g_2(z_0)$$

for $z_0 \in \partial\mathbb{B}_m$. Thus the proof is completed. □

References

1. Ü. Aksoy, Dirichlet Problem for Poisson and Bi-Poisson Equations in Clifford Analysis, in *Analysis as a Life*, Birkhäuser (2019), pp. 19–37
2. H. Begehr, Iterations of Pompeiu operators. Mem. Differ. Equ. Math. Phys. **12**, 3–21 (1997)
3. H. Begehr, Iterated integral operators in Clifford analysis. J. Analy. Appl. **18**, 361–377 (1999)
4. H. Begehr, Representation formulas in Clifford analysis, in *Acoustics, Mechanics, and the Related Topics of Mathematical Analysis*, ed. by A. Wirgin (World Scientific, Singapore, 2002), pp. 8–13
5. H. Begehr, Integral representation in complex, hypercomplex and Clifford analysis. Integr. Transf. Special Funct. **13**, 223–241 (2002)
6. H. Begehr, Ju. Dubinskii, Orthogonal decompositions of Sobolev spaces in Clifford analysis. Ann. Mat. Pura Appl. **181**, 55–71 (2002)
7. H. Begehr, Z.X. Zhang, J. Du, On Cauchy-Pompeiu formula for functions with values in a universal Clifford algebra. Acta Math. Sci. **23**, 95–103 (2003)
8. H. Begehr, H. Otto, Z.X. Zhang, Differential operators, their fundamental solutions and related integral representations in Clifford analysis. Complex Variables Elliptic Equ. **51**, 407–427 (2006)
9. F. Bracks, R. Delanghe, F. Sommen, *Clifford analysis* (Pitman, London, 1982)
10. J.E. Gilbert, M.A.M. Murray, *Clifford Algebras and Dirac Operators in Harmonic Analysis* (Cambridge University Press, Cambridge, 1991)
11. S. Huang, Y.Y. Qiao, G. Wen, *Real and Complex Clifford Analysis*. Advances in Complex Analysis and Its Applications, vol. 5 (Springer, Berlin, 2006)
12. K. Gürlebeck, U. Kähler, On a boundary value problem of the biharmonic equation. Math. Meth. Appl. Sci. **20**, 867–883 (1997)
13. K. Gürlebeck, W. Sprößig, *Quaternionic Analysis and Elliptic Boundary Value Problems* (Birkhäuser Verlag, Basel, 1990)
14. K. Gürlebeck, W. Sprößig, *Quaternionic and Clifford Calculus for Engineers and Physicists* (Wiley, Chichester, 1997)
15. E. Obolashvili, *Partial Differential Equations in Clifford Analysis* (Addison Wesley Longman, Harlow, 1998)
16. H. Otto, Cauchy-Pompeiusche Integraldarstellungen in der Clifford Analysis. Ph.D. Thesis, FU Berlin, 2006
17. J. Ryan, Cauchy-Green type formulae in Clifford analysis. Tran. Amer. Math. Soc. **347**, 1331–1341 (1995)
18. J. Ryan, Basic Clifford analysis. Cubo Math. Educ. **2**, 226–256 (2000)
19. I.N. Vekua, *Generalized Analytic Functions* (Pergamon Press, Oxford, 1962)
20. Z.X. Zhang, Integral representations and its applications in Clifford analysis. General Math. **13**, 81–98 (2005)

A Note on the Schwarz Problem in a Ring Domain

A. Okay Çelebi and Pelin Ayşe Gökgöz

Abstract In this presentation we discuss the Schwarz problem in a ring domain. After the preliminaries we have taken the inhomogeneous Cauchy–Riemann equation with revised boundary conditions. In the next section we give the unique solution of the Schwarz problem for generalized Beltrami equation in a ring domain using Fredholm alternative.

Keywords Schwarz problem · Ring domain · Integral representations

Mathematics Subject Classification (2010) Primary 32A26, 32W50; Secondary 30E25

1 Introduction

Many researchers have been attracted by the boundary value problems in $\mathbb{C}$ in various types of domains, see for example [1–5] and the references in them. In this note we revisit the Schwarz problem in a ring domain which was studied previously, [7].

In the next section we collect the relevant information for this subject. Section 3 is reserved for a Schwarz problem for inhomogeneous Cauchy–Riemann equation in a ring domain having slightly different boundary conditions than [7] has. In the final section the Schwarz problem for generalized Beltrami equations are investigated.

A. O. Çelebi (✉) · P. A. Gökgöz
Department of Mathematics, Yeditepe University, Istanbul, Turkey
e-mail: acelebi@yeditepe.edu.tr; pelin.gokgoz@yeditepe.edu.tr

P. Cerejeiras et al. (eds.), *Current Trends in Analysis, its Applications and Computation*, Research Perspectives,
https://doi.org/10.1007/978-3-030-87502-2_4

2 Preliminaries

One of the main tools to derive the solutions of boundary value problems is to construct a suitable integral representation for a function $w \in \mathbb{C}(D)$. Starting with Gauss Theorem

$$\int_D w_{\overline{z}}(z)dxdy = \frac{1}{2i}\int_{\partial D} w(z)dz$$

where $w \in C^1(D;\mathbb{C}) \cap C(\overline{D};\mathbb{C})$ in a regular domain D, we obtain the Cauchy–Pompeiu representation

$$w(z) = \frac{1}{2\pi i}\int_{\partial D} w(\zeta)\frac{d\zeta}{\zeta - z} - \frac{1}{\pi}\int_D w_{\overline{\zeta}}(\zeta)\frac{d\xi d\eta}{\zeta - z}, \qquad z \in D \tag{2.1}$$

which may be used to solve Dirichlet type boundary value problems.

From Eq. (2.1) we may derive

$$\begin{aligned} w(z) = &\frac{1}{2\pi i}\int_{\partial \mathbb{D}} Rew(\zeta)\frac{\zeta + z}{\zeta - z}\frac{d\zeta}{\zeta} + iImw(0) \\ &- \frac{1}{2\pi}\int_{\mathbb{D}} \Big[\frac{w_{\overline{\zeta}}(\zeta)}{\zeta}\frac{\zeta + z}{\zeta - z} + \frac{\overline{w_{\overline{\zeta}}(\zeta)}}{\overline{\zeta}}\frac{1 + z\overline{\zeta}}{1 - z\overline{\zeta}}\Big]d\xi d\eta \end{aligned} \tag{2.2}$$

for the unit disc $\mathbb{D} \subset \mathbb{C}$. Equation (2.2) is called as Cauchy–Schwarz - Pompeiu formula which may be employed to derive the solution of Schwarz problem stated as:

"Find the function $w(z)$ satisfying

$$w_{\overline{z}} = f(z) \text{ in } \mathbb{D},\ \ Rew(z) = \gamma(z) \text{ on } \partial\mathbb{D} \text{ and } \frac{1}{2\pi i}\int_{|z|=1} Imw(\zeta)\frac{d\zeta}{\zeta} = c$$

where $f \in L^p(\mathbb{D})$, $p \geq 1$, $\gamma \in C(\partial\mathbb{D};\mathbb{R})$ and c is a given arbitrary number."

Besides the boundary value problem defined above, a variety of related boundary value problems are investigated in various types of domains. Several people have treated the problem in a ring domain $R = \{z : r < |z| < 1\}$, see for example Vaitekhovich [7, 8].

Our aim in this paper is to discuss the Schwarz problem in R which have slightly revised conditions compared with Vaitekhovich [7].

3 Schwarz Problem for Inhomogeneous Cauchy–Riemann Equations

We start with the Schwarz problem for homogeneous Cauchy–Riemann equations in a ring domain, which was defined as

"Find an analytic function w in R satisfying

$$Re w(z) = \gamma(z) \text{ on } \partial R, \ \frac{1}{2\pi i} \int_{|\zeta|=\rho} Im\, w(\zeta) \frac{d\zeta}{\zeta} = c$$

for $\gamma \in C(\partial R; \mathbb{R})$, $c \in \mathbb{R}$ given and for arbitrary ρ, $R = \{z : r < \rho < 1\}$"

by Vaitekhovich [7]. The unique solution is

$$w(z) = \frac{1}{2\pi i} \int_{\partial R} \gamma(\zeta) \left[\frac{\zeta + z}{\zeta - z} + 2 \sum_{n=1}^{\infty} \left(\frac{r^{2n}\zeta}{r^{2n}\zeta - z} + \frac{r^{2n} z}{\zeta - r^{2n} z} \right) \right] \frac{d\zeta}{\zeta}$$
$$- \frac{1}{2\pi i} \int_{|\zeta|=r} \gamma(\zeta) \frac{d\zeta}{\zeta} + ic$$

if and only if the solvability condition

$$\frac{1}{2\pi i} \int_{\partial R} \gamma(\zeta) \frac{d\zeta}{\zeta} = 0$$

holds [7]. Later, the substitution $\varphi = w - Tf$ is used to determine the solution of the inhomogeneous Cauchy–Riemann equation.

In this presentation, we define the Schwarz problem in a ring domain as "Find the solution of $w_{\bar{z}} = f(z)$ in R satisfying

$$Re w(z) = \gamma(z) \text{ on } \partial R, \ \frac{1}{2\pi i} \int_{|\zeta|=r} Im w(\zeta) \frac{d\zeta}{\zeta} = c$$

for $\gamma(\zeta) \in C(\partial R; \mathbb{R})$, $c \in \mathbb{R}$ given."

Using similar computations employed in [7], we get the integral representation

$$w(z) = \frac{1}{2\pi i}\int_{\partial R} Rew(\zeta)\left[\frac{\zeta+z}{\zeta-z} + K_1(z,\zeta)\right]\frac{d\zeta}{\zeta} - \frac{1}{2\pi i}\int_{|\zeta|=r} Rew(\zeta)\frac{d\zeta}{\zeta}$$
$$+\frac{1}{2\pi}\int_{|\zeta|=r} Imw(\zeta)\frac{d\zeta}{\zeta} - \frac{1}{2\pi}\int_R \frac{w_{\bar\zeta}(\zeta)}{\zeta}\left[\frac{\zeta+z}{\zeta-z}+1+K_1(z,\zeta)\right]d\xi d\eta$$
$$-\frac{1}{2\pi}\int_R \frac{\overline{w_{\bar\zeta}(\zeta)}}{\bar\zeta}\left[\frac{1+z\bar\zeta}{1-z\bar\zeta}+3+K_2(z,\zeta)\right]d\xi d\eta$$

where

$$K_1(z,\zeta) = 2\sum_{n=1}^{\infty}\left(\frac{r^{2n}\zeta}{r^{2n}\zeta - z} + \frac{r^{2n}z}{\zeta - r^{2n}z}\right),$$
$$K_2(z,\zeta) = 2\sum_{n=1}^{\infty}\left(\frac{r^{2n}}{r^{2n} - z\bar\zeta} + \frac{r^{2n}z\bar\zeta}{1 - r^{2n}z\bar\zeta}\right).$$

The existence and uniqueness theorem for Schwarz problem in $R = \{z : r < |z| < 1\}$ is

Theorem 3.1 *The Schwarz problem defined in R has the unique solution*

$$w(z) = \frac{1}{2\pi i}\int_{\partial R} \gamma(\zeta)\left[\frac{\zeta+z}{\zeta-z} + K_1(z,\zeta)\right]\frac{d\zeta}{\zeta} - \frac{1}{2\pi i}\int_{|\zeta|=r} \gamma(\zeta)\frac{d\zeta}{\zeta}$$
$$+\frac{1}{2\pi}\int_{|\zeta|=r} Imw(\zeta)\frac{d\zeta}{\zeta} - \frac{1}{2\pi}\int_R \frac{f(\zeta)}{\zeta}\left[\frac{\zeta+z}{\zeta-z}+1+K_1(z,\zeta)\right]d\xi d\eta$$
$$-\frac{1}{2\pi}\int_R \frac{\overline{f(\zeta)}}{\bar\zeta}\left[\frac{1+z\bar\zeta}{1-z\bar\zeta}+3+K_2(z,\zeta)\right]d\xi d\eta \tag{3.1}$$

subject to the solvability condition

$$\frac{1}{2\pi i}\int_{|\zeta|=r} \gamma(\zeta)\frac{d\zeta}{\zeta} = -\frac{3}{2\pi}\int_R \left[\frac{f(\zeta)}{\zeta} + \frac{\overline{f(\zeta)}}{\bar\zeta}\right]d\xi d\eta.$$

Proof It is trivial that Eq. (3.1) satisfies the inhomogeneous Cauchy–Riemann equation. Thus we need to check the boundary conditions only. Hence we restrict

$$\begin{aligned}
2Rew(z) &= \frac{1}{2\pi i}\int_{\partial R}\gamma(\zeta)\left[\frac{\zeta+z}{\zeta-z}+\frac{\overline{\zeta}+\overline{z}}{\overline{\zeta}-\overline{z}}+K_1(z,\zeta)+\overline{K_1(z,\zeta)}\right]\frac{d\zeta}{\zeta}\\
&-\frac{1}{2\pi}\int_R\frac{f(\zeta)}{\zeta}\left[\frac{\zeta+z}{\zeta-z}+\frac{1+\overline{z}\zeta}{1-\overline{z}\zeta}+4+K_1(z,\zeta)+\overline{K_2(z,\zeta)}\right]d\xi d\eta\\
&-\frac{1}{2\pi}\int_R\frac{\overline{f(\zeta)}}{\overline{\zeta}}\left[\frac{\overline{\zeta}+\overline{z}}{\overline{\zeta}-\overline{z}}+\frac{1+z\overline{\zeta}}{1-z\overline{\zeta}}+4+\overline{K_1(z,\zeta)}+K_2(z,\zeta)\right]d\xi d\eta\\
&-\frac{2}{2\pi i}\int_{|\zeta|=r}\gamma(\zeta)\frac{d\zeta}{\zeta}
\end{aligned} \tag{3.2}$$

So Eq. (3.2) can be written as

$$\begin{aligned}
2Rew(z)\Big|_{\partial R} &= \left\{\frac{1}{2\pi i}\int_{\partial R}\gamma(\zeta)\left[\frac{2\zeta}{\zeta-z}+\frac{2\overline{\zeta}}{\overline{\zeta}-\overline{z}}-2+K_1(z,\zeta)+\overline{K_1(z,\zeta)}\right]\frac{d\zeta}{\zeta}\right.\\
&-\frac{1}{2\pi}\int_R\frac{f(\zeta)}{\zeta}\left[\frac{\zeta+z}{\zeta-z}+\frac{z+|z|^2\zeta}{z-|z|^2\zeta}+4\right.\\
&\left.+K_1(z,\zeta)+\overline{K_2(z,\zeta)}\right]d\xi d\eta\\
&-\frac{1}{2\pi}\int_R\frac{\overline{f(\zeta)}}{\overline{\zeta}}\left[\frac{\overline{\zeta}z+|z|^2}{\overline{\zeta}z-|z|^2}+\frac{1+\overline{\zeta}z}{1-\overline{\zeta}z}+4\right.\\
&\left.\left.+\overline{K_1(z,\zeta)}+K_2(z,\zeta)\right]d\xi d\eta\right\}\Big|_{\partial R}\\
&-\frac{2}{2\pi i}\int_{|\zeta|=r}\gamma(\zeta)\frac{d\zeta}{\zeta}
\end{aligned} \tag{3.3}$$

Recalling the argument given by Vaitekhovich [7] we get

$$2Rew(z)\Big|_{\partial R} = 2\gamma-\frac{2}{2\pi i}\int_{|\zeta|=r}\gamma(\zeta)\frac{d\zeta}{\zeta}-\frac{3}{\pi}\int_R\left(\frac{f(\zeta)}{\zeta}+\frac{\overline{f(\zeta)}}{\overline{\zeta}}\right)d\xi d\eta.$$

Hence the boundary condition $Re w(z) = \gamma(z)$ on ∂R holds if

$$\frac{1}{2\pi i}\int\limits_{|\zeta|=r} \gamma(\zeta)\frac{d\zeta}{\zeta} = -\frac{3}{2\pi}\int\limits_{R}\left(\frac{f(\zeta)}{\zeta} + \frac{\overline{f(\zeta)}}{\overline{\zeta}}\right) d\xi d\eta.$$

in which the boundary integral is in the clock-wise direction. □

4 Schwarz Problem for Generalized Beltrami Equations in a Ring Domain

In this section we derive the solution of the Schwarz problem for Beltrami equation

$$w_{\overline{z}} + A_1(z)w_z + A_2(z)\overline{w_z} + B_1(z)w + B_2(z)\overline{w} = f(z) \tag{4.1}$$

having homogeneous boundary conditions. Following the usual approach, we need to convert Eq. (4.1) into an integral equation. Now we point out two integral operators and their properties.

4.1 The Operators T_R and Π_R

Firstly we define the operator

$$T_R f(z) := -\frac{1}{2\pi}\int\limits_{R}\left\{\frac{f(\zeta)}{\zeta}\left[\frac{\zeta+z}{\zeta-z} + 1 + K_1(z,\zeta)\right]\right.$$
$$\left. + \frac{\overline{f(\zeta)}}{\overline{\zeta}}\left[\frac{1+z\overline{\zeta}}{1-z\overline{\zeta}} + 3 + K_2(z,\zeta)\right]\right\} d\xi d\eta$$

This operator satisfies the properties [2]

$$\partial_{\overline{z}} T_R f(z) = f(z),$$
$$Re T_R f(z) = 0, \quad z \in \partial R,$$

The operator $T_R f(z)$ may be decomposed as

$$T_R f(z) := T_{1R} f(z) + T_{2R} f(z)$$

where

$$T_{1R}f(z) = -\frac{1}{2\pi}\int\limits_R \left(\frac{f(\zeta)}{\zeta}\frac{\zeta+z}{\zeta-z} + \frac{\overline{f(\zeta)}}{\overline{\zeta}}\frac{1+z\overline{\zeta}}{1-z\overline{\zeta}}\right) d\xi d\eta$$

and

$$T_{2R}f(z) = -\frac{1}{2\pi}\int\limits_R \left(\frac{f(\zeta)}{\zeta}(1+K_1(z,\zeta)) + \frac{\overline{f(\zeta)}}{\overline{\zeta}}(3+K_2(z,\zeta))\right) d\xi d\eta.$$

We know that T_{1R} is a compact operator [2]. Since $K_1(z,\zeta)$ and $K_2(z,\zeta)$ are analytic in R, T_{2R} is also compact. Thus T_R is a compact operator.

Secondly we define

$$\Pi_R f(z) := \partial_z T_R f(z)$$

which may be written as

$$\Pi_R f(z) := \Pi_{1R} f(z) + \Pi_{2R} f(z)$$

where

$$\Pi_{1R} f(z) := \partial_z T_{1R} f(z),$$
$$\Pi_{2R} f(z) := \partial_z T_{2R} f(z).$$

Hence these operators are given by

$$\Pi_{1R} f(z) = -\frac{1}{\pi}\int\limits_R \left(\frac{f(\zeta)}{(\zeta-z)^2} + \frac{\overline{f(\zeta)}}{(1-z\overline{\zeta})^2}\right) d\xi d\eta$$

and

$$\Pi_{2R} f(z) = -\frac{1}{2\pi}\int\limits_R \left[\frac{f(\zeta)}{\zeta}\partial_z K_1(z,\zeta) + \frac{\overline{f(\zeta)}}{\overline{\zeta}}\partial_z K_2(z,\zeta)\right] d\xi d\eta$$

Π_{1R} is a strongly singular operator of Calderon - Zygmund type. It is known that [9], Π_{1R} is a bounded operator. On the other hand, $K_1(z,\zeta)$ and $K_2(z,\zeta)$ are bounded analytic functions in R by Montel theorem [[6], Thm. 9.12]. Hence the operarator Π_{2R} is also bounded. Hence Π_R is a bounded operator.

4.2 Solution of a Generalized Beltrami Equation

Now we may prove the existence and uniqueness theorem for the solutions of Schwarz problem.

Theorem 4.1 *There exits a unique solution $w \in W^{p,1}(R)$ of the Schwarz problem for Eq.* (4.1) *subject to the homogeneous boundary conditions*

$$Rew(z) = 0 \text{ on } \partial R, \quad \frac{1}{2\pi i} \int_{|\zeta|=r} Imw(\zeta)\frac{d\zeta}{\zeta} = 0$$

where $A_1(z)$, $A_2(z)$ are bounded measurable functions with

$$|A_1(z)| + |A_2(z)| \leq q_0 < 1;$$

satisfying

$$q_0\|\widehat{\Pi}_R\|_{L^p(R)} < 1$$

and $B_1, B_2, f \in L^p(R)$, $p > 2$.

Proof Assume that $w = T_R g$, $g \in L^p(R)$ is a solution of Eq. (4.1). Then we observe that g should be the solution of

$$(I + \widehat{\Pi}_R + \widehat{K}_R)g(z) = f(z) \tag{4.2}$$

where

$$\widehat{\Pi}_R g = A_1(z)\Pi_{1R} g + A_2(z)\Pi_{2R} g$$

$$\widehat{K}_R g = B_1(z)T_{1R} g + B_2(z)T_{2R} g$$

Thus we have converted Eq. (4.1) into an integral equation. As in [2] $\widehat{K}_R$ is a compact operator by Arzela–Ascoli theorem. $\widehat{\Pi}_R$ is a bounded strongly singular operator of Calderon–Zygmund type. Let us note that $I + \widehat{\Pi}_R$ is invertible if there exists q_0 such that

$$q_0\|\widehat{\Pi}_R\|_{L^p(R)} < 1 \text{ for } p > 2.$$

So $I + \widehat{\Pi}_R + \widehat{K}_R$ is a Fredholm operator with index zero. Hence Eq. (4.1) has a unique solution of the form $w = T_R g$ where $g \in L^p(R)$, $p > 2$ is a solution of Eq. (4.2). □

References

1. Ü. Aksoy, A.O. Çelebi, Dirichlet problem for a generalized inhomogeneous polyharmonic equation in an annular domain. Comp. Variab. Ellip. Equ. **57**, 229–241 (2012)
2. Ü. Aksoy, A.O. Çelebi, Schwarz problem for higher-order complex elliptic partial differential equations. Integ. Trans. Spec. Funct. **19**, 413–428 (2008)
3. Ü. Aksoy, H. Begehr, A.O. Çelebi, Schwarz problem for higher-order complex partial differential equations in the upper half plane. Math. Nachr **292**, 1183–1193 (2019)
4. H. Begehr, Integral representations in complex, hypercomplex and Clifford analysis. Integ. Trans. Spec. Funct. **13**, 223–241 (2002)
5. H. Begehr, T. Vaitekhovich, Schwarz problem in lens and lune. Comp. Variab. Ellip. Equ. **59**, 76–84 (2014)
6. J. Bruna, J. Cufi, *Complex Analysis*. EMS Textbooks in Mathematics (European Mathematical Society, Zürich, 2013)
7. T. Vaitekhovich, Boundary value problems to first-order complex partial differential equations. Integ. Trans. Spec. Funct. **19**, 211–233 (2008)
8. T. Vaitekhovich, Boundary value problems to second order complex partial differential equations in a ring domain. Siauliai Math. Semin. **2**, 117–146 (2007)
9. I.N. Vekua, *Generalized Analytic Functions* (Pergamon Press, Oxford, 1962)

On Hierarchical Models of Elastic Shallow Shells with Voids

Bakur Gulua

Abstract In the report the three-dimensional system of equations of equilibrium for solids with voids is considered. From this system of equations, using a reduction method of I. Vekua, we receive the equilibrium equations for the shallow shells. Further we consider the case of plates with constant thickness in more detail. Namely, the systems of equations corresponding to approximations $N = 0$ is written down in a complex form and we express the general solutions of these systems through analytic functions of complex variable and solutions of the Helmholtz equation. The received general representations give the opportunity to solve analytically boundary value problems.

Keywords Hierarchical models · Materials with voids · Shells

Mathematics Subject Classification (2010) 74K25, 74F05

1 Introduction

I. Vekua constructed hierarchical models for isotropic prismatic and standard shells in [1] and [2], respectively. All his results in this direction are summarized in [3, 4]. Vashakmadze [5] deals with hierarchical models for elastic anisotropic plates. In [6, 7] the results of I. Vekua are extended to the case of geometrically and physically nonlinear non-shallow elastic shells. To cusped prismatic shells (see [1], Chapter 2, Section 13; [2], Chapter 2, Section 2; [3], Section 1.9), in particular plates, are devoted [8, 9]. Jaiani [10] is devoted to hierarchical models of piezoelastic Kelvin-Voight cusped prismatic shells with voids, in particular these models contain

B. Gulua (✉)
Sokhumi State University, Tbilisi, Georgia

I. Vekua Institute of Applied Mathematics, I. Javakhishvili Tbilisi State University, Tbilisi, Georgia

P. Cerejeiras et al. (eds.), *Current Trends in Analysis, its Applications and Computation*, Research Perspectives,
https://doi.org/10.1007/978-3-030-87502-2_5

hierarchical models for elastic anisotropic and isotropic prismatic shells with voids. In Section 7 of [10] for N=0 approximation porous isotropic prismatic shells in particular plates are studied, included cusped ones and plates of constant thickness. In [11] vibration problem of a cusped elastic prismatic shell in case of the third model of Vekua's hierarchical model is studied. In [12], authors construction Vekua's hierarchical model for the shallow shells having double porosity. Relation of Vekua's hierarchical models to the corresponding 3D boundary value problems and their accuracy are studied in [13–16].

The physic-mathematical foundations of the linear theory of elastic materials with voids or empty pores and applications of this theory to some technological problems were originally proposed in the work Cowin and Nunziato [17]. Such materials include, in particular, rocks and soils, granulated and some other manufactured porous materials.

This theory differs essentially from the classical theory of elasticity in that the volume fraction corresponding to the void volume is considered as an independent variable. Voids have no mechanical or energetic meaning. In recent years, problems of elasticity for materials with voids were investigated by many authors [18–21].

The aim of the paper is to construct Hierarchical models for shallow standard shells with voids, applying I. Vekuas dimension reduction method. In particular, when the curvature $k = 0$, we get hierarchical models for plates with voids. In the $N = 0$ approximation of the plates of the constant thickness with voids we construct the complex representations of the general solution governing elliptic systems of equations.

2 Basic Three-Dimensional Relations

Let an elastic body with voids occupy the domain $\overline{\Omega} \in R^3$ Denote by x^1, x^2, x^3 a point of the domain $\overline{\Omega}$ in the arbitrary curvilinear system of coordinates. Let the domain $\overline{\Omega}$ be filled with an elastic isotropic homogenous medium with voids. The considered solid body is characterized by the displacement vector $\mathbf{u} = \left(u^1, u^2, u^3\right)$ and also by the function ϕ which is the change in volume fraction from the reference volume fraction.

Then a system of static equilibrium equations is written in the form [17, 18]

$$\overset{\circ}{\nabla}_i \sigma^{ij} + \Phi^j = 0, \tag{2.1}$$

$$\overset{\circ}{\nabla}_i h^i + g + F = 0, \tag{2.2}$$

where $\overset{\circ}{\nabla}_i$ are covariant derivatives with respect to the space coordinates x^i, σ^{ij} are contravariant components of stress tensor, Φ^j are contravariant components of volume forces $\mathbf{\Phi}$, h^i are is the component of the equilibrated stress vector, g and F are the intrinsic and extrinsic equilibrated volume forces.

Below we use the summation rule with respect to the dummy index, assuming that the Latin indices run through the values 1, 2, 3. It is also agreed that if the indices are denoted by Greek letters, they run through the values 1, 2.

Formulas that interrelate the stress components, the displacement vector components and the function ϕ have the form

$$\sigma^{ij} = \lambda e_k^k g^{ij} + 2\mu e^{ij} + \beta\phi g^{ij}, \tag{2.3}$$

$$h^i = \alpha \overset{\circ}{\nabla}{}^i \phi, \tag{2.4}$$

$$g = -\xi\phi - \beta e_k^k, \tag{2.5}$$

where λ and μ are the Lamé constants, $\alpha,\ \beta,\ \xi$ are the constants characterizing the body porosity, g^{ij} are the contravariant components of spatial metric tensor, e^{ij} are contravariant components of deformation tensor

$$e^{ij} = \frac{1}{2}\left(\overset{\circ}{\nabla}{}^i u^j + \overset{\circ}{\nabla}{}^j u^i\right),$$

$\overset{\circ}{\nabla}{}^i = g^{ij}\overset{\circ}{\nabla}_j$, the contravariant and covariant components of a vector of displacement are connected by a relation $u^i = g^{ij}u_j$.

3 Construction of Hierarchical Models for Elastic Shallow Shells with Voids

In 1955 Ilia Vekua published his models of elastic prismatic shells. In 1965 he offered analogous models for standard shells [1, 2].

Let Ω represent a shell with constant thickness $2h$, symmetric concerning the middle surface ω. ω are smooth bilateral surface. We will denote set of side surfaces of a shell through Γ. Surfaces of ω and Γ in each point are crossed at right angle. We assume that thickness $2h$ is much less in comparison with other sizes of a shell.

We will consider the coordinate system which is normal connected with a middle of surface. In this system the radius vector $\mathbf{R}$ of any point M of domain Ω is expressed by means of a formula (see Fig. 1)

$$\mathbf{R}(x^1, x^2, x^3) = \mathbf{r}(x^1, x^2) + x^3\mathbf{n}(x^1, x^2),$$

where $x^1,\ x^2$ are Gaussian parameters of surface ω; $\mathbf{r}$ and $\mathbf{n}$ are radius vector and normal of the point $\left(x^1, x^2\right) \in \omega$. x^3 is the relative length from point M to the surface ω.

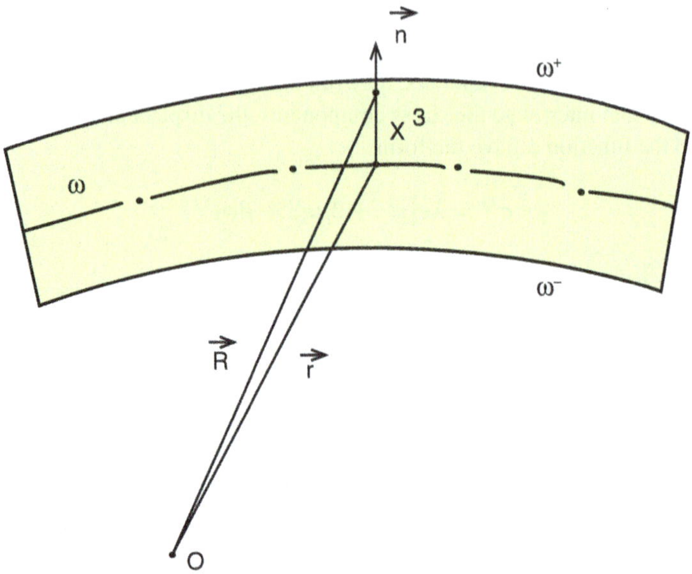

Fig. 1 Special coordinate system for shells

We apply I. Vekua's method to a reduction of the equations (2.1-2.5). We accept the following assumptions of geometrical character

$$1 - k_\alpha x^3 \cong 1, \quad -h \leq x^3 \leq h. \tag{3.1}$$

These requirements mean that or main curvature k_1 and k_2 of a surface are small (shallow shell), or thickness of shell is small (thin shell). If ω is the plane then $k_1 = k_2 = 0$ and conditions (3.1) are exactly satisfied. From assumptions (3.1) follows that spatial covariant $\mathbf{R}_\alpha$ and contravariant $\mathbf{R}^\alpha$ basis vectors are approximately equal to the corresponding basis vectors of a middle surface $\mathbf{r}_\alpha$, $\mathbf{r}^\alpha$. Therefore, also corresponding covariant and contravariant components and discriminants of metric tensors of space and a middle surface are approximately equal

$$\mathbf{R}_\alpha \cong \mathbf{r}_\alpha, \quad \mathbf{R}^\alpha \cong \mathbf{r}^\alpha, \quad \mathbf{R}^3 = \mathbf{r}_3 = \mathbf{n},$$

$$g_{\alpha\beta} \cong a_{\alpha\beta}, \quad g^{\alpha\beta} \cong a^{\alpha\beta}, \quad g \cong a, \quad \partial_3\sqrt{g} \cong -2H\sqrt{a},$$

where $a_{\alpha,\beta} = \mathbf{r}_\alpha \mathbf{r}_\beta$, $a^{\alpha,\beta} = \mathbf{r}^\alpha \mathbf{r}^\beta$, a is the discriminant of quadratic form of surface ω, $H = \frac{1}{2}(k_1 + k_2)$ are middle curvatures of the midsurface ω, $b_{\alpha\beta}$, b_α^β are covariant and mixed components of the tensor of curvature of the midsurface ω.

Further we carry out a reduction of system of the equations (2.1-2.5) using method of Vekua. At the same time we almost repeat verbatim reasonings in the monograph. Therefore we will present the reduced two-dimensional equations at once without details of derivation

$$\begin{cases} \nabla_\alpha \overset{(k)}{\sigma}{}^{\alpha\beta} - b^\beta_\alpha \overset{(k)}{\sigma}{}^{\alpha 3} - \dfrac{2k+1}{h}\left(\overset{(k-1)}{\sigma}{}^{3\beta} + \overset{(k-3)}{\sigma}{}^{3\beta} + \cdots\right) + \overset{(k)}{Q}{}^{\beta} = 0, \\ \nabla_\alpha \overset{(k)}{\sigma}{}^{\alpha 3} + b^\beta_{\alpha\beta} \overset{(k)}{\sigma}{}^{\alpha\beta} - \dfrac{2k+1}{h}\left(\overset{(k-1)}{\sigma}{}^{33} + \overset{(k-3)}{\sigma}{}^{33} + \cdots\right) + \overset{(k)}{Q}{}^{3} = 0, \\ \nabla_\alpha \overset{(k)}{h}{}^{\alpha} - \dfrac{2k+1}{h}\left(\overset{(k-1)}{h}{}^{3} + \overset{(k-3)}{h}{}^{3} + \cdots\right) + \overset{(k)}{g} + \overset{(k)}{Q} = 0, \end{cases} \tag{3.2}$$

where $k = 0, 1, \ldots$, ∇_α and ∇^α are symbols of a covariant and contravariant derivatives on the midsurface ω and

$$\overset{(k)}{\sigma}{}^{ij} = \frac{2k+1}{h} \int_{-h}^{h} \sigma^{ij}(x^1, x^2, x^3) P_k\left(\frac{x^3}{h}\right) dx^3,$$

$$\left(\overset{(k)}{h}{}^{i}, \overset{(k)}{g}\right) = \frac{2k+1}{h} \int_{-h}^{h} \left(h^i(x^1, x^2, x^3), g(x^1, x^2, x^3)\right) P_k\left(\frac{x^3}{h}\right) dx^3,$$

$$\overset{(k)}{Q}{}^{j} = \frac{2k+1}{h}\left(\sigma^{3j}(x^1, x^2, h) - (-1)^k \sigma^{3j}(x^1, x^2, -h)\right) + \overset{(k)}{\Phi}{}^{j},$$

$$\overset{(k)}{Q} = \frac{2k+1}{h}\left(h^3(x^1, x^2, h) - (-1)^k h^3(x^1, x^2, -h)\right) + \overset{(k)}{F},$$

$$\left(\overset{(k)}{\Phi}{}^{i}, \overset{(k)}{F}\right) = \frac{2k+1}{h} \int_{-h}^{h} \left(\Phi^i(x^1, x^2, x^3), F(x^1, x^2, x^3)\right) P_k\left(\frac{x^3}{h}\right) dx^3.$$

$\sigma^{3j}(x^1, x^2, \pm h)$ are components of the stress vectors acting on the upper and lower face surfaces and $h^3(x^1, x^2, \pm h)$ are components of the equilibrated stress vectors on the upper and lower face surfaces. Here $P_k\left(\dfrac{x^3}{h}\right)$ is the Legandre polynomials of order k. Then

$$\overset{(k)}{\sigma}{}^{\alpha\beta} = \lambda\left(\nabla_\gamma \overset{(k)}{u}{}^{\gamma} - 2H \overset{(k)}{u}{}^{3} + \overset{(k)}{u}{}'^{3}\right) a^{\alpha\beta}$$

$$+\mu\left(\nabla^\beta \overset{(k)}{u}{}^{\alpha} + \nabla^\alpha \overset{(k)}{u}{}^{\beta} - 2b^{\alpha\beta} \overset{(k)}{u}{}^{3}\right) + \beta \overset{(k)}{\phi} a^{\alpha\beta},$$

$$\overset{(k)}{\sigma}{}^{\alpha 3} = \mu\left(\nabla^\alpha \overset{(k)}{u}{}^{3} + b^\alpha_\gamma \overset{(k)}{u}{}^{\gamma} + \overset{(k)}{u}{}'^{\alpha}\right),$$

$$\overset{(k)}{\sigma}{}^{33} = \lambda \left(\nabla_\gamma \overset{(k)}{u}{}^\gamma - 2H\overset{(k)}{u}{}^3 + \overset{(k)}{u}{}'^3 \right) + 2\mu \overset{(k)}{u}{}'^3 + \beta \overset{(k)}{\phi},$$

$$\overset{(k)}{h}{}^\alpha = \alpha \nabla_\alpha \overset{(k)}{\phi}, \quad \overset{(k)}{h}{}^3 = \frac{\alpha}{h} \overset{(k)}{\phi}{}', \tag{3.3}$$

$$\overset{(k)}{g} = -\xi \overset{(k)}{\phi} - \beta \left(\nabla_\gamma \overset{(k)}{u}{}^\gamma - 2H\overset{(k)}{u}{}^3 + \overset{(k)}{u}{}'^3 \right),$$

where

$$\left(\overset{(k)}{u}{}^i, \overset{(k)}{\phi} \right) = \frac{2k+1}{h} \int\limits_{-h}^{h} \left(u^i(x^1, x^2, x^3), \phi(x^1, x^2, x^3) \right) P_k \left(\frac{x^3}{h} \right) dx^3,$$

$$\overset{(k)}{u}{}' = (2k+1) \left(\overset{(k+1)}{u} + \overset{(k+3)}{u} + \cdots \right).$$

If we substitute relations (3.3) into equations (3.2) we arrive at an infinite system of second-order equations with respect to components of vector $\overset{(k)}{\mathbf{u}}$ and $\overset{(k)}{\phi}$. As a result of simple computations the derived system of equations has the form

$$\nabla_\alpha \nabla^\alpha \overset{(k)}{u}{}^\beta + (\lambda + \mu) \nabla^\beta \nabla_\alpha \overset{(k)}{u}{}^\alpha + \nabla^\beta \overset{(k)}{\phi} + \overset{(k)}{M}{}^\beta + \overset{(k)}{Q}{}^\beta = 0,$$

$$\nabla_\alpha \nabla^\alpha \overset{(k)}{u}{}^3 + 2H \overset{(k)}{\phi} + \overset{(k)}{M}{}^3 + \overset{(k)}{Q}{}^3 = 0, \tag{3.4}$$

$$\nabla_\alpha \nabla^\alpha \overset{(k)}{\phi} - \xi \overset{(k)}{\phi} + \overset{(k)}{M} + \overset{(k)}{Q} = 0,$$

where $\overset{(k)}{M}{}^i$, $\overset{(k)}{M}$ are linear operators which contain unknown functions $\overset{(k)}{u}{}^i$, $\overset{(k)}{\phi}$. and their first-order derivatives.

An infinite system of equations (3.4) has the advantage that it contains two independent variables—Gaussian coordinates x^1, x^2 of the midsurface. But the decrease in the number of independent variables is achieved by increasing the number of equations to infinity, which, naturally, has an obvious practical inconvenience. Therefore it is necessary to make the next step for a further simplification of the problem.

In order to obtain the finite system of equations we accept the assumption

$$\left(\mathbf{u}\left(x^1, x^2, x^3\right), \phi\left(x^1, x^2, x^3\right) \right) = \sum_{k=0}^{N} \left(\overset{(k)}{\mathbf{u}}(x^1, x^2), \overset{(k)}{\boldsymbol{\phi}}(x^1, x^2) \right) P_k \left(\tfrac{x^3}{h} \right).$$

Thus, in all expressions (3.4) received above we will consider that $\overset{(k)}{\mathbf{u}} = 0$ and $\overset{(k)}{\phi} = 0$ when $k > N$. In addition, we obtain the Nth order approximation (hierarchical model) governing system consisting of $4N + 4$ equations with respect to $4N + 4$ unknown functions.

4 The $N = 0$ Approximation for the Plate

If we substitute $k = 0$ and assume all the mathematical moments of order greater than zero to be equal to zero, from (3.4) we obtain the basic relations of the $N = 0$ approximation of elastic isotropic plates with voids

$$\begin{aligned}
&\Delta \overset{(0)}{u_1} + (\lambda + \mu)\partial_1 \overset{(0)}{\theta} + \beta \partial_1 \overset{(0)}{\phi} = 0,\\
&\Delta \overset{(0)}{u_2} + (\lambda + \mu)\partial_2 \overset{(0)}{\theta} + \beta \partial_2 \overset{(0)}{\phi} = 0,\\
&\Delta \overset{(0)}{u_3} = 0,\\
&(\alpha\Delta - \xi)\Delta \overset{(0)}{\phi} - \beta \overset{(0)}{\theta} = 0,
\end{aligned} \tag{4.1}$$

where $\overset{(0)}{\theta} = \partial_1 \overset{(0)}{u_1} + \partial_1 \overset{(0)}{u_1}$, $\Delta \equiv \partial_{11} + \partial_{22}$ is the Laplace operator in two dimensions.

Assume: $\overset{(0)}{u_1} \equiv u_1$, $\overset{(0)}{u_2} \equiv u_2$, $\overset{(0)}{\phi} \equiv \phi$.

On the plane Ox_1x_2, we introduce the complex variable $z = x_1 + ix_2 = re^{i\vartheta}$, $(i^2 = -1)$ and the operators $\partial_z = 0.5(\partial_1 - i\partial_2)$, $\partial_{\bar{z}} = 0.5(\partial_1 + i\partial_2)$, $\bar{z} = x_1 - ix_2$, and $\Delta = 4\partial_z\partial_{\bar{z}}$.

To write system (4.1) in the complex form, the second equation of this system is multiplied by and summed up with the first equation

$$\begin{aligned}
&2\mu\partial_{\bar{z}}\partial_z u_+ + (\lambda + \mu)\partial_{\bar{z}}\theta + \beta\partial_{\bar{z}}\phi = 0,\\
&4\mu\partial_{\bar{z}}\partial_z u_3 = 0,\\
&(\alpha\Delta - \xi)\phi - \beta\theta = 0,
\end{aligned} \tag{4.2}$$

where $u_+ = u_1 + iu_2, \theta = \partial_z u_+ + \partial_{\bar{z}}\bar{u}_+$.

Theorem 4.1 *The general solution of the system of equations (4.2) is represented as follows:*

$$\begin{aligned}
&2\mu u_+ = \kappa\varphi(z) - z\overline{\varphi'(z)} - \overline{\psi(z)} - \frac{4\alpha\beta\mu}{\xi(\lambda + 2\mu) - \beta^2}\partial_{\bar{z}}\chi(z, \bar{z}),\\
&2\mu u_3 = f'(z) + \overline{f'(z)},\\
&\phi = \chi(z, \bar{z}) - \frac{\beta}{\xi(\lambda + \mu) - \beta^2}(\varphi'(z) + \overline{\varphi'(z)}),
\end{aligned}$$

where $\kappa = \frac{\xi(\lambda+3\mu)-\beta^2}{\xi(\lambda+\mu)-\beta^2}$, $\varphi(z)$ *and* $\psi(z)$ *and* $f(z)$ *are arbitrary analytic functions of a complex variable* z *in the domain* V, $\chi(z,\bar{z})$ *is an arbitrary solution of the Helmholtz equation*

$$\Delta\chi(z,\bar{z}) - \frac{\xi(\lambda+2\mu)-\beta^2}{\alpha(\lambda+2\mu)}\chi(z,\bar{z}) = 0.$$

The received general representations give the opportunity to solve analytically boundary value problems.

Acknowledgments This work is supported by the Ministry of Education and Culture of the Autonomous Republic of Abkhazia.

References

1. I. Vekua, On a way of calculating of prismatic shells, in *Proceedings of A. Razmadze Institute of Mathematics of Georgian Academy of Sciences*, vol. 21 (1995), pp. 191–259
2. I. Vekua, The theory of thin shallow shells of variable thickness, in *Proceedings of A. Razmadze Institute of Mathematics of Georgian Academy of Sciences*, vol. 30 (1965), pp. 5–103
3. I. Vekua, *Shell Theory: General Methods of Construction* (Pitman Advanced Publishing Program, Boston-London, 1985)
4. I. Vekua, Some general methods of constructing of various versions of theory of shells. Nauka, Moscow (Russian) (1982) 286 p.
5. T. Vashakmadze, The theory of anisotropic elastic plates, in *Mathematics and Its Applications* (Kluwer Academic Publishers, Dordrecht, 1999)
6. T. Meunargia, On one method of construction of geometrically and physically nonlinear theory of non-shallow shells. Proc. A. Razmadze Math. Inst. **119**, 133–154 (1999)
7. T. Meunargia, Some general methods for constructing the theory of shells. Differential equations and their applications. J. Math. Sci. (N.Y.) **157**(1), 1–15 (2009)
8. G. Jaiani, S. Kharibegashvili, D. Natroshvili, W.L. Wendland, Hierarchical models for elastic cusped plates and beams. Lect. Notes TICMI, **4**, 86 (2003)
9. N. Chinchaladze, On some nonclassical problems for differential equations and their applications to the theory of cusped prismatic shells. Lect. Notes TICMI, **9**, 92 (2008)
10. G. Jaiani, Piezoelectric viscoelastic Kelvin-Voigt cusped prismatic shells. Lect. Notes TICMI **19**, 83 (2018)
11. N. Chinchaladze, On one problem of a cusped elastic prismatic shells in case of the third model of Vekua's hierarchical model. Hacet. J. Math. Stat. **45**(6), 1665–1673 (2016)
12. B. Gulua, R. Janjgava, M. Narmania, Derivation of system of the equations of equilibrium for shallow shells and plates, having double porosity. Appl. Math. Inform. Mech. **21**(2), 16–37 (2016)
13. D.G. Gordeziani, The solvability of certain boundary value problems for a variant of the theory of thin shells. (Russian) Dokl. Akad. Nauk SSSR **215**, 1289–1292 (1974)
14. D.G. Gordeziani, The accuracy of a certain variant of the theory of thin shells. (Russian) Dokl. Akad. Nauk SSSR **216**, 751–754 (1974)
15. D. Gordeziani, M. Avalishvili, Investigation of dynamical two-dimensional model of prismatic shell. Bull. Georgian Acad. Sci. **166**(1), 16–19 (2002)
16. D. Gordeziani, M. Avalishvili, G. Avalishvili, Dynamical hierarchical models for elastic shells. Appl. Math. Inform. Mech. **10**(1), 19–38 (2005)

17. S.C. Cowin, J.W. Nunziato, Linear elastic materials with voids. J. Elasticity **13**, 125–147 (1983)
18. M. Ciarletta, A. Scalia, M. Svanadze, Fundamental solution in the theory of micropolar thermoelasticity for materials with voids. J. Thermal Stresses **30**(3), 213–229 (2007)
19. L. Bitsadze, The Neumann BVP of the linear theory of thermoelasticity for the sphere with voids and microtemperatures. Semin. I. Vekua Inst. Appl. Math. Rep. **44**, 15–29 (2018)
20. I. Tsagareli, Explicit solution of elastostatic boundary value problems for the elastic circle with voids. Adv. Math. Phys. **2018**, Art. ID 6275432, 6 (2018)
21. B. Gulua, R. Janjgava, On construction of general solutions of equations of elastostatic problems for the elastic bodies with voids. PAMM **18**(1), e201800306. https://doi.org/10.1002/pamm.201800306

Well-Posedness and Numerical Results to 3D Periodic Burgers' Equation in Lebesgue-Gevrey Class

Ridha Selmi and Abdelkerim Chaabani

Abstract We prove that a unique global in time solution to the three dimensional periodic Burger's equation exists, in the Lebesgue-Gevrey class. Also, we establish that the long time to this solution is determined by a finite number of Fourier modes; this is useful as a numerical result. Energy methods, compactness methods, maximum principle and Fourier analysis are the main tools.

Keywords Burgers equation · Well-posedness · Determining modes

Mathematics Subject Classification (2010) Primary 35A01, 35A02; Secondary 35B05, 35B10

1 Introduction

The three dimensional diffusive Burgers' equation is given by

$$\partial_t u - \nu \Delta u + (u \cdot \nabla) u = 0, \quad (t, x) \in \mathbb{R}_+ \times \mathbb{T}^3 \tag{1.1}$$

$$u|_{t=0} = u_0(x), \quad x \in \mathbb{T}^3, \tag{1.2}$$

This work was completed with the support of our TEX-pert.

R. Selmi (✉)
Department of Mathematics, Northern Border University, Arar, KSA
e-mail: ridha.selmi@nbu.edu.sa

A. Chaabani
Laboratory of PDEs and Applications, Faculty of Mathematical, Physical and Natural Sciences of Tunis, University of Tunis El Manar, Tunis, Tunisia
e-mail: abdelkerim.chaabani@fst.utm.tn

P. Cerejeiras et al. (eds.), *Current Trends in Analysis, its Applications and Computation*, Research Perspectives,
https://doi.org/10.1007/978-3-030-87502-2_6

where ν is the kinematic viscosity and u_0 is the initial data. Burgers' equation is one among the reliable nonlinear models describing diffusive waves in fluid dynamics. As stated in [4], "Burgers [2] first developed this equation primarily to shed light on the study of turbulence described by the interaction of the two opposite effects of convection and diffusion". Three-dimensional Burgers equation is used as an adhesive model for the large scale structures formation in the universe. Motivated by the structure of (1.1)–(1.2), one may try to perform a mathematical study parallel to the L^2-Navier-Stokes theory [6]. However, as explained in [8], the lack of the divergence free condition prevents to do so. Authors in [8] overcome this difficulty and proved global well-posedness, in the critical Sobolev space $H^{1/2}(\mathbb{T}^3)$; they used a bootstrap argument and a technical lemma to control the lack of the equivalence between homogeneous and nonhomogeneous Sobolev norms. In [10], we proved the global well-posedness, in the critical Sobolev-Gevrey space $H^{1/2}_{a,\sigma}(\mathbb{T}^3)$, without such bootstrap argument. In this article, we are improving our earlier result. Firstly, we will establish well-posedness in the Lebesgue-Gevrey space. Secondly, we will give estimates of the contribution from higher order Fourier modes in terms of lower order modes, that is modes of order above and below a critical value depending only and only on $\|u_0\|_{L^2_{a,\sigma}}$.

The following three inequalities will be useful throughout this paper

$$\|f\|_{L^\infty} \leq c_1\|f\|_{H^r}, \quad \text{for all } r > 3/2, \tag{1.3}$$

where c_1 is a positive constant that depends only on r (see [9]),

$$\|f\|_{H^r} \leq c_2\|f\|_{L^2_{a,\sigma}}, \quad \text{for all } f \in L^2_{a,\sigma}, \tag{1.4}$$

where $r \geq 0$ and c_2 is a positive constant that depends only on σ, a and r (see [7]),

$$\|\nabla f\|_{L^\infty} \leq c_3\|f\|_{\dot{H}^r}, \quad \text{for all } r > 5/2, \tag{1.5}$$

where c_3 is a positive constant that depends only on r (see [9]). Also, we recall that Lebesgue-Gevrey space is defined by

$$L^2_{a,\sigma} = \{f \in L^2(\mathbb{T}^3);\ \ e^{a\Lambda^{1/\sigma}} f \in L^2(\mathbb{T}^3)\},$$

and endowed by the norm

$$\|f\|_{L^2_{a,\sigma}} = \|e^{a\Lambda^{1/\sigma}} f\|_{L^2} = \Big(\sum_{k\in\mathbb{Z}^3} e^{2a|k|^{1/\sigma}} |\hat{f}_k|^2\Big)^{1/2}.$$

For the sake of completeness, the non-homogeneous Sobolev-Gevrey space is defined, for all $\sigma \geq 1$, $r \in \mathbb{R}$ and the radius of Gevrey class regularity $a \in (0, 1)$, by

$$H^r_{a,\sigma} = \{f \in L^2(\mathbb{T}^3);\ \ e^{a\Lambda^{1/\sigma}} f \in H^r(\mathbb{T}^3)\},$$

where by Λ is the operator $\sqrt{-\Delta}$. Naturally, the homogeneous one is given by

$$\dot{H}^r_{a,\sigma} = \{f \in L^2(\mathbb{T}^3);\ \ e^{a\Lambda^{1/\sigma}} f \in \dot{H}^r(\mathbb{T}^3)\}$$

and endowed by the norm

$$\|f\|_{\dot{H}^r_{a,\sigma}} = \|\Lambda^r e^{a\Lambda^{1/\sigma}} f\|_{L^2} = \Big(\sum_{k\in\mathbb{Z}^3} |k|^{2r} e^{2a|k|^{1/\sigma}} |\hat{f}_k|^2\Big)^{1/2},$$

where $\hat{f}_k$ is the kth-Fourier coefficient of f. Similar notations and definitions can be found in [1, 3, 7, 11].

Our first result is the following theorem.

Theorem 1.1 *Given u_0 in $L^2_{a,\sigma}(\mathbb{T}^3)$, where $\sigma \geq 1$ and $a \in (0,1)$, then there exists a unique global in time solution $u \in \mathcal{C}^0([0,\infty); L^2_{a,\sigma}(\mathbb{T}^3)) \cap L^2(0,\infty; H^1_{a,\sigma}(\mathbb{T}^3))$, to the Burgers' equation (1.1)–(1.2). Moreover, if $\|u_0\|_{L^2_{a,\sigma}} \leq \nu/c_1c_2$, then the solution u is uniformly bounded with respect to time and for all $t \geq 0$*

$$\|u(t)\|^2_{L^2_{a,\sigma}} + (\nu - c_1c_2\|u_0\|_{L^2_{a,\sigma}}) \int_0^t \|u(t)\|^2_{\dot{H}^1_{a,\sigma}} \leq \|u_0\|^2_{L^2_{a,\sigma}}, \tag{1.6}$$

where c_1 and c_2 are given by inequalities (1.3)–(1.4).

It is worthwhile to note that the regularity provided by Gevrey class analyticity is stronger even than that provided by $\mathcal{C}^\infty$ regularity [7]. This fact allows to control the nonlinear part. Here, we prove a theorem similar to our Theorem 1.1 in [10], but this time without using lemma 1.1 from [8]. In fact, the control of $L^2_{a,\sigma}$-norm of the solution allows to control its mean value. We make use of the maximum principle to establish that the solution is global in time.

Our second result is the following theorem.

Theorem 1.2 *Let $u \in \mathcal{C}^0([0,\infty); L^2_{a,\sigma}(\mathbb{T}^3)) \cap L^2(0,\infty; H^1_{a,\sigma}(\mathbb{T}^3))$ be a global solution to (1.1)–(1.2). There exists $m \sim \frac{\|u_0\|^{3/4}_{L^2_{a,\sigma}}}{(\nu\lambda_1^2)^{3/4}}$ such that the solution's $L^\infty_T(L^2_{a,\sigma}(\mathbb{T}^3))$ and $L^2_T(\dot{H}^1_{a,\sigma}(\mathbb{T}^3))$ norms are uniformly bounded with respect to time. Moreover,*

$$\|Q_m u(t)\|^2_{L^2_{a,\sigma}} + (\nu - c_1c_2\|Q_m u_0\|_{L^2_{a,\sigma}}) \int_0^t \|Q_m u(t)\|^2_{\dot{H}^1_{a,\sigma}} \leq \|Q_m u_0\|^2_{L^2_{a,\sigma}},$$

where $Q_m u(t)$ represents the part of the velocity field with modes higher than m, and so on for $Q_m u_0$.

To prove this theorem, we show that the Galerkin approximation of the solution converges strongly to u in $L^\infty_T(L^2_{a,\sigma}(\mathbb{T}^3)) \cap L^2_T(H^1_{a,\sigma}(\mathbb{T}^3))$, i.e. the spaces to which the solution belongs. Then, we use such strong convergence to identify $Q_m u$ (the part of the solution u with modes higher than m) with the unique solution w associated with the high frequencies part of the initial data u_0. The low modes associated with the Galerkin operator P_m of u inherit the properties of the solution. Finally, by proving these strong convergence, we deduce that the problem can be fairly split into a dissipative higher frequencies system and low frequencies system which inherit the properties of the original system (1.1)–(1.2).

The remainder of the paper is divided into two sections; the first is assigned to prove Theorem 1.1 and the second is devoted to prove Theorem 1.2.

2 Well-Posedness Result

We will use the Galerkin approximation. For $n \in \mathbb{N}$, let P_n the projection onto the Fourier modes of order up to n, that is $P_n(\sum_{k\in\mathbb{Z}^3} \hat{u}_k e^{ixk}) = \sum_{|k|\leq n} \hat{u}_k e^{ixk}$. Let $u_n = P_n u$ be the solution to

$$\partial_t u_n + P_n[(u_n \cdot \nabla)u_n] - \nu\Delta u_n = 0 \tag{2.1}$$

$$u_n(0) = P_n u_0. \tag{2.2}$$

For some T_n, there exists a solution $u_n \in C^\infty([0, T_n)\times\mathbb{T}^3)$ to this finite-dimensional locally-Lipschitz system (2.1)–(2.2). Taking the $L^2_{a,\sigma}$ inner product of (2.1) against u_n yields

$$\tfrac{1}{2}\tfrac{d}{dt}\|e^{a\Lambda^{1/\sigma}} u_n\|^2_{L^2} + \nu\|e^{a\Lambda^{1/\sigma}} u_n\|^2_{\dot{H}^1} \leq |\textstyle\sum_{k\in\mathbb{Z}^3} \mathcal{F}(u\cdot\nabla u)(k)\cdot\overline{\hat{\omega}_k}|,$$

where $\omega = \mathcal{F}^{-1}(e^{2a|k|^{1/\sigma}}\mathcal{F}(u_n)(k))$. Using Parseval's identity, Hölder's and Young's inequalities, it follows that

$$\frac{1}{2}\frac{d}{dt}\|u_n(t)\|^2_{L^2_{a,\sigma}} + \nu\|u_n(t)\|^2_{\dot{H}^1_{a,\sigma}} \leq c\|u_n(t)\|^2_{L^\infty(\mathbb{T}^3)}\|u_n(t)\|^2_{L^2_{a,\sigma}} + \nu/2\|u_n(t)\|^2_{\dot{H}^1_{a,\sigma}}.$$

Here and throughout the paper, we will use c to denote a generic positive constant which may depend on a, σ and the Sobolev index $r \geq 0$. Combining the Sobolev embedding $H^r(\mathbb{T}^3) \subset L^\infty(\mathbb{T}^3)$, for $r > 3/2$ and (1.4), it turns out that

$$\frac{d}{dt}\|u_n(t)\|^2_{L^2_{a,\sigma}} + \nu\|u_n(t)\|^2_{\dot{H}^1_{a,\sigma}} \leq c\|u_n(t)\|^4_{L^2_{a,\sigma}}. \tag{2.3}$$

Comparing $\|u_n\|^2_{L^2_{a,\sigma}}$ with the solution of the problem $\frac{dX}{dt} = cX^2, \quad X(0) = \|u_0\|^2_{L^2_{a,\sigma}}$, we deduce, for all t such that $0 \leq ct\|u_0\|^2_{L^2_{a,\sigma}} < 1$, that

$$\|u_n\|^2_{L^2_{a,\sigma}} \leq \frac{\|u_0\|^2_{L^2_{a,\sigma}}}{1 - ct\|u_0\|^2_{L^2_{a,\sigma}}}.$$

Let $T = \frac{1}{2c\|u_0\|^2_{L^2_{a,\sigma}}}$. Thus, for all $t \in [0, T]$, we have the uniform upper bound

$$\sup_{t\in[0,T]} \|u_n(t)\|^2_{L^2_{a,\sigma}} \leq X(T) = 2\|u_0\|^2_{L^2_{a,\sigma}}. \tag{2.4}$$

Integrating (2.3) over $(0, T)$ and dropping $\|u_n\|^2_{L^2_{a,\sigma}}$ from the left-hand side, we obtain

$$\nu \int_0^T \|u_n(t)\|^2_{\dot{H}^1_{a,\sigma}} dt \leq 3\|u_0\|^2_{L^2_{a,\sigma}}. \tag{2.5}$$

These are uniform bounds on the approximate solution u_n in $L^\infty(0, T; L^2_{a,\sigma})$ and $L^2(0, T; H^1_{a,\sigma})$. Aubin-Lions lemma allows to extract a strongly convergent subsequence and to construct a local in time solution.

Theorem 2.1 *If u is a classical solution of Burgers' equations (1.1)–(1.2) on a time interval $[\varepsilon, E]$ then*

$$\sup_{t\in[\varepsilon,E]} \|u(t)\|_{L^\infty} \leq \|u(\varepsilon)\|_{L^\infty}. \tag{2.6}$$

Proof (See [8]). □

Theorem 2.1 immediately yields

$$\|u(t)\|_{L^\infty} \leq \|u_0\|_{L^\infty}. \tag{2.7}$$

Using the same technicalities already used to prove local existence, namely, we apply the function's transformation $\vartheta = \mathcal{F}^{-1}(e^{2a|k|^{1/\sigma}}\mathcal{F}(u)(k))$, then Hölder's and finally Young's inequalities and (2.7) to obtain

$$\begin{aligned}\frac{1}{2}\frac{d}{dt}\|u(t)\|^2_{L^2_{a,\sigma}} + \nu\|u(t)\|^2_{\dot{H}^1_{a,\sigma}} &\leq |\langle(u\cdot\nabla u), \vartheta\rangle_{L^2(\mathbb{T}^3)}| \\ &\leq c\|u_0\|^2_{L^\infty(\mathbb{T}^3)}\|u(t)\|^2_{L^2_{a,\sigma}} + \nu/2\|u(t)\|^2_{\dot{H}^1_{a,\sigma}}.\end{aligned}$$

Thus, inequalities (1.3)–(1.4) and the Gronwall's lemma yield

$$\sup_{t\in[0,T]} \|u(t)\|_{L^2_{a,\sigma}} \leq \|u_0\|_{L^2_{a,\sigma}} \exp\left(ct\|u_0\|^2_{L^2_{a,\sigma}}\right). \tag{2.8}$$

In particular, if $\|u_0\|_{L^2_{a,\sigma}} \leq \nu/c_1c_2$, then the $L^2_{a,\sigma}$-norm of the solution is decreasing with respect to time t as it holds that

$$\begin{aligned}\frac{1}{2}\frac{d}{dt}\|u(t)\|^2_{L^2_{a,\sigma}} + \nu\|u(t)\|^2_{\dot{H}^1_{a,\sigma}} &\leq c_1c_2\|u_0\|_{L^2_{a,\sigma}}\|u(t)\|^2_{\dot{H}^{1/2}_{a,\sigma}}\\ &\leq c_1c_2\|u_0\|_{L^2_{a,\sigma}}\|u(t)\|^2_{\dot{H}^1_{a,\sigma}},\end{aligned}$$

since $\|u(t)\|_{\dot{H}^{1/2}_{a,\sigma}} \leq \|u(t)\|_{\dot{H}^1_{a,\sigma}}$. By integrating over $(0,\infty)$, inequality (1.6) follows. Thanks to the energy inequality (1.6), the upper bound on the $L^\infty_{\mathbb{R}_+}(L^2_{a,\sigma})$ and $L^2_{\mathbb{R}_+}(\dot{H}^1_{a,\sigma})$ norms of u are independent of time t unlike the one in (2.8). This finishes the proof of Theorem 1.1.

3 Numerical Result

In the higher modes and far away from $k = (0,0,0)$, we identify the operator $-\Delta$ with the Stokes operator A in the space-periodic case with vanishing space-average (see e.g. [5]). We recall that the Stokes operator A is self-adjoint, and its eigenvalues are of the form $\left(\frac{2\pi}{L}\right)^2 |k|^2$, where $k \in \mathbb{Z}^3$. We denote these eigenvalues by $0 < \lambda_1 = (2\pi/L)^2 \leq \lambda_2 \leq \lambda_3 \leq \ldots$ arranged in increasing order and counted according to their multiplicities. In dimension three, the asymptotic behavior of the eigenvalues is given by $\lambda_m \sim \lambda_1 m^{2/3}$, (see [5]).

Let v_n^0 be the projection onto the Fourier modes up to the order n of an initial data u_0. Consequently, v_n^0 converges in $L^2_{a,\sigma}$ to u_0. It is possible to prove that $u_0 \in B(v_n^0, \varepsilon)$ (the ball centered at v_n^0 of radius ε) for sufficiently large n (under the assumption that u_0 indeed gives rise to a solution u of (1.1)–(1.2) in the class of functional spaces stated in theorem 1.1). Thus, $w_n^0 = u_0 - v_n^0$ gives rise to a global solution w of

$$\partial_t w - \nu\Delta w + (w\cdot\nabla)w = 0, \qquad w|_{t=0} = u_0 - v_n^0 = w_n^0.$$

It is worth mentioning that the solution w to the system above does not agree with the higher frequencies modes of the solution $Q_n u = u - P_n u$ in general. However, by proving a convergence result in $L^\infty_T(L^2_{a,\sigma}(\mathbb{T}^3)) \cap L^2_T(\dot{H}^1_{a,\sigma}(\mathbb{T}^3))$ of $P_n u$ to u as $n \to \infty$, we can determine a wavenumber m for which $Q_m u$ can be identified with w.

Let u be the solution of the Burgers' equations (1.1)–(1.2) on $[0, T]$ and u_n be the sequence of Galerkin approximations of u. We want to show that the $L^2_{a,\sigma}$-norm of the difference $w_n = u - u_n$ tends uniformly to zero, on time interval $[0, T]$, as n tends to infinity. Since u is a solution, on the time interval $[0, T]$, we have

$$\langle e^{a\Lambda^{1/\sigma}}\partial_t u, e^{a\Lambda^{1/\sigma}}w_n\rangle - \nu\langle e^{a\Lambda^{1/\sigma}}\Delta u, e^{a\Lambda^{1/\sigma}}w_n\rangle_{L^2} + \langle e^{a\Lambda^{1/\sigma}}(u\cdot\nabla)u, e^{a\Lambda^{1/\sigma}}w_n\rangle_{L^2} = 0.$$

We introduce Q_n by $Id = P_n + Q_n$. We take the inner product of (2.1) against w_n, we subtract the equation satisfied by u_n from the one satisfied by u. Then, we use the identity $\langle e^{a\Lambda^{1/\sigma}}P_n f, e^{a\Lambda^{1/\sigma}}g\rangle_{L^2} = \langle e^{a\Lambda^{1/\sigma}}f, e^{a\Lambda^{1/\sigma}}P_n g\rangle_{L^2}$ to obtain

$$\begin{aligned}\frac{1}{2}\frac{d}{dt}\|w_n(t)\|_{L^2_{a,\sigma}} + \nu\|w_n(t)\|_{\dot H^1_{a,\sigma}} \le\ & |\langle e^{a\Lambda^{1/\sigma}}(u\cdot\nabla)w_n, e^{a\Lambda^{1/\sigma}}\underbrace{P_n w_n}_{0}\rangle_{L^2}| \\ &+ |\langle e^{a\Lambda^{1/\sigma}}(w_n\cdot\nabla)u, e^{a\Lambda^{1/\sigma}}\underbrace{P_n w_n}_{0}\rangle_{L^2} \\ &+ |\langle e^{a\Lambda^{1/\sigma}}(w_n\cdot\nabla)w_n, e^{a\Lambda^{1/\sigma}}\underbrace{P_n w_n}_{0}\rangle_{L^2}| \\ &+ |\langle e^{a\Lambda^{1/\sigma}}Q_n[(u\cdot\nabla)u], e^{a\Lambda^{1/\sigma}}w_n\rangle_{L^2}|.\end{aligned}$$

In fact, $P_n w_n = 0$ is due to $P_n w_n = P_n(Q_n(\sum_{k\in\mathbb{Z}^3}\hat u_k e^{ikx})) = P_n(\sum_{|k|>n}\hat u_k e^{ikx}) = 0$. Consequently, by integrating over $[0, T]$, we deduce that

$$\begin{aligned}\frac{1}{2}\|w_n(T)\|^2_{L^2_{a,\sigma}} + \nu\int_0^T \|w_n(t)\|^2_{\dot H^1_{a,\sigma}}\,dt \le\ & \frac{1}{2}\|w_n^0\|^2_{L^2_{a,\sigma}} \\ &+ \int_0^T \underbrace{|\langle e^{a\Lambda^{1/\sigma}}Q_n[(u\cdot\nabla)u], e^{a\Lambda^{1/\sigma}}w_n\rangle_{L^2}|}_{f_n(t)}\,dt.\end{aligned}$$

According to the above, strong uniform convergence of the $\|w_n\|_{L^2_{a,\sigma}}$ to zero follows provided that $\lim_{n\to\infty}\left(\|w_n^0\|^2_{L^2_{,\sigma}} + \int_0^T f_n(\tau)d\tau\right) = 0$. Clearly, $\lim_{n\to\infty}\|w_n^0\|^2_{L^2_{a,\sigma}} = 0$. To prove that $\lim_{n\to\infty}\int_0^T f_n(\tau)d\tau$, we use dominated convergence theorem. We have

$$\begin{aligned}f_n(t) &= |\langle e^{a\Lambda^{1/\sigma}}(u\cdot\nabla)u - P_n[(u\cdot\nabla)u], e^{a\Lambda^{1/\sigma}}w_n\rangle_{L^2}| \\ &\le 2\sum_{k\in\mathbb{Z}^3}\mathcal{F}(\vartheta_1\cdot\vartheta_2)(k)\overline{\mathcal{F}(\vartheta_3)(k)},\end{aligned}$$

where $\vartheta_1 = \mathcal{F}^{-1}(|\hat u_k|)$, $\vartheta_2 = \mathcal{F}^{-1}(|k|\cdot|\hat u_k|)$ and $\vartheta_3 = \mathcal{F}^{-1}(e^{2a|k|^{1/\sigma}}|\hat u_k|)$. So,

$$\begin{aligned}f_n(t) &\le 2\|\vartheta_1\|_{L^\infty}|\langle\vartheta_2,\vartheta_3\rangle_{L^2}| \\ &\le 2c_1c_2\|u(t)\|_{L^2_{a,\sigma}}\|u(t)\|^2_{\dot H^1_{a,\sigma}},\end{aligned}$$

where we used Hölder's inequality and (1.3) and (1.4). Now, since u belongs to $L^\infty_T(L^2_{a,\sigma}) \cap L^2_T(\dot{H}^1_{a,\sigma})$ then $f_n(t) \in L^1([0,T])$. It turns out that there exists m for which

$$\|Q_m u(t)\|^2_{L^2_{a,\sigma}} + 2\nu \int_0^t \|Q_m u(\tau)\|^2_{\dot{H}^1_{a,\sigma}} d\tau < \frac{\nu^2}{4(c_1 c_2)^2}, \quad \text{for all } 0 \le t < \infty,$$

where Q_m is the projection onto the modes higher than m. Moreover, it holds that $\|Q_m u(t)\|_{L^2_{a,\sigma}} \underset{t\to\infty}{\longrightarrow} 0$ under the assumption $\|P_m u(t)\|_{L^2_{a,\sigma}} \underset{t\to\infty}{\longrightarrow} 0$. This implies that the first m modes associated with P_m are in fact the determining modes. Consequently, there exists a critical m such that by splitting the initial data u_0 into $P_m u_0$ and $Q_m u_0$, the latter gives rise to the system

$$\partial_t w_m - \nu \Delta w_m + (w_m \cdot \nabla) w_m = 0, \quad (t,x) \in \mathbb{R}_+ \times \mathbb{T}^3 \tag{3.1}$$

$$w_m|_{t=0} = Q_m u_0, \quad x \in \mathbb{T}^3, \tag{3.2}$$

where w_m can be identified with $Q_m u$, i.e. the projection of u onto the modes higher than m. It is assumed that $\|Q_m u_0\|_{L^2_{a,\sigma}} < \frac{\nu}{2c_1c_2}$. Thus, $\|w_m(t)\|_{L^2_{a,\sigma}} \le \|Q_m u_0\|_{L^2_{a,\sigma}}$ for all $t \ge 0$, according to (1.6).

It remains to prove that $\|w_m(t)\|_{L^2_{a,\sigma}} \to 0$ as $t \to \infty$, as well as to obtain an upper bound for the number of determining modes m. We take the inner product of (3.1) against w_m to obtain

$$\frac{1}{2}\frac{d}{dt}\|e^{a\Lambda^{1/\sigma}} w_m(t)\|^2_{L^2} + \nu \|e^{a\Lambda^{1/\sigma}} w_m(t)\|^2_{\dot{H}^1} \le |\langle (w_m \cdot \nabla) w_m, \Xi \rangle_{L^2(\mathbb{T}^3)}|,$$

where $\Xi = \mathcal{F}^{-1}(e^{2a|k|^{1/\sigma}} \hat{w}_k)$. By using inequalities (1.5) and (1.4), it turns out

$$\frac{1}{2}\frac{d}{dt}\|w_m(t)\|^2_{L^2_{a,\sigma}} + \nu \|w_m(t)\|^2_{\dot{H}^1_{a,\sigma}} \le c_3 c_2 \|Q_m u_0\|_{L^2_{a,\sigma}} \|w_m(t)\|^2_{L^2_{a,\sigma}}.$$

We use the inequality $\lambda_{m+1}\|w_m\|_{L^2_{a,\sigma}} \le \|w_m\|_{\dot{H}^1_{a,\sigma}}$ (see e.g. [5]) to obtain

$$\frac{d}{d\tau}\|w_m(\tau)\|^2_{L^2_{a,\sigma}} + (\nu\lambda^2_{m+1} - c_3c_2\|Q_m u_0\|_{L^2_{a,\sigma}})\|w_m(\tau)\|^2_{L^2_{a,\sigma}} \le 0.$$

Multiplying by $\exp(\tau(\nu\lambda^2_{m+1} - c_3c_2\|Q_m u_0\|_{L^2_{a,\sigma}}))$ and integrating over $[0,t)$, we obtain

$$\|w_m(t)\|^2_{L^2_{a,\sigma}} \le \|Q_m u_0\|^2_{L^2_{a,\sigma}} e^{t(c_3c_2\|Q_m u_0\|_{L^2_{a,\sigma}} - \nu\lambda^2_{m+1})}. \tag{3.3}$$

So, we deduce that λ_{m+1} (the $(m+1)$th eigenvalue of A) must be as large as

$$\lambda^2_{m+1} > c_3c_2\|Q_m u_0\|_{L^2_{a,\sigma}}/\nu. \tag{3.4}$$

Therefore, inequality (3.3) immediately implies that $\limsup_{t\to\infty} \|w_m(t)\|_{L^2_{a,\sigma}} = 0$. Moreover, since $\lambda_m \sim \lambda_1 m^{2/3}$, we deduce that inequality (3.4) is satisfied provided m is such that $m \geq c\|u_0\|^{3/4}_{L^2_{a,\sigma}}/(\nu\lambda_1^2)^{3/4}$. This gives the wavenumber m beyond which the $L^\infty_{\mathbb{R}_+}(L^2_{a,\sigma})$ and $L^2_{\mathbb{R}_+}(\dot{H}^1_{a,\sigma})$ norms of the solution's high frequency modes are uniformly bounded with respect to time.

This result is useful as it solves the problem: how many modes should be retained if we want to be certain that the truncated system will have the same behavior for t large as the original one.

References

1. J. Benameur, On the exponential type explosion of Navier-Stokes equations. Nonlinear Anal. **103**, 87–97 (2016)
2. J.M. Burgers, A mathematical model illustrating the theory of turbulence. Adv. Appl. Mech. **1**, 171–191 (1948)
3. A. Chaabani, On well-posedness of the Cauchy problem for 3D MHD system in critical Sobolev–Gevrey space. SN Partial Differ. Equ. Appl. **2**, 24 (2021). https://doi.org/10.1007/s42985-021-00081-z
4. L. Debnath, *Nonlinear Partial Differential Equations for Scientist and Engineers* (Birkhauser, Boston, 1997)
5. C. Foias, O. Manley, R. Rosa, R. Temam, *Navier-Stokes Equations and Turbulence, Encyclopedia of Mathematics and its Applications*, vol. 83 (Cambridge University Press, Cambridge, 2001)
6. J. Leray, Sur le mouvement d'un liquide visqueux emplissant l'espace. Acta Math. **63**, 193–248 (1934)
7. M. Oliver, A Mathematical Investigation of Models of Shallow Water with a Varying Bottom. Ph.D. Dissertation, University of Arizona, Tucson, Arizona, 1996
8. B.C. Pooley, J.C. Robinson, Well-posedness for the diffusive 3D Burgers equations with initial data in $H^{1/2}$. London Math. Soc. Lecture Note Ser. **430**, 137–153 (2016)
9. J. Robinson, J. Rodrigo, W. Sadowski, *The Three-Dimensional Navier-Stokes Equations: Classical Theory (Cambridge Studies in Advanced Mathematics)* (Cambridge University Press, Cambridge, 2016)
10. R. Selmi, A. Chaabani, Well-posedness to 3D Burgers' equation in critical Gevrey Sobolev spaces. Arch. Math. **112**, 661–672 (2019)
11. R. Selmi, A. Châabani, Well-posedness, stability and determining modes to 3D Burgers equation in Gevrey class. Z. Angew. Math. Phys. **71**, 162 (2020). https://doi.org/10.1007/s00033-020-01389-3

BVP for the First Order Elliptic Systems in the Plane

Nino Manjavidze and Giorgi Akhalaia

Abstract In this paper the Riemann-Hilbert type boundary value problem for generalized analytic vectors in plane domains bounded by smooth curves is considered. In some cases Noethericity conditions of the problem are given.

Keywords Boundary value problem · Riemann-Hilbert type problem · Generalized analytic vector · Noethericity conditions

Mathematics Subject Classification (2010) Primary 30E25; Secondary 35E05

The linear system of partial differential equations

$$\frac{\partial u}{\partial x} = a(x, y)\frac{\partial u}{\partial y} + b(x, y)u(x, y) + f(x, y),$$

where $u = (u_1, u_2, \ldots, u_{2n})$ is the desired real vector with $2n$ components, a, b are given $2n \times 2n$ real matrices depending on two real variables (x, y), f is a given vector with $2n$ components is called elliptic in some domain D if the matrix a has no real characteristic numbers in this domain.It is well-known that in one dimensional case this system can be reduced to one complex equation:

$$\partial_{\bar{z}}w(z) + A(z)w(z) + B(z)\overline{w(z)} = F(z), \quad z \in D, \tag{1}$$

where $A,\ B,\ F$ are complex functions expressed by $a,\ b,\ f$ (see [1–3]).

N. Manjavidze (✉)
Ilia State University, Tbilisi, Georgia
e-mail: nino.manjavidze@iliauni.edu.ge

G. Akhalaia
I.Vekua Institute of Applied Mathematics, Tbilisi, Georgia

P. Cerejeiras et al. (eds.), *Current Trends in Analysis, its Applications and Computation*, Research Perspectives,
https://doi.org/10.1007/978-3-030-87502-2_7

In one dimensional case the system of such kind is called the Carleman-Vekua equation. They were investigated by Picard E., Beltrami E., Teodorescu N., Carleman T. and others. In the works of Polozhy G. N., Shabat B. several classes of this equations were studied. Using generalizations of the concepts of complex differentiation and integration Bers L. constructed the so-called theory of pseudo-analytic functions which are the solutions of the Eq. (1). Another theory of the solutions of the Eq. (1), the theory of generalized analytic functions was constructed by Vekua I. Further development of this theory by Vekua and his disciples and followers [3] is summarized in his fundamental monograph [4].

As a model problem of the boundary value problems of the theory of generalized analytic vectors we consider the following Riemann-Hilbert problem:

$$\mathrm{Re}\,[G(t)w(t)] = g(t), \quad t \in \Gamma, \tag{2}$$

for the Eq. (1). A, B are bounded measurable matrices, D is a bounded domain in the complex plane with the smooth boundary Γ.

We investigate the existence of the solutions of the problem (1) and (2). In particular, we prove the following proposition.

Proposition 1 *There exists a solution of the problem (1) and (2) in the form* $w(z) = w_0(z)+w_1(z)$*, where* $w_1(z) = R(F(z))$*,* R *is some linear bounded operator mapping the space* $L_s(D)$ *in the space of Hölder-continuous vectors,* $w_0(z)$ *is the solution of the homogeneous equation (1) in the class* $E_p(D, A, B, \varrho)$*,* $p > 1$.

In order to introduce the class from Proposition 1 we need some terms and notations [5] (see also [1]).

Let the matrix $V(t, z)$ be the generalized Cauchy kernel for the homogeneous equation (1). The equation

$$\partial_{\bar{z}}\Psi - A'(z)\Psi - \overline{B'(z)\Psi} = 0 \tag{3}$$

is called conjugate to the system (1). Using holomorphic vectors, generalized analytic vectors $w(z)$ can be represented as

$$w(z) = \phi(z) + \int\limits_D \Gamma_1(Z,t)\phi(t)d\sigma_t + \int\limits_D \Gamma_2(Z,t)\overline{\phi(t)}d\sigma_t + \sum_{k=1}^{N} c_k W_k(z), \tag{4}$$

where $\phi(z)$ is a holomorphic vector and $W_k(z)$ $(k = 1, \dots, N)$ is a complete system of linearly independent solutions of the Fredholm equation

$$Kw \equiv w(z) - \frac{1}{\pi}\int\limits_D V(t,z)\left[A(t)w(t) + B(t)\overline{w(t)}\right]d\sigma_t = 0,$$

$W_k(z)$ turn out to be continuous vectors in the whole plane, vanishing at infinity and c_k—are arbitrary real constants; the kernels $\Gamma_1(z,t)$ and $\Gamma_2(z,t)$ satisfy the system of the integral equations

$$\Gamma_1(z,t) + \frac{1}{\pi} V(t,z) A(t) + \frac{1}{\pi} \int\limits_D V(t,z) \left[A(t)\Gamma_1(\tau,t) + B(t)\overline{\Gamma_2(\tau,t)} \right] d\sigma_t$$

$$= -\frac{1}{2} \sum_{k=1}^{N} \{v_k(z), \bar{v}_k(t)\},$$

$$\Gamma_2(z,t) + \frac{1}{\pi} V(t,z) A(t) + \frac{1}{\pi} \int\limits_D V(t,z) \left[A(t)\Gamma_1(\tau,t) + B(t)\overline{\Gamma_2(\tau,t)} \right] d\sigma_t$$

$$= -\frac{1}{2} \sum_{k=1}^{N} \{v_k(z), \bar{v}_k(t)\},$$

where $v_k(z) \in L_p(\bar{D})$, $(k = 1, \ldots, N)$ form a system of linearly independent solutions of the Fredholm integral equation

$$v(z) + \frac{\overline{A'(z)}}{\pi} \int\limits_D V'(z,t) v(t) d\sigma_t + \frac{\overline{B'(z)}}{\pi} \int\limits_D V'(z,t) \bar{v}(t) d\sigma_t = 0$$

and v, w is the diagonal product of the vectors v and w. Notice that in formula (4) $\phi(z)$ is not an arbitrary holomorphic vector, it has to satisfy the following conditions

$$\operatorname{Re} \int\limits_D \phi(z) v_k(z) d\sigma_t = 0, \quad k = 1, \ldots, N. \tag{5}$$

In general, the Liouville theorem is not true for the solutions of (1). This is why the constants c_k appear in (4) and the conditions (5) have to be satisfied.

Denote by $E_p(D, A, B, \varrho)$ the class of the solutions of the homogeneous equation of (1), representable by the generalized Cauchy type integrals

$$w(z) = \frac{1}{2\pi i} \int\limits_\Gamma \left[\Omega_1(z,t)\varphi(t)dt - \Omega_2(z,t)\overline{\varphi(t)dt} \right] + \sum_{k=1}^{N} c_k W_k(z),$$

where $\varphi(t) \in L_p(\Gamma, \varrho)$ satisfies the condition

$$\operatorname{Im} \int\limits_\Gamma \big(\varphi(t), \Psi_j(t)\big)\, dt = 0, \quad (j = 1, \ldots, N),$$

where Ψ_j form a similar system for the conjugate equation (3). The kernels Ω_1 and Ω_2 are representable by the resolvents Γ_1 and Γ_2 according to the formulas

$$\Omega_1(z,t) = V(t,z) + \int_D \Gamma_1(z,\tau)V(t,\tau)d\sigma_\tau,$$

$$\Omega_2(z,t) = \int_D \Gamma_2(z,\tau)\overline{V(t,\tau)}d\sigma_\tau.$$

Let us also introduce the class

$$E_q\left(D, -A', -\overline{B'}, \varrho^{1-q}\right), q = \frac{p}{p-1}$$

of the solutions of the conjugate equation, representable in the form

$$\Psi(z) = \frac{1}{2\pi i}\int_\Gamma \left[\Omega_1'(t,z)h(t)dt - \overline{\Omega_2'}(t,z)\overline{h(t)dt}\right] + \sum_{k=1}^{N} c_k\Psi_k(z),$$

where the density $h(t) \in L_q(\Gamma, \varrho^{1-q})$ satisfies the condition

$$\operatorname{Im}\int_\Gamma \left(h(t), W_j(t)\right)dt = 0, \quad (j = 1,\ldots,N).$$

Therefore, for $w(z)$ satisfying (1), we get the following boundary condition

$$\operatorname{Re}\left[G(t)w_0(t)\right] = H^1(t), \tag{6}$$

where $H^1(t) = f(t) - \operatorname{Re}\left[G(t)w_1(t)\right]$. From (6) it follows

$$w_0(t) = G^{-1}(t)\left[H^1(t) + i\xi(t)\right],$$

where $\xi(t)$ is a desired real vector of the class $L_p(\Gamma, \varrho)$. For $\xi(t)$ we obtain the real system of singular integral equations

$$\begin{aligned}\xi(t_0) = &\int_\Gamma \left[G(t_0)G^{-1}(t) + \frac{t_0}{t}\overline{G(t_0)G^{-1}(t)}\right]\frac{\xi(t)}{t_0 - t} + \int_\Gamma K(t_0,t)\xi(t)ds \\ &= H^2(t_0) - 4\pi\sum_{k=1}^{N} c_k\operatorname{Re}\left[G(t_0)W_k(t_0)\right]\end{aligned} \tag{7}$$

and the additional conditions

$$\operatorname{Im} \int_{\Gamma} G^{-1}(t) \left(H^1(t) + i\xi(t) \right) \Psi_k(t) dt = 0, \quad (k = 1, \ldots, N),$$

where $K(t_0, t)$ is a real kernel with the weak singularity, $H^2(t)$ is a real vector which can be linearly expressed by means of $H^1(t)$.

From (7) we get

$$\xi(t_0) = \left(K H^3 \right)(t_0) + \sum_{k=1}^{l} l_k \xi_k(t_0),$$

where

$$H^3(t_0) = H^2(t_0) - 4\pi \sum_{k=1}^{N} \operatorname{Re}\left[G(t_0) W_k(t_0) \right]$$

and $\sum\limits_{k=1}^{l} l_k \xi_k$ is a general solution of the homogeneous equation of (7), K is a linear bounded operator on the space $L_p(\Gamma, \varrho)$.

We should take into the consideration the solvability condition of the Eq. (7)

$$\int_{\Gamma} \left(H^3(\tau), g^k(\tau) \right) d\tau = 0, \quad k = 1, \ldots, l^*,$$

where $g^k(t),\ k = 1, \ldots, l^*$ is a complete system of linearly independent solutions of the conjugate homogeneous equation of (7) in the class $L_q(\Gamma, \varrho^{1-q}), q = \frac{p}{p-1}$. If the Eq. (7) is Noetherian then the Riemann-Hilbert boundary value problem (7) is Noetherian in $E_p(D, A, B, \varrho)$ and the necessary and sufficient solvability conditions are the following

$$\operatorname{Im} \int_{\Gamma} \left(H^1(\tau), G'^{-1}(\tau) \eta_k(\tau) \right) d\tau = 0, \quad k = 1, \ldots, l',$$

where η_k is a complete system of linearly independent solutions of the Riemann-Hilbert problem

$$\operatorname{Re}\left[G'^{-1}(\tau) \eta(\tau) \right] = 0,$$

in the class $E_q(D, -A', -\overline{B'}, \varrho^{1-q}),\ q = \frac{p}{p-1}$.

Due to [5] the Eq. (7) is Noetherian in $L_p(\Gamma, \varrho)$ if

$$\frac{1+\varrho_k}{p} \neq \omega_{kj}, \quad k = 1, \ldots, r, \; j = 1, \ldots, n, \tag{8}$$

where $\omega_{kj} = \frac{1}{2\pi} \arg\lambda_{kj}, \; 0 \leq \arg\lambda_{kj} < 2\pi, \lambda_{kj}$ are the roots of the equation

$$\det\left[G^{-1}(t_k + 0)G(t_k - 0) - \lambda I\right] = 0.$$

If $G(t)$ is continuous on Γ then the condition (8) is not required.

Finally we remark that differential boundary value problem for the system of second order differential equations of elliptic type in plane domains bounded by smooth curves can be reduced to the problem for generalized analytic vectors considered above.

Acknowledgement The present work is supported by the Shota Rustaveli National Science Foundation (SRNSF grant FR 17-96).

References

1. G. Akhalaia, G. Giorgadze, V. Jikia, N. Kaldani, G. Makatsaria, N. Manjavidze, Elliptic systems on Riemann surface. Lect. Notes TICMI **13**, 1–167 (2012)
2. G. Akhalaia, Discontinuous boundary value problem for generalized analytic vectors. Proc. Seminar I. Vekua Inst. Appl. Math. **15**, 67–86 (1981)
3. B. Bojarski, Theory of generalized analytic vectors. Ann. Polon. Math. **17**, 281–320 (1966)
4. I. Vekua, *Generalized Analytic Functions* (Pergamon, Oxford, 1962)
5. G. Manjavidze, G. Akhalaia, Boundary value problems of the theory of generalized analytic vectors, in *Complex Methods for Partial Differential Equations*, ed. by H. Begehr, O. Celebi, W. Tutschke (Kluwer Academic Publishers, Dordrecht, 1999), pp. 57–97

Applications of Zalcman's Lemma in C^N

P. V. Dovbush

Abstract The aim of this paper is to give some applications of Marty's Criterion and Zalcman's Rescalling Lemma.

Keywords Marty's Criterion · Zalcman's lemma · Zalcman-Pang's lemma · Normal families · Holomorphic functions of several complex variables

Mathematics Subject Classification (2010) Primary 32A19

1 Introduction

Lemma 1.1 *A family $\mathcal{F}$ of functions meromorphic [analytic] on the unit disc Δ is not normal if and only if there exist (a) a number $0 < r < 1$ (b) points z_n, $|z_n| < r$ (c) functions $f_n \in \mathcal{F}$ (d) numbers $\rho_n \to 0+$ such that*

$$\frac{f_n(z_n + \rho_n \xi)}{\rho_n^\alpha} \quad (0 \leq \alpha < 1 \text{ arbitrary}).$$

spherically uniformly [uniformly] on compact subsets of C, where g is a nonconstant meromorphic [entire] function on C.

This is famous Pang's Lemma [6, Lemma 2, p. 787]. The case $\alpha = 0$ is due to Zalcman [13, p. 814] and is known as Zalcman's Rescalling Lemma. Zalcman's Lemma has numerous applications in the theory of entire and meromorphic functions, normality criteria, ordinary differential equations, complex dynamics, value distribution theory, quasiregular mappings in space, and minimal surfaces, for the list of some of these application see, for example, the papers [11, 12] and the book [9]. Zalcman's Lemma was generalised by Pang [6, 7], using in a way

---

P. V. Dovbush (✉)
Institute of Mathematics and Computer Science of Moldova, Chişinău, Republic of Moldova

P. Cerejeiras et al. (eds.), *Current Trends in Analysis, its Applications and Computation*, Research Perspectives,
https://doi.org/10.1007/978-3-030-87502-2_8

which makes it more flexible, and, in particular, applicable to algebraic differential equations.

Much attention has been given to find an appropriate generalization of Zalcman's Lemma to several complex variables, and more generally to complex manifolds (see, for example, [1–4, 10]).

Using the ideas of Zalcman and Pang we can prove the most of the theorems in this paper, but Montel's Criterion was used in all proofs!

2 Marty's Criterion

A family $\mathcal{F}$ of holomorphic functions on a domain $\Omega \subset C^n$ is normal in Ω if every sequence of functions $\{f_j\} \subseteq \mathcal{F}$ contains either a subsequence which converges to a limit function $f \not\equiv \infty$ uniformly on each compact subset of Ω, or a subsequence which converges uniformly to ∞ on each compact subset.

A family $\mathcal{F}$ is said to be normal at a point $z_0 \in \Omega$ if it is normal in some neighborhood of z_0. A family of holomorphic functions $\mathcal{F}$ is normal in a domain Ω if and only if $\mathcal{F}$ is normal at each point of Ω.

For every function φ of class $C^2(\Omega)$ we define at each point $z \in \Omega$ an Hermitian form

$$L_z(\varphi, v) := \sum_{k,l=1}^{n} \frac{\partial^2 \varphi}{\partial z_k \partial \overline{z}_l}(z) v_k \overline{v}_l,$$

(where $z_1, \ldots, z_n$ is the natural coordinate system in C^n) and call it the Levi form of the function φ at z.

For a holomorphic function f in Ω, set

$$f^\sharp(z) := \max_{|v|=1} \sqrt{L_z(\log(1 + |f|^2), v)} \tag{2.1}$$

and

$$_\sharp f(z) := \min_{|v|=1} \sqrt{L_z(\log(1 + |f|^2), v)}. \tag{2.2}$$

This quantities is well defined since the Levi form $L_z(\log(1 + |f|^2), v)$ is nonnegative for all $z \in \Omega$.

In (2.1) resp. (2.2) the supremum resp. infinum is at the same time a maximum/minimum, since by Weiershtrass Theorem the continuous real-valued function $L_z(\varphi, v)/(v, v)$ achieving a least upper bound resp. greatest lover bound on the unit sphere in C^n, a compact set.

In particular, for $n = 1$ the formula (2.1) takes the form

$$f^{\sharp}(z) := \frac{|f'(z)|}{1 + |f(z)|^2}$$

and $z^{\sharp}$ coincides with the spherical metric on C.

We have Marty's characterization of normal families in terms of the spherical metric.

Theorem 2.1 (Marty's Criterion, See [3]) *A family $\mathcal{F}$ of functions holomorphic on $\Omega \subset C^n$, is normal on Ω if and only if for each compact subset $K \subset \Omega$ there exists a constant $M(K)$ such that at each point $z \in K$*

$$f^{\sharp}(z) \leq M(K) \tag{2.3}$$

for all $f \in \mathcal{F}$.

Montel's criterion for normality, one of the more important results in function theory, and which will be used over and over again. Montel's criterion has a consequence that many might find astounding.

3 Non-normal Families

The following two theorems are the cornerstone of this chapter. Its are a powerful tools which allows to extract informations from non-normal holomorphic families.

The proof of the Zalcman's Rescaling Lemma is fairly short and elementary; it uses only Marty's Criterion (Theorem 2.1).

Using the ideas of Zalcman and Pang I derived the following important characterizations of non-normality.

Theorem 3.1 (Zalcman's Lemma, See [3]) *Suppose that a family $\mathcal{F}$ of functions holomorphic on $\Omega \subset C^n$ is not normal at some point $z_0 \in \Omega$. Then there exist sequences $f_j \in \mathcal{F}$, $z_j \to z_0$, $\rho_j = 1/f_j^{\sharp}(z_j) \to 0$, such that the sequence*

$$g_j(z) = f_j(z_j + \rho_j z)$$

converges locally uniformly in C^n to a non-constant entire function g satisfying $g^{\sharp}(z) \leq g^{\sharp}(0) = 1$.

Theorem 3.2 (Pang's Lemma) *The simultaneous proof of this theorems. The statement of Theorem 3.1 remains valid if the sequence* $g_j(z) = f_j(z_j + \rho_j z)$ *is replaced with*

$$g_j(z) := \frac{f_j(z_j + r_j z)}{r_j^\alpha} \quad (0 \leq \alpha < 1 \ arbitrary).$$

For the simultaneous proof of Theorems 3.1 and 3.2 we need the following lemma.

Lemma 3.3 *Let* f *be a holomorphic function on the unit ball* $B = \{z \in C^n : |z| < 1\}$, *and* α *be a real number with* $0 \leq \alpha < 1$. *Suppose* $j \geq 3$ *and*

$$\max_{|z| \leq 1/j} \frac{(1 - j|z|)^{1+\alpha}(1 + |f(z)|^2)f^\sharp(z)}{(1 - j|z|)^{2\alpha} + |f(z)|^2} > 1.$$

Then there exists a point ξ^*, $|\xi^*| < 1/j$, *and a real number* ρ, $0 < \rho < 1$, *such that*

$$\max_{|z| \leq 1/j} \frac{(1 - j|z|)^{1+\alpha}\rho^{1+\alpha}(1 + |f(z)|^2)f^\sharp(z)}{(1 - j|z|)^{2\alpha}\rho^{2\alpha} + |f(z)|^2} =$$

$$\frac{(1 - j|\xi^*|)^{1+\alpha}\rho^{1+\alpha}(1 + |f(\xi^*)|^2)f^\sharp(\xi^*)}{(1 - j|\xi^*|)^{2\alpha}\rho^{2\alpha} + |f(\xi^*)|^2} = 1.$$

In the proof of Theorem 3.2 the expression

$$\frac{(1 - j|z|)^{1+\alpha}(1 + |f_j(z)|^2)f_j^\sharp(z)}{(1 - j|z|)^{2\alpha} + |f_j(z)|^2}$$

takes the part of $(1 - j|z|)f_j^\sharp(z)$. The last function is continuous!

4 Montel's Theorem in C^N

Let us illustrate the use of Zalcman's Rescalling Lemma [3] by showing how it can be used to derive Montel's Theorem in several complex variables (see also [12] for one dimensional case).

Just to get a glimpse of a power and beauty of Zalcman's Lemma, we will give two proof of Montel's theorem as an application of this lemma. Judicious application of Zalcman's Lemma often leads to proofs which seem almost magical in their brevity.

Theorem 4.1 (Montel's Theorem) *Let $\mathcal{F}$ be a family of holomorphic functions on an open set $\Omega \subseteq C^n$ that omit two fixed complex values. Then, each sequence of functions in $\mathcal{F}$ has a subsequence which converges uniformly on compact subsets.*

First of all we shall need one auxiliary proposition for the proof of which see, for example, [5, Corollary, p.80].

Theorem 4.2 (Hurwitz's Theorem) *Let Ω be an open connected set in C^n and $\{g_j(z)\}$ a sequence of holomorphic functions on Ω, converging uniformly on compact sets to a holomorphic function g. Then if $g_j(z) \neq 0$ for all j, and all z, and g is nonconstant, we have $g(z) \neq 0$ for all $z \in \Omega$.*

First Proof of Montel's Theorem Composing the functions of $\mathcal{F}$ with a linear fractional transformation, we may also assume that the omitted values are 0 and 1. Suppose $\mathcal{F}$ is not normal on Ω. Then by Zalcman's Rescalling Lemma [3], there exist $f_j \in \mathcal{F}$, $z_j \in \Omega$ and $\rho_j \to 0+$ such that $f_j(z_j + \rho_j\zeta) = g_j(\zeta) \to g(\zeta)$ uniformly on compact subsets of C^n, where g is a nonconstant entire function. By Hurwitz's Theorem (see, for example, [5, Corollary, p. 80]), g does not take on the values 0 and 1, since no f_j does. Let $b \in C^n$. The function

$$g_b(\lambda) := g(\lambda \cdot b)$$

is entire function on C, satisfies

$$g_b(0) = g(0), \quad g_b(1) = g(b)$$

and

$$g_b(C) \subset g(C^N) \subseteq C \setminus \{0, 1\}.$$

But then, by one-dimensional version of Picard's Little Theorem, g_b is constant, hence $g(0) = g(b)$ for all $b \in C^n$, a contradiction. □

We also give a simple proof of the theorem of Montel, based on the idea of A. Ros [8] (see [11, p. 218]).

Second Proof of Montel's Theorem Since normality is a local notion, we may suppose that $\Omega = B$, the unit ball. So let $\mathcal{F}$ be as in the statement of Montel's Theorem and suppose that $\mathcal{F}$ is not normal. Composing with a linear fractional transformation, we may also assume that the omitted values are 0 and 1. This implies that if $f \in \mathcal{F}$ and $m \in N$ there exists a function g holomorphic in Ω such that $g^{2^m} = f$. Let $\mathcal{F}_m$ be the family of all such functions g. Note that

$$\left|\frac{1}{2^m}\right|^2 \left(\frac{|f|^{-1} + |f|}{|f|^{-1/2^m} + |f|^{1/2^m}}\right)^2 L_z(\log(1 + |f|^2, v) = L_z(\log(1 + |g|^2, v)$$

for all $(z, v) \in \Omega \times C^n$. This implies that

$$g^\sharp(z) = \frac{1}{2^m} \frac{|f|^{-1} + |f|}{|f|^{-1/2^m} + |f|^{1/2^m}} f^\sharp(z) \geq \frac{1}{2^m} f^\sharp(z) \ \ (z \in \Omega),$$

where we have used the inequality $a^{-1}+a \geq a^{-t}+a^t$ valid for $a > 0$ and $0 < t < 1$. By Marty's Criterion, the family $\{f^\sharp : f \in \mathcal{F}\}$ is not locally bounded. We deduce that, for fixed $m \in N$, the family $\{g^\sharp : g \in \mathcal{F}_m\}$ is not locally bounded. Using Marty's Criterion again we find that $\mathcal{F}_m$ is not normal, for all $m \in N$. Note that if $g \in \mathcal{F}_m$, then g omits the values $e^{2\pi ik/2m}$ for $k, m \in Z$. From the Zalcman Rescalling Lemma we thus deduce that there exists an entire function g_m omitting the values $e^{2\pi ik/2m}$ and satisfying $g_m(z) \leq g^\sharp_m(0) = 1$. The g_m thus form a normal family and we have $g_{m_j} \to G$ for some subsequence $\{g_{m_j}\}$ and some nonconstant entire function G. By Hurwitz's Theorem, G omits the values $e^{2\pi ik/2m}$ for all $k, m \in N$. Since $G(C^n)$ is open this implies that $|G(z)| \neq 1$ for all $z \in C^n$. Thus either $|G(z)| < 1$ for all $z \in C^n$ or $|G(z)| > 1$ for all $z \in C^n$. In the first case G is bounded and thus constant by Liouville's Theorem. In the second case $1/G$ is bounded. Again $1/G$ and thus G is constant. Thus we get a contradiction in both cases. □

We shall now show that the following result, in which the values omitted are allowed to vary with the function, as long as they do not approach one another too closely, is also true.

Theorem 4.3 (Carathéodory's Theorem) *Let $\mathcal{F}$ be a family of functions holomorphic on $\Omega \subset C^n$. Suppose that for some $\varepsilon > 0$, there exist for each $f \in \mathcal{F}$ distinct points $a_f, b_f \in C$ such that for all $z \in \Omega$, $f(z) \neq a_f, b_f$ and*

$$s(a_f, \infty)s(a_f, b_f)s(\infty, b_f) > \varepsilon.$$

Then $\mathcal{F}$ is normal on Ω.

Proof Otherwise, there exists some ball B in Ω, which we may assume to be unit ball, on which $\mathcal{F}$ fails to be normal. Then by Zalcman's Rescalling Lemma [3], there exist $f_j \in \mathcal{F}$, $z_j \in B$ and $\rho_j \to 0+$ such that $f_j(z_j + \rho_j\zeta) = g_j(\zeta) \to g(\zeta)$ uniformly on compact subsets of C^n, where g is a nonconstant entire function. Taking successive subsequences and renumbering, we can assume that $a_{f_j} \to a$ and $b_{f_j} \to b$, where a and b are distinct points in C. Since $g_j(\zeta) - a_{f_j} \neq 0$ and g is nonconstant, it follows from Theorem 4.2 that $g(\zeta) \neq a$. Similarly, $g(\zeta) \neq b$. Again, we may assume that the omitted values are 0 and 1. The reasoning used at the end of the first proof of Theorem 4.1 shows that g is a constant, a contradiction. □

Montel's theorem remains valid if the omitted values are replaced by omitted functions, so long as the omitted functions never take on the same value at points of Ω.

Theorem 4.4 (Fatou's Theorem) *Let $a(z)$ and $b(z)$ be functions holomorphic on $\Omega \subset C^n$ such that $a(z) \neq b(z)$ for each $z \in \Omega$. Let $\mathcal{F}$ be a family of functions holomorphic on Ω such that for each $z \in \Omega$*

$$f(z) \neq a(z) \qquad f(z) \neq b(z)$$

for all $f \in \mathcal{F}$. Then $\mathcal{F}$ is normal on Ω.

Proof Consider the family of functions

$$\mathcal{G} = \left\{ \frac{f(z) - a(z)}{f(z) - b(z)} \text{ for all } f \in \mathcal{F} \right\}.$$

Then each $g \in \mathcal{G}$ is holomorphic on Ω; and if $g \in \mathcal{G}$, then $g(z) \neq 0, 1$ for $z \in \Omega$. Thus $\mathcal{G}$ is normal on Ω by Theorem 4.1. But then, as is easily seen, $\mathcal{F}$ is normal on Ω as well. □

5 Criterion of Normality

Unfortunately, in practice Marty's criterion almost useless, as verification of the condition (2.3) in cases when normality is not already evident is generally extremely difficult. For example, given a family of holomorphic functions $\mathcal{F}$ such that on every compact $K \subset \Omega$, say,

$$\sharp f(z) \geq M(K), \quad z \in K,$$

we see that the Marty criterion is insufficient to establish normality, and a stronger version is required. The following has been

Theorem 5.1 *Let $\mathcal{F}$ be a family of holomorphic functions on $\Omega \subset C^n$ and assume that for each compact subset $K \subset \Omega$ there exists a constant $M(K) > 0$ such that*

$$\sharp f(z) \geq M(K), \quad z \in K,$$

holds for every $f \in \mathcal{F}$. Then $\mathcal{F}$ is normal in Ω.

The main tool in the proof of Theorem 5.1 are Montel's Theorem and Montel's Criterium.

References

1. G. Aladro, S.G. Krantz, A criterion for normality in C^n. J. Math. Anal. Appl. **161**, 1–8 (1991)
2. R. Brody, Compact manifolds and hyperbolicity. Trans. Am. Math. Soc. **235**, 213–219 (1978)

3. P.V. Dovbush, Zalcman's lemma in C^n. Complex Var Elliptic Equ. **65**, 796–800 (2019). https://doi.org/10.1080/17476933.2019.1627529
4. J.E. Joseph, M.H. Kwack, Some classical theorems and families of normal maps in several complex variables. Complex Variables, Theory and Application: An International Journal **29**(4), 343–362 (1996)
5. R. Narasimhan, *Several Complex Variables* (University of Chicago Press, Chicago, 1971)
6. X. Pang, Bloch's principle and normal criterion. Sci. China Ser. A **32**, 782–791 (1989)
7. X. Pang, On normal criterion of meromorphic functions. Sci. China Ser. A **33**, 521–527 (1990)
8. A. Ros, The Gauss map of minimal surfaces, in *Differential Geometry, Valencia 2001* (World Scientific, River Edge, 2002), pp. 235–252
9. N. Steinmetz, *Nevanlinna Theory, Normal Families, and Algebraic Differential Equations*. Universitext. ix+235 (Springer, Berlin, 2017). ISBN 978-3-319-59799-7
10. D.D. Thai, P.N.T. Trang, P.D. Huong, Families of normal maps in several complex variables and hyperbolicity of complex spaces. Complex Var. Theory Appl. Int. J. **48**(6), 469–482 (2003)
11. L. Zalcman, Normal families: new perspectives. Bull. Am. Math. Soc. **35**, 215–230 (1998)
12. L. Zalcman, Variations on Montel's theorem. Bull. Soc. Sci. Lett. de Łódź, Sér. Déform. **LIX**(1), 25–36 (2009)
13. L. Zalcman, A heuristic principle in complex function theory. Am. Math. Mon. **82**, 813–818 (1975)

To the Theory of One Class of Three-Dimensional Integral Equation with Super-Singular Kernels by Tube Domain

Nusrat Rajabov

Abstract In this paper, we study one new class of three-dimensional integral equations with super-singular kernels in a cylindrical domain, when the kernel has super-singularity on the lower base and the lateral surface of the cylinder. Depending on the roots of the characteristic equations (1.2) and (1.3), the integral representations of the solutions of equation (1.1) are found in explicit form. In the case where the parameters are present in the kernels, such that the general solution of the integral equation contains arbitrary functions, inversion formulas are found. On the basis of integral representations and their inversion formulas, in cases where the general solutions of the integral equation contain arbitrary functions, the correct formulation of a Dirichlet type problem is clarified and its solution is found.

Keywords Integral representation · Super-singular kernels · Inversion formula · Three-dimensional integral equations · Dirichlet type boundary value problem

Mathematics Subject Classification (2010). Primary 44A15; Secondary 35C10, 45E10

1 Integral Representations of a Variety of Solutions

Start the article with the command

Let Ω denote the cylindrical domain $\Omega = \{(t, z) : a < t < b, |z| < R\}$. The base of this cylinder will be denoted by $D = \{t = a, |z| < R\}$ and the lateral surface will be denoted by $S = \{a < t < b, |z| = R\}$, $z = x + iy$. We consider the

N. Rajabov (✉)
Tajik National University, Research Institute, Dushanbe, Tajikistan
e-mail: nusrat38@mail.ru

P. Cerejeiras et al. (eds.), *Current Trends in Analysis, its Applications and Computation*, Research Perspectives,
https://doi.org/10.1007/978-3-030-87502-2_9

integral equation in the domain Ω of the form

$$\varphi(t,z)+\int_a^t \frac{K_1(t,\tau)}{(\tau-a)^\alpha}\varphi(\tau,z)d\tau+\frac{1}{\pi}\iint_D \frac{exp[i\theta]K_2(r,\rho)}{(R-\rho)^\beta(s-z)}\varphi(t,s)ds+$$
$$+\frac{1}{\pi}\int_a^t \frac{d\tau}{(\tau-a)^\alpha}\iint_D \frac{exp[i\theta]K_3(t,\tau;r,\rho)}{(R-\rho)^\beta(s-z)}\varphi(\tau,s)ds=f(x,z), \tag{1.1}$$

where $\theta = args, s = \zeta + i\eta, ds = d\zeta d\eta, \rho^2 = \zeta^2+\eta^2, r^2 = x^2 + y^2$, $K_1(t,\tau) = \sum_{j=1}^{n} A_j(w_a^\alpha(t) - w_a^\alpha(\tau))^{j-1}$, $K_2(r,\rho) = \sum_{l=1}^{m} B_l(w_R^\beta(r) - w_R^\beta(\rho))^{l-1}$, $K_3(t,\tau;r,\rho;) = K_1(t,\tau).K_2(r,\rho)$, $A_j(1 \le j \le n)$, $B_l(1 \le l \le m)$—given constants , $f(t,z)$—are given functions and $\varphi(t,z)$—unknown functions, $w_a^\alpha(t) = [(\alpha-1)(t-a)^{\alpha-1}]^{-1}$, $\alpha = const > 1$, $w_R^\beta(r) = [(\beta-1)(R-r)^{\beta-1}]^{-1}$, $\beta = const > 1$.

The solution to the integral equation (1.1) will be sought in the class of functions $\varphi(t,z) \in C(\overline{\Omega})$, $\varphi(a,z) = 0$, $\varphi(t,Re^{i\theta}) = 0$, $\theta = argz$, and its asymptotic behavior, which for $t \to a$ and $r \to R$ are determined by the formulas

$$\varphi(t,z) = o[(t-a)^{\delta_1}],\ \delta_1 > (n+1)(\alpha-1) \text{ at } t \to a,$$

$$\varphi(t,z) = o[(R-r)^{\delta_2}],\ \delta_2 > (m+1)(\beta-1) \text{ at } r \to R.$$

In this paper, depending on the roots of the characteristic equations

$$\lambda^n + \sum_{j=1}^{n} A_j(j-1)!\lambda^{n-j} = o \tag{1.2}$$

and

$$\mu^m + \sum_{j=1}^{m} B_j(j-1)!\mu^{m-j} = o, \tag{1.3}$$

integral representations of the variety of solutions are obtained. A study of particular cases of Eq. (1.1) is devoted to the work [2–12] Let in the integral equation (1.1) $K_1(t,\tau)$, $K_2(r,\rho)$, $K_3(t,\tau;r,\rho)$ be related by the formula $K_3(t,\tau;r,\rho) = K_1(t,\tau)K_2(r,\rho)$. Then introducing the new $\Psi(t,z)$ function by the formula

$$\Psi(t,z) = \varphi(t,z) + \int_a^t \frac{K_1(t,\tau)}{(t-a)^\alpha}\varphi(\tau,z)d\tau \equiv K_\alpha(\varphi), \tag{1.4}$$

we arrive at the solution of the following Integral equation of the type I.N. Vekua [13]

$$\Psi(t,z)+\frac{1}{\pi}\iint_D \frac{exp[i\theta]K_2(r,\rho)}{(R-\rho)^\beta(s-z)}\psi(t,s)ds = f(t,z). \tag{1.5}$$

Thus, in this case, the problem of finding a solution to integral equation (1.1) has been reduced to solving an extended system of integral equations (1.4) and (1.5). The study of particular cases of integral equations (1.4) and (1.5) is the subject of [2–12]. In the case when the roots of the characteristic equation (1.2) are different, real, positive, and the solution of the integral equation (1.4) exists, then according to [7] the general solution of the integral equation (1.4) is representable in the form

$$\begin{aligned}\varphi(t,z) = &\sum_{j=1}^{n} C_j(z)exp[-\lambda_j w_a^\alpha(t)] + \psi(t,z)+\\ &\frac{1}{\Delta_0}\int_a^t\Big\{\sum_{j=1}^{n}(-1)^{n+j}\Delta_{jn}exp[\lambda_j(w_a^\alpha(\tau)-w_a^\alpha(t))]\Big\}\frac{\psi(\tau,z)}{(\tau-a)^\alpha}d\tau \equiv \\ &\sum_{j=1}^{n} C_j(z)exp[-\lambda_j w_a^\alpha(t)] + (K_\alpha)^{-1}(\psi),\end{aligned} \tag{1.6}$$

where Δ_0 is a Vandermond determinant, corresponding parameters $\lambda_j(1 \le j \le n)$, Δ_{jn} is minor of $(n-1)$-order, which is obtained from Δ_0 by dividing n-th lines and j-th column, $C_j(z)(1 \le j \le n)$ are arbitrary function of the domain D. Integral in the right part of the formula (1.6) converges, if $\psi(t,z) \in C(\overline{\Omega})$, $\psi(a,z) = 0$ with asymptotic behavior

$$\psi(t,z) = o[exp[-\lambda w_a^\alpha(t)](t-a)]^\gamma], \gamma > \alpha - 1 \text{ at } t \to a, \tag{1.7}$$

where $\lambda = \max(\lambda_1, \lambda_2, \ldots, \lambda_n)$. The function $\psi(t,z)$ have property(1.7), if $f(t,z) \in C(\overline{\Omega})$, $f(a,z) = 0$ with asymptotic behavior

$$f(t,z) = o[exp[-\lambda w_a^\alpha(t)](t-a)]^\gamma], \gamma > \alpha - 1 \text{ at } t \to a, \tag{1.8}$$

where $\lambda = \max(\lambda_1, \lambda_2, \ldots, \lambda_n)$. Note that

$$\begin{aligned}\frac{\partial}{\partial\overline{\zeta}}\Big[\frac{K_2(r,\rho)\psi(t,\rho)}{(R-\rho)^\beta}\Big] &= -\frac{\partial}{\partial\rho}\Big[\int_\rho^R \frac{K_2(r,\rho_1)\psi(t,\rho_1)}{(R-\rho_1)^\beta}d\rho_1\Big]e^{i\theta}\\ &= \frac{K_2(r,\rho)\psi(t,\rho)}{(R-\rho)^\beta}e^{i\theta}.\end{aligned}$$

From here

$$\frac{K_2(r,\rho)\psi(t,\rho)}{(R-\rho)^\beta}e^{i\theta} = \frac{\partial}{\partial\overline{\zeta}}\Big[\frac{K_2(r,\rho)\psi(t,\rho)}{(R-\rho)^\beta}\Big]$$

Then in the integral equation (1.5) $\Psi(t,z) = \Psi(t,r)$, we have

$$\frac{1}{\pi}\iint_D \frac{exp[i\theta]K_2(r,\rho)}{(R-\rho)^\beta(s-z)}\psi(t,s)ds = \frac{1}{\pi}\iint_D \frac{exp[i\theta]K_2(r,\rho)}{(R-\rho)^\beta(s-z)}\psi(t,\rho)ds =$$
$$= \frac{1}{\pi}\iint_D \frac{\partial}{\partial\overline{\zeta}}\Big[\frac{K_2(r,\rho_1)\psi(t,\rho_1)}{(R-\rho_1)^\beta}d\rho_1\Big]\frac{ds}{s-z} = -\int_r^R \frac{K_2(r,\rho)\psi(t,\rho)}{(R-\rho)^\beta}d\rho.$$

Then the integral equation (1.5) has the following form

$$\psi(t,r) - \int_r^R \frac{K_2(r,\rho)\psi(t,\rho)}{(R-\rho)^\beta}d\rho = f(t,r), \tag{1.9}$$

if $f(t,z) = f(t,r)$.

In addition, suppose that in Eq. (1.9) $\psi(t,r) \in C^{(m)}(D)$ respect to the variable r. Then, differentiating the integral equation (1.9) m times with respect to the variable r and each time multiplying by $(R-r)^\beta$, we arrive at the solution of the following expressed m-order differential equation

$$\begin{aligned}&(D_r^\beta)^m\psi(t,r) + B_1(D_r^\beta)^{m-1}\psi(t,r) + B_2(D_r^\beta)^{m-2}\psi(t,r)+\\&+2!B_3(D_r^\beta)^{m-3} + \ldots + (m-1)!B_m\psi(t,r) = 0,\end{aligned} \tag{1.10}$$

where $D_r^\beta = (R-r)^\beta\frac{\partial}{\partial r}$.

The homogeneous differential equation (1.10) corresponds to the characteristic equation (1.3). In the case when in the Eq. (1.9) the parameters $B_j(1 \le j \le m)$ are such that the roots of characteristic equation (1.3) are real, different, and negative, then the solution of the homogeneous differential equation (1.10) is given by the formula

$$\psi(t,r) = \sum_{j=1}^{n} exp[\mu_j w_R^\beta(r)]\psi_j(t), \tag{1.11}$$

where ψ_j $(t)(1 \le j \le m)$—arbitrary function variable t, μ_j—roots of characteristic equation (1.3). A function of the form (1.11) will also be a solution to the inhomogeneous integral equation (1.9). To find a general solution to the inhomogeneous integral equation (1.9), it is necessary to find a particular solution to the inhomogeneous integral equation (1.9).

In the case when the parameters $B_j(1 \leq j \leq m)$ in the integral equation (1.9) are such that the roots of the characteristic equation (1.3) are real, different, and negative, the following statement holds

Lemma 1 *Let the parameters $B_j(1 \leq j \leq m)$ in integral equation (1.9) be such that the roots of characteristic equation (1.3) are real, different, and negative, the function $f(t,r) \in C(\overline{\Omega})$, $f(t,R) = 0$ with asymptotic behavior*

$$f(t,r) = o[exp[-\mu w_R^\beta(r)](R-r)]^\delta], \delta > \alpha - 1 \ at \ r \to R, \tag{1.12}$$

where $\mu > max(|\mu_1|, |\mu_2|, \ldots, |\mu_m|)$.

Then the integral equation (1.9) in the class of functions $\Psi(t,z) \in C(\overline{\Omega})$, vanishes on the line $r = R$ is always solvable and its solution is given by the formula

$$\Psi(t,z) = \sum_{j=1}^{n} \Phi_j(t,z) exp[\mu_j w_R^\beta(r)] + f(t,r) +$$

$$\frac{1}{\Delta_0^1} \int_r^R \Big\{ \sum_{j=1}^{m} (-1)^{m+j} \Delta_{jm}^1 exp[\mu_j(w_R^\beta(r) - w_R^\beta(\rho))] \Big\} \frac{f(t,\rho)}{(R-\rho)^\beta} d\rho$$

where Δ_0^1 is Vandermond determinant for parameters $\mu_j(1 \leq j \leq m)$, Δ_{jm}^1 is minor of $(m-1)$-order, which obtained from Δ_0^1 by dividing m-th lines and j-th column, $\Phi_j(t,z)$ $(1 \leq j \leq m)$ are arbitrary function of two variables which are continuously by variables t and analytically by variables z.

Note that

$$\int_r^R \frac{exp[\mu_j w_R^\beta(\rho)] f(t,\rho)}{(R-\rho)^\beta} d\rho = -\frac{1}{\pi} \iint_D \frac{exp[i\theta + \mu_j w_R^\beta(\rho)]}{(R-\rho)^\beta (s-z)} f(t,\rho) ds,$$

we have

$$\Psi(t,z) = \sum_{j=1}^{n} \Phi_j(t,z) exp[-\mu_j w_R^\beta(r)] + f(t,r) -$$

$$-\frac{1}{\Delta_0^1} \frac{1}{\pi} \iint_D \Big\{ \sum_{j=1}^{m} (-1)^{m+j} \Delta_{jm}^1 exp[\mu_j(w_R^\beta(\rho) - w_R^\beta(r))] \Big\} \frac{exp[i\theta] f(t,\rho)}{(R-\rho)^\beta (s-z)} ds \equiv$$

$$\equiv \sum_{j=1}^{n} \Phi_j(t,z) exp[-\mu_j w_R^\beta(r)] + T_\beta(f). \tag{1.13}$$

For $\mu_j > 0$ $(1 \le j \le m)$ a solution of the form (1.3) exists if $f(t,r) \in C(\overline{D})$, $f(t,R) = 0$ with asymptotic behavior

$$f(t,r) = o[exp[-\mu w_R^\beta(r)](R-r)]^{\delta_1}], \delta_1 > \alpha - 1 \text{ at} r \to R, \tag{1.14}$$

where $\mu = max(\mu_1\mu_2, \ldots, \mu_m)$. Substituting the obtained value $\Psi(t,z)$ from expression (1.3) into formula (1.6), we have

$$\varphi(t,z) = \sum_{j=1}^{n} C_j(z) exp[-\lambda_j w_a^\alpha(t)] + \\ + \sum_{j=1}^{m} exp[\mu_j w_R^\beta(r)](K_\alpha)^{-1}(\Phi_j(t,z)) + (K_\alpha)^{-1} T_\beta(f). \tag{1.15}$$

The integrals in expression (1.15) converge if $f(a,r) = 0$ with asymptotic behavior (1.8). Thus, the following sentence is proved

Theorem 1 *Suppose that in the integral equation (1.1) the functions present in the nuclei are interconnected by the formula* $K_3(t,\tau;r,\rho) = K_1(t,\tau)K_2(r,\rho)$.

In $K_1(t,\tau)$ *the parameters* A_j $(1 \le j \le n)$ *are such that the roots of the characteristic equation (1.2)* λ_j $(1 \le j \le n)$ *are real, different, and positive and parameters* B_j $(1 \le j \le m)$ *such that the roots of characteristic equation (1.3) are real, different, and negative, the function* $f(t,r) \in C(\overline{\Omega})$, $f(t,R) = 0$ *with asymptotic behavior (1.12),* $f(a,r) = 0$ *with asymptotic behavior (1.8).*

Then any solution of the integral equation (1.1) from the class $C(\overline{\Omega})$ *is representable in the form (1.15), where* $C_j(z)(1 \le j \le n)$ *are arbitrary continuous functions in the domain* $\overline{D}$, *with* $C_j(Re^{i\theta}) = 0$ $(1 \le j \le n)$, *with asymptotic behaviors*

$$C_j(z) = o[(R-r)]^\gamma](1 \le j \le n), \gamma > \beta - 1 \, at \, r \to R. \tag{A}$$

$\Phi_j(t,z)(1 \le j \le m)$ *are arbitrary function of two variables which are continuously by variables t and analytically by variables z. Moreover* $\Phi_j(a,z) = 0$ *with asymptotic behavior*

$$\Phi_l(t,z) = o[(t-\alpha)]^\gamma], \gamma > \alpha - 1 \, at \, t \to a. \tag{B}$$

Depending on the sign of the roots of characteristic equations (1.2) and (1.3), various integral representations of the solution manifold are obtained containing a finite number of arbitrary functions $C_j(z)$ of the domain $\overline{D}$ and a finite number of arbitrary functions of two variables $\Phi_j(t,z)$ are continuous in the variable t and analytic in the variable z.

In particular, if the parameters $A_j (1 \leq j \leq n)$ are such that the roots of the characteristic equation (1.2) λ_j $(1 \leq j \leq n)$ are real, different, and negative, and the parameters B_j $(1 \leq j \leq m)$ such that the roots of characteristic equation (1.3) are real, different, and negative, the following statement holds:

Theorem 2 *Let in the integral equation (1.1) the functions present in the nuclei are interconnected by the formula* $K_3(t, \tau; r, \rho) = K_1(t, \tau)K_2(r, \rho)$.

In $K_1(t, \tau)$, *the parameters* A_j $(1 \leq j \leq n)$ *are such that the roots of the characteristic equation (1.2)* λ_j $(1 \leq j \leq n)$ *are real, different, and negative and in* $K_2(r, \rho)$ *the parameters* $B_j (1 \leq j \leq m)$ *are such that the roots of characteristic equation (1.3) are real, different, and positive, the function* $f(t, z) = f(t, r) \in C(\overline{\Omega})$, $f(a, z) = 0$ *with asymptotic behavior (1.8),* $f(t, R) = 0$ *with asymptotic behavior*

$$f(t, r) = o[(R - r)^{\delta_1}], \; \delta_1 > \beta - 1 \, at \, r \to R. \tag{1.16}$$

Then the integral equation (1.1) in the class $C(\overline{\Omega})$ *has a unique solution, which is given by the formula*

$$\varphi(t, z) = (K_\alpha)^{-1} T_\beta(f).$$

2 Inversion Formula

Now suppose that in the integral representation (1.13) the functions $\Psi(t, z) = K_\alpha(\varphi)$ are known, then according to [1] we find

$$\Phi_k(t, z) = \frac{exp[\mu_k w_R^\beta(r)]}{\Delta_0^1} \sum_{j=1}^{n} (-1)^{k+j} \Delta_{jk}^1 D_{\overline{z}}^{j-1} [K_\alpha(\varphi) - T_\beta(f)], \; (1 \leq k \leq m), \tag{2.1}$$

where $D_{\overline{z}} = 2exp[-i\theta]\frac{\partial}{\partial \overline{z}}$, Δ_0^1 is Vandermond determinant for parameters μ_j $(1 \leq j \leq m)$, Δ_{jm}^1 is minor of $(m-1)$-order, which obtained from Δ_0^1 by dividing m-th lines and j-th column.

Theorem 3 *Let all the conditions of Theorem 1 be fulfilled, the function* $f(t, z)$ *and the unknown function* $\Phi(t, z)$ *be differentiable* $(m-1)$ *times. Then in the integral equation (1.13) the functions* $\Phi_k(t, z)$ $(1 \leq k \leq m)$ *through* $\varphi(t, z)$, $f(t, z)$ *and their derivatives* $(m-1)$*th orders are found by formulas (2.1).*

Now, in the integral representation (1.15), suppose that the functions $f(t, z)$, $\Phi_k(t, z)$ $(1 \leq k \leq m)$, $\varphi(t, z)$ *are known and it is necessary to find* $C_j(z) (1 \leq j \leq n)$. *For this purpose, we represent formula (1.15) in the following form*

$$\sum_{j=1}^{n} C_j(z)exp[-\lambda_j w_a^\alpha(t)] = \varphi(t,z)-$$
$$-\sum_{j=1}^{m} exp[\mu_j w_R^\beta(r)](K_\alpha)^{-1}(\Phi_j(t,z)) - (K_\alpha)^{-1}T_\beta\beta(f) \equiv \tag{2.2}$$
$$\equiv T[f(t,z), \varphi(t,z), \Phi_1(t,z), \ldots, \Phi_m(t,z)].$$

Suppose that in (2.2) the well-known function $f(t,z)$, is the unknown function $\varphi(t,z)$ and the functions $\Phi_j(t,z)$ $(1 \le j \le m)$ by variable t differentiable $(n-1)$-times. Each time after differentiation, multiplying both sides of the resulting expression by $(t-a)^\alpha$, to find the function $C_j(z)$ $(1 \le j \le n)$, we obtain the following algebraic system

$$\sum_{j=1}^{n} C_j(z)exp[-\lambda_j w_a^\alpha(t)] = T[f(t,z), \varphi(t,z), \Phi_1(t,z), \ldots, \Phi_m(t,z)]$$

$$\sum_{j=1}^{n} C_j(z)exp[-\lambda_j w_a^\alpha(t)]\lambda_j = D_\alpha^t[T[f(t,z), \varphi(t,z), \Phi_1(t,z), \ldots, \Phi_m(t,z)]]$$

$$\sum_{j=1}^{n} C_j(z)exp[-\lambda_j w_a^\alpha(t)]\lambda_j^2 = (D_\alpha^t)^2[T[f(t,z), \varphi(t,z), \Phi_1(t,z), \ldots, \Phi_m(t,z)]]$$

....

....

$$\sum_{j=1}^{n} C_j(z)exp[-\lambda_j w_a^\alpha(t)]\lambda_j^{n-1} = (D_\alpha^t)^{n-1}$$

$$[T[f(t,z), \varphi(t,z), \Phi_1(t,z), \ldots, \Phi_m(t,z)]].$$

where $D_\alpha^t = (t-a)^\alpha \frac{d}{dt}$. Solving this system, we found

$$C_j(z) = \frac{exp[\lambda_j w_a^\alpha(t)]}{\Delta_0}$$
$$\sum_{k=1}^{n}(-1)^{j+k}\Delta_k^j((D_\alpha^t))^{k-1}[T[f(t,z), \varphi(t,z), \Phi_1(t,z), \ldots, \Phi_m(t,z)]] \tag{2.3}$$

$1 \le j \le n$, where Δ_0 is Vander mond determinant for parameters λ_j $(1 \le j \le n)$, Δ_k^j is minor of $(n-1)$-order, which obtained by dividing k-th lines and j-th column.

Theorem 4 *Let all the conditions of Theorem 1 be fulfilled, the function* $f(t, z)$, *the unknown function* $\varphi(t, z)$ *be differentiable* $(n - 1)$*-times with respect to the variable* t. *Then, in the integral representation (1.5) of the function* $C_k(z)$ $(1 \leq k \leq n)$ *through the values* $\varphi(t, z)$, $f(t, z)$, $\Phi_j(t, z)$ $(1 \leq k \leq n)$ *and their derivatives of order* $(m - 1)$ *with respect to variable* t *are found by formula (2.3).*

The integral representations and those obtained in Theorems 1 and their inversion formulas obtained in Theorems 5 and 6 make it possible for the integral equation (1.1) to pose and study various boundary value problems.

Problem D It is required to find a solution to the integral equation (1.1) from the class $C(\overline{\Omega})$ when all the conditions of Theorems 1, 3, and 4 are satisfied with respect to the boundary conditions

$$Re\Big[exp[\mu_k w_R^{\beta}(r) D_{\overline{z}}^{j-1}[K_\alpha(\varphi)]]\Big]_{r=R} = E_k^j(t, \theta)(1 \leq j, k \leq m), 0 \leq \theta \leq 2\pi \tag{2.4}$$

on the boundary of the region D, the condition

$$\Big[exp[\mu_k w_R^{\beta}(r) D_{\overline{z}}^{j-1}[K_\alpha(\varphi)]]\Big]_{z=0} = F_k^j(t) \quad (1 \leq j, k \leq m) \tag{2.5}$$

on the axis of the cylinder, condition

$$\Big[exp[\lambda_k w_R^{\alpha}(t)(D_\alpha^t)^{j-1}[(\varphi)]]\Big]_{t=a} = W_k^j(z) \quad (1 \leq j, k \leq n) \tag{2.6}$$

to the lateral surface of the cylinder, where $E_k^j(t, \theta)$ $(1 \leq j, k \leq m)$ are the given functions of the boundary of the lower base of the cylinder, $F_k^j(t)$ $(1 \leq j, k \leq m)$ are given functions of the cylinder axis and $W_k^j(z)$ $(1 \leq j, k \leq n)$ are given functions of the lower base of the cylinder.

Solution Problem D Let all the conditions of Theorems 3 and 4 be fulfilled. Then from formula (2.1) we find

$$\begin{aligned} &\Big[Re\Phi_k(t, z)\Big]_{r=R} = \\ &= \frac{1}{\Delta_0^1} \sum_{j=1}^{m} (-1)^{k+j} \Delta_{jk}^1 K_a\Big[Re\Big[exp\Big[\mu_k \omega_R^{\beta}(r)\Big] D_{\overline{z}}^{j-1}(\varphi)\Big]_{r=R}\Big] = \\ &= \frac{1}{\Delta_0^1} \sum_{j=1}^{m} (-1)^{k+j} \Delta_{jk}^1 K_a\Big[E_k^j(t, \theta)\Big] \equiv W_k(t, \theta) \quad (1 \leq k \leq m). \end{aligned} \tag{2.7}$$

Thus, to determine the unknown functions $\Phi_k(t, z)(1 \leq k \leq m)$, it is necessary to solve m Schwarz-type problems of the theory of analytic functions [13]. According to [1], the solution of type problems (2.7) are given by the formulas

$$\Phi_k(t,z)=\frac{1}{2\pi}\int_\gamma \frac{\tau+z}{\tau(\tau-z)}W_k(t,\theta)d\tau+i\Phi_k(t)\ (1\le k\le m), \tag{2.8}$$

where $\Phi_k(t)$—arbitrary functions point t. From formula (2.1) we find

$$\Phi_k(t,0)=$$
$$=\frac{1}{\Delta_0^1}\sum_{j=1}^{m}(-1)^{k+j}\Delta_{jk}^1\Big[K_a[exp[\mu_k\omega_R^\beta(r)]D_{\bar z}^{j-1}(\varphi)_{z=0}]\Big]=$$
$$=\frac{1}{\Delta_0^1}\sum_{j=1}^{m}(-1)^{k+j}\Delta_{jk}^1\Big[K_a[F_k^j(t)]\Big]\equiv E_k(t),\quad (1\le k\le m). \tag{2.9}$$

From formula (2.8) we find

$$\Phi_k(t,0)=\frac{1}{2\pi}\int_\gamma \frac{1}{\tau}W_k(t,\theta)d\tau+i\Phi_k(t),\ (1\le k\le m), \tag{2.10}$$

Comparing formulas (2.9) and (2.10), we have

$$\frac{1}{2\pi}\int_\gamma \frac{1}{\tau}W_k(t,\theta)d\tau+i\Phi_k(t)=E_k(t),\ \ (1\le k\le m),$$

From here we find

$$i\Phi_k(t)=E_k(t)-\frac{1}{2\pi}\int_\gamma \frac{1}{\tau}W_k(t,\theta)d\tau.$$

Representing the found values of $i\Phi_k(t)$ in formula (2.8), we find the explicit form $\Phi_k(t,z)$ in the following form

$$\begin{aligned}\Phi_k(t,z)=\frac{1}{2\pi}\int_\gamma \frac{\tau+z}{\tau(\tau-z)}W_k(t,\theta)d\tau+E_k(t)-\\ -\frac{1}{2\pi}\int_\gamma \frac{1}{\tau}W_k(t,\theta)d\tau,\ (1\le k\le m)\end{aligned} \tag{2.11}$$

To find $C_j(z)$ $(1 \leq j \leq n)$, we use formula (2.3) and condition (2.6). From these formulas we find

$$C_j(z) = \frac{1}{\Delta_0} \sum_{k=1}^{n} (-1)^{j+k} \Delta_k^j \Big[exp[\lambda_j \omega_a^\alpha(t)](D_a^t)^{k-1}(\varphi) \Big]_{t=a} =$$
$$= \frac{1}{\Delta_0} \sum_{k=1}^{n} (-1)^{j+k} \Delta_k^j W_k^j(z) \equiv \Psi_j(z) \ (1 \leq j \leq n) \tag{2.12}$$

Thus, if a solution to problem D exists, then it can be represented in the form (1.15), (2.11) and (2.12). If we take into account properties (A) and (B), we see that the functions $\Psi_j(z)(1 \leq j \leq n)$ vanish on the boundary of the region D with the following asymptotic behaviors

$$\Psi_j(z) = o[(R-r)^\gamma], \ (1 \leq j \leq n), \ \gamma_1 > \beta - 1 \quad \text{at} \quad r \to \alpha \tag{2.13}$$

and the functions $E_k(t)$, $W_k(t, \theta)$ vanish on the side surface of the cylinder with the following asymptotic behavior

$$E_k(t) = o[(t-\alpha)^{\gamma_1}], \ \gamma_1 > \alpha - 1 \quad \text{at} \quad t \to a \tag{2.14}$$

Thus, the following statement is proved

Theorem 5 *Let the conditions of Theorem 1, 3, and 4 be satisfied. In addition, under the conditions of problem D, let the functions* $E_k^j(t, \theta)$, $F_k^j(t)$ $(1 \leq j, k \leq m)$, $W_k^j(z)$ $(1 \leq j, k \leq n)$, *such that* $\Psi_j(Re^{i\theta}) = 0$ $(1 \leq j \leq n)$ *with asymptotic behavior (2.13) and* $E_k(a) = 0$ $(1 \leq k \leq m)$ *with asymptotic behavior (2.14). Then problem D has a unique solution, which is given by formulas (1.15), (2.11) and (2.12).*

References

1. F.D. Gakhov, *Boundary Value Problems*. Nauka, M. (Pergamon, Oxford, 1977) 640 p.
2. N. Rajabov, *Volterra Type Integral Equations with Boundary and Interior Fixed Singularity and Super-singularity Kernels and Their Application* (LAP LAMBERT Academic Publishing, 2011), 252 p.
3. N. Rajabov, To theory one class of two-dimensional complex integral equation for which all boundary are singular lines. Report Adigs (Cherces) International Academy of Science (2014), pp. 16–23
4. N. Rajabov, Rieman-Hilbert boundary value problems for one class of two-dimensional complex integral equation. Bulletin Chujand University of Name Academician Bobojon Gafurova "Modern Mathematics Problems and its Instruction, vol. 2, no. 29, p1, Chujand (2014), pp. 227–229

5. N. Rajabov, Integral representation of the manifold solution for new class of the volterra type integral equation with a boundary singularity in the case, when kernel contain logarithmic singularity and its power. J. Math. Syst. Sci. **6**, 23–37 (2016)
6. N. Rajabov, *New Methods for Volterra Type Integral Equation with Boundary Singular Point*. New Trends in Analysis and Interdisciplinary Applications, Select Contributions of the 10th ISAAC Congress, Macau 2015 (Springer, Berlin, 2017), pp. 121–127
7. N. Rajabov, *Volterra Type Integral Equation with Super-singular Kernels*. Functional Analysis in Interdisciplinary Applications, Astana, Kazakhstan. Springer Proceedings in Mathematics and Statistics, vol. 216 (Springer, Berlin, 2017), pp. 333–340
8. N. Rajabov, To theory one class of three- dimensional singular integral equation by laver domain. Bull. Tajik Natl. Univ. Ser. Nat. Sci. **2**, 6–10 (2018)
9. N. Rajabov, To theory of one class of three- dimensional singular integral equation. Bull. Tajik Natl. Univ. Ser. Nat. Sci. **3**, 5–11 (2018)
10. N. Rajabov, *To Theory One Class of Three-Dimensional Integral Equation by Tube Domain with Singular Kernels*. Modern problems explicit science and its role in forming scientific world outlook society, dedicate to 30-years independent States the Republic of Tajikistan, Chujand , 26–27 October 2018 (2018), pp. 147–150
11. N. Rajabov, *Well-Posed Boundary Value Problems for New Classes of Singular Integral Equations in Cylindrical Domains*. Analysis and Partial Differential Equations: Perspectives from Developing Countries. Springer Proceedings in Mathematics and Statistics, vol. 275 (Springer, Berlin, 2019), pp. 1–23
12. N. Rajabov, L. Rajaboba, *Introduction to Theory Many-Dimensional Volterra Type Integral Equation with Fixed Singular and Super-Singular Kernels and Applications* (LAP LAMBERT Academic Publishing, 2012), 502 p.
13. I.N. Vekua, *Generalized Analytic Functions* (Pergamum, Oxford, 1962)

Part III
Complex Geometry

Principal Higgs Bundles and Schottky Representations

Ana Casimiro and Susana Ferreira

Abstract Schottky representations are shown to be related to (A, B, A) branes in the moduli space of Principal Higgs bundles over a compact Riemann surface.

Keywords Representations of the fundamental group · Character varieties · Principal Higgs bundles · Branes · Riemann surfaces

Mathematics Subject Classification (2010) Primary 14H70 · 30F10; Secondary 32CXX

1 Introduction

The moduli space of G-Higgs bundles over a compact Riemann surface X, where G is a reductive complex algebraic group, was constructed by Hitchin [10], it has a structure of Hyperkähler manifold and it is related with mirror symmetry [9] and to the geometric Langlands correspondence [11]. It is well known by the Non Abelian Hodge theory [4, 5, 10, 15] that the moduli space of Higgs bundles is homeomorphic to the moduli space of solutions to the Hitchin equations and to the moduli space of representations of the fundamental group of X, $\pi_1(X)$, in G.

On the other hand, to parameterize all Riemann surfaces X of genus $g \geq 2$, a less well-known result, the so-called "retrosection theorem", or Schottky uniformization, asserts that we can write $X \cong \Omega/\Sigma$, for a certain *free group* of Möbius transformations $\Sigma \subset PSL_2\mathbb{C}$ of rank g (called, in this context, a *Schottky*

A. Casimiro (✉)
Departamento Matemática, Centro de Matemática e Aplicações, Faculdade de Ciências e Tecnologia, Universidade Nova de Lisboa, Caparica, Portugal
e-mail: amc@fct.unl.pt

S. Ferreira
Escola Superior de Tecnologia e Gestão, Instituto Politécnico de Leiria, Leiria, Portugal
e-mail: susfer@ipleiria.pt

P. Cerejeiras et al. (eds.), *Current Trends in Analysis, its Applications and Computation*, Research Perspectives,
https://doi.org/10.1007/978-3-030-87502-2_10

group) and region of discontinuity (for the Σ-action) $\Omega \subset \mathbb{CP}^1$ (see [2, 6]). This uniformization is defined on a less explicit subset of $\mathrm{Hom}(\Sigma, PSL_2\mathbb{C})/PSL_2\mathbb{C}$, having the advantage of providing holomorphic coordinates, when comparing for instance with the Fuchsian parameterization. Motivated by the mentioned Schottky uniformization, consider the presentation of the fundamental group of a fixed comapct Riemann surface X, of genus $g \geq 1$ (we are implicitly choosing a base point $x_0 \in X$, but this is irrelevant when considering isomorphism classes of representations) $\pi_1(X) = \left\langle \alpha_1, \cdots, \alpha_g, \beta_1, \cdots, \beta_g \mid \prod_{i=1}^{g} \alpha_i \beta_i \alpha_i^{-1} \beta_i^{-1} = 1 \right\rangle$. A representation $\rho : \pi_1(X) \to G$ is said to be *Schottky* (with respect to our choice of generators above) if $\rho(\alpha_i) = e$ for all $i = 1, \cdots, g$, with e the identity of G. Although the definitions require a choice of generators for $\pi_1(X)$, our results are independent of such choices. Thus, from an algebro-geometric perspective, Schottky representations (up to conjugation) are naturally parametrized by the affine *geometric invariant theory* (GIT) quotient

$$\mathbb{S} := \mathrm{Hom}\left(F_g, G\right) /\!\!/ G \subset \mathbb{B} := \mathrm{Hom}\left(\pi_1(X), G\right) /\!\!/ G$$

where F_g denotes a fixed free group of rank g. This affine algebraic variety was introduced and studied in [3] where was given its relation with the so called Schottky principal bundles. Schottky representations have the following natural topological interpretation. Suppose that M is a 3-manifold whose boundary is X, and the natural morphism $i_* : \pi_1(X) \to \pi_1(M)$ induced by the inclusion $i : X \hookrightarrow M$, has all the α_i in its kernel and the β_i are free, $i = 1, \cdots, g$. Then it is easy to see that Schottky representations are the representations of $\pi_1(X)$ which "extend to M", meaning that they factor through i_* (note that $\pi_1(M)$ is indeed a free group of rank g).

In addition to its relation to the uniformization problems for holomorphic G-bundles, Schottky representations also appear in a different context, related to the above mentioned non-abelian Hodge theory: recently, Baraglia and Schaposnik considered G-Higgs bundles over a Riemann surface equipped with an anti-holomorphic involution and showed that, inside the moduli space of G-Higgs bundles, the locus of those which are fixed by an associated involution define what is called an (A, B, A)-brane [1]. The study of branes is of great interest in connection with mirror symmetry and the geometric Langlands correspondence (see [11]).

In Sect. 2, we recall the definition of a Schottky respresentation introduced in [3], endow the set of Schottky representations, $\mathcal{S}$, with the structure of an affine algebraic variety, we consider the conjugation action of G on the variety and construct the GIT quotient, the Schottky space $\mathbb{S} := \mathcal{S} /\!\!/ G$, which is an irreducible affine algebraic variety but with singularities. In order to compute the dimension of the Schottky space we consider its tangent spaces in smooth representations (good representations). We describe them in terms of the first cohomology group of F_g in certain F_g-modules, and compute the dimension of the Schottky space. We also prove that the good locus of the Schottky space is a Lagrangian submanifold of the complex manifold of the smooth points of $\mathbb{B}$ and relate it with some moduli space of flat connections. In Sect. 3 we recall the moduli space of G-Higgs bundles,

$\mathcal{H}$, and the construction of branes inside the moduli space. More particulary, we consider anti-holomorphic involutions of a compact Rieman surface which will give an involution on $\mathcal{H}$ and on $\mathbb{B}$. Baraglia and Schaposnik [1] proved that the set of fixed smooth points of the involution is a Lagrangian submanifold of $\mathbb{B}$ and an (A, B, A)-brane inside $\mathcal{H}$. In Sect. 4, we identify all Schottky representations as elements of this brane (see [1, Proposition 43] and Proposition 4.1).

2 Schottky Representations

In this section, following [3], we give the definition of a Schottky representation of F_g into a general complex reductive algebraic group G, and some properties of the corresponding algebraic variety. Denote by $\pi_1 = \pi_1(X)$ the fundamental group of a compact Riemann surface X, with genus $g \geq 2$, and fix generators α_i, β_i, $i = 1, \cdots, g$, of π_1 giving the usual presentation $\pi_1 = \left\langle \alpha_1, \cdots, \alpha_g, \beta_1, \cdots, \beta_g \mid \prod_{i=1}^{g} \alpha_i \beta_i \alpha_i^{-1} \beta_i^{-1} = 1 \right\rangle$. Let G be a complex connected reductive algebraic group and denote by F_g a fixed free group of rank g, with g fixed generators $\gamma_1, \cdots, \gamma_g$. Since G is algebraic, and π_1 and F_g are finitely presented, both $\mathrm{Hom}(\pi_1, G)$ and $\mathrm{Hom}(F_g, G)$ are affine algebraic varieties. The reductive group G acts by conjugation on $\mathrm{Hom}(\pi_1, G)$ and hence, one can define a geometric invariant theory (GIT) quotient, called the G-character variety of π_1 (also called the *Betti space* in the context of the non-abelian Hodge theory, see [15]), as $\mathbb{B} := \mathrm{Hom}(\pi_1, G) /\!\!/ G$. This is a categorical quotient which, as an affine algebraic variety, is the maximal spectrum of the $\mathbb{C}$-algebra of G-invariant regular functions in $\mathbb{C}[\mathrm{Hom}(\pi_1, G)]$ (see, for example [13, Theorem 3.5]). Denote by $e \in G$, the unit element of G, and consider the short exact sequence of groups, $1 \to \ker\varphi \hookrightarrow \pi_1 \overset{\varphi}{\to} F_g \to 1$, where φ is the natural epimorphism given, in terms of the generators, by

$$\varphi(\alpha_i) = 1, \quad \text{and } \varphi(\beta_i) = \gamma_i, \quad \forall i = 1, \cdots, g, \tag{2.1}$$

so that $\ker\varphi$ is the normal subgroup of π_1 generated by all α_i.

Definition 2.1 A representation $\rho : \pi_1 \to G$ is called a **Schottky representation** if $\rho(\ker\varphi) = \{e\}$, for all $i \in \{1, \cdots, g\}$.

Let $\mathcal{S}$ denote the set of **Schottky representations**, it is easy to see that

$$\mathcal{S} \cong \mathrm{Hom}(F_g, \{e\} \times G) \cong \mathrm{Hom}(F_g, G) \cong G^g \subset \mathrm{Hom}(\pi_1, G)$$

where the last isomorphism is the evaluation map: $(\sigma : F_g \to G) \mapsto \left(\sigma(\gamma_1), \cdots, \sigma(\gamma_g)\right)$. Thus, $\mathcal{S}$ is a **smooth and irreducible** affine algebraic variety. The conjugation action of the reductive group G on $\mathrm{Hom}(\pi_1, G)$ restricts to an action on $\mathcal{S}$ thus it can be constructed the affine GIT quotient and also we have the homeomorphisms: $\mathbb{S} := \mathcal{S}/\!\!/G \cong G^g/\!\!/G \subset \mathbb{B} = \mathrm{Hom}(\pi_1, G)/\!\!/G$ The affine algebraic variety $\mathbb{S}$ is also **irreducible** but **singular** in general ([3, Proposition 2.4]).

The notion of a good representation allows us to consider smooth points of the GIT quotient, as we will see in this subsection. Let Γ be a finitely generated group, for example the fundamental group of a compact manifold. Denote by Z the center of G and given a representation $\rho : \Gamma \to G$ we denote by $Z(\rho) = \{h \in G : \rho(\gamma)h = h\rho(\gamma)\ \forall\gamma \in \Gamma\}$ its stabilizer in G, and denote by $G \cdot \rho$ its G-orbit in the algebraic variety $\mathrm{Hom}(\Gamma, G)$. Recall the following standard definitions.

Definition 2.2 Let $\rho : \Gamma \to G$ be a representation. We say that ρ is:

(a) *reducible* if $\rho(\Gamma)$ is contained in a proper parabolic subgroup of G,
(b) *irreducible* if it is not reducible,
(c) *good* if ρ is irreducible and $Z(\rho) = Z$.

In the case of Schottky representations.

Definition 2.3 A representation $\rho \in \mathcal{S} \subset \mathrm{Hom}\,(\pi_1, G)$ is said to be *good* if ρ is good as an element of $\mathrm{Hom}\,(\pi_1, G)$.

Denote the set of all good (resp. good Schottky) representations by $\mathrm{Hom}^{\mathsf{gd}}\,(\pi_1, G)$ (resp. $\mathcal{S}^{\mathsf{gd}}$). Since these notions are well defined under conjugation, we can define the corresponding quotient spaces: $\mathbb{B}^{\mathsf{gd}} := \mathrm{Hom}^{\mathsf{gd}}\,(\pi_1,\ G)/\!\!/G$ and $\mathbb{S}^{\mathsf{gd}} := \mathcal{S}^{\mathsf{gd}}/\!\!/G$, and, we have the inclusion $\mathbb{S}^{\mathsf{gd}} \subset \mathbb{B}^{\mathsf{gd}}$. The set of good representations is Zariski open in $\mathcal{S}$ (see for example [14]). By Martin [12, Lemma 4.6] there exists a good representation in $\mathrm{Hom}\,(\pi_1, G)$, that is, $\mathrm{Hom}^{\mathsf{gd}}\,(\pi_1,\ G) \neq \emptyset$, if X has genus $g \geq 2$. The case $g = 1$ is slightly different (see Section 9 of [3]).

Proposition 2.4 ([3, Proposition 2.13]) *Let $g \geq 2$. Then, there is always a good Schottky representation $\rho : \pi_1 \to G$. Moreover, such a representation can be defined to take values in a maximal compact subgroup of G.*

Theorem 2.5 *Let $g \geq 2$. The subsets of good representations* $\mathrm{Hom}^{\mathsf{gd}}\,(\pi_1,\ G)$ *and* $\mathcal{S}^{\mathsf{gd}}$ *are Zariski open in* $\mathrm{Hom}\,(\pi_1,\ G)$ *and* $\mathcal{S}$*, respectively. A good representation defines a smooth point in the corresponding geometric quotient. Thus, the geometric quotients* $\mathbb{B}^{\mathsf{gd}}$ *and* $\mathbb{S}^{\mathsf{gd}}$ *are complex manifolds, and* $\mathbb{S}^{\mathsf{gd}}$ *is a complex submanifold of* $\mathbb{B}^{\mathsf{gd}}$.

Proof By Proposition 2.4 there is a good Schottky representation, for $g \geq 2$. By Sikora [14, Proposition 33], the subspaces of good representations in $\mathrm{Hom}(\pi_1, G)$ and $\mathcal{S}$ are Zariski open. Thus, $\mathrm{Hom}^{\mathsf{gd}}\,(\pi_1,\ G)$ and $\mathcal{S}^{\mathsf{gd}}$ are open. Since we are considering either surface groups or free groups, [14, Corollary 50] shows that if $\rho \in \mathrm{Hom}^{\mathsf{gd}}\,(\pi_1,\ G)$, respectively $\rho \in \mathcal{S}^{\mathsf{gd}}$, then its class $[\rho]$ is a smooth point of $\mathbb{B}$, respectively $\mathbb{S}$. □

We begin by describing the tangent space of $\mathbb{B}$, at a good representation, in terms of the group cohomology of π_1. More generally, let Γ denote a finitely generated group and fix $\rho \in \mathrm{Hom}\,(\Gamma, G)$. The adjoint representation on the Lie algebra of G, $\mathfrak{g} = \mathsf{Lie}(G)$, composed with ρ, that is $\mathrm{Ad}_\rho : \Gamma \to G \to GL\,(\mathfrak{g})$, induces on $\mathfrak{g}$ a Γ-module structure, which we denote by $\mathfrak{g}_{\mathrm{Ad}_\rho}$. The following result giving an isomorphism between the Zariski tangent space of the character variety at a good representation ρ, and the first cohomology group $H^1\left(\Gamma, \mathfrak{g}_{\mathrm{Ad}_\rho}\right)$, was proved by Goldman [7] and Martin [12].

Theorem 2.6 *For a good representation* $\rho \in \mathrm{Hom}\,(\Gamma, G)$ *we have,*

$$T_{[\rho]}\left(\mathrm{Hom}\,(\Gamma, G) /\!\!/ G\right) \cong H^1\left(\Gamma, \mathfrak{g}_{\mathrm{Ad}_\rho}\right).$$

The identification between tangent spaces to character varieties and group cohomology spaces is very useful in many situations. In particular, we can use it to compute the dimension of the complex manifolds $\mathbb{B}^{\mathsf{gd}} = \mathrm{Hom}\,(\pi_1,\ G)^{\mathsf{gd}} /\!\!/ G$ and $\mathbb{S}^{\mathsf{gd}} \subset \mathbb{B}^{\mathsf{gd}}$, consisting of classes of good representations, when Γ is the fundamental group π_1 of a surface of genus g. In fact, by Martin [12, Lemma 6.2], we have, for $\rho \in \mathbb{B}^{\mathsf{gd}}$:

$$\dim Z^1\left(\pi_1, \mathfrak{g}_{\mathrm{Ad}_\rho}\right) = (2g-1)\dim G + \dim Z, \quad \dim B^1\left(\pi_1, \mathfrak{g}_{\mathrm{Ad}_\rho}\right) = \dim G - \dim Z,$$

and also if $[\rho] \in \mathbb{B}^{\mathsf{gd}}$, then $T_{[\rho]}\mathbb{B} \cong H^1\left(\pi_1, \mathfrak{g}_{\mathrm{Ad}_\rho}\right)$ and

$$\dim T_{[\rho]}\mathbb{B} = (2g-2)\dim G + 2\dim Z \tag{2.2}$$

We now compute the dimension of $\mathbb{S}$, using the techniques of group cohomology. By the density result (Theorem 2.5), the computations can be carried out at good representations.

Proposition 2.7 ([3, Proposition 7.1]) *Let* $g \geq 2$, *the dimension of* $\mathbb{S}$ *is given by* $\dim \mathbb{S} = (g-1)\dim G + \dim Z$.

Recall that a Lagrangian submanifold $L \subset M$ of a symplectic manifold M is a half dimensional submanifold such that the symplectic form vanishes on any tangent vectors to L. It is well known that character varieties of surface group representations have a natural symplectic structure [7], which can be constructed as follows. Consider an Ad-invariant bilinear form $\langle\, ,\, \rangle$ on $\mathfrak{g}$. Then, using the cup product on group cohomology

$$\cup : H^1\left(\pi_1, \mathfrak{g}_{\mathrm{Ad}_\rho}\right) \otimes H^1\left(\pi_1, \mathfrak{g}_{\mathrm{Ad}_\rho}\right) \to H^2\left(\pi_1, \mathfrak{g}_{\mathrm{Ad}_\rho}\right), \tag{2.3}$$

and composing it with the contraction with $\langle\, ,\, \rangle$ and with the evaluation on the fundamental 2-cycle, we obtain a non-degenerate bilinear pairing:

$$H^1\left(\pi_1, \mathfrak{g}_{\mathrm{Ad}_\rho}\right) \otimes H^1\left(\pi_1, \mathfrak{g}_{\mathrm{Ad}_\rho}\right) \xrightarrow{\cup} H^2\left(\pi_1,\ \mathfrak{g}_{\mathrm{Ad}_\rho}\right) \xrightarrow{\langle,\rangle} H^2\left(\pi_1,\ \mathbb{C}\right) \cong \mathbb{C} \tag{2.4}$$

Under the identification of the first cohomology group $H^1\left(\pi_1, \mathfrak{g}_{\mathrm{Ad}_\rho}\right)$ with the tangent space at a good representation $\rho \in \mathbb{B}^{\mathsf{gd}}$, this pairing defines a complex sympletic form on the complex manifold $\mathbb{B}^{\mathsf{gd}}$. This symplectic form is complex analytic with respect to the complex structure on $\mathbb{B}^{\mathsf{gd}}$ coming from the complex structure on G, and $\mathbb{S}^{\mathsf{gd}} \subset \mathbb{B}^{\mathsf{gd}}$ is Lagrangian.[1]

Theorem 2.8 *The good locus of the Schottky space $\mathbb{S}^{\mathsf{gd}}$ is a Lagrangian submanifold of $\mathbb{B}^{\mathsf{gd}}$.*

Proof The restriction of the map (2.3) to $H^1\left(F_g, \mathfrak{g}_{\mathrm{Ad}_\rho}\right)$ is a vanishing map:

$$\cup : H^1\left(F_g, \mathfrak{g}_{\mathrm{Ad}_\rho}\right) \otimes H^1\left(F_g, \mathfrak{g}_{\mathrm{Ad}_\rho}\right) \to H^2\left(F_g, \mathfrak{g}_{\mathrm{Ad}_\rho}\right) = 0,$$

because free groups have vanishing higher cohomology groups (see [2]). Since the tangent space, at a good point, to the strict Schottky locus $\mathbb{S}$ is identified with $H^1\left(F_g, \mathfrak{g}_{\mathrm{Ad}_\rho}\right)$ (see Theorem 2.6), this means that the symplectic form, defined above on $\mathbb{B}^{\mathsf{gd}}$, vanishes on any two tangent vectors to $\mathbb{S}^{\mathsf{gd}}$. Since the dimension of $\mathbb{B}^{\mathsf{gd}}$ is twice the dimension of $\mathbb{S}^{\mathsf{gd}}$ (see (2.2) and Proposition 2.7), we conclude the result. □

Let X be a compact Riemann surface of genus g, and let M be a **compact 3-handlebody** of genus g with boundary $\partial M \cong X$ such that $\pi_1(M, x_0) = F_g$ and let $x_0 \in X \subset M$. Thus, the inclusion $(X, x_0) \hookrightarrow (M, x_0)$ implies the surjective map $\varphi : \pi_1 = \pi_1(X, x_0) \to \pi_1(M, x_0)$ which asigns $\alpha_i \to 1$, $\beta_i \to \gamma_i$. Let $\mathbb{F}_M(G)$ denote the moduli space of flat G-connections over M.

Theorem 2.9 *The moduli space $\mathbb{S}$, of Schottky representations with respect to φ, coincides with the moduli space $\mathbb{F}_M(G)$. That is, $\mathbb{S} = \mathrm{Hom}(F_g, G)/\!\!/G \cong \mathbb{F}_M(G)$.*

Proof By hypothesis $\pi_1(M, x_0)$ is a free group of rank g, and π_1 has a "symplectic presentation" in terms of generators α_i and β_i, $i = 1, \cdots, g$, as in Eq. (2), so that $\varphi(\alpha_i) = 1, \quad \varphi(\beta_i) = \gamma_i, \quad i = 1, \cdots, g$, where $\gamma_1, \cdots, \gamma_g$ form a free basis of $\pi_1(M, x_0)$. Thus, a Schottky representation $\rho : \pi_1 \to G$ with respect to φ factors through a representation of $\pi_1(M, x_0) \cong F_g$ via φ. This is precisely the same as saying that the corresponding flat connection ∇_ρ on X extends, as a flat connection, to the 3-manifold M. Conversely, a flat G-connection on M induces a representation $\rho : \pi_1 \to G$ satisfying $\rho(\ker \varphi) = \{e\}$, and thus it is a Schottky representation of π_1 (with respect to φ). This correspondence is well defined up to conjugation by G, and so, we have a natural identification: $\mathbb{S} = \mathrm{Hom}(F_g, G)/\!\!/G \cong \mathbb{F}_M(G)$. □

[1] For a general real Lie group, the analogous pairing defines a smooth (C^∞) symplectic structure, see [7].

3 Principal Higgs Bundles and Branes

Let X be a compact Riemann surface of genus $g \geq 2$ and G a connected complex reductive group

Definition 3.1 A pair (E, ϕ) is a G-Higgs bundle on X: if E is a G-bundle on X and ϕ, the Higgs field, is a holomorphic section of $Ad(E)\otimes K$. ($Ad(E)$ is the adjoint bundle and K the canonical bundle of X)

Considering a notion of stability it can be constructed $\mathcal{H}$, the moduli space of G-Higgs bundles which has a hyperkähler structure (Hitchin, [10]). Denoting by (I, J, K) the choice of the three Hyperkähler complex structures we can consider submanifolds of $\mathcal{H}$ that are Lagrangian (type A) or complex (type B) with respect to each of the hyperkähler structures. Kasputin and Witten [11] called these submanifolds branes, more specifically, (B, A, A), (A, B, A) or (A, A, B)-branes. They have connection with the geometric Langlands program and mirror symmetry. In order to obtain branes we can consider anti-holomorphic involutions of X.

Let X be a compact Riemann surface of genus $g \geq 2$ and $f : X \to X$ an **anti-holomorphic involution**. This induces an involution on $\mathbb{B}$, indeed, fixing $x_0 \in X$, f induces an isomorphism between $\pi_1(X, x_0)$ and $\pi_1(X, f(x_0))$ and fixing γ, a path from x_0 to $f(x_0)$, the composition of the isomorphism with the conjugation by γ gives an automorphism of $\pi_1(X, x_0)$. Changing γ, the automorphism changes by composing with a inner automorphism. If we consider the good locus, $\mathbb{B}^{\mathsf{gd}}$, the involution preserves this subvariety. This can be identified with the moduli space of gauge equivalence classes of flat G-connections on X with reductive holonomy, so we get an involution of this one by pullback of connections. Now, the moduli space of gauge equivalence classes of flat G-connections on X is isomorphic to the moduli space of solutions to the Hitchin equations and this last one is isomorphic to $\mathcal{H}$. (Non-Abelian Hodge Theorem [4, 5, 10, 15]). In [1], it is denoted by $\mathcal{L}_G$ the set of **fixed points** of the involution in $\mathbb{B}^{\mathsf{gd}}$ (or in $\mathcal{H}$) and proved that

Proposition 3.2 ([1, Proposition 10]) *If non-empty, the set of good points of $\mathcal{L}_G$ is a smooth Lagrangian submanifold of $\mathbb{B}^{\mathsf{gd}}$.*

Following the ideas of [11], Baraglia and Schaposnik proved that

Theorem 3.3 ([1, Theorem 14]) *$\mathcal{L}_G$ is an (A, B, A)-brane defined on $\mathcal{H}$.*

Consider now the 3-manifold with boundary $\hat{X} := X \times [-1, 1]$, such that f defines an orientation preserving involution $\sigma : \hat{X} \to \hat{X}$ given by $\sigma(x, t) = (f(x), -t)$. The boundary of $\hat{X}$ consists of two copies of X and the boundary of the compact 3-manifold $M := \hat{X}/\sigma$, is homeomorphic to X.

Proposition 3.4 ([1, Proposition 43]) *The representations of X in G, which extend to M, belong to the (A, B, A)-brane $\mathcal{L}_G$.*

This subspace of representations can be viewed as flat G-connections on X that extend to flat G-connections over M, that is, as $\mathbb{F}_M(G)$.

4 Schottky Representations and Branes

Suppose now that we have X a compact Riemann surface with an anti-holomorphic involution $f : X \to X$, defining a real structure on X. Using the construction and notations of the Sect. 3, let M be the compact 3-manifold whose boundary is homeomorphic to X. Then,

Theorem 4.1 *Let $f : X \to X$ be an anti-holomorphic involution such that M is a handlebody of genus g, and let $x_0 \in X \subset M$ be fixed by f. Then, the moduli space $\mathbb{S}$ of Schottky representations with respect to the map φ in* (2.1) *is included in the Baraglia-Schaposnik brane $\mathcal{L}_G$.*

Proof In Proposition 3.4 it is proved the existence of an inclusion: $\mathbb{F}_M(G) \to \mathcal{L}_G \subset \mathcal{H}$. Since, by Theorem 2.9, $\mathbb{S}$ can be identified with $\mathbb{F}_M(G)$ the result follows. □

Remark 4.2 The assumption of the previous proposition is verified when the anti-holomorphic involution f has as fixed point locus the union of $g + 1$ disjoint loops and disconnected orientation double cover (see [8]). In this case, Proposition 3.2 says that the set of smooth points of $\mathcal{L}_G$ is a non-empty Lagrangian submanifold of $\mathcal{H}$. In a future work, we plan to further address this construction.

Conclusions

- Under our approach, since there are good Schottky representations for every $g \geq 2$, this furnishes a proof that the set of smooth points of the Baraglia-Schaposnik brane is non-empty.
- As $\mathbb{S} \subset \mathcal{L}_G$, in a future work, we plan to study the conditions under which this inclusion is actually a bijection.

Acknowledgments We thank C. Florentino for his support and comments and to A. Schmitt for organizing the session "Complex Geometry" in the 12th ISAAC Congress, where this work was presented.

This work was supported by Fundação para a Ciência e a Tecnologia (Portuguese Foundation for Science and Technology) through the project UID/MAT/00297/2019.

References

1. D. Baraglia, L. Schaposnik, Higgs bundles and (A,B,A)-branes. Commun. Math. Phys. **331**, 1271–1300 (2014)
2. K. Brown, *Cohomology of Groups*. Graduate Texts in Mathematics, vol. 87 (Springer, New York, 1994)
3. A. Casimiro, S. Ferreira, C. Florentino, Principal Schottky bundles over Riemann surfaces. Geom. Dedicata **201**, 379–409 (2019). http://dx.doi.org/10.1007/s10711-018-0398-2
4. K. Corlette, Flat G-bundles with canonical metrics. J. Differ. Geom. **28**, 361–382 (1988)
5. S.K. Donaldson, Twisted harmonic maps and the self-duality equations. Proc. Lond. Math. Soc. **55**(1), 127–131 (1988)
6. L. Ford, *Automorphic Functions*, 2nd edn. (Chelsea Publishing, New York, 1951)

7. W. Goldman, The symplectic nature of fundamental groups of surfaces. Adv. Math. **54**(2), 200–225 (1984)
8. B.H. Gross, J. Harris, Real algebraic curves. Ann. Sci. École Norm. Sup. **14**(2), 157–182 (1981)
9. T. Hausel, M. Thaddeus, Mirror symmetry, Langlands duality, and the Hitchin system. Invent. Math. **153**, 197–229 (2003)
10. N.J. Hitchin, The self-duality equations on a Riemann surface. Proc. Lond. Math. Soc. **55**(1), 59–126 (1987)
11. A. Kapustin, E. Witten, Electric-magnetic duality and the geometric Langlands program. Commun. Number Theory Phys. **1**, 1–236 (2007)
12. B. Martin, Restrictions of Representaions of surface group to a pair of free subgroups. J. Algebra **225**(1), 231–249 (2000)
13. P. Newstead, *Introduction to Moduli Problems and Orbit Spaces*. Tata Institute of Fundamental Research, Lectures on Mathematics and Physics, vol. 51 (Narosa Publishing House, New Delhi, 1978)
14. A. Sikora, Character varieties. Trans. Am. Math. Soc. **364**(10), 5173–5208 (2012)
15. C.T. Simpson, Constructing variations of Hodge structure using Yang-Mills theory and applications to uniformisation. J. Am. Math. Soc. **1**, 867–918 (1988)

On Hodge Polynomials of Singular Character Varieties

Carlos Florentino, Azizeh Nozad, Jaime Silva, and Alfonso Zamora

Abstract Let $\mathcal{X}_\Gamma G := \mathrm{Hom}(\Gamma, G)/\!\!/G$ be the G-character variety of Γ, where G is a complex reductive group and Γ a finitely presented group. We introduce new techniques for computing Hodge-Deligne and Serre polynomials of $\mathcal{X}_\Gamma G$, and present some applications, focusing on the cases when Γ is a free or free abelian group. Detailed constructions and proofs of the main results will appear elsewhere.

Keywords Hodge-Deligne polynomials · E-polynomials · Character varieties · Free group representations

Mathematics Subject Classification (2010) Primary 14L30; Secondary 32S35, 14D20

C. Florentino (✉)
Departamento de Matemática, Faculdade de Ciências, Universidade de Lisboa, Lisbon, Portugal
e-mail: caflorentino@ciencias.ulisboa.pt

A. Nozad
School of Mathematics, Institute for Research in Fundamental Sciences (IPM), Tehran, Iran
e-mail: anozad@ipm.ir

J. Silva
Área Departamental de Matemática, ISEL - Instituto Superior de Engenharia de Lisboa, Lisboa, Portugal
e-mail: jaime.silva@isel.pt

A. Zamora
Departamento de Matemática Aplicada a las TIC, ETSI Informáticos, Universidad Politécnica de Madrid, Madrid, Spain
e-mail: alfonso.zamora@upm.es

P. Cerejeiras et al. (eds.), *Current Trends in Analysis, its Applications and Computation*, Research Perspectives,
https://doi.org/10.1007/978-3-030-87502-2_11

1 Introduction

Let G be a connected reductive complex algebraic group, and Γ be a finitely presented group. The G-character variety of Γ is defined to be the (affine) geometric invariant theory (GIT) quotient

$$\mathcal{X}_\Gamma G = \mathrm{Hom}(\Gamma, G) /\!\!/ G.$$

The most well studied families of character varieties include the cases when the group Γ is the fundamental group of a Riemann surface Σ, and its "twisted" variants. In these cases, the non-abelian Hodge correspondence (see, for example [20]) shows that (components of) $\mathcal{X}_\Gamma G$ are homeomorphic to certain moduli spaces of G-Higgs bundles which appear in connection to important problems in Mathematical-Physics: for example, these spaces play an important role in the quantum field theory interpretation of the geometric Langlands correspondence, in the context of mirror symmetry [13].

The study of geometric and topological properties of character varieties is an active topic and there are many recent advances in the computation of their Poincaré polynomials and other invariants. For the surface group case ($\Gamma = \pi_1(\Sigma)$ and related groups) the calculations of Poincaré polynomials started with Hitchin and Gothen, and have been pursued more recently by Hausel, Lettelier, Mellit, Rodriguez-Villegas, Schiffmann and others, who also considered the parabolic version of these character varieties (see [12, 16, 19]). Those recent results use arithmetic methods: it is shown that the number of points of the corresponding moduli space over finite fields is given by a polynomial, which turns out to coincide with the E-polynomial of $\mathcal{X}_\Gamma G$ ([12, Appendix]). Then, in the smooth case, the pure nature of the cohomology of Higgs bundles moduli spaces allows the derivation of the Poincaré polyomial from the E-polynomial.

On the other hand, for many important classes of *singular* character varieties, explicitly computable formulas for the E-polynomials (also called Serre polynomials) are very hard to obtain. In the articles of Logares, Muñoz, Newstead and Lawton [14, 15] (using geometric methods) and of Baraglia and Hekmati [1] (using arithmetic methods), the E-polynomials are computed for several character varieties, with $G = GL(n, \mathbb{C})$, $SL(n, \mathbb{C})$ and $PGL(n, \mathbb{C})$ for small values of n, but the computations quickly become intractable for n higher than 3.

In this short article, we describe some of the techniques and constructions that we have recently developed for computations of E-polynomials of singular character varieties, and present some of their main applications.

The outline of the article is as follows. Section 1 covers notations and preliminaries on mixed Hodge and E-polynomials and on character varieties in the context of GIT. In Sect. 2, we explain how to use equivariant mixed Hodge structures to study (the identity component of) $\mathcal{X}_\Gamma G$ when Γ is a free abelian group and G a classical group. These character vareities have orbifold singularities and we can obtain their full mixed Hodge polynomials. In Sect. 3, for arbitrary Γ, we define a stratification of $GL(n, \mathbb{C})$-character varieties (which also exists for $G = SL(n, \mathbb{C})$ or $PGL(n, \mathbb{C})$) which allows writing down an explicit plethystic exponential relation between generating functions of the E-polynomials of $\mathcal{X}_\Gamma GL(n, \mathbb{C})$ and of its locus of irreducible representations $\mathcal{X}_\Gamma^{irr} GL(n, \mathbb{C})$. Finally, in Sect. 4, we consider the free group $\Gamma = F_r$ of rank r, and announce the solution of a conjecture of Lawton and Muñoz: the E-polynomials of $\mathcal{X}_{F_r} SL(n, \mathbb{C})$ and of $\mathcal{X}_{F_r} PGL(n, \mathbb{C})$ coincide, for every $n \in \mathbb{N}$. For lack of space, the proofs are omitted and will be published elsewhere.

2 Preliminaries on Hodge-Deligne Polynomials, Affine GIT and Character Varieties

In this article, all algebraic varieties are defined over $\mathbb{C}$, G is a connected reductive algebraic group, and Γ is a finitely presented group.

Let X be a quasi-projective variety (not necessarily irreducible), of complex dimension $\leq d$. Deligne showed that the compactly supported cohomology $H_c^*(X) := H_c^*(X, \mathbb{C})$ can be endowed with a mixed Hodge structure whose mixed Hodge numbers are given by

$$h^{k,p,q}(X) := \dim_{\mathbb{C}} H_c^{k,p,q}(X) \in \mathbb{N}_0,$$

for $k, p, q \in \{0, \cdots, 2d\}$, and we call (p, q) the k-weights of X, if $h^{k,p,q} \neq 0$ (c.f. [3, 18]).

Mixed Hodge numbers are symmetric in the weights, $h^{k,p,q} = h^{k,q,p}$, and $\dim_{\mathbb{C}} H_c^k(X) = \sum_{p,q} h^{k,p,q}$. Therefore, they provide the (compactly supported) Betti numbers, yielding the usual Betti numbers, by Poincaré duality, in the non-singular case. They are also the coefficients of the mixed Hodge polynomial of X on three variables,

$$\mu(X;\, t, u, v) := \sum_{k,p,q} h^{k,p,q}(X)\, t^k u^p v^q \in \mathbb{N}_0[t, u, v], \tag{2.1}$$

which specializes to the (compactly supported) Poincaré polynomial by setting $u = v = 1$, $P_t^c(X) := \mu(X;\, t, 1, 1)$ (and provides the usual Poincaré polynomial in the smooth situation). Plugging $t = -1$, mixed Hodge polynomials convert into the E-polynomial of X, or the *Serre polynomial* of X, given by

$$E(X;\, u, v) = \sum_{k,p,q} (-1)^k h^{k,p,q}(X)\, u^p v^q \in \mathbb{Z}[u, v].$$

From the E-polynomial we can compute the (compactly supported) Euler characteristic of X as $\chi^c(X) = E(X;\, 1, 1) = \mu(X;\, -1, 1, 1)$.

Serre polynomials satisfy an *additive property with respect to stratifications* by locally closed (in the Zariski topology) strata: if X has a closed subvariety $Z \subset X$ we have (see, eg. [18]),

$$E(X) = E(Z) + E(X \setminus Z).$$

The E-polynomial also satisfies (c.f. [4, 15]) a *multiplicative property for fibrations*. Namely, for a given algebraic fibration $F \hookrightarrow X \to B$, we have

$$E(X) = E(F) \cdot E(B)$$

in any of the following three situations:

(i) the fibration is locally trivial in the Zariski topology of B,
(ii) F, X and B are smooth, the fibration is locally trivial in the complex analytic topology, and $\pi_1(B)$ acts trivially on $H_c^*(F)$, or
(iii) X, B are smooth and F is a complex connected Lie group.

We say X is of *Hodge-Tate type* (also called *balanced* type) if all the k-weights are of the form (p, p) with $p \in \{0, \cdots, k\}$, in which case the sum in $\mu(X)$ reduces to a one-variable sum. In particular, the E-polynomials of Hodge-Tate type varieties depend only the product uv, so we write $x = uv$ and use the notation $E(X; x) := E(X; \sqrt{x}, \sqrt{x}) \in \mathbb{Z}[x]$.

Now let X be an affine algebraic variety, and let the reductive group G act algebraically on X. The induced action of G on the ring $\mathbb{C}[X]$ of regular functions on X defines the (affine) GIT quotient

$$X /\!\!/ G := Spec\left(\mathbb{C}[X]^G\right),$$

where $\mathbb{C}[X]^G$ is the subring of G-invariants in $\mathbb{C}[X]$. This quotient identifies G-orbits whose closures intersect, such that each point in the quotient classifies an equivalence class of orbits, leading to a stability condition. Let $G_x \subset G$ be the stabilizer of $x \in X$ and consider the orbit map through x, $\psi_x : G \to X;\ g \mapsto g \cdot x$. We define $x \in X$ to be *stable* if ψ_x is a proper map and *polystable* if the orbit $G \cdot x$ is closed in X. Stability implies polystability, but not conversely.

GIT shows that the stable locus $X^s \subset X$ is a Zariski open set (hence dense, when non-empty) and that the restriction of the affine quotient map $\Phi : X \to X /\!\!/ G$ to the stable locus, $X^s \to X^s/G$, is a geometric quotient (or an orbit space), where $\Phi(X^s)$ is Zariski open in $X /\!\!/ G$.

Now, consider a finitely presented group Γ. The (generally singular) algebraic variety of representations of Γ in G is

$$\mathcal{R}_\Gamma G = \mathrm{Hom}(\Gamma, G).$$

Each $\rho \in \mathcal{R}_\Gamma G$ is determined by $\rho(\gamma)$, for each generator $\gamma \in \Gamma$, and satisfying the relations of the group Γ. There is an algebraic action of G on the variety $\mathcal{R}_\Gamma G$ by conjugation of representations, $g^{-1}\rho g$, yielding the G-character variety of Γ,

$$\mathcal{X}_\Gamma G := \mathrm{Hom}(\Gamma, G)/\!\!/G,$$

as the GIT quotient.

By definition, polystable representations are representations $\rho : \Gamma \to G$ whose orbits $G \cdot \rho := \{g\rho g^{-1} : g \in G\}$ are Zariski closed in $\mathcal{R}_\Gamma G$. Alternatively, a representation ρ is polystable if and only if it is completely reducible (i.e, if $\rho(\Gamma) \subset P \subset G$ for some proper parabolic P of G, then $\rho(\Gamma)$ is contained in a Levi subgroup of P). Denote the subset of polystable representations in $\mathcal{R}_\Gamma G$ by $\mathcal{R}_\Gamma^{ps} G \subset \mathcal{R}_\Gamma G$, which is a Zariski locally-closed subvariety containing the stable locus $\mathcal{R}_\Gamma^{s} G \subset \mathcal{R}_\Gamma G$.

Proposition 2.1 ([7]) *There is a bijective correspondence:*

$$\mathcal{X}_\Gamma G = \mathcal{R}_\Gamma G/\!\!/G \cong \mathcal{R}_\Gamma^{ps} G/G,$$

where the right hand side is called the polystable quotient.

We say that ρ is *irreducible* if $\rho(\Gamma)$ is not contained in a proper parabolic subgroup of G. Alternatively, ρ is irreducible if it is polystable and Z_ρ, the centralizer of $\rho(\Gamma)$ inside G, is a finite extension of the center $ZG \subset G$. Denote by $\mathcal{R}_\Gamma^{irr} G \subset \mathcal{R}_\Gamma^{ps} G$ the subset of irreducible representations (being a Zariski open subset of $\mathcal{R}_\Gamma G$, $\mathcal{R}_\Gamma^{irr} G$ is a quasi-projective variety), and since irreducibility is well defined on G-orbits, denote by

$$\mathcal{X}_\Gamma^{irr} G := \mathcal{R}_\Gamma^{irr} G/G \tag{2.2}$$

the *G-irreducible character variety of* Γ, which is a geometric quotient, as it happens with the stable locus. In fact, it can be proved that irreducibility is equivalent to GIT stability for character varieties (see [2, Thm. 1.3(1)]).

3 The Free Abelian Case

In this section, we are concerned with the determination of the mixed Hodge polynomials of character varieties $\mathcal{X}_\Gamma G$ of the free abelian group of rank r, $\Gamma \cong \mathbb{Z}^r$. As we always work over $\mathbb{C}$, we abbreviate the notation of the classical groups such as the linear group, special linear, special orthogonal and symplectic to GL_n, SL_n, SO_n and Sp_n, respectively (instead of $GL(n, \mathbb{C})$, etc).

The topology and geometry of the character varieties $\mathcal{X}_{\mathbb{Z}^r}G$ was studied in [8, 21], among others. Most important for us are the following facts:

(i) there is only one irreducible component containing the trivial representation, that we denote by $\mathcal{X}^0_{\mathbb{Z}^r}G$ [21, Theorem 2.1],
(ii) if the semisimple part of G is a classical group (ie, one of SL_n, SO_n and Sp_n), there exists an algebraic isomorphism

$$\mathcal{X}^0_{\mathbb{Z}^r}G \cong (T_G)^r / W_G \tag{3.1}$$

where T_G is a maximal torus of G, and W_G its Weyl group [21, Theorem 2.1],
(iii) the irreducibility of the free abelian character varieties $\mathcal{X}_{\mathbb{Z}^r}G$ can be characterized, in terms of G: for example, if the semisimple part of G is a product of SL_n's and Sp_n's then $\mathcal{X}_{\mathbb{Z}^r}G$ is irreducible, so that $\mathcal{X}_{\mathbb{Z}^r}G \cong \mathcal{X}^0_{\mathbb{Z}^r}G$ [8, Theorem 1.2].

We now focus on the determination of the mixed Hodge numbers of $\mathcal{X}_{\mathbb{Z}^r}G$ when it is irreducible, or of $\mathcal{X}^0_{\mathbb{Z}^r}G$ when the algebraic isomorphism (3.1) applies. We start by explaining how mixed Hodge numbers transform under finite quotients.

Let X be a complex quasi-projective variety and F a finite group acting algebraically on it. The action of F on X induces an action on its cohomology. Since F acts by algebraic isomorphisms, it also induces an action on the mixed Hodge components. Then we can regard $H^{k,p,q}(X)$ as F-modules, that we denote by $\left[H^{k,p,q}(X)\right]_F$. As in Eq. (2.1) for the mixed Hodge polynomial, we codify these in the *equivariant mixed Hodge polynomial*, defined by

$$\mu_F(X;\, t, u, v) := \sum_{k,p,q} \left[H^{k,p,q}(X)\right] t^k u^p v^q \in R(F)[t, u, v]$$

whose coefficients belong to $R(F)$, the representation ring of F. The polynomial $\mu_F(X;\, t, u, v)$ may also be seen as a polynomial-weighted representation. For instance, one can consider equivariant cohomology to obtain an isomorphism

$$H^*(X/F) \cong H^*(X)^F \tag{3.2}$$

that respects mixed Hodge structures. In particular, this isomorphism allows us to identify the mixed Hodge polynomial of the quotient X/F as the coefficient of the trivial representation of $\mu_F(X;\, t, u, v)$ when written on a basis of irreducible representations of F. Another important consequence for us is the inequality $h^{k,p,q}(X) \geq h^{k,p,q}(X/F)$, which holds since $H^{k,p,q}(X/F)$ is given by the F-invariant part of $H^{k,p,q}(X)$. We conclude that if X is, for instance, a balanced variety, or if its mixed Hodge structure is actually *pure* (that is, if $h^{k,p,q} \neq 0$ then $k = p + q$), then the same holds for X/F.

We now summarize our strategy to obtain the mixed Hodge polynomials of $\mathcal{X}^0_{\mathbb{Z}^r}G$, in the cases when the isomorphism (3.1) holds (so, these character varieties are isomorphic to finite quotients of algebraic tori). The only non-zero Hodge numbers of the maximal torus $T_G \cong (\mathbb{C}^*)^n$ are $h^{k,k,k}(T_G)$. Moreover, its natural mixed Hodge structure satisfies:

$$H^{k,k,k}(T_G) \cong \bigwedge^k H^{1,1,1}(T_G).$$

So, the action of W_G on the cohomology ring can be understood from the one on the mixed Hodge component $H^{1,1,1}(T_G)$. The next three theorems are proved in [11].

Theorem 3.1 *For a reductive group G satisfying* (3.1), *we have*

$$\mu(\mathcal{X}^0_{\mathbb{Z}^r}G;\, t,u,v) = \frac{1}{|W_G|}\sum_{g\in W_G}\left[\det\left(I + tuv\, A_g\right)\right]^r$$

where A_g is the automorphism of $H^{1,1,1}(T_G)$ induced by the action of $g \in W_G$.

The proof starts by establishing the $r = 1$ case, and using the diagonal action for higher r as well as the isomorphism (3.1), together with the multiplicative relation for the equivariant polynomials $\mu_{W_G}\left(T^r_G\right) = \mu_{W_G}(T_G)^{\otimes r}$. We remark that Theorem 3.1 generalizes a formula for the Poincaré polynomial of $\mathcal{X}^0_{\mathbb{Z}^r}G$, recently obtained in [22].

To further work with Theorem 3.1, we examine the induced action of W_G on $H^{1,1,1}(T_G)$ for some classical groups. In the case $G = GL_n$, the Weyl group is the symmetric group S_n on n letters, which acts on $H^{1,1,1}(T_G) \cong \mathbb{C}^n$ by permutation of coordinates, and we obtain a general formula in terms of partitions of n.

A partition of $n \in \mathbb{N}$ is denoted by $[k] = [1^{k_1}\cdots j^{k_j}\cdots n^{k_n}]$ where the exponent $k_j \geq 0$ is the number of parts of size $j \in \{1,\cdots,n\}$, so that $n = \sum_{j=1}^n j\cdot k_j$. Let $\mathcal{P}_n$ denote the finite set of partitions of n.

Theorem 3.2 *The mixed Hodge polynomials of $\mathcal{X}_{\mathbb{Z}^r}GL_n$ and of $\mathcal{X}_{\mathbb{Z}^r}SL_n$ satisfy*

$$\mu(\mathcal{X}_{\mathbb{Z}^r}GL_n;\, t,u,v) = \mu(\mathcal{X}_{\mathbb{Z}^r}SL_n;\, t,x)(1+tuv)^r = \sum_{[k]\in\mathcal{P}_n}\prod_{j=1}^n\frac{(1-(-tuv)^j)^{k_j r}}{k_j!\, j^{k_j}},$$

By using similar considerations as for the GL_n case, we can also deduce a concrete formula for Sp_n in terms of *bipartitions*. A bipartition of n, denoted $[a,b] \in \mathcal{B}_n$ consists of two partitions $[a] \in \mathcal{P}_k$ and $[b] \in \mathcal{P}_l$, such that $0 \leq k,l \leq n$ with $k + l = n$. One can show that bipartitions of n are in one-to-one correspondence with conjugacy classes in W_{Sp_n}, the Weyl group of Sp_n.

Theorem 3.3 *The mixed Hodge polynomial of $\mathcal{X}_{\mathbb{Z}^r} Sp_n\mathbb{C}$ is given by*

$$\mu(\mathcal{X}_{\mathbb{Z}^r} Sp_n;\ t, u, v) = \frac{1}{2^n n!} \sum_{[a,b]\in\mathcal{B}_n} c_{[a,b]} \prod_{i=1}^{k} (1 - (-tuv)^i)^{a_i r} \prod_{j=1}^{l} (1 + (-tuv)^j)^{b_j r}$$

where $c_{[a,b]}$ is the size of the conjugacy class in W_{Sp_n}, corresponding to $[a, b] \in \mathcal{B}_n$.

The same method allows to obtain explicit expressions for $\mu(\mathcal{X}^0_{\mathbb{Z}^r} G)$ in the case of other reductive G; the special orthogonal groups SO_n will be addressed in a future work.

4 Generating Functions for E-Polynomials

In this section we consider character varieties with arbitrarily bad singularities. In this case, there are formidable difficulties in computing the corresponding Poincaré polynomials in general, and previous explicit methods have dealt with the E-polynomials for low dimensional groups such as SL_2 and SL_3 [1, 14, 15].

By using the additive and multiplicative properties of E-polynomial, for $G = GL_n$ we now address our new approach on E-polynomial computations based on a stratification of $\mathcal{X}_\Gamma G$ that we term *by partition type*, and which works for arbitrary Γ.

Using standard arguments in GIT, any character variety admits a stratification by the dimension of the stabilizer of a given representation. When G is the general linear group GL_n (as well as the related groups SL_n and PGL_n), there is a more convenient refined stratification that gives a lot of information on the corresponding character varieties $\mathcal{X}_\Gamma G$ which we call *stratification by partition type*.

Definition 4.1 Let $G = GL_n$ and $[k] \in \mathcal{P}_n$. We say that $\rho \in \mathcal{R}_\Gamma G = \mathrm{Hom}(\Gamma, G)$ is $[k]$-polystable if ρ is conjugated to $\bigoplus_{j=1}^n \rho_j$ where each ρ_j is, in turn, a direct sum of $k_j > 0$ *irreducible* representations of $\mathcal{R}_\Gamma(GL_j)$, for $j = 1, \cdots, n$ (by convention, if some $k_j = 0$, then ρ_j is not present in the direct sum).

We denote $[k]$-polystable representations by $\mathcal{R}^{[k]}_\Gamma G$ and use similar terminology/notation for equivalence classes under conjugation $\mathcal{X}^{[k]}_\Gamma G \subset \mathcal{X}_\Gamma G$. It is to be noted that the trivial partition $[n] = [n^1] \in \mathcal{P}_n$ corresponds exactly to the irreducible (or stable) locus: $\mathcal{X}^{[n]}_\Gamma G = \mathcal{X}^{irr}_\Gamma G$.

Proposition 4.2 *Fix $n \in \mathbb{N}$, and let $G = GL_n$. Then $\mathcal{X}_\Gamma G = \bigsqcup_{[k]\in\mathcal{P}_n} \mathcal{X}^{[k]}_\Gamma G$, as a disjoint union of locally closed quasi-projective varieties.*

The next result relates, by the plethystic exponential, the generating functions of the E-polynomials $E(\mathcal{X}_\Gamma GL_n)$ to the corresponding generating functions of the E-polynomials of the irreducible character varieties $E(\mathcal{X}^{irr}_\Gamma GL_n)$.

The *plethystic exponential* of a formal power series $f(x, y, z) = \sum_{n\geq 0} f_n(x, y) z^n \in \mathbb{Q}[x, y][[z]]$ is denoted by $\mathrm{PExp}(f)$, and defined formally (in terms of the usual exponential) as $\mathrm{PExp}(f) := e^{\Psi(f)} \in \mathbb{Q}[x, y][[z]]$, where Ψ acts on monomials as: $\Psi(x^i y^j z^k) = \sum_{l\geq 1} \frac{x^{li} y^{lj} z^{lk}}{l}$, where $(i, j, k) \in \mathbb{N}_0^3 \setminus \{(0, 0, 0)\}$, and is $\mathbb{Q}$-linear on $\mathbb{Q}[x, y][[z]]$. This exponential plays a prominent role in the combinatorics of symmetric functions, and has applications in counting of gauge invariant operators in supersymmetric quantum field theories (see eg. [5]).

Theorem 4.3 *Let Γ be any finitely presented group. Then:*

$$\sum_{n\geq 0} E(\mathcal{X}_\Gamma GL_n; u, v)\, t^n = \mathrm{PExp}\left(\sum_{n\geq 1} E(\mathcal{X}_\Gamma^{irr} GL_n; u, v)\, t^n\right).$$

The proofs of Theorem 4.3 and Proposition 4.2 are detailed in [9]; they allow to write explicit expressions for $E(\mathcal{X}_\Gamma GL_n)$, for any group Γ, for which we have a formula for $E(\mathcal{X}_\Gamma^{irr} GL_m)$, for all $m \leq n$, by a simple finite algorithm (and vice-versa). The formula of Theorem 4.3 generalizes a formula of [17] to an arbitrary group Γ, even if the corresponding GL_n-character variety is not of polynomial type.

5 The Free Group Case

In this last section, we describe applications of the above methods to the case of the free group of rank r, $\Gamma = F_r$; for simplicity we adopt the notations $\mathcal{X}_r GL_n$, $\mathcal{X}_r SL_n$, etc, for the corresponding character varieties. In [17], it was shown that $\mathcal{X}_r^{irr} GL_n$ and $\mathcal{X}_r GL_n$ are of polynomial type. Moreover, by counting points over finite fields and using a theorem of Katz ([12, Appendix]), Mozgovoy and Reineke found a formula for the E-polynomial of $\mathcal{X}_r^{irr} GL_n$ that can be written as follows (dropping the x variable in $E(X;\ x)$, and using $||[k]|| := k_1 + \cdots + k_d$ for the length of a partition $[k] \in \mathcal{P}_d$).

Proposition 5.1 ([9, 17]) *For $r, n \geq 2$, we have:*

$$E(\mathcal{X}_r^{irr} GL_n) = (x-1) \sum_{d|n} \frac{\mu(n/d)}{n/d} \sum_{[k]\in\mathcal{P}_d} \frac{(-1)^{||[k]||}}{||[k]||} \binom{||[k]||}{k_1, \cdots, k_d}$$

$$\prod_{j=1}^{d} b_j (x^{n/d})^{k_j} x^{\frac{n(r-1)k_j}{d} \binom{j}{2}},$$

where μ is the Möbius function, and the $b_j(x)$ are polynomials defined by:

$$\left(1+\sum_{n\geq 1} b_n(x)\,t^n\right)\left(1+\sum_{n\geq 1}\left((x-1)(x^2-1)\dots(x^n-1)\right)^{r-1}t^n\right)=1. \tag{5.1}$$

Using Propositions 4.2 and 5.1 and Theorem 4.3, we are able to write down very explicit expressions for $E(\mathcal{X}_r^{[k]}GL_n)$, the E-polynomials of all polystable strata of $\mathcal{X}_r GL_n$ (see [9, Secs. 5 and 6], where we also compute $E(\mathcal{X}_\Gamma^{irr}GL_n)$ for other Γ and low n).

We now provide a few lines on a forthcoming proof of the equality between the E-polynomials of $\mathcal{X}_r SL_n$ and of $\mathcal{X}_r PGL_n$ for all $n \in \mathbb{N}$. This has been conjectured in Lawton-Muñoz in [14], who proved by explicit computation the cases $n = 2$ and 3.

In a analogous way as for GL_n (see Sect. 4), we can define the $[k]$-polystable loci $\mathcal{X}_r^{[k]}SL_n$ and $\mathcal{X}_r^{[k]}PGL_n$ as follows. For a partition $[k] \in \mathcal{P}_n$, the $[k]$-stratum of $\mathcal{X}_r SL_n$ is defined by restriction of the corresponding one for GL_n:

$$\mathcal{X}_r^{[k]}SL_n := \{\rho \in \mathcal{X}_r^{[k]}GL_n \mid \det\rho = 1\},$$

where the determinant of a representation is an element of $\mathcal{R}_r\mathbb{C}^*$. By considering the action $\mathcal{R}_r\mathbb{C}^* \times \mathcal{X}_r GL_n \to \mathcal{X}_r GL_n$ given by multiplication of (conjugacy classes of) representations, which is well defined on the GIT quotients and preserves the stratification of GL_n, we can define

$$\mathcal{X}_r^{[k]}PGL_n := \mathcal{X}_r^{[k]}GL_n/\mathcal{R}_r\mathbb{C}^* = \mathcal{X}_r^{[k]}GL_n/(\mathbb{C}^*)^r. \tag{5.2}$$

Theorem 5.2 ([10]) *For the free group F_r, we have the equalities:*

$$E(\mathcal{X}_r SL_n) = E(\mathcal{X}_r PGL_n) = E(\mathcal{X}_r GL_n)(x-1)^{-r}$$
$$E(\mathcal{X}_r^{[k]}SL_n) = E(\mathcal{X}_r^{[k]}PGL_n) = E(\mathcal{X}_r^{[k]}GL_n)(x-1)^{-r},$$

for every r, n and partition $[k] \in \mathcal{P}_n$.

The proof of Theorem 5.2 uses geometric methods and has two parts. The easy part is the relation between the E-polynomials of $\mathcal{X}_r^{[k]}PGL_n$ and of $\mathcal{X}_r^{[k]}GL_n$, which follows from the locally trivial (in the Zariski topology) fibration corresponding to the quotient (5.2). The difficult part is the relation between the strata $\mathcal{X}_r^{[k]}PGL_n$ and $\mathcal{X}_r^{[k]}SL_n$ which involves finite quotients: it requires the proof of the triviality of the action of the center $\mathbb{Z}_n \subset SL_n$ on the cohomology (with compact support) of all the strata $\mathcal{X}_r^{[k]}SL_n$; for this we use equivariant cohomology and a deformation retraction between $\mathcal{X}_r^{irr}SL_n$ and the smooth part of the semialgebraic set $\mathrm{Hom}(F_r, SU(n))/SU(n)$ (see [6]).

Acknowledgments We would like to thank S. Lawton, M. Logares, S. Mozgovoy, V. Muñoz, A. Oliveira and F. Rodriguez-Villegas for several interesting and very useful conversations on topics around mixed Hodge structures and E-polynomials. We also thank Alexander Schmitt for the invitation to deliver a talk at the Special Session on Complex Geometry of the conference ISAAC 2019, where these results were presented.

This work was partially supported by CAMGSD and CMAFcIO of the University of Lisbon, the projects PTDC/MAT-GEO/2823/2014 and PTDC/MAT-PUR/30234/2017, FCT Portugal, a grant from IPM (Iran) and ICTP (Italy), and the project MTM2016-79400-P by the Spanish government.

References

1. D. Baraglia, P. Hekmati, Arithmetic of singular character varieties and their E-polynomials. P. Lond. Math. Soc. **114**(2), 293–332 (2017)
2. A.C. Casimiro, C. Florentino, Stability of affine G-varieties and irreducibility in reductive groups. Int. J. Math. **23**(8), 1250082 (2012)
3. P. Deligne, Théorie de Hodge, II. Inst. Hautes études Sci. Publ. Math. **40**(1), 5–57 (1971)
4. A. Dimca, G.I. Lehrer, Purity and equivariant weight polynomials, in *Algebraic Groups and Lie Groups* (Cambridge University Press, Cambridge, 1997)
5. B. Feng, A. Hanany, Y.-H. He, Counting gauge invariants: the plethystic program. J. High Energy Phys. **03**, 090 (2007)
6. C. Florentino, S. Lawton, The topology of moduli spaces of free group representations. Math. Ann. **345**, 453–489 (2009)
7. C. Florentino, S. Lawton, Singularities of free group character varieties. Pac. J. Math. **260**, 149–179 (2012)
8. C. Florentino, S. Lawton, Topology of character varieties of abelian groups. Topol. Appl. **173**, 32–58 (2014)
9. C. Florentino, A. Nozad, A. Zamora, Generating series for the E-polynomials of $GL(n, \mathbb{C})$-character Varieties. Math. Nachrichten. (preprint, 2019). arXiv:1902.06837
10. C. Florentino, A. Nozad, A. Zamora, Serre polynomials of SL_n and PGL_n-character Varieties of the free group. J. Geom. Phys. **161**, 104008 (2021). https://doi.org/10.1016/j.geomphys.2020.104008
11. C. Florentino, J. Silva, Hodge-Deligne polynomials of character varieties of abelian groups. Open Math. (to appear, 2017). arXiv:1711.07909
12. T. Hausel, F. Rodriguez-Villegas, Mixed Hodge polynomials of character varieties, with an appendix by N. M. Katz. Invent. Math. **174**(3), 555–624 (2008)
13. A. Kapustin, E. Witten, Electric-magnetic duality and the geometric Langlands program. Commun. Num. Theor. Phys. **1** , 1–236 (2007)
14. S. Lawton, V. Muñoz, E-polynomial of the SL(3, C)-character variety of free groups. Pac. J. Math. **282**(1), 173–202 (2016)
15. M. Logares, V. Muñoz, P.E. Newstead, Hodge polynomials of $SL(2, \mathbb{C})$-character varieties for curves of small genus. Rev. Mat. Comput. **26**, 635–703 (2013)
16. A. Mellit, Poincaré polynomials of moduli spaces of Higgs bundles and character varieties (no punctures). Inv. Math. **221**, 301–327 (2020)
17. S. Mozgovoy, M. Reineke, Arithmetic of character varieties of free groups. Int. J. Math. **26**, 12 (2015)
18. C. Peters, J. Steenbrink, *Mixed Hodge Structures*. Ergebnisse der Mathematik, vol. 52 (Springer, Berlin, 2008)
19. O. Schiffmann, Indecomposable vector bundles and stable Higgs bundles over smooth projective curves. Ann. Math. **183**, 297–362 (2016)

20. C.T. Simpson, Constructing variations of Hodge structure using Yang-Mills theory and applications to uniformization. J. Am. Math. Soc. **1**, 867–918 (1988)
21. A.S. Sikora, Character varieties of Abelian groups. Math. Z. **277**, 241–256 (2014)
22. M. Stafa, Poincaré series of character varieties for nilpotent groups. J. Group Theory **22**(3), 419–440 (2019)

Diagonal Double Kodaira Structures on Finite Groups

Francesco Polizzi

Abstract We introduce some special system of generators on finite groups, that we call *diagonal double Kodaira structures* and whose existence is equivalent to the existence of some special Kodaira fibred surfaces, that we call *diagonal double Kodaira fibrations*. This allows us to rephrase in purely algebraic terms some results about finite Heisenberg groups, previously obtained in Causin and Polizzi (Ann. Sc. Norm. Super. Pisa Cl. Sci. (5) XXII:1309–1352, 2021), and makes possible to extend them to the case of arbitrary extra-special p-groups.

Keywords Surface braid groups · Extra-special p-groups · Kodaira fibrations

Mathematics Subject Classification (2010). 14J29, 14J25, 20D15

1 Introduction

A *Kodaira fibration* is a smooth, connected holomorphic fibration $f_1 \colon S \longrightarrow B_1$, where S is a compact complex surface and B_1 is a compact complex curve, which is not isotrivial (this means that not all its fibres are biholomorphic to each others). The genus $b_1 := g(B_1)$ is called the *base genus* of the fibration, whereas the genus $g := g(F)$, where F is any fibre, is called the *fibre genus*. If a surface S is the total space of a Kodaira fibration, we will call it a *Kodaira fibred surface*; it is possible to prove that every such a surface is minimal and of general type.

Examples of Kodaira fibrations were originally constructed in [1, 11] in order to show that, unlike the topological Euler characteristic, the signature σ of a real manifold is not multiplicative for fibre bundles. In fact, every Kodaira fibred surface S satisfies $\sigma(S) > 0$, see for example the introduction of [12], whereas $\sigma(B_1) = \sigma(F) = 0$, and so $\sigma(S) \neq \sigma(B_1)\sigma(F)$.

F. Polizzi (✉)
Dipartimento di Matematica e Informatica, Università della Calabria, Cosenza, Italy
e-mail: francesco.polizzi@unical.it

P. Cerejeiras et al. (eds.), *Current Trends in Analysis, its Applications and Computation*, Research Perspectives,
https://doi.org/10.1007/978-3-030-87502-2_12

A *double Kodaira surface* is a compact complex surface S, endowed with a *double Kodaira fibration*, namely a surjective, holomorphic map $f\colon S \longrightarrow B_1 \times B_2$ yielding, by composition with the natural projections, two Kodaira fibrations $f_i\colon S \longrightarrow B_i$, $i = 1, 2$.

In [5] the author (in collaboration with A. Causin) introduced a new topological method to construct double Kodaira fibrations, based on the so-called *Heisenberg covers* of $\Sigma_b \times \Sigma_b$, where Σ_b denotes a real, closed, connected, orientable surface of genus b (from now on, we will simply write "a real surface of genus b"). These are finite Galois covers $S \longrightarrow \Sigma_b \times \Sigma_b$, whose branch locus is the diagonal $\Delta \subset \Sigma_b \times \Sigma_b$ and whose Galois group is isomorphic to a finite Heisenberg group. In this note we rephrase the group-cohomological methods of [5] in a purely algebraic way, by introducing the so-called *diagonal double Kodaira structures* on a finite group G, see Definition 2.1. These are special systems of generators of G, whose existence is equivalent to the fact that G is a good quotient of some higher genus pure braid group on two strands, where "good" means that the natural braid called A_{12} in [6] has non-trivial image under the quotient map. The existence of diagonal double Kodaira structures yields in turn the existence of some special double Kodaira fibrations, that we call *of diagonal type*, see Definition 4.2.

With this new and compact terminology, we give a short account of some of the main results contained in [5], namely

- the existence of (double) Kodaira fibrations over every curve of genus b (and not only over special curves with extra automorphisms) and the proof that the number of such fibrations over a fixed base can be arbitrarily large, see Theorem 4.5;
- the first "double solution" to a problem, posed by Geoff Mess, from Kirby's problem list in low-dimensional topology, see Theorem 4.6;
- the existence of an infinite family of (double) Kodaira fibrations with slope strictly higher than $3 + 1/3$, see Theorem 4.8 and Remark 4.9.

This paper also contains some new results, namely

- the construction of diagonal double Kodaira structures on extra-special p-groups of any exponent, see Theorems 3.7 and 3.10. This extends the equivalent statements for extra-special p-groups of exponent p proved in [5];
- an explicit upper bound for the slope of a diagonal double Kodaira fibration, see Proposition 4.12 and Remark 4.13.

An intriguing problem is the existence of diagonal double Kodaira structures on finite groups that are not extra-special or, more generally, on finite groups whose nilpotency class is at least 3, cf. Remark 2.4. However, we will not develop this point here, hoping to come back on it in a sequel to this paper.

Notation and Conventions The order of a finite group G is denoted by $|G|$. If $x \in G$, the order of x is denoted by $o(x)$. The subgroup generated by $x_1, \ldots, x_n \in G$ is denoted by $\langle x_1, \ldots, x_n \rangle$. The center of G is denoted by $Z(G)$. If $x, y \in G$, their commutator is defined as $[x, y] = xyx^{-1}y^{-1}$. We denote both the cyclic group of order p and the field with p elements by $\mathbb{Z}_p$.

2 Diagonal Double Kodaira Structures

Let G be a finite group and let $b,\ n \geq 2$ be two positive integers.

Definition 2.1 A *diagonal double Kodaira structure of type* $(b,\ n)$ on G is an ordered set of $4b+1$ generators

$$\mathfrak{S} = (\mathsf{r}_{11},\ \mathsf{t}_{11}, \ldots, \mathsf{r}_{1b},\ \mathsf{t}_{1b},\ \mathsf{r}_{21},\ \mathsf{t}_{21}, \ldots, \mathsf{r}_{2b},\ \mathsf{t}_{2b},\ \mathsf{z}),$$

with $o(\mathsf{z}) = n$, such that the following relations are satisfied. We systematically use the commutator notation in order to indicate the conjugacy action, writing for instance $[x,\ y] = zy^{-1}$ instead of $xyx^{-1} = z$.

- Surface relations

$$[\mathsf{r}_{1b}^{-1},\ \mathsf{t}_{1b}^{-1}]\,\mathsf{t}_{1b}^{-1}\,[\mathsf{r}_{1\,b-1}^{-1},\ \mathsf{t}_{1\,b-1}^{-1}]\,\mathsf{t}_{1\,b-1}^{-1}\cdots[\mathsf{r}_{11}^{-1},\ \mathsf{t}_{11}^{-1}]\,\mathsf{t}_{11}^{-1}\,(\mathsf{t}_{11}\,\mathsf{t}_{12}\cdots\mathsf{t}_{1b}) = \mathsf{z}$$

$$[\mathsf{r}_{21}^{-1},\ \mathsf{t}_{21}]\,\mathsf{t}_{21}\,[\mathsf{r}_{22}^{-1},\ \mathsf{t}_{22}]\,\mathsf{t}_{22}\cdots[\mathsf{r}_{2b}^{-1},\ \mathsf{t}_{2b}]\,\mathsf{t}_{2b}\,(\mathsf{t}_{2b}^{-1}\,\mathsf{t}_{2\,b-1}^{-1}\cdots\mathsf{t}_{21}^{-1}) = \mathsf{z}^{-1}$$

- Conjugacy action of r_{1j}

$$\begin{aligned}
&[\mathsf{r}_{1j},\ \mathsf{r}_{2k}] = 1 && \text{if } j < k \\
&[\mathsf{r}_{1j},\ \mathsf{r}_{2j}] = 1 && \\
&[\mathsf{r}_{1j},\ \mathsf{r}_{2k}] = \mathsf{z}^{-1}\,\mathsf{r}_{2k}\,\mathsf{r}_{2j}^{-1}\,\mathsf{z}\,\mathsf{r}_{2j}\,\mathsf{r}_{2k}^{-1} && \text{if } j > k \\
&[\mathsf{r}_{1j},\ \mathsf{t}_{2k}] = 1 && \text{if } j < k \\
&[\mathsf{r}_{1j},\ \mathsf{t}_{2j}] = \mathsf{z}^{-1} && \\
&[\mathsf{r}_{1j},\ \mathsf{t}_{2k}] = [\mathsf{z}^{-1},\ \mathsf{t}_{2k}] && \text{if } j > k \\
&[\mathsf{r}_{1j},\ \mathsf{z}] = [\mathsf{r}_{2j}^{-1},\ \mathsf{z}] &&
\end{aligned} \tag{2.1}$$

- Conjugacy action of t_{1j}

$$\begin{aligned}
&[\mathsf{t}_{1j},\ \mathsf{r}_{2k}] = 1 && \text{if } j < k \\
&[\mathsf{t}_{1j},\ \mathsf{r}_{2j}] = \mathsf{t}_{2j}^{-1}\,\mathsf{z}\,\mathsf{t}_{2j} && \\
&[\mathsf{t}_{1j},\ \mathsf{r}_{2k}] = [\mathsf{t}_{2j}^{-1},\ \mathsf{z}] && \text{if } j > k \\
&[\mathsf{t}_{1j},\ \mathsf{t}_{2k}] = 1 && \text{if } j < k \\
&[\mathsf{t}_{1j},\ \mathsf{t}_{2j}] = [\mathsf{t}_{2j}^{-1},\ \mathsf{z}] && \\
&[\mathsf{t}_{1j},\ \mathsf{t}_{2k}] = \mathsf{t}_{2j}^{-1}\,\mathsf{z}\,\mathsf{t}_{2j}\,\mathsf{z}^{-1}\,\mathsf{t}_{2k}\,\mathsf{z}\,\mathsf{t}_{2j}^{-1}\,\mathsf{z}^{-1}\,\mathsf{t}_{2j}\,\mathsf{t}_{2k}^{-1} && \text{if } j > k \\
&[\mathsf{t}_{1j},\ \mathsf{z}] = [\mathsf{t}_{2j}^{-1},\ \mathsf{z}] &&
\end{aligned} \tag{2.2}$$

Remark 2.2 From (2.1) and (2.2) we can deduce the corresponding conjugacy actions of r_{1j}^{-1} and t_{1j}^{-1}. We leave the cumbersome but standard computations to the reader.

Remark 2.3 Abelian groups admit no diagonal double Kodaira structures. Indeed, the relation $[\mathsf{r}_{1j}, \mathsf{t}_{2j}] = \mathsf{z}^{-1}$ in (2.1) provides a non-trivial commutator in G, because $o(\mathsf{z}) = n$.

Remark 2.4 Assume that the commutator subgroup $[G, G]$ is contained in the center $Z(G)$, i.e., that $G/Z(G)$ is abelian (being G non-abelian, this is equivalent to the fact that G has nilpotency class 2, see [9, p. 22]). Then the relations defining a diagonal double Kodaira structure on G assume the following simplified form.

- Relations expressing the centrality of z

$$[\mathsf{r}_{1j}, \mathsf{z}] = [\mathsf{t}_{1j}, \mathsf{z}] = [\mathsf{r}_{2j}, \mathsf{z}] = [\mathsf{t}_{2j}, \mathsf{z}] = 1 \tag{2.3}$$

- Surface relations

$$\begin{aligned} &[\mathsf{r}_{1b}^{-1}, \mathsf{t}_{1b}^{-1}]\,[\mathsf{r}_{1\,b-1}^{-1}, \mathsf{t}_{1\,b-1}^{-1}] \cdots [\mathsf{r}_{11}^{-1}, \mathsf{t}_{11}^{-1}] = \mathsf{z} \\ &[\mathsf{r}_{21}^{-1}, \mathsf{t}_{21}]\,[\mathsf{r}_{22}^{-1}, \mathsf{t}_{22}] \cdots [\mathsf{r}_{2b}^{-1}, \mathsf{t}_{2b}] = \mathsf{z}^{-1} \end{aligned} \tag{2.4}$$

- Conjugacy action of r_{1j}

$$\begin{aligned} &[\mathsf{r}_{1j}, \mathsf{r}_{2k}] = 1 && \text{for all } j, k \\ &[\mathsf{r}_{1j}, \mathsf{t}_{2k}] = \mathsf{z}^{-\delta_{jk}} \end{aligned} \tag{2.5}$$

- Conjugacy action of t_{1j}

$$\begin{aligned} &[\mathsf{t}_{1j}, \mathsf{r}_{2k}] = \mathsf{z}^{\delta_{jk}} \\ &[\mathsf{t}_{1j}, \mathsf{t}_{2k}] = 1 && \text{for all } j, k \end{aligned} \tag{2.6}$$

where δ_{jk} stands for the Kronecker symbol.

If $\mathfrak{S}$ is a diagonal double Kodaira structure of type (b, n) on G, then the subgroup

$$K_2 := \langle \mathsf{r}_{21}, \mathsf{t}_{21}, \ldots, \mathsf{r}_{2b}, \mathsf{t}_{2b}, \mathsf{z} \rangle$$

is normal in G and so there is a short exact sequence

$$1 \longrightarrow K_2 \longrightarrow G \longrightarrow Q_1 \longrightarrow 1,$$

where the elements $\mathsf{r}_{11}, \mathsf{t}_{11}, \ldots, \mathsf{r}_{1b}, \mathsf{t}_{1b}$ yield a complete set of representatives for Q_1. On the other hand, the set of relations defining $\mathfrak{S}$ is invariant under the substitutions

$$\mathsf{z} \longleftrightarrow \mathsf{z}^{-1}, \quad \mathsf{t}_{1j} \longleftrightarrow \mathsf{t}_{2\ b+1-j}^{-1}, \quad \mathsf{r}_{1j} \longleftrightarrow \mathsf{r}_{2\ b+1-j},$$

hence we can also see G as the middle term of a short exact sequence

$$1 \longrightarrow K_1 \longrightarrow G \longrightarrow Q_2 \longrightarrow 1,$$

where

$$K_1 := \langle \mathsf{r}_{11}, \mathsf{t}_{11}, \ldots, \mathsf{r}_{1b}, \mathsf{t}_{1b}, \mathsf{z} \rangle$$

and $\mathsf{r}_{21}, \mathsf{t}_{21}, \ldots, \mathsf{r}_{2b}, \mathsf{t}_{2b}$ yield a complete set of representatives for Q_2.

Definition 2.5 A diagonal double Kodaira structure $\mathfrak{S}$ as above will be called *of strong type* (b, n) if $K_1 = K_2 = G$. Otherwise, it will be called *of non-strong type* (b, n).

Sometimes we will not specify the pair (b, n), and we will simply say that $\mathfrak{S}$ is "of strong type" or "of non-strong type", respectively.

3 The Case of Extra-Special p-Groups

The following classical definition can be found, for instance, in [7, p. 183] and [9, p. 123].

Definition 3.1 Let p be a prime number. A finite p-group G is called *extra-special* if its center $Z(G)$ is cyclic of order p and the quotient $V = G/Z(G)$ is a non-trivial, elementary abelian p-group.

An elementary abelian p-group is a finite-dimensional vector space over the field $\mathbb{Z}_p$, hence it is of the form $V = (\mathbb{Z}_p)^{\dim V}$ and G fits into a short exact sequence

$$1 \longrightarrow \mathbb{Z}_p \longrightarrow G \longrightarrow V \longrightarrow 1. \tag{3.1}$$

Note that, V being abelian, we must have $[G, G] = \mathbb{Z}_p$, namely the commutator subgroup of G coincides with its center. Furthermore, since the extension (3.1) is central, it cannot be split, otherwise G would be isomorphic to the direct product of the two abelian groups $\mathbb{Z}_p$ and V, which is impossible because G is non-abelian. It can be also proved that, if G is extra-special, then $\dim V$ is even, so $|G| = p^{\dim V+1}$ is an odd power of p.

For every prime number p, there are precisely two isomorphism classes $M(p)$, $N(p)$ of non-abelian groups of order p^3, namely

$$M(p) = \langle \mathsf{r},\ \mathsf{t},\ \mathsf{z} \mid \mathsf{r}^p = \mathsf{t}^p = 1,\ \mathsf{z}^p = 1, [\mathsf{r},\ \mathsf{z}] = [\mathsf{t},\ \mathsf{z}] = 1,\ [\mathsf{r},\ \mathsf{t}] = \mathsf{z}^{-1} \rangle$$

$$N(p) = \langle \mathsf{r},\ \mathsf{t},\ \mathsf{z} \mid \mathsf{r}^p = \mathsf{t}^p = \mathsf{z},\ \mathsf{z}^p = 1, [\mathsf{r},\ \mathsf{z}] = [\mathsf{t},\ \mathsf{z}] = 1,\ [\mathsf{r},\ \mathsf{t}] = \mathsf{z}^{-1} \rangle$$

and both of them are in fact extra-special, see [7, Theorem 5.1 of Chapter 5].

If p is odd, then the groups $M(p)$ and $N(p)$ are distinguished by their exponent, which equals p and p^2, respectively. If $p = 2$, the group $M(p)$ is isomorphic to the dihedral group D_8, whereas $N(p)$ is isomorphic to the quaternion group Q_8.

The classification of extra-special p-groups is provided by the result below, see [7, Section 5 of Chapter 5].

Proposition 3.2 *If $b \geq 2$ is a positive integer and p is a prime number, there are exactly two isomorphism classes of extra-special p-groups of order p^{2b+1}, that can be described as follows.*

- *The central product $\mathsf{H}_{2b+1}(\mathbb{Z}_p)$ of b copies of $M(p)$, having presentation*

$$\begin{aligned} \mathsf{H}_{2b+1}(\mathbb{Z}_p) = \langle\, & \mathsf{r}_1,\ \mathsf{t}_1, \ldots, \mathsf{r}_b,\ \mathsf{t}_b,\ \mathsf{z} \mid \mathsf{r}_j^p = \mathsf{t}_j^p = \mathsf{z}^p = 1, \\ & [\mathsf{r}_j,\ \mathsf{z}] = [\mathsf{t}_j,\ \mathsf{z}] = 1, \\ & [\mathsf{r}_j, \mathsf{r}_k] = [\mathsf{t}_j, \mathsf{t}_k] = 1, \\ & [\mathsf{r}_j,\ \mathsf{t}_k] = \mathsf{z}^{-\delta_{jk}}\, \rangle. \end{aligned}$$

  If p is odd, this group has exponent p.
- *The central product $\mathsf{G}_{2b+1}(\mathbb{Z}_p)$ of $b-1$ copies of $M(p)$ and one copy of $N(p)$, having presentation*

$$\begin{aligned} \mathsf{G}_{2b+1}(\mathbb{Z}_p) = \langle\, & \mathsf{r}_1,\ \mathsf{t}_1, \ldots, \mathsf{r}_b,\ \mathsf{t}_b,\ \mathsf{z} \mid \mathsf{r}_b^p = \mathsf{t}_b^p = \mathsf{z}, \\ & \mathsf{r}_1^p = \mathsf{t}_1^p = \ldots = \mathsf{r}_{b-1}^p = \mathsf{t}_{b-1}^p = \mathsf{z}^p = 1, \\ & [\mathsf{r}_j,\ \mathsf{z}] = [\mathsf{t}_j,\ \mathsf{z}] = 1, \\ & [\mathsf{r}_j, \mathsf{r}_k] = [\mathsf{t}_j, \mathsf{t}_k] = 1, \\ & [\mathsf{r}_j,\ \mathsf{t}_k] = \mathsf{z}^{-\delta_{jk}}\, \rangle. \end{aligned}$$

  If p is odd, this group has exponent p^2.

Remark 3.3 In both cases, from the relations above we deduce

$$[\mathsf{r}_j^{-1},\ \mathsf{t}_k] = \mathsf{z}^{\delta_{jk}}, \quad [\mathsf{r}_j^{-1},\ \mathsf{t}_k^{-1}] = \mathsf{z}^{-\delta_{jk}}. \tag{3.2}$$

Remark 3.4 For both groups $\mathsf{H}_{2b+1}(\mathbb{Z}_p)$ and $\mathsf{G}_{2b+1}(\mathbb{Z}_p)$, the center is $\langle \mathsf{z} \rangle \simeq \mathbb{Z}_p$.

Remark 3.5 If $p = 2$, we can distinguish the two groups $\mathsf{H}_{2b+1}(\mathbb{Z}_p)$ and $\mathsf{G}_{2b+1}(\mathbb{Z}_p)$ by counting the number of elements of order 4.

Remark 3.6 The group $\mathsf{H}_{2b+1}(\mathbb{Z}_p)$ is isomorphic to the *matrix Heisenberg group* of order p^{2b+1}, that is, the subgroup of $\mathrm{GL}_{b+2}(\mathbb{Z}_p)$ consisting of matrices with 1 along the diagonal and 0 elsewhere, except for the top row and rightmost column, namely

$$\mathsf{H}_{2b+1}(\mathbb{Z}_p) = \left\{ \begin{pmatrix} 1 & \mathbf{x} & z \\ {}^{t}\mathbf{0} & I_b & {}^{t}\mathbf{y} \\ 0 & \mathbf{0} & 1 \end{pmatrix} \;\middle|\; \mathbf{x}, \mathbf{y} \in (\mathbb{Z}_p)^b, \; z \in \mathbb{Z}_p \right\}.$$

With this identification, calling $\{\mathbf{e}_1, \ldots, \mathbf{e}_b\}$ the standard basis of $(\mathbb{Z}_p)^b$, we have that:

- r_j corresponds to the matrix with $\mathbf{x} = \mathbf{0}$, $\mathbf{y} = \mathbf{e}_j$, $z = 0$;
- t_j corresponds to the matrix with $\mathbf{x} = \mathbf{e}_j$, $\mathbf{y} = \mathbf{0}$, $z = 0$;
- z corresponds to the matrix with $\mathbf{x} = \mathbf{0}$, $\mathbf{y} = \mathbf{0}$, $z = 1$.

Here is our first main result, cf. [5, Section 3].

Theorem 3.7 *Let $b \geq 2$ be a positive integer and let p be a prime number. If p divides $b+1$, then every extra-special p-group G of order p^{2b+1} admits a diagonal double Kodaira structure of strong type (b, p).*

Proof In both cases $G = \mathsf{H}_{2b+1}(\mathbb{Z}_p)$ and $G = \mathsf{G}_{2b+1}(\mathbb{Z}_p)$, set

$$\mathsf{r}_{1j} = \mathsf{r}_{2j} := \mathsf{r}_j, \quad \mathsf{t}_{1j} = \mathsf{t}_{2j} := \mathsf{t}_j$$

and define

$$\mathfrak{S} = (\mathsf{r}_{11}, \mathsf{t}_{11}, \ldots, \mathsf{r}_{1b}, \mathsf{t}_{1b}, \mathsf{r}_{21}, \mathsf{t}_{21}, \ldots, \mathsf{r}_{2b}, \mathsf{t}_{2b}, \mathsf{z}).$$

Since every extra-special p-group G satisfies $[G, G] = Z(G)$, it suffices to check the simplified set of relations given in Remark 2.4. Verifying (2.3), (2.5) and (2.6) is immediate from the presentation of G (see Proposition 3.2), whereas the surface relations (2.4) follow from (3.2) because, by assumption, we have $b = -1$ in $\mathbb{Z}_p$. Thus $\mathfrak{S}$ provides a diagonal double Kodaira structure on G, that is of strong type by construction. □

Our next goal is to show that, if in Theorem (3.7) we drop the condition that p divides $b+1$, we can still obtain some diagonal double Kodaira structures *of non-strong type* on extra-special p-groups of (bigger) order p^{4b+1}. Let us first show a couple of technical lemmas.

Lemma 3.8 *If $b \geq 2$ is an integer and $p \geq 5$ is a prime number, we can find non-zero elements*

$$\lambda_1, \ldots, \lambda_b, \mu_1, \ldots, \mu_b \in \mathbb{Z}_p$$

such that

$$\sum_{j=1}^{b} \lambda_j = \sum_{j=1}^{b} \mu_j = 1 \tag{3.3}$$

and $\lambda_j \mu_j \neq 1$ *for all* $j \in \{1, \ldots, b\}$.

Proof The following simple argument is borrowed from [5, proof of Proposition 2.16]. Choose arbitrarily λ_j, with $j \in \{1, \ldots, b-1\}$, and μ_j, with $j \in \{1, \ldots, b-2\}$, such that $\lambda_j \mu_j \neq 1$ for all $j \in \{1, \ldots, b-2\}$. Then λ_b is uniquely determined by $\lambda_b = 1 - \sum_{j=1}^{b-1} \lambda_j$, whereas μ_{b-1} and μ_b are subject to the following conditions:

- $\mu_{b-1} + \mu_b$ is equal to a constant $c = 1 - \sum_{j=1}^{b-2} \mu_j$
- $\mu_{b-1} \neq \lambda_{b-1}^{-1}$, $\mu_b \neq \lambda_b^{-1}$.

These requirements are in turn equivalent to $\mu_{b-1} \notin \{\lambda_{b-1}^{-1},\, c - \lambda_b^{-1}\}$. If $p \geq 5$ this can be clearly satisfied, because there are more than two non-zero elements in $\mathbb{Z}_p$. □

Now, take any anti-symmetrix matrix $A = (a_{jk})$ of order $2n$ over $\mathbb{Z}_p$, and consider the finitely presented groups

$$\begin{aligned} \mathsf{H}(A) = \langle \mathsf{x}_1, \ldots, \mathsf{x}_{2n}, \mathsf{z} \mid \mathsf{x}_1^p = \ldots = \mathsf{x}_{2n}^p = \mathsf{z}^p = 1,& \\ [\mathsf{x}_1, \mathsf{z}] = \ldots = [\mathsf{x}_{2n}, \mathsf{z}] = 1,& \\ [\mathsf{x}_j, \mathsf{x}_k] = \mathsf{z}^{a_{jk}} \rangle,& \end{aligned} \tag{3.4}$$

$$\begin{aligned} \mathsf{G}(A) = \langle \mathsf{x}_1, \ldots, \mathsf{x}_{2n}, \mathsf{z} \mid \mathsf{x}_1^p = \ldots = \mathsf{x}_{2n-2}^p = \mathsf{z}^p = 1,& \\ \mathsf{x}_{2n-1}^p = \mathsf{x}_{2n}^p = \mathsf{z},& \\ [\mathsf{x}_1, \mathsf{z}] = \ldots = [\mathsf{x}_{2n}, \mathsf{z}] = 1,& \\ [\mathsf{x}_j, \mathsf{x}_k] = \mathsf{z}^{a_{jk}} \rangle,& \end{aligned}$$

where the exponent in $\mathsf{z}^{a_{jk}}$ stands for any representative in $\mathbb{Z}$ of $a_{jk} \in \mathbb{Z}_p$.

Recall that, given three elements a, b, c in a group G, we have the commutator relation $[a, bc] = [a, b]b[a, c]b^{-1}$. Since all commutators in $\mathsf{H}(A)$ are central, we get

$$[a, bc] = [a, b][a, c] \quad \text{for all } a, b, c \in \mathsf{H}(A), \tag{3.5}$$

and similarly for $\mathsf{G}(A)$.

Lemma 3.9 *If* $\det A \neq 0$*, then the following holds*:

- $\mathsf{H}(A)$ *is an extra-special* p*-group of order* p^{2n+1} *and exponent* p*. In particular, it is isomorphic to* $\mathsf{H}_{2n+1}(\mathbb{Z}_p)$*;*
- $\mathsf{G}(A)$ *is an extra-special* p*-group of order* p^{2n+1} *and exponent* p^2*. In particular, it is isomorphic to* $\mathsf{G}_{2n+1}(\mathbb{Z}_p)$*.*

Proof We prove only the first point, the second being similar. The commutator relations in (3.4) show that every element of $\mathsf{H}(A)$ can be written in the form $\mathsf{x}_1^{t_1} \ldots \mathsf{x}_{2n}^{t_{2n}}$, with $t_1, \ldots, t_{2n} \in \mathbb{Z}$. Since x_j has order p and $[\mathsf{x}_j, \mathsf{x}_k]$ is central, it follows that $\mathsf{H}(A)$ has exponent p.

The quotient of $\mathsf{H}(A)$ by the central subgroup $\langle \mathsf{z} \rangle$ is an elementary abelian p-group of order p^{2n}. Therefore the only remaining issue is check that the center of $\mathsf{H}(A)$ is precisely $\langle \mathsf{z} \rangle$, and no larger. To this purpose, it suffices to check that an element of the form $\mathsf{x}_1^{t_1} \ldots \mathsf{x}_{2n}^{t_{2n}}$ is central if and only if all the t_j are zero. By using (3.4) and (3.5), we get

$$\begin{aligned} \mathsf{x}_k, \mathsf{x}_1^{t_1} \ldots \mathsf{x}_{2n}^{t_{2n}} t &= [\mathsf{x}_k, \mathsf{x}_1]^{t_1} \ldots [\mathsf{x}_k, \mathsf{x}_{2n}]^{t_{2n}} \\ &= \mathsf{z}^{a_{k1}t_1 + \ldots + a_{k2n}t_{2n}}. \end{aligned}$$

It follows that $\mathsf{x}_1^{t_1} \ldots \mathsf{x}_{2n}^{t_{2n}}$ is central if and only if we have

$$a_{k1}t_1 + \ldots + a_{k2n}t_{2n} = 0, \quad k = 1, \ldots, 2n.$$

This is a homogeneous system of linear equations in the variables $t_1, \ldots, t_{2n}$ and whose coefficient matrix is A. Being A non-singular by assumption, there is only the trivial solution $t_1 = \ldots = t_{2n} = 0$. □

We are now in a position to prove our second main result, cf. [5, Section 2].

Theorem 3.10 *If* $b \geq 2$ *is a positive integer and* $p \geq 5$ *is a prime number, then every extra-special* p*-group* G *of order* p^{4b+1} *admits a diagonal double Kodaira structure of non-strong type* (b, p)*.*

Proof Again, we treat in detail the case $G = \mathsf{H}_{4b+1}(\mathbb{Z}_p)$; the proof for $G = \mathsf{G}_{4b+1}(\mathbb{Z}_p)$ is similar. Let us consider the anti-symmetric matrix

$$\Omega_b = \begin{pmatrix} L_b & J_b \\ J_b & M_b \end{pmatrix} \in \mathrm{Mat}_{4b}(\mathbb{Z}_p),$$

where the blocks are the elements of $\mathrm{Mat}_{2b}(\mathbb{Z}_p)$ given by

$$L_b = \begin{pmatrix} 0 & \lambda_1 & & & \\ -\lambda_1 & 0 & & & 0 \\ & & \ddots & & \\ & & & 0 & \lambda_b \\ 0 & & & -\lambda_b & 0 \end{pmatrix} \quad M_b = \begin{pmatrix} 0 & \mu_1 & & & \\ -\mu_1 & 0 & & & 0 \\ & & \ddots & & \\ & & & 0 & \mu_b \\ 0 & & & -\mu_b & 0 \end{pmatrix}$$

$$J_b = \begin{pmatrix} 0 & -1 & & & \\ 1 & 0 & & & 0 \\ & & \ddots & & \\ & & & 0 & -1 \\ 0 & & & 1 & 0 \end{pmatrix}$$

and $\lambda_1, \ldots, \lambda_b, \mu_1, \ldots, \mu_b$ are as in Lemma 3.8. We have

$$\det \Omega_b = (1 - \lambda_1\mu_1)^2(1 - \lambda_2\mu_2)^2 \cdots (1 - \lambda_b\mu_b)^2 \neq 0$$

and so, by Lemma 3.9, we infer that $\mathsf{H}(\Omega_b)$ is isomorphic to $\mathsf{H}_{4b+1}(\mathbb{Z}_p)$. By definition, the group $\mathsf{H}(\Omega_b)$ is generated by a set of $4b + 1$ elements

$$\mathfrak{S} = \{\mathsf{r}_{11}, \mathsf{t}_{11}, \ldots, \mathsf{r}_{1b}, \mathsf{t}_{1b}, \mathsf{r}_{21}, \mathsf{t}_{21}, \ldots, \mathsf{r}_{2b}, \mathsf{t}_{2b}, \mathsf{z}\}$$

subject to the relations

$$\begin{aligned} &\mathsf{r}_{1j}^p = \mathsf{t}_{1j}^p = \mathsf{r}_{2j}^p = \mathsf{t}_{2j}^p = \mathsf{z}^p = 1, \\ &[\mathsf{r}_{1j}, \mathsf{z}] = [\mathsf{t}_{1j}, \mathsf{z}] = [\mathsf{r}_{2j}, \mathsf{z}] = [\mathsf{t}_{2j}, \mathsf{z}] = 1, \\ &[\mathsf{r}_{1j}, \mathsf{r}_{1k}] = [\mathsf{t}_{1j}, \mathsf{t}_{1k}] = 1, \\ &[\mathsf{r}_{1j}, \mathsf{r}_{2k}] = [\mathsf{t}_{1j}, \mathsf{t}_{2k}] = 1, \\ &[\mathsf{r}_{2j}, \mathsf{r}_{2k}] = [\mathsf{t}_{2j}, \mathsf{t}_{2k}] = 1, \\ &[\mathsf{r}_{1j}, \mathsf{t}_{1k}] = \mathsf{z}^{\delta_{jk}\lambda_j}, \\ &[\mathsf{r}_{2j}, \mathsf{t}_{2k}] = \mathsf{z}^{\delta_{jk}\mu_j}, \\ &[\mathsf{r}_{1j}, \mathsf{t}_{2k}] = [\mathsf{r}_{2j}, \mathsf{t}_{1k}] = \mathsf{z}^{-\delta_{jk}}. \end{aligned}$$

Using (3.3), we can check that the two surface relations (2.4) are satisfied. Since the remaining relations (2.3), (2.5) and (2.6) clearly hold, it follows that $\mathfrak{S}$ provides a diagonal double Kodaira structure of type (b, p) on $\mathsf{H}(\Omega_b)$, and so a diagonal double Kodaira structure of the same type on the isomorphic group $\mathsf{H}_{4b+1}(\mathbb{Z}_p)$. Such a structure is not strong, because the two subgroups $K_1 = \langle \mathsf{r}_{11}, \mathsf{t}_{11}, \ldots, \mathsf{r}_{1b}, \mathsf{t}_{1b}, \mathsf{z} \rangle$

and $K_2 = \langle \mathsf{r}_{21}, \mathsf{t}_{21}, \ldots, \mathsf{r}_{2b}, \mathsf{t}_{2b}, \mathsf{z} \rangle$ are isomorphic to $\mathsf{H}_{2b+1}(\mathbb{Z}_p)$, hence they both have index p^{2b} in G. □

Remark 3.11 The conclusion of Lemma 3.8 is false when $p \leq 3$. If $p = 2$, this follows immediately from the fact that there exists a unique non-zero element in $\mathbb{Z}_2$. If $p = 3$, two non-zero elements $\lambda_i, \mu_i \in \mathbb{Z}_3$ satisfy $\lambda_i \mu_i \neq 1$ if and only if $\lambda_i = -\mu_i$, so (3.3) cannot hold. This shows that, if Theorem 3.10 is also true for $p \leq 3$, then it must be proved in a different way.

Remark 3.12 The existence of a diagonal double Kodaira structure of non-strong type on $\mathsf{H}_{4b+1}(\mathbb{Z}_p)$ was first showed in [5, Section 2], although we did not use this terminology; the original proof relies on some group-cohomological results related to the structure the cohomology algebra $H^*(\Sigma_b \times \Sigma_b - \Delta, \mathbb{Z}_p)$, where Σ_b is a real surface of genus b and $\Delta \subset \Sigma_b \times \Sigma_b$ is the diagonal. Besides, such a proof does not use Lemma 3.9, but an equivalent statement coming from the identification of $\mathsf{H}_{4b+1}(\mathbb{Z}_p)$ with the so-called *symplectic Heisenberg group* $\mathsf{Heis}(V, \omega)$, where $V = H_1(\Sigma_b \times \Sigma_b - \Delta, \mathbb{Z}_p) \simeq (\mathbb{Z}_p)^{4b}$ and ω is any symplectic form on V.

Then, assuming that p divides $b + 1$, in [5, Section 3] we deduced the existence of a diagonal double Kodaira structure of strong type on $\mathsf{H}_{2b+1}(\mathbb{Z}_p)$ by setting

$$\lambda_1 = \ldots = \lambda_b = \mu_1 = \ldots = \mu_b = -1.$$

Indeed, this yields a diagonal double Kodaira structure on a "degenerate" Heisenberg group of order p^{4b+1} (in this case $\det \Omega_b = 0$), admitting the group $\mathsf{H}_{2b+1}(\mathbb{Z}_p)$ as a quotient.

In this note we adopted, instead, a purely group-theoretical approach; it is less geometric but shorter than the original one and it naturally yields new results, namely the existence of diagonal double Kodaira structures on the extra-special p-groups of exponent p^2.

4 Geometric Interpretation: From Diagonal Double Kodaira Structures to Diagonal Double Kodaira Fibrations

For more details on the basic definitions and results of this section, we refer the reader to the Introduction and to [5], especially Sects. 1 and 3. Recall that a *Kodaira fibration* is a smooth, connected holomorphic fibration $f_1 \colon S \longrightarrow B_1$, where S is a compact complex surface and B_1 is a compact complex curve, which is not isotrivial. The genus $b_1 := g(B_1)$ is called the *base genus* of the fibration, whereas the genus $g := g(F)$, where F is any fibre, is called the *fibre genus*.

Definition 4.1 A *double Kodaira surface* is a compact complex surface S, endowed with a *double Kodaira fibration*, namely a surjective, holomorphic map $f \colon S \longrightarrow B_1 \times B_2$ yielding, by composition with the natural projections, two Kodaira fibrations $f_i \colon S \longrightarrow B_i$, $i = 1, 2$.

The aim of this section is to show how the existence of diagonal double Kodaira structures is equivalent to the existence of some special double Kodaira fibrations, that we call *diagonal double Kodaira fibrations*. Looking at Gonçalves-Guaschi's presentation of surface pure braid groups, see [6, Theorem 7] and [5, Theorem 1.7], we see that a finite group G admits a diagonal double Kodaira structure $\mathfrak{S}$ of type (b, n) if and only if there is a surjective group homomorphism

$$\varphi\colon \mathsf{P}_2(\Sigma_b) \longrightarrow G, \tag{4.1}$$

such that $\mathsf{z} := \varphi(A_{12})$ has order n. Here $\mathsf{P}_2(\Sigma_b)$ is the pure braid group of genus b on two strands, which is isomorphic to the fundamental group $\pi_1(\Sigma_b \times \Sigma_b - \Delta, (p_1, p_2))$ of the configuration space of two ordered points on a real surface of genus b, and the generator A_{12} is the homotopy class in $\Sigma_b \times \Sigma_b - \Delta$ of a loop in $\Sigma_b \times \Sigma_b$ that "winds once" around the diagonal Δ.

With a slight abuse of notation, in the sequel we will use the symbol Σ_b to indicate both a smooth complex curve of genus b and its underlying real surface. By using Grauert-Remmert's extension theorem together with Serre's GAGA, the group epimorphism φ gives the existence of a smooth, complex, projective surface S endowed with a Galois cover

$$\mathbf{f}\colon S \longrightarrow \Sigma_b \times \Sigma_b,$$

with Galois group G and branched precisely over Δ with branching order n, see [5, Proposition 3.4].

The braid group $\mathsf{P}_2(\Sigma_b)$ is the middle term of two short exact sequences

$$1 \longrightarrow \pi_1(\Sigma_b - \{p_i\}, p_j) \longrightarrow \mathsf{P}_2(\Sigma_b) \longrightarrow \pi_1(\Sigma_b, p_i) \longrightarrow 1, \tag{4.2}$$

where $\{i, j\} = \{1, 2\}$, induced by the two natural projections of pointed topological spaces $(\Sigma_b \times \Sigma_b - \Delta, (p_1, p_2)) \longrightarrow (\Sigma_b, p_i)$. Composing the left homomorphism in (4.2) with $\varphi\colon \mathsf{P}_2(\Sigma_b) \longrightarrow G$, we get two homomorphisms

$$\varphi_1\colon \pi_1(\Sigma_b - \{p_2\}, p_1) \longrightarrow G, \quad \varphi_2\colon \pi_1(\Sigma_b - \{p_1\}, p_2) \longrightarrow G,$$

whose image equals K_1 and K_2, respectively. By construction, these are the homomorphisms induced by the restrictions $\mathbf{f}_i\colon \Gamma_i \longrightarrow \Sigma_b$ of the Galois cover $\mathbf{f}\colon S \longrightarrow \Sigma_b \times \Sigma_b$ to the fibres of the two natural projections $\pi_i\colon \Sigma_b \times \Sigma_b \longrightarrow \Sigma_b$. Since Δ intersects transversally at a single point all the fibres of the natural projections, it follows that both such restrictions are branched at precisely one point, and the number of connected components of the smooth curve $\Gamma_i \subset S$ equals the index $m_i := [G : K_i]$ of K_i in G.

So, taking the Stein factorizations of the compositions $\pi_i \circ \mathbf{f} \colon S \longrightarrow \Sigma_b$ as in the diagram below

$$\begin{array}{ccc} S & \xrightarrow{\quad \pi_i \circ \mathbf{f} \quad} & \Sigma_b \\ & {\scriptstyle f_i} \searrow \quad \nearrow {\scriptstyle \theta_i} & \\ & \Sigma_{b_i} & \end{array} \tag{4.3}$$

we obtain two distinct Kodaira fibrations $f_i \colon S \longrightarrow \Sigma_{b_i}$, hence a double Kodaira fibration by considering the product morphism

$$f = f_1 \times f_2 \colon S \longrightarrow \Sigma_{b_1} \times \Sigma_{b_2}.$$

Definition 4.2 We call $f \colon S \longrightarrow \Sigma_{b_1} \times \Sigma_{b_2}$ the *diagonal double Kodaira fibration* associated with the diagonal double Kodaira structure $\mathfrak{S}$ on the finite group G. Conversely, we will say that a double Kodaira fibration $f \colon S \longrightarrow \Sigma_{b_1} \times \Sigma_{b_2}$ is *of diagonal type* $(b,\, n)$ if there exists a finite group G and a diagonal double Kodaira structure $\mathfrak{S}$ of type $(b,\, n)$ on it such that f is associated with $\mathfrak{S}$.

One can wonder whether all double Kodaira fibrations are of diagonal type; the answer is negative, as we will show in Example 4.11, see also Proposition 4.12 and Remark 4.13.

Since the morphism $\theta_i \colon \Sigma_{b_i} \longrightarrow \Sigma_b$ is étale of degree m_i, by using the Hurwitz formula we obtain

$$b_1 - 1 = m_1(b-1), \quad b_2 - 1 = m_2(b-1). \tag{4.4}$$

Moreover, the fibre genera $g_1,\, g_2$ of the Kodaira fibrations $f_1 \colon S \longrightarrow \Sigma_{b_1}$, $f_2 \colon S \longrightarrow \Sigma_{b_2}$ are computed by the formulae

$$2g_1 - 2 = \frac{|G|}{m_1}(2b - 2 + \mathfrak{n}), \quad 2g_2 - 2 = \frac{|G|}{m_2}\,(2b - 2 + \mathfrak{n})\,,$$

where $\mathfrak{n} := 1 - 1/n$. Finally, the surface S fits into a diagram

$$\begin{array}{ccc} S & \xrightarrow{\qquad \mathbf{f} \qquad} & \Sigma_b \times \Sigma_b \\ & {\scriptstyle f} \searrow \quad \nearrow {\scriptstyle \theta_1 \times \theta_2} & \\ & \Sigma_{b_1} \times \Sigma_{b_2} & \end{array}$$

so that the diagonal double Kodaira fibration $f \colon S \longrightarrow \Sigma_{b_1} \times \Sigma_{b_2}$ is a finite cover of degree $\frac{|G|}{m_1 m_2}$, branched precisely over the curve

$$(\theta_1 \times \theta_2)^{-1}(\Delta) = \Sigma_{b_1} \times_{\Sigma_b} \Sigma_{b_2}.$$

Such a curve is always smooth, being the preimage of a smooth divisor via an étale morphism. However, it is reducible in general, see [5, Proposition 3.11]. The invariants of S can be now computed as follows, see [5, Proposition 3.8].

Proposition 4.3 *Let $f\colon S \longrightarrow \Sigma_{b_1} \times \Sigma_{b_2}$ be the diagonal double Kodaira fibration associated with a diagonal double Kodaira structure $\mathfrak{S}$ of type $(b,\ n)$ on a finite group G. Then we have*

$$c_1^2(S) = |G|\,(2b-2)(4b-4+4\mathfrak{n}-\mathfrak{n}^2)$$
$$c_2(S) = |G|\,(2b-2)(2b-2+\mathfrak{n}).$$

As a consequence, the slope and the signature of S can be expressed as

$$\nu(S) = \frac{c_1^2(S)}{c_2(S)} = 2 + \frac{2\mathfrak{n}-\mathfrak{n}^2}{2b-2+\mathfrak{n}}$$
$$\sigma(S) = \frac{1}{3}\left(c_1^2(S) - 2c_2(S)\right) = \frac{1}{3}\,|G|\,(2b-2)(2\mathfrak{n}-\mathfrak{n}^2),$$

where $\mathfrak{n} = 1 - 1/n$.

Remark 4.4 By definition, $\mathfrak{S}$ is a diagonal double Kodaira structure of strong type if and only if $m_1 = m_2 = 1$, that in turn implies $b_1 = b_2 = b$, i.e., $f = \mathbf{f}$. In other words, $\mathfrak{S}$ is of strong type if and only if no Stein factorization as in (4.3) is needed or, equivalently, if and only if the Galois cover $\mathbf{f} : S \longrightarrow \Sigma_b \times \Sigma_b$ induced by (4.1) is already a double Kodaira fibration, branched on the diagonal $\Delta \subset \Sigma_b \times \Sigma_b$.

We can now specialize the previous results, by taking as G an extra-special p-group and using what we have proved in Sect. 3. Let $\omega : \mathbb{N} \longrightarrow \mathbb{N}$ be the arithmetic function counting the number of distinct prime factors of a positive integer, see [8, p.335]. The following is [5, Corollary 3.18].

Theorem 4.5 *Let Σ_b be any smooth curve of genus b. Then there exists a double Kodaira fibration $f : S \longrightarrow \Sigma_b \times \Sigma_b$. Moreover, denoting by $\kappa(b)$ the number of such fibrations, we have*

$$\kappa(b) \geq \omega(b+1).$$

In particular,

$$\limsup_{b \to +\infty} \kappa(b) = +\infty.$$

Proof Given a prime number p dividing $b+1$, every extra-special p-group G of order p^{2b+1} admits a diagonal double Kodaira structure of strong type $(b,\ p)$, see Theorem 3.7, and this gives in turn a diagonal double Kodaira fibration $f\colon S \longrightarrow \Sigma_b \times \Sigma_b$, see Remark 4.4. Two different prime divisors of $b+1$ give rise to two non-homeomorphic double Kodaira surfaces, because the corresponding signatures are different (use the last equality in Proposition (4.3) with $n = p$ and note that, for fixed b, the function expressing $\sigma(S)$ is strictly increasing in p). Since the number of distinct prime factors of $b + 1$ can be arbitrarily large when b goes to infinity, the last statement follows. □

The case $b = 2$, $p = 3$ is particularly interesting. In fact, it provides (to our knowledge) the first "double solution" to a problem (posed by Geoff Mess) from Kirby's problem list in low-dimensional topology ([10, Problem 2.18A]), asking what is the smallest number b for which there exists a real surface bundle over a surface with base genus b and non-zero signature, see [5, Proposition 3.19].

Theorem 4.6 *Let S be the diagonal double Kodaira surface associated with a diagonal double Kodaira structure of strong type* $(2,\ 3)$ *on an extra-special* 3*-group G of order* 3^5*. Then the real manifold X underlying S is a closed, orientable* 4*-manifold of signature* 144 *that can be realized as a real surface bundle over a surface of genus* 2*, with fibre genus* 325*, in two different ways.*

This naturally leads to the following interesting problem, see [5, Question 3.20].

Question 4.7 What are the minimal possible fibre genus f_{min} and the minimum possible signature σ_{min} for a double Kodaira fibration $S \longrightarrow \Sigma_2 \times \Sigma_2$?

Note that Theorem 4.6 implies $f_{\mathrm{min}} \leq 325$ and $\sigma_{\mathrm{min}} \leq 144$.

Let us show now how to use our methods in order to obtain double Kodaira fibrations with slope strictly higher than $2 + 1/3$. Fix $b = 2$ and let $p \geq 5$ be a prime number. Then every extra-special p-group G of order $p^{4b+1} = p^9$ admits a diagonal double Kodaira structure $\mathfrak{S}$ of non-strong type $(2,\ p)$ and such that $m_1 = m_2 = p^{2b}$, see Theorem 3.10. Setting $b' := p^4 + 1$, cf. Eq. (4.4), and using also Proposition 4.3, we obtain the following particular case of [5, Proposition 3.12].

Theorem 4.8 *Let $f \colon S_{2,\,p} \longrightarrow \Sigma_{b'} \times \Sigma_{b'}$ be the diagonal double Kodaira fibration associated with a diagonal double Kodaira structure of non-strong type* $(2,\ p)$ *on an extra-special p-group G of order p^9. Then the maximum slope $\nu(S_{2,\,p})$ is attained for precisely two values of p, namely*

$$\nu(S_{2,\,5}) = \nu(S_{2,\,7}) = 2 + \frac{12}{35}.$$

Furthermore, $\nu(S_{2,\,p}) > 2+1/3$ for all $p \geq 5$. More precisely, if $p \geq 7$ the function $\nu(S_{2,\,p})$ is strictly decreasing and

$$\lim_{p \to +\infty} \nu(S_{2,\,p}) = 2 + \frac{1}{3}.$$

Remark 4.9 The original examples by Atiyah, Hirzebruch and Kodaira have slope lying in the interval $(2, 2+1/3]$, see [3, p. 221]. Our construction provides an infinite family of Kodaira fibred surfaces such that $2+1/3 < \nu(S) \leq 2+12/35$, maintaining at the same time a complete control on both the base genus and the signature. By contrast, the "tautological construction" used in [4] yields a higher slope than ours, namely $2 + 2/3$, but it involves an étale pullback "of sufficiently large degree", that completely loses control on the other quantities.

Remark 4.10 By Liu's inequality (see [13]), every Kodaira fibred surface S satisfies $\nu(S) < 3$. The value $\nu = 2 + 2/3$ is the current record for the slope, in particular it is unknown whether the slope of a Kodaira fibred surface can be arbitrarily close to 3.

Finally, let us show that there exist double Kodaira fibrations that are not of diagonal type.

Example 4.11 Take any double Kodaira fibration $f: S \longrightarrow \Sigma_{b_1} \times \Sigma_{b_2}$ with $b_2 = 2$ and $\nu(S) = 2 + 1/2$, see for instance [12, Examples 6.3 and 6.6 of Table 3]. We claim that such a f cannot be of diagonal type. In fact, assume by contradiction that f is associated with a diagonal double Kodaira structure of type $(b, \mathfrak{n})$ on a finite group G. Then, by using the second equation in (4.4), we obtain $2 - 1 = m_2(b - 1)$, hence $b = 2$. Substituting in the slope expression provided by Proposition 4.3, we get

$$\frac{1}{2} = \frac{2\mathfrak{n} - \mathfrak{n}^2}{2 + \mathfrak{n}},$$

or, equivalently, $n^2 - n + 2 = 0$, that has no integer solutions.

In fact, Example 4.11 is an instance of the following, more general result.

Proposition 4.12 *Let $f: S \longrightarrow \Sigma_{b_1} \times \Sigma_{b_2}$ be a double Kodaira fibration of diagonal type $(b, \mathfrak{n})$. Then we have $\nu(S) = 2 + s$, where s is a strictly positive rational number such that $(s+2)^2 - 8bs$ is a perfect square in $\mathbb{Q}$. As a consequence, we obtain $s < 6 - 4\sqrt{2}$.*

Proof By definition $\nu(S)$ is a rational number, and moreover $\nu(S) > 2$ because of Arakelov inequality, see [2] . So we can write $\nu(S) = 2 + s$, with $s > 0$. Since we are assuming that S is associated with a diagonal double Kodaira structure of type $(b, \mathfrak{n})$, the slope identity in Proposition 4.3 yields

$$s = \frac{2\mathfrak{n} - \mathfrak{n}^2}{2b - 2 + \mathfrak{n}},$$

or, equivalently,

$$(2bs - s - 1)n^2 - sn + 1 = 0.$$

The discriminant of this quadratic equation is $(s+2)^2 - 8bs$, and this quantity must be a perfect square in $\mathbb{Q}$ because n is an integer number. In particular, we have $(s+2)^2 \geq 8bs$, that is,

$$2 \leq b \leq \frac{(s+2)^2}{8s}.$$

From this we deduce the inequality $(s+2)^2 - 16s \geq 0$; since Remark 4.10 gives $s < 1$, we infer $s < 6 - 4\sqrt{2}$. □

Remark 4.13 Since $6 - 4\sqrt{2} = 0.3431...$ and $12/35 = 0.3428...$, we see that the surfaces $S_{2,5}$ and $S_{2,7}$, described in Theorem 4.8, "almost maximize" the slope of a double Kodaira fibration of diagonal type. In fact, the upper bound $s < 6 - 4\sqrt{2}$ shows that high slope examples, like Catanese-Rollenske's one for which $s = 2/3$, are out of reach of the methods of this paper.

Acknowledgments The author was partially supported by GNSAGA-INdAM. He thanks Andrea Causin for useful comments and remarks. He is also grateful to the organizers of the 12*th ISAAC Congress*, Universidade de Aveiro (2019), and in particular to Alexander Schmitt, for the invitation and the hospitality. Finally, he is indebted with Geoff Robinson and Derek Holt for their generous help with the proof of Lemma 3.9, see the MathOverflow thread https://mathoverflow.net/questions/351163, and to the anonymous referee for constructive criticism and precise suggestions that considerably improved the presentation of these results.

References

1. M.F. Atiyah, The signature of fibre bundles, in *Global Analysis* (Papers in honor of K. Kodaira) (University of Tokyo Press, Tokyo, 1969), pp. 73–84
2. A. Beauville, L'inegalité $p_g \geq 2q - 4$ pour les surfaces de type générale. Bull. Soc. Math. de France **110**, 343–346 (1982)
3. W. Barth, K. Hulek, C.A.M. Peters, A. Van de Ven, *Compact Complex Surfaces*. Grundlehren der Mathematischen Wissenschaften, vol. 4, Second enlarged edition (Springer, Berlin, 2003)
4. F. Catanese, S. Rollenske, Double Kodaira fibrations. J. Reine Angew. Math. **628**, 205–233 (2009)
5. A. Causin, F. Polizzi, Surface braid groups, finite Heisenberg covers and double Kodaira fibrations. Ann. Sc. Norm. Super. Pisa, Cl. Sci. (5) **XXII**, 1309–1352 (2021)
6. D.L. Gonçalves, J. Guaschi, On the structure of surface pure braid groups. J. Pure Appl. Algebra **186**, 187–218 (2004)
7. D. Gorenstein, *Finite Groups*, reprinted edition (AMS Chelsea Publishing, Rochester, 2007)
8. G.H. Hardy, E.M. Wright, *An Introduction to the Theory of Numbers*, 6th edn. (Oxford University Press, Oxford, 2008)
9. M. Isaacs, *Finite Groups Theory*. Graduate Studies in Mathematics, vol. 92 (American Mathematical Society, Providence, 2008)
10. R. Kirby (ed.), *Problems in Low-Dimensional Topology*. AMS/IP Studies in Advanced Mathematics, 2.2, Geometric Topology (Athens, GA, 1993), (American Mathematical Society, Providence, 1997), pp. 35–473

11. K. Kodaira, A certain type of irregular, algebraic surfaces. J. Anal. Math. **19**, 207–215 (1967)
12. J.A. Lee, M. Lönne, S. Rollenske, Double Kodaira fibrations with small signature. Int. J. Math. **31**(7), 2050052 (2020)
13. K. Liu, Geometric height inequalities. Math. Res. Lett. **3**(5), 693–702 (1996)

Symmetric Differentials and the Dimension of Hitchin Components for Orbi-Curves

Florent Schaffhauser

Abstract This note is based on a talk given at the 2019 ISAAC Congress in Aveiro. We give an expository account of joint work with Daniele Alessandrini and Gye-Seon Lee on Hitchin components for orbifold groups, recasting part of it in the language of analytic orbi-curves. This reduces the computation of the dimension of the Hitchin component for orbifold groups to an application of the orbifold Riemann-Roch theorem.

Keywords Fuchsian groups · Symmetric differentials · Hitchin components

Mathematics Subject Classification (2010) Primary 30F35, 30F30; Secondary 53C07.

1 Hitchin Components for Orbifold Fundamental Groups

1.1 Compact Orbi-Surfaces of Negative Euler Characteristic

An *orbifold* is a kind of space that generalises the notion of a manifold (be it a topological, differentiable or analytic one). For instance, a differentiable orbifold is a type of space that locally looks like the quotient of an open set $U \subset \mathbb{R}^n$ by a finite group of diffeomorphisms $\Gamma \subset \mathrm{Diff}(U)$. What is meant here by *quotient* depends a lot on how one understands the expression *a type of space*. For us, it will be sufficient to consider (topological, differentiable or analytic) *stacks* as our notion of space. Such a stack is then called an orbifold if it admits a *covering* by open substacks of the form $[U/\Gamma]$, parameterising families of Γ-orbits in U, where U is

F. Schaffhauser (✉)
Universidad de Los Andes, Departamento de Matemáticas, Bogotá, Colombia

Université de Strasbourg, IRMA UMR 7501, Strasbourg, France
e-mail: florent@uniandes.edu.co

P. Cerejeiras et al. (eds.), *Current Trends in Analysis, its Applications and Computation*, Research Perspectives,
https://doi.org/10.1007/978-3-030-87502-2_13

the local model for representable stacks (i.e. manifolds) and Γ is a finite subgroup of the automorphism group of U. A fundamental example of orbifold is the stack $\mathcal{X} := [M/\pi]$, where π is a discrete group acting (effectively and) properly on a manifold M.

A *coarse moduli space* (CMS) for an orbifold $\mathcal{X}$ is a manifold X equipped with a morphism $p : \mathcal{X} \longrightarrow X$ that satisfies the following universal property for all manifolds M:

$$\begin{array}{ccc} \mathcal{X} & \longrightarrow & M \\ {\scriptstyle p}\downarrow & \nearrow_{\exists !} & \\ X & & \end{array}$$

In the first part of the paper, we will work only with (effective) *differentiable* orbifolds. Then, up to real dimension 2, it suffices to enlarge the category of manifolds slightly and accommodate manifolds with corners, to ensure that coarse moduli spaces always exist. This is convenient because it allows us to think of an orbi-surface (or even an orbi-surface with boundary) as an ordinary surface with an extra structure, namely some "special points", all of whose open neighbourhoods are of the form $[U/\,\Gamma]$ with non-trivial Γ. As a matter of fact, since we are in the $\mathcal{C}^\infty$ setting and Γ is finite, we can always assume that it acts on the open set $U \subset \mathbb{R}^2$ preserving a positive-definite metric. The classification of linear isometries of the Euclidean plane then tells that a point in the coarse moduli space $U/\,\Gamma$ is of one of the following three types:

1. A *cone point*, which admits an open neighbourhood of the form $D(0;\varepsilon)/\mathrm{C}_m$, where $\mathrm{C}_m \simeq \mathbb{Z}/m\mathbb{Z}$ is a finite cyclic group of order m, acting on the open disk $D(0;\varepsilon)$ by rotation. Such a cone point is said to have *order* m.
2. A *dihedral point* (or *corner reflector*), which admits an open neighbourhood of the form $D(0;\varepsilon)/\mathrm{D}_m$, where $\mathrm{D}_m \simeq \mathrm{C}_m \rtimes \mathbb{Z}/2\mathbb{Z}$ is the dihedral group of order $2m$. Such a dihedral point is said to have *order* m.
3. A *mirror point*, which admits an open neighbourhood of the from $D(0;\varepsilon)/\mathbb{Z}/2\mathbb{Z}$, where $\mathbb{Z}/2\mathbb{Z}$ acts on $D(0;\varepsilon)$ by reflection through a diameter.

For instance, a 2-dimensional orbifold could have a triangle for a coarse moduli space: the edges are mirror points, while the vertices are dihedral points. Another good thing about (compact) orbi-surfaces is that they admit an *orbifold Euler characteristic*, computable explicitly from the coarse moduli space through the following formula (in which k is the number of cone points, ℓ the number of dihedral points, m_i is the order of the i-th cone point and n_j is the order of the j-th dihedral point):

$$\chi(\mathcal{X}) = \chi(X) - \textstyle\sum_{i=1}^{k}\left(1 - \frac{1}{m_i}\right) - \frac{1}{2}\sum_{j=1}^{\ell}\left(1 - \frac{1}{n_j}\right) \in \mathbb{Q}. \qquad (1.1)$$

In the so-called orientable case, the quantity $\chi(\mathcal{X})$ is negative if and only if the coarse moduli space X is a closed surface of genus at least 2, or a torus with at least one cone point, or a sphere with at least three cone points. For a more complete treatment of the fundamental properties of orbifolds, we refer for instance to [4, 5, 15, 17] and for the stacky point of view, we refer to [3, 9].

1.2 Fundamental Group and Hyperbolic Structures

A *cover* of an orbifold $\mathcal{X}$ is a morphism $\mathcal{Y} \longrightarrow \mathcal{X}$ which, in orbifold charts, is conjugate to a morphism of the form $\amalg_{i \in I}[U/\Gamma_i] \longrightarrow [U/\Gamma]$, where each Γ_i is a subgroup of Γ. In particular, the canonical map $U \longrightarrow [U/\Gamma]$ is an orbifold cover. A more concrete example is given as follows: the "flattening" of a sphere with 3 cone points is a 2-to-1 cover of a triangle, the cone points upstairs being mapped to dihedral points of the same order downstairs. There is an orbifold structure on the inverse limit of all connected covers and the latter is called the *universal cover* of $\mathcal{X}$. The fundamental group of $\mathcal{X}$ is the automorphism group of the universal cover (whose total space may or may not be a manifold). We denote by $\pi_1\mathcal{X}$ the fundamental group of $\mathcal{X}$. For instance, if M is a simply connected manifold and π is a discrete group acting properly on M, then $\pi_1([M/\pi]) \simeq \pi$. If one is careful about base points, connected covers of an orbifold $\mathcal{X}$ correspond bijectively to subgroups of $\pi_1\mathcal{X}$.

A *hyperbolic structure* on a differentiable orbifold $\mathcal{X}$ is a covering by open substacks of the form $[U/\Gamma]$ in which $U \subset \mathbf{H}^2$ is an open subspace *of the (real) hyperbolic plane* and $\Gamma \subset \mathrm{Isom}(\mathbf{H}^2) \simeq \mathbf{PGL}(2;\mathbb{R})$ is a finite subgroup of the isometry group of $\mathbf{H}^2$ that leaves U invariant. If $\chi(\mathcal{X}) < 0$, then $\mathcal{X}$ admits hyperbolic structures and its universal cover is isomorphic to $\mathbf{H}^2$. The deformation space of hyperbolic structures on $\mathcal{X}$ is identified, via the space of holonomy representations of such structures, to a connected component of the topological space

$$\mathrm{Hom}\big(\pi_1\mathcal{X};\mathbf{PGL}(2;\mathbb{R})\big)/\mathbf{PGL}(2;\mathbb{R}).$$

Namely, it is the space of *discrete and faithful* representations $\varrho : \pi_1\mathcal{X} \longrightarrow \mathbf{PGL}(2;\mathbb{R})$. Thus, if $\chi(\mathcal{X}) < 0$, it is always possible to identify $\pi_1\mathcal{X}$ with a discrete subgroup of $\mathbf{PGL}(2;\mathbb{R})$, i.e. a *Fuchsian group*. If the orbifold $\mathcal{X}$ is orientable (which in dimension 2 amounts to saying that the CMS X is an orientable surface and that all groups Γ appearing in the orbifold charts contain only orientation-preserving transformations), then the fundamental group of $\mathcal{X}$ admits the following presentation:

$$\pi_1\mathcal{X} \simeq \Big\langle (a_i,b_i)_{1\leqslant i\leqslant g},(c_j)_{1\leqslant j\leqslant k} \mid \textstyle\prod_{1\leqslant i\leqslant g}[a_i,b_i]\prod_{1\leqslant j\leqslant k} c_j = 1$$
$$= c_1^{m_1} = \ldots = c_k^{m_k}\Big\rangle =: \pi_{g,(m_1,\ldots,m_k)}. \tag{1.2}$$

The space of discrete and faithful representations of $\pi_1\mathcal{X}$ in $\mathbf{PGL}(2;\mathbb{R})$ will be called the *Teichmüller space* of $\mathcal{X}$ and denoted by $\mathcal{T}(\mathcal{X})$. It is homeomorphic to a real vector space of dimension $-3\chi(X) + 2k + \ell$. In particular, it is reduced to a point if $\mathcal{X}$ is a (quotient of a) sphere with three cone points (see for instance [4] for a full account on this, including for the more refined notion of orbifold with boundary).

1.3 Hitchin Components

Let $\mathfrak{g}_{\mathbb{C}}$ be a simple complex Lie algebra. The *adjoint group* $G_{\mathbb{C}} := \mathrm{Int}(\mathfrak{g}_{\mathbb{C}})$ is the neutral component, in the Lie group topology, of $\mathrm{Aut}(\mathfrak{g}_{\mathbb{C}})$, and it is a complex Lie group with trivial centre, whose Lie algebra is isomorphic to $\mathfrak{g}_{\mathbb{C}}$. Given a real form $\mathfrak{g}$ of $\mathfrak{g}_{\mathbb{C}}$, there is an associated anti-holomorphic involution θ of $G_{\mathbb{C}}$, whose fixed-point set we denote by G. It consists of interior automorphisms of $\mathfrak{g}$ that commute with θ. The neutral component of G is $\mathrm{Int}(\mathfrak{g})$. In particular, G is not necessarily connected. For instance, if we choose $\mathfrak{g} = \mathfrak{sl}(n;\mathbb{R})$, then $G \simeq \mathbf{PGL}(n;\mathbb{R})$, which is connected if n is odd and has two connected components if n is even. In what follows, we shall always assume that $\mathfrak{g}$ is the *split real form* of the Lie algebra $\mathfrak{g}_{\mathbb{C}}$.

In [8], N. Hitchin studies representations of surface groups into G and shows that the representation space $\mathrm{Hom}(\pi_1 X; G)/G$ has a contractible connected component. His definition of that component rests on the notion of *Fuchsian representation*, which itself depends on the choice of a so-called *principal morphism* $\kappa : \mathbf{PGL}(2;\mathbb{R}) \longrightarrow G$, first introduced by B. Kostant [11]. When $G = \mathbf{PGL}(n;\mathbb{R})$, this morphism is induced by the linear action of $\mathbf{GL}(2;\mathbb{R})$ on the space V_n of homogeneous polynomials of degree $n-1$ in two variables x, y. Hitchin's definition, extended to the orbifold case, is then the following. Given an orbi-surface $\mathcal{X}$ of negative Euler characteristic and a principal morphism κ from $\mathbf{PGL}(2;\mathbb{R})$ to the split real form G of $\mathrm{Int}(\mathfrak{g}_{\mathbb{C}})$, a representation $\varrho : \pi_1\mathcal{X} \longrightarrow G$ is called *Fuchsian* if it lifts to a discrete and faithful representation $h : \pi_1\mathcal{X} \longrightarrow \mathbf{PGL}(2;\mathbb{R})$, in the sense that the following diagram becomes commutative:

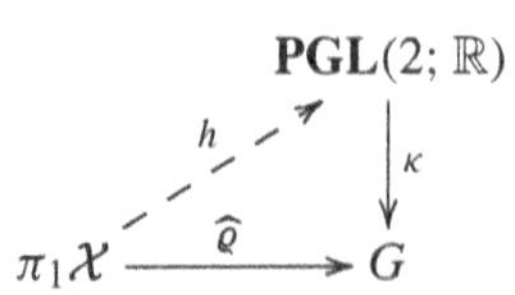

This defines a map $\mathcal{T}(\mathcal{X}) \longrightarrow \mathrm{Hom}(\pi_1\mathcal{X}; G)/G$ whose image is called the *Fuchsian locus*. As the Teichmüller space $\mathcal{T}(\mathcal{X})$ is connected, this map picks out a single connected component of the representation space $\mathrm{Hom}(\pi_1\mathcal{X}; G)/G$, called the *Hitchin component* and denoted by $\mathrm{Hit}(\pi_1\mathcal{X}; G)$. When $G = \mathbf{PGL}(2;\mathbb{R})$, we have $\mathrm{Hit}(\pi_1\mathcal{X}; \mathbf{PGL}(2; R)) \simeq \mathcal{T}(\mathcal{X})$, by definition. For split real groups G of *higher rank*, Hitchin components form a family of so-called *Higher Teichmüller*

spaces [18]. Indeed, Hitchin representations are discrete and faithful [6, 12]. In the surface group case, Hitchin has proved that $\mathrm{Hit}(\pi_1 X, G)$ has a trivial topology:

Theorem 1.1 (Hitchin, [8]) *Let X be a closed orientable surface of negative Euler characteristic and let G be the split real form of $\mathrm{Int}(\mathfrak{g}_{\mathbb{C}})$, where $\mathfrak{g}_{\mathbb{C}}$ is a simple complex Lie algebra. Then $\mathrm{Hit}(\pi_1 X; G)$ is homeomorphic to a real vector space of dimension $-\chi(X)\dim G$.*

This formula cannot be generalised directly to the orbifold case, as $\chi(\mathcal{X})$ is not an integer in general. However, for $G = \mathbf{PGL}(3;\mathbb{R})$, S. Choi and W. Goldman have proved the following formula.

Theorem 1.2 (Choi and Goldman, [4]) *Let $\mathcal{X}$ be a closed orbi-surface of negative Euler characteristic and with coarse moduli space X. Then $\mathrm{Hit}(\pi_1\mathcal{X}; \mathbf{PGL}(3;\mathbb{R}))$ is homeomorphic to a real vector space of dimension $-8\chi(X) + (6k - 2k_2) + (3\ell - \ell_2)$, where k_2 (respectively, ℓ_2) is the number of cone points (respectively, dihedral points) of order 2 of $\mathcal{X}$.*

In collaboration with D. Alessandrini and G.S. Lee, we have been looking at Hitchin components for orbifold groups and we have obtained the following common generalisation of the two results above. For the sake of clarity, we will present it here for the group $G = \mathbf{PGL}(n;\mathbb{R})$ only, but our results hold for split real forms of all adjoint groups of simple complex Lie algebras (e.g. $\mathbf{PO}(m, m+1)$ or $\mathbf{PO}^{\pm}(m, m)$).

Theorem 1.3 ([2]) *Let $\mathcal{X}$ be a closed orbi-surface of negative Euler characteristic and with coarse moduli space X. Then $\mathrm{Hit}(\pi_1\mathcal{X}; \mathbf{PGL}(n;\mathbb{R}))$ is homeomorphic to a real vector space of dimension*

$$-(n^2-1)\chi(X) + \sum_{d=2}^{n}\left(2\sum_{i=1}^{k} R(d, m_i) + \sum_{j=1}^{\ell} R(d, n_j)\right)$$

where m_i (respectively, n_j) is the order of the i-th cone point (respectively, the j-th dihedral point) of $\mathcal{X}$ and $R(d, m) := \left\lfloor d - \frac{d}{m}\right\rfloor$ is the integral part of the real number $\left(d - \frac{d}{m}\right)$.

As a matter of fact, like Choi and Goldman in [4], we can also deal with the case of orbifolds with boundary. We also note that, when Y has only mirror points as orbifold singularities (no cone or dihedral points), then $\chi(\mathcal{X}) = \chi(X)$ and Hitchin's formula holds without modifications. There is another way of writing the formula in Theorem 1.3, which resembles more that of Theorem 1.2, and we refer to [2] for it. We get for instance

$$\dim \mathrm{Hit}\big(\pi_1\mathcal{X}; \mathbf{PGL}(4;\mathbb{R})\big) = -15\chi(X) + (12k - 4k_2 - 2k_3) + (6\ell - 2\ell_2 - \ell_3),$$

where again k_i (respectively, ℓ_i) is the number of cone points (respectively, dihedral points) of order i of $\mathcal{X}$. We see that this dimension may vanish for certain orbifolds $\mathcal{X}$ and that such orbifolds form an infinite family, containing for instance all spheres with three cone points of order $(2, 3, r)$ for all $r \geqslant 7$. This has applications to the rigidity of projective structures on Seifert-fibered spaces with base $\mathcal{X}$ (see [2] for details).

The methods of proof for Theorems 1.1 and 1.2 are quite different. Hitchin uses tools from analytic and differential geometry (namely, Higgs bundles and the Non-Abelian Hodge Correspondence), while Choi and Goldman's methods are based on the interpretation of $\mathrm{Hit}(\pi_1\mathcal{X}; G)$ as the deformation space of *convex projective structures* on $\mathcal{X}$. In the absence of such a geometric interpretation for general G, our approach in [2] consists in adapting Hitchin's method to our setting. Thanks to an orbifold version of the Non-Abelian Hodge Correspondence, we show that the Hitchin component is homeomorphic to a space of symmetric differentials on an analytic orbi-curve, the dimension of which we can compute using the orbifold Riemann-Roch theorem, similarly to Hitchin's proof in the surface group case. We explain this in greater detail in the next section.

2 Analytic Parameterisation of Hitchin Components

2.1 Analytic Orbi-Curves

A *complex analytic orbifold* is an analytic stack $\mathcal{X}$ (over complex analytic manifolds) that admits a covering by open substacks of the form $[U/\Gamma]$, where $U \subset \mathbb{C}^n$ is an open subset and $\Gamma \subset \mathrm{Aut}(U)$ is a finite group of holomorphic transformations of U. If the open sets U are all of complex dimension 1, we say that $\mathcal{X}$ is an *orbi-curve* or an *orbi-Riemann surface*. The only possible orbifold points in this case are cone points and it is a remarkable fact that there always exist coarse moduli spaces: an orbi-Riemann surface always has an "underlying" Riemann surface, because if $\Gamma \simeq \mathrm{C}_m$ acts by rotation of angle $\frac{2\pi}{m}$ on the open disk $D(0; \varepsilon)$ then the map $z \longmapsto z^m$ induces a holomorphic chart $D(0; \varepsilon)/\mathrm{C}_m \simeq D(0; \varepsilon^m)$. In fact, the whole theory of complex analytic orbi-curves can be phrased in terms of *Riemann surfaces with signature*, where the signature is the map $X \longrightarrow \mathbb{N}$ taking a point to its order (so the map is constant equal to 1, except possibly over a finite set of points in X). We prefer to work, however, in the orbifold setting. In particular, subgroups of the orbifold fundamental group (1.2) correspond to connected analytic covers of the compact orbi-curve $\mathcal{X} := [\mathbf{H}^2/\pi_{g,(m_1, \dots, m_k)}]$.

To prove Theorem 1.3, complex analytic orbi-curves will not be quite enough if we want to include the case of non-orientable differentiable orbi-surfaces. To deal with those, we need to consider also orbi-curves which are *defined over the real numbers*. This essentially means complex analytic orbi-curves $\mathcal{X}^+$ equipped with an anti-analytic involution $\sigma : \mathcal{X}^+ \longrightarrow \mathcal{X}^+$ given, in local charts, by a Γ-equivariant anti-holomorphic involution $\sigma : U \longrightarrow U'$. In particular, the orders of the points x and $\sigma(x)$ have to coincide for all x. More intrinsically perhaps, one could consider *dianalytic orbifolds*, for which local models are quotient stacks

$[U/\Gamma]$, where $U \subset \mathbb{C}^n$ is an open subset but the finite group $\Gamma \subset \mathrm{Aut}^{\pm}(U)$ is now allowed to also contain anti-holomorphic transformations of U. If we consider such a dianalytic orbifold $\mathcal{X}$, its fundamental group $\pi := \pi_1\mathcal{X}$ has a subgroup π^+, of index at most 2, consisting of transformations that preserve the orientation of the universal cover $\widetilde{\mathcal{X}}$, the latter being necessarily complex analytic: the quotient orbifold $\mathcal{X}^+ := [\widetilde{\mathcal{X}}/\pi^+]$ is a complex analytic orbifold which is a cover, of degree at most 2, of $\mathcal{X}$. If $\pi^+ \neq \pi$, then π/π^+ acts on $\mathcal{X}^+$ via an anti-holomorphic involution σ and $\mathcal{X} \simeq [\mathcal{X}^+/\langle\sigma\rangle]$. In this case, there is a short exact sequence

$$1 \longrightarrow \pi_1\mathcal{X}^+ \longrightarrow \pi_1\mathcal{X} \longrightarrow \{\pm 1\} \longrightarrow 1.$$

Note that $\mathcal{X}^+$ has two cone points x and $\sigma(x)$ (of the same order) for each cone point of $\mathcal{X}$, and one cone point which is fixed by σ for each dihedral point of $\mathcal{X}$. Consider for instance the fundamental group of a triangle $\mathcal{X}$ with vertices of respective orders (p, q, r). The double cover $\mathcal{X}^+$ is a sphere with three cone points, of respective orders (p, q, r). The fundamental group of $\mathcal{X}^+$ is the Von Dyck group

$$\pi_{0,(p,q,r)} \simeq \langle a, b, c \mid a^p = b^q = c^r = abc = 1\rangle$$

of (1.2), while that of $\mathcal{X}$ is the Coxeter (triangle) group with presentation

$$T_{(p,q,r)} := \Big\langle x, y, z \mid x^2 = y^2 = z^2 = (xy)^p = (yz)^q = (zx)^r = 1\Big\rangle. \tag{2.1}$$

The covering (flattening map) $\mathcal{X}^+ \longrightarrow \mathcal{X}$ induces the injective group morphism $\pi_{0,(p,q,r)} \longrightarrow T_{(p,q,r)}$ defined by $a \longmapsto xy$, $b \longmapsto yz$, $c \longmapsto zx$, and the quotient map $T_{(p,q,r)} \longrightarrow \{\pm 1\}$ is given by the reduced word length modulo 2.

When $\mathcal{X}$ is a compact orbi-curve of negative Euler characteristic, the fundamental group $\pi_1\mathcal{X}$ is a finitely generated group that embeds onto a discrete subgroup of $\mathbf{PGL}(2;\mathbb{R})$. Therefore, by Selberg's lemma, it contains a finite index normal subgroup which is torsion-free [16]. Geometrically, this means that there exists a compact Riemann surface Y and a finite Galois cover $Y \longrightarrow \mathcal{X}$. If we denote by π the automorphism group of that cover, we therefore have an isomorphism of orbifolds $[Y/\pi] \simeq \mathcal{X}$, and a short exact sequence

$$1 \longrightarrow \pi_1 Y \longrightarrow \pi_1\mathcal{X} \longrightarrow \pi \longrightarrow 1.$$

2.2 The Riemann-Roch Formula

An *orbifold line bundle* $\mathfrak{L}$ over $\mathcal{X}$ is a morphism of stacks $\mathfrak{L} \longrightarrow \mathcal{X}$ which is locally conjugate, in the orbifold chart $[U/\Gamma]$ about x, to the orbifold $[(U \times \mathbb{C})/\Gamma]$, where Γ acts on $U \times \mathbb{C}$ via a linear representation $\varrho_\Gamma : \Gamma \longrightarrow \mathbf{GL}(1, \mathbb{C})$. When the finite group Γ is cyclic of order m, the morphism ϱ_Γ sends a generator of Γ to an m-th root of unity. If we choose a generator γ of Γ and a primitive m-th root of unity ζ, then $\varrho_\Gamma(\gamma) = \zeta^a$ for a certain $a \in \{0; \ldots ; m-1\}$ which does not depend on the choices just made and is sometimes called the *isotropy* at the point x. When Γ is a dihedral group, we write $\Gamma \simeq \mathrm{C}_m \rtimes \mathbb{Z}/2\mathbb{Z}$, where $\mathbb{Z}/2\mathbb{Z}$ acts on the cyclic group C_m by inversion, and think of $\varrho_\Gamma : \Gamma \longrightarrow \mathbb{C}^*$ as a morphism $\varrho_{\mathrm{C}_m} : \mathrm{C}_m \longrightarrow \mathbb{C}^*$ as before, which in addition is $\mathbb{Z}/2\mathbb{Z}$-equivariant with respect to complex conjugation on $\mathbb{C}^*$. In particular, the number $a \in \{0; \ldots ; m-1\}$ again completely determines the morphism ϱ_Γ. Given a cone or dihedral point x of order m, the quantity $\frac{a}{m}$, where $a \in \{0; \ldots ; m-1\}$ is defined as above, will be called the *age* of the orbifold line bundle $\mathfrak{L}$ at x and denoted by $\mathrm{age}_x(\mathfrak{L})$. Consider for instance the canonical line bundle $K_{\mathcal{X}}$ of an analytic orbi-curve $\mathcal{X}$. The age of the tangent bundle at a cone point of order m is $\frac{1}{m}$ (the action of C_m on tangent vectors being multiplication by a primitive root of unity) and, since the canonical bundle is the dual of the tangent bundle in this case, the group $\Gamma \simeq \mathrm{C}_m$ acts on tangent covectors at a point via multiplication by ζ^{-1}, so the age of $K_{\mathcal{X}}$ at a cone point of order m is $\frac{m-1}{m}$. If we now look at tensor powers of $K_{\mathcal{X}}$, then the action of C_m on homogeneous polynomial functions of degree d over the tangent space at a cone point is given by multiplication by $\zeta^{d(m-1)}$, so the age of $K_{\mathcal{X}}^d$ at a cone point is

$$\frac{d(m-1) \bmod m}{m} = \frac{d(m-1)}{m} - \left\lfloor \frac{d(m-1)}{m} \right\rfloor .$$

We will see in Sect. 2.3 below that this is the origin of the term $R(d,m) := \left\lfloor \frac{d(m-1)}{m} \right\rfloor$ in Theorem 1.3.

Let us denote by $\underline{\mathfrak{L}}$ the sheaf of local sections of $\mathfrak{L}$. There are associated cohomology groups $\mathrm{H}^0(\mathcal{X}; \underline{\mathfrak{L}})$ and $\mathrm{H}^1(\mathcal{X}; \underline{\mathfrak{L}})$, which are finite-dimensional complex or real vector spaces (depending on the field of definition of $\mathcal{X}$). The *Euler characteristic* of $\underline{\mathfrak{L}}$ is the integer $\chi(\mathcal{X}; \underline{\mathfrak{L}}) := \dim \mathrm{H}^0(\mathcal{X}; \underline{\mathfrak{L}}) - \dim \mathrm{H}^1(\mathcal{X}; \underline{\mathfrak{L}})$. The Riemann-Roch formula computes this quantity by comparing it to the Euler characteristic of the structure sheaf $\mathcal{O}_{\mathcal{X}}$. To state the result, we still need the notion of *degree* of an orbifold line bundle, of which we recall the following two definitions (in the complex case). When $\mathcal{X} \simeq [Y/\pi]$, where Y is a compact Riemann surface and π is a finite group of analytic transformations of Y, an orbifold line bundle $\mathfrak{L} \longrightarrow \mathcal{X}$ pulls back to a Γ-equivariant analytic line bundle $\mathcal{E} \longrightarrow Y$ and we can define the degree of $\mathfrak{L}$ as $\frac{\deg(\mathcal{E})}{|\pi|} \in \mathbb{Q}$, since this quantity is independent of the choice of the finite Galois cover $Y \longrightarrow \mathcal{X}$. Equivalently, if we denote by $p : \mathcal{X} \longrightarrow X$ the coarse moduli space of $\mathcal{X}$, then, given an orbifold line bundle $\mathfrak{L} \longrightarrow \mathcal{X}$, there exists a unique analytic line bundle $L \longrightarrow X$ and for each cone point x_i of $\mathcal{X}$ a well-defined integer $a_i \in \{0; \ldots ; m_i - 1\}$ such that

$$\mathfrak{L} \simeq p^*L \otimes \mathcal{O}_{\mathcal{X}}\big(\textstyle\sum_{i=1}^k a_i x_i\big).$$

We then have $\mathrm{age}_{x_i}(\mathfrak{L}) = \frac{a_i}{m_i}$ and $\deg(\mathfrak{L}) := \deg(L) + \sum_{i=1}^{k} \frac{a_i}{m_i}$, where m_i is the order of the cone point x_i. For instance, when $\mathfrak{L} = K_{\mathcal{X}}^d \simeq [K_Y^d/\pi]$, one can check that

$$K_{\mathcal{X}}^d \simeq p^*\left[K_X^d \otimes O_X\left(\sum_{i=1}^{k} R(d, m_i)p(x_i)\right)\right] \otimes O_{\mathcal{X}}\left(\sum_{i=1}^{k}(d(m_i - 1) \bmod m_i)x_i\right)$$

so, using (1.1),

$$\begin{aligned}\deg(K_{\mathcal{X}}^d) &= d(2g-2) + \sum_{i=1}^{k} R(d, m_i) + \sum_{i=1}^{k} \frac{d(m_i-1) \bmod m_i}{m_i} = -d\chi(\mathcal{X}) \\ &= -d\frac{\chi(Y)}{|\pi|} = \frac{\deg(K_Y^d)}{|\pi|}, \qquad (2.2)\end{aligned}$$

where $g := \dim \mathrm{H}^1(\mathcal{X}; O_{\mathcal{X}})$ is the genus of $\mathcal{X}$. Indeed, π-invariant holomorphic sections of K_Y^d correspond bijectively to meromorphic sections of K_X^d with poles of order at most $R(d, m_i)$ at x_i, for all $i \in \{1; \ldots; k\}$. More generally, for $j = 0, 1$, there are isomorphisms $\mathrm{H}^j(\mathcal{X}; \underline{\mathfrak{L}}) \simeq \mathrm{Fix}_\pi \mathrm{H}^j(Y; \underline{\mathcal{E}}) \simeq \mathrm{H}^j(X; \underline{L})$, from which one can deduce the following orbifold Riemann-Roch formula (see for instance [1, 14] or [13] for an exposition; the theorem itself is due to Kawasaki, [10]).

Theorem 2.1 (Orbifold Riemann-Roch, [10]) *Let $\mathcal{X}$ be a compact complex analytic orbi-curve and denote its cone points by $(x_i)_{1 \leqslant i \leqslant k}$. Let $\mathfrak{L}$ be an analytic line bundle over $\mathcal{X}$. Then*

$$\chi(\mathcal{X}; \underline{\mathfrak{L}}) = \chi(\mathcal{X}; O_{\mathcal{X}}) + \deg \mathfrak{L} - \sum_{i=1}^{k} \mathrm{age}_{x_i}(\mathfrak{L}).$$

For instance, $\chi(\mathcal{X}; \underline{K_{\mathcal{X}}}) = g - 1$ and $\chi(\mathcal{X}; \underline{K_{\mathcal{X}}^2}) = 3(g-1) + k$. When $\mathcal{X}$ is defined over $\mathbb{R}$, we can deduce the appropriate version of the Riemann-Roch formula from the complex case, by applying Theorem 2.1 to the complex analytic orbifold $\mathcal{X}^+$. Indeed, the real structure $\sigma : \mathcal{X}^+ \longrightarrow \mathcal{X}^+$ induces a $\mathbb{C}$-antilinear involution σ of the complex vector spaces $\mathrm{H}^j(\mathcal{X}^+; \underline{\mathfrak{L}^+})$, in such a way that $\mathrm{H}^j(\mathcal{X}; \underline{\mathfrak{L}}) \simeq \mathrm{Fix}_\sigma \mathrm{H}^j(\mathcal{X}^+; \underline{\mathfrak{L}^+})$, so $\dim_{\mathbb{R}} \mathrm{H}^j(\mathcal{X}; \underline{\mathfrak{L}}) = \dim_{\mathbb{C}} \mathrm{H}^j(\mathcal{X}^+; \underline{\mathfrak{L}^+})$ and $\chi(\mathcal{X}; \underline{\mathfrak{L}}) = \chi(\mathcal{X}^+; \underline{\mathfrak{L}^+})$. Setting $\deg \mathfrak{L} := \deg \mathfrak{L}^+$, one gets:

$$\chi(\mathcal{X}; \underline{\mathfrak{L}}) = \chi(\mathcal{X}; O_{\mathcal{X}}) + \deg(\mathfrak{L}) - 2\sum_{i=1}^{k} \mathrm{age}_{x_i}(\mathfrak{L}) - \sum_{j=1}^{\ell} \mathrm{age}_{y_j}(\mathfrak{L})$$

where the $(x_i)_{1 \leqslant i \leqslant k}$ and the $(y_j)_{1 \leqslant j \leqslant \ell}$ are respectively the cone points and dihedral points of $\mathcal{X}$. In particular, $\chi(\mathcal{X}; K_{\mathcal{X}}^2) = 3(g_X - 1) + 2k + \ell$, where again $g := \dim \mathrm{H}^1(\mathcal{X}; O_{\mathcal{X}})$ is the genus of $\mathcal{X}$.

2.3 *Spaces of Symmetric Differentials*

As we saw in Sect. 2.2, if $\mathcal{X}$ is a compact analytic orbi-curve, then $\chi(\mathcal{X}; K_{\mathcal{X}}^2) = \dim \mathcal{T}(\mathcal{X})$ (this is a complex dimension if $\mathcal{X}$ is defined over $\mathbb{C}$ and a real dimension if $\mathcal{X}$ is defined over $\mathbb{R}$). While this result is well-known, it is also the $n = 2$ case of the *Hitchin parameterisation* of $\mathrm{Hit}(\pi_1\mathcal{X}; \mathbf{PGL}(n;\mathbb{R}))$, as we shall see momentarily. Let us first recall Hitchin's result in the surface group case [8]: If Y is a closed orientable surface of negative Euler characteristic, the choice of a complex analytic structure on Y induces a homeomorphism

$$\mathrm{Hit}\big(\pi_1 Y; \mathbf{PGL}(n;\mathbb{R})\big) \simeq \bigoplus_{d=2}^{n} \mathrm{H}^0(Y; K_Y^d).$$

The main result of [2] is the following extension of Hitchin's result to the orbifold case.

Theorem 2.2 ([2]) *Let $\mathcal{X}$ be a compact differentiable orbi-surface of negative Euler characteristic. Then the choice of an analytic structure on $\mathcal{X}$ induces a homeomorphism*

$$\mathrm{Hit}\big(\pi_1\mathcal{X}; \mathbf{PGL}(n;\mathbb{R})\big) \simeq \bigoplus_{d=2}^{n} \mathrm{H}^0(\mathcal{X}; K_{\mathcal{X}}^d).$$

Here, choosing an analytic structure on $\mathcal{X}$ reduces to choosing a finite Galois cover by a closed orientable surface $Y \longrightarrow \mathcal{X}$ and a complex analytic structure on Y which is preserved by the automorphism group of that cover. As we have seen, the fact that such a cover always exists is a consequence of Selberg's lemma. Note that we are considering at the same time the case where the differentiable orbifold $\mathcal{X}$ is orientable (so admits a complex analytic structure, i.e. the finite group $\pi := \mathrm{Aut}_X(Y)$ acts holomorphically on Y) and the case where it is not (here Y is still a closed orientable surface but π will contain orientation-reversing transformation; as a consequence, the coarse moduli space X of $\mathcal{X} \simeq [Y/\pi]$ will be a differentiable surface with corners that has non-empty boundary or is non-orientable or both).

The proof of Theorem 2.2 consists in adapting Hitchin's proof to the orbifold case. The main tool is the orbifold version of the Non-Abelian Hodge Correspondence (NAHC). In [2], we took a largely equivariant approach to the latter, making the resulting formulation of the NAHC dependent on the choice of a presentation $\mathcal{X} \simeq [Y/\pi]$. Equivalently, we can rephrase this in terms of $\mathcal{G}$-Higgs bundles on $\mathcal{X}$, where G is a real reductive group and $\mathcal{G}$ is the orbifold group bundle $[(\widetilde{\mathcal{X}} \times G)/\pi_1\mathcal{X}]$. But in any case, the point is that, if G is the split real form of the adjoint group $\mathrm{Int}(\mathfrak{g}_{\mathbb{C}})$, where $\mathfrak{g}_{\mathbb{C}}$ is a simple complex Lie algebra, then the Hitchin component $\mathrm{Hit}(\pi_1 X; G)$ embeds into the moduli space of $\mathcal{G}$-Higgs bundles, denoted by $\mathcal{M}_{\mathcal{X}}(G)$.

$$\begin{array}{ccc} \mathrm{Hit}\big(\pi_1\mathcal{X}, \mathbf{PGL}(n,\mathbb{R})\big) & \overset{\text{NAHC}}{\hookrightarrow} & \mathcal{M}_{\mathcal{X}}\big(\mathbf{PGL}(n;\mathbb{R})\big) \\ & \nwarrow_{\text{Hitchin section}} & \downarrow{\scriptstyle \text{Hitchin fibration}} \\ & & \bigoplus_{d=2}^{n} \mathrm{H}^0(\mathcal{X}, K_{\mathcal{X}}^d) \end{array} \tag{2.3}$$

In [2], we showed that the *Hitchin fibration*, which is a morphism from the moduli space $\mathcal{M}_{\mathcal{X}}(G)$ to a vector space $\mathcal{B}_{\mathcal{X}}(\mathfrak{g})$ called the *Hitchin base*, first constructed by Hitchin in the surface group case [7], was well-defined in the orbifold case. For $\mathfrak{g} = \mathfrak{sl}(n;\mathbb{R})$, the Hitchin base is $\bigoplus_{d=2}^{n} \mathrm{H}^0(\mathcal{X}; K_{\mathcal{X}}^d)$, as in Diagram (2.3). Then we extended Hitchin's construction of a section of that fibration: The image of that section being exactly the embedded copy of $\mathrm{Hit}(\pi_1\mathcal{X}; G)$ in $\mathcal{M}_{\mathcal{X}}(G)$, thus proving Theorem 2.2.

This shows that $\mathrm{Hit}(\pi_1\mathcal{X}; \mathbf{PGL}(n;\mathbb{R}))$ is homeomorphic to the vector space $\mathcal{B}_{\mathcal{X}}(\mathfrak{sl}(n;\mathbb{R}))$, which is a complex vector space if $\mathcal{X}$ is complex and a real vector space if $\mathcal{X}$ is real. Using Theorem 2.1, we can compute the dimension of that vector space. Since we already know how to deduce the result in the real case from the result in the complex case, we will present the proof in the latter case only. From (2.2), we get that, for all $d \in \{2; \ldots; n\}$,

$$\chi(\mathcal{X}; K_{\mathcal{X}}^d) = (2d-1)(g-1) + \textstyle\sum_{i=1}^{k} R(d, m_i).$$

But for $d \geqslant 2$, one has $\deg K_{\mathcal{X}}^d = d \deg K_{\mathcal{X}} > \deg K_{\mathcal{X}}$, so $\mathrm{H}^1(\mathcal{X}; K_{\mathcal{X}}^d) = 0$ and $\chi(\mathcal{X}; K_{\mathcal{X}}) = \dim \mathrm{H}^0(\mathcal{X}; K_{\mathcal{X}}^d)$. Thus, when $\mathcal{X}$ is complex analytic, $\mathcal{B}_{\mathcal{X}}(\mathfrak{g})$ is a complex vector space of dimension

$$\begin{aligned}(g-1)\textstyle\sum_{d=2}^{n}(2d-1) + \sum_{d=2}^{n}\sum_{i=1}^{k} R(d, m_i) &= (g-1)(n^2-1) \\ &\quad + \textstyle\sum_{d=2}^{n}\sum_{i=1}^{k} R(d, m_i).\end{aligned}$$

The real dimension is twice as much, which indeed coincides with the formula in Theorem 1.3 (for $\ell = 0$).

Let us denote $\mathbf{PGL}(n;\mathbb{R})$ simply by G. A consequence of Theorem 2.2 is that, given an analytic orbi-curve $\mathcal{X}$, we can embed the Hitchin component $\mathrm{Hit}(\pi_1\mathcal{X}; G)$ into the Hitchin component $\mathrm{Hit}(\pi_1\mathcal{Y}; G)$ associated to any Galois cover $\mathcal{Y} \longrightarrow \mathcal{X}$. More precisely, given a Galois cover $\mathcal{Y} \longrightarrow \mathcal{X}$ with automorphism group π, consider the short exact sequence

$$1 \longrightarrow \pi_1\mathcal{Y} \longrightarrow \pi_1\mathcal{X} \longrightarrow \pi \longrightarrow 1,$$

the induced morphism $\pi \longrightarrow \mathrm{Out}(\pi_1\mathcal{Y})$ and the associated action of π on $\mathrm{Hit}(\pi_1\mathcal{Y}; G)$. Then, the map taking a representation $\varrho : \pi_1\mathcal{X} \longrightarrow G$ to its restriction $\varrho|_{\pi_1\mathcal{Y}}$ induces a homeomorphism $\mathrm{Hit}(\pi_1\mathcal{X}; G) \simeq \mathrm{Fix}_\pi\, \mathrm{Hit}(\pi_1\mathcal{Y}; G)$, since $\mathcal{X} \simeq [\mathcal{Y}/\pi]$ implies that $\mathrm{H}^0(\mathcal{X}; K_{\mathcal{X}}^d) = \mathrm{Fix}_\pi\, \mathrm{H}^0(\mathcal{Y}; K_{\mathcal{Y}}^d)$. As an example of this, consider the Coxeter triangle group $T_{(2,3,7)}$ of (2.1). It is the orbifold fundamental group of a hyperbolic triangle with vertices of respective orders 2, 3 and 7, which can be obtained as the quotient of the Klein quartic $\mathcal{K}$ by its full automorphism group. As $\mathrm{Hit}(T_{(2,3,7)}; \mathbf{PGL}(6;\mathbb{R}))$ is of (real) dimension 1 by Theorem 1.3, it defines a one-parameter family of Hitchin representations in $\mathrm{Hit}(\pi_1\mathcal{K}; \mathbf{PGL}(6;\mathbb{R}))$, the latter being, by Hitchin's result for the closed orientable surface $\mathcal{K}$ (of genus 3), of real dimension 140.

Acknowledgments It is a pleasure to thank Alexander Schmitt and all participants of the Session on Complex Geometry at ISAAC 2019, as well as Georgios Kydonakis, for feedback on the topics discussed in this paper. I also thank the referee for careful reading and valuable suggestions.

References

1. D. Abramovic, Lectures on Gromov-Witten invariants of orbifolds (2005). https://arxiv.org/abs/math/0512372
2. D. Alessandrini, G.S. Lee, F. Schaffhauser, Hitchin components for orbifolds (2018, submitted). arxiv 1811.05366v3
3. K. Behrend, G. Ginot, B. Noohi, P. Xu, String topology for stacks. Astérisque **343**, xiv+169 (2012)
4. S. Choi, W.M. Goldman, The deformation spaces of convex $\mathbb{RP}^2$-structures on 2-orbifolds. Am. J. Math. **127**(5), 1019–1102 (2005)
5. D. Cooper, C.D. Hodgson, S.P. Kerckhoff, *Three-Dimensional Orbifolds and Cone-Manifolds*. MSJ Memoirs, vol. 5 (Mathematical Society of Japan, Tokyo, 2000). With a postface by S. Kojima
6. V. Fock, A. Goncharov, Moduli spaces of local systems and higher Teichmüller theory. Publ. Math. Inst. Hautes Études Sci. **103**, 1–211 (2006)
7. N.J. Hitchin, Stable bundles and integrable systems. Duke Math. J. **54**(1), 91–114 (1987)
8. N.J. Hitchin, Lie groups and Teichmüller space. Topology **31**(3), 449–473 (1992)
9. V. Hinich, A. Vaintrob, Augmented Teichmüller spaces and orbifolds. Sel. Math. **16**(3), 533–629 (2010)
10. T. Kawasaki, The Riemann-Roch theorem for complex V-manifolds. Osaka Math. J. **16**(1), 151–159 (1979)
11. B. Kostant, The principal three-dimensional subgroup and the Betti numbers of a complex simple Lie group. Am. J. Math. **81**, 973–1032 (1959)
12. F. Labourie, Anosov flows, surface groups and curves in projective space. Invent. Math. **165**(1), 51–114 (2006)
13. C.-C.M. Liu, Localization in Gromov-Witten theory and orbifold Gromov-Witten theory, in *Handbook of Moduli. Vol. II*. Advanced Lectures in Mathematics, vol. 25 (International Press, Somerville, 2013), pp. 353–425
14. E.B. Nasatyr, B. Steer, Orbifold Riemann surfaces and the Yang-Mills-Higgs equations. Ann. Scuola Norm. Sup. Pisa Cl. Sci. **22**(4), 595–643 (1995)
15. P. Scott, The geometries of 3-manifolds. Bull. Lond. Math. Soc. **15**(5), 401–487 (1983)
16. A. Selberg, On discontinuous groups in higher-dimensional symmetric spaces, in *Contributions to Function Theory (International Colloquium on Function Theory, Bombay, 1960)* (Tata Institute of Fundamental Research, Bombay, 1960), pp. 147–164
17. W. Thurston, *Geometry and Topology of 3-Manifolds*. Lecture Notes (Princeton University, Princeton, 1979)
18. A. Wienhard, An invitation Higher Teichmüller Theory, in *Proceedings of the International Congress of Mathematicians—Rio de Janeiro 2018*, vol. 1 (World Scientific, Singapore, 2018), pp. 1007–1034

Part IV
Complex Variables and Potential Theory

Extremal Decomposition of a Multidimensional Complex Space with Poles on the Boundary of a Polydisk

Iryna Denega

Abstract In the paper we obtain estimates of the maximums of products of generalized inner radii of mutually non-overlapping polycylindrical domains in $\mathbb{C}^n$. The main theorems of the paper generalize and strengthening known results in the theory of non-overlapping domains with free poles on the unit circle onto the case of n-dimensional complex space.

Keywords Inner radius of the domain · Non-overlapping polycylindrical domain · The Green function · Transfinite diameter · Theorem on minimizing of area · The Cauchy inequality

Mathematics Subject Classification (2010). Primary 32A30; Secondary 30C75

1 Preliminaries

The goal of the present work is the study of the problems of a products of the generalized inner radii of polycylindrical non-overlapping domains with poles on the boundary of a polydisk. The spatial analogs of a number of known results concerning the non-overlapping domains on a plane were obtained in [1], where a generalization of the notion of inner radius was given. Namely, the notion of harmonic radius of the spatial domain $B \subset \mathbb{R}^n$ relative to some internal point was introduced. Work [1] was the essential break-through in the consideration of non-overlapping domains in the spatial case. Then, work [2] advanced an approach that allowed the transfer of some results known in the case of a complex plane onto $\mathbb{C}^n$. At the same time, the problems of non-overlapping domains in the case of a complex plane represent a sufficiently well-developed trend of the geometric theory of functions of complex variable (see, e.g., [1–14]).

I. Denega (✉)
Institute of Mathematics of the National Academy of Sciences of Ukraine, Department of Complex Analysis and Potential Theory, Kyiv, Ukraine

P. Cerejeiras et al. (eds.), *Current Trends in Analysis, its Applications and Computation*, Research Perspectives,
https://doi.org/10.1007/978-3-030-87502-2_14

In this paper, we have obtained analogs of estimates of the maximums of products of inner radii of domains for the case of multidimensional complex spaces, which can be applied to coverage theorems, distortion theorems, and estimates of the coefficients of univalent functions, also in holomorphic dynamics for study of the number of critical points in parabolic basins.

Let $\mathbb{N}$, $\mathbb{R}$, and $\mathbb{C}$ be the sets of natural, real, and complex numbers, respectively, and $\mathbb{R}^+ = (0, \infty)$. Let $\overline{\mathbb{C}}$ be a Riemann sphere (extended complex plane). It is well known that $\mathbb{C}^n = \underbrace{(\mathbb{C} \times \mathbb{C} \times \ldots \times \mathbb{C})}_{n-\text{times}}$, $n \in \mathbb{N}$. $\overline{\mathbb{C}}^n = \underbrace{(\overline{\mathbb{C}} \times \overline{\mathbb{C}} \times \ldots \times \overline{\mathbb{C}})}_{n-\text{times}}$ is a compactification of the space $\mathbb{C}^n$ (see, e.g., [3–5]), where the set of infinitely remote points has the complex dimension $n - 1$. Let $[D]^n$ (Cartesian degree of a domain $D \subset \overline{\mathbb{C}}$) denote the Cartesian product $\underbrace{D \times D \times \ldots \times D}_{n-\text{times}}$, and let $[d]^n$ (Cartesian degree of a point $d \in \overline{\mathbb{C}}$) denote the point with $\overline{\mathbb{C}}^n$, which have the coordinates $\underbrace{(d, \ldots, d)}_{n-\text{times}}$. It is clear that $\mathbb{C}^1 = \mathbb{C}$, $\overline{\mathbb{C}}^1 = \overline{\mathbb{C}}$. The topology in $\overline{\mathbb{C}}^n$ is introduced like in a Cartesian product of topological spaces. In this topology, $\overline{\mathbb{C}}^n$ is compact (see [3–5]).

Definition 1.1 The domain $\mathbb{B} = B_1 \times B_2 \times \ldots \times B_n \subset \overline{\mathbb{C}}^n$, where each domain $B_k \subset \overline{\mathbb{C}}$, $k = \overline{1, n}$, is called a polycylindrical domain in $\overline{\mathbb{C}}^n$ (see, e.g., [3]). The domains B_k, $k = \overline{1, n}$, are called coordinate domains of the domain $\mathbb{B}$.

Definition 1.2 Let B be a domain from $\overline{\mathbb{C}}$. Let

$$g_B(z, a) = h_{B,a}(z) + \log \frac{1}{|z - a|}$$

be a generalized Green's function of the domain B relative to the point $a \in B$. If $a \to \infty$, then

$$g_B(z, \infty) = h_{B,\infty}(z) + \log \frac{1}{|z|}.$$

The quantity $r(B, a) := \exp(h_{B,a}(a))$ means the inner radius of the domain $B \subset \overline{\mathbb{C}}$ relative to the point $a \in B$ (see [6–11]).

Definition 1.3 The generalized inner radius of the polycylindrical domain $\mathbb{B}$ relative to the point $\mathbb{A} = (a_1, a_2, \ldots, a_n) \in \mathbb{B}$, $a_k \in B_k$, $k = \overline{1, n}$, is

$$R(\mathbb{B}, \mathbb{A}) := \prod_{k=1}^{n} r(B_k, a_k),$$

where the quantities $r(B_k, a_k)$, $k = \overline{1, n}$, mean the inner radii of the coordinate domains B_k relative to a_k. For $n = 1$, the quantity $R(\mathbb{B}, \mathbb{A})$ is the ordinary inner radius of the domain $\mathbb{B} \subset \overline{\mathbb{C}}$ relative to the point $\mathbb{A}$.

Let $\mathbb{U}^n = [U]^n$, where $U = \{z \in \mathbb{C} : |z| < 1\}$ (unit disk in the complex plane $\mathbb{C}$). By Γ_n we denote the skeleton (distinguished boundary) of the polydisk $\mathbb{U}^n$ (see [7, 8]), i.e., the set of points $\mathbb{A} = (a_1, a_2, \ldots, a_n) \subset \mathbb{C}^n$, $|a_s| = 1$, $s = \overline{1, n}$.

Definition 1.4 The system $\{\mathbb{B}_k\}_{k=1}^m$ $\left(\mathbb{B}_k = B_1^{(k)} \times \ldots \times B_n^{(k)},\ k = \overline{1, m}\right)$ is called a system of non-overlapping polycylindrical domains, if, for every fixed p_0, $p_0 = \overline{1, n}$, the system of domains $\left\{B_{p_0}^{(k)}\right\}$, $k = \overline{1, m}$, is a system of non-overlapping domains on $\overline{\mathbb{C}}$.

Further, we will consider the systems of points in the space $\mathbb{C}^n$ of the form

$$\{\mathbb{A}_k\}_{k=1}^m, \quad \mathbb{A}_k = \left(a_1^{(k)}, a_2^{(k)}, \ldots, a_n^{(k)}\right) \in \mathbb{C}^n,$$

$$k = \overline{1, m}, \quad a_{p_0}^{(1)} > 0, \quad p_0 = \overline{1, n}, \tag{1.1}$$

$$\arg a_{p_0}^{(k)} < \arg a_{p_0}^{(k+1)}, \quad k = \overline{1, m-1}, \quad \arg a_{p_0}^{(m)} < 2\pi.$$

2 Main Results

In the above posed notations, we establish the following results.

Theorem 2.1 *Let $m, n \in \mathbb{N}$, $m \geq 2$, $\gamma \in (0, m]$. Then, for any system of different points of the form (1.1) $\{\mathbb{A}_k\}_{k=1}^m = \left\{a_p^{(k)}\right\}_{k=1}^m \in \overline{\mathbb{C}}^n$, $p = \overline{1, n}$, such that $\mathbb{A}_k \in \Gamma_n$, $k = \overline{1, m}$, and for any collection of mutually non-overlapping polycylindrical domains $\mathbb{B}_0$, $\mathbb{B}_k$, $\mathbb{A}_0 = [0]^n \in \mathbb{B}_0 \subset \overline{\mathbb{C}}^n$, $\mathbb{A}_k \in \mathbb{B}_k \subset \overline{\mathbb{C}}^n$, $k = \overline{1, m}$, the inequality*

$$R^\gamma(\mathbb{B}_0, \mathbb{A}_0) \prod_{k=1}^m R(\mathbb{B}_k, \mathbb{A}_k) \leq m^{-\frac{\gamma n}{2}} \left(\frac{4}{m}\right)^{n(m-\gamma)} \tag{2.1}$$

holds.

Proof We make the transformation

$$R^\gamma(\mathbb{B}_0, \mathbb{A}_0) \prod_{k=1}^m R(\mathbb{B}_k, \mathbb{A}_k) = \left[\prod_{p=1}^n r\left(B_p^{(0)}, 0\right)\right]^\gamma \prod_{k=1}^m \left[\prod_{p=1}^n r\left(B_p^{(k)}, a_p^{(k)}\right)\right] =$$

$$= \prod_{p=1}^n \left[r^\gamma\left(B_p^{(0)}, 0\right) \prod_{k=1}^m r\left(B_p^{(k)}, a_p^{(k)}\right)\right]. \tag{2.2}$$

Then, for a fixed $p = \overline{1, n}$ the domains $B_p^{(k)}$, $k = \overline{0, m}$, form a system of pairwise non-overlapping domains on the complex plane $\overline{\mathbb{C}}$. Further, we will consider the following product

$$r^{\gamma}(D_0, 0) \prod_{k=1}^{m} r(D_k, a_k),$$

where $\gamma \in (0, m]$, $D_0, D_1, D_2, \ldots, D_m$, $m \geq 2$, are mutually non-overlapping domains in $\overline{\mathbb{C}}$, $a_0 = 0$, $|a_k| = 1$, $k = \overline{1, m}$, $a_k \in D_k \subset \overline{\mathbb{C}}$, $k = \overline{0, m}$.

Let $d(E)$ be the transfinite diameter of a compact set $E \subset \mathbb{C}$. Then the following relation holds

$$r(D_0, 0) = r\left(D_0^+, \infty\right) = \frac{1}{d\left(\overline{\mathbb{C}} \setminus D_0^+\right)} \leq \frac{1}{d\left(\bigcup\limits_{k=1}^{m} \overline{D_k^+}\right)}, \tag{2.3}$$

where $D^+ = \{z : \frac{1}{z} \in D\}$. Using the well-known Polya theorem [8, p. 34] and [13, p. 28] the inequality

$$\mu E \leq \pi d^2(E),$$

where μE denotes the Lebesgue measure of a compact set E, is valid. From whence, we get

$$d(E) \geq \left(\frac{1}{\pi}\mu E\right)^{\frac{1}{2}}.$$

Then from (2.3) we have

$$r(D_0, 0) \leq \frac{1}{d\left(\bigcup\limits_{k=1}^{m} \overline{D_k^+}\right)} \leq \frac{1}{\sqrt{\frac{1}{\pi}\mu\left(\bigcup\limits_{k=1}^{m} \overline{D_k^+}\right)}} = \left(\frac{1}{\pi}\sum_{k=1}^{m} \mu \overline{D_k^+}\right)^{-\frac{1}{2}}. \tag{2.4}$$

From the theorem of minimization of an area [8, p. 34], we obtain

$$\mu(D) \geq \pi r^2(D, a).$$

Inequality (2.4) implies directly that

$$r(D_0, 0) \leq \left(\frac{1}{\pi}\sum_{k=1}^{m} \mu \overline{D_k^+}\right)^{-\frac{1}{2}} \leq \left(\frac{1}{\pi}\sum_{k=1}^{m} \mu D_k^+\right)^{-\frac{1}{2}} \leq \left(\sum_{k=1}^{m} r^2\left(D_k^+, a_k^+\right)\right)^{-\frac{1}{2}}.$$

Using conformal invariance of the Green function, we have

$$g_{D_k}(z, a_k) = g_{D_k^+}(w^+, a_k^+), \quad w^+ = \frac{1}{z}.$$

Then, using relation

$$g_{D_k^+}(w^+, a_k^+) = g_{D_k^+}\left(\frac{1}{z}, \frac{1}{a_k}\right) = \ln \frac{1}{|\frac{1}{z} - a_k^+|} + \ln r(D_k^+, a_k^+) + o(1),$$

we obtain

$$r\left(D_k^+, a_k^+\right) = \frac{r\left(D_k, a_k\right)}{|a_k|^2}.$$

Thus,

$$r\left(D_0, 0\right) \leq \left(\sum_{k=1}^{m} \frac{r^2\left(D_k, a_k\right)}{|a_k|^4}\right)^{-\frac{1}{2}}.$$

This result yields the relation

$$r^{\gamma}\left(D_0, 0\right) \prod_{k=1}^{m} r\left(D_k, a_k\right) \leq \frac{\prod\limits_{k=1}^{m} r\left(D_k, a_k\right)}{\left(\sum\limits_{k=1}^{m} \frac{r^2(D_k, a_k)}{|a_k|^4}\right)^{\frac{\gamma}{2}}}.$$

The Cauchy inequality yields automatically the inequality

$$\frac{1}{m} \sum_{k=1}^{m} \frac{r^2\left(D_k, a_k\right)}{|a_k|^4} \geq \left(\prod_{k=1}^{m} \frac{r^2\left(D_k, a_k\right)}{|a_k|^4}\right)^{\frac{1}{m}}.$$

Then we get easily

$$\left(\sum_{k=1}^{m} \frac{r^2\left(D_k, a_k\right)}{|a_k|^4}\right)^{\frac{\gamma}{2}} \geq \left(m\left(\prod_{k=1}^{m} \frac{r^2\left(D_k, a_k\right)}{|a_k|^4}\right)^{\frac{1}{m}}\right)^{\frac{\gamma}{2}} \geq m^{\frac{\gamma}{2}}\left(\prod_{k=1}^{m} \frac{r\left(D_k, a_k\right)}{|a_k|^2}\right)^{\frac{\gamma}{m}}.$$

In this way, using condition $|a_k| = 1$, $k = \overline{1, m}$, we obtain

$$r^{\gamma}(D_0, 0) \prod_{k=1}^{m} r(D_k, a_k) \le \frac{\prod_{k=1}^{m} r(D_k, a_k)}{m^{\frac{\gamma}{2}} \left(\prod_{k=1}^{m} r(D_k, a_k) \right)^{\frac{\gamma}{m}}} = m^{-\frac{\gamma}{2}} \left(\prod_{k=1}^{m} r(D_k, a_k) \right)^{1-\frac{\gamma}{m}}.$$

In [11, Theorem 6.11] for any different points a_k on the circle $|a_k| = 1$, $k = \overline{1, m}$ $(m \ge 2)$, and any pairwise non-overlapping domains $D_k \subset \overline{\mathbb{C}}$ such that $a_k \in D_k$, $k = \overline{1, m}$, the inequality

$$\prod_{k=1}^{m} r(D_k, a_k) \le \left(\frac{4}{m} \right)^{m}$$

is proved. Thus,

$$r^{\gamma}(D_0, 0) \prod_{k=1}^{m} r(D_k, a_k) \le m^{-\frac{\gamma}{2}} \left(\frac{4}{m} \right)^{m-\gamma}.$$

Finally, upon combining (2.2) and the last inequality, we easily see that, the following relation

$$R^{\gamma}(\mathbb{B}_0, \mathbb{A}_0) \prod_{k=1}^{m} R(\mathbb{B}_k, \mathbb{A}_k) \le \prod_{p=1}^{n} \left[m^{-\frac{\gamma}{2}} \left(\frac{4}{m} \right)^{m-\gamma} \right] = m^{-\frac{\gamma n}{2}} \left(\frac{4}{m} \right)^{n(m-\gamma)}$$

holds. Thus, Theorem 2.1 is proved. □

Remark 2.2 If $\gamma = m$, then under all conditions of above posed Theorem 2.1, the inequality

$$R^{m}(\mathbb{B}_0, \mathbb{A}_0) \prod_{k=1}^{m} R(\mathbb{B}_k, \mathbb{A}_k) \le m^{-\frac{mn}{2}}$$

holds.

Theorem 2.3 *Let $m, n \in \mathbb{N}$, $m \ge 2$, $\gamma \in (0, m]$. Then, for any system of different points of the form (1.1) $\{\mathbb{A}_k\}_{k=1}^{m} = \left\{ a_p^{(k)} \right\}_{k=1}^{m} \in \overline{\mathbb{C}}^n$, $p = \overline{1, n}$, such that $\mathbb{A}_k \in \Gamma_n$, $k = \overline{1, m}$, and for any collection of mutually non-overlapping polycylindrical domains $\mathbb{B}_0$, $\mathbb{B}_k$, $\mathbb{A}_0 = [0]^n \in \mathbb{B}_0 \subset \overline{\mathbb{C}}^n$, $\mathbb{A}_k \in \mathbb{B}_k \subset \overline{\mathbb{C}}^n$, $k = \overline{1, m}$, and $\mathbb{B}_k$, $k = \overline{1, m}$, are mirror-symmetric relative to Γ_n, the inequality (2.1) holds.*

The proof of Theorem 2.3 is similar to that of Theorem 2.1, so we have chosen to omit the analogous details.

Theorem 2.4 *Let* $m, n \in \mathbb{N}$, $m \geq 2$, $\gamma \in (0, \frac{m+2}{2}]$. *Then, for any system of different points of the form* (1.1) $\{\mathbb{A}_k\}_{k=1}^m = \left\{a_p^{(k)}\right\}_{k=1}^m \in \overline{\mathbb{C}}^n$, $p = \overline{1,n}$, *such that* $\mathbb{A}_k \in \Gamma_n$, $k = \overline{1,m}$, *and for any collection of mutually non-overlapping polycylindrical domains* $\mathbb{B}_0$, $\mathbb{B}_\infty$, $\mathbb{B}_k$, $\mathbb{A}_0 = [0]^n \in \mathbb{B}_0 \subset \overline{\mathbb{C}}^n$, $\mathbb{A}_\infty = [\infty]^n \in \mathbb{B}_\infty \subset \overline{\mathbb{C}}^n$, $\mathbb{A}_k \in \mathbb{B}_k \subset \overline{\mathbb{C}}^n$, $k = \overline{1,m}$, *the inequality*

$$(R(\mathbb{B}_0, \mathbb{A}_0) R(\mathbb{B}_\infty, \mathbb{A}_\infty))^\gamma \prod_{k=1}^m R(\mathbb{B}_k, \mathbb{A}_k) \leq (m+1)^{-\gamma \frac{n(m+1)}{m+2}} \left(\frac{4}{m}\right)^{mn\left(1-\frac{2\gamma}{m+2}\right)}$$

holds.

Proof The proof of Theorem 2.4 is based on constructions given in proof of the Theorem 2.1. First, let us consider the following product

$$J_m(\gamma) = (r\,(D_0, 0)\, r\,(D_\infty, \infty))^\gamma \prod_{k=1}^m r\,(D_k, a_k)\,,$$

where $\gamma \in (0, \frac{m+2}{2}]$, D_0, D_∞, D_1, $D_2, \ldots, D_m$, $m \geq 2$, are mutually non-overlapping domains in $\overline{\mathbb{C}}$, $a_0 = 0$, $|a_k| = 1$, $k = \overline{1,m}$, $0 \in D_0 \subset \overline{\mathbb{C}}$, $\infty \in D_\infty \subset \overline{\mathbb{C}}$, $a_k \in D_k \subset \overline{\mathbb{C}}$, $k = \overline{1,m}$. Using inequalities (2.3) and (2.4), we have

$$r\,(D_0, 0) \leq \left[r^2\,(D_\infty, \infty) + \sum_{k=1}^m \frac{r^2\,(D_k, a_k)}{|a_k|^4}\right]^{-\frac{1}{2}},$$

$$r\,(D_\infty, \infty) \leq \left[r^2\,(D_0, 0) + \sum_{k=1}^m r^2\,(D_k, a_k)\right]^{-\frac{1}{2}}.$$

Taking into account the Cauchy inequality

$$\left(r^2\,(D_\infty, \infty) + \sum_{k=1}^m \frac{r^2\,(D_k, a_k)}{|a_k|^4}\right)^{\frac{1}{2}} \geq (m+1)^{\frac{1}{2}} \left[r\,(D_\infty, \infty) \prod_{k=1}^m \frac{r\,(D_k, a_k)}{|a_k|^2}\right]^{\frac{1}{m+1}},$$

and

$$\left(r^2\,(D_0, 0) + \sum_{k=1}^m r^2\,(D_k, a_k)\right)^{\frac{1}{2}} \geq (m+1)^{\frac{1}{2}} \left[r\,(D_0, 0) \prod_{k=1}^m r\,(D_k, a_k)\right]^{\frac{1}{m+1}}.$$

Upon combining two previous inequalities we easily see that

$$r\left(D_0,0\right)r\left(D_\infty,\infty\right)\le\frac{\left(\prod\limits_{k=1}^{m}|a_k|\right)^{\frac{2}{m+2}}}{(m+1)^{\frac{m+1}{m+2}}\left(\prod\limits_{k=1}^{m}r\left(D_k,a_k\right)\right)^{\frac{2}{m+2}}}.$$

And we thus find that

$$J_m(\gamma)\le(m+1)^{-\gamma\frac{m+1}{m+2}}\left(\prod_{k=1}^{m}r\left(D_k,a_k\right)\right)^{1-\frac{2\gamma}{m+2}}\left(\prod_{k=1}^{m}|a_k|\right)^{\frac{2\gamma}{m+2}}.$$

By the virtue of [11, Theorem 6.11] and condition $|a_k|=1$, $k=\overline{1,m}$, we conclude that

$$J_m(\gamma)\le(m+1)^{-\gamma\frac{m+1}{m+2}}\left(\frac{4}{m}\right)^{m\left(1-\frac{2\gamma}{m+2}\right)}.$$

Thus, using the fact that

$$(R(\mathbb{B}_0,\mathbb{A}_0)R(\mathbb{B}_\infty,\mathbb{A}_\infty))^\gamma\prod_{k=1}^{m}R(\mathbb{B}_k,\mathbb{A}_k)=$$

$$=\prod_{p=1}^{n}\left[\left(r\left(B_p^{(0)},0\right)r\left(B_p^{(\infty)},\infty\right)\right)^\gamma\prod_{k=1}^{m}r\left(B_p^{(k)},a_p^{(k)}\right)\right]$$

and for a fixed $p=\overline{1,n}$ the domains $B_p^{(0)}$, $B_p^{(\infty)}$, $B_p^{(k)}$, $k=\overline{1,m}$, form a system of pairwise non-overlapping domains on the complex plane $\overline{\mathbb{C}}$, it is easy to see that

$$(R(\mathbb{B}_0,\mathbb{A}_0)R(\mathbb{B}_\infty,\mathbb{A}_\infty))^\gamma\prod_{k=1}^{m}R(\mathbb{B}_k,\mathbb{A}_k)\le(m+1)^{-\gamma\frac{n(m+1)}{m+2}}\left(\frac{4}{m}\right)^{mn\left(1-\frac{2\gamma}{m+2}\right)}.$$

Theorem 2.4 is proved. □

Remark 2.5 If $\gamma=\frac{m+2}{2}$, then under all conditions of above posed Theorem 2.4, the inequality holds

$$(R(\mathbb{B}_0,\mathbb{A}_0)R(\mathbb{B}_\infty,\mathbb{A}_\infty))^{\frac{m+2}{2}}\prod_{k=1}^{m}R(\mathbb{B}_k,\mathbb{A}_k)\le(m+1)^{-\frac{n(m+1)}{2}}.$$

From Theorem 2.4 (see also, [11, p. 176]), we obtain the following result.

Theorem 2.6 *Let* $m, n \in \mathbb{N}$, $m \geq 2$, $\gamma \in (0, \frac{m+2}{2}]$ *and* $\mathbb{B}_0 \subset \mathbb{U}^n$. *Then, for any system of different points of the form* (1.1) $\{\mathbb{A}_k\}_{k=1}^m = \left\{a_p^{(k)}\right\}_{k=1}^m \in \overline{\mathbb{C}}^n$, $p = \overline{1,n}$, *such that* $\mathbb{A}_k \in \Gamma_n$, $k = \overline{1,m}$, *and for any collection of mutually non-overlapping polycylindrical domains* $\mathbb{B}_0$, $\mathbb{B}_k$, $\mathbb{A}_0 = [0]^n \in \mathbb{B}_0 \subset \overline{\mathbb{C}}^n$, $\mathbb{A}_k \in \mathbb{B}_k \subset \overline{\mathbb{C}}^n$, $k = \overline{1,m}$, *and* $\mathbb{B}_k$, $k = \overline{1,m}$, *are mirror-symmetric relative to* Γ_n, *the following inequality holds*

$$R^{2\gamma}(\mathbb{B}_0, \mathbb{A}_0) \prod_{k=1}^{m} R(\mathbb{B}_k, \mathbb{A}_k) \leq (m+1)^{-\gamma \frac{n(m+1)}{m+2}} \left(\frac{4}{m}\right)^{mn\left(1-\frac{2\gamma}{m+2}\right)}.$$

Acknowledgments The present work is financially held up by the budget program "Support of the development of priority trends of scientific researches" (KPKVK 6541230).

References

1. V.N. Dubinin, E.G. Prilepkina, On the extreme partition of spatial domains (in Russian). Zap. Nauch. Sem. POMI **254**, 95–107 (1998)
2. A.K. Bakhtin, A generalization of some results of the theory of one-sheeted functions onto multidimensional complex spaces (in Russian). Dop. NAN Ukr. **3**, 7–11 (2011)
3. B.V. Shabat, *Introduction to Complex Analysis. Part II. Functions of Several Variables* (in Russian), 2nd edn., revised and augmented, Izdat (Nauka, Moscow, 1976)
4. E.M. Chirka, *Complex Analytic Sets* (in Russian). (Nauka, Moscow, 1985)
5. B. A. Fuks, *Introduction to the Theory of Analytic Functions of Several Complex Variables* (Fizmatgiz, Moscow, 1962); English transl., Amer. Math. Soc., Providence, R. I., 1963
6. M.A. Lavrent'ev, On the theory of conformal mappings (in Russian). Tr. Sci. Inst An USSR **5**, 159–245 (1934)
7. V. Hayman, *Multivalent Functions* (Cambridge University Press, Cambridge, 1958)
8. G.M. Goluzin, *Geometric Theory of Functions of a Complex Variable* (American Mathematical Society, Providence, 1969)
9. N.A. Lebedev, *The Area Principle in the Theory of Univalent Functions* (in Russian) (Science, Moscow, 1975)
10. A.K. Bakhtin, G.P. Bakhtina, Y.B. Zelinskii, *Topological-Algebraic Structures and Geometric Methods in Complex Analysis* (in Russian). Zb. Prats of the Inst. of Math. of NASU (2008)
11. V.N. Dubinin, *Condenser Capacities and Symmetrization in Geometric Function Theory* (Birkhäuser, Basel, 2014)
12. W. Bergweiler, On the number of critical points in parabolic basins. Ergod. Theory Dyn. Syst. **22**, 655–669 (2002)
13. G. Polya, G. Szego, *Isoperimetric Inequalities in Mathematical Physics* (in Russian). M:Fizmatgiz (Princeton University Press, Princeton, 1962)
14. G.P. Bakhtina, *On Conformal Radii of Symmetric Non-overlapping Domains* (in Russian). Modern Problems of the Real and Complex Analysis I(nstitute of Mathematics of the Academy of Sciences of USSR, Kiev, 1984), pp. 21–27

Some Properties of the Solutions Space of Irregular Elliptic Systems

Grigori Giorgadze and Giorgi Makatsaria

Abstract In this paper we prove Liouville theorem for the irregular nonhomogeneous Cauchy-Riemann equation depended on parameters and we show that qualitative properties of generalized analytic vectors strongly depend on the asymptotic parameters. We give an explicit formula for the solutions of the special type elliptic system of two unknown functions by the spectrum of the corresponding matrix. This result is a revision of the similarity principle for the elliptic system in a whole complex plane.

Keywords Elliptic system · Nonhomogeneous Cauchy-Riemann equation · Liouville theorem · Similarity principle

Mathematics Subject Classification (2010) Primary 30G20; Secondary 30C55

1 Introduction and Motivation

A matrix elliptic system of the form

$$\partial_{\bar{z}} W(z, \bar{z}) = V(z, \bar{z}) W(z, \bar{z}) + U(z, \bar{z}) \overline{W(z, \bar{z})} \tag{1.1}$$

on the domain $D \subset \mathbf{C}_{(z,\bar{z})}$, where $V(z, \bar{z})$ and $U(z, \bar{z})$ are the matrix functions, given on D and $W(z, \bar{z})$ is an unknown vector function, is the generalization of the Carleman–Bers–Vekua equation

$$\partial_{\bar{z}} \omega + A\omega + B\bar{\omega} = 0, \tag{1.2}$$

G. Giorgadze (✉)
Tbilisi State University, Tbilisi, Georgia
e-mail: gia.giorgadze@tsu.ge

G. Makatsaria
BTU, Tbilisi, Georgia

P. Cerejeiras et al. (eds.), *Current Trends in Analysis, its Applications and Computation*, Research Perspectives,
https://doi.org/10.1007/978-3-030-87502-2_15

where A and B is a pair of regular coefficients (functions) on a complex plane $\mathbf{C}$ and $\partial_{\bar{z}}$ is an operator of partial derivation with respect to the independent variable $\bar{z}$ in generalized (Sobolev) sense [16]. We call the Eq. (1.2) *regular* on $\mathbf{C}$ if the coefficients A and B are regular. It means that both A and B belong to the space $L_{p,2},\ p > 2$. By the definition $L_{p,2},\ p > 2$ consists of all functions g defined on the whole plane satisfying the conditions

$$\int\int_D |g(\xi)|^p dD < \infty, \quad \int\int_D \frac{1}{|z|^{2p}} \left| g\left(\frac{1}{z}\right) \right|^p dD < \infty,$$

where $D = \{|z| \leq 1\}$ is a unit disc (see [1, 3, 6]). Similarly, the system (1.1) is called regular if the entries of the matrix functions V and W are regular. Hence, we call (1.1) or (1.2) *irregular*, if at least one coefficient is not regular.

The generalized solutions of the system (1.1) and Eq. (1.2) are called the generalized analytic vectors [3] and generalized analytic functions respectively [16]. General representation of generalized analytic functions by the analytic functions is the strongest tool in order to investigate the solutions of the Eq. (1.2). In particular, the following takes place

$$\omega = f exp(T), \tag{1.3}$$

where f is an arbitrary entire function and the function $T = T(z)$ is evaluated by the formula

$$T(z) = -\frac{1}{\pi} \int\int_D \left(A(\xi) + B(\xi) \frac{\overline{\omega(\xi)}}{\omega(\xi)} \right) \frac{1}{\xi - z} dD_\xi .$$

Obtained representation of the solution by means of the entire function is crucial for the solutions of (1.2) in order to get the analogue of the Liouville classical theorem. Due to the regularity of the coefficients of equation, the factor $\exp(T)$ is continuous on the whole plane, never equals zero and $\exp(T) \to 1,\ z \to \infty$. Such representation of the solution of the Eq. (1.2) is known as the main lemma [16] or the similarity principle [2] in theory of generalized analytic or pseudo-analytic functions theory.

Respectively, if the coefficients of the Eq. (1.2) are regular, solution ω is bounded and is zero at some point of the plane then it is identically zero. Thus, every bounded solution of the Eq. (1.2) with regular coefficients has exactly the same property as the entire function of a complex variable; in particular, there is an alternative: either the solution is not zero anywhere or is identically zero. At the same time one principal difference should be mentioned. The bounded entire function is constant, whereas the bounded solution of the Eq. (1.2) except of some special cases, is not constant. This makes difficult describe effectively the solutions of the Eq. (1.2) with $O(z^N)$ asymptotic at infinity, in particular to obtain representation of type (1.3).

The Liouville theorem, which we will use below, has the following form:

Theorem 1.1 (J. Liouville) *Let w be classical (of class C^1) solution of the equation $\partial_{\bar{z}}\omega = 0$ on the whole complex plane* $\mathbf{C}$, *for which there exists nonnegative integer number N, such that*

$$\omega(z) = O(z^N), \quad z \to \infty. \tag{1.4}$$

Then

$$w(z) = a_0 + a_1 z + \ldots + a_N z^N,$$

where $a_0, a_1, \ldots, a_N$ are complex numbers.

From this it follows that if we denote by $\Omega(N)$ the solutions space of equation $\partial_{\bar{z}}\omega = 0$ with properties (1.4), then $\dim_{\mathbf{C}} \Omega(N) = N + 1$. Finiteness of the dimensions of the solution space of the system of type (1.1), which follows from the Liouville theorem, guarantees successful application of algebraic-topological methods for the investigation of the space of holomorphic sections or deformation of complex structures of the vector bundle [4, 7, 8].

Many fundamental works and monographs (see [3, 5, 6, 11, 14–17]) are dedicated to the generalization of Theorem 1.1 for the solutions spaces of the Eqs. (1.1) and (1.2). As it was mentioned above for the solution of the Eq. (1.2), obtained analogue of Liouville classical theorem is essentially based on the principal limitation for the coefficients A and B of the Eq. (1.2)—they must be regular. Theoretically (also for the analysis of applied problems) the largest interest is attracted to find the analogues of Liouville classical theorem for the Eq. (1.2) when these coefficients are not regular. Recent research [10, 12] shows that irregular systems of type (1.1) arise from the problems of mathematical physics. For such equations the Liouville type theorems are obtained in the works of several authors [11, 17]. It seems to be most notable the simplest case of the coefficients $A = const \neq 0$, $B = const \neq 0$. Exactly, for these coefficients the most important results related with the Liouville type theorem has been obtained in [17]. In general, for the regular system of type (1.1) the Liouville theorem is not valid (see [6]).

In the next section we consider irregular elliptic system of type $\partial_{\bar{z}} W(z, \bar{z}) = V(z, \bar{z}) W(z, \bar{z})$ and investigate solutions space of such system in the case of two equations in detail.

2 Main Theorem

Let $m \in \mathbf{N}$ and a_{kp}, b_{kp} $1 \leq k, p \leq m$ be given real valued functions on the domain $D \subset \mathbf{R}^2_{(x,y)}$. Consider the first order elliptic system

$$\frac{\partial u_k}{\partial x} + \sum_{p=1}^{m} a_{kp} \frac{\partial u_p}{\partial y} = \sum_{p=1}^{m} b_{kp} u_p \tag{2.1}$$

with unknown vector function $(u_1, \ldots, u_m)$. From the ellipticity if follows that the equation $\det\left((a_{jk}(x,y))_{j,k=1}^{m} - \lambda\mathbf{I}\right) = 0$ has only complex solutions for every $(x, y) \in D$, where $\mathbf{I}$ is identity matrix.

Below we consider special case of the system (2.1) which in complex notation $\omega_k = u_{2k-1} + iu_{2k}$, $k = 1, 2, \ldots, n$, $n = \frac{m}{2}$, has the form

$$\partial_{\bar{z}}\omega_k = \sum_{p=1}^{n} \tau_{kp}\omega_p, \quad 1 \le k \le n, \tag{2.2}$$

and

$$\tau_{kp} = \gamma\alpha_{kp}, \quad 1 \le k, p \le n, \tag{2.3}$$

where α_{kp} are complex numbers and γ is a complex valued continous function on the whole complex plane.

We call that the continous function $(\omega_1, \ldots, \omega_n)$ is the solution of the system (2.2), if it satisfies (2.2) at all points of $\mathbf{C}$.

In the following lemma we give an effective formula for the solutions of the system (2.2) with assumption (2.3) by the spectrum of constant matrix $(\alpha_{ij})_{i,j=1}^{n}$. We give the formulation and the proof of this lemma for the system of two equations. Note that, the theorem is true for an arbitrary n.

Consider now the elliptic system of the equations

$$\partial_{\bar{z}}\omega_1 = \tau_{11}\omega_1 + \tau_{12}\omega_2, \quad \partial_{\bar{z}}\omega_2 = \tau_{21}\omega_1 + \tau_{22}\omega_2, \tag{2.4}$$

where the functions τ_{kp} satisfy the conditions (2.3). Suppose λ_1 and λ_2 are eigenvalues of the matrix $(\alpha_{ij})_{i,j=1}^{2}$ and $H = (h_1, h_2)$, $G = (g_1, g_2)$ are the corresponding eigenvectors.

Lemma 2.1 *The general solutions of the system (2.4) are*

$$\omega_1 = \Phi_1 h_1 e^{\lambda_1\Gamma} + \Phi_2 g_1 e^{\lambda_2\Gamma}, \quad \omega_2 = \Phi_1 h_2 e^{\lambda_1\Gamma} + \Phi_2 g_2 e^{\lambda_2\Gamma}. \tag{2.5}$$

Here Φ_1 and Φ_2 are arbitrary entire functions, Γ is some $\partial_{\bar{z}}$-primitive of the continuous function γ.

Proof Indeed, direct computations show that (2.5) satisfy the Eq. (2.4). Assume that ω_1 and ω_2 are defined on the whole complex plane and satisfy (2.4). We have to prove that they have the form (2.5). Fix some point z on complex plane and for the pair of complex numbers $\Phi_1(z)$, $\Phi_2(z)$ consider the system of algebraic equations

$$h_1 e^{\lambda_1\Gamma(z)}\Phi_1(z) + g_1 e^{\lambda_2\Gamma(z)}\Phi_2(z) = \omega_1(z),$$

$$h_2 e^{\lambda_1\Gamma(z)}\Phi_1(z) + g_2 e^{\lambda_2\Gamma(z)}\Phi_2(z) = \omega_2(z).$$

The matrix of this system

$$K = \begin{pmatrix} h_1 e^{\lambda_1 \Gamma(z)} & g_1 e^{\lambda_2 \Gamma(z)} \\ h_2 e^{\lambda_1 \Gamma(z)} & g_2 e^{\lambda_2 \Gamma(z)} \end{pmatrix}$$

is nondegenerate, because the vectors H and G are linearly independent and therefore $s = h_1 g_2 - h_2 g_1 \neq 0$. From the identity

$$s \begin{pmatrix} \Phi_1(z) \\ \Phi_2(z) \end{pmatrix} = K^{-1} \begin{pmatrix} \omega_1(z) \\ \omega_2(z) \end{pmatrix}$$

it follows that

$$s \partial_{\bar{z}} \begin{pmatrix} \Phi_1 \\ \Phi_2 \end{pmatrix} = \partial_{\bar{z}} \left(K^{-1} \begin{pmatrix} \omega_1 \\ \omega_2 \end{pmatrix} \right) = \gamma T,$$

where

$$T = \begin{pmatrix} e^{-\lambda_1 \Gamma(z)}(-\lambda_1 g_2 + \alpha_{11} g_2 - \alpha_{21} g_1) & e^{-\lambda_2 \Gamma(z)}(\lambda_1 g_1 + \alpha_{12} g_2 - \alpha_{22} g_1) \\ e^{-\lambda_1 \Gamma(z)}(-\lambda_2 h_2 - \alpha_{11} h_2 - \alpha_{21} h_1) & e^{\lambda_2 \Gamma(z)}(-\lambda_2 h_1 - \alpha_{12} h_2 + \alpha_{22} h_1) \end{pmatrix}.$$

Since $\lambda_1 + \lambda_2 = \alpha_{11} + \alpha_{22}$, we obtain that

$$\partial_{\bar{z}} \Phi_1 \equiv \partial_{\bar{z}} \Phi_2 \equiv 0.$$

The lemma is proved.

Remark 2.2 Here we assume the existence of $\partial_{\bar{z}}$-primitive of continous function γ. For the irregular equations of type (1.2), difference from regular case, existence of primitive distinct non-trivial problem (see [9]).

Get back to (2.3) and suppose that the function γ has a specific form:

$$\gamma_{\nu,\mu} = \begin{cases} 0, & if \quad z = 0, \\ |z|^{\nu} \exp(i\mu\varphi), & if \quad z \neq 0. \end{cases}$$

Here $\varphi = \arg z$, $\nu > 0$ is a real number and μ is a nonnegative integer.

To find explicitly $\partial_{\bar{z}}$-primitive of $\gamma_{\nu,\mu}$ rewrite it in the form

$$\gamma_{\nu,\mu} = z^{\frac{\nu+\mu}{2}} \bar{z}^{\frac{\nu-\mu}{2}}.$$

After the formal integration with respect to $\bar{z}$ we obtain

$$\int \gamma_{\nu,\mu} d\bar{z} = \frac{2\bar{z}}{\nu - \mu + 2} \gamma_{\nu,\mu}.$$

It means that, when the parameters ν, μ satisfy the inequality $\nu - \mu + 2 \neq 0$, all $\partial_{\bar{z}}$-primitives $\Gamma_{\mu,\nu}$ of the continuous function $\gamma_{\mu,\nu}$ have the form

$$\Gamma_{\nu,\mu} = \frac{2\bar{z}}{\nu - \mu + 2}\gamma_{\nu,\mu} + \Phi(z),$$

where Φ is an arbitrary entire function. Indeed, to prove the identity $\partial_{\bar{z}}\Gamma_{\nu,\mu} = \gamma_{\nu,\mu}(z), \quad z \in \mathbf{C}$ it is sufficient to rewrite $\partial_{\bar{z}}$ operator in polar coordinates.

Let N be a nonnegative integer and δ be a nonnegative real number. Denote by $\Omega(N, \delta)$ such (ω_1, ω_2) solutions of system (2.4), with $\gamma = \gamma_{\nu,\mu}$, which at infinity satisfy the condition

$$\max_{1 \leq k \leq 2} |\omega_k(z)| = O(z^N \exp(\delta |z|^{\nu+1}), \quad z \to \infty. \tag{2.6}$$

It is clear that $\Omega(N, \delta)$ is the vector space over $\mathbf{C}$.

Theorem 2.3 (The Liouville Theorem with Parameters) *If the parameters* ν, μ *of the irregular equation*

$$\partial_{\bar{z}}\omega = \lambda |z|^{\nu} \exp(i\mu\varphi)\omega \tag{2.7}$$

satisfy the condition

$$|\mu - 1| > 2(\nu + 1), \tag{2.8}$$

then the solution space $\Omega(N, \delta)$ *of (2.7) with condition (2.6) has finite dimension and* $\dim_{\mathbf{C}} \Omega(N, \delta) = N+1$, *where* $\delta = \frac{|\lambda|}{|\mu - \nu + 2|}$ *and* $\lambda \in \mathbf{C}$ *defined form (2.3),when* $n = 1$.

Proof The proof of the theorem bases on the explicit representation of the solutions of Eq. (2.7) and the application of Phragmén–Lindelöf principle for such solution.

The solution of (2.7) has form

$$\omega(z) = \Phi(z) \exp\left\{\frac{2\lambda}{\nu - \mu + 2}|z|^{\nu+1}\right\} \exp\{i(\mu - 1)\varphi\}, \quad \varphi = \arg z, \tag{2.9}$$

where $\Phi(z)$ is an arbitrary entire function. From this follows that along all ray

$$\Gamma_{\varphi*} = \{z : z = re^{i\varphi_*}\}, \quad r > 0, (\mu - 1)\varphi_* = 2\pi k, k = 0, \pm 1, \pm 2, \dots,$$

$\Phi(z)$ satisfies the condition $\Phi(z) = O(z^N), \ z \to \infty, \ z \in \Gamma_{\varphi*}$. Therefore, $\Phi(z) = O(z^N), \ z \to \infty$. It means that $\dim_{\mathbf{C}} \Omega(N, \delta) = N + 1$. The theorem is proved.

In the Lemma 2.1 we additionally required that the eigenvalues λ_1 and λ_2 of elliptic system (2.4) satisfy the condition $|\lambda_1| = |\lambda_2|$.

Proposition 2.4 *Let $\gamma = \gamma_{\nu,\mu}$ and $|\lambda_1| = |\lambda_2| = \lambda_0$ in the system (2.4) satisfy the inequality (2.8), then*

$$\dim_{\mathbb{C}} \Omega\left(N, \frac{2\lambda_0}{|\mu - \nu + 2|}\right) = 2(N+1) \tag{2.10}$$

for every nonnegative integer N.

The basis of $\Omega\left(N, \frac{2\lambda_0}{|\nu-\mu+2|}\right)$ is the system of vectors

$$\tilde{H}, z\tilde{H}, z^2\tilde{H}, \ldots, z^N\tilde{H}, \quad \tilde{G}, z\tilde{G}, z^2\tilde{G}, \ldots, z^N\tilde{G},$$

where $\tilde{H} = e^{\lambda_1 \Gamma_{\nu,\mu}} H$, $\tilde{G} = e^{\lambda_2 \Gamma_{\nu,\mu}} G$ and $H = (h_1, h_2)$, $G = (g_1, g_2)$ are the eigenvectors corresponding to the eigenvalues λ_1, λ_2.

Proof By Lemma 2.1 the general solutions of the system (2.4) are expressed by the pair of eigenvalues and arbitrary entire functions. On the other hand by the Theorem 2.3 numbers of independent solutions of scalar equation satisfying condition (2.8) equal to $N+1$. Hence for the system of two equations the dimension of the vector space of the solutions of the system (2.4) satisfying the condition (2.8) will be $2(N+1)$. The proposition is proved.

Note that if the inequality (2.8) is not fulfilled, then in general the equality (2.10) is not valid. In particular, if

$$|\mu - 1| < 2(\nu + 1), \tag{2.11}$$

then

$$\dim_{\mathbb{C}} \Omega\left(N, \frac{2\lambda_0}{|\mu - \nu + 2|}\right) = \infty.$$

Example Let $\mu = -\nu$, then for such μ and ν the relation (2.11) is valid and $|\mu - \nu + 2| \neq 0$. The corresponding function $\gamma_{\nu,-\nu}$ has the form $\gamma_{\nu,-\nu} = \bar{z}^\nu$. Consider the equation

$$\partial_{\bar{z}}\omega = \lambda_0 \bar{z}^\nu \omega, \tag{2.12}$$

where $\lambda_0 \neq 0$ and ν nonnegative integer as above. The solutions of (2.12) are the functions

$$\omega = \Psi(z) \exp\left(\lambda_0 \frac{\bar{z}^{\nu+1}}{\nu+1}\right), \tag{2.13}$$

where $\Psi(z)$ is arbitrary entire function. $\omega \in \Omega(N, \delta)$ if and only if

$$|\omega| = |\Psi(z)||\exp\left(\frac{\lambda_0}{\nu+1}\bar{z}^{\nu+1}\right)| \leq M|z|^N \exp\left(\frac{|\lambda_0|}{\nu+1}|z|^{\nu+1}\right). \tag{2.14}$$

Denote by $f(z) = \Psi(z)e^{\frac{\bar{\lambda}_0}{\nu+1}z^{\nu+1}}$. Then from (2.14) we have the estimation

$$|\Psi(z)| \left|\exp\left(\frac{\bar{\lambda}_0}{\nu+1}z^{\nu+1}\right)\exp\left(\frac{\lambda_0}{\nu+1}\bar{z}^{\nu+1} - \frac{\bar{\lambda}_0}{\nu+1}z^{\nu+1}\right)\right| =$$

$$|f(z)| \left|\exp\left(\frac{\lambda_0}{\nu+1}\bar{z}^{\nu+1} - \frac{\bar{\lambda}_0}{\nu+1}z^{\nu+1}\right)\right| \leq M|z|^N \exp\left(\frac{|\lambda_0|}{\nu+1}|z|^{\nu+1}\right).$$

Therefore

$$|f(z)| \leq M|z|^N \exp\left(\frac{|\lambda_0|}{\nu+1}|z|^{\nu+1}\right).$$

The space of entire functions satisfying the last inequality have infinite dimension.

Remark 2.5 Denote $W = (\omega_1, \omega_2)$ and by $T = (\tau_{i,j})^2_{i,j=1}$. Then the system (2.4) obtains the form $\partial_{\bar{z}} W = TW$. Let C be an invertible analytic matrix function. Then the systems $\partial_{\bar{z}} W = TW$ and $\partial_{\bar{z}} W_1 = T_1 W_1$, where $T_1 = CTC^{-1}$ are equivalent (see Sect. 1) and therefore Lemma 2.1 and Proposition 2.4 are true for such systems also.

Acknowledgments This work is supported by Shota Rustaveli National Science Foundation grant N FR17-96.

References

1. G. Akhalaia, G. Giorgadze, V. Jikia, N. Kaldani, G. Makatsaria, N. Manjavidze, Elliptic systems on Riemann surface. Lect. Notes of TICMI, **13**, (2012)
2. L. Bers, *Theory of Pseudo-Analytic Functions* (New York University of Institute Mathematics Mechanics, New York, 1953)
3. B. Bojarski, Theory of generalized analytic vectors. Ann. Polon. Math. **17**, 281–320 (1966)
4. B. Bojarski, G. Giorgadze, Some analytical and geometric aspect of the stable partial indices. Proc. I. Vekua Inst. Appl. Math. **61–62**, 14–32 (2012)
5. A. Douglis, A function-theoretic approach to elliptic systems of equations in two variables. Commun. Pure Appl. Math. **6**, 259–289 (1953)
6. R. Gilbert, J. Buchanan, *First Order Elliptic Systems: A Function Theoretic Approach* (Academic Press, Cambridge, 1983)
7. G. Giorgadze, Geometric aspects of generalized analytic functions, in *Topics in Analysis and its Applications*, ed. by H. Begehr, G. Barsegian. NATO Science Series, vol. 147 (Springer, Berlin, 2004), pp. 69–81

8. G. Giorgadze, Moduli space of complex structures. J. Math. Sci. **160**, 697–716 (2009)
9. G. Giorgadze, V. Jikia, On some properties of generalized analytic functions induced from irregular Carleman–Bers–Vekua equations. J. Complex Variables Elliptic Equ. Int. J. **58**(9), 1183–194 (2013)
10. P. Grinevich, R. Novikov, Moutard transform for generalized analytic functions. J. Geom. Anal. **26**(4), 2984–2995 (2016)
11. G. Makatsaria, Liouville-type theorems for generalized analytic functions. Bull. Acad. Sci. Georgian SSR **113**(3), 485–488 (1984)
12. A. Meziani, Representation of solutions of a singular Cauchy-Riemann equation in the plane. J. Complex Var. Elliptic Equ. Int. J. **53**(12), 1111–1130 (2008)
13. R. Novikov, I. Taimanov, Mourand type transformation for matrix generalized analytic functions and gauge transformations. Russ. Math. Surv. **71**(5), 179–180 (2016)
14. A. Soldatov, Hyperanalytic functions and their applications. J. Math. Sci. **132**, 827–882 (2006)
15. Z.D. Usmanov, *Generalized Cauchy-Riemann Systems with a Singular Point* (Longman, Harlow, 1997)
16. I.N. Vekua, *Generalized Analytic Functions* (Pergamon, Oxford, 1962)
17. V.S. Vinogradov, Liouville's theorem for generalized analytic functions. Dokl. Akad. Nauk SSSR **183**, 503–508 (1968)

Bases in Commutative Algebras of the Second Rank and Monogenic Functions Related to Some Cases of Plane Orthotropy

S. V. Gryshchuk

Abstract Among all two-dimensional commutative associative algebras of the second rank with unity over the field of complex numbers we want to find all pairs $(\mathbb{B}_*, \{e_1, e_2\})$, where $\mathbb{B}_*$ is an algebra and $\{e_1, e_2\}$ are its bases such that $e_1^4 + 2pe_1^2e_2^2 + e_2^4 = 0$ for every fixed p, $-1 < p < 1$. This problem is solved in an explicit form. An approach of $\mathbb{B}_*$-valued "analytic" functions $\Phi(xe_1 + ye_2)$ ($\{e_1, e_2\}$ is fixed, x and y are real variables), such that their real-valued functions-components satisfy the equation on finding the stress function in certain cases of orthotropic plane deformations, is developing.

Keywords Anisotropic (orthotropic) media · Hooke's generalized law · Stress function · Lamé equilibrium system with respect to displacements · Commutative and associative algebras · Monogenic functions

Mathematics Subject Classification (2010) Primary 30G35; Secondary 74B05

1 Statement of the Problem

Let p be an arbitrary fixed number such that $-1 < p < 1$. We assume that a model of an elastic anisotropic medium occupied a bounded domain D of the Cartesian plane xOy is a homogeneous (cf., e.g., [1, p. 25]) plane orthotropic (cf., e.g., [1, p. 35]) body, therefore, it physically obeys Hooke's generalized law (to be more exact, an element of the class of Hooke's generalized laws including the

S. V. Gryshchuk (✉)
Institute of Mathematics, National Academy of Sciences of Ukraine, Kyiv, Ukraine
e-mail: gryshchuk@imath.kiev.ua; serhii.gryshchuk@gmail.com

P. Cerejeiras et al. (eds.), *Current Trends in Analysis, its Applications and Computation*, Research Perspectives,
https://doi.org/10.1007/978-3-030-87502-2_16

corresponding parameters p and a_{12}) of the form

$$\begin{pmatrix} \sigma_x \\ \tau_{xy} \\ \sigma_y \end{pmatrix} = \begin{pmatrix} \frac{1}{1-(a_{12})^2} & 0 & -\frac{a_{12}}{1-(a_{12})^2} \\ 0 & \frac{1}{2(p-a_{12})} & 0 \\ -\frac{a_{12}}{1-(a_{12})^2} & 0 & \frac{1}{1-(a_{12})^2} \end{pmatrix} \begin{pmatrix} \varepsilon_x \\ \gamma_{xy} \\ \varepsilon_y \end{pmatrix}, \tag{1.1}$$

or, the inverse equivalent form:

$$\varepsilon_x = \sigma_x + a_{12}\sigma_y,\ \gamma_{xy} = 2\,(p - a_{12})\,\tau_{xy},\ \varepsilon_y = a_{12}\sigma_x + \sigma_y, \tag{1.2}$$

where σ_x, τ_{xy}, σ_y and ε_x, $\frac{\gamma_{xy}}{2}$, ε_y are components of the stress tensor [1] and the strain tensor [1], respectively, and the number a_{12} satisfies the relation $-1 < a_{12} < p$. The physical meaning of the parameters p and a_{12} is given in [2], and the case $p > 1$ is considered in [3, 4] (in this case, $-1 < a_{12} < 1$).

The equation of the stress function $u(x, y)$ ($\sigma_x(x_0, y_0) = \frac{\partial^2 u}{\partial y^2}(x_0, y_0)$, $\sigma_y(x_0, y_0) = \frac{\partial^2 u}{\partial x^2}(x_0, y_0)$, $\tau_{xy}(x_0, y_0) = -\frac{\partial^2 u}{\partial x \partial y}(x_0, y_0)$ for all $(x_0, y_0) \in D$) in the absence of body forces has a form (cf., e.g., [1, 5])

$$\widetilde{l}_p u(x, y) := \left(\frac{\partial^4}{\partial x^4} + 2p\frac{\partial^4}{\partial x^2 \partial y^2} + \frac{\partial^4}{\partial y^4}\right) u(x, y) = 0\ \forall (x, y) \in D. \tag{1.3}$$

By the same conditions a system of equilibrium equations for displacements known also as the "Lamé system of equilibrium equations" has the following form (see, e.g., [1, 5, 6]) for all $(x, y) \in D$:

$$\begin{cases} B_{11}\frac{\partial^2 u(x,y)}{\partial x^2} + B_{12}\frac{\partial^2 u(x,y)}{\partial y^2} + \frac{\partial^2 v(x,y)}{\partial x \partial y} = 0, \\ B_{21}\frac{\partial^2 v(x,y)}{\partial^2 x^2} + B_{22}\frac{\partial^2 v(x,y)}{\partial y^2} + \frac{\partial^2 u(x,y)}{\partial x \partial y} = 0, \end{cases} \tag{1.4}$$

where

$$B_{11} = B_{22} := \frac{2\,(p - a_{12})}{(a_{12})^2 - 2pa_{12} + 1},\ B_{12} = B_{21} := \frac{1 - (a_{12})^2}{(a_{12})^2 - 2pa_{12} + 1}. \tag{1.5}$$

Let B_* denote a two-dimensional commutative associative algebra with unity e over the complex field $\mathbb{C}$ which has a basis $\{e_1, e_2\}$ satisfying the condition

$$\mathcal{L}_p(e_1, e_2) := e_1^4 + 2pe_1^2e_2^2 + e_2^4 = 0. \tag{1.6}$$

We assume also that every nonzero element $h \in \mu_{e_1,e_2} := \{xe_1 + ye_2 : (x, y) \in \mathbb{R}\}$ is invertible (i.e., there exists the inverse element $h^{-1} \in \mathbb{B}_*$ such that $hh^{-1} = e$), here $\mathbb{R}$ is a field of real numbers.

For any complex numbers c_1 and c_2 we introduce the notation

$$l_p(c_1, c_2) := c_1^4 + 2pc_1^2c_2^2 + c_2^4. \tag{1.7}$$

The characteristic equation of the Eq. (1.3) has a form

$$l_p(s, 1) \equiv s^4 + 2ps^2 + 1 = 0, \ s \in \mathbb{C}, \tag{1.8}$$

a set of its roots is

$$\{s_1, s_2, \overline{s_1}, \overline{s_2}\} =: \ker l_p(s, 1), \tag{1.9}$$

where

$$s_1 = \frac{\sqrt{2(1-p)}}{2} - \frac{\sqrt{2(1+p)}}{2}i, \ s_2 = -\frac{\sqrt{2(1-p)}}{2} + \frac{\sqrt{2(1+p)}}{2}i, \tag{1.10}$$

$\overline{x+iy} := x - iy \equiv \mathrm{Re}z - i\mathrm{Im}z$, $x, y \in \mathbb{R}$, $z = x + iy$, i is the imaginary complex unity. Thus, the relation (1.6) is generated by the Eq. (1.8).

In what follows, $(x, y) \in D$, $\zeta = xe_1 + ye_2 \in D_\zeta := \{\zeta = xe_1 + ye_2 : (x, y) \in D\} \subset \mu_{e_1,e_2}$.

An arbitrary function $\Phi\colon D_\zeta \longrightarrow \mathbb{B}$ can be presented in the form

$$\Phi(\zeta) = U_1(x, y)\, e_1 + U_2(x, y)\, ie_1 + U_3(x, y)\, e_2 + U_4(x, y)\, ie_2 \,\forall (x, y) \in D, \tag{1.11}$$

where $U_k\colon D \longrightarrow \mathbb{R}$, $k = \overline{1, 4}$, are real-valued functions.

We call the function Φ monogenic if there exists a finite (in the sense of a norm) limit at every point $\zeta \in D_\zeta$:

$$\Phi'(\zeta) := \lim_{h \to 0,\, h \in \mu_{e_1,e_2}} \left(\Phi(\zeta + h) - \Phi(\zeta)\right) h^{-1}. \tag{1.12}$$

We call $\Phi'(\zeta)$ the *derivative* of a function Φ at the point $\zeta \in D_\zeta$.

A *problem of our consideration* in the present work is to find all pairs $(\mathbb{B}_*, \mathcal{B}_p)$, where $\mathcal{B}_p$ is a totality of all required bases $\{e_1, e_2\}$, and, to deliver a procedures of finding solutions of Eq. (1.3) and the system (1.4) via monogenic functions $\Phi\colon D_\zeta \longrightarrow \mathbb{B}_*$.

Note, that a similar problem is solved in [7–9] for $p = 1$, and in [3, 4] for $p > 1$.

2 Two-Dimensional Algebras Over the Field of Complex Numbers and Their Bases Related to Plane Orthotropy

As it is well known (see, e.g., [10]), there exist (to within an isomorphism) two associative commutative algebras of the second rank with unity e over the field of complex numbers. Those algebras are generated by the bases $\{e, \rho\}$ and $\{e, \omega\}$, respectively:

$$\mathbb{B} := \{c_1 e + c_2 \rho : c_k \in \mathbb{C}, k = 1, 2\},\ \rho^2 = 0, \tag{2.1}$$

$$\mathbb{B}_0 := \{c_1 e + c_2 \omega : c_k \in \mathbb{C}, k = 1, 2\},\ \omega^2 = e. \tag{2.2}$$

It is obvious that the algebra $\mathbb{B}_0$ is semisimple and has the basis composed of orthogonal idempotents $\{\mathcal{I}_1, \mathcal{I}_2\}$, where

$$\mathcal{I}_1 = \frac{1}{2}(e + \omega),\ \mathcal{I}_2 = \frac{1}{2}(e - \omega),\ \mathcal{I}_1\mathcal{I}_2 = 0. \tag{2.3}$$

It is obvious that

$$\mathcal{I}_1 + \mathcal{I}_2 = e,\ \mathcal{I}_1 - \mathcal{I}_2 = \omega. \tag{2.4}$$

It is clear that if

$$e_1 = \alpha_1\mathcal{I}_1 + \alpha_2\mathcal{I}_2,\ e_2 = \beta_1\mathcal{I}_1 + \beta_2\mathcal{I}_2,\ \alpha_k, \beta_k \in \mathbb{C}, k = 1, 2, \tag{2.5}$$

are basis elements of the algebra (2.2) satisfying condition (1.6), then

$$e_1 = \beta_1\mathcal{I}_1 + \beta_2\mathcal{I}_2,\ e_2 = \alpha_1\mathcal{I}_1 + \alpha_2\mathcal{I}_2$$

are also basis elements of the algebra (2.2) satisfying condition (1.6).Combining these two cases, we say that relation (2.5) specifies the basis $\{e_1, e_2\}$ of $\mathbb{B}_0$ *satisfying condition* (1.6) *to within permutations.*

Lemma 2.1 *The algebra $\mathbb{B}$ does not contain any basis $\{e_1, e_2\}$ satisfying condition* (1.6).

All bases of the algebra $\mathbb{B}_0$ satisfying condition (1.6) *to within permutations can be represented in the form*

$$e_1 = \alpha_1\mathcal{I}_1 + \alpha_2\mathcal{I}_2,\ e_2 = \beta_1\mathcal{I}_1 + \beta_2\mathcal{I}_2,$$

where the complex numbers $\alpha_k \neq 0$*,* $\beta_k \neq 0$*,* $k = 1, 2$*, satisfy one of the following two conditions:*

(a) $\beta_k = \widetilde{s}_k\, \alpha_k, k = 1, 2;$

(b) $\beta_1 = \widehat{s}_1\, \alpha_1,\ \beta_2 = \frac{1}{\widehat{s}_2}\, \alpha_2,$

where $\widetilde{s}_1$ *and* $\widetilde{s}_2$ *are arbitrary distinct elements from* $\ker l_p(s, 1)$; $\widehat{s}_k \in \ker l_p(s, 1)$, $k = 1, 2$*, such that*

$$\overline{\widehat{s}_2} \neq \widehat{s}_1. \tag{2.6}$$

Proof Seeking a basis $\{e_1, e_2\}$ of the algebra $\mathbb{B}_0$ of the form (2.5), we get the condition: $\mathcal{L}_p(e_1, e_2) = l_p(\alpha_1, \beta_1)\mathcal{I}_1 + l_p(\alpha_2, \beta_1)\mathcal{I}_2$. Modifying a method of the prove of [3, Theorem 1] to our case $-1 < p < 1$, we obtain the validity of all statements of Lemma 2.1. □

Using (1.10) we get relations $\frac{1}{s_k} = \overline{s_k}$, $k = 1, 2$, which yield a fact, that a set of pairs $(\widetilde{s}_1, \widetilde{s}_2)$ is equal to the set of pairs $(\widehat{s}_1, \frac{1}{\widehat{s}_2})$. Thus, a set of bases $\{e_1, e_2\}$ generated by the case (b) of Lemma 2.1 is a subset of the totality of bases $\{e_1, e_2\}$ generated by the case (a) of Lemma 2.1, and Lemma 2.1 turns into the following form.

Theorem 2.2 *The algebra* $\mathbb{B}$ *does not contain any basis* $\{e_1, e_2\}$ *satisfying condition* (1.6).

All bases of the algebra $\mathbb{B}_0$ *satisfying condition* (1.6) *can be represented in the form*

$$e_1 = \alpha\,\mathcal{I}_1 + \beta\,\mathcal{I}_2,\ e_2 = \widetilde{s}_1\alpha\,\mathcal{I}_1 + \widetilde{s}_2\beta\,\mathcal{I}_2, \tag{2.7}$$

where α *and* β *are arbitrary complex numbers such that* $\alpha \neq 0$*,* $\beta \neq 0$*,* $\widetilde{s}_1$ *and* $\widetilde{s}_2$ *are arbitrary distinct elements from* $\ker l_p(s, 1)$.

By $\mathcal{B}_{p,1}$ we denote the totality of bases (2.7) with $e_1 = e$. The symbol $\mathcal{E}$ we use for a set of all invertible elements $\{e_1 = a_1\mathcal{I}_1 + a_2\mathcal{I}_2 \in \mathbb{B}_0 : a_k \in \mathbb{C} \setminus \{0\}, k = 1, 2\}$. The product of sets $E_k \subset B_0$, $k = 1, 2$, is defined as the set $E \equiv E_1 E_2 := \{x_1 x_2 : x_k \in E_k, k = 1, 2\}$.

A relationship between the sets $\mathcal{B}_p$ and $\mathcal{B}_{p,1}$ is obtained similar to [3, Lemma 1].

Lemma 2.3 *The equality of sets* $\mathcal{B}_p = \mathcal{E}\,\mathcal{B}_{p,1}$ *is true.*

3 Monogenic Functions in the Plane Generated by a Basis Satisfying Condition (1.6)

Consider monogenic functions $\Phi\colon D_\zeta \longrightarrow \mathbb{B}_0$, $D_\zeta \subset \mu_{e_1,e_2}$, where a basis $\{e_1, e_2\}$ is chosen in (2.7). It easy to verify that any element e_k, $k = 1, 2$, in (2.7) is invertible. Every non-zero element in μ_{e_1,e_2} is invertible too.

Analogously to the corresponding result in [11], we can prove that the a function of the type $\Phi\colon D_\zeta \longrightarrow \mathbb{B}_0$ is monogenic iff its components U_k, $k = \overline{1,4}$, (in (1.11)) are differentiable in the domain D and the following analog of the Cauchy–Riemann conditions holds:

$$\frac{\partial \Phi(\zeta)}{\partial y}\, e_1 = \frac{\partial \Phi(\zeta)}{\partial x}\, e_2 \equiv \Phi'(\zeta) e_1 e_2 \quad \forall \zeta \in D_\zeta. \tag{3.1}$$

In a similar way to corresponding result of the paragraph 5 in [3], we obtain that the functional algebra of monogenic functions of the variable $\zeta_* = x e_1 + y e_2$ is isomorphic to the functional algebra of monogenic functions of the variable $\zeta = xe + ye_2$.

Thus, the investigation of monogenic functions in domains lying in μ_{e_1,e_2} with $e_1 \neq e$, is equivalent to the investigation of monogenic functions in domains lying in μ_{e_1,e_2} with $e_1 = e$.

By this reason, here and below we assume that $e_1 = e$ in (2.7), i.e.,

$$e_1 = \mathcal{I}_1 + \mathcal{I}_2,\ e_2 = \widetilde{s}_1\, \mathcal{I}_1 + \widetilde{s}_2\, \mathcal{I}_2. \tag{3.2}$$

For the variable $\zeta = x e_1 + y e_2$ we introduce complex variables $Z_k \subset \mathbb{C}$ and their domains $D_{Z_k} \subset \mathbb{C}$, $k = 1, 2$, by the formulas

$$Z_k := x + \widetilde{s}_k\, y,\ D_{Z_k} := \{Z_k = x + \widetilde{s}_k\, y : x e_1 + y e_2 \in D_\zeta\}, k = 1, 2,$$

here $\widetilde{s}_k$, $k = 1, 2$, are the same as in (3.2). Then we can represent the variable ζ in the form $\zeta = Z_1 \mathcal{I}_1 + Z_2 \mathcal{I}_2$.

Analogous to [3, Theorem 3] we obtain the following statement: *the function* $\Phi\colon D_\zeta \longrightarrow \mathbb{B}_0$ *is monogenic in* D_ζ *if and only if the following equality is true*

$$\Phi(\zeta) = F_1(Z_1)\mathcal{I}_1 + F_2(Z_2)\mathcal{I}_2\ \forall \zeta \in D_\zeta, \tag{3.3}$$

where $F_k\colon D_{Z_k} \longrightarrow \mathbb{C}$, $k \in \{1, 2\}$, *is a holomorphic function.*

Note that a relation between continuous and differentiable in the sense of Gateaux functions considered in [12, 13] and monogenic functions (3.3) is the same as the similar relation for $p > 1$ (see remarks after Theorem 3 in [3]), i.e., these two notions coincide.

The equalities (3.2) yield relations

$$\mathcal{I}_1 = \frac{\widetilde{s}_2}{\widetilde{s}_2 - \widetilde{s}_1}\, e - \frac{e_2}{\widetilde{s}_2 - \widetilde{s}_1}, \quad \mathcal{I}_2 = -\frac{\widetilde{s}_1}{\widetilde{s}_2 - \widetilde{s}_1}\, e + \frac{e_2}{\widetilde{s}_2 - \widetilde{s}_1}.$$

Without loss of generality we replace $F_1(Z_1)$ onto $\frac{\widetilde{s}_2}{\widetilde{s}_2-\widetilde{s}_1} F_1(Z_1)$ and $F_2(Z_2)$ onto $\left(-\frac{\widetilde{s}_1}{\widetilde{s}_2-\widetilde{s}_1}\right) F_2(Z_1)$, then (3.3) turns onto

$$\Phi(\zeta) = (F_1(Z_1) + F_2(Z_2))\, e_1 - \left(\frac{1}{\widetilde{s}_2} F_1(Z_1) + \frac{1}{\widetilde{s}_1} F_2(Z_2)\right) e_2 \;\forall \zeta \in D_\zeta, e_1 = e. \tag{3.4}$$

4 Monogenic Functions Related to Eq. (1.3) and System (1.4)

It follows from (3.1) and (3.3) that every monogenic functions has derivatives $\Phi^{(n)}$ of any order $n = 1, 2, \ldots$. Then each component

$$\mathrm{U}_k\,[\Phi(\zeta)] := U_k(x, y) \;\forall \zeta \in D_\zeta,\; k = \overline{1, 4},$$

of monogenic function Φ in (1.11) satisfies Eq. (1.3) due to equalities

$$\widetilde{l}_p \Phi(\zeta) = \mathcal{L}_p(e_1, e_2)\Phi^{(4)(\zeta)} \equiv 0.$$

Here and below we restrict our attention on the case $\widetilde{s}_k$, $k = 1, 2$, ($\widetilde{s}_1 \neq \widetilde{s}_2$) such that

$$\overline{\widetilde{s}_2} \neq \widetilde{s}_1.$$

Let us assume now that the domain D is bounded and simply connected.

It is known (see, e.g., [1, 5, 14]) that the general solution of Eq. (1.3) takes the form

$$u(x, y) = \mathrm{Re}\; (F_1\,(Z_1) + F_2\,(Z_2)) \;\; \forall (x, y) \in D, \tag{4.1}$$

here $F_k \colon D_{Z_k} \longrightarrow \mathbb{C}$, $k = 1, 2$, are any holomorphic functions of the corresponding complex variables.

A fixed solution u of (1.3) satisfies the equality

$$u(x, y) = \mathrm{U}_1\,[\Phi_u(\zeta)] \;\; \forall \zeta \in D_\zeta, \tag{4.2}$$

where $\Phi_u\colon D_\zeta \longrightarrow \mathbb{B}_0$ is a monogenic function which has the same F_k, $k = 1, 2$, in the equality (3.4) for $\Phi := \Phi_u$ as in (4.1) for u. Denote by $\Phi_{1,0}$ a function which is monogenic in D_ζ such that $\mathrm{U}_1\left[\Phi_{1,0}\right] \equiv 0$.

Lemma 4.1 *Let D be a bounded and simply connected domain of the Cartesian plane xOy. Then all monogenic functions $\Phi\colon D_\zeta \longrightarrow \mathbb{B}_0$ satisfying the equality* (4.2) *can be presented in the form*

$$\Phi(\zeta) = \Phi_u(\zeta) + \Phi_{1,0}(\zeta)\,\forall\zeta \in D_\zeta. \tag{4.3}$$

Note that the equality (4.3) is found in the explicit form (the explicit formula for $\Phi_{1,0}$) in [2] for a case when $\widetilde{s}_k := s_k$, $k = 1, 2$.

Now, let us assume that $\widetilde{s}_k = s_k$, $k = 1, 2$. In this case a procedure for building solutions of the system (1.4) by means of components $U_k = \mathrm{U}_k[\Phi]$ of the monogenic in D_ζ function Φ is found in [6, Theorem 5.1]

Theorem 4.2 *Let $\Phi\colon D_\zeta \longrightarrow \mathbb{B}_0$ be a monogenic function, α_k, $k = \overline{1,4}$, be any real numbers, and let the numbers β_k, $k = \overline{1,4}$, be connected with them by the linear relations*

$$\begin{cases} \beta_1 = (pB_{11} - B_{12})\,\alpha_3 - \sqrt{1-p^2}B_{11}\alpha_4, \\ \beta_2 = \sqrt{1-p^2}B_{11}\alpha_3 + (pB_{11} - B_{12})\,\alpha_4, \\ \beta_3 = (pB_{12} - B_{11})\,\alpha_1 + \sqrt{1-p^2}B_{12}\alpha_2, \\ \beta_4 = -\sqrt{1-p^2}B_{12}\alpha_1 + (pB_{12} - B_{11})\,\alpha_2. \end{cases} \tag{4.4}$$

In addition, let the functions u and v be linear combinations of the components $U_k = U_k[\Phi]$, $k = \overline{1,4}$, of the form

$$u(x, y) = \sum_{k=1}^{4} \alpha_k U_k(x, y),\; v(x, y) = \sum_{k=1}^{4} \beta_k U_k(x, y)\;\forall(x, y) \in D. \tag{4.5}$$

Then the pair of functions (u, v) satisfies the system (1.4) *in D.*

Acknowledgments The research is partially supported by the Fundamental Research Programme funded by the Ministry of Education and Science of Ukraine (project No. 0116U001528) and a common grant of the National Academy of Sciences of Ukraine and the Poland Academy of Sciences (No. 32/2018).

References

1. S.G. Lekhnitskii, *Anisotropic Plates* (Engl. transl. from the 2nd Russ. edition by S. W. Tsai, T. Cheron) (Gordon and Breach, New York, 1984)
2. S.V. Gryshchuk, Monogenic Functions in Two Dimensional Commutative Algebras to Equations of Plane Orthotropy. Pratsi Inst. Prikl. Mat. Mikh. NANU **32**, 18–29 (2018)

3. S.V. Gryshchuk, Commutative complex algebras of the second rank with unity and some cases of plane orthotropy I (in Ukrainian). Ukr. Mat. Zh. **70**(8), 1058–1071 (2018); **English translation:** Ukr. Math. J. **70**(8), 1221–1236 (2019)
4. S.V. Gryshchuk, Commutative complex algebras of the second rank with unity and some cases of plane orthotropy II (in Ukrainian). Ukr. Mat. Zh. **70**(10), 1382–1389 (2018); **English translation:** Ukr. Math. J. **70**(10), 1594–1603 (2019)
5. Y.A. Bogan, Regular integral equations for the second boundary-value problem in the anisotropic twodimensional elasticity theory (in Russian). Izv. RAN. Mekh Tver. Tela **4**, 17–26 (2005)
6. S.V. Gryshchuk, Monogenic functions in commutative complex algebras of the second rank and the Lamé equilibrium system for some class of plane orthotropy (in Russian, English summary). Ukr. Math. Bull. **16**(3), 345–356 (2019); **English translation:** J. Math. Sci. **246**(1), (2020)
7. I.P. Mel'nichenko, Biharmonic bases in algebras of the second rank. Ukr. Mat. Zh. **38**(2), 224–226 (1986); **English translation:** Ukr. Math. J. **38**(2), 252–254 (1986)
8. S.V. Gryshchuk, Hypercomplex monogenic functions of the biharmonic variable in some problems of the plane elasticity (in Ukrainian, English summary). Rep. Acad. Sci. Ukraine A **6**, 7–12 (2015)
9. S.V. Gryshchuk $\mathbb{B}$-valued monogenic functions and their applications to boundary value problems in displacements of 2-D Elasticity, in *Analytic Methods of Analysis and Differential Equations: AMADE 2015*, ed. by S.V. Rogosin, M.V. Dubatovskaya, Belarusian State University, Minsk, Belarus (Cambridge Scientic, Cambridge, 2016), pp. 37–47. ISBN 978-1-908106-56-8
10. E. Study, Uber systeme complexer zahlen und ihre anwendungen in der theorie der transformationsgruppen. Monatshefte für Mathematik **1**(1), 283–354 (1890)
11. V.F Kovalev, I.P. Mel'nichenko, Biharmonic functions on the biharmonic plane (in Russian). Rep. Acad. Sci. USSR A **8**, 25–27 (1981)
12. S.A. Plaksa, R.P. Pukhtaievych, Constructive description of monogenic functions in a finite-dimensional semisimple commutative algebra (in Ukrainian). Rep. Natl. Acad. Sci. Ukr. **1**, 14–21 (2014)
13. S.A. Plaksa, R.P. Pukhtaievych, Monogenic functions in a finite-dimensional semi-simple commutative algebra. An. St. Univ. Ovidius Constanta **22**(1), 221–235 (2014)
14. N.I. Muskhelishvili *Some Basic Problems of the Mathematical Theory of Elasticity. Fundamental Equations, Plane Theory of Elasticity, Torsion and Bending* (English transl. from the 4th Russian edition by R.M. Radok) (Noordhoff International, Leiden, 1977)

On the Behavior at Infinity of One Class of Homeomorphisms

Ruslan Salimov and Bogdan Klishchuk

Abstract We study the behavior at infinity of ring Q-homeomorphisms with respect to p-modulus for $p > n$.

Keywords Ring Q-homeomorphisms · p-modulus of a family of curves · Quasiconformal mappings · Condenser · p-capacity of a condenser

Mathematics Subject Classification (2010) 30C65

1 Introduction

Let us recall some definitions, see [1]. Let Γ be a family of curves γ in $\mathbb{R}^n$, $n \geqslant 2$. A Borel measurable function $\rho : \mathbb{R}^n \to [0, \infty]$ is called *admissible* for Γ, (abbr. $\rho \in \operatorname{adm} \Gamma$), if

$$\int\limits_{\gamma} \rho(x)\, ds \geqslant 1$$

for any curve $\gamma \in \Gamma$. Let $p \in (1, \infty)$.

The quantity

$$M_p(\Gamma) = \inf_{\rho \in \operatorname{adm} \Gamma} \int\limits_{\mathbb{R}^n} \rho^p(x)\, dm(x)$$

is called *p–modulus* of the family Γ.

R. Salimov (✉) · B. Klishchuk
Institute of Mathematics, National Academy of Sciences of Ukraine, Kyiv, Ukraine

P. Cerejeiras et al. (eds.), *Current Trends in Analysis, its Applications and Computation*, Research Perspectives,
https://doi.org/10.1007/978-3-030-87502-2_17

For arbitrary sets E, F and G of $\mathbb{R}^n$ we denote by $\Delta(E, F, G)$ a set of all continuous curves $\gamma : [a, b] \to \mathbb{R}^n$ that connect E and F in G, i.e., such that $\gamma(a) \in E$, $\gamma(b) \in F$ and $\gamma(t) \in G$ for $a < t < b$.

Let D be a domain in $\mathbb{R}^n$, $n \geqslant 2$, $x_0 \in D$ and $d_0 = \mathrm{dist}(x_0, \partial D)$. Set

$$\mathbb{A}(x_0, r_1, r_2) = \{x \in \mathbb{R}^n : r_1 < |x - x_0| < r_2\},$$

$$S_i = S(x_0, r_i) = \{x \in \mathbb{R}^n : |x - x_0| = r_i\}, \quad i = 1, 2.$$

Let a function $Q : D \to [0, \infty]$ be Lebesgue measurable. We say that a homeomorphism $f : D \to \mathbb{R}^n$ is ring Q-homeomorphism with respect to p-modulus at $x_0 \in D$ if the inequality

$$M_p(\Delta(fS_1, fS_2, fD)) \leqslant \int\limits_{\mathbb{A}} Q(x)\, \eta^p(|x - x_0|)\, dm(x)$$

holds for any ring $\mathbb{A} = \mathbb{A}(x_0, r_1, r_2)$, $0 < r_1 < r_2 < d_0$, $d_0 = \mathrm{dist}(x_0, \partial D)$, and for any measurable function $\eta : (r_1, r_2) \to [0, \infty]$ such that

$$\int\limits_{r_1}^{r_2} \eta(r)\, dr = 1.$$

The theory of Q-homeomorphisms for $p = n$ was studied in works [2–6] , for $1 < p < n$ in works [7–14] and for $p > n$ in works [15–19], see also [20, 21].

Denote by ω_{n-1} the area of the unit sphere $\mathbb{S}^{n-1} = \{x \in \mathbb{R}^n : |x| = 1\}$ in $\mathbb{R}^n$ and by $q_{x_0}(r) = \frac{1}{\omega_{n-1} r^{n-1}} \int\limits_{S(x_0,r)} Q(x)\, d\mathcal{A}$ the integral mean over the sphere $S(x_0, r) = \{x \in \mathbb{R}^n : |x - x_0| = r\}$, here $d\mathcal{A}$ is the element of the surface area.

Now we formulate a criterion which guarantees for a homeomorphism to be ring Q-homeomorphism with respect to p-modulus for $p > 1$ in $\mathbb{R}^n$, $n \geqslant 2$.

Proposition 1 *Let D be a domain in $\mathbb{R}^n$, $n \geqslant 2$, and let $Q : D \to [0, \infty]$ be a Lebesgue measurable function such that $q_{x_0}(r) \neq \infty$ for a.e. $r \in (0, d_0)$, $d_0 = \mathrm{dist}(x_0, \partial D)$. A homeomorphism $f : D \to \mathbb{R}^n$ is ring Q-homeomorphism with respect to p-modulus at a point $x_0 \in D$ if and only if the inequality*

$$M_p\left(\Delta(fS_1, fS_2, f\mathbb{A})\right) \leqslant \frac{\omega_{n-1}}{\left(\int\limits_{r_1}^{r_2} \frac{dr}{r^{\frac{n-1}{p-1}}\, q_{x_0}^{\frac{1}{p-1}}(r)}\right)^{p-1}}$$

holds for any $0 < r_1 < r_2 < d_0$ (see [12], Theorem 2.3).

Following the paper [22], a pair $\mathcal{E} = (A, C)$ where $A \subset \mathbb{R}^n$ is an open set and C is a nonempty compact set contained in A, is called *condenser*. We say that a condenser $\mathcal{E} = (A, C)$ lies in a domain D if $A \subset D$. Clearly, if $f : D \to \mathbb{R}^n$ is a homeomorphism and $\mathcal{E} = (A, C)$ is a condenser in D then (fA, fC) is also condenser in fD. Further, we denote $f\mathcal{E} = (fA, fC)$.

Let $\mathcal{E} = (A, C)$ be a condenser. Denote by $\mathcal{C}_0(A)$ a set of continuous functions $u : A \to \mathbb{R}^1$ with compact support. Let $\mathcal{W}_0(\mathcal{E}) = \mathcal{W}_0(A, C)$ be a family of nonnegative functions $u : A \to \mathbb{R}^1$ such that 1) $u \in \mathcal{C}_0(A)$, 2) $u(x) \geqslant 1$ for $x \in C$ and 3) u belongs to the class ACL and

$$|\nabla u| = \left(\sum_{i=1}^{n}\left(\frac{\partial u}{\partial x_i}\right)^2\right)^{\frac{1}{2}}.$$

For $p \geqslant 1$ the quantity

$$\operatorname{cap}_p \mathcal{E} = \operatorname{cap}_p (A, C) = \inf_{u \in \mathcal{W}_0(\mathcal{E})} \int_A |\nabla u|^p \, dm(x)$$

is called *p-capacity* of the condenser $\mathcal{E}$. It is known that for $p > 1$

$$\operatorname{cap}_p \mathcal{E} = M_p(\Delta(\partial A, \partial C; A \setminus C)), \tag{1.1}$$

see in ([23] ,Theorem 1). For $p > n$ the inequality

$$\operatorname{cap}_p (A, C) \geqslant n\, \Omega_n^{\frac{p}{n}} \left(\frac{p-n}{p-1}\right)^{p-1} \left[m^{\frac{p-n}{n(p-1)}}(A) - m^{\frac{p-n}{n(p-1)}}(C)\right]^{1-p} \tag{1.2}$$

holds where Ω_n is a volume of the unit ball in $\mathbb{R}^n$ (see, e.g., the inequality 8.7 in [24]).

2 Main Results

Now we present the main result of our paper on the behavior at infinity of ring Q-homeomorphisms with respect to p-modulus for $p > n$. The case $p = n$ was studied in the work [25]. Let

$$L(x_0, f, R) = \sup_{|x - x_0| \leqslant R} |f(x) - f(x_0)| \, .$$

Theorem 2.1 (Main Theorem) *Suppose that $f : \mathbb{R}^n \to \mathbb{R}^n$ is a ring Q-homeomorphism with respect to p-modulus at a point x_0 with $p > n$ where x_0*

is some point in $\mathbb{R}^n$ *and for some numbers* $r_0 > 0$, $K > 0$ *the condition*

$$q_{x_0}(t) \leqslant K\,t^{\alpha} \tag{2.1}$$

holds for a.a. $t \in [r_0, +\infty)$. *If* $\alpha \in [0, p-n)$ *then*

$$\varliminf_{R\to\infty} \frac{L(x_0, f, R)}{R^{\frac{p-n-\alpha}{p-n}}} \geqslant K^{\frac{1}{n-p}} \left(\frac{p-n}{p-n-\alpha}\right)^{\frac{p-1}{p-n}} > 0\,.$$

If $\alpha = p - n$ *then*

$$\varliminf_{R\to\infty} \frac{L(x_0, f, R)}{(\ln R)^{\frac{p-1}{p-n}}} \geqslant K^{\frac{1}{n-p}} \left(\frac{p-n}{p-1}\right)^{\frac{p-1}{p-n}} > 0\,.$$

Proof Consider a condenser $\mathcal{E} = (A, C)$ in $\mathbb{R}^n$, where $A = \{x \in \mathbb{R}^n : |x - x_0| < R\}$, $C = \{x \in \mathbb{R}^n : |x - x_0| \leqslant r_0\}$, $0 < R < r_0 < \infty$. Then $f\mathcal{E} = (fA, fC)$ is a ringlike condenser in $\mathbb{R}^n$, and by (1.1) we have equality

$$\operatorname{cap}_p f\mathcal{E} = \mathrm{M}_p\left(\Delta(\partial fA, \partial fC; f(A \setminus C))\right).$$

Due to the inequality (1.2),

$$\operatorname{cap}_p(fA, fC) \geqslant n\,\Omega_n^{\frac{p}{n}} \left(\frac{p-n}{p-1}\right)^{p-1} \left[m^{\frac{p-n}{n(p-1)}}(fA) - m^{\frac{p-n}{n(p-1)}}(fC)\right]^{1-p},$$

we obtain

$$\operatorname{cap}_p(fA, fC) \geqslant n\,\Omega_n^{\frac{p}{n}} \left(\frac{p-n}{p-1}\right)^{p-1} [m(fA)]^{\frac{n-p}{n}}\,. \tag{2.2}$$

On the other hand, by Proposition 1, one gets

$$\operatorname{cap}_p(fA, fC) \leqslant \frac{\omega_{n-1}}{\left(\int\limits_{r_0}^{R} \frac{dt}{t^{\frac{n-1}{p-1}}\, q_{x_0}^{\frac{1}{p-1}}(t)}\right)^{p-1}}\,. \tag{2.3}$$

Combining the inequalities (2.2) and (2.3), we obtain

$$n\,\Omega_n^{\frac{p}{n}} \left(\frac{p-n}{p-1}\right)^{p-1} [m(fA)]^{\frac{n-p}{n}} \leqslant \frac{\omega_{n-1}}{\left(\int\limits_{r_0}^{R} \frac{dt}{t^{\frac{n-1}{p-1}}\, q_{x_0}^{\frac{1}{p-1}}(t)}\right)^{p-1}}\,.$$

Due to $\omega_{n-1} = n\,\Omega_n$, the last inequality can be rewritten as

$$\Omega_n^{\frac{p}{n}-1}\left(\frac{p-n}{p-1}\right)^{p-1}[m(fA)]^{\frac{n-p}{n}} \leqslant \left(\int\limits_{r_0}^{R}\frac{dt}{t^{\frac{n-1}{p-1}}q_{x_0}^{\frac{1}{p-1}}(t)}\right)^{1-p}. \tag{2.4}$$

Consider the case $\alpha \in [0, p-n)$. Then from the condition (2.1) the estimate

$$\Omega_n^{\frac{p}{n}-1}\left(\frac{p-n}{p-1}\right)^{p-1}[m(fA)]^{\frac{n-p}{n}} \leqslant K\left(\frac{p-n-\alpha}{p-1}\right)^{p-1}\left(R^{\frac{p-n-\alpha}{p-1}} - r_0^{\frac{p-n-\alpha}{p-1}}\right)^{1-p}$$

holds. Therefore,

$$m(fB(x_0, R)) \geqslant \Omega_n\,K^{\frac{n}{n-p}}\left(\frac{p-n}{p-n-\alpha}\right)^{\frac{n(p-1)}{p-n}}\left(R^{\frac{p-n-\alpha}{p-1}} - r_0^{\frac{p-n-\alpha}{p-1}}\right)^{\frac{n(p-1)}{p-n}}. \tag{2.5}$$

Due to

$$m(fB(x_0, R)) \leqslant \Omega_n\,L^n(x_0, f, R)\,, \tag{2.6}$$

from the inequality (2.5) we have

$$L(x_0, f, R) \geqslant K^{\frac{1}{n-p}}\left(\frac{p-n}{p-n-\alpha}\right)^{\frac{p-1}{p-n}}\left(R^{\frac{p-n-\alpha}{p-1}} - r_0^{\frac{p-n-\alpha}{p-1}}\right)^{\frac{p-1}{p-n}}.$$

Dividing the last inequality by $R^{\frac{p-n-\alpha}{p-n}}$ and taking the lower limit for $R \to \infty$, we conclude

$$\varliminf_{R\to\infty}\frac{L(x_0, f, R)}{R^{\frac{p-n-\alpha}{p-n}}} \geqslant K^{\frac{1}{n-p}}\left(\frac{p-n}{p-n-\alpha}\right)^{\frac{p-1}{p-n}}.$$

Now we consider the case $\alpha = p-n$. Then from (2.4) we get

$$\Omega_n^{\frac{p}{n}-1}\left(\frac{p-n}{p-1}\right)^{p-1}[m(fA)]^{\frac{n-p}{n}} \leqslant K\left(\ln\frac{R}{r_0}\right)^{1-p}.$$

Therefore,

$$m(fB(x_0, R)) \geqslant \Omega_n\,K^{\frac{n}{n-p}}\left(\frac{p-n}{p-1}\right)^{\frac{n(p-1)}{p-n}}\left(\ln\frac{R}{r_0}\right)^{\frac{n(p-1)}{p-n}}.$$

Due to the estimate (2.6) we obtain

$$L(x_0, f, R) \geqslant K^{\frac{1}{n-p}} \left(\frac{p-n}{p-1}\right)^{\frac{p-1}{p-n}} \left(\ln \frac{R}{r_0}\right)^{\frac{p-1}{p-n}}.$$

Finally, dividing the last inequality by $(\ln R)^{\frac{p-1}{p-n}}$ and taking the lower limit for $R \to \infty$, we conclude

$$\varliminf_{R\to\infty} \frac{L(x_0, f, R)}{(\ln R)^{\frac{p-1}{p-n}}} \geqslant K^{\frac{1}{n-p}} \left(\frac{p-n}{p-1}\right)^{\frac{p-1}{p-n}}.$$

This completes the proof of Main Theorem.

Let us consider some examples.

Example 2.1 Let $f_1 : \mathbb{R}^n \to \mathbb{R}^n$, where

$$f_1(x) = \begin{cases} K^{\frac{1}{n-p}} \left(\frac{p-n}{p-n-\alpha}\right)^{\frac{p-1}{p-n}} |x|^{\frac{p-n-\alpha}{p-n}} \frac{x}{|x|}, & x \neq 0 \\ 0, & x = 0. \end{cases}$$

It can be easily seen that $\lim\limits_{x\to\infty} \frac{|f(x)|}{|x|^{\frac{p-n-\alpha}{p-n}}} = K^{\frac{1}{n-p}} \left(\frac{p-n}{p-n-\alpha}\right)^{\frac{p-1}{p-n}}$. Let us show that the mapping f_1 is a ring Q-homeomorphism with respect to p-modulus with the function $Q(x) = K\,|x|^{\alpha}$ at the point $x_0 = 0$. Clearly, $q_{x_0}(t) = K\,t^{\alpha}$. Consider a ring $\mathbb{A}(0, r_1, r_2)$, $0 < r_1 < r_2 < \infty$. Note that the mapping f_1 maps the ring $\mathbb{A}(0, r_1, r_2)$ onto the ring $\widetilde{\mathbb{A}}(0, \widetilde{r}_1, \widetilde{r}_2)$, where

$$\widetilde{r}_i = K^{\frac{1}{n-p}} \left(\frac{p-n}{p-n-\alpha}\right)^{\frac{p-1}{p-n}} r_i^{\frac{p-n-\alpha}{p-n}}, \quad i = 1, 2.$$

Denote by Γ a set of all curves that join the spheres $S(0, r_1)$ and $S(0, r_2)$ in the ring $\mathbb{A}(0, r_1, r_2)$. Then one can calculate p-modulus of the family of curves $f_1\Gamma$ in an implicit form:

$$\mathrm{M}_p(f_1\Gamma) = \omega_{n-1} \left(\frac{p-n}{p-1}\right)^{p-1} \left(\widetilde{r}_2^{\frac{p-n}{p-1}} - \widetilde{r}_1^{\frac{p-n}{p-1}}\right)^{1-p}$$

(see, e.g., the relation (2) in [26]). Substituting in the above equality the values $\widetilde{r}_1$ and $\widetilde{r}_2$, defined above, one gets

$$\mathrm{M}_p(f_1\Gamma) = \omega_{n-1}\, K \left(\frac{p-n-\alpha}{p-1}\right)^{p-1} \left(r_2^{\frac{p-n-\alpha}{p-1}} - r_1^{\frac{p-n-\alpha}{p-1}}\right)^{1-p}.$$

Note that the last equality can be written by

$$M_p(f_1\Gamma) = \frac{\omega_{n-1}}{\left(\int\limits_{r_1}^{r_2} \frac{dt}{t^{\frac{n-1}{p-1}} q_{x_0}^{\frac{1}{p-1}}(t)}\right)^{p-1}},$$

where $q_{x_0}(t) = K\, t^{\alpha}$.

Hence, by Proposition 1, the homeomorphism f_1 is a ring Q-homeomorphism with respect to p-modulus for $p > n$ with the function $Q(x) = K\,|x|^{\alpha}$ at the point $x_0 = 0$.

Example 2.2 Let $\alpha = p - n$ and $f_2 : \mathbb{R}^n \to \mathbb{R}^n$, where

$$f_2(x) = \begin{cases} K^{\frac{1}{n-p}} \left(\frac{p-n}{p-1}\right)^{\frac{p-1}{p-n}} (\ln|x|)^{\frac{p-1}{p-n}} \frac{x}{|x|}, & x \neq 0 \\ 0, & x = 0\,. \end{cases}$$

It can be easily seen that $\lim\limits_{x\to\infty} \frac{|f(x)|}{(\ln|x|)^{\frac{p-1}{p-n}}} = K^{\frac{1}{n-p}} \left(\frac{p-n}{p-1}\right)^{\frac{p-1}{p-n}}$. By analogy to Example 2.1, we can show that the mapping f_2 is a ring Q-homeomorphism with respect to p-modulus with the function $Q(x) = K\,|x|^{p-n}$.

Remark 2.1 Examples 2.1 and 2.2 show that the estimates in Main Theorem are sharp, i.e. the bounds are attained on the above mappings.

Acknowledgments This work was supported by the budget program "Support of the development of priority trends of scientific researches" (KPKVK 6541230).

References

1. J. Väisälä, *Lectures on N-Dimensional Quasiconformal Mappings*. Lecture Notes in Mathematics, vol. 229 (Springer, Berlin, 1971)
2. V.I. Ryazanov, E.A. Sevost'yanov, Equicontinuous classes of ring Q-homeomorphisms. Siberian Math. J. **48**(6), 1093–1105 (2007)
3. O. Martio, V. Ryazanov, U. Srebro, E. Yakubov, *Q-Homeomorphisms*. Complex Analysis and Dynamical Systems (Contemporary Mathematics), vol. 364 (American Mathematical Society, Providence, 2004), pp. 193–203
4. O. Martio, V. Ryazanov, U. Srebro, E. Yakubov, On Q-homeomorphisms. Ann. Acad. Sci. Fenn. Math. **30**(1), 49–69 (2005)
5. O. Martio, V. Ryazanov, U. Srebro, E. Yakubov, *Moduli in Modern Mapping Theory*. Springer Monographs in Mathematics (Springer, New York, 2009)
6. R. Salimov, ACL and differentiability of a generalization of quasiconformal maps. Izv. Math. **72**(5), 977–984 (2008)
7. A. Golberg, Differential properties of (α, Q)-homeomorphisms, in *Further Progress in Analysis, Proceedings of the 6th ISAAC Congress* (2009), pp. 218–228

8. A. Golberg, Integrally quasiconformal mappings in space. Trans. Inst. Math. NAS Ukr. **7**(2), 53–64 (2010)
9. A. Golberg, R. Salimov, Logarithmic Holder continuity of ring homeomorphisms with controlled p-module. Complex Var. Elliptic Equ. **59**(1), 91–98 (2014)
10. A. Golberg, R. Salimov, E. Sevost'yanov, Distortion estimates under mappings with controlled p-module. Ann. Univ. Bucharest Ser. Math. **5**(LXIII), 95–114 (2014)
11. R. Salimov, On finitely Lipschitz space mappings. Siberian Electron. Math. Rep. **8**, 284–295 (2011)
12. R. Salimov, Estimation of the measure of the image of the ball. Siberian Math. J. **53**(4), 920–930 (2012)
13. R. Salimov, To a theory of ring Q-homeomorphisms with respect to a p-modulus. Ukr. Math. Bull. **10**(3), 379–396 (2013)
14. R. Salimov, One property of ring$Q$-homeomorphisms with respect to a p-module. Ukr. Math. J. **65**(5), 728–733 (2013)
15. R. Salimov, B. Klishchuk, The extremal problem for the area of an image of a disc. Rep. NAS Ukr. **10**, 22–27 (2016)
16. B. Klishchuk, R. Salimov, Lower bounds for the area of the image of a circle. Ufa. Math. J. **9**(2), 55–61 (2017)
17. R. Salimov, B. Klishchuk, Extremal problem for the area of the image of a disk. Zapiski Nauchnykh Seminarov POMI **456**, 160–171 (2017)
18. R. Salimov, B. Klishchuk, An extremal problem for the volume functional. Matematychni Studii **50**(1), 36–43 (2018)
19. B. Klishchuk, R. Salimov, Lower bounds for the volume of the image of a ball. Ukr. Math. J. **71**(6), 774–785 (2019)
20. M. Cristea, Some properties of open, discrete generalized ring mappings. Complex Var. Elliptic Equ. **61**(5), 623–643 (2016)
21. M. Cristea, The limit mapping of generalized ring homeomorphisms. Complex Var. Elliptic Equ. **61**(5), 608–622 (2016)
22. O. Martio, S. Rickman, J. Väisälä, Definitions for quasiregular mappings. Ann. Acad. Sci. Fenn. Ser. A1. Math. **448**, 1–40 (1969)
23. V.A. Shlyk, The equality between p-capacity and p-modulus. Siberian Math. J. **34**(6), 216–221 (1993)
24. V. Maz'ya, Lectures on isoperimetric and isocapacitary inequalities in the theory of Sobolev spaces. Contemp. Math. **338**, 307–340 (2003)
25. R. Salimov, E. Smolovaya, On the order of growth of ring Q-homeomorphisms at infinity. Ukr. Math. J. **62**(6), 829–836 (2010)
26. F.W. Gehring, Lipschitz mappings and the p-capacity of ring in n-space. Advances in the theory of Riemann surfaces (Proceedings of the Conference Stonybrook, 1969). Ann. Math. Stud. **66**, 175–193 (1971)

Domain Perturbation for the Solution of a Periodic Dirichlet Problem

Paolo Luzzini and Paolo Musolino

Abstract We prove that the solution of the periodic Dirichlet problem for the Laplace equation depends real analytically on a suitable parametrization of the shape of the domain, on the periodicity parameters, and on the Dirichlet datum.

Keywords Laplace operator · Periodically perforated domains · Domain perturbation · Real analyticity · Shape analysis · Integral equations method

Mathematics Subject Classification (2010). Primary 35J25; Secondary 45A05 31B10 35J05 35B20

1 Introduction

In this paper we study the dependence of the solution of the periodic Dirichlet problem for the Laplace equation in $\mathbb{R}^n$ upon joint perturbation of the shape of the domain, of the periodicity structure, and of the Dirichlet datum. The shape of the domain is determined by the image of a fixed domain through a map ϕ in a suitable class of diffeomorphisms and the periodicity cell is a box of edges of length $q_{11}, \ldots, q_{nn}$. As a main result, we prove that the solution of the problem depends real analytically on the 'periodicity-domain-Dirichlet datum' triple $((q_{11}, \ldots, q_{nn}), \phi, g)$. Our method is based on a periodic version of potential theory which has already revealed to be a powerful tool to analyze boundary value problems for elliptic differential equations in periodic domains.

P. Luzzini (✉)
Dipartimento di Matematica 'Tullio Levi-Civita', Università degli Studi di Padova, Padova, Italy
e-mail: pluzzini@math.unipd.it

P. Musolino
Dipartimento di Scienze Molecolari e Nanosistemi, Università Ca' Foscari Venezia, Venezia Mestre, Italy
e-mail: paolo.musolino@unive.it

P. Cerejeiras et al. (eds.), *Current Trends in Analysis, its Applications and Computation*, Research Perspectives,
https://doi.org/10.1007/978-3-030-87502-2_18

Many authors have exploited potential theory to analyze perturbation problems. In the non-periodic setting, Potthast [20] and Potthast and Stratis [21] have proved a Fréchet differentiability result for layer potentials associated to the Helmholtz operator, with an application to inverse problems in scattering theory. Lanza de Cristoforis and Preciso [15] have shown that the Cauchy integral depends real analytically on domain perturbations. Lanza de Cristoforis and Rossi [16] have considered the case of layer potentials associated to the Laplace operator and have obtained real analyticity results. Later on, Lanza de Cristoforis [11, 12] has exploited these results to prove that the solutions of boundary value problems for the Laplace and Poisson equations depend real analytically upon domain perturbation. Then, these results have been extended to singular perturbation problems and to systems of partial differential equations (see, e.g., Dalla Riva and Lanza de Cristoforis [4] for the Lamé equations and Dalla Riva [3] for the Stokes' system). Moreover, analyticity results for domain perturbation problems for eigenvalues and eigenfunctions have been obtained for example for the Laplace equation by Lanza de Cristoforis and Lamberti [10] and for the biharmonic operator by Buoso [2]. We mention also Keldysh [9], Henry [8] and Sokolowski and Zolésio [22] for elliptic domain perturbation problems.

Now, we introduce our problem. We fix once for all $n \in \mathbb{N} \setminus \{0, 1\}$. If $(q_{11}, \dots, q_{nn}) \in]0, +\infty[^n$ we introduce a periodicity cell Q and a matrix $q \in \mathbb{D}_n^+(\mathbb{R})$ by setting

$$Q \equiv \prod_{j=1}^{n}]0, q_{jj}[, \qquad q \equiv \begin{pmatrix} q_{11} & 0 & \cdots & 0 \\ 0 & q_{22} & \cdots & 0 \\ \vdots & \vdots & \ddots & \vdots \\ 0 & 0 & \cdots & q_{nn} \end{pmatrix},$$

where $\mathbb{D}_n(\mathbb{R})$ is the space of $n \times n$ diagonal matrices with real entries and $\mathbb{D}_n^+(\mathbb{R})$ is the set of elements of $\mathbb{D}_n(\mathbb{R})$ with diagonal entries in $]0, +\infty[$. We also denote by $|Q|_n$ the n-dimensional measure of the cell Q, by ν_Q the outward unit normal to ∂Q, where it exists, and by q^{-1} the inverse matrix of q. Clearly $q\mathbb{Z}^n$ is the set of vertices of a periodic subdivision of $\mathbb{R}^n$ corresponding to the fundamental cell Q. Moreover, we find convenient to set

$$\tilde{Q} \equiv]0, 1[^n, \qquad \tilde{q} \equiv I_n \equiv \begin{pmatrix} 1 & 0 & \cdots & 0 \\ 0 & 1 & \cdots & 0 \\ \vdots & \vdots & \ddots & \vdots \\ 0 & 0 & \cdots & 1 \end{pmatrix}.$$

Then we take

$$\begin{aligned} &\alpha \in]0, 1[\text{ and a bounded open connected subset } \Omega \text{ of } \mathbb{R}^n \\ &\text{of class } C^{1,\alpha} \text{ such that } \mathbb{R}^n \setminus \overline{\Omega} \text{ is connected.} \end{aligned} \tag{1.1}$$

The symbol '‾' denotes the closure of a set. For the definition of sets and functions of the Schauder class $C^{1,\alpha}$ we refer, e.g., to Gilbarg and Trudinger [7]. Then we consider a class of diffeomorphisms $\mathcal{A}_{\partial\Omega}^{\widetilde{Q}}$ from $\partial\Omega$ into their images contained in $\widetilde{Q}$ (see (2.1) below). If $\phi \in \mathcal{A}_{\partial\Omega}^{\widetilde{Q}}$, the Jordan-Leray separation theorem ensures that $\mathbb{R}^n \setminus \phi(\partial\Omega)$ has exactly two open connected components (see, e.g, Deimling [6, Thm. 5.2, p. 26]), and we denote by $\mathbb{I}[\phi]$ the bounded open connected component of $\mathbb{R}^n \setminus \phi(\partial\Omega)$. Since $\phi(\partial\Omega) \subseteq \widetilde{Q}$, a simple topological argument shows that $\widetilde{Q} \setminus \overline{\mathbb{I}[\phi]}$ is also connected. Then we consider the following two periodic domains:

$$\mathbb{S}_q[q\mathbb{I}[\phi]] \equiv \bigcup_{z \in \mathbb{Z}^n} (qz + q\mathbb{I}[\phi]), \qquad \mathbb{S}_q[q\mathbb{I}[\phi]]^- \equiv \mathbb{R}^n \setminus \overline{\mathbb{S}_q[q\mathbb{I}[\phi]]}.$$

Now, we take $g \in C^{1,\alpha}(\partial\Omega)$ and we consider the following periodic Dirichlet problem for the Laplace equation:

$$\begin{cases} \Delta u = 0 & \text{in } \mathbb{S}_q[q\mathbb{I}[\phi]]^-, \\ u(x+qz) = u(x) & \forall x \in \overline{\mathbb{S}_q[q\mathbb{I}[\phi]]^-}, \forall z \in \mathbb{Z}^n, \\ u(x) = g \circ \phi^{(-1)}(q^{-1}x) & \forall x \in \partial q\mathbb{I}[\phi]. \end{cases} \tag{1.2}$$

If $\phi \in C^{1,\alpha}(\partial\Omega, \mathbb{R}^n) \cap \mathcal{A}_{\partial\Omega}^{\widetilde{Q}}$, then the solution of problem (1.2) in the space $C_q^{1,\alpha}(\overline{\mathbb{S}_q[q\mathbb{I}[\phi]]^-})$ of q-periodic functions in $\overline{\mathbb{S}_q[q\mathbb{I}[\phi]]^-}$ of class $C^{1,\alpha}$ exists and is unique and we denote it by $u[q,\phi,g]$. Then we pose the following question:

What can be said on the regularity of the map $(q,\phi,g) \mapsto u[q,\phi,g]$? (1.3)

Our work stems from Lanza de Cristoforis [11, 12] where the author proved the real analytic dependence of the solution of the Dirichlet problem for the Laplace and Poisson equations upon domain perturbations. Moreover, it can be thought as a continuation of [18] where the authors proved a real analyticity result for the periodic layer potentials upon variation of the periodicity, of the shape of the support of integration, and of the density. We note that this paper generalizes a part of [17] where the authors proved an analyticity result for the longitudinal flow along a periodic array of cylinders.

In this work, we answer to the question in (1.3) by proving that the map $(q,\phi,g) \mapsto u[q,\phi,g]$ is real analytic between suitable Banach spaces (see Theorem 3.6). Such a result implies that if $\delta_0 > 0$ and we have a family of triples $\{(q_\delta,\phi_\delta,g_\delta)\}_{\delta\in]-\delta_0,\delta_0[}$ in a suitable Banach space such that the map $\delta \mapsto (q_\delta,\phi_\delta,g_\delta)$ is real analytic, then, if x belongs to the domain of $u[q_\delta,\phi_\delta,g_\delta]$ for all $\delta \in]-\delta_0,\delta_0[$, we can deduce the possibility to expand $u[q_\delta,\phi_\delta,g_\delta](x)$ as a power series in δ, i.e.,

$$u[q_\delta,\phi_\delta,g_\delta](x) = \sum_{k=0}^{\infty} c_k \delta^k \tag{1.4}$$

for δ close to zero. Moreover, the coefficients $(c_k)_{k\in\mathbb{N}}$ in (1.4) can be constructively determined by exploiting the method developed in [5].

2 Preliminary Results

In order to consider shape perturbations, we introduce a class of diffeomorphisms. Let Ω be as in (1.1). We denote by $\mathcal{A}_{\partial\Omega}$ the set of functions of class $C^1(\partial\Omega, \mathbb{R}^n)$ which are injective and whose differential is injective at all points of $\partial\Omega$. One can verify that $\mathcal{A}_{\partial\Omega}$ is open in $C^1(\partial\Omega, \mathbb{R}^n)$ (see, e.g., Lanza de Cristoforis and Rossi [16, Lem. 2.5, p. 143]). Then we set

$$\mathcal{A}_{\partial\Omega}^{\tilde{Q}} \equiv \{\phi \in \mathcal{A}_{\partial\Omega} : \phi(\partial\Omega) \subseteq \tilde{Q}\}. \tag{2.1}$$

Our method is based on a periodic version of classical potential theory. Therefore, to introduce layer potentials, we replace the fundamental solution of the Laplace operator by a q-periodic tempered distribution $S_{q,n}$ such that $\Delta S_{q,n} = \sum_{z\in\mathbb{Z}^n} \delta_{qz} - \frac{1}{|Q|_n}$, where δ_{qz} is the Dirac measure with mass in qz (see e.g., [13, p. 84]). The distribution $S_{q,n}$ is determined up to an additive constant, and we can take

$$S_{q,n}(x) = -\sum_{z\in\mathbb{Z}^n\setminus\{0\}} \frac{1}{|Q|_n 4\pi^2 |q^{-1}z|^2} e^{2\pi i (q^{-1}z)\cdot x}$$

in the sense of distributions in $\mathbb{R}^n$ (see e.g., Ammari and Kang [1, p. 53] and [13, §3]). Moreover, $S_{q,n}$ is even, real analytic in $\mathbb{R}^n \setminus q\mathbb{Z}^n$, and locally integrable in $\mathbb{R}^n$ (see e.g., [13, §3]). We now introduce the periodic double layer potential. Let Ω_Q be a bounded open subset of $\mathbb{R}^n$ of class $C^{1,\alpha}$ for some $\alpha \in]0, 1[$ such that $\overline{\Omega_Q} \subseteq Q$. Then we consider the following two periodic domains:

$$\mathbb{S}_q[\Omega_Q] \equiv \bigcup_{z\in\mathbb{Z}^n} (qz + \Omega_Q), \qquad \mathbb{S}_q[\Omega_Q]^- \equiv \mathbb{R}^n \setminus \overline{\mathbb{S}_q[\Omega_Q]}.$$

We set

$$w_q[\partial\Omega_Q, \mu](x) \equiv -\int_{\partial\Omega_Q} \nu_{\Omega_Q}(y) \cdot DS_{q,n}(x-y)\mu(y)\, d\sigma_y \qquad \forall x \in \mathbb{R}^n$$

for all $\mu \in L^2(\partial\Omega_Q)$. The symbol ν_{Ω_Q} denotes the outward unit normal field to $\partial\Omega_Q$, $d\sigma$ denotes the area element on $\partial\Omega_Q$, and $DS_{q,n}(\xi)$ denotes the gradient of $S_{q,n}$ computed at the point $\xi \in \mathbb{R}^n \setminus q\mathbb{Z}^n$. The function $w_q[\partial\Omega_Q, \mu]$ is called the q-periodic double layer potential. As is well known, if $\mu \in C^0(\partial\Omega_Q)$ then $w_q[\partial\Omega_Q, \mu]_{|\mathbb{S}_q[\Omega_Q]}$ admits a continuous extension to $\overline{\mathbb{S}_q[\Omega_Q]}$, which we denote by $w_q^+[\partial\Omega_Q, \mu]$ and $w_q[\partial\Omega_Q, \mu]_{|\mathbb{S}_q[\Omega_Q]^-}$ admits a continuous extension to $\overline{\mathbb{S}_q[\Omega_Q]^-}$, which we denote by $w_q^-[\partial\Omega_Q, \mu]$ (cf. e.g., [13, §3]). We also need the following lemma about the real analyticity upon the diffeomorphism ϕ of some maps related to the change of variables in the integrals and to the outer normal field (for a proof, see Lanza de Cristoforis and Rossi [16, p. 166]).

Lemma 2.1 *Let α, Ω be as in* (1.1)*. Then the following statements hold.*

(i) For each $\phi \in C^{1,\alpha}(\partial\Omega, \mathbb{R}^n) \cap \mathcal{A}_{\partial\Omega}$, there exists a unique $\tilde{\sigma}[\phi] \in C^{0,\alpha}(\partial\Omega)$ such that $\tilde{\sigma}[\phi] > 0$ and

$$\int_{\phi(\partial\Omega)} w(s)\, d\sigma_s = \int_{\partial\Omega} w \circ \phi(y)\tilde{\sigma}[\phi](y)\, d\sigma_y, \qquad \forall w \in L^1(\phi(\partial\Omega)).$$

Moreover, the map $\tilde{\sigma}[\cdot]$ from $C^{1,\alpha}(\partial\Omega, \mathbb{R}^n) \cap \mathcal{A}_{\partial\Omega}$ to $C^{0,\alpha}(\partial\Omega)$ is real analytic.

(ii) The map from $C^{1,\alpha}(\partial\Omega, \mathbb{R}^n) \cap \mathcal{A}_{\partial\Omega}$ to $C^{0,\alpha}(\partial\Omega, \mathbb{R}^n)$ which takes ϕ to $\nu_{\mathbb{I}[\phi]} \circ \phi$ is real analytic.

3 Analyticity of the Solution

As we shall see, we will reduce the analysis of the solution $u[q, \phi, g]$ of problem (1.2) to that of a related integral equation. To do so, we start with a result on a boundary integral operator, which is proved in [19, Prop. A.3].

Lemma 3.1 *Let $q \in \mathbb{D}_n^+(\mathbb{R})$. Let α, Ω be as in* (1.1)*. Let $\phi \in C^{1,\alpha}(\partial\Omega, \mathbb{R}^n) \cap \mathcal{A}_{\partial\Omega}^{\tilde{Q}}$. Let N be the map from $C^{1,\alpha}(\partial q\mathbb{I}[\phi])$ to itself, defined by*

$$N[\mu] \equiv -\frac{1}{2}\mu + w_q[\partial q\mathbb{I}[\phi], \mu] \qquad \forall \mu \in C^{1,\alpha}(\partial q\mathbb{I}[\phi]).$$

Then N is a linear homeomorphism from $C^{1,\alpha}(\partial q\mathbb{I}[\phi])$ to $C^{1,\alpha}(\partial q\mathbb{I}[\phi])$.

Now we are able to establish a correspondence between the solution of our Dirichlet problem and the solution of an integral equation in the proposition below, whose proof follows from a straightforward modification of the proof of [18, Prop. 5.2].

Proposition 3.2 *Let $q \in \mathbb{D}_n^+(\mathbb{R})$. Let α, Ω be as in* (1.1)*. Let $\phi \in C^{1,\alpha}(\partial\Omega, \mathbb{R}^n) \cap \mathcal{A}_{\partial\Omega}^{\tilde{Q}}$. Let $g \in C^{1,\alpha}(\partial\Omega)$. Then the boundary value problem*

$$\begin{cases} \Delta u = 0 & \text{in } \mathbb{S}_q[q\mathbb{I}[\phi]]^-, \\ u(x + qz) = u(x) & \forall x \in \overline{\mathbb{S}_q[q\mathbb{I}[\phi]]^-}, \forall z \in \mathbb{Z}^n, \\ u(x) = g \circ \phi^{(-1)}(q^{-1}x) & \forall x \in \partial q\mathbb{I}[\phi] \end{cases}$$

has a unique solution $u[q, \phi, g]$ *in* $C_q^{1,\alpha}(\overline{\mathbb{S}_q[q\mathbb{I}[\phi]]^-})$. *Moreover,*

$$u[q, \phi, g](x) = w_q^-[\partial q\mathbb{I}[\phi], \mu](x) \qquad \forall x \in \overline{\mathbb{S}_q[q\mathbb{I}[\phi]]^-},$$

where μ *is the unique solution in* $C^{1,\alpha}(\partial q\mathbb{I}[\phi])$ *of the integral equation*

$$-\frac{1}{2}\mu(x) + w_q[\partial q\mathbb{I}[\phi], \mu](x) = g \circ \phi^{(-1)}(q^{-1}x) \qquad \forall x \in \partial q\mathbb{I}[\phi]. \tag{3.1}$$

Next, we analyze the dependence of the solution of (3.1) upon (q, ϕ, g). Since Eq. (3.1) is defined on the (q, ϕ)-dependent domain $\partial q\mathbb{I}[\phi]$, the first step is to provide a reformulation on a fixed domain. More precisely, we have the following lemma. The proof follows by a change of variable and by Lemma 3.1 (cf. [19, Lem. 3.4]).

Lemma 3.3 *Let* $q \in \mathbb{D}_n^+(\mathbb{R})$. *Let* α, Ω *be as in* (1.1). *Let* $\phi \in C^{1,\alpha}(\partial\Omega, \mathbb{R}^n) \cap \mathcal{A}_{\partial\Omega}^{\widetilde{Q}}$. *Let* $g \in C^{1,\alpha}(\partial\Omega)$. *Then the function* $\theta \in C^{1,\alpha}(\partial\Omega)$ *solves the equation*

$$-\frac{1}{2}\theta(t) - \int_{q\phi(\partial\Omega)} \nu_{q\mathbb{I}[\phi]}(s) \cdot DS_{q,n}(q\phi(t) - s)(\theta \circ \phi^{(-1)})(q^{-1}s)\,d\sigma_s = g(t) \qquad \forall t \in \partial\Omega, \tag{3.2}$$

if and only if the function $\mu \in C^{1,\alpha}(\partial q\mathbb{I}[\phi])$, *with* μ *delivered by*

$$\mu(x) = (\theta \circ \phi^{(-1)})(q^{-1}x) \qquad \forall x \in \partial q\mathbb{I}[\phi],$$

solves the equation

$$-\frac{1}{2}\mu(x) + w_q[\partial q\mathbb{I}[\phi], \mu](x) = g \circ \phi^{(-1)}(q^{-1}x) \qquad \forall x \in \partial q\mathbb{I}[\phi].$$

Moreover, Eq. (3.2) *has a unique solution* θ *in* $C^{1,\alpha}(\partial\Omega)$.

Now, our aim is to prove the analyticity upon (q, ϕ, g) of the function θ which solves Eq. (3.2). Inspired by Lemma 3.3, we introduce the map Λ from $\mathbb{D}_n^+(\mathbb{R}) \times \left(C^{1,\alpha}(\partial\Omega, \mathbb{R}^n) \cap \mathcal{A}_{\partial\Omega}^{\widetilde{Q}}\right) \times \left(C^{1,\alpha}(\partial\Omega)\right)^2$ to $C^{1,\alpha}(\partial\Omega)$ by setting

$$\begin{aligned}\Lambda[q, \phi, g, \theta](t) \equiv &-\frac{1}{2}\theta(t)\\ &- \int_{q\phi(\partial\Omega)} \nu_{q\mathbb{I}[\phi]}(s) \cdot DS_{q,n}(q\phi(t) - s)(\theta \circ \phi^{(-1)})(q^{-1}s)\,d\sigma_s - g(t) \quad \forall t \in \partial\Omega,\end{aligned}$$

for all $(q,\phi,g,\theta) \in \mathbb{D}_n^+(\mathbb{R}) \times \left(C^{1,\alpha}(\partial\Omega,\mathbb{R}^n) \cap \mathcal{A}_{\partial\Omega}^{\widetilde{Q}}\right) \times \left(C^{1,\alpha}(\partial\Omega)\right)^2$. Next, we apply the implicit function theorem to the equation $\Lambda[q,\phi,g,\theta]=0$.

Proposition 3.4 *Let α, Ω be as in* (1.1). *Then the following statements hold.*

(i) *Λ is real analytic.*

(ii) *For each $(q,\phi,g) \in \mathbb{D}_n^+(\mathbb{R}) \times \left(C^{1,\alpha}(\partial\Omega,\mathbb{R}^n) \cap \mathcal{A}_{\partial\Omega}^{\widetilde{Q}}\right) \times C^{1,\alpha}(\partial\Omega)$, there exists a unique θ in $C^{1,\alpha}(\partial\Omega)$ such that*

$$\Lambda[q,\phi,g,\theta]=0 \qquad \text{on } \partial\Omega,$$

and we denote such a function by $\theta[q,\phi,g]$.

(iii) *The map $\theta[\cdot,\cdot,\cdot]$ from $\mathbb{D}_n^+(\mathbb{R}) \times \left(C^{1,\alpha}(\partial\Omega,\mathbb{R}^n) \cap \mathcal{A}_{\partial\Omega}^{\widetilde{Q}}\right) \times C^{1,\alpha}(\partial\Omega)$ to $C^{1,\alpha}(\partial\Omega)$ which takes (q,ϕ,g) to $\theta[q,\phi,g]$ is real analytic.*

Proof Statement (i) follows from [18, Thm. 3.2 (ii)], while (ii) is a consequence of Lemma 3.3. Next we consider (iii). Since the analyticity is a local property, we fix (q_0,ϕ_0,g_0) in $\mathbb{D}_n^+(\mathbb{R}) \times \left(C^{1,\alpha}(\partial\Omega,\mathbb{R}^n) \cap \mathcal{A}_{\partial\Omega}^{\widetilde{Q}}\right) \times C^{1,\alpha}(\partial\Omega)$ and we show that $\theta[\cdot,\cdot,\cdot]$ is real analytic in a neighborhood of (q_0,ϕ_0,g_0) in the product space $\mathbb{D}_n^+(\mathbb{R}) \times \left(C^{1,\alpha}(\partial\Omega,\mathbb{R}^n) \cap \mathcal{A}_{\partial\Omega}^{\widetilde{Q}}\right) \times C^{1,\alpha}(\partial\Omega)$. By standard calculus in normed spaces, the partial differential $\partial_\theta\Lambda[q_0,\phi_0,g_0,\theta[q_0,\phi_0,g_0]]$ of Λ at $(q_0,\phi_0,g_0,\theta[q_0,\phi_0,g_0])$ with respect to the variable θ is delivered by

$$\begin{aligned}
&\partial_\theta\Lambda[q_0,\phi_0,g_0,\theta[q_0,\phi_0,g_0]](\psi)(t)\\
&= -\frac{1}{2}\psi(t) - \int_{q_0\phi_0(\partial\Omega)} \nu_{q_0\mathbb{I}[\phi_0]}(s)\cdot DS_{q_0,n}(q_0\phi_0(t)-s)(\psi\circ\phi_0^{(-1)})(q_0^{-1}s)\,d\sigma_s\\
&\qquad\qquad \forall t\in\partial\Omega,
\end{aligned}$$

for all $\psi \in C^{1,\alpha}(\partial\Omega)$. By Lemma 3.1, $\partial_\theta\Lambda[q_0,\phi_0,g_0,\theta[q_0,\phi_0,g_0]]$ is a linear homeomorphism from $C^{1,\alpha}(\partial\Omega)$ onto $C^{1,\alpha}(\partial\Omega)$. Then the implicit function theorem for real analytic maps in Banach spaces (see, e.g., Deimling [6, Thm. 15.3]) implies that $\theta[\cdot,\cdot,\cdot]$ is real analytic in a neighborhood of (q_0,ϕ_0,g_0) in $\mathbb{D}_n^+(\mathbb{R}) \times \left(C^{1,\alpha}(\partial\Omega,\mathbb{R}^n) \cap \mathcal{A}_{\partial\Omega}^{\widetilde{Q}}\right) \times C^{1,\alpha}(\partial\Omega)$. □

Remark 3.5 By Lemma 2.1, Propositions 3.2 and 3.4, the solution $u[q,\phi,g]$ of problem (1.2) can be written as

$$\begin{aligned}
u[q,\phi,g](x) &= -\int_{\partial\Omega} \nu_{q\mathbb{I}[\phi]}(q\phi(s))\cdot DS_{q,n}(x-q\phi(s))\theta[q,\phi,g](s)\tilde{\sigma}[q\phi](s)\,d\sigma_s\\
&\qquad\qquad \forall x\in\mathbb{S}_q[q\mathbb{I}[\phi]]^-,
\end{aligned}$$

for all $(q,\phi,g) \in \mathbb{D}_n^+(\mathbb{R}) \times \left(C^{1,\alpha}(\partial\Omega,\mathbb{R}^n) \cap \mathcal{A}_{\partial\Omega}^{\widetilde{Q}}\right) \times C^{1,\alpha}(\partial\Omega)$.

We are now able to deduce our main result, which answers to (1.3).

Theorem 3.6 *Let α, Ω be as in* (1.1). *Let*

$$(q_0, \phi_0, g_0) \in \mathbb{D}_n^+(\mathbb{R}) \times \left(C^{1,\alpha}(\partial\Omega, \mathbb{R}^n) \cap \mathcal{A}_{\partial\Omega}^{\widetilde{Q}}\right) \times C^{1,\alpha}(\partial\Omega).$$

Let U be a bounded open subset of $\mathbb{R}^n$ such that $\overline{U} \subseteq \mathbb{S}_{q_0}[q_0\mathbb{I}[\phi_0]]^-$. Then there exists an open neighborhood $\mathcal{U}$ of (q_0, ϕ_0, g_0) in

$$\mathbb{D}_n^+(\mathbb{R}) \times \left(C^{1,\alpha}(\partial\Omega, \mathbb{R}^n) \cap \mathcal{A}_{\partial\Omega}^{\widetilde{Q}}\right) \times C^{1,\alpha}(\partial\Omega)$$

such that the following statements hold.

(i) $\overline{U} \subseteq \mathbb{S}_q[q\mathbb{I}[\phi]]^-$ *for all* $(q, \phi, g) \in \mathcal{U}$.
(ii) Let $k \in \mathbb{N}$. Then the map of $\mathcal{U}$ to $C^k(\overline{U})$ which takes (q, ϕ, g) to the restriction $u[q, \phi, g]_{|\overline{U}}$ of $u[q, \phi, g]$ to $\overline{U}$ is real analytic.

Proof By taking $\mathcal{U}$ small enough, we can deduce the validity of statement (i). Statement (ii) follows from the representation formula of Remark 3.5 together with Lemma 2.1, Proposition 3.4 and standard properties of integral operators with real analytic kernels and with no singularity (cf. [14]). □

Remark 3.7 We considered the periodic Dirichlet problem for the Laplace equation. Our method can be used for other periodic problems. For example, one can consider the Dirichlet problem

$$\begin{cases} \Delta v = 1 & \text{in } \mathbb{S}_q[q\mathbb{I}[\phi]]^-, \\ v(x + qz) = v(x) & \forall x \in \overline{\mathbb{S}_q[q\mathbb{I}[\phi]]^-}, \forall z \in \mathbb{Z}^n, \\ v(x) = 0 & \forall x \in \partial q\mathbb{I}[\phi], \end{cases} \tag{3.3}$$

which generalizes the one considered in [17]. Then, if we denote by $v[q, \phi]$ the solution to problem (3.3), by exploiting the periodic volume potential we can prove that the map from $\mathbb{D}_n^+(\mathbb{R}) \times \left(C^{1,\alpha}(\partial\Omega, \mathbb{R}^n) \cap \mathcal{A}_{\partial\Omega}^{\widetilde{Q}}\right)$ to $\mathbb{R}$

$$(q, \phi) \mapsto \int_{Q \setminus q\mathbb{I}[\phi]} v[q, \phi](x)\, dx$$

is real analytic. Moreover, one can replace the right-hand side in the first equation of problem (3.3) by a more general sufficiently regular periodic function.

Acknowledgments The authors are members of the 'Gruppo Nazionale per l'Analisi Matematica, la Probabilità e le loro Applicazioni' (GNAMPA) of the 'Istituto Nazionale di Alta Matematica' (INdAM) and acknowledge the support of the Project BIRD191739/19 'Sensitivity analysis of partial differential equations in the mathematical theory of electromagnetism' of the University of Padova. P.M. also acknowledges the support of the grant 'Challenges in Asymptotic and Shape Analysis—CASA' of the Ca' Foscari University of Venice.

References

1. H. Ammari, H. Kang, *Polarization and Moment Tensors. With Applications to Inverse Problems and Effective Medium Theory* (Springer, Berlin, 2007)
2. D. Buoso, Analyticity and criticality results for the eigenvalues of the biharmonic operator, in *Geometric Properties for Parabolic and Elliptic PDE's*. Springer Proceedings in Mathematics & Statistics, vol. 176 (Springer, Cham, 2016), pp. 65–85
3. M. Dalla Riva, Stokes flow in a singularly perturbed exterior domain. Complex Var. Elliptic Equ. **58**(2), 231–257 (2013)
4. M. Dalla Riva, M. Lanza de Cristoforis, Microscopically weakly singularly perturbed loads for a nonlinear traction boundary value problem: a functional analytic approach. Complex Var. Elliptic Equ. **55**(8–10), 771–794 (2010)
5. M. Dalla Riva, P. Musolino, S.V. Rogosin, Series expansions for the solution of the Dirichlet problem in a planar domain with a small hole. Asymptot. Anal. **92**(3–4), 339–361 (2015)
6. K. Deimling, *Nonlinear Functional Analysis* (Springer, Berlin, 1985)
7. D. Gilbarg, N.S. Trudinger, *Elliptic Partial Differential Equations of Second Order*, 2nd edn. (Springer, Berlin, 1983)
8. D. Henry, *Topics in Nonlinear Analysis*, vol. 192. Trabalho de Matematica (Universidade de Brasilia, Brasilia, 1982)
9. M.V. Keldysh, On the solubility and the stability of Dirichlet's problem. Uspekhi Matem. Nauk **8**, 171–231 (1941)
10. P.D. Lamberti, M. Lanza de Cristoforis, A real analyticity result for symmetric functions of the eigenvalues of a domain dependent Dirichlet problem for the Laplace operator. J. Nonlinear Convex Anal. **5**(1), 19–42 (2004)
11. M. Lanza de Cristoforis, A domain perturbation problem for the Poisson equation. Complex Var. Theory Appl. **50**(7–11), 851–867 (2005)
12. M. Lanza de Cristoforis, Perturbation problems in potential theory, a functional analytic approach. J. Appl. Funct. Anal. **2**(3), 197–222 (2007)
13. M. Lanza de Cristoforis, P. Musolino, A perturbation result for periodic layer potentials of general second order differential operators with constant coefficients. Far East J. Math. Sci. **52**(1), 75–120 (2011)
14. M. Lanza de Cristoforis, P. Musolino, A real analyticity result for a nonlinear integral operator. J. Integral Equ. Appl. **25**(1), 21–46 (2013)
15. M. Lanza de Cristoforis, L. Preciso, On the analyticity of the Cauchy integral in Schauder spaces. J. Integral Equ. Appl. **11**, 363–391 (1999)
16. M. Lanza de Cristoforis, L. Rossi, Real analytic dependence of simple and double layer potentials upon perturbation of the support and of the density. J. Integral Equ. Appl. **16**, 137–174 (2004)
17. P. Luzzini, P. Musolino, R. Pukhtaievych, Shape analysis of the longitudinal flow along a periodic array of cylinders. J. Math. Anal. Appl. **477**(2), 1369–1395 (2019)
18. P. Luzzini, P. Musolino, R. Pukhtaievych, Real analyticity of periodic layer potentials upon perturbation of the periodicity parameters and of the support, in *Proceedings of the 12th ISAAC Congress (Aveiro, 2019)*. Research Perspectives (Birkhäuser, Basel, 2022)
19. P. Musolino, A singularly perturbed Dirichlet problem for the Laplace operator in a periodically perforated domain. A functional analytic approach. Math. Methods Appl. Sci. **35**(3), 334–349 (2012)
20. R. Potthast, Domain derivatives in electromagnetic scattering. Math. Methods Appl. Sci. **19**(15), 1157–1175 (1996)
21. R. Potthast, I.G. Stratis, On the domain derivative for scattering by impenetrable obstacles in chiral media. IMA J. Appl. Math. **68**(3), 621–635 (2003)
22. J. Sokolowski, J.P. Zolésio, *Introduction to Shape Optimization. Shape Sensitivity Analysis* (Springer, Berlin, 1992)

Real Analyticity of Periodic Layer Potentials Upon Perturbation of the Periodicity Parameters and of the Support

Paolo Luzzini, Paolo Musolino, and Roman Pukhtaievych

Abstract We prove that the periodic layer potentials for the Laplace operator depend real analytically on the density function, on the supporting hypersurface, and on the periodicity parameters.

Keywords Periodic simple layer potential · Periodic double layer potential · Laplace operator · Domain perturbation · Special nonlinear operators

Mathematics Subject Classification (2010) Primary 31B10; Secondary 45A05, 35J05, 35J25

1 Introduction

This paper is devoted to the study of the dependence of the periodic simple and double layer potentials upon perturbation of the periodicity cell and of the support of the integration. A periodic version of potential theory has revealed to be a powerful tool to analyze boundary value problems for elliptic differential equations in spatially periodic domains. If one is interested into studying the behavior of the solutions of boundary value problems for the Laplacian in a periodic domain upon

P. Luzzini
Dipartimento di Matematica 'Tullio Levi-Civita', Università degli Studi di Padova, Padova, Italy
e-mail: pluzzini@math.unipd.it

P. Musolino (✉)
Dipartimento di Scienze Molecolari e Nanosistemi, Università Ca' Foscari Venezia, Venezia Mestre, Italy
e-mail: paolo.musolino@unive.it

R. Pukhtaievych
Department of Complex Analysis and Potential Theory, Institute of Mathematics of the National Academy of Sciences of Ukraine, Kyiv, Ukraine
e-mail: p_r_p@imath.kiev.ua

P. Cerejeiras et al. (eds.), *Current Trends in Analysis, its Applications and Computation*, Research Perspectives,
https://doi.org/10.1007/978-3-030-87502-2_19

perturbation of the periodicity cell and of the shape of the domain, then one faces the problem of studying the behavior of the corresponding layer potentials upon the same perturbations.

In view of such an application, several authors have studied the dependence of the layer potentials upon domain perturbations. Potthast [15, 16] has obtained Fréchet differentiability results for the dependence of the layer potentials. Costabel and Le Louër [2] have analyzed the Fréchet differentiability of a class of boundary integral operators with pseudohomogeneous hypersingular and weakly singular kernels in the framework of Sobolev spaces. For elastic obstacle scattering, we mention Le Louër [13]. Also, Lanza de Cristoforis and collaborators have developed a method based on potential theory with the aim of proving real analyticity results in the framework of Schauder spaces for the dependence of the solutions of boundary value problems upon domain perturbations. In order to apply such a method, one has to verify the real analytic dependence of the layer potentials on both variation of the support of integration and on data. In [11, 12], Lanza de Cristoforis and Rossi have considered the layer potentials associated with the Laplace and the Helmholtz operators. In [5], instead, Dalla Riva and Lanza de Cristoforis have studied the case of layer potentials associated to a family of second order differential operators with constant coefficients. In Dalla Riva [3, 4] the author has considered the single layer potential corresponding to the fundamental solution of a given elliptic partial differential operator of order $2k$ with constant coefficients.

In order to introduce the problem, we fix $n \in \mathbb{N} \setminus \{0, 1\}$. If $(q_{11}, \dots, q_{nn}) \in]0, +\infty[^n$ we introduce a periodicity cell Q and a matrix q by setting

$$Q \equiv \prod_{j=1}^{n}]0, q_{jj}[, \qquad q \equiv \begin{pmatrix} q_{11} & 0 & \dots & 0 \\ 0 & q_{22} & \dots & 0 \\ \dots & \dots & \dots & \dots \\ 0 & 0 & \dots & q_{nn} \end{pmatrix} .$$

We also denote by $|Q|_n$ the n-dimensional measure of the fundamental cell Q, by ν_Q the outward unit normal to ∂Q, where it exists, and by q^{-1} the inverse matrix of q. Clearly, $q\mathbb{Z}^n \equiv \{qz : z \in \mathbb{Z}^n\}$ is the set of vertices of a periodic subdivision of $\mathbb{R}^n$ corresponding to the fundamental cell Q. In order to construct periodic layer potentials, we replace the fundamental solution of the Laplace operator by a q-periodic tempered distribution $S_{q,n}$ such that

$$\Delta S_{q,n} = \sum_{z \in \mathbb{Z}^n} \delta_{qz} - \frac{1}{|Q|_n} ,$$

where δ_{qz} denotes the Dirac measure with mass in qz (see e.g., [8, p. 84]). The distribution $S_{q,n}$ is determined up to an additive constant, and we can take

$$S_{q,n}(x) = - \sum_{z \in \mathbb{Z}^n \setminus \{0\}} \frac{1}{|Q|_n 4\pi^2 |q^{-1}z|^2} e^{2\pi i (q^{-1}z) \cdot x}$$

in the sense of distributions in $\mathbb{R}^n$ (see e.g., Ammari and Kang [1, p. 53], [8, §3]). Moreover, $S_{q,n}$ is even, real analytic in $\mathbb{R}^n \setminus q\mathbb{Z}^n$, and locally integrable in $\mathbb{R}^n$ (see e.g., [8, §3]). We now introduce the periodic layer potentials. We take a bounded open subset Ω_Q of $\mathbb{R}^n$ of class $C^{1,\alpha}$ for some $\alpha \in]0, 1[$ such that $\overline{\Omega_Q} \subseteq Q$. For the definition of sets and functions of the Schauder class $C^{k,\alpha}$ ($k \in \mathbb{N}$) we refer, e.g., to Gilbarg and Trudinger [7]. We set

$$v_q[\partial\Omega_Q, \mu](x) \equiv \int_{\partial\Omega_Q} S_{q,n}(x-y)\mu(y)\,d\sigma_y \qquad \forall x \in \mathbb{R}^n\,,$$

$$w_q[\partial\Omega_Q, \mu](x) \equiv -\int_{\partial\Omega_Q} DS_{q,n}(x-y)\cdot \nu_{\Omega_Q}(y)\mu(y)\,d\sigma_y \qquad \forall x \in \mathbb{R}^n\,,$$

$$w_{q,*}[\partial\Omega_Q, \mu](x) \equiv \int_{\partial\Omega_Q} DS_{q,n}(x-y)\cdot \nu_{\Omega_Q}(x)\mu(y)\,d\sigma_y \qquad \forall x \in \partial\Omega_Q\,,$$

for all $\mu \in L^2(\partial\Omega_Q)$. Here above, the symbol ν_{Ω_Q} denotes the outward unit normal field to $\partial\Omega_Q$, $d\sigma$ denotes the area element on $\partial\Omega_Q$, and $DS_{q,n}(\xi)$ denotes the gradient of $S_{q,n}$ computed at the point $\xi \in \mathbb{R}^n \setminus q\mathbb{Z}^n$. The functions $v_q[\partial\Omega_Q, \mu]$ and $w_q[\partial\Omega_Q, \mu]$ are called the q-periodic simple (or single) and double layer potentials, respectively.

In order to consider the dependence of periodic layer potentials under shape perturbations, we need to introduce some notation. First, we find convenient to set $\widetilde{Q} \equiv]0, 1[^n$ and to set $\tilde{q}$ equal to the $n \times n$ identity matrix. Then we take

$$\begin{aligned}&\alpha \in]0, 1[\text{ and a bounded open connected subset } \Omega \text{ of } \mathbb{R}^n \text{ of class } C^{1,\alpha}\\ &\text{such that } \mathbb{R}^n \setminus \overline{\Omega} \text{ is connected.}\end{aligned} \tag{1.1}$$

The symbol '$\overline{\cdot}$' denotes the closure. Then we consider a class of diffeomorphisms $\mathcal{A}_{\partial\Omega}^{\widetilde{Q}}$ from $\partial\Omega$ into their images contained in $\widetilde{Q}$ (see (1.2)). To define such a class, we take Ω as in (1.1) and a bounded open connected subset Ω' of $\mathbb{R}^n$ of class $C^{1,\alpha}$. We denote by $\mathcal{A}_{\partial\Omega}$ and by $\mathcal{A}_{\overline{\Omega'}}$ the sets of functions of class $C^1(\partial\Omega, \mathbb{R}^n)$ and of class $C^1(\overline{\Omega'}, \mathbb{R}^n)$ which are injective and whose differential is injective at all points of $\partial\Omega$ and of $\overline{\Omega'}$, respectively. One can verify that $\mathcal{A}_{\partial\Omega}$ and $\mathcal{A}_{\overline{\Omega'}}$ are open in $C^1(\partial\Omega, \mathbb{R}^n)$ and $C^1(\overline{\Omega'}, \mathbb{R}^n)$, respectively (see, e.g., Lanza de Cristoforis and Rossi [12, Lem. 2.2, p. 197] and [11, Lem. 2.5, p. 143]). Then we set

$$\mathcal{A}_{\partial\Omega}^{\widetilde{Q}} \equiv \{\phi \in \mathcal{A}_{\partial\Omega} : \phi(\partial\Omega) \subseteq \widetilde{Q}\}, \quad \mathcal{A}_{\overline{\Omega'}}^{\widetilde{Q}} \equiv \{\Phi \in \mathcal{A}_{\overline{\Omega'}} : \Phi(\overline{\Omega'}) \subseteq \widetilde{Q}\}. \tag{1.2}$$

If $\phi \in \mathcal{A}_{\partial\Omega}^{\widetilde{Q}}$, the Jordan-Leray separation theorem ensures that $\mathbb{R}^n \setminus \phi(\partial\Omega)$ has exactly two open connected components (see, *e.g.*, Deimling [6, Thm. 5.2, p. 26]), and we denote by $\mathbb{I}[\phi]$ the bounded open connected component of $\mathbb{R}^n \setminus \phi(\partial\Omega)$.

We denote by $\mathbb{D}_n(\mathbb{R})$ the space of $n \times n$ diagonal matrices with real entries and by $\mathbb{D}_n^+(\mathbb{R})$ the set of elements of $\mathbb{D}_n(\mathbb{R})$ with diagonal entries in $]0, +\infty[$.

Then for each triple (q, ϕ, θ) in $\mathbb{D}_n^+(\mathbb{R}) \times (C^{1,\alpha}(\partial\Omega, \mathbb{R}^n) \cap \mathcal{A}_{\partial\Omega}^{\widetilde{Q}}) \times C^{0,\alpha}(\partial\Omega)$ we denote by $V[q, \phi, \theta]$ the function in $C^{1,\alpha}(\partial\Omega)$ defined by

$$V[q, \phi, \theta](x) \equiv \int_{q\phi(\partial\Omega)} S_{q,n}(q\phi(x) - s)(\theta \circ \phi^{(-1)})(q^{-1}s)d\sigma_s \quad \forall x \in \partial\Omega \,,$$

and by $W_*[q, \phi, \theta]$ the function in $C^{0,\alpha}(\partial\Omega)$ defined by

$$W_*[q, \phi, \theta](x) \equiv \int_{q\phi(\partial\Omega)} DS_{q,n}(q\phi(x) - s) \cdot \nu_{q\mathbb{I}[\phi]}(q\phi(x))(\theta \circ \phi^{(-1)})(q^{-1}s)d\sigma_s \qquad \forall x \in \partial\Omega \,.$$

Similarly, for each triple (q, ϕ, θ) in $\mathbb{D}_n^+(\mathbb{R}) \times (C^{1,\alpha}(\partial\Omega, \mathbb{R}^n) \cap \mathcal{A}_{\partial\Omega}^{\widetilde{Q}}) \times C^{1,\alpha}(\partial\Omega)$ we denote by $W[q, \phi, \theta]$ the function in $C^{1,\alpha}(\partial\Omega)$ defined by

$$W[q, \phi, \theta](x) \equiv -\int_{q\phi(\partial\Omega)} DS_{q,n}(q\phi(x) - s) \cdot \nu_{q\mathbb{I}[\phi]}(s)(\theta \circ \phi^{(-1)})(q^{-1}s)d\sigma_s \qquad \forall x \in \partial\Omega \,.$$

The functions $V[q, \phi, \theta]$ and $W[q, \phi, \theta]$ are associated with the $q\phi$-pull backs on $\partial\Omega$ of the periodic simple layer potential and of the periodic double layer potential, respectively. The function $W_*[q, \phi, \theta]$, instead, is associated with the $q\phi$-pull back on $\partial\Omega$ of the normal derivative of the periodic simple layer potential. These functions are well known to intervene in the integral equations associated with periodic boundary value problems. We are interested in understanding the dependence of $V[q, \phi, \theta]$, $W[q, \phi, \theta]$, and $W_*[q, \phi, \theta]$ upon perturbation of (q, ϕ, θ), i.e., of the periodicity matrix, the support of integration, and the density function. Hence, we pose the following question:

What can be said on the regularity of the maps $(q, \phi, \theta) \mapsto V[q, \phi, \theta]$, $(q, \phi, \theta) \mapsto W[q, \phi, \theta]$, and $(q, \phi, \theta) \mapsto W_*[q, \phi, \theta]$? (1.3)

Our work stems from that of Lanza de Cristoforis and Preciso [10] for the Cauchy integral operator, from that of Lanza de Cristoforis and Rossi [11, 12] for the Laplace and for the Helmholtz operator, and from that of Dalla Riva and Lanza de Cristoforis [5] for second order elliptic operators. Moreover this work can be seen as complement of [8], where it has been shown that periodic layer potentials associated with parameter dependent analytic families of fundamental solutions of second order differential operators with constant coefficients depend real analytically upon the density function and on a suitable parametrization of the supporting hypersurface and on the parameter. Furthermore, it generalizes a part of [14] where the authors have proven analyticity results for the double layer potential in dimension two for a specific perturbation of the periodicity cell, in order to study the longitudinal flow through a periodic array of cylinders.

In this paper, we answer to the question (1.3) by proving that the maps in (1.3) are real analytic (see Theorem 3.2).

2 Preliminary Technical Results

To prove the analyticity of the operators $V[\cdot,\cdot,\cdot]$, $W[\cdot,\cdot,\cdot]$, and $W_*[\cdot,\cdot,\cdot]$, we need the following results from Lanza de Cristoforis and Rossi [12, §2].

Lemma 2.1 *Let α, Ω be as in* (1.1). *Then there exists $\beta \in C^{1,\alpha}(\partial\Omega, \mathbb{R}^n)$ such that $|\beta(x)| = 1$ and $\beta(x) \cdot \nu_\Omega(x) > 1/2$ for all $x \in \partial\Omega$.*

Lemma 2.2 *Let α, Ω be as in* (1.1). *Let β be as in Lemma 2.1. Then the following statements hold.*

(i) There exists $\delta_\Omega \in]0, +\infty[$ such that the sets

$$\Omega_{\beta,\delta} \equiv \{x + t\beta(x) : x \in \partial\Omega,\ t \in]-\delta, \delta[\},$$

$$\Omega^+_{\beta,\delta} \equiv \{x + t\beta(x) : x \in \partial\Omega,\ t \in]-\delta, 0[\},$$

are connected and of class $C^{1,\alpha}$, and

$$\partial\Omega_{\beta,\delta} = \{x + t\beta(x) : x \in \partial\Omega,\ t \in \{-\delta, \delta\}\},$$

$$\partial\Omega^+_{\beta,\delta} = \{x + t\beta(x) : x \in \partial\Omega,\ t \in \{-\delta, 0\}\},$$

and $\Omega^+_{\beta,\delta} \subseteq \Omega$ for all $\delta \in]0, \delta_\Omega[$.

(ii) Let $\delta \in]0, \delta_\Omega[$. If $\Phi \in \mathcal{A}_{\overline{\Omega_{\beta,\delta}}}$, then $\Phi_{|\partial\Omega} \in \mathcal{A}_{\partial\Omega}$.

(iii) If $\delta \in]0, \delta_\Omega[$, then the set

$$\mathcal{A}'_{\overline{\Omega_{\beta,\delta}}} \equiv \{\Phi \in \mathcal{A}_{\overline{\Omega_{\beta,\delta}}} : \Phi(\Omega^+_{\beta,\delta}) \subseteq \mathbb{I}[\Phi_{|\partial\Omega}]\}$$

is open in $\mathcal{A}_{\overline{\Omega_{\beta,\delta}}}$.

(iv) If $\delta \in]0, \delta_\Omega[$ and $\Phi \in C^{1,\alpha}(\overline{\Omega_{\beta,\delta}}, \mathbb{R}^n) \cap \mathcal{A}'_{\overline{\Omega_{\beta,\delta}}}$, then $\Phi(\Omega^+_{\beta,\delta})$ is an open set of class $C^{1,\alpha}$ and $\partial\Phi(\Omega^+_{\beta,\delta}) = \Phi(\partial\Omega^+_{\beta,\delta})$.

3 Analyticity of the Integral Operators Associated with Layer Potentials

In this section, we prove our main result on the analyticity of the maps in (1.3). We first need to prove the following lemma which represents an intermediate step.

Lemma 3.1 *Let* α, Ω *be as in* (1.1). *Let* β *and* δ_Ω *be as in Lemma 2.2. Let*

$$\mathcal{A}'^{\widetilde{Q}}_{\overline{\Omega_{\beta,\delta}}} \equiv \mathcal{A}'_{\overline{\Omega_{\beta,\delta}}} \cap \mathcal{A}^{\widetilde{Q}}_{\overline{\Omega_{\beta,\delta}}} \quad \forall \delta \in]0, \delta_\Omega[.$$

Let $\eta \in]0, 1[$. *Then there exists* $\delta_\eta \in]0, \delta_\Omega[$ *such that for all* $\delta \in]0, \delta_\eta[$ *the map which takes*

$$(q, \Phi, \theta) \in \mathbb{D}^+_n(\mathbb{R}) \times \left(C^{1,\alpha}(\overline{\Omega_{\beta,\delta}}, \mathbb{R}^n) \cap \mathcal{A}'^{\widetilde{Q}}_{\overline{\Omega_{\beta,\delta}}}\right) \times C^{0,\alpha}(\partial\Omega)$$

to the function $V^+[q, \Phi, \theta]$, *which is defined as*

$$V^+[q, \Phi, \mu](x) \equiv \int_{q\Phi(\partial\Omega)} S_{q,n}(q\Phi(x) - s)\left(\mu \circ \Phi^{(-1)}\right)(q^{-1}s)d\sigma_s \quad \forall x \in \overline{\Omega^+_{\beta,\delta}},$$

is real analytic from $\mathcal{O}(\eta) \times \mathcal{U}_{\eta,\delta} \times C^{0,\alpha}(\partial\Omega)$ *to* $C^{1,\alpha}(\overline{\Omega^+_{\beta,\delta}})$, *where*

$$\mathcal{O}(\eta) \equiv \left\{ q \in \mathbb{D}^+_n(\mathbb{R}) : \inf_{\xi\in\mathbb{R}^n, |\xi|=1} \left\{ \sum_{j=1}^n (q_{jj})^{-2}\xi_j^2 \right\} > \eta,\ \max\{(q_{jj})^{-2} : j = 1, \dots, n\} < \eta^{-1} \right\},$$

$$\mathcal{U}_{\eta,\delta} \equiv \left\{ \Phi \in \mathcal{A}'^{\widetilde{Q}}_{\overline{\Omega_{\beta,\delta}}} \cap C^{1,\alpha}(\overline{\Omega_{\beta,\delta}}, \mathbb{R}^n) : \sup_{\overline{\Omega_{\beta,\delta}}} |\det(D\Phi)| < \eta^{-1} \right\}.$$

Proof Let $\delta \in]0, \delta_\Omega[$. Next, we note that if

$$(q, \Phi) \in \mathbb{D}^+_n(\mathbb{R}) \times \left(C^{1,\alpha}(\overline{\Omega_{\beta,\delta}}, \mathbb{R}^n) \cap \mathcal{A}'^{\widetilde{Q}}_{\overline{\Omega_{\beta,\delta}}}\right)$$

then

$$V^+[q, \Phi, \mu](x) = \int_{q\Phi(\partial\Omega)} S_{q,n}(q\Phi(x) - s)\left(\mu \circ \Phi^{(-1)}\right)(q^{-1}s)d\sigma_s$$

$$= \int_{\Phi(\partial\Omega)} S_{q,n}(q(\Phi(x) - y))\left(\mu \circ \Phi^{(-1)}\right)(y)|\det q||q^{-1} \cdot \nu_{\mathbb{I}[\Phi_{|\partial\Omega}]}(y)|d\sigma_y$$

for all $\mu \in C^{0,\alpha}(\partial\Omega)$ and for all $x \in \Omega^+_{\beta,\delta}$. Then we set

$$\tilde{S}_n(q,x) \equiv \det q S_{q,n}(qx) \qquad \forall x \in \mathbb{R}^n \setminus \mathbb{Z}^n .$$

We note that the $\tilde{q}$-periodic function $\tilde{S}_n(q,\cdot)$ is a $\tilde{q}$-periodic $\{0\}$-analog of the fundamental solution of the operator $\sum_{j=1}^n \frac{1}{(q_{jj})^2}\frac{\partial^2}{\partial x_j^2}$, i.e., a $\tilde{q}$-periodic tempered distribution such that

$$\sum_{j=1}^{n} \frac{1}{(q_{jj})^2}\frac{\partial^2}{\partial x_j^2}\tilde{S}_n(q,\cdot) = \sum_{z\in\mathbb{Z}^n} \delta_z - 1 ,$$

in the sense of distributions (see [8, §1]). Then, if we set

$$\sigma_\#[q,\Phi](s) \equiv |q^{-1}\cdot(\nu_{\mathbb{I}[\Phi_{|\partial\Omega}]}\circ\Phi)(s)| \qquad \forall s \in \partial\Omega ,$$

for all $(q,\Phi) \in \mathbb{D}_n^+(\mathbb{R}) \times \mathcal{U}_{\eta,\delta}$, we can write

$$\begin{aligned}
&\int_{\Phi(\partial\Omega)} S_{q,n}(q(\Phi(x)-y))\left(\mu\circ\Phi^{(-1)}\right)(y)|\det q||q^{-1}\cdot\nu_{\mathbb{I}[\Phi_{|\partial\Omega}]}(y)|d\sigma_y \\
&= \int_{\Phi(\partial\Omega)} \tilde{S}_n(q,\Phi(x)-y)\left(\mu\circ\Phi^{(-1)}\right)(y)\sigma_\#[q,\Phi]\circ\Phi^{(-1)}(y)d\sigma_y \\
&\equiv \widetilde{V}^+_{\tilde{q}}[q,\Phi,\mu](x) \qquad \forall x \in \overline{\Omega^+_{\beta,\delta}},
\end{aligned}$$

for all $(q,\Phi,\mu) \in \mathbb{D}_n^+(\mathbb{R}) \times \mathcal{U}_{\eta,\delta} \times C^{0,\alpha}(\partial\Omega)$. By Lanza de Cristoforis and Rossi [11, Lem. 3.3 and p. 166] and standard calculus in Banach spaces, we have that the map from $\mathbb{D}_n^+(\mathbb{R}) \times \mathcal{U}_{\eta,\delta}$ to $C^{0,\alpha}(\partial\Omega)$ which takes a pair (q,Φ) to $\sigma_\#[q,\Phi]$ is real analytic. Now we note that by Lanza de Cristoforis and Musolino [9, Thm. 7] and [8, §3] the map from $\mathbb{D}_n^+(\mathbb{R}) \times (\mathbb{R}^n \setminus \mathbb{Z}^n)$ to $\mathbb{R}$ which takes the pair (q,x) to $\tilde{S}_n(q,x)$ is real analytic. Moreover, as noted above, for all $q \in \mathbb{D}_n^+(\mathbb{R})$, the map $\tilde{S}_n(q,\cdot)$ is a $\tilde{q}$-periodic function in $L^1_{\mathrm{loc}}(\mathbb{R}^n)$ such that $\sum_{j=1}^n \frac{1}{(q_{jj})^2}\frac{\partial^2}{\partial x_j^2}\tilde{S}_n(q,\cdot) = \sum_{z\in\mathbb{Z}^n}\delta_z - 1$ in the sense of distributions. Accordingly, one can readily verify that the assumptions (1.8) of [8, pp. 78, 79] are satisfied and thus we can apply the results of [8]. Hence, [8, Prop. 5.6, pp. 105, 106] implies that there exists $\delta_\eta \in]0,\delta_\Omega[$ such that for all $\delta \in]0,\delta_\eta[$ the map $\widetilde{V}^+_{\tilde{q}}[\cdot,\cdot,\cdot]$ is real analytic from $\mathcal{O}(\eta) \times \mathcal{U}_{\eta,\delta} \times C^{0,\alpha}(\partial\Omega)$ to $C^{1,\alpha}(\overline{\Omega^+_{\beta,\delta}})$, and thus the proof is complete. □

We can now deduce our main theorem on the analyticity of the periodic layer potentials upon the periodicity parameter, the shape, and the density. The proof follows the strategy exploited in Lanza de Cristoforis and Rossi [11, Thm. 3.12] and in [8, Thm. 5.10].

Theorem 3.2 *Let α, Ω be as in* (1.1)*. Then the following statements hold.*

(i) The map from $\mathbb{D}_n^+(\mathbb{R}) \times (C^{1,\alpha}(\partial\Omega, \mathbb{R}^n) \cap \mathcal{A}_{\partial\Omega}^{\widetilde{Q}}) \times C^{0,\alpha}(\partial\Omega)$ *to* $C^{1,\alpha}(\partial\Omega)$ *which takes a triple* (q, ϕ, θ) *to the function* $V[q, \phi, \theta]$ *is real analytic.*

(ii) The map from $\mathbb{D}_n^+(\mathbb{R}) \times (C^{1,\alpha}(\partial\Omega, \mathbb{R}^n) \cap \mathcal{A}_{\partial\Omega}^{\widetilde{Q}}) \times C^{1,\alpha}(\partial\Omega)$ *to* $C^{1,\alpha}(\partial\Omega)$ *which takes a triple* (q, ϕ, θ) *to the function* $W[q, \phi, \theta]$ *is real analytic.*

(iii) The map from $\mathbb{D}_n^+(\mathbb{R}) \times (C^{1,\alpha}(\partial\Omega, \mathbb{R}^n) \cap \mathcal{A}_{\partial\Omega}^{\widetilde{Q}}) \times C^{0,\alpha}(\partial\Omega)$ *to* $C^{0,\alpha}(\partial\Omega)$ *which takes a triple* (q, ϕ, θ) *to the function* $W_*[q, \phi, \theta]$ *is real analytic.*

Proof To prove statements (i) and (iii), it suffices to argue as in the proof of [8, Thm. 5.10] and to replace [8, Prop. 5.6] by Lemma 3.1. Instead, the proof of statement (ii), follows by the same argument as the one of the proof of [14, Lem. 4.2], with the combination of the proof of [14, Lem. 4.1] and Lemma 3.1. □

Acknowledgments The authors are members of the 'Gruppo Nazionale per l'Analisi Matematica, la Probabilità e le loro Applicazioni' (GNAMPA) of the 'Istituto Nazionale di Alta Matematica' (INdAM). P.L. and P.M. acknowledge the support of the Project BIRD191739/19 'Sensitivity analysis of partial differential equations in the mathematical theory of electromagnetism' of the University of Padova. P.M. acknowledges the support of the grant 'Challenges in Asymptotic and Shape Analysis - CASA' of the Ca' Foscari University of Venice. R.P. was supported by the budget program 'Support for the Development of Priority Areas of Scientific Research' (КПКВК 6541230) of the National Academy of Sciences of Ukraine.

References

1. H. Ammari, H. Kang, *Polarization and Moment Tensors. With Applications to Inverse Problems and Effective Medium Theory* (Springer, Berlin, 2007)
2. M. Costabel, F. Le Louër, Shape derivatives of boundary integral operators in electromagnetic scattering. Part I: shape differentiability of pseudo-homogeneous boundary integral operators. Integr. Equ. Oper. Theory **72**(4), 509–535 (2012)
3. M. Dalla Riva, The layer potentials of some partial differential operators: real analytic dependence upon perturbations, in *Further Progress in Analysis* (World Scientific Publishing, Hackensack, 2009), pp. 208–217
4. M. Dalla Riva, A family of fundamental solutions of elliptic partial differential operators with real constant coefficients. Integr. Equ. Oper. Theory **76**(1), 1–23 (2013)
5. M. Dalla Riva, M. Lanza de Cristoforis, A perturbation result for the layer potentials of general second order differential operators with constant coefficients. J. Appl. Funct. Anal. **5**(1), 10–30 (2010)
6. K. Deimling, *Nonlinear Functional Analysis* (Springer, Berlin, 1985)
7. D. Gilbarg, N.S. Trudinger, *Elliptic Partial Differential Equations of Second Order*, 2nd edn (Springer, Berlin, 1983)
8. M. Lanza de Cristoforis, P. Musolino, A perturbation result for periodic layer potentials of general second order differential operators with constant coefficients. Far East J. Math. Sci. **52**(1), 75–120 (2011)
9. M. Lanza de Cristoforis, P. Musolino, Analytic dependence of a periodic analog of a fundamental solution upon the periodicity parameters. Ann. Mat. Pura Appl. **197**(4), 1089–1116 (2018)

10. M. Lanza de Cristoforis, L. Preciso, On the analyticity of the Cauchy integral in Schauder spaces. J. Integral Equ. Appl. **11**, 363–391 (1999)
11. M. Lanza de Cristoforis, L. Rossi, Real analytic dependence of simple and double layer potentials upon perturbation of the support and of the density. J. Integral Equ. Appl. **16**, 137–174 (2004)
12. M. Lanza de Cristoforis, L. Rossi, Real analytic dependence of simple and double layer potentials for the Helmholtz equation upon perturbation of the support and of the density, in *Analytic Methods of Analysis and Differential Equations, AMADE 2006*, ed. by A.A. Kilbas, S.V. Rogosin (Cambridge Scientific Publishers, Cambridge, 2008), pp. 193–220
13. F. Le Louër, On the Fréchet derivative in elastic obstacle scattering. SIAM J. Appl. Math. **72**(5), 1493–1507 (2012)
14. P. Luzzini, P. Musolino, R. Pukhtaievych, Shape analysis of the longitudinal flow along a periodic array of cylinders. J. Math. Anal. Appl. **477**(2), 1369–1395 (2019)
15. R. Potthast, Fréchet differentiability of boundary integral operators in inverse acoustic scattering. Inverse Probl. **10**(2), 431–447 (1994)
16. R. Potthast, Domain derivatives in electromagnetic scattering. Math. Methods Appl. Sci. **19**(15), 1157–1175 (1996)

Directional Derivatives and Stolz Condition for Bicomplex Holomorphic Functions

Lino F. Reséndis O., Gamaliel Y. Téllez S., and Luis M. Tovar S.

Abstract The focus of this paper is to present a characterization of bicomplex holomorphic function in terms of the directional derivatives over the idempotent planes. Moreover we give an equivalence between bicomplex holomorphy and the Stolz's condition. We have considered different representation of bicomplex numbers and applying tools of complex analysis.

Keywords Bicomplex Holomorphic functions · Stolz condition

Mathematics Subject Classification (2010) 30G30, 32A10

1 Introduction

Bicomplex numbers are generated by two imaginary units in a similar way that quaternions but bicomplex numbers are commutative respect to the product and with zero divisors. The study of this set was started by J. Cockle in [1] under the name of tessarines and recently there has been an increasing interest about this.

The bicomplex analysis has a hybrid behavior between the one complex analysis and the analysis of several complex variables. So, many results from the theory of one complex variable—that can be consulted in [2, 3]—, have a natural extension into the bicomplex space. In the other hand, several results in several complex variables specifically in two variables,—where the fundamental reference is [4]—have a corresponding result in Bicomplex Holomorphic theory.

L. F. Reséndis O. (✉)
Universidad Autónoma Metropolitana - Unidad Azcapotzalco, Area de Análisis Matemático y sus Aplicaciones, Mexico City, Mexico
e-mail: lfro@azc.uam.mx

G. Y. Téllez S. · L. M. Tovar S.
Escuela Superior de Fisica y Matemáticas, Instituto Politécnico Nacional, Mexico City, Mexico
e-mail: tovar@esfm.ipn.mx

P. Cerejeiras et al. (eds.), *Current Trends in Analysis, its Applications and Computation*, Research Perspectives,
https://doi.org/10.1007/978-3-030-87502-2_20

Section 2, contains a brief compiled of the main tools of bicomplex holomorphic functions analysis that we applied in this work, for a complete development of this subject the featured references are [5–8].

Observe that the definition of bicomplex holomorphy evade the zero divisors. So the main results are presented in Sect. 3, where we give the relationships between bicomplex holomorphy with the directional derivatives over the idempotent planes **e** and $\mathbf{e}^{\dagger}$. Then we show that the Stolz condition for a complex holomorphic function is extended for the bicomplex case.

2 Some Basic Results on Bicomplex Holomorphic Theory

We present several common facts about bicomplex numbers and bicomplex holomorphic functions. We will be free to use results and notation of [6].

The set of bicomplex numbers $\mathbb{BC}$ is defined as

$$\mathbb{BC} := \{\, z_1 + \mathbf{j} z_2 \;:\; z_1,\ z_2 \in \mathbb{C}(i),\ \mathbf{j}^2 = -1\}.$$

Sum and product of bicomplex numbers are made in the expected way. We write all the bicomplex numbers $Z = z_1 + \mathbf{j} z_2$, with $z_l = x_l + i y_l \in \mathbb{C}(\mathbf{i})$, in theirs $\mathbb{C}(\mathbf{i})$-idempotent form, that is

$$Z = \beta_1\, \mathbf{e} + \beta_2 \mathbf{e}^{\dagger} \tag{2.1}$$

where

$$\beta_1 = z_1 - \mathbf{i} z_2 \qquad \text{and} \qquad \beta_2 = z_1 + \mathbf{i} z_2$$

and

$$\mathbf{e} := \frac{1 + \mathbf{ij}}{2} \qquad \text{and} \qquad \mathbf{e}^{\dagger} := \frac{1 - \mathbf{ij}}{2}.$$

Observe that $\mathbf{e}\,\mathbf{e}^{\dagger} = 0$; $1 = \mathbf{e} + \mathbf{e}^{\dagger}$ or more general $\lambda = \lambda(\,\mathbf{e} + \mathbf{e}^{\dagger})$ with $\lambda \in \mathbb{C}(\mathbf{i})$.

A bicomplex valued function $F : X \to \mathbb{BC}$ is completely determined by its component functions $F_1, F_2 : X \to \mathbb{C}(i)$ such that for every $x \in X$,

$$F(x) = F_1(x) + \mathbf{j} F_2(x)$$

Or we can consider the idempotent representation of F as

$$F(x) = G_1(x)\, \mathbf{e} + G_2(x) \mathbf{e}^{\dagger},$$

where the functions $G_1, G_2 : X \to \mathbb{BC}$ are the idempotent components and these are related with F_1 and F_2 by 2.1.

Let Ω be a domain in $\mathbb{BC}$ and $F : \Omega \to \mathbb{BC}$ a $\mathbb{BC}$-holomorphic function in Ω, that means that for every $Z_0 \in \Omega$ exists the limit

$$F'(Z_0) = \lim_{Z \to Z_0} \frac{F(Z) - F(Z_0)}{Z - Z_0}, \qquad Z \in \Omega$$

and such that $H = Z - Z_0$ is an invertible bicomplex number, that is

$$Z - Z_0 \notin \{\beta\,\mathbf{e} + 0\,\mathbf{e}^{\dagger}\} \cup \{0\,\mathbf{e} + \beta\,\mathbf{e}^{\dagger}\}, \qquad \beta \in \mathbb{C}(i).$$

If F is derivable for all $Z \in \Omega$ we say that is a bicomplex holomorphic function in Ω.

Like in the one complex variable case, the bicomplex holomorphic functions fulfill with a Cauchy-Riemann type system

$$\frac{\partial F_1}{\partial z_1} = \frac{\partial F_2}{\partial z_2} \quad \text{and} \quad \frac{\partial F_1}{\partial z_2} = -\frac{\partial F_2}{\partial z_1}. \tag{2.2}$$

As a consequence of this system we have the following essential result in the theory of bicomplex holomorphic functions [5].

Theorem 2.1 *Let $\Omega \subset \mathbb{BC}$ be a domain. A bicomplex function $F : \Omega \to \mathbb{BC}$ of class C^1 and idempotent decomposition*

$$F = G_1\,\mathbf{e} + G_2\,\mathbf{e}^{\dagger}$$

is $\mathbb{BC}$-holomorphic if and only if, the following two conditions hold:

(a) *The component G_1, seen as a $\mathbb{C}(\mathbf{i})$-valued function of the complex variables (β_1, β_2) is holomorphic; moreover does not depend on the variable β_2 and thus G_1 is a holomorphic function of the variable β_1.*
(b) *The component G_2, seen as a $\mathbb{C}(\mathbf{i})$-valued function of the complex variables (β_1, β_2) is holomorhic; moreover does not depend on the variable β_1 and thus G_2 is a holomorphic function of the variable β_2.*

Therefore applying this result, the rules of derivability are the same like usual ones and the derivatives of higher order are given by

$$F^{(n)}(Z) = G_1^{(n)}(\beta_1)\,\mathbf{e} + G_2^{(n)}(\beta_2)\mathbf{e}^{\dagger}, \qquad n = 0,\ 1,\ 2\ldots \tag{2.3}$$

Observe that as a straightforward result of this idempotent representation, we get the following inverse mapping theorem.

Theorem 2.2 *Let Ω be a domain in $\mathbb{BC}$ and $F = F_1 + \mathbf{j}F_2 : \Omega \to \mathbb{BC}$ a $\mathbb{BC}$-holomorphic function in Ω. Then if $F'(Z_0)$ is not a zero divisor, then the function*

F is locally invertible and the inverse function F^{-1} is $\mathbb{BC}$-holomorphic at $W_0 = F(Z_0)$ with

$$(F^{-1})'(W_0) = \frac{1}{F'(Z_0)}.$$

Proof If we consider the idempotent decomposition

$$F(Z) = G_1(\beta_1)\,\mathbf{e} + G_2(\beta)\mathbf{e}^{\dagger}$$

As $G_1'(\beta_1^0) \neq 0 \neq G_2'(\beta_2^0)$, then G_1 and G_2 are locally invertibles. Thus F results locally invertible in a neighboorhood of Z_0. If $W_0 = F(Z_0)$ then

$$F'(W_0) = \frac{1}{G_1'(\beta_1^0)}\,\mathbf{e} + \frac{1}{G_2'(\beta^0)}\mathbf{e}^{\dagger} = \frac{1}{F'(Z_0)}.$$

□

Suppose that $\Omega_1,\ \Omega_2 \subset \mathbb{C}(\mathbf{i})$ are domains of holomorphy for $G_1,\ G_2$ respectively, that is, G_1 and G_2 can not be holomorphically extended to any bigger open set in $\mathbb{C}(\mathbf{i})$. Then the bicomplex function $F = G_1\,\mathbf{e}{+}G_2\,\mathbf{e}^{\dagger}$ defined on $\Omega = \Omega_1\,\mathbf{e}{+}\Omega_2\,\mathbf{e}^{\dagger}$ can not be $\mathbb{BC}$-holomorphically extended to any open set containing Ω, thus Ω is a domain of bicomplex holomorphy for F.This fact represent an important difference between the concept of Domains of Holomorphy in Several Complex Variables and the Bicomplex Holomorphic Theory: Domains of holomorphy in Bicomplex Theory are just product domains. While, for instance, the euclidian ball in C^2 is a domain of holomorphy in Two Complex Variables Theory but not in Bicomplex Holomorphic Theory.

3 The Stolz Condition

Observe that $\{\beta\,\mathbf{e} + 0\,\mathbf{e}^{\dagger}\}$ and $\{0\,\mathbf{e} + \beta\,\mathbf{e}^{\dagger}\}$ are fields isomorphic to $\mathbb{C}$, where the inverse of $\beta\,\mathbf{e}$ and $\beta\,\mathbf{e}^{\dagger}$ are $\beta^{-1}\,\mathbf{e}$ and $\beta^{-1}\,\mathbf{e}^{\dagger}$, respectively, for $\beta \neq 0$.

Let $Z^0 = \beta_1^0\,\mathbf{e}+\beta_2^0\,\mathbf{e}^{\dagger},\ Z = \beta_1^0\,\mathbf{e}+\beta_2\,\mathbf{e}^{\dagger} \in \Omega$ and $\Delta Z = Z - Z_0 = 0\,\mathbf{e}+(\beta_2 - \beta_2^0)\,\mathbf{e}^{\dagger}$. Note that ΔZ is an increment in the $\mathbf{e}^{\dagger}$ direction. If F is $\mathbb{BC}$-holomorphic in Z^0, then

$$\begin{aligned} F(Z) &= G_1(\beta_1^0)\,\mathbf{e} + G_2(\beta_2)\,\mathbf{e}^{\dagger} \\ F(Z^0) &= G_1(\beta_1^0)\,\mathbf{e} + G_2(\beta_2^0)\,\mathbf{e}^{\dagger} \\ F'(Z^0) &= G_1'(\beta_1^0)\,\mathbf{e} + G_2'(\beta_2^0)\,\mathbf{e}^{\dagger}. \end{aligned}$$

As $G_2(\beta_2)$ is a holomorphic function in β_2,and if $\Delta Z \to 0$, then $\beta_2 \to \beta_2^0$. By Stolz condition in the complex case

$$\begin{aligned}\Delta F(Z) = F(Z) - F(Z^0) &= G_1(\beta_1^0)\,\mathbf{e} + G_2(\beta_2)\,\mathbf{e}^\dagger - (G_1(\beta_1^0)\,\mathbf{e} + G_2(\beta_2^0)\,\mathbf{e}^\dagger)\\ &= 0\,\mathbf{e} + (G_2(\beta_2) - G_2(\beta_2^0))\,\mathbf{e}^\dagger\\ &= 0\,\mathbf{e} + (G_2'(\beta_2^0)(\beta_2 - \beta_2^0) + \alpha_2(\beta_2, \beta_2^0)(\beta_2 - \beta_2^0)\,)\mathbf{e}^\dagger\\ &= 0\,\mathbf{e} + (G_2'(\beta_2^0) + \alpha_2(\beta_2, \beta_2^0))(\beta_2 - \beta_2^0)\,\mathbf{e}^\dagger\end{aligned}$$

where

$$\lim_{\beta_2 \to \beta_2^0} \alpha_2(\beta_2, \beta_2^0) = 0$$

The complex number $\beta_2 - \beta_2^0 \in \mathbb{C}(\mathbf{i})$ is invertible in $\mathbb{C}(\mathbf{i})$.If we multiply the previous expression by the bicomplex number $0\,\mathbf{e} + (\beta_2 - \beta_2^0)^{-1}\,\mathbf{e}^\dagger$, thus

$$0\,\mathbf{e} + (G_2(\beta_2) - G_2(\beta_2^0))(\beta_2 - \beta_2^0)^{-1}\,\mathbf{e}^\dagger = 0\,\mathbf{e} + (G_2'(\beta_2^0) + \alpha_2(\beta_2, \beta_2^0))\,\mathbf{e}^\dagger$$

As G_2 is holomorphic at β_2^0 , if $\beta_2 \to \beta_2^0$ we get

$$0\,\mathbf{e} + G_2'(\beta_2^0)\,\mathbf{e}^\dagger = 0\,\mathbf{e} + (G_2'(\beta_2^0) + \lim_{\beta_2 \to \beta_2^0} \alpha_2(\beta_2, \beta_2^0))\,\mathbf{e}^\dagger.$$

Thus, by Stolz criterium for the one complex case, there exists the directional derivative $F'_{\mathbf{e}^\dagger}(Z^0)$ and

$$F'_{\mathbf{e}^\dagger}(Z^0) = 0\,\mathbf{e} + G_2'(\beta_2^0)\,\mathbf{e}^\dagger.$$

In the same way, we obtain

$$F'_{\mathbf{e}}(Z^0) = G_1'(\beta_1^0)\,\mathbf{e} + 0\,\mathbf{e}^\dagger.$$

Thus we have obtained the next result

Theorem 3.1 *Let $F(Z) = G_1(\beta_1)\,\mathbf{e} + G_2(\beta_2)\,\mathbf{e}^\dagger$ be a $\mathbb{BC}$-holomorphic function in a domain $\Omega \subset \mathbb{BC}$. If $Z^0 = \beta_1^0\,\mathbf{e} + \beta_2^0\,\mathbf{e}^\dagger$ then*

- *The directional derivatives $F'_{\mathbf{e}}(Z^0)$, $F'_{\mathbf{e}^\dagger}(Z^0)$ exist and*

$$\begin{aligned}F'_{\mathbf{e}}(Z^0) &= G_1'(\beta_1^0)\,\mathbf{e} + 0\,\mathbf{e}^\dagger.\\ F'_{\mathbf{e}^\dagger}(Z^0) &= 0\,\mathbf{e} + G_2'(\beta_2^0)\,\mathbf{e}^\dagger\end{aligned}$$

•

$$F'(Z^0) = F'_{\mathbf{e}}(Z^0)\,\mathbf{e} + F'_{\mathbf{e}^\dagger}(Z^0)\,\mathbf{e}^\dagger.$$

• *The tangent space at the point* $(Z^0, F(Z^0))$ *is given by*

$$W = F(Z^0) + F'(Z^0)(Z - Z^0).$$

As Corollary of the previous proof and Theorem 7.6.3 in [5], we obtain the following result.

Corollary 3.2 *Let* $\Omega \subset \mathbb{BC}$ *be a domain and let* $F : \Omega \to \mathbb{BC}$ *be a bicomplex function given by* $F(Z) = G_1(\beta_1)\,\mathbf{e} + G_2(\beta_2)\,\mathbf{e}^\dagger$. *If the directional derivatives* $F'_{\mathbf{e}}(Z^0)$, $F'_{\mathbf{e}^\dagger}(Z^0)$ *exist for all* $Z^0 \in \Omega$ *then* F *is a bicomplex holomorphic function with derivative* $F'(Z^0) = F'_{\mathbf{e}}(Z^0)\,\mathbf{e} + F'_{\mathbf{e}^\dagger}(Z^0)\,\mathbf{e}^\dagger$.

Theorem 3.3 *Let* $\Omega \subset \mathbb{BC}$ *be a domain and let* $F : \Omega \to \mathbb{BC}$ *be a bicomplex function. Then* F *is a bicomplex holomorphic function on* Ω *if and only if,* F *is* $\mathbb{BC}$ *Stolz differentiable on* Ω*, that is*

$$F(Z + H) = F(Z) + A_Z H + \alpha(H)H, \qquad \text{with } A_Z \in \mathbb{BC} \quad \text{and} \quad \lim_{H\to 0} \alpha(H) = 0. \tag{3.1}$$

Proof If F satisfies the Stolz condition (3.1) and if H is a non zero divisor, we have for $Z \in \Omega$

$$F'(Z) = \lim_{H\to 0} \frac{F(Z + H) - F(Z)}{H} = \lim_{H\to 0} (A_Z + \alpha(H)) = A_Z.$$

Reciprocally, if $F = G_1\,\mathbf{e} + G_2\,\mathbf{e}^\dagger$ is a $\mathbb{BC}$ holomorphic function at $Z^0 \in \Omega$, $Z^0 = \beta_1^0\,\mathbf{e} + \beta_2^0\,\mathbf{e}^\dagger$, by the Stolz condition for G_1, G_2 we have

$$\begin{aligned}\Delta F(Z) &= F(Z) - F(Z^0)\\ &= (G_1(\beta_1) - G_1(\beta_1^0))\,\mathbf{e} + (G_2(\beta_2) - (G_2(\beta_2^0))\,\mathbf{e}^\dagger\\ &= (G_1'(\beta_1^0)(\beta_1 - \beta_1^0) + \alpha_1(\beta_1, \beta_1^0)(\beta_1 - \beta_1^0))\,\mathbf{e}\\ &\quad + (\,G_2'(\beta_2^0)(\beta_2 - \beta_2^0) + \alpha_2(\beta_2, \beta_2^0)(\beta_2 - \beta_2^0))\mathbf{e}^\dagger\end{aligned}$$

with

$$\lim_{\beta_1\to\beta_1^0} \alpha_1(\beta_1, \beta_1^0) = 0 \quad \text{and} \quad \lim_{\beta_2\to\beta_2^0} \alpha_2(\beta_2, \beta_2^0) = 0.$$

Define

$$\alpha(Z, Z^0) = \alpha_1(\beta_1, \beta_1^0)\,\mathbf{e} + \alpha_2(\beta_2, \beta_2^0)\,\mathbf{e}^\dagger$$

then it follows $\lim_{Z \to Z^0} \alpha(Z, Z^0) = 0$. □

Acknowledgments This research was partially supported by Consejo Nacional de Ciencia y Tecnología (CONACYT) and by Instituto Politécnico Nacional in the framework of SIP programs and with Universidad Autónoma Metropolitana.

References

1. J. Cockle, On certain functions resembling quaternions and on a new imaginary in algebra. London-Dublin-Edinburgh Philos. Mag. Sci. J. **32**(3), 435–439 (1848)
2. R.A. Silverman, *Introductory Complex Analysis* (Dover Publications, New York, 1972)
3. J.B. Conway, *Functions of One Complex Variable* (Springer, New York, 1978)
4. H. Poincaré, Sur les fonctions de deux variables. Acta Math. **2**, 97–113 (1883)
5. D. Alpay, M.E. Luna-Elizarrarás, M. Shapiro, D.C. Struppa, *Basics on Functional Analysis with Bicomplex Scalars, and Bicomplex Schur Analysis* (Springer International Publishing, Berlin, 2014)
6. M.E. Luna-Elizarrarás, M. Shapiro, D.C. Struppa, A. Vajiac, *Bicomplex Holomorphic Functions: The Algebra, Geometry and Analysis* (Birkhäuser, Basel, 2015)
7. F. Colombo, I. Sabadini, D. Struppa, Bicomplex holomorphic functional calculus. Math. Nachrichten. **257**(10), 1093–1105 (2013)
8. K.S. Charak, D. Rochon, N. Sharma, Normal families of bicomplex meromorphic functions. Ann. Polonici Math. **103**(3), 303–317 (2012)

Part V
Constructive Methods in the Theory of Composite and Porous Media

Fast Method for 2D Dirichlet Problem for Circular Multiply Connected Domains

Olaf Bar and Krzysztof Wójcik

Abstract This paper is devoted the optimization of application to determine the flux around closely spaced nonoverlapping disks on the conductive plane. This method is based on successive approximations applied to the functional equations. This paper is concerned on influence of checking diagrams on convergence fast Poincaré series method. This can be used to solve Laplace's equation on a conductive plane with nonoverlapping inclusions. The initial stream is composed from set of two-point functions which are dependent on the graph which represents connection between nearest neighbours circles.

Keywords Fast Poincaré series · Laplace's equation · Voronoi diagram · Delaunay triangulation

Mathematics Subject Classification (2010) Primary 99Z99; Secondary 00A00

1 Introduction

The steady state heat distribution or electrical flux on the plane can be modelled by two dimensional Laplace's equation. Such a model can describe the phenomenon of the flow perpendicular to the fibres embedded in matrix [1, 2].

O. Bar (✉)
Institute of Computer Science and Computational Methods, Pedagogical University of Kraków, Cracow, Poland
e-mail: olaf.bar@up.krakow.pl

K. Wójcik
Production Engineering Institute, Faculty of Mechanical Engineering, Cracow University of Technology, Cracow, Poland
e-mail: krzysztof.wojcik@mech.pk.edu.pl

P. Cerejeiras et al. (eds.), *Current Trends in Analysis, its Applications and Computation*, Research Perspectives,
https://doi.org/10.1007/978-3-030-87502-2_21

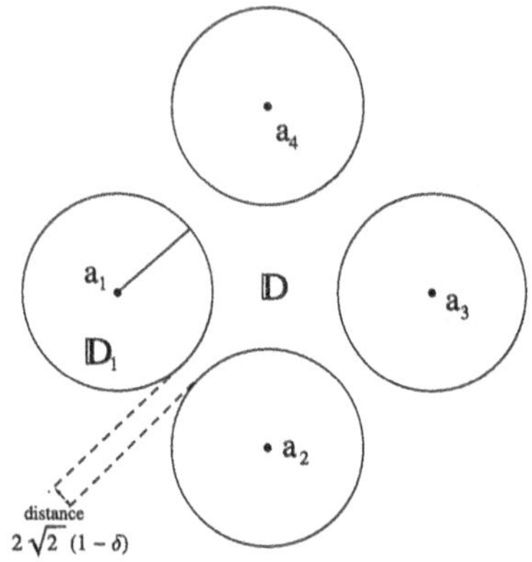

Fig. 1 Four disks and the multiply connected region $\mathbb{D}$

In many case we can compute the satisfied solution by the standard numerical methods. One of most recently used method is finite element method FEM. Numerical solutions are accurate but provides solutions of the local fields.

One can find papers devoted to the issue of boundary conditions with a large number of inclusions. They usually relate to cases when the distances between inclusions are relatively large [3, 4]

On the other hand the specific geometry of boundary conditions causes that the numerical algorithm sometimes deliver poor results.

This paper is devoted to the analytical method of solution Laplace's equation with Dirichlet boundary conditions. The geometry of the system shown in Fig. 1.

It is convenient to go to the complex variables $z = x + iy$. Then introduce proper definitions:

Definition 1.1

$\mathbb{D}_k = \{z \in \mathbb{C} : |z - a_k| < r_k\}(k = 1, 2, \ldots)$ nonoverlapping open disks, $\partial\mathbb{D}_k\{z \in \mathbb{C} : |z - a_k| = r_k\}(k = 1, 2, \ldots)$,

$\mathbb{D}$ complement of the closest disks to the extended complex plane $\mathbb{C} \cup \{\infty\}$

Physical problem is defined as follows: we are looking the flux ∇u which is a gradient of potential $u(x, y)$. Function $u(x, y)$ is defined on plane $\mathbb{D}$ and satisfies Laplace's equation $\Delta u = 0$ with Dirichlet boundary conditions on the closure $\partial\mathbb{D}$.

In this paper we restrict the Dirichlet boundary condition to the constants defined on the $\partial\mathbb{D}_k$:

$$u(t) = u_k, \quad |t - a_k| = r_k, k = 1, 2, \ldots, 4. \tag{1.1}$$

This problem can be reduced to the Riemann–Hilbert problem [5] by introduce the complex potential function $\psi(z) = u_x - iu_y$. Then we can rewrite the boundary condition (1.1) in the form:

$$\operatorname{Im} \psi(t) := \operatorname{Im}\left[\frac{t - a_k}{r_k}\psi(t)\right] = 0, \quad |t - a_k| = r_k, \ k = 1 \ldots, 4, \tag{1.2}$$

The function $\varphi(z)$ which is a primitive integral of $\psi(z)$ satisfies the equation (1.3)

$$Re\ \varphi(t) = u_k, \quad |t - a_k| = r_k,\ k = 1, 2, \ldots, \mathcal{N}, \tag{1.3}$$

where the function $\varphi(t)$ is analytic in $\mathbb{D}$.

2 Description of the Fast Poincaré Series Method

The fast series method is based on algorithm published in [6]

Introduce the inersion with respect to the k-circle:

$$z^*_{(k)} = \frac{r_k^2}{z - a_k} + a_k \tag{2.1}$$

As we know, the analytical solution for the flux between two circles is known [7]:

$$\Psi(z) = \frac{1}{z - z_{12}} - \frac{1}{z - z_{21}} \tag{2.2}$$

where $\Psi(z) = \varphi'(z)$.

The function $\Psi(z)$ describes the flux for the known difference $u_1 - u_2$ and the z_{12}, z_{21} satisfy the quadratic equation $z^*_{(1)} = z^*_{(2)}$. This function can be used as the zero-th approximation for the fast algorithm.

Define the analytic function [6]:

$$f_{km}(z) := \begin{cases} 0, & k = m, \\ \sum_{\ell \in J_m; \ell \neq k} \Psi(z; m, \ell)(z), & k \in J_m, \\ \sum_{\ell \in J_m} \Psi(z; m, \ell)(z), & k \in J^*_m, \end{cases} \tag{2.3}$$

where J^*_m is the complement of $J_m \cup \{m\}$ to $\{1, 2, \ldots, n\}$.

The following algorithm can be applied. First, we compute auxiliary functions $\psi_k(z)$ by the following iterations:

$$\psi_k^{(0)}(z) = f_{km}(z), \tag{2.4}$$

$$\psi_k^{(p)}(z) = \sum_{m \neq k} \left(\frac{r_m}{z - a_m}\right)^2 \overline{\psi_m^{(p-1)}\left(z^*_{(m)}\right)} + f_{km}(z),\ p = 1, 2, \ldots. \tag{2.5}$$

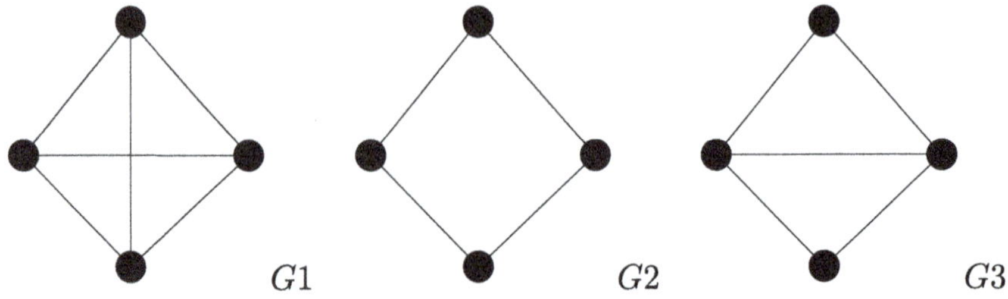

Fig. 2 Diagrams for settings J_m and J_m^* files

The p-th approximation of the complex flux $\psi(z) = \varphi'(z)$ is calculated by formula

$$\psi^{(p)}(z) = \sum_{m=1}^{n} \left(\frac{r_m}{z - a_m} \right)^2 \overline{\psi_m^{(p)}\left(z_{(m)}^*\right)} + \psi_\delta(z), \quad z \in \mathbb{D}, \tag{2.6}$$

where $\psi_\delta(z) = \sum_{\ell \in J_m} \Psi(z; m, \ell)$.

The potential $\varphi(z)$ is obtained by integration of $\psi(z)$.

The main goal of this paper is studying the effect the selection of set J_m on convergence speed of algorithm (2.6). The crucial point is that to determine which circles are closest neighbours. The well known method of solutions this problem is Delaunay triangulation. Disadvantage of this approach is that the Delaunay triangulation takes into account only the centres of the circles. In our case the most important parameter is the gap between closest circles (see the δ in the Fig. 1).

In this article three types of diagram ($G1$, $G2$, $G3$) were taken into account (see Fig. 2). The first of them represents the nonplanar graph where interaction between each of vertex with all other were allowed. The second graph were constructed only from the least gaps. Third is the classic Delaunay triangulation (Figs. 3, 4 and 5).

Calculations were made for the following geometry properties:

[-1,$-i$, 1.1, 1.2 i]—the vector of the circle centres (see Fig. 1)

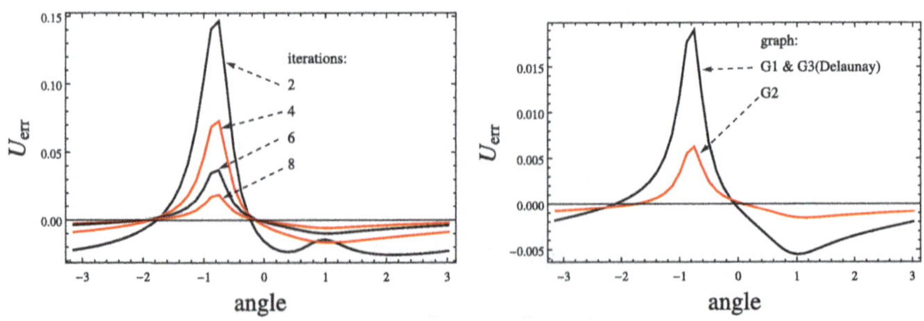

Fig. 3 Dependence of U_{err} on number of iterations(left) and on graph (right) for parameters: $\delta = 0.98$, $b = 1$, $k = 1$

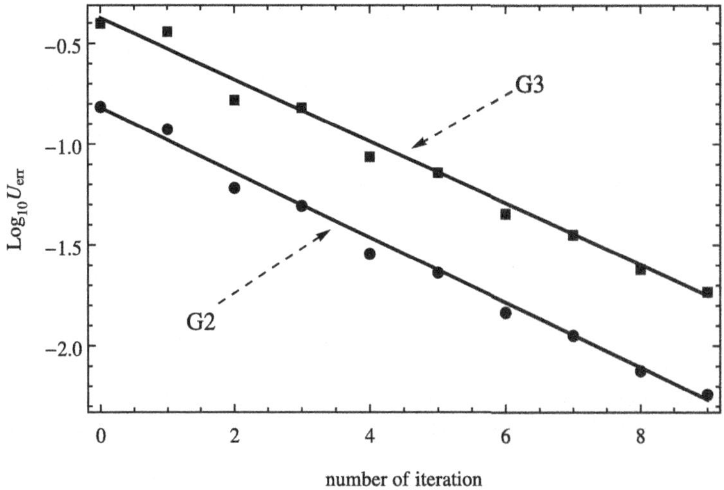

Fig. 4 Dependence of U_{err} on number of iterations for two graph G2 and G3. Parameters: $\delta = 0.98$, $b = 1$, $k = 1$

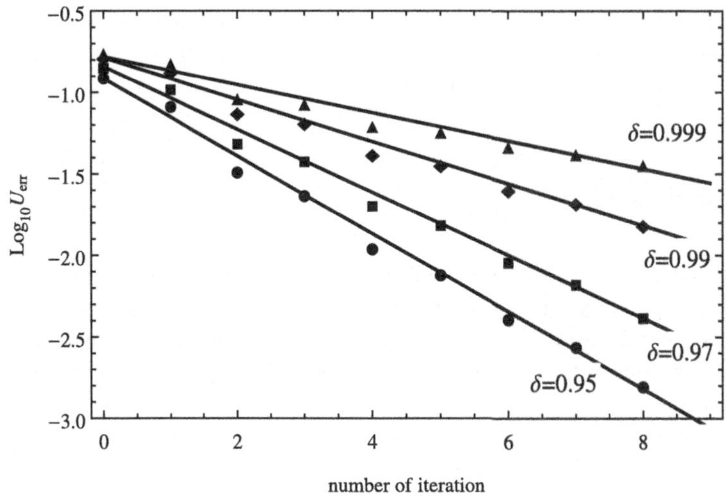

Fig. 5 Dependence of U_{err} on number of iterations for different δ. Parameters: graph $G2$, $b = 1$, $k = 1$

$r = \frac{\sqrt{2}}{2}\delta$—the radius (all of the circles has the same r)

The limit case $\delta = 1$ yields tangent circles $\mathbb{D}_1$ and $\mathbb{D}_2$.

As is known [8] there is three ($n-1$ in general case) linear independent functions $\psi^{(p)}(z)$ (2.6). These functions form three-dimensional basis. In previous formulas basis index were omitted for transparency.

Table 1 Calculation for $\delta = 0.98$

Iterations	U_{err} $G2$	U_{err} $G3$
0	0.159	0.413
1	0.123	0.376
2	0.063	0.172
3	0.0513	0.158
4	0.0297	0.090
5	0.0239	0.075
6	0.0152	0.047
7	0.0117	0.037
8	0.00781	0.025

Thus, the procedure of error computation is defined as follows:

$U_{err}^{b,k} = max(\mathrm{Re}\,\varphi_b(t)) - min(\mathrm{Re}\,\varphi_b(t))$ where $t \in \partial\mathbb{D}_k$ and b denotes the basis number.

Then as the total error U_{err} assumed:

$$U_{err} = Max\{U_{err}^{b,k} \quad b = 1\ldots 3, k = 1\ldots 4\} \tag{2.7}$$

The potential $\varphi_b(z)$ was obtained by analytical integration of the flux $\psi(z)$.

During the computation it turn out that the maximum error exists on the circles 1 and 2 for basis number 1 and 2 (Table 1).

3 Conclusions

The procedure to obtain the potential from the flux requires integration of the long algebraic expressions. The analytical integration procedure requires a lot of time and it is possible at most for the 8–9 iterations. Accordingly, optimal way to set the starting point functions for fast series method is the most important point of computation.

The results presented in this paper show that the best choice to construct the J_m files is diagram $G2$. In this case the files J_m are equal:

$$J_1 = \{2, 4\} \quad J_1^* = \{3\} \quad , J_2 = \{1, 3\} \quad J_2^* = \{4\} \quad ,$$

$$J_3 = \{2, 4\} \quad J_3^* = \{1\} \quad , J_4 = \{1, 3\} \quad J_4^* = \{2\}$$

The proper choice of diagram decreases the boundary condition error about three times in comparison with the other.

References

1. G.W. Milton, *The Theory of Composites* (Cambridge University Press, Cambridge, 2002)
2. S. Gluzman, V. Mityushev, W. Nawalaniec, G. Sokal, Random composite: stirred or shaken?, Archives Mech. **68**(3), 229–241 (2016)
3. W. Baran, K. Kurnik, W. Nawalaniec, A. Shareif, Local stationary heat fields in fibrous composites. Silesian J. Pure Appl. Math. **9**(1), 1–8 (2019)
4. N. Rylko, Edge effects for heat flux in fibrous composites. Appl. Math. Comput. **70**, 2283–2291 (2015)
5. V. Mityushev, S. Rogosin, Constructive methods to linear and non-linear boundary value problems of the analytic function, in *Theory and Applications*. Pure and Applied Mathematics (Chapman & Hall/CRC, Boca Raton, 2000)
6. V. Mityushev, N. Rylko, A fast algorithm for computing the flux around non-overlapping disks on the plane. Math. Comput. Modell. **57**, 1350–1359 (2013)
7. L.D. Landau, E.M. Lifshitz, *Electrodynamics of Continuous Media*. Course of Theoretical Physics, vol. 8 (Pergamon Press, Oxford, 1960)
8. V. Mityushev, Riemann-Hilbert problems for multiply connected domains and circular slit maps. Comput. Methods Funct. Theory **11**, 575–590 (2011)

Local Stationary Heat Fields in Fibrous Composites

Wojciech Baran, Krystian Kurnik, and Shareif Albasher

Abstract This work covers an analysis of the complex flux along a domain surface D. The flux function is analytic in D surface and determined by different boundary values. The boundary components of the surface are determined by a method of functional equations. The flux distribution behaviours for different conditions were investigated. This work is a follow up for the paper: local stationary heat fields in fibrous composites.

Keywords Complex potential · Fibrous composite · Heat flux · Functional equation · Schwarz problem

Mathematics Subject Classification (2010) 74A40

1 Introduction

The complex flux along a domain D filled by non-overlapping discs depend on many factors that affect the flux flow and therefore the distribution of temperature. The goal of this work is to estimate the complex flux. In the present paper, we applied analytical approximate formula of complex potentials [2] to describe the flux in two-

---

Sincere words of gratitude are extended to Professor Vladimir Mityushev for his valuable advice.

W. Baran (✉) · K. Kurnik
Institute of Computer Science, Pedagogical University of Krakow, Materialica+ Research Group, Krakow, Poland
e-mail: wojciech.baran@up.krakow.pl; krystian.kurnik@up.krakow.pl

S. Albasher
Institute of Mathematics, Pedagogical University of Krakow, Materialica+ Research Group, Krakow, Poland
e-mail: shareif.albasher@up.krakow.pl

P. Cerejeiras et al. (eds.), *Current Trends in Analysis, its Applications and Computation*, Research Perspectives,
https://doi.org/10.1007/978-3-030-87502-2_22

dimensional composites. As any function in a connected surface, the distribution of the flow could be expressed as the real part of the flux function in the domain [3, 4], the distribution is expressed through the real part of the complex potential $u(z) = \text{Re}\,\varphi(z)$

2 Methodology

Let $z = x_1 + ix_2$ denote a complex variable in the extended complex plane $\widehat{\mathbb{C}} = \mathbb{C} \cup \{\infty\}$. Consider non-overlapping disks $|z - a_k| < r$ $(k = 1, 2, \ldots, n)$, and the domain D, the complement of all the disks $|z - a_k| \leq r$ to $\widehat{\mathbb{C}}$. The potentials $u(z)$ is harmonic in D except at infinity where $u(z) \sim x_1 = \text{Re}\, z$ and continuously differentiable in the closures of the considered domain. The singularity of $u(z)$ determine the external flux applied at infinity.

The distribution of temperature $u(z) \equiv u(x_1, x_2)$ is expressed through the real part of the complex potential [3, 4]

$$u(z) = \text{Re}\,\varphi(z), \quad z \in D, \tag{2.1}$$

where $\varphi(z) = u(z) + iv(z)$ is analytic in D except at infinity where $\varphi(z) \sim z$, and continuously differentiable in the closures of the considered domain. For definiteness, we assume that the disks $|z - a_k| < r$ $(k = 1, 2, \ldots, n)$ are filled by a conductor with non-vanishing conductivity. It is worth noting that in this case the function $\varphi(z)$ is single-valued in the multiply connected domain D and does not contain logarithmic terms [4]. It follows from the fact that the divergence of the normal flux through every boundary component vanishes

$$\int_{|z-a_k|=r} \frac{\partial u}{\partial \mathbf{n}}(z)\, ds = \int_{|z-a_k|=r} \frac{\partial v}{\partial \mathbf{s}}(z)\, ds = [v]_{|z-a_k|=r} = 0, \; k = 1, 2, \ldots, n. \tag{2.2}$$

Here, $\frac{\partial}{\partial \mathbf{n}}$ denotes the outward unit normal derivative $\frac{\partial}{\partial \mathbf{s}}$ the tangent derivative to $|z - a_k| = r$, respectively, $[v]_{|z-a_k|=r}$ the increment of the function $v(z)$ along the circle $|z - a_k| = r$.

The gradient of $u(z)$ is related to the heat flux [3] and can be calculated by formula

$$\psi(z) = \frac{\partial u}{\partial x} - i\frac{\partial u}{\partial y}, \tag{2.3}$$

where $\psi(z) = \varphi'(z)$ in the closure of D.

The perfect contact condition between the components is expressed by two real relations

$$u_k(z) = u(z), \quad \lambda \frac{\partial u_k}{\partial \mathbf{n}}(z) = \frac{\partial u}{\partial \mathbf{n}}(z), \quad |z - a_k| = r \ (k = 1, 2, \ldots, n), \tag{2.4}$$

where the conductivity of matrix is normalized to unity and the conductivity of inclusions is equal to λ. Introduce the contrast parameter

$$\varrho = \frac{\lambda - 1}{\lambda + 1}. \tag{2.5}$$

Two real equations (2.4) are reduced to the $\mathbb{R}$-linear complex condition [3]

$$\varphi(z) = \varphi_k(z) - \varrho\, \overline{\varphi_k(z)}, \quad |z - a_k| = r \ (k = 1, 2, \ldots, n) \tag{2.6}$$

where $\varphi_k(z)$ are analytic in $|z - a_k| < r$, respectively, and continuously differentiable in the closures of the considered disks. The harmonic and analytic functions are related by the equalities

$$u_k(z) = \frac{2}{\lambda + 1} \operatorname{Re} \varphi_k(z), \quad |z - a_k| \le r. \tag{2.7}$$

Consider Schottky group of inversions and their compositions with respect to the circles $|z - a_k| = r$, $k = 1, 2, \ldots, n$ (plus the identity element)

$$z^*_{(k)} = \frac{r^2}{\overline{z - a_k}} + a_k, \quad z^*_{(k_1 k_2 \ldots, k_m)} := (z^*_{(k_2 \ldots, k_{m-1})})^*_{k_1}, \quad (k_{j+1} \ne k_j). \tag{2.8}$$

Exact solution of the considered problem for any $|\varrho| < 1$ was found in the form of the absolutely and uniformly convergent Poincaré type series [3, formula (2.3.100)] up to an additive constant

$$\varphi_k(z) = z + \varrho \sum_{k_1 \ne k} \overline{z^*_{(k_1)}} + \varrho^2 \sum_{k_1 \ne k} \sum_{k_2 \ne k_1} z^*_{(k_2 k_1)} + \varrho^3 \sum_{k_1 \ne k} \sum_{k_2 \ne k_1} \sum_{k_3 \ne k_2} \overline{z^*_{(k_3 k_2 k_1)}} + \cdots \tag{2.9}$$

and

$$\varphi(z) = z + \varrho \sum_{k=1}^{n} \overline{z^*_{(k)}} + \varrho^2 \sum_{k=1}^{n} \sum_{k_1 \ne k} z^*_{(k_1 k)} + \varrho^3 \sum_{k=1}^{n} \sum_{k_1 \ne k} \sum_{k_2 \ne k_1} \overline{z^*_{(k_2 k_1 k)}} + \cdots. \tag{2.10}$$

It is worth noting that our formulae (2.9)–(2.10) have simpler form than [3, formula (2.3.100)] because the restriction $|\varrho| < 1$ is supposed. The functions $\varphi_k(z)$ and $\varphi(z)$ in the limit case $\varrho = 1$ have more complicated structure, since it is represented by a uniformly and not necessary absolutely convergent Poincaré type series [3].

Below, we consider this, the most difficult in computations, case. The relation $\varrho = 1$ means that inclusions are filled by a perfectly conducting materials when λ tends to infinity. Formally, the $\mathbb{R}$-linear problem (2.6) does not hold in this case and has to be written as the modified Dirichlet problem [3]

$$u_k(z) = c_k, \quad |z - a_k| = r \ (k = 1, 2, \ldots, n), \tag{2.11}$$

where c_k are undetermined constants.

In the present paper, we do not consider the boundary value problem (2.11). We use the limit $\varrho \to 1$ in the final formulae for the local flux justified by uniform convergence of the corresponding series for $\varrho \leq 1$

Therefore, we may differentiate the corresponding uniformly convergent Poincaré type series (2.9)–(2.10) term by term and arrive at the uniformly convergent series

$$\psi_k(z) = f(z) + \varrho \sum_{k_1 \neq k} \frac{d}{dz}\overline{z^*_{(k_1)}} + \varrho^2 \sum_{k_1 \neq k} \sum_{k_2 \neq k_1} \frac{d}{dz}\overline{z^*_{(k_2 k_1)}} \tag{2.12}$$

$$+\varrho^3 \sum_{k_1 \neq k} \sum_{k_2 \neq k_1} \sum_{k_3 \neq k_2} \frac{d}{dz}\overline{z^*_{(k_3 k_2 k_1)}} + \cdots, \quad z \in D_k \quad (k = 1, 2, \ldots, n)$$

and

$$\psi(z) = f(z) + \varrho \sum_{k=1}^{n} \frac{d}{dz}\overline{z^*_{(k)}} + \varrho^2 \sum_{k=1}^{n} \sum_{k_1 \neq k} \frac{d}{dz}\overline{z^*_{(k_1 k)}} \tag{2.13}$$

$$+\varrho^3 \sum_{k=1}^{n} \sum_{k_1 \neq k} \sum_{k_2 \neq k_1} \frac{d}{dz}\overline{z^*_{(k_2 k_1 k)}} + \cdots, \quad z \in D.$$

It is worth noting that formula (2.13) is universal and takes place for $\varrho \leq 1$.

3 Numerical Examples

We can now compare a number of flux behaviours which shows the distribution of the flux in different conditions.

As any function in a connected surface, the distribution of the flow could be expressed as the real part of the flux function in the domain [3, 4], the distribution is expressed through the real part of the complex potential $u(z) = \text{Re}\,\varphi(z)$.

In the first cases we considered the function $f(z) = 1$ and applied that on a surface contains.

140 non-overlapped discs firstly with 0.828402 concentration and radiuses of 0.027448063 for all the discs, and secondly with reducing disc radius and concentration to 0.00178412 and 0.2 respectively.

The computation was executed and showed a linear flux in both cases. The results represented the complex flux in a standard way of a heat flowing and leaving through the surface.

With changing the function to an exponential function and applying that $f(z) = z^2$, the flux flows in a continuous whirlpool towards the centre. We investigated a three cases of a surface with also 140 discs (firstly with a concentration of 0.828402 and radiuses: 0.027448063 for all discs. Secondly with a concentration of 0.3 and radiuses: 0.021851, and the third with reducing the concentration to 0.1 and the radiuses: 0.0126157.

In these cases, a big amount of the heat flux flows inside the surface instead of leaving the surface. And because we used the approximate functional equations for calculating the flux, then in reality, the flow leaving through the surface should be more than the shown whirlpool limits as well as more intertwined in the centre.

Even in the cases of using a various properties and keeping the exponential function is maintained, the bulk of the heat flow remains in vortex inward, as some heat flow out of the surface, and accordingly, the models of the fibrous composites are designed to reflect the heat flow functionality which serves the required technology.

Finally, with applying an inverse functionality $f(z) = 1/z + 07$ (0.0126157 radius, 0.1 concentration), and $f(z) = 1/z + 0.7$ (0.021851 radius, 0.3 concentration) and also with (radius of 0.027448063 and concentration: 0.828402). All the cases were applied on the surface of 100 symmetric discs.

Although, the flux distribution is not perfectly anisotropic due to the distribution of the random discs and the boundary conditions, but when using an inverse values, the thermal behaviour will still reverses the flow towards the centre of the discs. However, choosing the linear or non-linear functions of $f(z)$ strongly affect the flux distribution and it is a main factor in heat flow technology of these composites (Figs. 1, 2, 3, 4, 5, 6, 7, and 8).

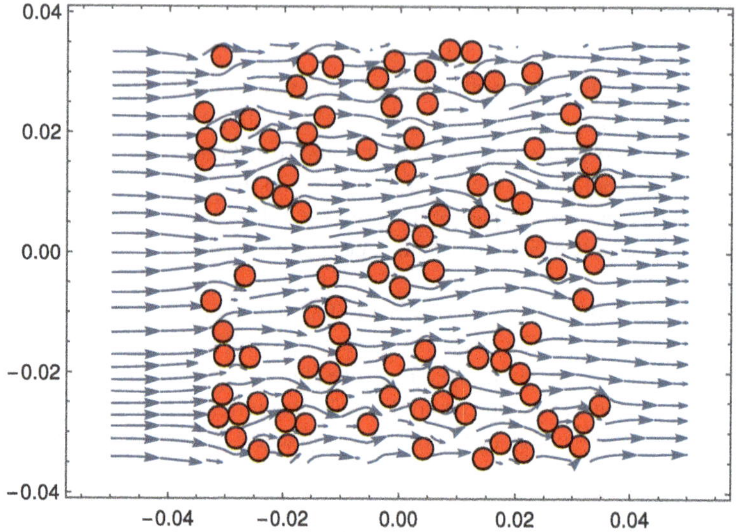

Fig. 1 Nawalaniec [5] Function: $f(z) = 1$ Number of disks: 140 radius: 0.00178412 concentration: 0.2

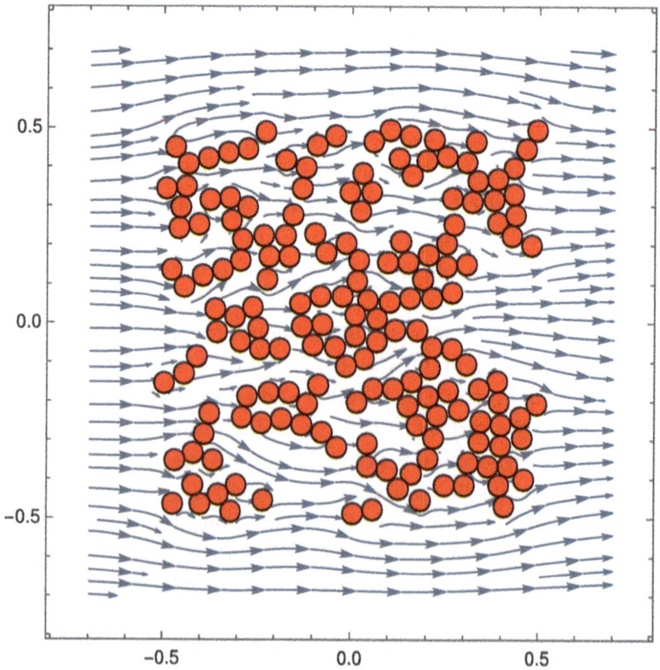

Fig. 2 Nawalaniec [5] Function: $f(z) = 1$ Number of disks: 140 radius: 0.027448063 concentration: 0.828402

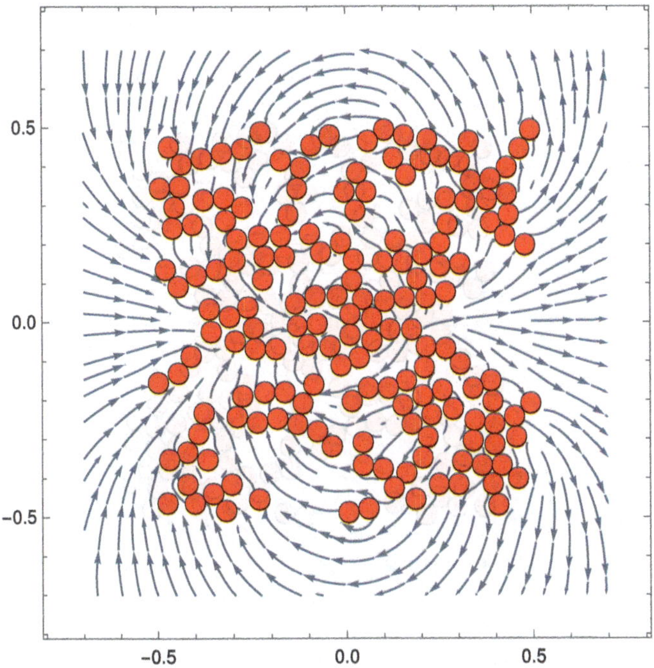

Fig. 3 Nawalaniec [5] Function: $f(z) = z^2$ Number of disks: 140 radius: 0.027448063 concentration: 0.828402

4 Conclusion

The considered samples are a cluster of non-overlapping disks simulated randomly along a domain surface D. Numerical approximation solution was computed to discover the complex flux along a domain surface filled by $non-overlapping$ discs. The simulation had been repeated several times with different complex conditions. It was clearly shown that changing the heat flow control for these models is highly dependent on the complex variable functions $f(z)$ and whether it is linear or non-linear. The simulation was applied for different conditions and different multiple disks. On the basis of the result, it is possible to adjust and determine the flow performance characteristics to fulfil the required heat transfer technology and its applications on fibrous composites.

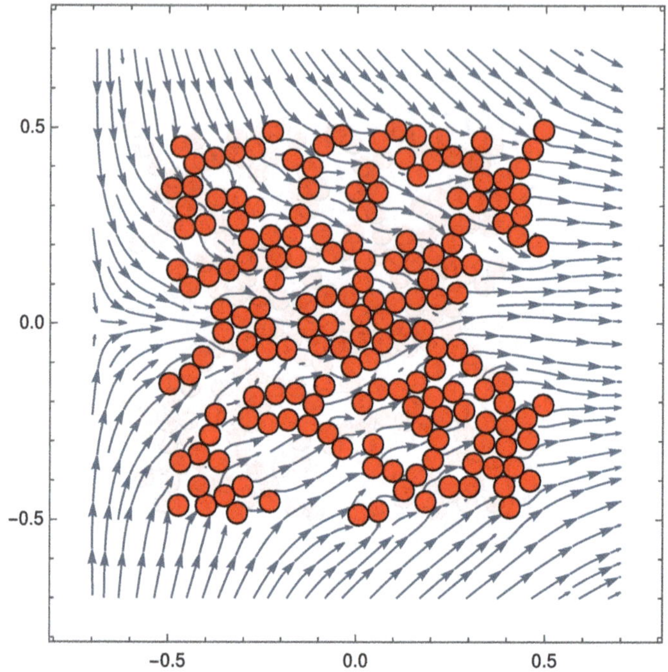

Fig. 4 Nawalaniec [5] Function: $f(z) = 1/z + 0,7$ Number of disks: 140 radius: 0.027448063 concentration: 0.828402

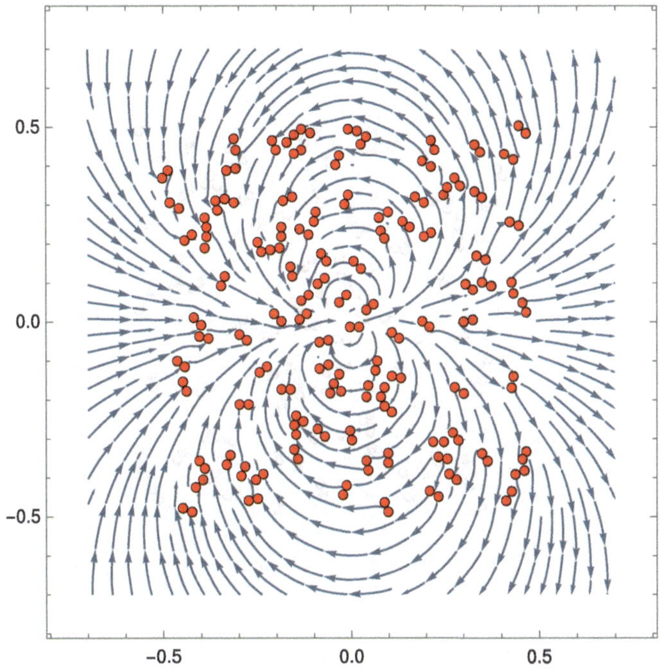

Fig. 5 Czapla [1] Function: $f(z) = z^2$ Number of disks: 100 radius: 0.0126157 concentration: 0.1

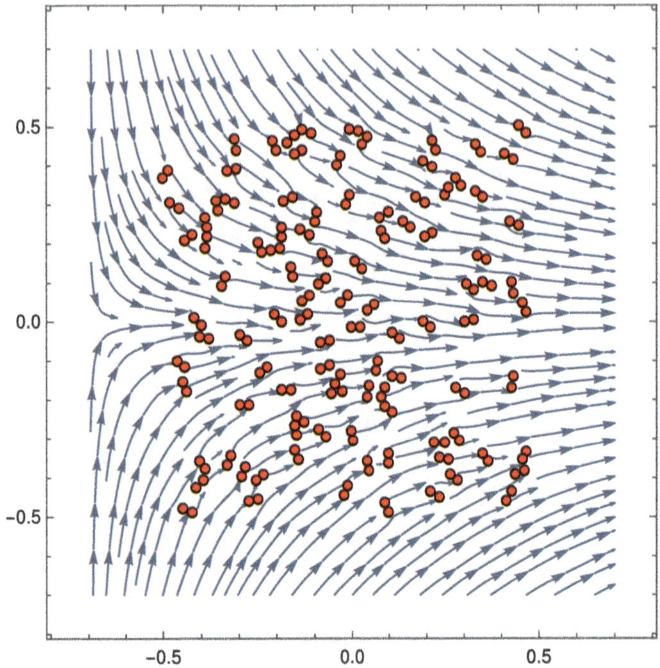

Fig. 6 Czapla [1] Function: $f(z) = 1/z + 07$ Number of disks: 100 radius: 0.0126157 concentration: 0.1

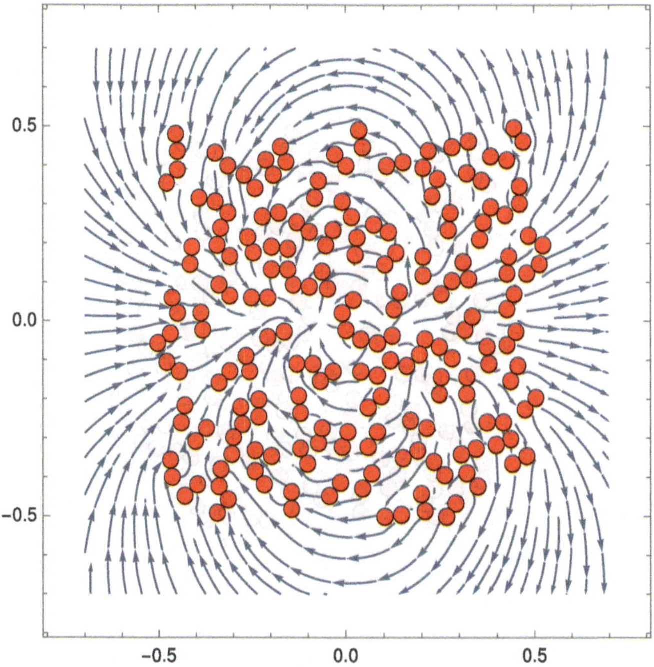

Fig. 7 Czapla [1] Function: $f(z) = z^2$ Number of disks: 100 radius: 0.021851 concentration: 0.3

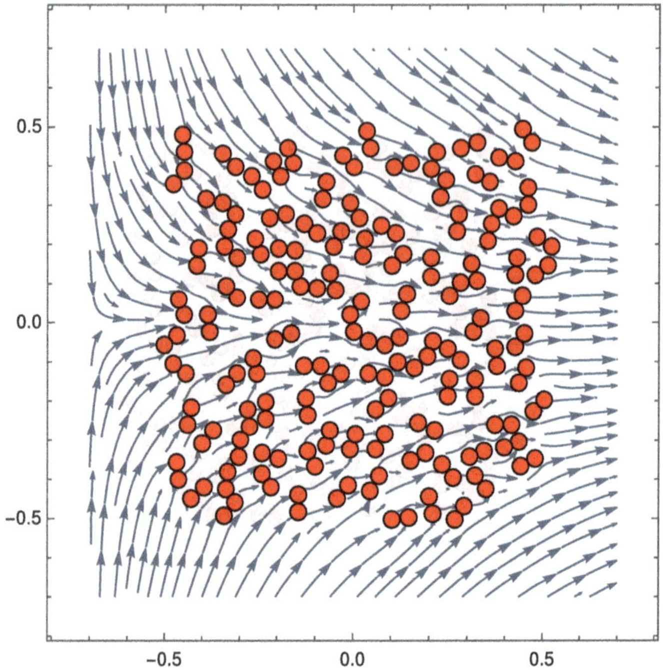

Fig. 8 Czapla [1] Function: $f(z) = 1/z + 0,7$ Number of disks: 100 radius: 0.021851 concentration: 0.3

References

1. R. Czapla, Random sets of stadiums in square and collective behavior of bacteria. IEEE/ACM Trans. Computat. Biol. Bioinf. **15**, 251–256 (2018)
2. H. Gliński, R. Grzymkowski, A. Kapusta, D. Słota, *Mathematica*, vol. 8 (Jacek Skalmierski Publ., Gliwice, 2012)
3. S. Gluzman, V. Mityushev, W. Nawalaniec, *Computational Analysis of Structured Media* (Academic, Elsevier, 2017)
4. V.V. Mityushev, S.V. Rogosin, *Constructive Methods for Linear and Non-linear Boundary Value Problems of the Analytic Function. Theory and Applications* (Chapman & Hall/CRC, Boca Raton, 2000)
5. W. Nawalaniec, *Random Non-overlapping Walks of Disks on the Plane*. Trends in Mathematics, vol. 2 (Springer, Berlin, 2015), pp. 769–775

The Robin Problem for Quasi-Linear Elliptic Equation $p(x)$-Laplacian in a Domain with Conical Boundary Point

Mikhail Borsuk

Abstract This paper is a survey of our last results about bounded weak solutions to the Robin boundary and the Robin transmission problems for an elliptic quasi-linear second-order equation with the variable $p(x)$-Laplacian in a conical bounded $n-$dimensional domain.

Keywords $p(x)$-Laplacian · Robin problem · Conical points

Mathematics Subject Classification (2010) Primary 99Z99; Secondary 00A00

1 Introduction

In this paper I present a survey of our last results [5–8] about bounded weak solutions to the Robin boundary and the Robin transmission problems for an elliptic quasi-linear second-order equation with the variable $p(x)$-Laplacian in a conical bounded $n-$dimensional domain. We consider the next problems:

(1) L_∞—estimate of bounded weak solutions. (2) The behavior of weak solutions near the angular or conical point of the boundary. (3) Existence of bounded weak solutions. (4) Transmission Robin problem.

Boundary value problems for elliptic second order equations with a non-standard growth in function spaces with variable exponents have been **an object of active investigation** in recent years. Differential equations with variable exponents-growth conditions arise from the nonlinear elasticity theory, electrorheological fluids, etc. There are many essential differences between the variable exponent problems and the constant exponent problems. In the variable exponent problems, many singular phenomena occurred and many special questions were raised. V. Zhikov [14] has

M. Borsuk (✉)
Faculty of Mathematics and Computer Science, University of Warmia and Mazury in Olsztyn, Olsztyn, Poland
e-mail: borsuk@uwm.edu.pl

P. Cerejeiras et al. (eds.), *Current Trends in Analysis, its Applications and Computation*, Research Perspectives,
https://doi.org/10.1007/978-3-030-87502-2_23

given examples of the Lavrentiev phenomenon for the variational problems with variable exponent.

Most of the works devoted to the quasi-linear elliptic second-order equations with the variable $p(x)$-Laplacian refers to the Dirichlet problem (see [1, 11]). In [2, 3, 10] the Robin problem for such equations has been considered, but in smooth domains only. What is more, in these works the lower order terms depend only on (x, u) and do not depend on $|\nabla u|$. A problem with a lower order term that does not depend on $|\nabla u|$ in a non-smooth domain has been recently studied in [9]. Our recent works [5–8] are devoted to the Robin boundary and the Robin transmission problems **in a cone** for equations with **a singular** $p(x)-$ **power gradient lower order term**. We use the following standard notations:

- $\mathbb{C}$: an open cone in $\mathbb{R}^n$, $n \geq 2$, with the vertex at the origin $\mathcal{O}$ and the angular opening of cone $\omega_0 \in (0, \pi)$; B_r : an open ball with radius r centered at $\mathcal{O}$;
- S^{n-1} : a unit sphere in $\mathbb{R}^n$ centered at $\mathcal{O}$; $(r, \omega), \omega = (\omega_1, \omega_2, \ldots, \omega_{n-1})$: the spherical coordinates of $x \in \mathbb{R}^n$ with pole $\mathcal{O}$:
- Ω : a domain on the unit sphere S^{n-1} with the smooth boundary $\partial\Omega$ obtained by the intersection of the cone $\mathbb{C}$ with the sphere S^{n-1}; $\partial\Omega = \partial\mathbb{C} \cap S^{n-1}$;
- $G_0^d \equiv \mathbb{C} \cap B_d = \{(r, \omega) \mid 0 < r < d;\ \omega \in \Omega\}$; $G_0^d = G_+^d \cup G_-^d$ is divided into two subdomains $G_\pm^d := \{(r, \omega) : 0 < r < d,\ \omega \in \Omega_\pm\}$ by a $\Sigma_0^d := G_0^d \cap \{x_n = 0\}$, where $\mathcal{O} \in \overline{\Sigma_0^d}$; $\Omega_+ = \Omega \cap \{x_n > 0\}$, $\Omega_- = \Omega \cap \{x_n < 0\} \Longrightarrow \Omega = \Omega_+ \cup \Omega_-$;
- $\Gamma_0^d \equiv \partial\mathbb{C} \cap B_d = \{(r, \omega) \mid 0 < r < d;\ \omega \in \partial\Omega\}$; $\Gamma_0^d = \Gamma_+^d \cup \Gamma_-^d$, $\Gamma_\pm^d := \{(r, \omega) : 0 < r < d,\ \omega \in \partial_\pm\Omega\}$; $\partial_\pm\Omega = \overline{\Omega_\pm} \cap \partial C$; $\partial\Omega_\pm = \partial_\pm\Omega \cup \sigma_0$; $\sigma_0 = \overline{\Sigma_0^d} \cap \Omega_d$; $\Omega_d = \overline{G_0^d} \cap \{|x| = d\}$.
- $u(x) = \begin{cases} u_+(x), & x \in G_+^d, \\ u_-(x), & x \in G_-^d; \end{cases} \quad f(x) = \begin{cases} f_+(x), & x \in G_+^d, \\ f_-(x), & x \in G_-^d \end{cases}$ etc.;
- $[u]_{\Sigma_0^d}$ denotes the saltus of the function $u(x)$ on crossing Σ_0^d, i.e. $[u]_{\Sigma_0^d} = u_+(x)\Big|_{\Sigma_0^d} - u_-(x)\Big|_{\Sigma_0^d}$, where $u_\pm(x)\Big|_{\Sigma_0^d} = \lim_{G_\pm^d \ni y \to x \in \Sigma_0^d} u_\pm(y)$;
- $n_i = \cos(\vec{n}, x_i)$, $i = 1, 2$, where $\vec{n}$ denotes the unit outward vector with respect to G_+^d (or G_0^d) normal to Σ_0^d (respectively $\partial G_0^d \setminus \mathcal{O}$).

We shall investigate the bounded weak solutions of the Robin boundary problem:

$$\begin{cases} -\Delta_{p(x)}u + a(x)u|u|^{p(x)-1} + b(u, \nabla u) = f(x), & x \in G, \\ |\nabla u|^{p(x)-2}\frac{\partial u}{\partial \vec{n}} + \frac{\gamma}{|x|^{p(x)-1}}u|u|^{p(x)-2} = g(x), & x \in \partial G, \end{cases} \tag{RQL}$$

(here $G \in C^{0,1}$ is a bounded domain in $\mathbb{R}^n$ with the boundary ∂G, containing **a conical point in the origin** $\mathcal{O}$, and near $\mathcal{O}$ it is a conical surface) and of the Robin transmission problem:

$$\begin{cases} -\Delta_{p(x)}u + a(x)u|u|^{p(x)-1} + b(u,\nabla u) = f(x), & x \in G_0^{d_0}, \\ [u]_{\Sigma_0^{d_0}} = 0, & \\ \left[|\nabla u|^{p(x)-2}\frac{\partial u}{\partial \vec{n}}\right]_{\Sigma_0^{d_0}} + \frac{\beta}{|x|^{p(x)-1}}u|u|^{p(x)-2} = h(x,u), & x \in \Sigma_0^{d_0}, \\ |\nabla u|^{p(x)-2}\frac{\partial u}{\partial \vec{n}} + \frac{\gamma}{|x|^{p(x)-1}}u|u|^{p(x)-2} = g(x,u), & x \in \Gamma_0^{d_0}, \end{cases} \qquad (TRQL)$$

where $0 < d_0 \ll 1$ (d_0 is fixed); $\quad \Delta_{p(x)}u \equiv \operatorname{div}\left(|\nabla u|^{p(x)-2}\nabla u\right)$;

$$1 < p_- \le p(x) \le p_+ = p(0) < n, \ \forall x \in \overline{G}. \qquad (1.1)$$

We define the functions class $\mathfrak{N}^{1,p(x)}_{-1,\infty}(G) = \left\{u \;\middle|\; u(x) \in L_\infty(G) \text{ and } \int_G \left(|x|^{-p(x)}|u|^{p(x)} + |u|^{-1}|\nabla u|^{p(x)}\right)dx < \infty\right\}$. It is obvious that $\mathfrak{N}^{1,p(x)}_{-1,\infty}(G) \subset W^{1,p(x)}(G)$.

Remark 1.1 If $p(x) > n$, by the Sobolev imbedding theorem, we have $u \in C^{1-\frac{n}{p(0)}}(G)$. Therefore we investigate only $p(x) \in (1,n)$.

2 L_∞: Estimate of Bounded Weak Solutions

Definition 2.1 The function u is called a weak bounded solution of problem (RQL) provided that $u(x) \in \mathfrak{N}^{1,p(x)}_{-1,\infty}(G)$ and u satisfies the integral identity

$$Q(u,\eta) :\equiv \int_G \left(|\nabla u|^{p(x)-2}u_{x_i}\eta_{x_i} + a(x)u|u|^{p(x)-1}\eta(x) + b(u,\nabla u)\eta(x)\right)dx$$

$$+\gamma \int_{\partial G} |x|^{1-p(x)}u|u|^{p(x)-2}\eta(x)ds = \int_{\partial G} g(x)\eta(x)ds + \int_G f(x)\eta(x)dx. \qquad (II)$$

for all $\eta(x) \in \mathfrak{N}^{1,p(x)}_{-1,\infty}(G)$.

Theorem 2.2 *Let $u(x)$ be a weak bounded solution of (RQL). Assume that*

(i) *$p(x) \in C^{(0)}(\overline{G})$ and (1.1);*
(ii) *$0 \le a_0 \le a(x) \le const \cdot |x|^{-p(x)}$; $a_0 = const > 0$, $\forall x \in \overline{G}$; $b(u,\xi) : \mathbb{R}^{n+1} \Rightarrow \mathbb{R}$ is a Carathéodory function satisfying for all $(u,\xi) \in \mathbb{R}^{n+1}$ the following inequalities:*

$$|b(u,\xi)| \le \mu|u|^{-1}|\xi|^{p(x)}, \quad 0 \le \mu < 1, \ \forall x \in \overline{G};$$

(iii)

$$|f(x)| \leq f_0 |x|^{\beta(x)}, \qquad \beta(x) \geq \beta_0 - \frac{n}{s};\ s > \frac{n}{p_-};\ f_0 \geq 0,\ \beta_0 > 0, \quad \forall x \in \overline{G};$$

$$|g(x)| \leq g_0 |x|^{1-p(x)}, \quad g_0 \geq 0,\ \forall x \in \partial G.$$

Then there exists a constant $M_0 > 0$ depending only on meas $G, n, p_\pm, s, \mu, f_0, g_0, a_0, \beta_0, \gamma$ such that $\|u\|_{L_\infty(G)} \leq M_0$.

Remark 2.3 It is easy to verify that the assumptions **(i)–(iii)**) guarantee the existence of integrals over G and ∂G in the integral identity (II). Therefore, $Q(u, \eta)$ is well defined.

The proof of this Theorem is based on the well-known level method and the Stampacchia Lemma. Namely, we consider the set
$A(k) = \{x \in \overline{G}, \quad |u(x)| > k\}$. New is our presentation
$A(k) = A_-(k) \cup A_+(k)$, where $A_-(k) = A(k) \cap \{|\nabla u| \leq 1\}$,
$A_+(k) = A(k) \cap \{|\nabla u| \geq 1\}$ and the application of the inequalities
$|\nabla u|^{p_+} \leq |\nabla u|^{p(x)} \leq |\nabla u|^{p_-}$ on $A_-(k)$; $|\nabla u|^{p_-} \leq |\nabla u|^{p(x)} \leq |\nabla u|^{p_+}$ on $A_+(k)$.

We note that prior to that the L_∞—regularity of weak solutions for quasi-linear equations with $p(x)$—Laplacian was studied:

- in [1] for $b(u, \xi) \equiv 0$ (the Dirichlet problem),
- in [2, 3] for $b(u, \xi)$ not depending on ξ (the Dirichlet and the Robin problems),
- in [12] for $|b(x, u, \xi)| \leq c_1 |\xi|^{\alpha(x)} + c_2 |u|^{r(x)-1} + c_3$; $\alpha(x) = \frac{r(x)-1}{r(x)} p(x)$, $p(x) \leq r(x) < p^*(x)$, where $p^*(x)$ is the Sobolev embedding exponent of $p(x)$ (the Dirichlet problem).

3 The Behavior of Weak Solutions Near the Angular or Conical Point of the Boundary

Here we describe qualitatively the behavior of the weak solution near a conical point, namely, we derive the sharp estimate of the type $|u(x)| = O(|x|^\varkappa)$ for the weak solution modulus (for the solution decrease rate) of problem (RQL) near a conical boundary point. As well, we establish the comparison principle for weak solutions.

Theorem 3.1 *Let $u(x)$ be a weak bounded solution of (RQL) in a cone $G_0^d \subset G$, $0 < d \ll 1$ with the boundary condition with $g(x, u)$ instead of $g(x)$ on the lateral surface Γ_0^d of the cone G_0^d and $M_0 = \sup_{x \in G_0^d} |u(x)|$. Assume that*

(v) $p(x) \in C^{(0,1)}(\overline{G_0^d})$ *and (1.1);*

(iv) *$b(u,\xi):\mathbb{R}^{n+1}\Rightarrow\mathbb{R}$ is a Carathéodory function satisfying for all $(u,\xi)\in\mathbb{R}^{n+1}$ the following inequalities:*

$$\textbf{(iv)}_{\mathbf{a}}\qquad |b(u,\xi)|\le\delta|u|^{-1}|\xi|^{p(x)}+b_0|u|^{p(x)-1},\ \begin{cases}0\le\delta<\mu, & \textit{if } \mu>0;\\ \delta\ge 0, & \textit{if } \mu=0;\end{cases}$$

$$\textbf{(iv)}_{\mathbf{b}}\ b(u,\xi)\operatorname{sign}u\ge\nu|u|^{-1}|\xi|^{p(x)}-b_0|u|^{p(x)-1},\quad \nu>0;\quad \textit{if } \mu=0;$$

$$\textbf{(iv)}_{\mathbf{c}}\qquad \sqrt{\sum_{i=1}^{n}\left|\frac{\partial b(u,\xi)}{\partial\xi_i}\right|^2}\le b_1|u|^{-1}|\xi|^{p(x)-1};\qquad \frac{\partial b(u,\xi)}{\partial u}\ge b_2|u|^{-2}|\xi|^{p(x)};$$

$$b_0\ge 0,\ b_1\ge 0,\ b_2\ge 0;$$

(vi) $0\le a_0\le a(x)\le const\cdot|x|^{-p(x)}$;
(vii) $|f(x)|\le f_0|x|^{\beta(x)},\quad f_0\ge 0,\ \beta(x)\ge\frac{p_+-1}{p_+-1+\mu}\,(p(x)-1)\,\lambda-p(x)$;
$\forall x\in\overline{G_0^{d_0}}$; $0\le\mu<1$, $\gamma=const\ge 1$ *and λ is the least positive eigenvalue of problem*

$$\begin{cases}\qquad -div_\omega\left((\lambda^2\psi^2+|\nabla_\omega\psi|^2)^{(p_+-2)/2}\nabla_\omega\psi\right)= & \\ \lambda\left(\lambda(p_+-1)+n-p_+\right)(\lambda^2\psi^2+|\nabla_\omega\psi|^2)^{(p_+-2)/2}\psi, & \omega\in\Omega,\\ (\lambda^2\psi^2+|\nabla_\omega\psi|^2)^{(p_+-2)/2}\frac{\partial\psi}{\partial\vec{\nu}}+\gamma\left(\frac{p_+-1+\mu}{p_+-1}\right)^{p_+-1}\cdot\psi|\psi|^{p_+-2}=0, & \omega\in\partial\Omega,\end{cases}$$

for $\psi(\omega)\in C^2(\Omega)\cap C^1(\overline{\Omega})$, where $|\nabla_\omega\psi|$ denotes the projection of the vector $\nabla\psi$ onto the tangent plane to the unit sphere at the point ω and $\vec{\nu}$ denotes the exterior normal to $\partial\mathbb{C}$ at points of $\partial\Omega$;
(vv) $|g(x,u)|\le g_0|x|^{1+\beta(x)}$; $\forall u\in L_\infty$; $\frac{\partial g(x,u)}{\partial u}\le 0,\ g(x,0)\equiv 0,\ x\in\Gamma_0^{d_0}$;
(vvv) *the spherical region $\Omega\subset S^{n-1}$ is invariant with respect to rotations in S^{n-2}.*

Then there exist $\tilde{d}\in(0,d)$ and a constant $C_0>0$ depending only on $\lambda, d, M_0, p_+, p_-, L, n, (\mu-\delta), \nu, b_0, f_0$ and such that

$$|u(x)|\le C_0|x|^{\varkappa},\quad \varkappa=\frac{p_+-1}{p_+-1+\mu}\lambda;\quad \forall x\in G_0^{\tilde{d}}.$$

The proof of this Theorem is based on the our **new comparison principle**, adapted to the problem, and the construction of **a barrier function**

$$w=w(r,\omega)=r^{\varkappa}\psi^{\varkappa/\lambda}(\omega),\quad \varkappa=\frac{p_+-1}{p_+-1+\mu}\lambda$$

as a solution of the auxiliary problem:

$$\begin{cases} -\Delta_{p_+} w = \mu w^{-1} |\nabla w|^{p_+}, & x \in G_0^d, \\ |\nabla w|^{p_+-2} \frac{\partial w}{\partial \vec{n}} + \frac{\gamma}{|x|^{p_+-1}} w|w|^{p_+-2} = 0, & x \in \Gamma_0^d, \end{cases}$$

where $(\lambda, \psi(\omega))$ is the solution of the above eigenvalue problem.

4 Existence of Bounded Weak Solutions

Here we study the existence of bounded weak solutions of (RQL). We need the space $M(G)$: it is the set of all measurable and bounded almost everywhere in $\overline{G}$ functions $u(x)$ with the norm

$$\|u\| = vrai \max_{x \in \overline{G}} |u(x)| = \inf_{meas\, E=0} \left\{ \sup_{x \in \overline{G} \setminus E} |u(x)| \right\}.$$

The convergence in $M(G)$ is the uniform convergence almost everywhere.

Theorem 4.1 *Let suppositions* (**v**), (**vi**) *be satisfy and assume that*

(**w**) $b(u, \xi) : \mathbb{R}^{n+1} \Rightarrow \mathbb{R}$ *is a Carathéodory function satisfying for all* $(u, \xi) \in \mathbb{R}^{n+1}$ *the following inequalities:*

$$(\mathbf{w})_{\mathbf{a}} \quad |b(u, \xi)| \le b_1 \left(|u(x)|^{q_0(x)} + |\xi|^{q_1(x)} \right), \quad \textit{where} \quad b_1 = const \ge 0,$$

$$q_0(x) < p^*(x) - 1, \quad q_1(x) < p(x) - 1 + \frac{p(x)}{n}, \quad p^*(x) = \frac{np(x)}{n - p(x)};$$

$$(\mathbf{w})_{\mathbf{b}} \quad ub(x, u, \xi) \ge |u|^{p(x)} \quad \textit{for} \quad |u| > 1;$$

(**ww**) $\quad f \in L_{p'(x)}(G), \quad \frac{1}{p'(x)} + \frac{1}{p(x)} = 1; \quad g(x) \equiv 0.$

Then (RQL) *has at least one bounded weak solution*

$$u \in V_{p(x)}(G) \equiv W^{1,p(x)}(G) \cap M(G).$$

Proof For all $u, \eta \in V_{p(x)}(G)$ we define nonlinear operators

$$J, B, \Gamma : V_{p(x)}(G) \to V^*_{p(x)}(G) \text{ and an element } f^* \in V^*_{p(x)}(G)$$

by

$$\langle J(u), \eta\rangle = \int_G |\nabla u(x)|^{p(x)-2}\nabla u(x)\nabla\eta(x)dx,$$

$$\langle B(u), \eta\rangle = \int_G \widetilde{b}\,(x, u(x), \nabla u(x))\,\eta(x)dx,$$

$$\langle \Gamma(u), \eta\rangle = \int_{\partial G} |x|^{1-p(x)}u(x)|u(x)|^{p(x)-2}\eta(x)dS, \quad \langle f^*, \eta\rangle = \int_G f(x)\eta(x)dx.$$

At first, we verify that operators J, B, Γ, f^* are well defined. Next we put $T := J + B + \Gamma$. Then the operator equation $T(u) = f^*$ is equivalent to validity of the integral identity (II). This fact means that the solutions of this operator equation correspond one-to-one to the weak solutions of (RQL). Further, to prove that there is a solution of the operator equation we verify the assumptions of the Leray-Lions Theorem (see [13]). □

5 Transmission Robin Problem

We investigate the behavior in a neighborhood of the origin $\mathcal{O}$ of solutions to the transmission Robin problem $(TRQL)$. The same problems for $p(x) = p = const$ were studied in our monograph [4]. We suppose from Sect. 3 **(iv)**, **(vi)**, **(vvv)** and **(vii)** with λ being the least positive eigenvalue of problem

$$\begin{cases} -\mathrm{div}_\omega\left((\lambda^2\psi^2 + |\nabla_\omega\psi|^2)^{(p_+-2)/2}\nabla_\omega\psi\right) \\ \quad = \lambda\left(\lambda(p_+ - 1) + n - p_+\right)(\lambda^2\psi^2 + |\nabla_\omega\psi|^2)^{(p_+-2)/2}\psi, \qquad \omega\in\Omega, \\ [\psi]_{\sigma_0} = 0; \\ \left[(\lambda^2\psi^2 + |\nabla_\omega\psi|^2)^{(p_+-2)/2}\frac{\partial\psi}{\partial\vec{\nu}}\right]_{\sigma_0} + \beta\left(\frac{p_+-1+\mu}{p_+-1}\right)^{p_+-1}\cdot\psi|\psi|^{p_+-2}\Big|_{\sigma_0} = 0, \\ (\lambda^2\psi^2 + |\nabla_\omega\psi|^2)^{(p_+-2)/2}\frac{\partial\psi}{\partial\vec{\nu}} + \gamma\left(\frac{p_+-1+\mu}{p_+-1}\right)^{p_+-1}\cdot\psi|\psi|^{p_+-2} = 0, \ \omega\in\partial\Omega, \end{cases}$$

and the following:

(1) $1 < p_- \le p(x) \le p_+ = p(0) < n;\ \forall x \in \overline{G_0^{d_0}};\quad p(x) \in C^{0,1}(\overline{G_0^{d_0}}) \implies$ $0 \le p_+ - p(x) \le L|x|,\ \forall x \in \overline{G_0^{d_0}};$ where L is the Lipschitz constant for $p(x)$;

(2) constants $\beta,\ \gamma$, are such that $\begin{cases} \beta \ge 2 \ \text{ and } \ \gamma \ge \frac{1}{2}\beta, & \text{if } \ p_+ \ge 2; \\ 1 \le \gamma \le \frac{1}{2}\beta, & \text{if } \ 1 < p_+ \le 2; \end{cases}$

(3) $|h(x,u)| \le h_0|x|^{1+\beta(x)};\ \forall u \in L_\infty;\ \frac{\partial h(x,u)}{\partial u} \le 0,\ h(x,0) \equiv 0,\ x \in \Sigma_0^{d_0};$
$|g(x,u)| \le g_0|x|^{1+\beta(x)};\ \forall u \in L_\infty;\ \frac{\partial g(x,u)}{\partial u} \le 0,\ g(x,0) \equiv 0,\ x \in \Gamma_0^{d_0}.$

The main result is the following statement:

Theorem 5.1 *Let u be a weak bounded solution of problem $(TRQL)$, $M_0 = \sup_{x\in G_0^{d_0}} |u(x)|$ and above assumptions hold. Then there exist $\widetilde{d} \in (0, d_0)$ and a constant $C_0 > 0$ depending only on $\lambda, d_0, M_0, p_+, p_-, L, n, (\mu-\delta), \nu, b_0, f_0$ such that*

$$|u(x)| \le C_0|x|^{\varkappa},\quad \varkappa = \frac{p_+ - 1}{p_+ - 1 + \mu}\lambda;\quad \forall x \in G_0^{\widetilde{d}}.$$

Proof As in the proof of the Theorem 3.1 we prove maximum and comparison principles and construct a barrier function

$$w(r,\omega) = r^{\varkappa}\psi^{\varkappa/\lambda}(\omega),\qquad \varkappa = \frac{p_+ - 1}{p_+ - 1 + \mu}\lambda,$$

where $(\lambda, \psi(\omega))$ is the solution of the above eigenvalue problem. For details we refer to [6, 8]. □

References

1. Yu. Alkhutov, M. Borsuk, The behavior of solutions to the Dirichlet problem for second order elliptic equations with variable non-linearity exponent in a neighborhood of a conical boundary point. J. Math. Sci. **210**(4), 341–370 (2015)
2. S. Antontsev, L. Consiglieri, Elliptic boundary value problems with nonstandard growth conditions. Nonlinear Analy. **71**, 891–902 (2009)
3. S. Antontsev, S. Shmarev, Elliptic equations and systems with nonstandard growth conditions: existence, uniqueness and localization properties of solutions. Nonlinear Analy. **65**, 728–761 (2006)
4. M. Borsuk, *Transmission Problems for Elliptic Second-Order Equations in Non-smoooth Domains*. Frontiers in Mathematics (Birkhäuser, Basel, 2010), 218 p. https://doi.org/10.1007/978-3-0346-0477-2
5. M. Borsuk, L_∞ -estimate for the Robin problem of a singular variable p - Laplacian equation in a conical domain. Electr. J. Differ. Equ. **2018**(49), 1–9 (2018)
6. M. Borsuk, The Robin problem for singular $p(x)$-Laplacian equation in a cone. Electr. J. Qualitat. Theory Differ. Equ. **93**, 1–14 (2018). https://doi.org/10.14232/ejqtde.20218.1.93
7. M. Borsuk, Existence of bounded weak solutions of the Robin problem for quasi-linear elliptic equation with p(x)-Laplacian. Electr. J. Qualitat. Theory Differ. Equ. **16**,1–11 (2019). https://doi.org/10.14232/ejqtde.2019.1.16
8. M. Borsuk, Transmission Robin problem for singular $p(x)$-Laplacian equation in a cone. Electr. J. Qualitat. Theory Differ. Equ. **93**, 1–17 (2019). https://doi.org/10.14232/ejqtde.2019.1.93

9. M.M. Boureanu, A. Vélez-Santiago, Fine regularity for elliptic and parabolic anisotropic Robin problems with variable exponents. J. Differ. Equ. (2019). https://doi.org/10.1016/j.jde.2018.12.026
10. Sh.-G. Deng, Positive solutions for Robin problem involving the $p(x)-$Laplacian. J. Math. Anlys. Appli. **360**, 548–560 (2009)
11. X. Fan, Existence of solutions for $p(x)-$Laplacian Dirichlet problem. Nonlinear Analys. **52**, 1843–1852 (2003)
12. X. Fan, D. Zhao, A class of De Giorgi type and Hölder continuity. Nonlinear Analy. **36**, 295–318 (1999)
13. J. Leray, J.L Lions, Quelques résultats de Višik sur les problémes elliptiques non linéaires par les méthodes de Minty-Browder. Bull. Soc. Math. France **93**, 97–107 (1965)
14. V.V. Zhikov, On Lavrentiev's phenomenon. Russian J. Math. Phys. **13**(2), 249–269 (1994)

Analysis of Distributions of Stadiums on the Plane Using e-Sums

Roman Czapla

Abstract The main goal of this paper is to discuss the method of characterization of distributions of geometric objects on the plane. The method is based on the structural sums performing the same role in description of image as the n-point correlation functions. In this paper we present applications of this theory to description of distributions of non-circular objects (stadiums) on the plane approximated by disks.

Keywords Structural sums · Distributions of object on the plane · Stadium shape

Mathematics Subject Classification (2010) Primary 62-07; Secondary 60G55

1 Introduction

Description of geometry of images[1] has fundamental meaning in different branches of science i.e. engineering materials, biology, medicine, astronomy etc. In addition to analyzing visible features from studied image it is very interesting to extract the hidden features—invisible to the human, but significant (for example isotropy of material). This problem can be reduced to the construction of set G containing parameters describing geometry. More precisely, we can build structural sums feature vectors [13, 22]. Usually we use this representation for images containing non-overlapping disks, but in this paper we will extend the method to other shapes. Structural sums constitute a crucial part of the framework of modern computational material science [13, 14] forming a coherent whole with other contemporary results and applications in the field [1, 2, 6–12, 16, 18, 19, 22, 25–28, 30].

[1] Without loss of generality, we can consider only two-phase images [29, p. 24].

R. Czapla (✉)
Faculty of Mathematics, Physics and Technology, Pedagogical University of Cracow, Krakow, Poland

P. Cerejeiras et al. (eds.), *Current Trends in Analysis, its Applications and Computation*, Research Perspectives,
https://doi.org/10.1007/978-3-030-87502-2_24

2 Structural Sums

Consider a lattice $\mathcal{Q}$ which is determined on the complex plane by two vectors ω_1 i ω_2 $\left(\text{for definiteness, it is assumed that } \mathrm{Im}\left[\omega_2/\omega_1\right] > 0\right)$.We introduce the cell $(0, 0)$ as the parallelogram:

$$\mathcal{Q}_{(0,0)} := \left\{ z = s_1\omega_1 + s_2\omega_2 : \ -\frac{1}{2} < s_k < \frac{1}{2} \ (k = 1, 2) \right\}.$$

The lattice $\mathcal{Q}$ consists of the cells

$$\mathcal{Q}_{(m_1,m_2)} := \left\{ z \in \mathbb{C}: \ z - m_1\omega_1 - m_2\omega_2 \in \mathcal{Q}_{(0,0)} \right\}$$

where m_1, m_2 run over integer numbers. We will consider unit cell which include N non-overlapping disks of radii r_k $(k = 1, 2, \ldots, N)$ (see Fig. 1). This distribution will be realized in the torus topology. We define concentration of disks by equals $\nu = \pi \sum_{k=1}^{N} r_k^2$. Let R be the largest of the radii r_k $(k = 1, 2, \ldots, N)$ and introduce constants $\upsilon_k = \left(\frac{r_k}{R}\right)^2$, $k = 1, 2, 3, \ldots, N$ describing polydispersity of disks.

Consider points $a_k = x_k + iy_k$ $(k = 1, 2, \ldots, N)$ in the cell $\mathcal{Q}_{(0,0)}$. Let q be a natural number, $k_0, k_1, \ldots, k_q$ run over integer numbers from 1 to N, and $p_j > 1$ $(j = 1, 2, \ldots, q)$ are integer numbers. The following sum was introduced by Mityushev [15]:

$$\begin{aligned} e_{p_1,p_2,\ldots,p_q}^{\upsilon_1,\upsilon_2,\ldots,\upsilon_q} &= \frac{1}{\eta^{\delta+1}} \textstyle\sum_{k_0,k_1,\ldots,k_q} \upsilon_{k_0}^{t_0} \upsilon_{k_1}^{t_1} \upsilon_{k_2}^{t_2} \cdots \upsilon_{k_q}^{t_q} E_{p_1}(a_{k_0} - a_{k_1}) \\ &\times \overline{E_{p_2}(a_{k_1} - a_{k_2})} E_{p_3}(a_{k_2} - a_{k_3}) \cdots \mathbf{C}^{q+1} E_{p_q}(a_{k_{q-1}} - a_{k_q}), \end{aligned} \tag{2.1}$$

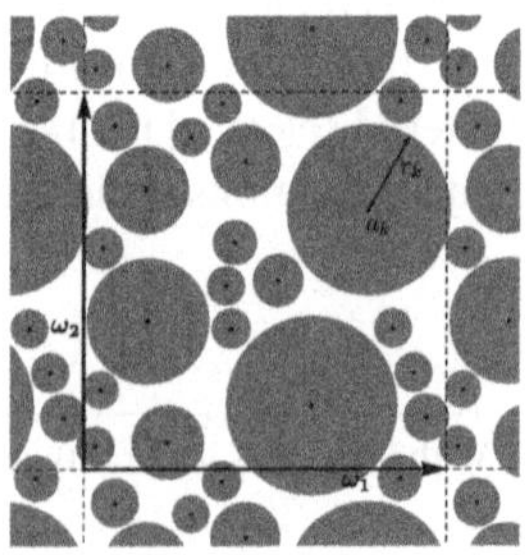

Fig. 1 Doubly periodic cell $\mathcal{Q}_{(0,0)}$ with a configuration of non-overlapping disks

where $\eta = \sum_{j=1}^{n} v_j$ and $\delta = \frac{1}{2}\sum_{j=1}^{n} p_j$. Symbol $\boldsymbol{C}$ denote operator of complex conjugation. Superscripts t_j $(j = 0, 1, \ldots, n)$ are given by recurrence relations: $t_0 = 1$, $t_j = p_j - t_{j-1}$, $j = 1, 2, \ldots, n$. Functions E_m $(m = 2, 3, \ldots)$ are the *Eisenstein functions* corresponding to double periodic cell $Q_{(0,0)}$ (see [13] for more information). The sum (2.1) is called *structural sum* (or *e-sum*) *of the multi-index* $\boldsymbol{p} = (p_1, p_2, \ldots, p_q)$ and number δ is called *the order of the structural sum*.

Now we describe the connection of structural sums with the geometry of a set of disks distributed on a plane. It turns out that we can choose geometric parameters as the following set of structural sums:

$$G = \{e_{\mathbf{m}},\ \mathbf{m} \in \mathcal{M}_e\},$$

where the set $\mathcal{M}_e$ is defined recursively in [20]. The justifications for this fact show latest research by Mityushev and Nawalaniec [17], in which structural sums were used in systematic studies of dynamically changing structures. We can consider the set G as the parameters describe *structural features* of data represented by non-overlapping disks. Because the set $\mathcal{M}_e$ is infinite, therefore practical applications require finite approximations in form of the *structural sums feature vector X_q of order q* defined in [22]. For example, the feature vector of order 5 has the following form:

$$X_5 = \{e_2, e_{2,2}, e_{2,2,2}, e_{3,3}, e_{2,2,2,2}, \\ e_{3,3,2}, e_{4,4}, e_{2,2,2,2,2}, e_{2,3,3,2}, e_{3,3,2,2}, e_{3,4,3}, e_{4,4,2}, e_{5,5}\}.$$

More detailed construction of X_q is presented in [22] as well as in [13]. In next part of this paper we will consider selected sums from the vector X_5 i.e.: $e_{2,2}$, $e_{3,3}$, $e_{4,4}$, $e_{5,5}$.

3 Applications of Structural Sums in Description of the Geometry of Distributions of Stadiums

As we know application of basis sums in description of the geometry of two-phase image is justified for geometric objects being disks. However, any geometric object can be approximated by clusters of disks. In this section we focus on the study of the impact of choosing an approximation of a given geometric object by disk on the ability of distinguishing distributions of objects. We will study distributions of geometric objects called *stadiums*. The study of such distributions is related to the collective behavior of bacteria [3–5].

First, we introduce a formal definition of stadium. Let $R > 0$ be a real number and $a, b \in \mathbb{C}$ be such that $a \neq b$. The *stadium* $\mathcal{S}\big((a,b), r\big) \subset \mathbb{C}$ is defined as follows

$$\mathcal{S}\big((a,b), R\big) := \Big\{ z \in \mathbb{C}: \; |z-a| \le r \\ \vee |z-b| \le r \vee \Big(\mathrm{Im}[z-a] \cdot \mathrm{Im}[a-b] \\ + \mathrm{Re}[z-a] \cdot \mathrm{Re}[a-b] \le 0 \wedge \mathrm{Im}[z-b] \cdot \mathrm{Im}[a-b] \\ + \mathrm{Re}[z-b] \cdot \mathrm{Re}[a-b] \ge 0 \wedge \Big| \mathrm{Im}\big[(b-a) \cdot \mathrm{Re}[z] \\ + (z-b) \cdot \mathrm{Re}[a] + (a-z) \cdot \mathrm{Re}[b]\big]\Big| \le R \cdot |a-b| \Big) \Big\},$$

where R is called its *radius* and the (a, b) its *centers* (see Fig. 2).

Consider a lattice $\mathcal{Q}$ which is defined by two fundamental translation vectors $\omega_1 = 1$ and $\omega_2 = i$ forming a square cell $Q_{(0,0)}$ (see Fig. 3). Consider N non-overlapping stadiums $\mathcal{S}_k = \mathcal{S}_k\big((a_k, b_k), r\big)$ of radius r with centers $a_k, b_k \in Q_{(0,0)}$ and such that $|a_k - b_k| = \mu = 6r$. Let $\nu_{\mathcal{S}}$ stands for the concentration of stadiums in the unit cell, i.e.

$$\nu_{\mathcal{S}} = N(r^2\pi + 2r\mu).$$

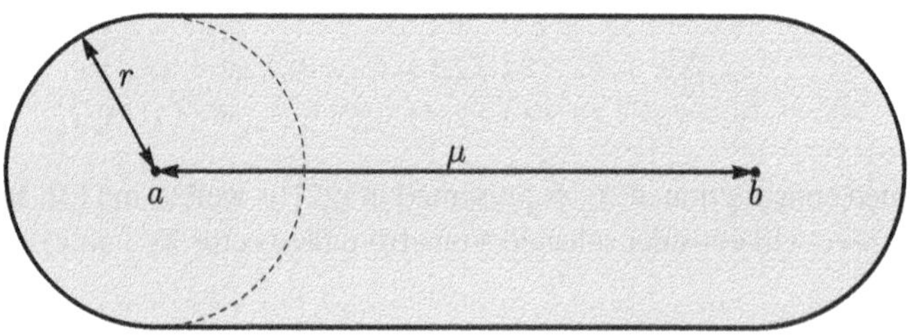

Fig. 2 Stadium with centers a and b, radius r and distance between centers equal to μ

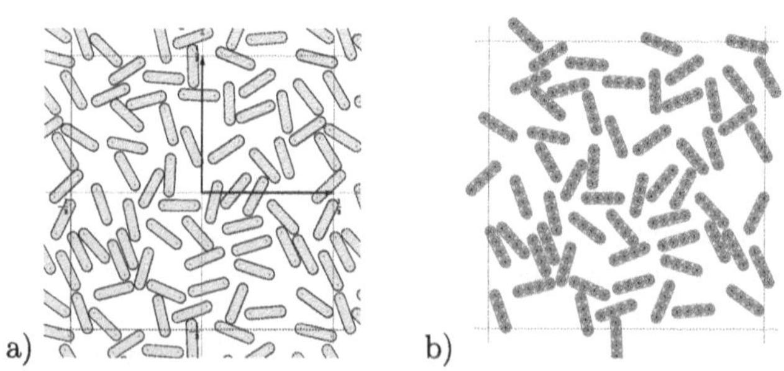

Fig. 3 Doubly periodic unit cell $Q_{(0,0)}$ with a configuration of non-overlapping identical stadiums (**a**) and their approximations (**b**)

We will consider distributions of stadiums generated by two algorithms. First algorithm (A) is based on the RSA method - this algorithm is described in details in [4]. Second algorithm (B) generate distributions of stadiums with deterministically fixed orientations of objects (description of this algorithm can be found in [3]). In order to study the distributions obtained by algorithms A and B, we approximate stadiums by disks. Let's consider four types of such approximations with radii equal to the radius of the stadium:

- approximation by one disk with center in geometric center of stadium;
- approximation by two disks with centers in centers of stadium;
- approximation by three disks with centers in geometric center and centers of stadium;
- approximation by the chain of four disks (see Fig. 3b).

Using algorithms A and B we generated $M = 50$ distributions containing $N = 300$ stadiums for $\nu_S \in \{0.15, 0.25, 0.35\}$. In the next step, we calculated the feature vector X_5 considered approximations. Selected results are presented in Figs. 4, 5, 6 and 7 show comparison of real parts of values $e_{2,2}$, $e_{3,3}$, $e_{4,4}$ and $e_{5,5}$ for two sets of 50 samples (M) drawn from distributions of stadiums. Black points denote distributions generated by algorithm A, gray points denote distributions generated by algorithm B.

4 Conclusions

By analysis of the obtained results one can notice the following remarks:

- approximation of stadiums (such that $\mu = 6r$) by one disk is insufficient to study their distributions;
- approximation of stadiums using two disks seems to be the best; moreover, for higher concentrations, approximation of stadiums by three or four disks may lead to less expressive distinction between the considered distributions;
- as the concentration increases, the distributions generated by the A and B algorithms are less and less distinguishable;
- the parameter $e_{4,4}$ seems to be the best for analyzing the considered distributions.

In order to calculate structural features, one can use the Python software package `basicsums` [23, 24] providing high level of abstraction in the computation of structural sums using algorithms reported in [20] and [21].

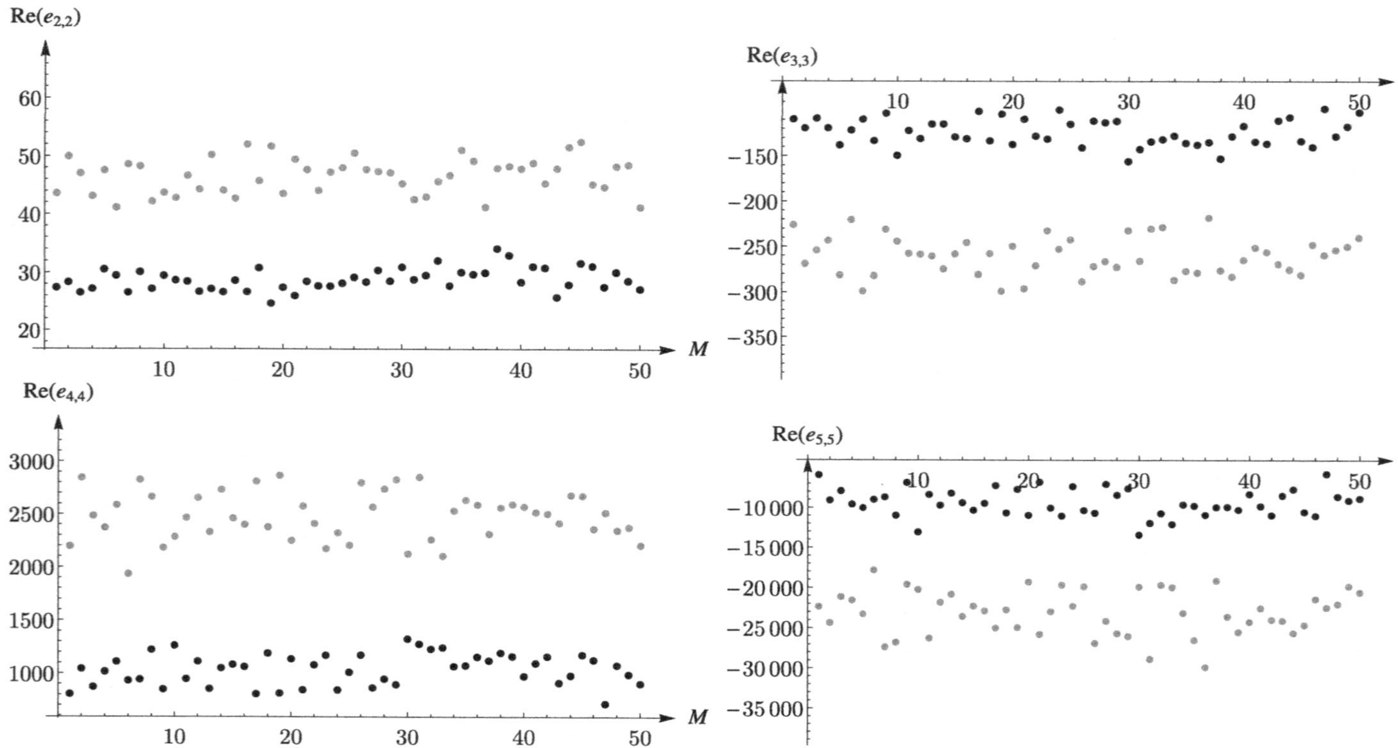

Fig. 4 Real parts of structural sums for stadiums approximated by two disks; $\nu_S = 0.15$

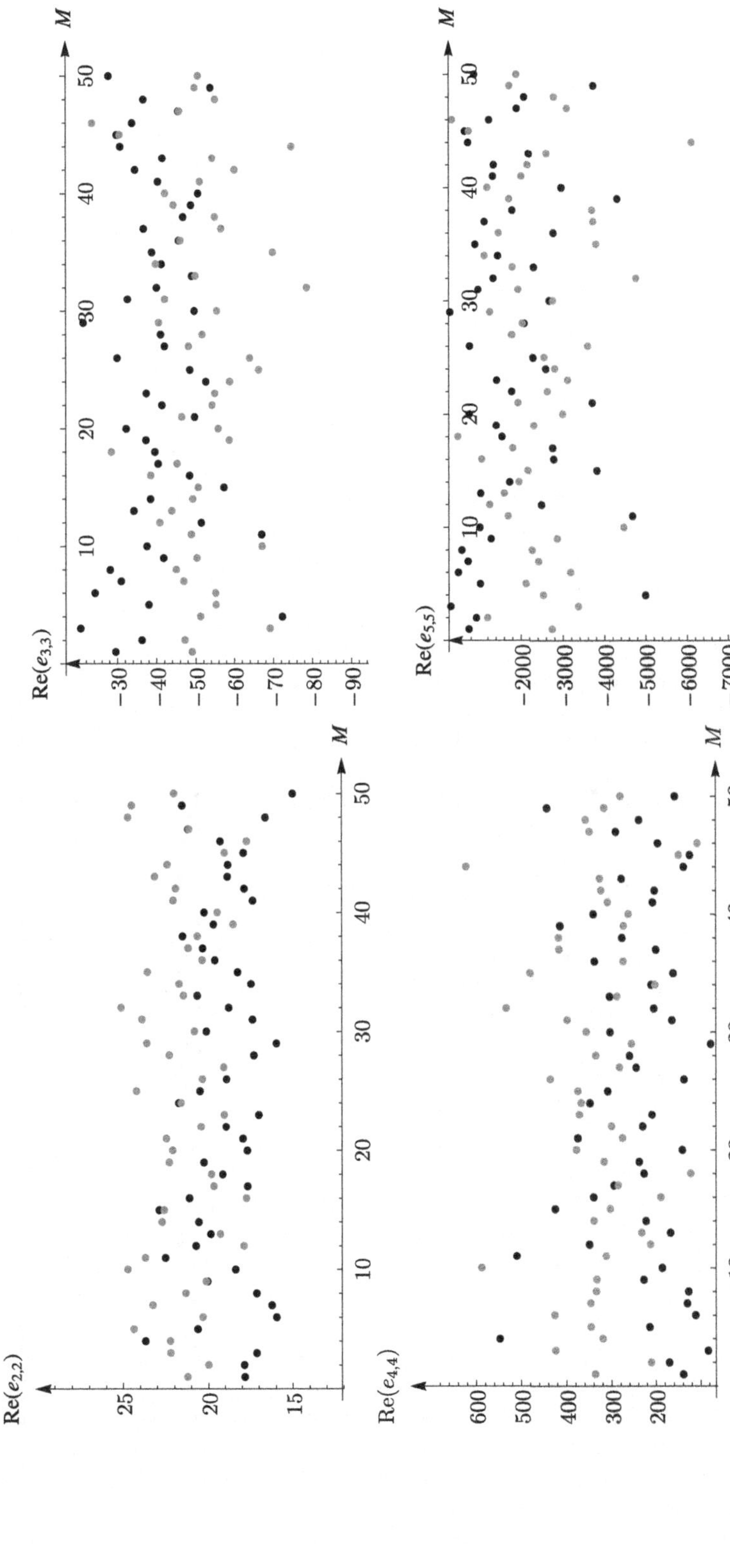

Fig. 5 Real parts of structural sums for stadiums approximated by one disk; $\nu_S = 0.25$

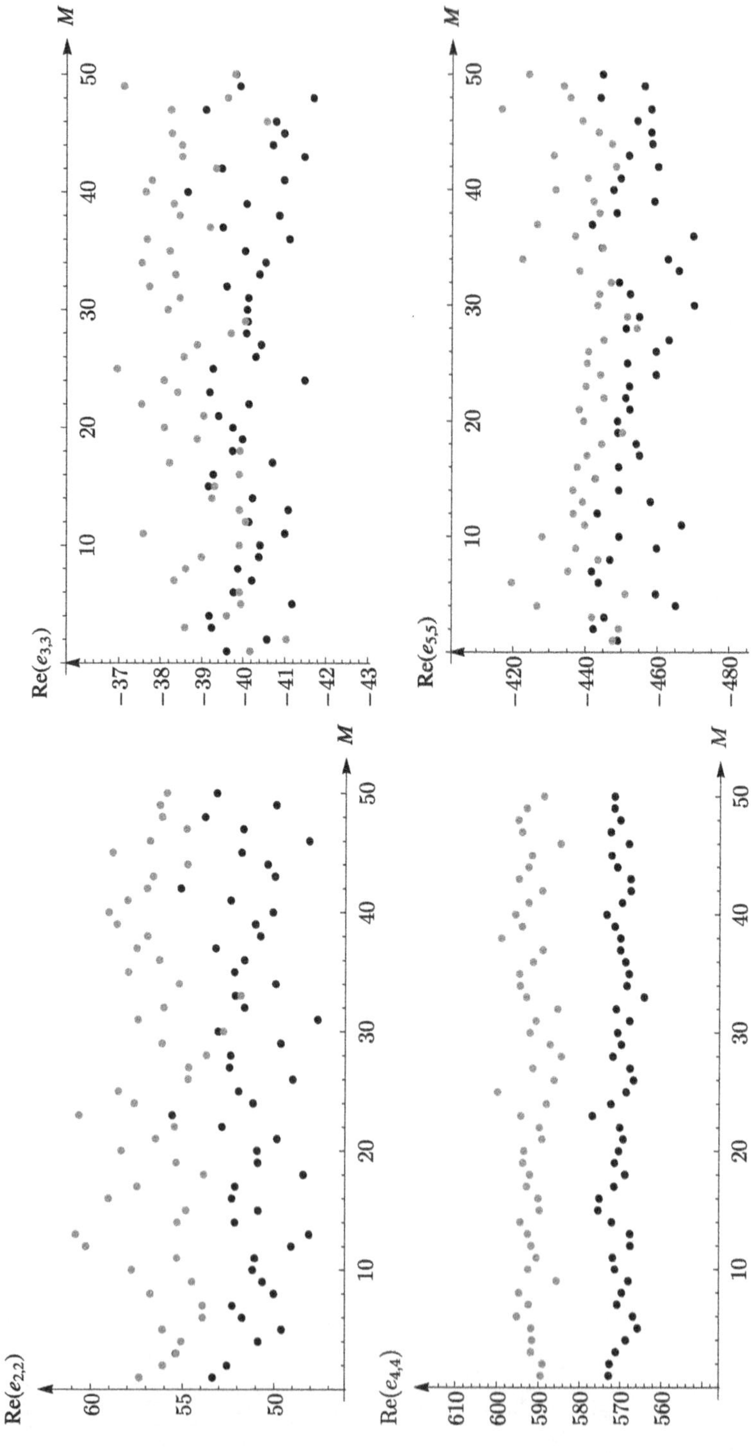

Fig. 6 Real parts of structural sums for stadiums approximated by four disks; $\nu_S = 0.25$

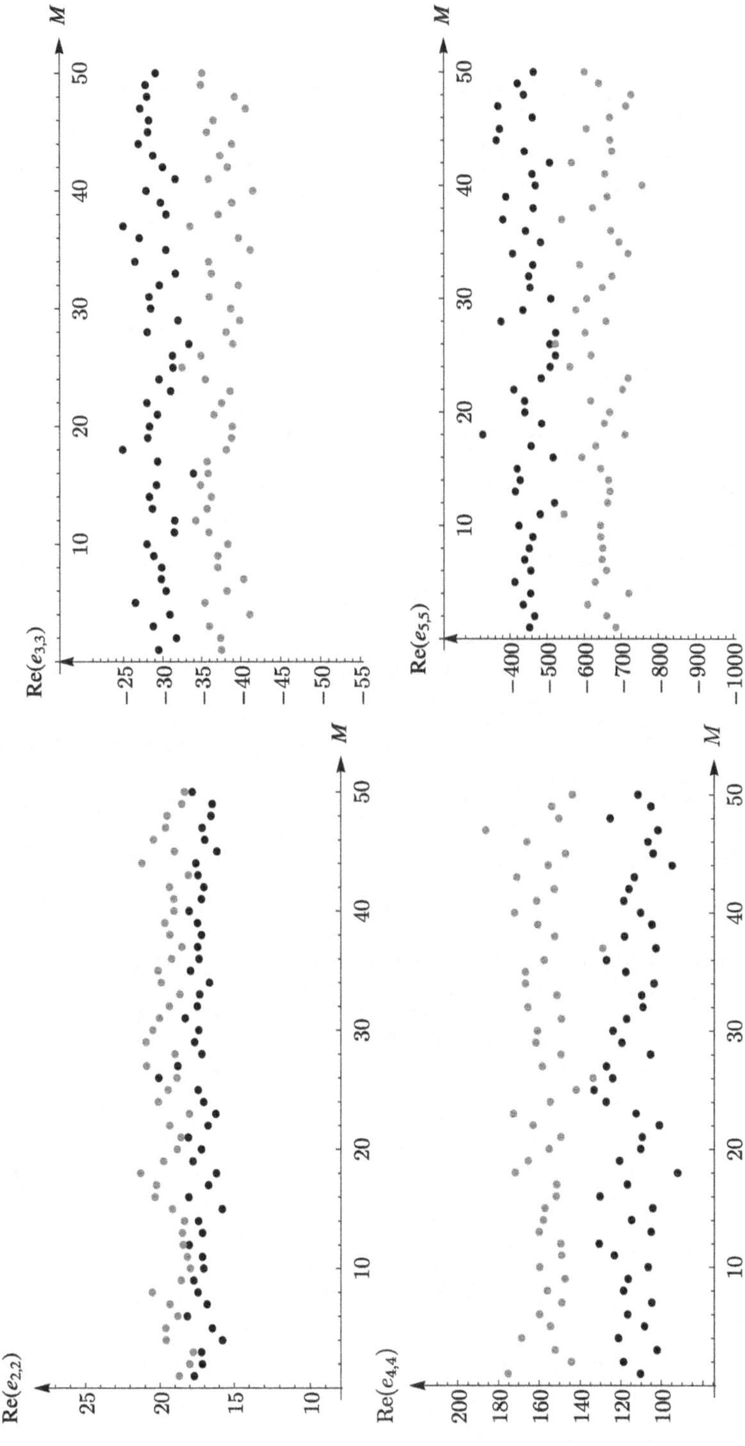

Fig. 7 Real parts of structural sums for stadiums approximated by two disks; $\nu_S = 0.35$

References

1. O. Bar, Fast algorithm to determine the flux around closely spaced non-overlapping disks, in *New Trends in Analysis and Interdisciplinary Applications, Selected Contributions of the 10th ISAAC Congress* (Springer International Publishing, Cham, 2017)
2. W. Baran, K. Kurnik, W. Nawalaniec, A. Shareif, Local stationary heat fields in fibrous composites. Silesian J. Pure Appl. Math. **9(1)**, 1–8 (2019)
3. R. Czapla, Basic sums as parameters characterizing. Silesian J. Pure Appl. Math. **6**(1), 85–96 (2016)
4. R. Czapla, Random sets of stadiums in square and collective behavior of bacteria. IEEE/ACM Trans. Computat. Biol. Bioinf. **15**, 251–256 (2018)
5. R. Czapla, V.V. Mityushev, A criterion of collective behavior of bacteria. Math. Biosci. Eng. **14**, 277–287 (2017)
6. R. Czapla, V.V. Mityushev, E. Pesetskaya, An analytical formula for the effective conductivity of 2D domains with cracks of high density. Appl. Math. Modell. **53**, 214–222 (2018)
7. P. Drygaś, A functional-differential equation in a class of analytic functions and its application. Aequationes Math. **73**(3), 222–232 (2007)
8. P. Drygaś, Functional-differential equations in a class of analytic functions and its application to elastic composites. Complex Var. Elliptic Eq. **61**(8), 1145–1156 (2016)
9. P. Drygaś, Generalized Eisenstein functions. J. Math. Analy. Appl. **444**(2), 1321–1331 (2016)
10. P. Drygaś, V. Mityushev, Effective elastic properties of random two-dimensional composites. Int. J. Solids Struct. **97–98**, 543–553 (2016)
11. P. Drygaś, V. Mityushev, Contrast expansion method for elastic incompressible fibrous composites. Adv. Math. Phys. **2017** art. no. 4780928 (2017)
12. P. Drygaś, S. Gluzman, V. Mityushev, W. Nawalaniec, Effective elastic constants of hexagonal array of soft fibers. Computat. Mat. Sci. **139**, 395–405 (2017)
13. P. Drygaś, S. Gluzman, V. Mityushev, W. Nawalaniec, *Applied Analysis of Composite Media: Analytical and Computational Results for Materials Scientists and Engineers* (Woodhead Publishing, Sawston, 2019)
14. S. Gluzman, V. Mityushev, W. Nawalaniec, *Computational Analysis of Structured Media* (Academic Press, Cambridge, 2017)
15. V. Mityushev, Representative cell in mechanics of composites and generalized Eisenstein - Rayleigh sums. Complex Variables **51**(8–11), 1033–1045 (2006)
16. V. Mityushev, Cluster method in composites and its convergence. Appl. Math. Lett. **77**, 44–48 (2018)
17. V. Mityushev, W. Nawalaniec, Basic sums and their random dynamic changes in description of microstructure of 2D composites. Computat. Mat. Sci. **97**, 64–74 (2015)
18. V. Mityushev, N. Rylko, Optimal distribution of the nonoverlapping conducting disks. Multiscale Model. Simul. SIAM **10**(1), 180–190 (2012)
19. V. Mityushev, N. Rylko, M. Bryła, Conductivity of two-dimensional composites with randomly distributed elliptical inclusions. ZAMM **98**, 512–516 (2018)
20. W. Nawalaniec, Algorithms for computing symbolic representations of basic *e*–sums and their application to composites. J. Symb. Computat. **74**, 328–345 (2016)
21. W. Nawalaniec, Comp. Appl. Math. (2017). https://doi.org/10.1007/s40314-017-0449-6
22. W. Nawalaniec, Classifying and analysis of random composites using structural sums feature vector. Proc. Roy. Soc. Math. Phys. Eng. Sci. **475**(2225), 20180698 (2019). https://doi.org/10.1098/rspa.2018.0698
23. W. Nawalaniec, Basicsums: A Python package for computing structural sums and the effective conductivity of random composites. J. Open Source Softw. **4**(39), 1327 (2019). https://doi.org/10.21105/joss.01327
24. W. Nawalaniec. The `basicsums` Python package documentation. https://basicsums.bitbucket.io

25. P. Kurtyka, N. Rylko, Quantitative analysis of the particles distributions in reinforced composites. Composite Struct. **182**, 412–419 (2017)
26. N. Rylko, Effective anti-plane properties of piezoelectric fibrous composites. Acta Mechanica **224**, 2719–2734 (2013)
27. N. Rylko, Edge effects for heat flux in fibrous composites. Appl. Math. Comput. **70**, 2283–2291 (2015)
28. N. Rylko, Fractal local fields in random composites. Appl. Math. Comput. **69**, 247–254 (2015)
29. S. Torquato, *Random Hetrogeneous Materials: Microstructure and Macroscopic Properties* (Spinger, Berlin, 2002)
30. S. Yakubovich, P. Drygaś, V. Mityushev, Closed-form evaluation of two-dimensional static lattice sums. Proc. Roy. Soc. A Math. Phys. Eng. Sci. **472**(2195), 20160510 (2016)

Boundary Value Problems and Their Applications to 2D Composites Theory

Drygaś Piotr

Abstract Analytical effective formulas are derived separately for conductivity and for elasticity. We consider 2D material with circular inclusions with different radii and different properties (n-phase material). We derive new analytical formulas determining the effective properties of such materials. They are connected by *structural basic sum* expressed through the Eisenstein and Natanzon functions.

Keywords Elasticity · Conductivity · Complex potential · Random composite

Mathematics Subject Classification (2010) Primary 74B05; Secondary 76T99, 30E25

1 Introduction

Traditional computational methods for calculation of the effective properites of random composites are usually based on the repeated applications of the pure numerical schemes to simulated random structures [1]. Application of analytical methods reduces the computational expenses [2] and gives an effective possibility to study structures whose representative cells contains more than 1000 inclusions per cell [3–5].

In the present paper, we make the next step to reduce the computational expenses for a class of random composites. It was noted in [2, 6, 7] that application of the Schwarz alternating method to multi-phase dispersed composites leads to a representation of the effective constants as a series in the concentration f of inclusions with coefficients explicitly decomposed onto terms with physical properties of components and terms depending only on the location of inclusions. This yields a simple method for computation of the effective properties of random

D. Piotr (✉)
Institute of Mathematics, University of Rzeszow, Rzeszow, Poland
e-mail: drygaspi@ur.edu.pl

P. Cerejeiras et al. (eds.), *Current Trends in Analysis, its Applications and Computation*, Research Perspectives,
https://doi.org/10.1007/978-3-030-87502-2_25

composites [8–10]. It should be noted that research is also being conducted into analytical formulas for elliptic inclusions [11]. In the present paper, we apply this method to 2D conducting composites with circular inclusions with the imperfect contact between the components. An analytic symbolic algorithm to solve the problem for random composites was proposed in [12]. The symbolic algorithm from [12] is developed in the present paper to deduce a decomposition formula for the effective conductivity of random composites. This decomposition yields explicitly the terms including the contrast parameter ρ_k and the thermal resistance $\mathcal{R}_k$ on the kth component of composite. When the distribution of ρ_k and $\mathcal{R}_k$ does not depend on the geometry, one can replace the huge expressions with these parameters by their mathematical expectations. As a result, we obtain an analytical approximate formula for the effective conductivity of the considered class of random composites in terms of the series in f. The effective conductivti expression has many apllication to study heat flux and local field in composite materials [9, 10, 13–15]. Effective conductivity problems are also used for biological problems [16, 17] In these paper we use basic sums of the multi-orser p, which was hard study in [18–22]

It is considered that for random composite materials, it is imposible to obtain exact analytical solution for the elastic moduli. Exists only approximate expresions most of which are only for small concentyration [23]. Recently in works [24–26] this defect has been removed. In this paper, we present some results coincides effective conductivity and effective elastic properties.

Consider the square lattice $\mathcal{Q}$, the set of points $\mathbb{Z}[i] = \{m_1 + im_2 : m_1, m_2 \in \mathbb{Z}\}$, on the complex plane $\mathbb{C}$ determined by the pair of periods 1 and $i = \sqrt{-1}$. The zero-th square cell Q is defined as

$$Q = Q_{(0,0)} = \left\{ z = z_1 + iz_2 \in \mathbb{C} : -\frac{1}{2} < t_1, t_2 < \frac{1}{2} \right\}.$$

Here, $\mathbb{Z}$ stands for the set of integer numbers. Introduce the cells $Q_{(m_1,m_2)} = Q_{(0,0)} + m_1 + im_2$.

Consider N non-overlapping equal disks $D_k = \{z \in \mathbb{C} : |z - a_k| < r\}$ of radius r with the centers $a_k \in Q$ $(k = 1, 2, \ldots, N)$. Let each circle $L_k = \{z \in \mathbb{C} : |z - a_k| = r\}$ leave D_k to the left. Introduce the multiply connected domain $D = Q \setminus \cup_{k=1}^{N} (D_k \cup L_k)$, the complement of all the closures of D_k to Q.

Introduce

$$r^2 = \frac{1}{N} \sum_{k=1}^{N} r_k^2,$$

which is proportional to concentration f, i.e.,

$$r_k^2 = R_k r^2,$$

where R_k is the proportionality ratio.

2 General Formulae for Effective Conductivity

Consider a conducting infinite host medium of the normalized unit conductivity with the inclusions $D_k + m$ ($m \in \mathbb{Z}[i]$) of conductivity $\lambda_k > 0$, respectively.

The temperature distribution in the considered composite is expressed by the function

$$u(z) = \begin{cases} u_0(z), & z \in D, \\ u_k(z), & z \in D_k, \end{cases} \tag{2.1}$$

harmonic in every component of the composite, continuously differentiable in the closures of D and D_k. Let $\mathbf{n} = (n_1, n_2)$ denote the outward unit normal vector to the circle L_k and $\frac{\partial}{\partial \mathbf{n}}$ the normal derivative. Introduce the Kapitza interface resistance $\mathcal{R}_k$ on L_k [23]. The following conjugation conditions express the imperfect contact between the components

$$\frac{\partial u_0}{\partial \mathbf{n}}(t) = \lambda_k \frac{\partial u_k}{\partial \mathbf{n}}(t), \tag{2.2}$$

$$\mathcal{R}_k r_k \frac{\partial u_k}{\partial \mathbf{n}}(t) + u_k(t) - u_0(t) = 0, \quad t \in L_k \ (k = 1, 2, \ldots, N). \tag{2.3}$$

The relations (2.2)–(2.3) model the imperfect thermal contact between inclusion and matrix. Equation (2.2) expresses equality of the normal heat flux on L_k. Equation (2.3) describes a linear relation between the jump of the temperature across the boundary of inclusion and the normal heat flux on the interface. The case $\mathcal{R}_k = 0$ corresponds to the perfect thermal contact between fiber and matrix, while $\mathcal{R}_k = +\infty$ corresponds to the perfect thermal insulator.

It is assumed that the external flux is applied in the x–direction and it is normalized in such a way that the function $u_0(z) - x$ is doubly periodic in the perforated domain $\mathcal{D} = \cup_{m \in \mathbb{Z}[i]}(D + m)$. The function $u_0(z)$ has the unit jump per a cell along the x–axis and is periodic along the y–axis

$$u_0(z+1) - u_0(z) = 1, \quad u_0(z+i) - u_0(z) = 0. \tag{2.4}$$

Introduce the complex potentials $\varphi_k(z)$ analytic in D_k respectively, continuously differentiable in the closures of the considered domains. The harmonic and the analytic functions are related by the equalities

$$u_0(z) = \mathrm{Re}\ \varphi_0(z), \ z \in D, \qquad u_k = \mathrm{Re}\ \varphi_k(z), \ z \in D_k. \tag{2.5}$$

The heat flux $\nabla u(x, y)$ is determined by means of the complex potentials

$$\psi_k(z) := \frac{\partial \varphi_k}{\partial z} = \frac{\partial u_k}{\partial x} - i\frac{\partial u_k}{\partial y}, \quad z \in D_k,\ k = 0, 1, \ldots, n. \tag{2.6}$$

The problem (2.2)–(2.3) is reducing to the followign boundary problem

$$\psi^-(t) = \left(\frac{1}{1-\rho_k} + \frac{\mathcal{R}_k}{2}\right)\psi_k(t) + \left(\frac{\rho_k}{1-\rho_k} - \frac{\mathcal{R}_k}{2}\right)\left(\frac{r_k}{t-a_k}\right)^2 \overline{\psi_k(t)}$$
$$+ \frac{\mathcal{R}_k}{2}(t-a_k)\psi_k'(t) - \frac{\mathcal{R}_k}{2}\frac{r_k^4}{(t-a_k)^3}\overline{\psi_k'(t)}, \tag{2.7}$$

where $|t - a_k| = r_k$.

Introduce the contrast parameter ρ_k assigned to the kth inclusion

$$\rho_k = \frac{\lambda_k - 1}{\lambda_k + 1} \tag{2.8}$$

and the constant parameters

$$\Theta_k^{(l)} = \left(\frac{1}{1-\rho_k} + (l+1)\frac{\mathcal{R}_k}{2}\right)^{-1}, \quad \Omega_k^{(l)} = \frac{\rho_k}{1-\rho_k} - (l+1)\frac{\mathcal{R}_k}{2} \tag{2.9}$$

for $k = 1, 2, \ldots, N$ and $l = 0, 1, \ldots$. This problem (2.2)–(2.3) was solved in [12] by method of successive approximation for Taylor coefficients of potentials ψ.

In the case of macroscopically isotropic composites, the effective conductivity tensor becomes $\hat{\boldsymbol{\lambda}} = \hat{\lambda} I$, where $\hat{\lambda}$ stands for the scalar effective conductivity and I for the identity matrix. The following formula was deduced in [12]

$$\hat{\lambda} = 1 + 2\pi r^2 \sum_{k=1}^{N} \Omega_k^{(0)} \psi_k(a_k). \tag{2.10}$$

Such formulae were obtained and described in [12] for deterministic composites.

Let $f = N\pi r^2$ denote the concentration of inclusions in composite. Then, the effective conductivity (2.10) can be written in the form of series in f

$$\hat{\lambda} = 1 + 2f \sum_{p=0}^{\infty} \mathbf{A}_p f^p, \tag{2.11}$$

where the first few coefficients $\mathbf{A}_p$ have the form

$$\mathbf{A}_0 = \frac{1}{N}\sum_{k=1}^{N}\tilde{\rho}_k^{(0)}, \quad \mathbf{A}_1 = \frac{1}{\pi}g_2^{(0)}, \quad \mathbf{A}_2 = \frac{1}{\pi^2}g_{22}^{(00)},$$
$$\mathbf{A}_3 = \frac{1}{\pi^3}\left[-2g_{33}^{(10)} + g_{222}^{(000)}\right], \tag{2.12}$$
$$\mathbf{A}_4 = \frac{1}{\pi^4}\left[3g_{44}^{(20)} - 2(g_{332}^{(100)} + g_{233}^{(010)}) + g_{2222}^{(0000)}\right],$$

where the multiple convolution sums is defined by

$$g_{m_1,\dots,m_q}^{(l_1,\dots,l_q)} = \frac{1}{N^{1+\frac{1}{2}(m_1+\dots+m_q)}}\sum_{k_0,k_1,\dots,k_q}\tilde{\rho}_{k_0}^{(0)}\tilde{\rho}_{k_1}^{(l_1)}\cdots\tilde{\rho}_{k_q}^{(l_q)}\times$$
$$E_{m_1}(a_{k_0} - a_{k_1})\overline{E_{m_2}(a_{k_1} - a_{k_2})\dots\mathbf{C}^{q+1}E_{m_q}(a_{k_{q-1}} - a_{k_q})}, \tag{2.13}$$

where $\mathbf{C}$ denote the operator of complex conjugation,

$$\tilde{\rho}_k^{(l)} = R_k^{l+1}\Omega_k^{(l)}\Theta_k^{(l)} = R_k^{l+1}\frac{\rho_k - (1-\rho_k)(l+1)\frac{\mathcal{R}_k}{2}}{1+(1-\rho_k)(l+1)\frac{\mathcal{R}_k}{2}}, \tag{2.14}$$

$$E_2(z) = \wp(z) + S_2$$

and for $n > 2$

$$E_n(z) = \frac{(-1)^n}{(n-1)!}\frac{d^{n-2}}{dz^{n-2}}\wp(z).$$

Properties of the Eisenstein functions are described in [18, 20, 21, 27, 28].

The next coefficients can be explicitly written by using of the symbolic computation code. The structure of $\mathbf{A}_p$ as linear combinations of $g_{m_1,\dots,m_q}^{(l_1,\dots,l_q)}$ follows form the iterative scheme. The obtained expressions can be considered as a decomposition of the effective conductivity on the "geometrical" and "physical" terms as follows. First, the formula (2.11) contains the sum of terms with the powers of the concentration f. The coefficients $\mathbf{A}_p$ are presented in (2.12) as linear combinations of the sums (2.13). The latter sums $g_{m_1,\dots,m_q}^{(l_1,\dots,l_q)}$ are decomposed onto the sum of pure geometric objects expressed by means of the Eisenstein functions with the pure physical multipliers $\tilde{\rho}_{k_0}^{(0)}\tilde{\rho}_{k_1}^{(l_1)}\cdots\tilde{\rho}_{k_q}^{(l_q)}$. Therefore, we have the explicit decomposition of the effective conductivity on the powers of f, the physical parameters $\tilde{\rho}_k$ and the geometrical parameters $E_m(a_k - a_l)$ calculated by the centers of inclusions.

3 General Formulae for Effective Shear Moduli

3.1 Statement the Problem

The component of the stress tensor can be determined by the Kolosov-Muskhelishvili formulae [29]

$$\sigma_{xx} + \sigma_{yy} = \begin{cases} 4\mathrm{Re}\ \varphi_k'(z), \ z \in D_k, \\ 4\mathrm{Re}\ \varphi_0'(z), \ \ z \in D, \end{cases} \tag{3.1}$$

$$\sigma_{xx} - \sigma_{yy} + 2i\sigma_{xy} = \begin{cases} -2\left[z\overline{\varphi_k''(z)} + \overline{\psi_k'(z)}\right], \ z \in D_k, \\ -2\left[z\overline{\varphi_0''(z)} + \overline{\psi_0'(z)}\right], \ \ z \in D, \end{cases}$$

where Re denotes the real part and the bar the complex conjugation. Let $\begin{pmatrix} \sigma_{xx}^{\infty} & \sigma_{xy}^{\infty} \\ \sigma_{yx}^{\infty} & \sigma_{yy}^{\infty} \end{pmatrix}$ be the stress tensor applied at infinity. Following [29] introduce the constants

$$B_0 = \frac{\sigma_{xx}^{\infty} + \sigma_{yy}^{\infty}}{4}, \quad \Gamma_0 = \frac{\sigma_{yy}^{\infty} - \sigma_{xx}^{\infty} + 2i\sigma_{xy}^{\infty}}{2}. \tag{3.2}$$

Then,

$$\varphi_0(z) = B_0 z + \varphi(z), \quad \psi_0(z) = \Gamma_0 z + \psi(z), \tag{3.3}$$

where $\varphi(z)$ and $\psi(z)$ are analytical in D and bounded at infinity. The functions $\varphi_k(z)$ and $\psi_k(z)$ are analytical in D_k and twice differentiable in the closures of the considered domains. The displacement (u, v) are calculated by formulae [29]

$$u + iv = \begin{cases} \frac{1}{2G_k}\left(\kappa_k\varphi_k(t) - t\overline{\varphi_k'(t)} - \overline{\psi_k(t)}\right), \ z \in D_k, \\ \frac{1}{2G}\left(\kappa\varphi_0(t) - t\overline{\varphi_0'(t)} - \overline{\psi_0(t)}\right), \ \ z \in D. \end{cases} \tag{3.4}$$

The perfect bonding at the matrix-inclusion interface present displacement and traction vectors continuity across the interface, can be expressed by two equations [29]

$$\varphi_k(t) + t\overline{\varphi_k'(t)} + \overline{\psi_k(t)} = \varphi_0(t) + t\overline{\varphi_0'(t)} + \overline{\psi_0(t)}, \tag{3.5}$$

$$\kappa_k\varphi_k(t) - t\overline{\varphi_k'(t)} - \overline{\psi_k(t)} = \frac{G_k}{G}\left(\kappa\varphi_0(t) - t\overline{\varphi_0'(t)} - \overline{\psi_0(t)}\right). \tag{3.6}$$

The problem (3.5)–(3.6) is the classic boundary value problem of the plane elasticity. It was discussed in many works by various methods [29]. Below, we concentrate our attention to its analytical solution. Introduce the following nondimensional contrast parameters

$$\varrho_{1k,m} = \frac{G_m - G}{G_k + \kappa_k G} \cdot \frac{G_k}{G_m}, \; \varrho_{2k,m} = \frac{\kappa G_m - \kappa_m G}{\kappa G_k + G} \cdot \frac{G_k}{G_m}, \; \varrho_{3k,m} = \frac{G_m - G}{\kappa G_k + G} \cdot \frac{G_k}{G_m}.$$

In the case of $k = m$ for brevity we use notations $\varrho_{ik,k} = \varrho_{ik}$ $(i = 1, 2, 3)$.

The problem (3.5)–(3.6) was reduce to the system of functional equations and solved by method of successive approximation [26].

3.2 *Effective Constants Up to $O(f^3)$*

The effective shear moduli is calculted from formula [26]

$$\frac{G_e}{G} = \frac{1 + \operatorname{Re} A}{1 - \kappa \operatorname{Re} A}. \tag{3.7}$$

where

$$A = \sum_{s=1}^{\infty} A^{(s)} f^s,$$

and

$$A^{(s)} = \frac{i}{n^s \pi^{s-1}} \sum_{k=1}^{n} R_k \frac{\varrho_{3k}}{1 + \varrho_{3k}} \left(3 R_k \overline{\alpha}_{k,3}^{(s-2)} + 2 a_k \overline{\alpha}_{k,2}^{(s-1)} + \overline{\beta}_{k,1}^{(s-1)} \right). \tag{3.8}$$

The few initial coefficients o A has the form

$$A^{(1)} = \frac{1}{n} \sum_{k=1}^{n} R_k \varrho_{3k}, \; A^{(2)} = -\frac{2}{n^2 \pi} \sum_{k=1}^{n} \sum_{m=1}^{n} R_k R_m \varrho_{3k} \varrho_{3m} \overline{E_3^{(1)}(a_k - a_m)}$$

$$A^{(3)} = \frac{1}{n^3 \pi^2} \sum_{m_1=1}^{n} \sum_{m_2=1}^{n} \sum_{m_3=1}^{n} \Big[4 R_{m_1} R_{m_2} R_{m_3} \varrho_{3m_1} \varrho_{3m_2} \varrho_{3m_3}$$

$$\times \overline{E_3^{(1)}(a_{m_1} - a_{m_2})} E_3^{(1)}(a_{m_2} - a_{m_3})$$

$$+ R_{m_1} R_{m_2} R_{m_3} \frac{\varrho_{3m_1} (\varrho_{1m_2} - \varrho_{3m_2}) \varrho_{3m_3}}{(1 + \varrho_{1m_2}) \varrho_{3m_2}} \overline{E_2(a_{m_1} - a_{m_2})} E_2(a_{m_2} - a_{m_3})$$

$$-R_{m_1}R_{m_2}R_{m_3}\frac{\varrho_{3m_1}(\varrho_{1m_2}-\varrho_{3m_2})\varrho_{3m_3}}{(1+\varrho_{1m_2})\varrho_{3m_2}}\overline{E_2(a_{m_1}-a_{m_2})E_2(a_{m_2}-a_{m_3})}\Bigg]$$

$$+\frac{3}{n^3\pi^2}\sum_{m_1=1}^{n}\sum_{m_2=1}^{n}R_{m_1}R_{m_2}(R_{m_1}+R_{m_2})\varrho_{3m_1}\varrho_{3m_2}\overline{E_4(a_{m_1}-a_{m_2})}\,, \tag{3.9}$$

where the Natanzon functions are used

$$E_p^{(1)}(z)=\frac{(-1)^p}{(p-1)!}\left(\overline{z}\frac{d^{p-2}}{dz^{p-2}}\wp(z)-\frac{d^{p-2}}{dz^{p-2}}\wp_1(z)\right), \tag{3.10}$$

$$\wp_1'(z)=-2\sum_{\substack{m\in\mathbb{Z}[i]\\ m\neq 0}}\left(\frac{\overline{m}}{(z-m)^3}+\frac{\overline{m}}{m^3}\right). \tag{3.11}$$

References

1. I.I. Zohdi, P. Wriggers, *Introduction to Computational Micromechanics* (Springer, Berlin, 2005)
2. S. Gluzman, V. Mityushev, W. Nawalaniec, *Computational Analysis of Structured Media* (Academic/Elsevier, Cambridge/Amsterdam, 2018)
3. N. Rylko, Effective anti-plane properties of piezoelectric fibrous composites. Acta Mech. **224**, 2719–2734 (2013)
4. D. Legut, U. Wdowik, P. Kurtyka, Vibrational and dielectric properties of α-Si3N4 from density functional theory. Mat. Chem. Phys. **147**(1–2), 42–49 (2014)
5. P. Kurtyka, N. Rylko, Quantitative analysis of the particles distributions in reinforced composites. Composite Struct. **182**, 412–419 (2017)
6. V. Mityushev, Cluster method in composites and its convergence. Appl. Math. Lett. **77**, 44–48 (2018)
7. V. Mityushev, N. Rylko, Optimal distribution of the nonoverlapping conducting disks. Multiscale Model. Simul. **10**(1), 180–190 (2012)
8. R. Czapla, V.V. Mityushev, E. Pesetskaya, An analytical formula for the effective conductivity of 2D domains with cracks of high density. Appl. Math. Modell. **53**, 214–222 (2018)
9. N. Rylko, Edge effects for heat flux in fibrous composites. Appl. Math. Comput. **70**, 2283–2291 (2015)
10. N. Rylko, Fractal local fields in random composites. Appl. Math. Comput. **69**, 247–254 (2015)
11. V. Mityushev, N. Rylko, M. Bryła, Conductivity of two-dimensional composites with randomly distributed elliptical inclusions. ZAMM **98**, 512–516 (2018)
12. P. Drygaś, V. Mityushev, Effective conductivity of unidirectional cylinders with interfacial resistance. Quarterly J. Mech. Appl. **62**(3), 235–262 (2009)
13. S. Gluzman, V. Mityushev, W. Nawalaniec, G. Sokal, Random composite: stirred or shaken? Archives Mech. **68**(3), 229–241 (2016)
14. O. Bar, Fast algorithm to determine the flux around closely spaced non-overlapping disks, in *New Trends in Analysis and Interdisciplinary Applications, Selected Contributions of the 10th ISAAC Congress* (Springer International Publishing, Cham, 2017)
15. W. Baran, K. Kurnik, W. Nawalaniec, A. Shareif, Local stationary heat fields in fibrous composites. Silesian J. Pure Appl. Math. **9**(1), 1–8 (2019)

16. R. Czapla, V.V. Mityushev, A criterion of collective behavior of bacteria. Math. Biosci. Eng. **14**, 277–287 (2017)
17. R. Czapla, Random sets of stadiums in square and collective behavior of bacteria. IEEE/ACM Trans. Computat. Biol. Bioinf. **15**, 251–256 (2018)
18. W. Nawalaniec, Algorithms for computing symbolic representations of basic e-sums and their application to composites. J. Symb. Comput. **74**, 328–345 (2016)
19. W. Nawalaniec, Classifying and analysis of random composites using structural sums feature vector. Proc. Roy. Soc. A Math. Phys. Eng. Sci. **475**(2225), 20180698 (2019)
20. W. Nawalaniec, Efficient computation of basic sums for random polydispersed composites. Computat. Appl. Math. **37**(2), 2237–2259 (2017)
21. R. Czapla, Basic sums as parameters characterizing. Silesian J. Pure Appl. Math. **6**(1), 85–96 (2016)
22. V. Mityushev, W. Nawalaniec, N. Rylko, *Introduction to Mathematical Modeling and Computer Simulations* (CRC Taylor & Francis, Boca Raton, 2018)
23. G.W. Milton, *The Theory of Composites* (Cambridge University Press, Cambridge, 2002)
24. P. Drygaś, Functional-differential equations in a class of analytic functions and its application to elastic composites. Complex Variables Elliptic Equ. **61**(8), 1145–1156 (2016)
25. P. Drygaś, S. Gluzman, V. Mityushev, W. Nawalaniec, Effective elastic constants of hexagonal array of soft fibers. Computat. Mat. Sci. **139**, 395–405 (2017)
26. P. Drygaś, S. Gluzman, V. Mityushev, W. Nawalaniec, *Applied Analysis of Composite Media Analytical and Computational Results for Materials Scientists and Engineers* (Elsevier, Amsterdam, 2020)
27. P. Drygaś, Generalized Eisenstein functions. J. Math. Analy. Appl. **444**(2), 1321–1331 (2016)
28. S. Yakubovich, P. Drygaś, V. Mityushev, Closed-form evaluation of two-dimensional static lattice sums. Proc. Roy. Soc. A Math. Phys. Eng. Sci. **472**(2195), 20160510 (2016)
29. N.I. Muskhelishvili, *Some Mathematical Problems of the Plane Theory of Elasticity* (Nauka Moscow, 1966)

Selected Aspects of Visualization in Education of Functions

Ján Gunčaga

Abstract The aspect of visualization plays an important role in the calculus teaching. The notion of the function is basic notion at the lower secondary level and in the teacher training of future mathematics teachers. The notion of the graph of the function and his graphical and visual representation is important at upper secondary level and university level. If the student draws the graph of the function, then the work with graphical representation of the function can help to make his knowledge about notions such continuity, derivative more deep. The development of information and communication technologies (ICT) gives the possibility to use different tools (for example educational software) as a supporting aspect for mathematics education with better understanding. We would like to show some examples with the help of GeoGebra for motivational analysis teaching.

Keywords Function · Graph of the function · Calculus · Visualization

Mathematics Subject Classification (2010) Primary 97I20; Secondary 97I40

1 Introduction

Visual imagery has been an effective tool to communicate ideas connected with basic mathematics concepts since the dawn of mankind. The notion of the function belongs to these concepts. The development of educational visualization technology allows these ideas to be demonstrated with the help of some educational software. In this paper, we specifically consider the use of GeoGebra, a free, open-source educational application developed by an international consortium of mathematics and statistics educators, but other educational software could also be used for the same visualization tasks.

J. Gunčaga (✉)
Comenius University in Bratislava, Faculty of Education, Bratislava, Slovakia
e-mail: guncaga@fedu.uniba.sk

P. Cerejeiras et al. (eds.), *Current Trends in Analysis, its Applications and Computation*, Research Perspectives,
https://doi.org/10.1007/978-3-030-87502-2_26

Using ICT and educational software is effective tool for supporting visualization during the educational process in mathematics. These tools bring according [15] and [3] relevant contribution to:

- Effecting working processes and improving production, notably by increasing the speed and efficiency of such processes, and supporting visual presentation of results, so contributing to the pace and productivity of lessons;
- Supporting processes of checking, validating and refinement, notably with respect to checking and correcting elements of work, and testing and improving problem strategies and solutions; see also [8]
- Overcoming student difficulties and building assurance, notably by circumventing problems experienced by students when writing and drawing by hand, and easing correction of mistakes, so enhancing students sense of capability in their work;
- Enhancing the variety and appeal of classroom activity, notably by varying the format of lessons and altering their ambience by introducing elements of play, fun and excitement and supporting using real life problems;
- Using of different visual separate and universal models for understanding mathematic notions, their properties and relationships to other notions.

Visualization belongs according Brunner [1] to the stage of iconic forms and many notions in calculus teaching are taught without visual and graphical representation.

Weigand and Weth in [18] argue, that students need according the properties of mathematic notions adequate visual image of this notion, which alow graphically represent their main properties. We shows some examples for this aspect of mathematic teaching in the next part of the paper.

2 Threshold Concepts and Using ICT in the Education of Functions

A threshold concept according Meyer and Land [9] can be considered as akin to a portal, opening up a new and previously inaccessible way of thinking about something. It represents a transformed way of understanding, or interpreting, or viewing something without which the learner cannot progress. As a consequence of comprehending a threshold concept there may thus be a transformed internal view of subject matter, subject landscape, or even world view. This transformation may be sudden or it may be protracted over a considerable period of time, with the transition to understanding proving troublesome.

The use of ICT in mathematics education is possible to characterize on several levels (see also [10]):

Cognitive level ICT support students in various ways in the development of mathematical notions. Computer Algebra Systems (CAS) support functional thinking according Vollrath [17]. In the case of universal mathematical programs such GeoGebra—they integrate CAS, dynamic geometry systems (DGS) and spreadsheets—several different representations can be according [3] achieved in one teaching unit.

Affective level ICT increases the joy of the students in dealing with mathematics.

ICT tools They enable students to quickly and correctly solve problems with changed parameters (see also [4]).

We present now some threshold concepts used in education of functions [14]. The first notion will be a derivative. We use in our reasoning the approximate description of a real function f differentiable at x_0, when we use linear approximation (see [6]). We define the real function $f(x) = x^2 + 1.2|x - 2| - 2.2$.

The following figure show, how we can analyze with students graph of the function with educational software, if we can show via visualization, that some continuous function hasn't a derivative in some point (see Fig. 1).

Let us continue with the discussion of the local extremes of a real function. We define for this a new real function $g = x^4 - 0.02x^2 + 0.1$. We see in the next figure (see Fig. 2) the function g with three extremals, which we cannot see in the beginning (see [5]).

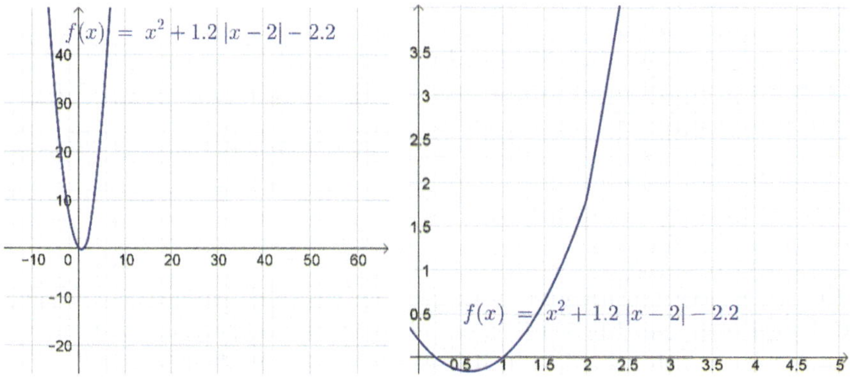

Fig. 1 The function f hasn't a derivative at the point 2

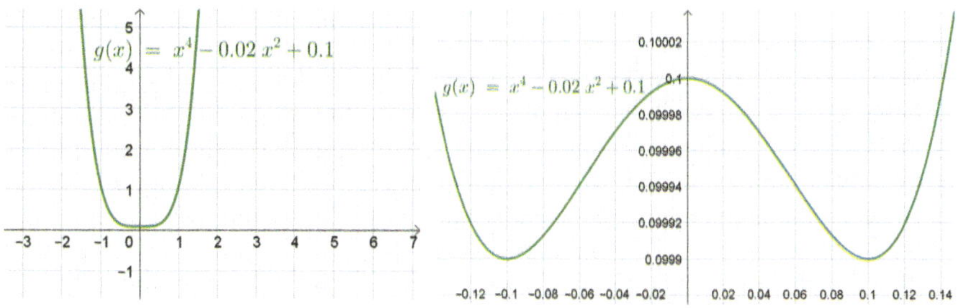

Fig. 2 The function g with the three extremals in two views

3 Visualization as a Helping Tool for Explanation of Misconceptions

Visualization with the graph of the function can help in the mathematics education, if we find some misconceptions connected with the notion of function. Many practical examples with concrete works prepared by students about different kind of functional properties show Sajka in [16]. Csachová and Jurečková in [2] gave attention to the analysis of the knowledge in understanding and correct solution of mathematical tasks with figures and graphs of the functions by students in secondary level.

The identification of problems in calculus teaching is good feedback for the preparation of the future mathematics teachers at the universities. We present now one situation from lesson with future teachers of mathematics at University Innsbruck in Austria. Student Simone solved following quadratic inequality in the wrong way in the end of solution:

$$x^2 + 7x + 10 > 0$$

$$\left(x + \frac{7}{2}\right)^2 - \frac{7^2}{4} + 10 > 0$$

$$\left(x + \frac{7}{2}\right)^2 > \frac{9}{4}$$

$$x + \frac{7}{2} > \pm\frac{3}{2}$$

$$x_1 > -2$$

$$x_2 > -5$$

This student forget connection between quadratic inequality with quadratic equation and graph of the quadratic function. We can start with quadratic equation $x^2 + 7x + 10 = 0$. It is possible to follow the style of solving strategy of this student.

$$x^2 + 7x + 10 = 0$$

$$\left(x + \frac{7}{2}\right)^2 - \frac{7^2}{4} + 10 = 0$$

$$\left(x + \frac{7}{2}\right)^2 = \frac{9}{4}$$

$$x + \frac{7}{2} = \pm\frac{3}{2}$$

$$x_1 = -2$$

$$x_2 = -5$$

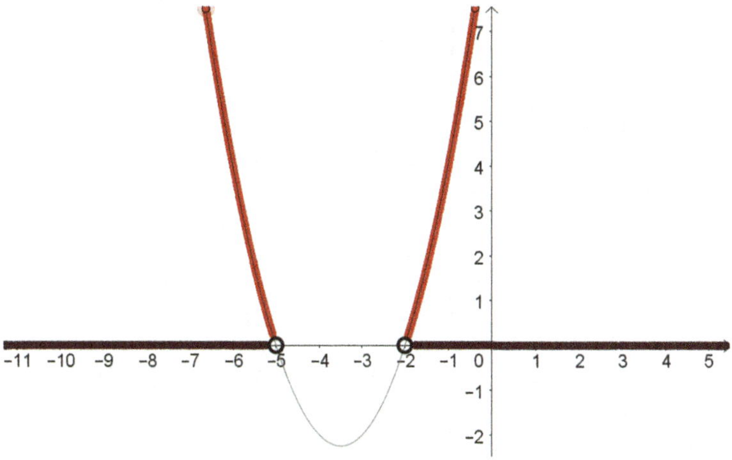

Fig. 3 The graphic solution for $x^2 + 7x + 10 > 0$

Now, we can draw the graph of the function $y = x^2 + 7x + 10$ and we can show graphically the solution for the quadratic inequality $x^2 + 7x + 10 > 0$. We see from the Fig. 3, that correct solution is the set $(-\infty, -5) \cup (-2, \infty)$. The notion of function belongs to some mathematical structure, many practical information is possible to find in Pauer and Stampfer [12].

4 The Notion of the Inverse Function

The notion of inverse function is interesting for discovery of graphs of function and its inverse function. Base property is, that these graphs are in the axial symmetry according axis $y = x$. We show this kind of discovery on example of the quadratic functions of type $x^2 + a$.

$f(x) = x^2 + a$, $x \geq 0$ is our function and the inverse function is $f^{-1}(x) = \sqrt{x - a}$, $x \geq a$. We can find common points of the graphs of these two functions:

$$f(x) = f^{-1}(x)$$
$$x^2 + a = \sqrt{x - a}$$
$$x^4 + 2ax^2 + a^2 = x - a$$
$$x^4 + 2ax^2 - x + a^2 + a = 0$$
$$x^4 + 2ax^2 + x^2 - x^2 - x + a^2 + a = 0$$
$$x^4 + (2a + 1)x^2 - x^2 - x + a^2 + a = 0$$

$$\left(x^2+\left(a+\frac{1}{2}\right)\right)^2-\left(a+\frac{1}{2}\right)^2-x^2-x+a^2+a=0$$

$$\left(x^2+\left(a+\frac{1}{2}\right)\right)^2-a^2-a-\frac{1}{4}-x^2-x+a^2+a=0$$

$$\left(x^2+\left(a+\frac{1}{2}\right)\right)^2-\left(x^2+x+\frac{1}{4}\right)=0$$

$$\left(x^2+\left(a+\frac{1}{2}\right)\right)^2-\left(x+\frac{1}{2}\right)^2=0$$

$$\left(x^2+x+(a+1)\right)\left(x^2-x+a\right)=0$$

We obtain following solutions:

$$x_{1,2}=\frac{-1\pm\sqrt{1-4(a+1)}}{2}\qquad x_{3,4}=\frac{1\pm\sqrt{1-4a}}{2}.$$

It is possible to conclude solutions according parameter a in following Table 1:

These solutions have also graphical interpretation. We present in the Fig. 4 the case, when the graphs have one common point.

5 Conclusions

It is possible according [13] to formulate some general impediments to the use of technology for mathematics teaching:

1. teachers should not prioritizing technological tools,
2. the curriculum supports in the few range the use of technology (the task is, how to use it),

Table 1 Solutions according parameter a

$a \in (-\infty, -1\rangle$	$\left\{\frac{-1-\sqrt{-3-4a}}{2}, \frac{1+\sqrt{1-4a}}{2}\right\}$
$a \in (-1, 0)$	$\left\{\frac{1+\sqrt{1-4a}}{2}\right\}$
$a \in \left\langle 0, \frac{1}{4}\right)$	$\left\{\frac{1-\sqrt{1-4a}}{2}, \frac{1+\sqrt{1-4a}}{2}\right\}$
$a = \frac{1}{4}$	$\left\{\frac{1}{2}\right\}$
$a \in \left(\frac{1}{4}, \infty\right)$	$\emptyset$

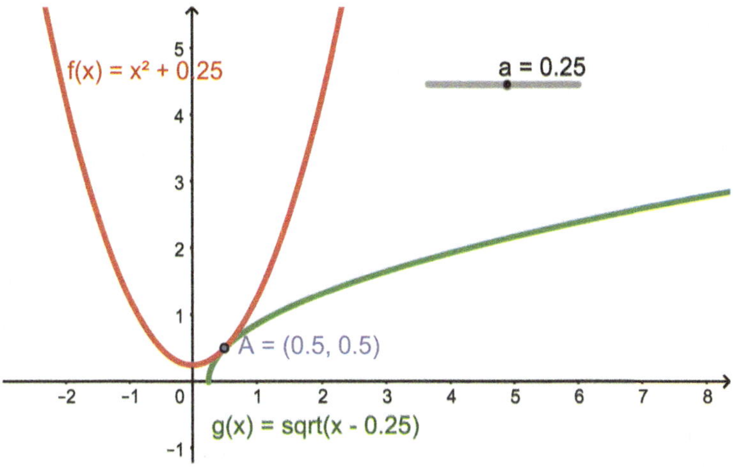

Fig. 4 The solution for the case $a = 0.25$. g is inverse function to f

3. assessment of teachers and students sometimes not encourage the use of technological tools,
4. teachers' unwillingness to attend professional development programs or up-skill on the latest technology developments, and
5. the use of technology reinforces other skills (e.g. computation) rather than the development of concepts,

One solution for these problems can be according [19] life-long learning courses for in-service teachers can support their teaching digital skills.

Visual representation in mathematics education plays according [7] an important role. We tried to present in our contribution some examples suitable for calculus teaching. It is needed for this topic to know and connect the knowledge of many parts of school mathematics. One part is arithmetic with operation (see more in [11]).

Graph of the function, pictures and schemas can help in the building of mathematic notions in the knowledge of students in secondary level. These tasks are important to present in appropriate way also for students-future teachers mathematics at universities, who can use them in their educational practise (see also [20]).

Acknowledgments This article is prepared with the help of the project VEGA 1/0079/19 "Analysis of critical points in school mathematics and identification of factors influencing pupils' attitude to mathematics".

References

1. S.J. Bruner, *Toward a Theory of Instruction* (Harvard University Press, Cambridge, 1966), 176 p.
2. L. Csachová, M. Jurečková, Notes to the problems with figures at school mathematics, in *17th Conference on Applied Mathematics (APLIMAT 2018)* (Slovak Technical University, Bratislava, 2018), pp. 270-279
3. M. Hejný, et al., *Creative Teaching in Mathematics*. Faculty of Education (Charles University, Prague, 2006), 320 p.
4. *Innovated Slovak Curriculum for Mathematics at Upper Secondary Level*. Retrieved from [30/12/2019]: http://www.statpedu.sk/files/articles/dokumenty/inovovany-statny-vzdelavaci-program/matematika_g_4_5_r.pdf
5. A. Kirsch, Eine "intellektuell ehrliche" Einführung des Integralbegriffs in Grundkursen [Some Intelectual Introduction to the Notion of Integral in the Basic Courses]. Didaktik der Mathematik **4**, 87–105 (1976)
6. N. Knoche, H. Wippermann, *Vorlesungen zur Methodik und Didaktik der Analysis [Lectures about Methodology and Didactics of Calculus]* (BI Wissenschaftsverlag, Mannheim, Wien, Z'rich, 1986), 350 p.
7. L. Koreňová, M. Kis, Z. Lavicza, P. Ostradický, New directions in nationwide technology integration into mathematics teaching: The geomatech project, in *DIVAI 2018: 12th International Scientific Conference on Distance Learning in Applied Informatics* (Wolters Kluwer, Prague, 2018), pp. 81–90
8. J. Medová, K. Páleniková, L. Rybanský, Z. Naštická, *Undergraduate Students' Solutions of Modeling Problems in Algorithmic Graph Theory*. Mathematics **7**(7), 572 (2019)
9. H.F.J. Meyer, R. Land, Threshold concepts and troublesome knowledge 1 - Linkages to ways of thinking and practising, in *Improving Student Learning - Ten Years On* (OCSLD, Oxford, 2003), pp. 412-424
10. P. Micheuz, K.J. Fuchs, C. Landerer, Mission possible - computers in "anyschool", in *Proceedings of IFIP-Conference on "Informatics, Mathematics and ICT: A Golden Triangle"*, ed. by D. Benzie, M. Iding, June 27-29 (College of Computer and Information Science Northeastern University, Boston, 2007). ISBN 13: 978-0-615-14623-2
11. E. Partová, T. Marcinek, Third Graders' representations of multiplication, in *International Symposium on Elementary Mathematics Teaching (SEMT)*. Faculty of Education (Charles University, Prague, 2015), pp. 267–275
12. F. Pauer, F. Stampfer, Mit Funktionen rechnen ein wichtiges Thema der Sekundarstufe 2 [Calculation with Function One Important Topic for Upper Secondary Level] . Schriftenreihe zur Didaktik der Mathematik der Österreichischen Mathematischen Gesellschaft (ÖMG) **47**, 62–67 (2015)
13. D. Pratt, N. Davies, D. Connor, The role of technology in teaching and learning statistics, in *Teaching Statistics in School Mathematics: Challenges for Teaching and Teacher Education: A Joint ICMI/IASE Study* (Springer, Dordrecht, 2011), pp. 97–107
14. D. Polcin, H. Petruskova, J. Stando, Support and Improvement of the Educational Process in Regional Education in Slovakia Through Tablet Classrooms. *Third International Conference on Computer Science, Computer Engineering and Education Technologies (CSCEET2016)*. (Lodz: Lodz University of Technology, 2016), pp. 71–78
15. K. Ruthven, R. Deaney, S. Hennessy, Using graphing software to teach about algebraic forms: a study of technology-supported practice in secondary-school mathematics. Educat. Stud. Math. **71**(3), 279–297 (2009)
16. M. Sajka, *Pojecie Funkcji [The Notion of Function]* (Wydawnictwo Naukowe UP, Kraków, 2019), 144 p.
17. H.J. Vollrath, Funktionales Denken [Functional thinking]. J. für Mathematikdidaktik **10**, 3–37 (1989)

18. H.G. Weigand, T. Weth, *Computer im Mathematikunterricht [Computer in Mathematics Education]* (Spektrum Akademischer Verlag, Berlin, 2002), 273 p.
19. J. Záhorec, A. Hašková, M. Munk, Teachers' professional digital literacy skills and their upgrade. Euro. J. Contemp. Edu. **8**(2), 378–393 (2019)
20. K. Žilková, J. Kopáčová, L. Rybanský, Similarities and differences in identification of rectangles of fourth-grade pupils, in *International Symposium on Elementary Mathematics Teaching (SEMT)*. Faculty of Education, (Charles University, Prague, 2017), pp. 466-475

Key Cell Method Based on the ARVE. How Not to Fall Into the Representative Cell "Trap"

Natalia Rylko, Pawel Kurtyka, and Michal Stawiarz

Abstract We develop novel Key Cell method and the theory of analytical representative volume element (ARVE) based on the Eisenstein-Rayleigh-Mityushev sums. ARVE is constructed by means of the discrete convolutions of elliptic functions. Examples of various types of particle-reinforced composite structures are given instead of discussed.

Keywords Key Cell · ARVE theory with Eisenstein–Rayleigh-Mityushev sums (ERM-sums) · Homogeneous and heterogeneous structure analysis · Multiscale modeling · Anisotropy

Mathematics Subject Classification (2010) Primary 99Z99; Secondary 00A00

1 Introduction

Composite structures (reinforcing phase distributions, shapes and sizes of reinforcing particles) largely determine various properties of the manufactured material [1–5]. This is the reason why issues related to the optimization of the composite structure are extremely important in the development of technology for their manufacture and subsequent use [6, 7]. Therefore, the selection of the composite structures analysis methodology becomes one of the basic issue in conducting research. The contemporary approach to the subject of composite structure analysis

---

Sincere words of gratitude are extended to Professor Vladimir Mityushev for his valuable advice.

N. Rylko (✉) · P. Kurtyka
Institute of Technology, Pedagogical University of Krakow, Materialica+ Research Group, Krakow, Poland

M. Stawiarz
Independent Researcher in Collaboration with Materialica+ Research Group, Krakow, Poland

P. Cerejeiras et al. (eds.), *Current Trends in Analysis, its Applications and Computation*, Research Perspectives,
https://doi.org/10.1007/978-3-030-87502-2_27

always includes the stage of image analysis using advanced 2D or 3D structure imaging techniques which is not the subject of the presented work [8, 9]. Photomicrograph, electron micrograph as the image of a field of the composite samples is the primary unit of the geometrical data collection. Data and parameters obtained in this way are used in further research by selecting appropriate statistical, numerical or analytical methods [10–15].

Methodology appropriate to the issue of structure optimization has to provide the possibility of an objective and not a subjective comparison of composites structures of the same content, obtained as a result of various technological processes, changes in process parameters, processing methods, etc. Statistical methods are commonly used in order to compare the quality of a structure. But we cannot consider them as the best solution to the problem. In this way, researchers are able to prove that e.g. homogenization of the composite structure occurs as a result of the technological process used. However, they are not able to estimate the effects of homogenization or compare the results obtained [16, 17].

An alternative analysis of changes in the distribution of reinforcing particles can be carried out by other conventional methods, among them Voronoi tessellation, examination of the function of covariance or radial distribution, etc. [10, 18–23].

Determining a representative cell is based on appropriate parameters—for example, it can be reinforcing phase concentration, anisotropy coefficient, etc.

2 Methodology of the Study Composite Structures

2.1 Eisenstein–Rayleigh-Mityushev Sums (ERM-Sums)

Methodology of ARVE theory is based on the application of fast computational algorithms to calculating subsequent terms of the ERM series [24]. The calculated values are an infinite set of parameters that precisely determine the structure. A truncated set is taken into account in practical applications.

The structural ERM-sums are defined by the following discrete convolutions [25, 29]:

$$\begin{aligned} e_{m_1\ldots m_q} := N^{-[1+\frac{1}{2}(m_1+\cdots+m_q)]} \sum_{k_0 k_1 \ldots k_q} E_{m_1}(a_{k_0}-a_{k_1}) \\ \times \overline{E_{m_2}(a_{k_1}-a_{k_2})} \ldots \mathbf{C}^{q+1} E_{m_q}(a_{k_{q-1}}-a_{k_q}). \end{aligned} \tag{2.1}$$

It is assumed for convenience that $\mathbf{C}$ stands for the operator of complex conjugation and $E_m(0) := S_m$, where S_m denotes the lattice sum of order m; a_k denotes the complex coordinate of the center of the kth inclusion and $E_m(z)$ the Eisenstein series described in [26, 27]; the bar denotes the complex conjugation, N the number of inclusions per cell.

The sum e_2 was used in practical applications to study anisotropy of random structures [28]. It yields the anisotropy factor [24, 27]

$$\kappa = \frac{1}{2\pi N^2} \left| \mathrm{Re} \sum_{k=1}^{N} \left[\sum_{m \neq k}^{N} \mathbb{E}_2(a_k - a_m) - \pi \right] \right|, \tag{2.2}$$

As $\kappa = 0$, the material is isotropic, and the degree of anisotropy increased as κ increases.

The method of structural sums was extended to the elastisity problems [30–33]

2.2 Determination of the Key RVE_c and RVE_a Cells

The input data needed for the analysis of the reinforcing particles distribution and for the calculations of the composites effective properties can be obtained with the use of variety of data sets. Typically, these data are established by image processing techniques applyed to the microstructure images obtained e.g. by the LM (Light microscopy), SEM (Scanning electron microscopy), TEM (Transmission electron microscopy), XTM (X-ray microtomograph) and by other methods.

As basis to the further analysis we use the geometrical parameters of structure, for instance:

(i) particles distribution,
(ii) particles size,
(iii) particles shape factor.

Often, the subject of such analysis is to determine whether the structure under study is homogeneous or heterogeneous. Heterogeneity can be described by means of the features introduced below in order to determine the composite structure in the context of its properties. We mean here such parameters as the concentration of reinforcing phase, the size of particles, etc. Below, we construct two types of the cells RVE_c and RVE_a formed from the concentration and from the anisotropy coefficient, respectively. These cells are called the Key Cells (KC).

For this purpose, the KC method was developed, which involves the implementation of subsequent stages of structure research:

(i) determination of RVE through the selected parameters, the KC to further.
(ii) creation of the data collection assotiated to the KC.
(iii) detailed analysis of data obtained and computation of the non-uniformity coefficient U [34].

The analyzed area contains at least several hundred thousand reinforcing phase particles with the largest possible presentation their typical distributions. In order to facilitate the presentation of the method, let us introduce the upper index in the cell

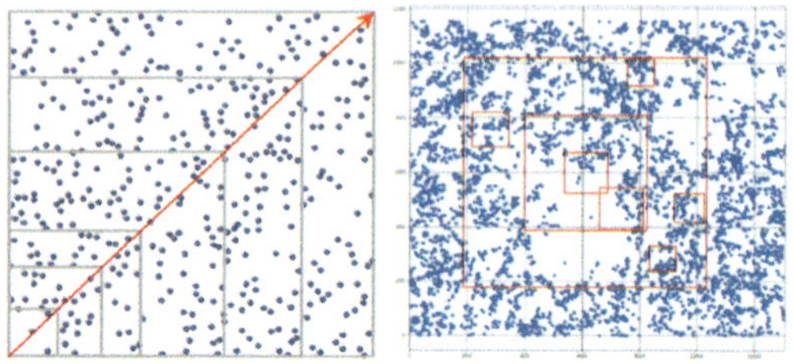

Fig. 1 Scheme of the RVE_c^N build

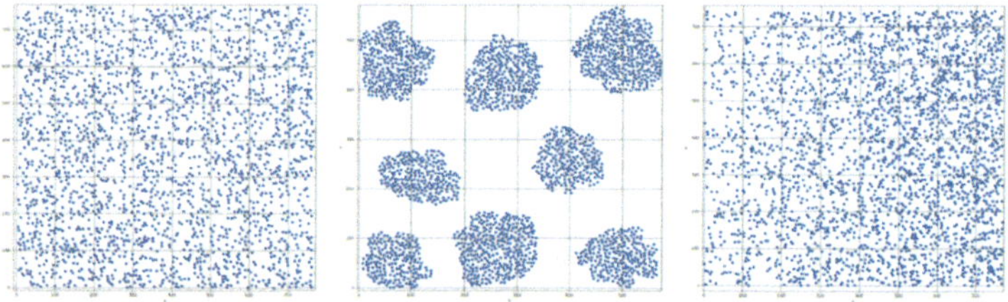

Fig. 2 The generated composite structures, from left: random, clusters, gradient

designations RVE_c^N where N, as mentioned above, gives us the number of inclusions per cell.

To determine the key cell RVE_c we started by measuring of the concentration in a randomly selected 2D section of composite. The cell has a square shape. The adopted scheme of cell expansion is shown in Fig. 1. Subsequent particles are included to the cell according to the specific algorithm [34].

To make the KC method presentation more complete we generated structures with the same concentration of the particle phase as for the real composite, i.e. $c = 10\%$. At the generated structures the particles are arranged randomly, in clusters, with a gradient, see Fig. 2. Total number particles in each generated structures equals to 2000. The concentration of the reinforcing phase is determined for each subsequent cell built by increasing number of particles up to $N = 2000$ for the generated structures and $N = 5000$ for the real composite structures.

2.3 Results of RVE_c and RVE_a Cells Analysis

The obtained results are shown in Fig. 3 as the dependence of concentration on the number of inclusions per cell. The red line color corresponds to the specially selected cell that contains a large area without particles. This yields the shape of the concentration curve gradually increasing from $c = 0.03$ for RVE_c^{50} through $c = 0.08$ for RVE_c^{400}, see Table 1.

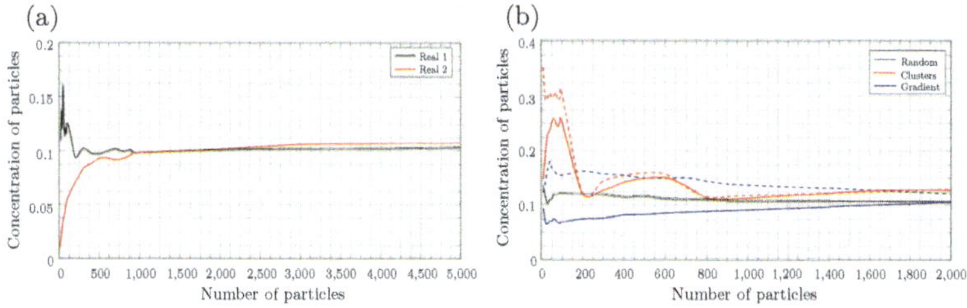

Fig. 3 The concentration stabilization for (**a**) the real composite structures and (**b**) the generated structures

Table 1 Examples of RVE_c^N cell data sets

Parameter	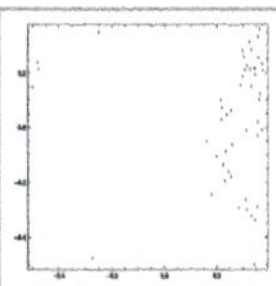	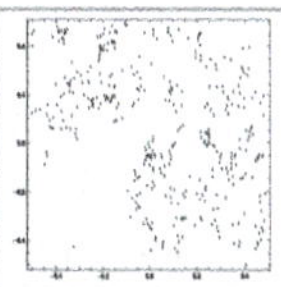	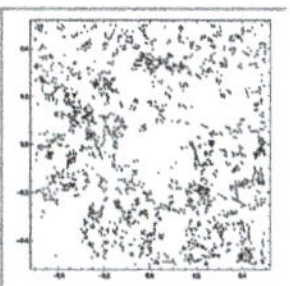
N	50	400	2000
Concentration	0.03	0.08	0.10
$Re\ e_2$	−14.01	−23.49	−14.89
$Im\ e_2$	63.92	1.14	0.42
κ	2.73	4.24	2.87

In order to determine the parameters of the key cell, we find the concentration stabilization area by comparing the obtained concentration with the one concentration of the composite manufacturer. For both real structure we observe stabilization of the reinforcement phase concentration at the level 1000 particles per cell.

The presented KC method seems to be quite general. We will try validate it by examining three types of generated structures shown in Fig. 2. Additionally, for cluster and gradient structures, the cell building process was carried out twice starting the cell building from the point belong to the cluster (dashed lines) or out of them, and from the area of the low or the high (dashed lines) concentration for the gradient structure.

The graph for the generated random structure has the same character as well as the key cell parameters for the previously analyzed real structure.

In the case of the gradient structure, it is clear that we cannot determine a RVE cell. But the shape of the related curves and the symmetry will help us to classify the structure by the gradient type.

In the case of cluster structures, concentration curves have a wave shape with decreasing amplitude and increasing wavelength. The same area between the curve and the line $c = 0.1$ for every wave indicates homogeneity in terms of the size of cluster surface. The distance between the maximum concentration values of the neighboring waves gives us the average distance between the nearest clusters. In this

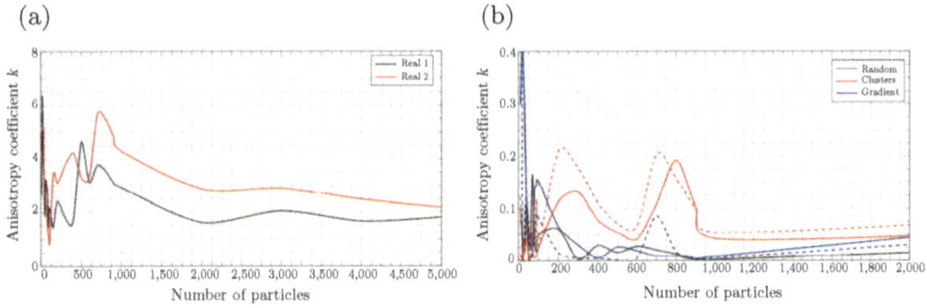

Fig. 4 The anisotropy coefficient stabilization for (**a**) the real composite structures and (**b**) the generated structures

case, the determination of the KC is still possible, but the generated structure does not contain sufficient number of particles to complete the investigation.

The same procedure is used to build the RVE_a^N. The results of anisotropy coefficient calculations are shown in Fig. 4 and can be analyzed by the KC method [2, 3, 34]. We pay attention to the comparison of key cells show the RVE_a cell is significantly larger than the RVE_c cell. However, the big surprise is the lack of anisotropy stabilization for the gradient material. For this type of materials when the concentration changes according to a prescribed rule, this procedure has to be developed yet.

2.4 Summary and Conclusion

The developed KC method based on ARVE theory and ERM-sums can be a powerful tool for optimization of technological processes. The presented methodology allows us:

(i) to carry out a quantitative analysis of composite structures,
(ii) to identify the parameters of technological processes,
(iii) to design materials within a strictly defined structure, etc.

KC method was presented for 2D structures. Research on its extension to 3D structures is currently underway. Hence, based on the 2KC method after its development the 3KC method will be created.

Acknowledgments This work has been funded by the Research Fund of the Institute of Technology of Pedagogical University of Krakow.

References

1. S.D. Ryan, V. Mityushev, V.M. Vinokur, L. Berlyand, Rayleigh approximation to ground state of the Bose and Coulomb glasses. Sci. Rep. **5**, 7821–7825 (2015)
2. P. Kurtyka, N. Rylko, Structure analysis of the modified cast metal matrix composites by use of the RVE theory. Arch. Metall. Mater. **58**, 357–360 (2013)
3. P. Kurtyka, N. Rylko, T. Tokarski, A. Wójcicka, A. Pietras, Cast aluminium matrix composites modified with using FSP process - Changing of the structure and mechanical properties. Compos. Struct. **133**, 959–967 (2015)
4. A.J. Dolata, M. Dyzia, J. Wieczorek, Tribological properties of single (AlSi7/SiCp, AlSi7/GCsf) and hybrid (AlSi7/SiCp + GCsf) composite layers formed in sleeves via centrifugal casting. Materials, **12**(17), Article Number 2803 (2019)
5. S. Boczkal, A.J. Dolata, M. Nowak, Effect of SiC and GR reinforcement particles on the structure and functional properties of composite casting. Arch. Metall. Mater. **61**(1), 399-404 (2016)
6. J. Luyan, Z. Jianbing, M. Yuqin, X. Shengyong, W. Wenlan, Z. Yong, Fabrication and characterization of ZrW2O8-Cf/E51 negative thermal expansion composite. Mater. Res. Express. **7** (2019). https://doi.org/10.1088/2053-1591/ab6638
7. A.V. Lima, J.L. Cardoso, C.J. Lobo, Research on hybrid sisal/glass composites: A review. J. Reinfor. Plastics Composites **38**(17), 789–821 (2019)
8. E. Olson, Particle shape factors and their use in image analysis-part 1: theory. J. GXP Compl. **15**(3), 85–96 (2011)
9. M. Pinz, G. Weber, S. Ghosh, Generating 3D virtual microstructures and statistically equivalent RVEs for subgranular gamma-gamma' microstructures of nickel-based superalloys. Computat. Mater. Sci. **167**, 198–214 (2019)
10. S. Torquato, *Random Heterogeneous Materials* (Springer, Berlin, 2002)
11. M.N. Nguyen, T.Q. Bui, T.T. Truong, S. Tanaka, S. Hirose, Numerical analysis of 3-D solids and composite structures by an enhanced 8-node hexahedral element. Finite Elements Analy. Design **131**, 1–16 (2017)
12. W. Nawalaniec, Classifying and analysis of random composites using structural sums feature vector. Proc. Roy. Soc. A Math. Phys. Eng. Sci. **475**(2225), 20180698 (2019)
13. R. Czapla, Basic sums as parameters characterizing. Siles. J. Pure Appl. Math. **6**(1), 85–96 (2016)
14. Z. Hashin, Analysis of composite materials - a survey. J. Appl. Mech. **50**, 481—504 (1983)
15. M. Stawiarz, P. Kurtyka, M. Rylko, S. Gluzman, Influence of FSP process modification on selected properties of Al-Si-Cu/SiCp composite surface layer. Compos. Theory Practice **19**(4), 161–168 (2019)
16. D.-H. Choi, Y.-I. Kim, D.-U. Kim, S.-B. Jung, Effect of SiC particles on microstructure and mechanical property of friction stir processed AA6061-T4. Trans. Nonferrous Metals Soc. China **22**, 614–618 (2013)
17. H. Nami, H. Adgi, M. Sharifitabar, H. Shamabadi, Microstructure and mechanical properties of friction stir welded Al/Mg2Si metal matrix cast composite. Trans. Nonferrous Metals Soc. China **32**, 976–983 (2011)
18. R. Babaei, A. Farrokhabadi, Prediction of debonding growth in two-dimensional RVEs using an extended interface element based on continuum damage mechanics concept. Composite Struct. **238**, 111981 (2020)
19. P. Karimi, A. Malyarenko, M. Ostoja-Starzewski, X. Zhang, RVE Problem: mathematical aspects and related stochastic mechanics. Int. J. Eng. Sci. **146**, 103169 (2020)
20. L. Jasinski, D. Sangaré, P.M. Adler, V.V. Mourzenko, J.-F. Thovert, N. Gland, S. Békri, Transport properties of a Bentheim sandstone under deformation. Phys. Rev. E **91**, 013304 (2015)

21. M. Dzikowski, L. Jasinski, M. Dabrowski, Depth-averaged Lattice Boltzmann and Finite Element methods for single-phase flows in fractures with obstacles. Comput. Math. Appl. **75**, 3453-3470 (2018)
22. A. Ayyar, G.A. Crawford, J.J. Williams, N. Chawla, Numerical simulation of the effect of particle spatial distribution and strength on tensile behavior of particle reinforced composites. Computat. Mat. Sci. **44**, 496–506 (2008)
23. P.M. Adler, J-F. Thovert, V.V. Mourzenko, *Fractured Porous Media* (Oxford University Press, Oxford, 2012)
24. V. Mityushev, Transport properties of doubly periodic arrays of circular cylinders and optimal design problems. Appl. Math. Optim. **44**, 17–31 (2001)
25. S. Gluzman, V. Mityushev, W. Nawalaniec, *Computational Analysis of Structured Media* (Academic/Cambridge, Elsevier/Amsterdam, 2018)
26. V. Mityushev, Representative cell in mechanics of composites and generalized Eisenstein-Rayleigh sums. Complex Variables Elliptic Eq. **51**, 1033–1045 (2006)
27. V. Mityushev, Effective conductivity of random two-dimensional composites with circular non-overlapping inclusions. Computat. Mat. Sci. **63**, 118–126 (2012)
28. P. Drygaś, S. Gluzman, V. Mityushev, W. Nawalaniec, *Applied Analysis of Composite Media. Analytical and Computational Results for Materials Scientists and Engineers* (Woodhead Publishing, Sawston, 2019)
29. V. Mityushev, W. Nawalaniec, Basic sums and their random dynamic changes in description of microstructure of 2D composites. Computat. Mater. Sci. **97**, 64–74 (2015)
30. P. Drygaś, Generalized Eisenstein functions. J. Math. Analy. Appl. **444**(2), 1321–1331 (2016)
31. S. Yakubovich, P. Drygaś, V. Mityushev, Closed-form evaluation of two-dimensional static lattice sums. Proc. Roy. Soc. Math. Phys. Eng. Sci. **472**(2195), 20160510 (2016)
32. P. Drygaś, S. Gluzman, V. Mityushev, W. Nawalaniec, Effective elastic constants of hexagonal array of soft fibers. Computat. Mater. Sci. **139**, 395–405 (2017)
33. P. Drygaś, V. Mityushev, Contrast expansion method for elastic incompressible fibrous composites. Adv. Math. Phys. **2017**, art. no. 4780928 (2017)
34. P. Kurtyka, N. Rylko, Quantitative analysis of the particles distributions in reinforced composites. Compos. Struct. **182**, 412–419 (2017)

Neutral Inclusion and Spectral $\mathbb{R}$-Linear Problem

Vladimir Mityushev

Abstract An eigenvalue $\mathbb{R}$-linear problem arisen in the theory of invisible and neutral inclusions is discussed by a method of complex potentials. A nodal domains conjecture is posed. Demonstration of the conjecture allows to justify that a set of inclusions can be made invisible by surrounding it with an appropriate coating.

Keywords $\mathbb{R}$-linear spectral problem · Invisible inclusion · Coated neutral inclusion · Composite · Metamaterial

Mathematics Subject Classification (2010) Primary 30E25; Secondary 39B32, 74Q15

1 Introduction

The concentric-spheres model introduced by Hashin [1] led to advanced models for neutral (invisible) inclusions in composites, see [2, 3] and the conference *etopim11.up.krakow.pl/*. Following these models in the last few years physicists and engineers create metamaterials which possess surprising cloaking properties.

The present paper is devoted to a 2D stationary model of neutral inclusions and the corresponding spectral boundary value problem problem discussed in [4–6] and works cited therein. It is suggested that the maximal eigenvalue and the corresponding eigenfunction determine the physical coefficient (conductivity) and the shape, respectively, of the coating which hides a given inclusion from the external field.

The spectral problems are sufficiently well studied when the spectral parameter λ is included into PDE in a domain D, e.g., $-\Delta u = \lambda u$ [7]. Another type of the spectral problem is the Steklov problem $\frac{\partial u}{\partial n} = \lambda u$ when the relation

V. Mityushev (✉)
Institute of Computer Science, Pedagogical University, Krakow, Poland
e-mail: mityu@up.krakow.pl

P. Cerejeiras et al. (eds.), *Current Trends in Analysis, its Applications and Computation*, Research Perspectives,
https://doi.org/10.1007/978-3-030-87502-2_28

takes place on the boundary ∂D of the domain [8], see also the workshop *events.math.unipd.it/spectralPD2019/*. The neutral problem leads to the spectral conjugation (transmission) problem discussed below.

2 Eigenvalue Problem

Introduce the complex variable $\zeta = x_1 + ix_2$ on the plane $\mathbb{R}^2$. Let a simply connected domain G_1 be bounded by a smooth curve Γ and the domain $G_1 \cup \Gamma \cup G$ be by a smooth curve γ as shown in Fig. 1. It is assumed for definiteness that $0 \in G_1$. Let an unknown function $u(\zeta)$ be harmonic in the doubly connected domain G and continuously differentiable in its closure. The function $u(\zeta)$ satisfies the Dirichlet boundary condition

$$u(\zeta) = 0, \quad \zeta \in \Gamma. \tag{2.1}$$

The condition (2.1) holds for the perfectly conducting inclusion. The perfect insulator is modeled by the Neumann condition $\frac{\partial u}{\partial n} = 0$.

Let σ and σ_2 be given positive constants which denote the coefficients of conductivity of materials occupied the domains G and the exterior domain $G_2 = \mathbb{R}^2 \backslash (G_1 \cup \Gamma \cup G)$. The following conjugation (transmission) condition holds

$$u(\zeta) = u_2(\zeta), \quad \sigma \frac{\partial u}{\partial n}(\zeta) = \sigma_2 \frac{\partial u_2}{\partial n}(\zeta), \quad \zeta \in \gamma, \tag{2.2}$$

where $\frac{\partial}{\partial n}$ stands for the normal derivative on γ.

Below, we consider the inverse problem when the function $u_2(\zeta)$ is given and the curve γ is unknown. In the theory of neutral inclusions, it is interesting to consider $u_2(\zeta)$ as an $\mathbb{R}$-linear function, for instance,

$$u_2(\zeta) = \frac{2\sigma}{\sigma + \sigma_2} \operatorname{Re} \zeta \equiv \frac{2\sigma}{\sigma + \sigma_2} x_1. \tag{2.3}$$

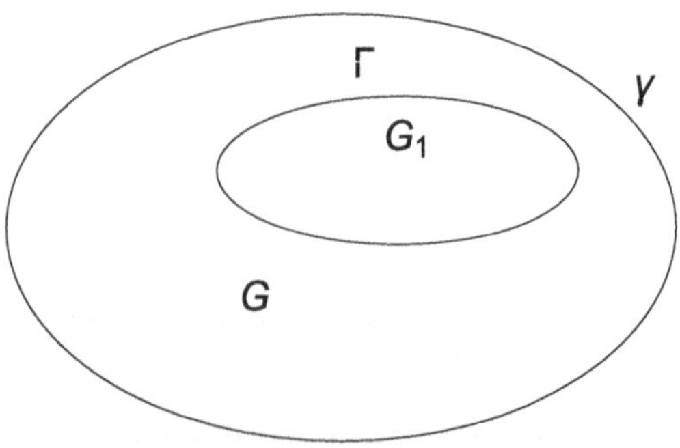

Fig. 1 The plane $\zeta = x_1 + ix_2$. Inclusion G_1 is bounded by Γ, coating G by $\Gamma \cup \gamma$

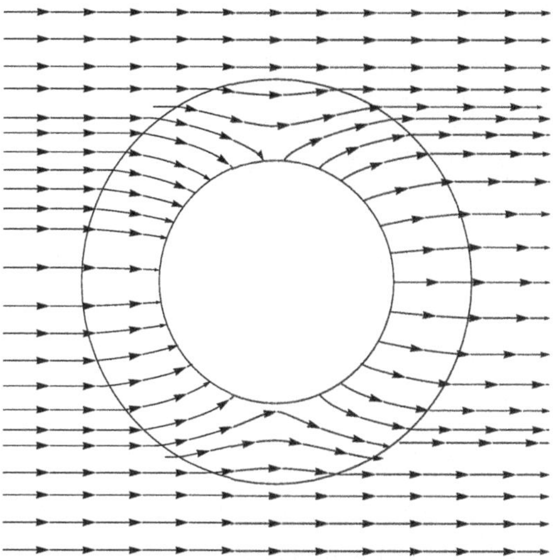

Fig. 2 Hashin's neutral perfectly conducting inclusion of the normalized radius $r < 1$ embedded in the circular coating of the exterior unit radius. The vector field is calculated by the gradients $\sigma \nabla u$ in $r < |\zeta| < 1$ and $\sigma_2 \nabla u_2$ in $|\zeta| > 1$ with formulas (4.12)–(4.13)

The potential (2.3) yields the constant gradient ∇u_2 in G_2, i.e., the inclusion G_1 does not perturb the constant external field and becomes invisible.

The multiplier $\frac{2\sigma}{\sigma+\sigma_2}$ is taken in (2.3) for convenience. Two real conditions (2.2) can be written in the complex form

$$\phi(\zeta) = \phi_2(\zeta) - \lambda\overline{\phi_2(\zeta)}, \quad \zeta \in \gamma, \tag{2.4}$$

where $u_2(\zeta) = \frac{2\sigma}{\sigma+\sigma_2}$ Re $\phi_2(\zeta)$ with $\phi_2(\zeta) = \zeta$ and $u(\zeta) =$ Re $\phi(\zeta)$ in the corresponding domains. The spectral parameter

$$\lambda = \frac{\sigma_2 - \sigma}{\sigma_2 + \sigma} \tag{2.5}$$

determines the unknown conductivity σ of the coating G.

In the case of the circular inclusion G_1, a solution of the problem is given by Hashin's fomulas (4.12)–(4.13) at the end of the paper. The corresponding flux with the invisible perfect inclusion is displayed in Fig. 2.

3 ℝ-Linear Eigenvalue Problem

Let $\zeta = \omega(z)$ be a conformal mapping of the unit disk $|z| < 1$ onto $G_1 \cup \Gamma \cup G$ normalized by the relations $\omega(0) = 0$ and $\omega'(0) > 0$. The unit circle $\partial\mathbb{U}$ corresponds to γ and a curve L to Γ. The doubly connected domain D is transformed onto G by the conformal mapping $\omega(z)$ (Fig. 3).

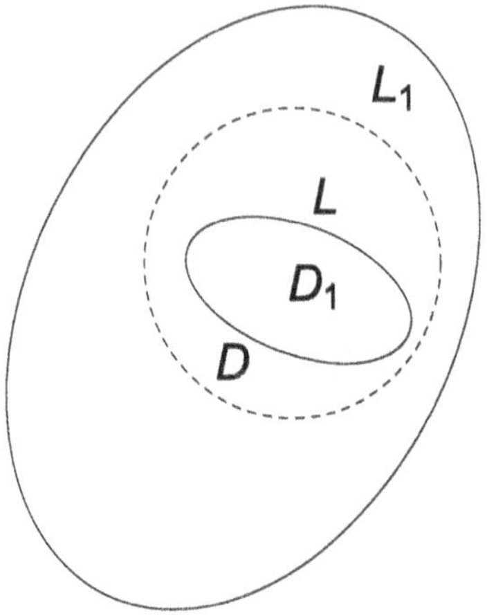

Fig. 3 The complex plane z. The unit circle $\partial\mathbb{U}$ is shown by dashed line, $\omega(D_1) = G_1$, $\omega(D) = G$, $\omega(L) = \Gamma$. The curve L_1, the inversion of L with respect to the unit circle, and L are shown by solid line

Introduce the function $\varphi(z) = \phi[\omega(z)]$ analytic in the domain D and continuously differentiable in its closure. Application of the conformal mapping transforms the problem (2.1) and (2.4) into the spectral $\mathbb{R}$-linear problem

$$\operatorname{Re} \varphi(z) = 0, \quad z \in L, \tag{3.1}$$

$$\varphi(z) = \omega(z) - \lambda\overline{\omega(z)}, \quad |z| = 1. \tag{3.2}$$

In the above statement of the problem, the boundary L of the domain D is given, the unknown functions $\varphi(z)$ and $\omega(z)$ are analytic in G and $|z| < 1$, respectively. The spectral parameter λ has to be positive. The corresponding eigenfunction $\omega_\lambda(z)$ has to be a conformal mapping. This is equivalent to the condition that $z = 0$ is the unique zero of $\omega_\lambda(z)$ in the unit disk or to the vanishing winding number (index) $\text{wind}_L \omega_\lambda = 0$.

The relation (3.2) can be written in the form

$$\varphi(z) = -\lambda\varphi_2(z) + \overline{\varphi_2(z)}, \quad |z| = 1, \tag{3.3}$$

where the function $\varphi_2(z) = \overline{\omega\left(\frac{1}{\overline{z}}\right)}$ is analytic in $|z| > 1$.

The spectral $\mathbb{R}$-linear problem (3.1) and (3.3) can be stated in the equivalent real form [9]

$$v(z) = 0, \quad z \in L, \tag{3.4}$$

$$v = v_2, \quad (-\sigma)\frac{\partial v}{\partial n} = \sigma_2\frac{\partial v_2}{\partial n}, \quad |z| = 1, \tag{3.5}$$

where

$$v(z) = \text{Re}\ \varphi(z),\ z \in D,\ \text{and}\ v_2(z) = \frac{2\sigma}{\sigma + \sigma_2}\text{Re}\ \varphi_2(z),\ |z| > 1. \tag{3.6}$$

The spectral parameter λ is related to σ by Eq. (2.5) where the constant $\sigma_2 > 0$ is supposed to be given as the conductivity of the surround medium. The relation (3.5) on the unknown constant σ differs from (2.2) by the minus sign on σ. It is worth noting that the direct problem (3.4)–(3.5) with negative $(-\sigma)$ and positive σ_2 models the field in metamaterials [2, 3].

The main spectral properties of the problems (3.1), (3.3) such as countability of real eigenvalues were established in [4], see the integral equation (3.12) with $\varrho = 1$. The condition (3.5) with $(-\sigma) < 0$ does not refer to the elliptic case introduced by Mikhajlov, see the most general discussion in [10] and [9]. An alternative way to study the spectral properties can be based on the reduction of (3.4), (3.5) to integral equations. In such a statement, the study can be extended to the Helmholtz equation important in applications.

The condition that the function $\omega(z)$ has the unique zero at the origin in the unit disk is equivalent to the unique zero at infinity of the function $\varphi_2(z) = \overline{\omega\left(\frac{1}{\bar{z}}\right)}$ in the domain $|z| > 1$. For the function $v_2(z)$ this condition transforms into the nodal domain condition. The nodal lines of the function $v_2(z)$ are determined by equation $v_2(z) = 0$. The nodal lines divide the domain $|z| > 1$ onto subdomains where the function $v_2(z)$ has the same sign. The numerical examples [4] suggest Courant's type theorem about nodal domains [11] though its proof has not been established yet.

4 Functional Equation

In the present section, we reduce the boundary value problem (3.1)–(3.2) to an iterative functional equation. Introduce the function $\Phi(z) = \varphi(z) - \omega(z)$ analytic in D. It follows from (3.2) that $\Phi(z)$ is analytically continued into $|z| > 1$ by the relation

$$\Phi(z) = -\lambda\ \overline{\omega\left(\frac{1}{\bar{z}}\right)}, \quad |z| > 1. \tag{4.1}$$

Therefore, $\Phi(z)$ is analytic in $D_1^- = (\mathbb{C} \cup \{\infty\}) \backslash (D_1 \cup L)$, the complement of the closure of D_1 to the extended complex plane. Then, (4.1) yields the analytic continuation of $\omega(z)$ through the unit circle up to the curve L_1 obtained from L by the inversion $z \mapsto \frac{1}{\bar{z}}$

$$\omega(z) = -\frac{1}{\lambda}\ \overline{\Phi\left(\frac{1}{\bar{z}}\right)}, \quad z \in D^*. \tag{4.2}$$

Here, D^* is the image of D after the inversion $z \mapsto \frac{1}{\overline{z}}$. Therefore, $\omega(z)$ is analytic in the domain $D^+ = D_1 \cup L \cup D \cup \partial\mathbb{U} \cup D^*$, the interior domain to the curve L_1, and continuously differentiable in its closure.

Substitute $\varphi(z) = \Phi(z) + \omega(z)$ in (3.1)

$$\text{Re}\,[\Phi(z) + \omega(z)] = 0, \quad z \in L. \tag{4.3}$$

Using (4.1) we obtain the boundary condition on the function $\omega(z)$ analytic in D^+ and continuously differentiable in its closure $D^+ \cup L_1$

$$\text{Re}\left[\omega(z) - \lambda\,\overline{\omega\left(\frac{1}{\overline{z}}\right)}\right] = 0, \quad z \in L. \tag{4.4}$$

Using the inversion we rewrite this boundary condition in the form

$$\text{Re}\left[\lambda\,\omega(z) - \overline{\omega\left(\frac{1}{\overline{z}}\right)}\right] = 0, \quad z \in L_1 \equiv \partial D^+. \tag{4.5}$$

Let $U(z)$ be the real part of $\omega(z)$. Hence, $U(z)$ is harmonic in D^+. Then, (4.5) can be written in the form of functional equation in a space of harmonic functions

$$\lambda\, U(z) = U\left(\frac{1}{\overline{z}}\right), \quad z \in \partial D^+. \tag{4.6}$$

The shift $\frac{1}{\overline{z}}$ maps $L_1 \equiv \partial D^+$ onto L, hence, it is a shift into domain.

The functional equation (4.6) yields the third way to investigate the spectral boundary value problems. Let $f : D^+ \to \mathbb{U}$ be the conformal mapping normalized by the relations $f(0) = 0$ and $f'(0) > 0$. Using this conformal mapping we transform the functional equation (4.6) to the following one

$$\lambda\, V(z) = V[\alpha(z)], \quad |z| = 1, \tag{4.7}$$

where $U = V \circ f$ and $\alpha(z) = f\left(\frac{1}{\overline{f^{-1}(z)}}\right)$ is a diffeomorphism of the unit circle onto a curve lying interior of the unit disk and having the same orientation as the unit circle. Using the Poisson formula for harmonic functions one can write (4.7) as the integral equation

$$\lambda\, V(z) = \frac{1}{2\pi}\int_0^{2\pi} V(e^{i\theta})\frac{1 - |\alpha(z)|^2}{|\alpha(z) - e^{i\theta}|^2}\, d\theta, \quad |z| = 1. \tag{4.8}$$

It is convenient to consider (4.8) as the Fredholm integral equation in the Hilbert space L^2. This yields the compactness of operators from the right part of (4.6) and of (4.7) in the corresponding Hardy type spaces.

Hypothesis

(i) All the eigenvalues of (4.7) are positive and can be arranged in a decreasing sequence $1 = \lambda_0 \geq \lambda_1 \geq \lambda_2 \geq \ldots$ where each eigenvalue is repeated the number of times equal to its multiplicity.

(ii) The number of nodal domains in the unit disk of the eigenfunction $V_k(z)$ corresponding to the eigenvalue λ_k is less than or equal to k ($k = 0, 1, 2, \ldots$).

In particular, the eigenfunction $V_1(z)$ yields the desired confomal mapping $\omega(z)$, hence, the shape of the coating and λ_1 determines its conductivity through (2.5).

Example Let L be the circle $|z| = r < 1$. Then, L_1 is the circle $|z| = r^{-1}$ on which $\frac{1}{\overline{z}} = r^2 z$. The functional equation (4.6) becomes

$$\lambda\, U(z) = U\left(r^2 z\right), \quad |z| = r^{-1}. \tag{4.9}$$

It is equivalent to the functional equation for analytic functions up to an arbitrary pure imaginary constant

$$\lambda\, \omega(z) = \omega\left(r^2 z\right), \quad |z| \leq r^{-1}, \tag{4.10}$$

since the function $\lambda\ U(z) - U\left(r^2 z\right)$ is harmonic in the unit disk and vanishes on its boundary. It follows from the theory of iterative functional equations [9] that the spectral problem (4.10) has the countable number of solutions

$$\lambda = r^{2k}, \quad \omega(z) = z^k, \quad k = 0, 1, \ldots. \tag{4.11}$$

Excluding the constant eigenvalue one can see that only the eigenfunction $\omega(z) = z$ is a conformal mapping. The corresponding eigenvalue $\lambda = r^2$ is the maximal one after 1.

The considered example represents Hashin's result [1] concerning the circular inclusion $|z| = r$, since the conformal mapping $\zeta = \omega(z) = z$ is identical. One can check that the analytic functions $\phi(\zeta) = \zeta - \frac{r^2}{\zeta}$ and $\phi_2(\zeta) = \zeta$ satisfy the boundary conditions (2.4) on $|\zeta| = 1$ with $\lambda = r^2$ and $\operatorname{Re} \varphi(\zeta) = 0$ on $|\zeta| = r$.

The conductivity of coating and the harmonic potentials have the form

$$\sigma = \sigma_2\, \frac{1 - r^2}{1 + r^2}, \tag{4.12}$$

$$u(\zeta) = x_1 \left(1 - \frac{r^2}{x_1^2 + x_2^2}\right), \; r \leq |\zeta| \leq 1, \; u_2(\zeta) = (1 - r^2) x_1, \; |\zeta| \geq 1, \tag{4.13}$$

where $\zeta = x_1 + i x_2$.

5 Conclusion

In the present paper, we discuss relations between 2D stationary conductivity problem governed by Laplace's equation and the spectral $\mathbb{R}$-linear problem.

One can find various extensions and applications of metamaterials in the presentation [2] and works cited therin. An extension of Hashin's result to the heat equation in a circular ring can be found in [12].

Application of the spectral theory lead to the conjecture that any smooth inclusion can be made invisible by appropriate coating. It is a mathematical result concerning existence of solutions. Physicists and engineers successfully discover new metamaterils and do not think too much about a mathematical justification of their existence. In this field, physicists overtake mathematicians and solve the problems existence of solutions of which has been not proved yet. However, the mathematical problem outlined in the present paper might be useful to discuss various shapes of neutral inclusions and to develop methods of their determination.

Acknowledgments This paper was supported by National Science Centre, Poland, Research Project No. 2016/21/B/ST8/01181.

References

1. Z. Hashin, The elastic moduli of heterogeneous materials. ASME J. Appl. Mech. **29**, 143–150 (1962)
2. R. McPhedran, Metamaterials, transform optics and cloaking, in *Mesoscopic Physics in Complex Media* (Springer, Berlin, 2010), p. 01016
3. *2D and Quasi-2D Composite and Nanocomposite Materials. Properties and Photonic Applications*, ed. by R. McPhedran, S. Gluzman, V. Mityushev, N. Rylko (Elsevier, 2020)
4. P. Jarczyk, V. Mityushev, Neutral coated inclusions of finite conductivity. Proc. Roy. Soc. London **468A**, 954–970 (2012)
5. V. Mityushev, Composites with invisible inclusions: Eigenvalues of $\mathbb{R}$-linear problems. Euro. J. Appl. Math. **27**, 1–11 (2016)
6. X. Wang, P. Schiavone, Neutrality of coated holes in the presence of screw dislocation dipoles or circular thermal inclusions. Arch. Mech. **70**, 1–11 (2018)
7. *Integral Methods in Science and Engineering*, ed. by C. Constanda, M. Dalla Riva, P.D. Lamberti, P. Musolino, v.1-2 (Springer, Birkhäuser, 2017)
8. P.D. Lamberti, L. Provenzano, Neumann to Steklov eigenvalues: asymptotic and monotonicity results. Proc. R. Soc. Edinb. Sect. A **147**, 429–447 (2017)
9. S. Gluzman, V. Mityushev, W. Nawalaniec, *Computational Analysis of Structured Media* (Elsevier, Amsterdam, 2018)
10. B. Bojarski, V. Mityushev, $\mathbb{R}$-linear problem for multiply connected domains and alternating method of Schwarz. J. Math. Sci. **189**, 68–77 (2013)
11. V. Mityushev, New boundary value problems and their applications to invisible materials, in *Analytic Methods of Analysis and Differential Equations: AMADE-2011: materials of the 6th International Conference, Dedicated to the Memory of prof. A.A. Kilbas*, ed. by S.V. Rogosin (Publishing House of BSU, Minsk, 2012), pp. 141–146
12. S. Guenneau, C. Amra, D. Veynante, Transformation thermodynamics: cloaking and concentrating heat flux. Optics Express **20**, 8207–8218 (2012)

Regularization Method for Stable Structural Features

Wojciech Nawalaniec

Abstract The main goal of this paper is to provide regularization method for structural sums in order to achieve stable features in analysis of data represented by distributions of non-overlapping disks on the plane as well as by point process patterns. The presented approach is illustrated by calculation of stable structural features characterizing the Poisson point process.

Keywords Structural sums · Structural features vector · Data analysis · Point pattern analysis · Distributions of disks on the plane

Mathematics Subject Classification (2010) Primary 62-07; Secondary 60G55

1 Introduction

Structural sums constitute a crucial part of the framework of modern computational material science [14, 15] forming a coherent whole with other contemporary results and applications in the field [1, 2, 4–13, 16, 18, 21, 24, 27–30].

Recent study brings yet another application of structural sums in construction of *structural sums feature vector* [24]. Such an approach enables the immediate application of machine learning tools and data analysis techniques to data represented by distributions of non-overlapping disks on the plane. In case of data represented by distributions of disks, we calculate basics sums directly. On the other hand, the analysed data can be represented by points (i.e. point process patterns). In such a case, a standard scenario involves straightforward application of the equal-disks form of basic sums assuming that the data points are centres of identical disks. Unfortunately, for a certain kind of data, where pairs of a very close points may appear, the calculation may lead to unstable results (with very high deviations). This

W. Nawalaniec (✉)
Faculty of Mathematics, Physics and Technology, Pedagogical University of Cracow, Krakow, Poland

P. Cerejeiras et al. (eds.), *Current Trends in Analysis, its Applications and Computation*, Research Perspectives,
https://doi.org/10.1007/978-3-030-87502-2_29

work presents a method of obtaining stable features for such systems. We propose a method of data transformation into the corresponding polydispersed disks case that may regularize the problem.

The present paper is organized as follows. In Sect. 2, we briefly review the background material on structural sums and structural features. Section 3.1 covers calculations of selected structural features for Poisson point process patterns yielding results with high deviations. In Sect. 3.2 we observe that the transformation of data into the corresponding configuration of disks with so-called *nearest neighbour radii*, defined therein, yields reliable results.

2 Structural Sums and Structural Features

2.1 Definition of Structural Sum

Consider a periodic two–dimensional lattice $\mathcal{Q}$, defined by complex numbers ω_1 and ω_2 on the complex plane $\mathbb{C}$. The (0, 0)-cell is introduced as the unit parallelogram $Q_{(0,0)} := \{z = t_1\omega_1 + t_2\omega_2 : -1/2 < t_j < 1/2 \ (j = 1, 2)\}$. The lattice $\mathcal{Q}$ consists of the cells $Q_{(m_1,m_2)} := \{z \in \mathbb{C} : z - m_1\omega_1 - m_2\omega_2 \in Q_{(0,0)}\}$, where m_1 and m_2 run over integer numbers. Consider N non-overlapping disks of radii r_j $(j = 1, 2, \ldots, N)$ distributed in the (0, 0)-cell (see Fig. 1). The total concentration of disks equals $\nu = \pi \sum_{j=1}^{N} r_j^2$. Let r be the largest of the radii r_j $(j = 1, 2, \ldots, N)$ and introduce constants

$$\nu_j = \left(r_j/r\right)^2, \quad j = 1, 2, 3, \ldots, N \tag{2.1}$$

describing polydispersity, i.e. heterogeneity of sizes of disks.

Consider a set of points a_k $(k = 1, 2, \ldots, N)$ being the centres of the disks. Let n be a natural number; $k_0, k_1 \ldots, k_n$ be integers from 1 to N; $p_j \geq 2$. Let $\mathbf{C}$ be

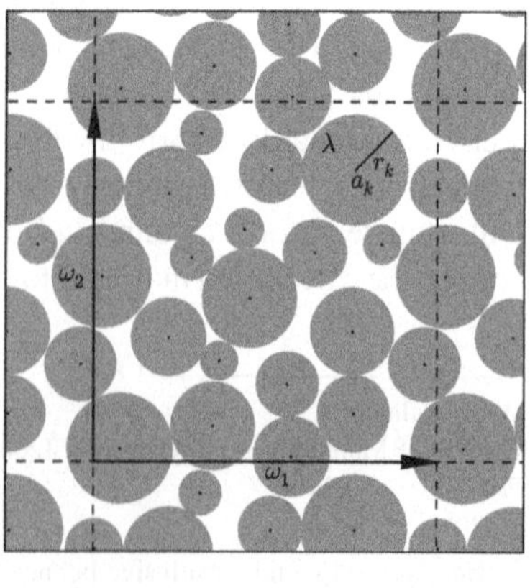

Fig. 1 Doubly periodic cell $Q_{(0,0)}$ with a configuration of non-overlapping disks

the operator of the complex conjugation. The following sums were introduced by Mityushev [17]:

$$e_{p_1,p_2,\ldots,p_n}^{\nu_0,\nu_1,\ldots,\nu_n} = \frac{1}{\eta^{\delta+1}} \sum_{k_0,k_1,\ldots,k_n} \nu_{k_0}^{t_0} \nu_{k_1}^{t_1} \nu_{k_2}^{t_2} \cdots \nu_{k_n}^{t_n} E_{p_1}(a_{k_0} - a_{k_1}) \\ \times \overline{E_{p_2}(a_{k_1} - a_{k_2})} E_{p_3}(a_{k_2} - a_{k_3}) \cdots \mathbf{C}^{n+1} E_{p_n}(a_{k_{n-1}} - a_{k_n}), \tag{2.2}$$

where $\eta = \sum_{j=1}^{N} \nu_j$ and $\delta = \frac{1}{2}\sum_{j=1}^{n} p_j$. Functions E_k $(k = 2, 3, \ldots)$ are Eisenstein functions corresponding to the doubly periodic cell $Q_{(0,0)}$ (see [14] for more details), and the superscripts t_j $(j = 0, 1, \ldots, n)$ are given by recurrence relations

$$t_0 = 1, \\ t_j = p_j - t_{j-1},\ j = 1, 2, \ldots, n. \tag{2.3}$$

The sum (2.2) is called the *structural sum of the multi-order* $\mathbf{p} = (p_1, \ldots, p_n)$. We also call δ the *order* of the structural sum. From this point, the superscripts for structural sums are omitted for the purpose of conciseness. In addition, following [22] we have $t_n = 1$.

For example, in case of data represented by N non-identical disks with radii r_j $(j = 1, 2, \ldots, N)$, structural sums e_2 and $e_{2,2}$ take the following forms:

$$e_2 = \frac{1}{N^2} \sum_{k_0=1}^{N} \sum_{k_1=1}^{N} \nu_{k_0} \nu_{k_1} E_2(a_{k_0} - a_{k_1}), \\ e_{2,2} = \frac{1}{N^3} \sum_{k_0=1}^{N} \sum_{k_1=1}^{N} \sum_{k_2=1}^{N} \nu_{k_0} \nu_{k_1} \nu_{k_2} E_2(a_{k_0} - a_{k_1}) \overline{E_2(a_{k_1} - a_{k_2})} \tag{2.4}$$

On the other hand, in case of system of identical disks we have $\nu_j = 1$ $(j = 1, 2, \ldots, N)$.

2.2 Structural Features

Let the properties of the composite be fixed. Then, the fundamental problem of composites consists in the construction of a homogenization operator $\mathcal{H} : G \to M$, where G stands for microstructure (geometry) and M for the macroscopic physical constants. The key point is a precise and convenient description of the geometrical set G which can be given, for instance, as a set of images. The recent research by Mityushev and Nawalaniec [19], where structural sums were applied in the systematic investigation of the dynamically changing structures, proposes a choice of the geometric parameters as the following set of structural sums:

$$G = \{e_{\mathbf{m}}, \ \mathbf{m} \in \mathcal{M}_e\},$$

where the set $\mathcal{M}_e$ is defined recursively in [22]. One can consider the set G as the base for *structural features* of data represented by non-overlapping disks as well as points distributed on the plane. The set $\mathcal{M}_e$ is infinite, therefore applications require finite approximations in form of the *structural sums feature vector of order q* X_q defined in [24]. For example, the feature vector of order 4 has the following form:

$$X_4 = \{e_2, e_{2,2}, e_{2,2,2}, e_{3,3}, e_{2,2,2,2}, e_{2,3,3}, e_{4,4}\}.$$

The detailed construction of X_q is presented in [24] as well as in [14]. Let us limit our considerations to sums e_2 and $e_{2,2}$.

3 Numerical Experiments

3.1 Structural Features of Poisson Point Process

As an application let us compute structural features of Poisson point processes considered to be completely random process [3, p. 36]. We generated 1000 patterns consisting of 256 points each. Figure 2 presents complex modulus of sums (2.4) calculated with no assumption about radii, i.e. the points are treated as centres of disks with identical radii ($\nu_j = 1$). One can observe a number of outliers, hence we cannot consider these values as reliable characteristics. Table 1 presents high standard deviations of obtained sums. Moreover, it is known from the theory that the values of e_2 for isotropic distributions oscillate around π which is not the case in considered example. We will tackle the above issues in the following

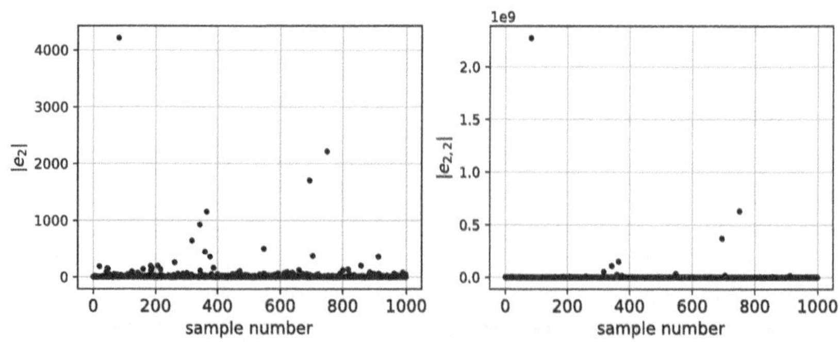

Fig. 2 Complex modulus of considered structural sums for 1000 Poisson process point patterns computed with application of identical radii representation

Table 1 Mean value and standard deviation of considered structural sums computed for identical radii

Sum	Mean	Stdev
e_2	$1.7851 - 8.5439i$	$1.7238 \cdot 10^2$
$e_{2,2}$	$3.7894 \cdot 10^6 - 2.2854 \cdot 10^{-10}i$	$7.5646 \cdot 10^7$

subsection introducing the method for transforming such *ill-conditioned* systems onto corresponding configurations of disks with so-called nearest neighbour radii.

3.2 Stable Features with the Nearest Neighbour Radii

Let us introduce *nearest neighbour* (NN) radii r_j yielding corresponding set of constants ν_j in (2.2). Let a_j $(j = 1, 2, 3, \ldots, N)$ be points distributed in the cell $Q_{(0,0)}$. Then r_j is defined as follows:

$$r_j = \frac{1}{2} \min_{k \neq j} dist(a_j, a_k), \quad j = 1, 2, 3, \ldots, N, \tag{3.1}$$

where $dist$ is the distance between points a_k and a_j in torus topology. Hence, we transform point process pattern into a system of disks with non-zero radii (see Fig. 3) and apply (2.4) with constants ν_j calculated via (2.1). Such an approach results in obtaining reliable structural features (see Fig. 4 and Table 2).

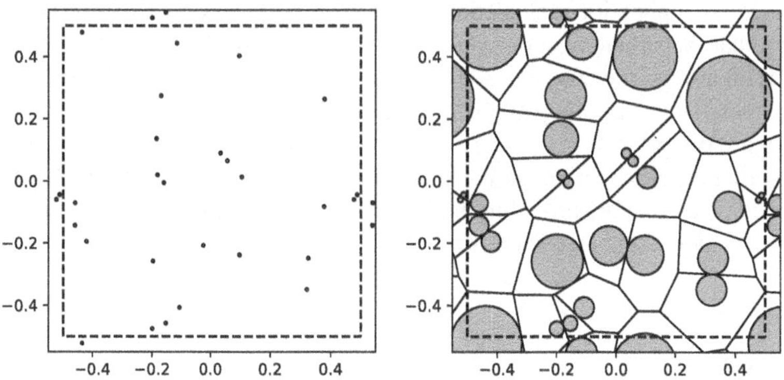

Fig. 3 Example of regularization of data: distribution of data points (left) and the corresponding system of disks with the nearest neighbour radii (right)

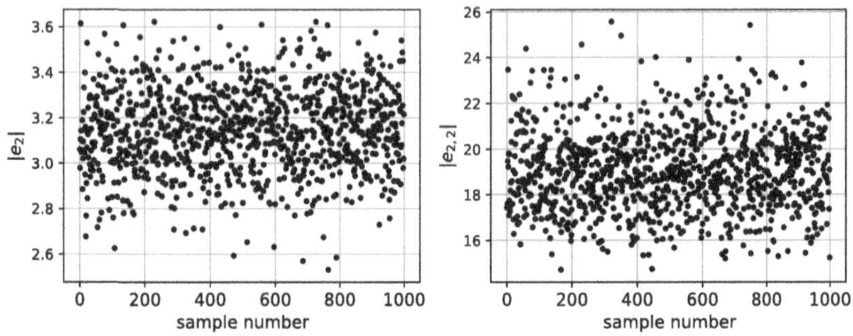

Fig. 4 Complex modulus of considered structural sums for 1000 Poisson process point patterns computed with application of nearest neighbour radii

Table 2 Mean value and standard deviation of considered structural sums computed for nearest neighbour radii

Sum	Mean	Stdev
e_2	$3.1424 - 1.5135 \cdot 10^{-3}i$	0.2602
$e_{2,2}$	$19.0998 + 9.5403 \cdot 10^{-20}i$	1.7613

4 Conclusions

We proposed a method of obtaining reliable structural features in case of data represented by distributions of circles or points on the plane. In order to present the method, we applied it to calculations of Poisson point process patterns. Our study shown that the structural sums can be applied as characteristics of point fields, an important area of statistics [3]. A detailed investigation of structural sums in a role of characteristics of point patterns is under the development and will be published in a future paper.

In order to calculate structural features, one can use the Python software package `basicsums` [25, 26] providing high level of abstraction in the computation of structural sums using algorithms reported in [22] and [23].

References

1. O. Bar, Fast algorithm to determine the flux around closely spaced non-overlapping disks, in *New Trends in Analysis and Interdisciplinary Applications, Selected Contributions of the 10th ISAAC Congress* (Springer International Publishing, Cham, 2017)
2. W. Baran, K. Kurnik, W. Nawalaniec, A. Shareif, Local stationary heat fields in fibrous composites. Silesian J. Pure Appl. Math. **9**(1), 1–8 (2019)
3. S.N. Chiu, D. Stoyan, W.S. Kendall, J. Mecke, *Stochastic Geometry and Its Applications*, 3rd edn. (Wiley, Hoboken, 2013)
4. R. Czapla, Basic sums as parameters characterizing. Silesian J. Pure Appl. Math. **6**(1), 85–96 (2016)

5. R. Czapla, Random sets of stadiums in square and collective behavior of bacteria. IEEE/ACM Trans. Computat. Biology Bioinf. **15**, 251–256 (2018)
6. R. Czapla, V.V. Mityushev, A criterion of collective behavior of bacteria. Math. Biosci. Eng. **14**, 277–287 (2017)
7. R. Czapla, V.V. Mityushev, E. Pesetskaya, An analytical formula for the effective conductivity of 2D domains with cracks of high density. Appl. Math. Modell. **53**, 214–222 (2018)
8. P. Drygaś, A functional-differential equation in a class of analytic functions and its application. Aequationes Math. **73**(3), 222–232 (2007)
9. P. Drygaś, Functional-differential equations in a class of analytic functions and its application to elastic composites. Complex Variables Elliptic Equ. **61**(8), 1145–1156 (2016)
10. P. Drygaś, Generalized Eisenstein functions. J. Math. Analy. Appl. **444**(2), 1321–1331 (2016)
11. P. Drygaś, S. Gluzman, V. Mityushev, W. Nawalaniec, Effective elastic constants of hexagonal array of soft fibers. Computat. Mater. Sci. **139**, 395–405 (2017)
12. P. Drygaś, V. Mityushev, Effective elastic properties of random two-dimensional composites. Int. J. Solids Struct. **97–98**, 543–553 (2016)
13. P. Drygaś, V. Mityushev, Contrast expansion method for elastic incompressible fibrous composites. Adv. Math. Phys. **2017**, Art. no. 4780928 (2017)
14. P. Drygaś, S. Gluzman, V. Mityushev, W. Nawalaniec, *Applied Analysis of Composite Media: Analytical and Computational Results for Materials Scientists and Engineers* (Woodhead Publishing, Sawston, 2019)
15. S. Gluzman, V. Mityushev, W. Nawalaniec, *Computational Analysis of Structured Media* (Academic, Cambridge, 2017)
16. P. Kurtyka, N. Rylko, Quantitative analysis of the particles distributions in reinforced composites. Composite Struct. **182**, 412–419 (2017)
17. V. Mityushev, Representative cell in mechanics of composites and generalized Eisenstein - Rayleigh sums. Complex Variables **51**(8–11), 1033–1045 (2006)
18. V. Mityushev, Cluster method in composites and its convergence. Appl. Math. Lett. **77**, 44–48 (2018)
19. V. Mityushev, W. Nawalaniec, Basic sums and their random dynamic changes in description of microstructure of 2D composites. Computat. Mat. Sci. **97**, 64–74 (2015)
20. V. Mityushev, N. Rylko, Optimal distribution of the nonoverlapping conducting disks. Multiscale Model. Simul. SIAM **10**(1), 180–190 (2012)
21. V. Mityushev, N. Rylko, M. Bryła, Conductivity of two-dimensional composites with randomly distributed elliptical inclusions. ZAMM **98**, 512–516 (2018)
22. W. Nawalaniec, Algorithms for computing symbolic representations of basic *e*–sums and their application to composites. J. Symb. Computat. **74**, 328–345 (2016)
23. W. Nawalaniec, Comp. Appl. Math. (2017). https://doi.org/10.1007/s40314-017-0449-6
24. W. Nawalaniec, Classifying and analysis of random composites using structural sums feature vector. Proc. Roy. Soc. A Math. Phys. Eng. Sci. **475**(2225), 20180698 (2019). https://doi.org/10.1098/rspa.2018.0698
25. W. Nawalaniec, Basicsums: A Python package for computing structural sums and the effective conductivity of random composites. J. Open Source Softw. **4**(39), 1327 (2019). https://doi.org/10.21105/joss.01327
26. W. Nawalaniec. The `basicsums` Python package documentation. https://basicsums.bitbucket.io
27. N. Rylko, Effective anti-plane properties of piezoelectric fibrous composites. Acta Mech. **224**, 2719–2734 (2013)
28. N. Rylko, Edge effects for heat flux in fibrous composites. Appl. Math. Comput. **70**, 2283–2291 (2015)
29. N. Rylko, Fractal local fields in random composites. Appl. Math. Comput. **69**, 247–254 (2015)
30. S. Yakubovich, P. Drygaś, V. Mityushev, Closed-form evaluation of two-dimensional static lattice sums. Proc. Roy. Soc. A Math. Phys. Eng. Sci. **472**(2195), 20160510 (2016)

Multidimensional Potential and Its Application to Social Networks

Natalia Rylko and Karolina Tytko

Abstract We consider application of the structural approximation method to research information flow in social networks. The notations of potential, inter-particle flux and energy are considered in terms of social networks. The basic notations of society functioning related to the communication and information transfer processes are modeled by means of multidimensional potentials and the corresponding interactions arisen in networks. A graph associated to the network is considered in the high dimensional space $\mathbb{R}^d$ when d is comparable with the number of vertices in graphs.

Keywords Social network · Structural approximation · Multidimensional potential

Mathematics Subject Classification (2010) Primary 94C15; Secondary 00A71, 05C21

1 Introduction

Communication and information transfer plays a great role in society [15]. It is supposed that existance of society relies on communication processes and information transfer [4]. Structural approximation is a power method used in Physics and Mechanics in order to study densely packed composites [1, 7, 11, 12, 17, 19, 21, 22]. A similar approach was applied to the study of collective behavior of bacteria [2, 3]. Previous models were created for two– and three– dimensional media. Analogous model can be created for social networks in the high dimensional space $\mathbb{R}^d$ when d

N. Rylko (✉)
Pedagogical University, Krakow, Poland

K. Tytko
Pontifical University of John Paul II, Krakow, Poland
e-mail: karolina.tytko@doktorant.upjp2.edu.pl

P. Cerejeiras et al. (eds.), *Current Trends in Analysis, its Applications and Computation*, Research Perspectives,
https://doi.org/10.1007/978-3-030-87502-2_30

is comparable with the number of participants n involved into a network. The high dimension of space is required by the following reasons. Every participant may connect with the $n-1$ others despite of their remoteness in the physical space $\mathbb{R}^3$. Structural approximation is strictly related to the dimension of the considered space. Hence, in order to properly describe all the connections in the geometry of social network we need to work in the higher dimensional space $\mathbb{R}^{n-1}$.

In this paper, we present application of the multidimensional potentials introduced in [18] to social networks. We use the following terminology associated to the graph theory and flows in the graphs. The term *vertex* is used for a participant of social network, edge for interpersonal relation. Instead of interparticle flux we say *intensity of interaction* between two participants of social network. Potential is understood as the information flow potential in the social network. Energy becomes the *amount of information* or simply information. Let we have the external potential in the form of information $u_0(\mathbf{x})$ determined on the boundary vertices of graph and the potential u_k prescribed to the kth participant.

The Voronoi tessellations of finite networks in bounded domains were used in [1, 11] to estimate the effective properties of composites. Following [1, 11, 18] we extend the theory to network communications.

2 Structural Approximation in $\mathbb{R}^d$

Let $D_k = \{\mathbf{x} \in \mathbb{R}^d : |\mathbf{x} - \mathbf{a}_k| < r\}$ denote a ball with the centers $\mathbf{a}_k$ and with radius r. Consider mutually disjoint balls D_k $(k = 1, 2, \ldots, n)$ in the space $\mathbb{R}^d$. Let D_0 be the complement of all closure balls $D_k \cup \partial D_k$ to $\mathbb{R}^d$. Following [1, 11, 18] and works cited therein we outline the results concerning closely spaced balls and flows in the domain D_0 governed by the p-Laplace equation

$$\nabla \cdot |\nabla u|^{p-2} \nabla u = 0, \quad \mathbf{x} \in D_0, \tag{2.1}$$

where $p \geq 2$. In this model [5, 8], the function $u(\mathbf{x})$ denotes the information flow potential, the vector function $\mathbf{J} = |\nabla u|^{p-2} \nabla u$ the intensity interaction between participant of social network. Let $u_0(\mathbf{x})$ denote the given external information.

The potential satisfies the boundary conditions

$$u(\mathbf{x}) = u_k, \quad |\mathbf{x} - \mathbf{a}_k| = r \ (k = 1, 2, \ldots, n), \tag{2.2}$$

where u_k are undetermined constants. The total normal flow through each sphere vanishes:

$$\int_{\partial D_k} \mathbf{J}(\mathbf{x}) \cdot \mathbf{n}\, ds = 0, \quad k = 1, 2, \ldots, n, \tag{2.3}$$

where $\mathbf{n}$ is the outward unit normal vector to the sphere ∂D_k [18].

Information passing through the whole network is calculated by formula [18]

$$E = \frac{1}{2}\int_{D_0} |\nabla u|^p d\mathbf{x}. \tag{2.4}$$

The quantitive of information E can be found as the minimum of the functional [1, 18]:

$$E = \min_{v \in V} \frac{1}{2}\int_{D_0} |\nabla v|^p d\mathbf{x}, \tag{2.5}$$

where the space V consists of the quasi-periodic functions from the Sobolev space $W^{1,p}(Q_0)$ [18]:

$$V = \{v \in W^{1,p}(Q_0) : v(\mathbf{x}) = t_k \text{ on } D_k \ (k = 1, 2, \ldots, n)\}. \tag{2.6}$$

The discrete social network is a graph Γ with the numbered vertices at $\mathbf{a}_k$ ($k = 1, 2, \ldots, n$) and the edges which show interpersonal relations between two participants of social network. Participants are defined as balls that divide a common interaction (edge) of the Voronoi tessellation. For each fixed $\mathbf{a}_k$, state the set J_k of numbers for neighbor vertices. Their total number $N_k = \#J_k$ is called the degree of the vertex $\mathbf{a}_k$.

The discrete social network model is based on the justification that the intensity interaction is concentrated in the relationships between people. In the linear case $p = 2$, two participants of social network D_k and D_j have the relative intensity interaction $g_{kj}^{(0)}$ calculated by the following asymptotic Keller formula [10]

$$g_{kj}^{(0)} = -\pi r \ln \delta_{kj} \text{ in } \mathbb{R}^3, \tag{2.7}$$

and

$$g_{kj}^{(0)} = \pi\sqrt{\frac{r}{\delta_{kj}}} \text{ in } \mathbb{R}^2. \tag{2.8}$$

Here, δ_{kj} denotes the gap between the balls—the communication gap D_k and D_j

$$\delta_{kj} = |\mathbf{a}_k - \mathbf{a}_j| - 2r. \tag{2.9}$$

The above formulas hold asymptotically as $\delta_{kj} \to 0$.

Let for shortness, $\mathbf{t} = (t_1, t_2, \ldots, t_n) \in \mathbb{R}^n$ and $\mathbf{a} = (\mathbf{a}_1, \mathbf{a}_2, \ldots, \mathbf{a}_n) \in \mathbb{R}^n \times \mathbb{R}^d$. Following [18], introduce the double sum

$$\sum_{k,j}{}' = \sum_{k=1}^{n}\sum_{j\sim k} \equiv \sum_{k=1}^{n}\sum_{j\in J_k}. \tag{2.10}$$

In order to formulate the main asymptotic result of [1] introduce the maximal length of edges $\delta = \max_k \max_{j\in J_k} \delta_{jk}$ of the graph Γ. Following [11, 18] we consider a class $\mathcal{D}$ of social networks. We assume the existence of a percolation chain for $\delta = 0$. This case $\delta = 0$ means that information transmitted instantly spread in the social network. Change of δ means that the centers of balls are fixed but their radii r change and remain equal, more precisely, participants can improve communication by the control parameter r. If $\delta = 0$ then there exists a chain of contacting balls which connect a set of balls. Such a chain is defined a percolation chain. Macroscopic isotropy of the graph representing the network is assumed in our consideration.

Any location of balls not belonging to the class $\mathcal{D}$ of participants having close enough relationships with each other, can be replaced by an element of $\mathcal{D}$ having higher communication intencity. It is possible to do it by parallel translations of non-touching groups of balls to make them touched [18].

3 Discrete Network in $\mathbb{R}^d$

Introduce the main term of the discrete information

$$\mathcal{E} = \min_{\mathbf{u}} \frac{1}{2} {\sum_{k,j}}' g_{kj}^{(0)} |u_k - u_j|^p. \tag{3.1}$$

Mityushev [18] estimated (3.1) following Keller [10] by the integral

$$g_{jk}^{(0)} = \int_B \frac{d\mathbf{x}_{d-1}}{\left(\delta_{jk} + \frac{R^2}{r}\right)^{p-1}}, \tag{3.2}$$

where $d\mathbf{x}_{d-1} = dx_1\ dx_2\ \cdots dx_{d-1}$. The integral (3.2) was calculated in [18] in terms of the hypergeometric function ${}_2F_1$

$$g_{jk} = \frac{2\pi^{\frac{d-1}{2}} r^{d-1}}{\Gamma\left(\frac{d-1}{2}\right)} \frac{{}_2F_1\left(\frac{d-1}{2}, p-1, \frac{d+1}{2}, -\frac{r}{\delta_{jk}}\right)}{\delta_{jk}^{p-1}(d-1)} \tag{3.3}$$

for natural $p > 2$ and

$$p > \frac{d+1}{2}. \tag{3.4}$$

The main asymptotic term of (3.3) as $\delta_{jk} \to 0$ was simplified in [18]. The final expressions are given below

$$g_{jk}^{(0)} = \frac{1}{\delta_{jk}^{p-\frac{d+1}{2}}} \frac{(\pi r)^{\frac{d-1}{2}} \left(p - \frac{d+3}{2}\right)!}{(p-2)!}, \tag{3.5}$$

when d is an odd number. If d is even,

$$g_{jk}^{(0)} = \frac{1}{\delta_{jk}^{p-\frac{d+1}{2}}} \frac{\sqrt{\pi}(\pi r)^{\frac{d-1}{2}}(2p-d-3)!!}{2^{p-\frac{d+1}{2}}(p-2)}. \tag{3.6}$$

Introduce the discrete information

$$\mathcal{E}(\mathbf{t}, \mathbf{a}) = \frac{1}{2} \sum_{k,j}{}' |u_k - u_j|^p f(\|\mathbf{a}_k - \mathbf{a}_j\|) \tag{3.7}$$

and its minimization

$$\mathcal{E}_0 = \min_{\mathbf{t}} \frac{1}{2} \sum_{k,j}{}' |u_k - u_j|^p f(\|\mathbf{a}_k - \mathbf{a}_j\|). \tag{3.8}$$

Following [18] we note that the main term of the intensity interaction $g_{jk}^{(0)}$ is written through the function $f(x) = c(x-2r)^{-\left(p-\frac{d+1}{2}\right)}$ of one variable $x \geq 0$. The constant c is expressed through r, p, d and can be exactly written by use of (3.5) and (3.6).

4 Discussion

In this article, we proposed an application of multidimensional potential to the social network model. We point out analogies between the structures of physics/mechanics and social networks. We considered the intensity interaction problem in the interpersonal relations. The intensity interaction $g_{jk}^{(0)}$ depends on the number of participants of the given social network n related to the dimension of the space d, and on p also related to d by (3.4). Intensity interaction depends also on r. It can be interpret as the dependence between intensity interaction and quality of interpersonal relation, see (2.9), and the dependence between intensity interaction and properties (weights) of social network participants expressed by means of the parameter r.

Interpersonal relation is a process which based on the communication and information transfer. Information theory investigate the physical properties of information and flow of information. Communication theory investigate the processes of human interaction [14]. If we accept the theory of emboddied mind, studied in [6, 13] and [25] we can argue, that social mechanisms are also embodied [9]. Therefore, the information flow, can also change a given social network. It should be noted that we do not take part in the discussion on the scope of determination of human behavior and, as a consequence, determination of social mechanisms.

The proposed model can be applied to participants and their relationschip in the steady state time regime. This is a simplification since the scientific research suggests that the internet [16] or kind of the information (for example gossip [24]) changes the quality of interpersonal relations in time.

In the future paper, we will minimize the information flow potential for the social network model including the time regimes. This will be related to resistance, or impediments in the communication theory. Thanks to this, it will be possible to examine more thoroughly difficulties in communication processes and in interactions of interpersonal relations.

References

1. L. Berlyand, A.G. Kolpakov, A. Novikov, *Introduction to the Network Approximation. Method for Materials Modeling* (Cambridge University Press, Cambridge 2013)
2. R. Czapla, Random sets of stadiums in square and collective behavior of bacteria. IEEE/ACM Trans. Comput. Biol. Bioinf. **15**, 251–256 (2018)
3. R. Czapla, V. Mityushev, A criterion of collective behavior of bacteria. Math. Biosci. Engin. **14**, 277–287 (2017)
4. J. Dewey, Human nature and conduct: an introduction to social psychology. J. Comparat. Psychol. **3**, 73–83 (1923)
5. P. Drygaś, S. Gluzman, V. Mityushev, W. Nawalaniec, *Applied Analysis of Composite Media Analytical and Computational Results for Materials Scientists and Engineers* (Woodhead Publishing Elsevier, Cambridge, 2020)
6. G.M. Edelman, Naturalizing consciousness: A theoretical framework. Proc. Nat. Acad. Sci. **100**, 5520–5524 (2003)
7. S. Gluzman, V. Mityushev, W. Nawalaniec, G. Şokal, Random composite: stirred or shaken? Arch. Mech. **68**, 229–241 (2016)
8. S. Gluzman, V. Mityushev, W. Nawalaniec, *Computational Analysis of Structured Media* (Academic/Cambridge, Elsevier/Amsterdam 2018)
9. M. Hohol, P. Urbańczyk, Some remarks on embodied-embedded social cognition, in *The Emotional Brain Revisited*, ed. by J. Ledoux, M. Heller, B. Brozek, J. Debiec (Copernicus Center Press, Kraków, 2014)
10. J.B. Keller, Conductivity of a medium containing a dense array of perfectly conducting spheres or cylinders or nonconducting cylinders. J. Appl. Phys. **34**, 991–993 (1963)
11. A.A. Kolpakov, A.G. Kolpakov, *Capacity and Transport in Contrast Composite Structures: Asymptotic Analysis and Applications* (CRC Press, Boca Raton 2010)
12. P. Kurtyka, N. Rylko, Quantitative analysis of the particles distributions in reinforced composites. Comp. Struct. **182**, 412–419 (2017)
13. G. Lakoff, Explaining embodied cognition results. Topics Cog. Sci. **4**, 773–785 (2012)
14. R.L Lanigan, Information theories, in *Theories and Models of Communication Vol. 1*, ed. by P. Cobley, P. Schulz (De Gruyter Mouton, Boston, 2013)
15. H. Lasswell, The structure and function of communication in society, in *The Communication of Ideas*, ed. by L. Bryson (Institute for Religious and Social Studies, New York 1948)
16. P.S.N. Lee, L. Leung, V. Lo, Ch. Xiong, T. Wu, Internet communication versus face-to-face interaction in quality of life. Soc. Ind. Res. **100**, 375–389 (2011)
17. D. Legut, U. Wdowik, P. Kurtyka, Vibrational and dielectric properties of α-Si3N4 from density functional theory. Mat. Chem. Phys. **147**, 42–49 (2014)
18. V. Mityushev, Optimal packing of spheres in R^d *and extremal effective conductivity* (2014). arXiv 1412.7527

19. V. Mityushev, N. Rylko, A fast algorithm for computing the flux around non-overlapping disks on the plane. Math. Comput. Model. **57**, 1350–1359 (2013)
20. V. Mityushev, N. Rylko, Optimal distribution of the non-overlapping conducting disks. Multiscale Model. Simul. **10**, 180–190 (2012)
21. V. Mityushev, W. Nawalaniec, N. Rylko, *Introduction to Mathematical Modeling and Computer Simulations* (CRC–Taylor & Francis, Boca Raton 2018)
22. W. Nawalaniec, Classifying and analysis of random composites using structural sums feature vector. Proc. Roy. Soc. A **475**, 1364–5021 (2019)
23. N. Rylko, Structure of the scalar field around unidirectional circular cylinders. Proc. R. Soc. A **464**, 391–407 (2008)
24. A. Shaw, M. Tsvetkova, R. Deneshvar, D. Brooks, Ch. Yu, The effect of gossip on social networks. Complexity **16**, 39–47 (2011)
25. C. Stafford, The remembered present: a biological theory of consciousness. JAMA **264**, 1758–1759 (1990)

Heat and Energy Consumption Management of a Public Object

S. Sh. Ixanov, Zh. Kh. Zhunussova, V. V. Nikulin, and L. Kh. Zhunussova

Abstract The effects of market and seasonal changes in the cost of heat and energy resources on the financial self-sufficiency of a public object. As an example, we take a college, the most important link in educational institutions of Kazakhstan. The necessity to calculate the share of the energy-saving budget compensator, the adjustments of which will reduce the loss of unplanned funds during the period of sharp cold snap and achieve financial sustainability of the college as a management object is justified by computer experiments in MathCad 15 and MatLab 6.5 packages. The calculated data make it possible to predict the amplitude-frequency characteristics of the control signal for smoothing jumps and disturbances in the adaptive control system at the optimal time. This allows to ultimately save college money and spend part of it on additional financial support for the educational process and increase teachers' salaries. It is shown, that the introduction of resource saving technologies (heat, electricity, utilities, staff) contributes to the sustainable development of the institution.

Keywords Adaptive management · Adaptive automated management system · Computer simulation modeling · Stability · The two-link adaptive control system · Heat consumption

The original version of the chapter has been revised. A correction to this chapter can be found at https://doi.org/10.1007/978-3-030-87502-2_70

S. Sh. Ixanov · Zh. Kh. Zhunussova (✉)
Al-Farabi Kazakh National University, Almaty, Republic of Kazakhstan

V. V. Nikulin
State University of New York, New York, NY, USA
e-mail: vnikulin@binghamton.edu

L. Kh. Zhunussova
Abai Kazakh National Pedagogical University, Almaty, Republic of Kazakhstan

P. Cerejeiras et al. (eds.), *Current Trends in Analysis, its Applications and Computation*, Research Perspectives,
https://doi.org/10.1007/978-3-030-87502-2_31

Mathematics Subject Classification (2010) Primary 93C85, 93C40; Secondary 90B50, 90B70

1 Introduction

Education in colleges in the Republic of Kazakhstan is an important and promising direction for improving the business and innovative education of the country's population. The management structure of colleges is based, in the majority, on private entrepreneurship. The competitive environment in the field of special pre-university education requires the effective use of all creative and material resources of the college whose ultimate goal is to train demanded middle-level specialists in the field of innovative economics and banking informational systems. The high quality training of specialists is the most important advantage in the work of the admission committee of a private college, and this, in many ways, determines its financial and economic sustainability and the prospects for the development of the educational institution. One of the components that negatively affects the effectiveness of college management is the consideration of unpredictable external factors: different levels of training for students, undifferentiated wages, spasmodic inflation of money, a reduction in the real incomes of teachers, higher tariffs for electricity, water, higher utility costs, etc. Computerization and informatization of educational and teaching and educational processes, creation of modern microcontroller automated management of heat, electricity and other material resources of the college set the new tasks for adapting management of technical and economic parameters and indicators of the educational process to the realities of modern society [1]. It should be noted, that among all colleges of the country there is no example of a fully computerized college that meets the standards of the digital society. Therefore, the study and application of mathematical and computer analysis methods of sustainability colleges in the Republic of Kazakhstan as an management object are an actual and poorly studied problem in the theory of control.

2 Materials and Methods

In the scientific and technical literature and in legislative norms, the concept of an adaptive educational institution (school, college, university) is actively discussed. In the city of Karaganda, on the basis of the secondary school 27 under the auspices of the Ministry of Education and Science of the Republic of Kazakhstan, an experimental program "Adaptive School - School for All" has been implemented for more than 5 years. It is aimed at the preparation and development of intellectual abilities for the various levels of training, mental and physical abilities of secondary school pupils [2, 3]. Creation of technologies of inclusive education from the viewpoint of control theory is interpreted as the management of education with significantly fuzzy regulators requiring the creation of costly corrective and development methods and the corresponding payment for teachers. Thus, tough budgetary financing without appropriate adjusting regulators led this new and necessary for the country project to the path of unsustainable development due to strong fluctuations in the financial and economic parameters of the school. Children with disabilities often can not obtain private elitist and specialized education in the school system. For compulsory secondary education the decrease in the probability of fuzziness in the input parameters in the management of an established institution is discussed at the legislative level and the state legislatively creates a system of additional funding for children with disabilities, then for secondary pre-university education these standards do not work yet. When students with disabilities enter the college, this institution must ensure, at the expense of its own resources, a quality education for all students, regardless of their abilities. Therefore, the definition of an adaptive college should be considered much broader and taking into account the characteristics of modern development of society. A more precise definition of adaptive college is reduced to three words—a digital adaptive college. The most important criteria for such educational institution are: availability of all conditions (ramps, elevators, special suites, places in canteens, parking lots and special means of electronic communication and transfer) for full-time education of students with limited opportunities, depending on medical indications on technical specialties it is necessary to ensure by individual ventilation systems and heating; availability of all information resources (portal, content, simulators of laboratory works, webinars, smart desk, etc.) for distance learning of students with limited opportunities; availability of special financial, material and pedagogical resources for training and consultations in educational and counseling centers close to the residence of students with disabilities. Thus, to create a digital adaptive college, it is necessary to develop and constantly improve modern adaptive control systems of information, material and pedagogical resources for colleges using fuzzy microcontroller PID regulators [4, 5]. The created mathematical model Smart-system should provide a stable, optimal and efficient management of all resources of the educational institution [6–10]. It should be noted, that with the availability of distance learning facilities in the college, it can gradually increase the number of students without entering new training areas, not only from the number of people with disabilities, but also those who temporarily do not have opportunity to study internally. It is known, that the most common adaptive control systems are automatic control system (ACS) with feedback [11], under the conditions that all parameters of the system are digitized and formalized [12, 13]. The principle of operation of such automatic control system is based on the fact that from the vector of input signals r(t) (a test of knowledge of applicants, current and final control of knowledge, over normative heat, power consumption, decline in academic performance, attendance, etc.) in order to form the error vector e(t), the response vector c(t) is extracted, on the basis of which the signal e(t) is calculated. This signal, using the control system (the college directorate), affects the managed controlled system—management object (college) until the error signal is zero. Figure 1 shows the ACS variants with feedback. This model and model with a compensator can be acceptable both to the technical elements of the college and to the collective of the college whose effectiveness can be quantified as the ratio of the real functionality of the organization vector to the vector of its functionality as an "ideal" functional system.

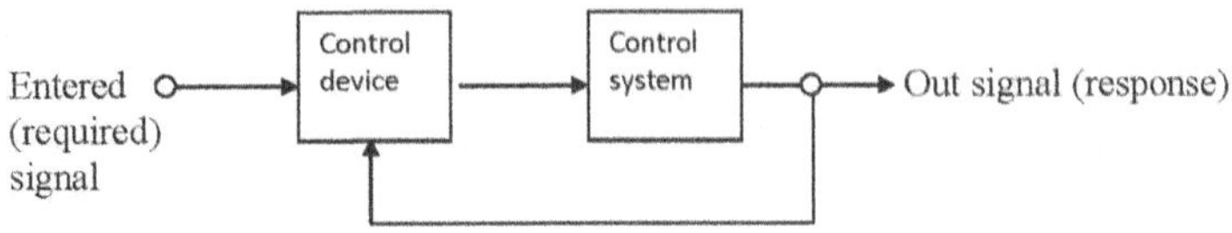

Fig. 1 Control system with feedback

This ratio N.N. Moiseyev designated "as the amount of dissipation (dissipation, volatilization) of energy, a certain analog of this ratio is the coefficient of efficiency" [13]. The objective function of the adaptation model should tend to minimum of the absolute difference between the vectors c(t) and r(t), provided that the poles of the transfer function of the entire system (college) in the stability zone. This definition has a geometric interpretation and is easily calculated using the sisotool tools of the MatLab 6.5 software package [14]. One of the criteria for the stability of the transfer function and the system as a whole is the character of the location of the points on the complex plane of the roots of the characteristic equation. There are three variants of the location of the roots: all the roots lie in the left semi-plane (for example, p1, p2, p3, p4, p5, p6). The system is stable; at least one root lies in the right semi-plane (for example, p7). The system is not stable; the roots lie on the imaginary axis Ym and in the left semi-plane. The system is at the stability bound (conditionally stable).

3 Results and Discussion

The control object (college) can not be studied as an adaptive system with fuzzy PID controllers unless the system is stable with feedback or with corrective elements [7]. We note, that the content of the educational building, as a technical object, is the most important and most costly component of the multi-linked management system of the college. Technological processes associated with the operation of the college's academic building are characterized by the presence of complex links such as "control-output", "perturbation-output". We note, that the algorithm for implementation the study of stability of college as an management object by means of the MathCad 15 software package have not been used anywhere. In particular, when applying a semi-empirical method based on the selection and evaluation of parameters of a controlled system by the least squares method, it is possible to restore the form of the transfer function. This opens the possibility to study all the complex processes which are characteristic for educational institution on the principle of "from a simple adaptive system to a complex multi-linked adaptive control system with fuzzy PID regulators." We consider, for example, the effect of overclocking curves of financial indicators of heat and power supply on the manageability and stability of the college as a two-link management system. Data on heat and electricity are obtained by processing the financial and statistical data of the financial and technical services of the college. It is known, that there is no possibility to select analytically differential equations for the adaptive management of the material resources of the college, therefore, the specific values of the coefficients of the transfer function is determined by the technical and economic indicators of the research object, the normative indicators of the material balance of heat and electricity flows, depending on the season and the time of year. The overclocking curves are taken from the financial and statistical data of the technical department of the college. When removing the acceleration curves on the "control-output" channel or processing the existing ones, it is necessary to monitor that the control and output values of the object are within the limits specified by the technical standards, the perturbations were constant and their values corresponded to the intervals normalized for the selected periods of the year. Accordingly, the acceleration curves for each perturbation through the channels "perturbation-output" must be removed with a fixed control action and constant other perturbations. We give an algorithm for obtaining a transfer function. The curve for the acceleration of the discharge characteristics of the college is derived from the origin of coordinates as a function of time by training calendar.

Step 1. According to the overclocking curve, the tabulated values of the heat and power consumption data are compiled by the integral regulatory indicators, depending on the time of the year and day.
Step 2. By the form of the curve, as well as by the physical nature of the object, the possible orders of the left and right parts of the differential equation describing the object are selected.
Step 3. We describe the differential equation of the object in general form and the solution of the equation for a given input action.
Step 4. By the method of the least squares with respect to the reduced acceleration curve and the solution of the differential equation determines the coefficients of the differential equation. The graphs of the initial curve and the calculated curve are plotted.
Step 5. The transfer function is written by the differential equation.

y	t	y	t	y	t	y
0.00	10.0	0.00	15.0	0.02	20.0	0.90
1.67	45.0	1.68	50.0	2.01	55.0	2.26
2.85	85.0	3.00	90.0	3.08	95	3.1
3.64	140.0	3.79	150.0	3.93	160.	4.05

It can be seen, that the process of consumption of heat and electricity by the college is delayed due to the fact that in the first months of the school year consumption increases through a certain time interval τ, which is measured from the moment of the input signal change to the resource saving before the change of the output control signal begins. The delayed link is a link in which the output quantity y exactly repeats the input quantity x with some delay t: $y(t) = x(t - \tau)$. Transfer function of this link: $W(s) = e - \tau s$. Cutting out the first eight constant values corresponding to the pure delay with $\tau = 40$ we introduce the following values, starting the countdown of the time and the output quantity from zero. Assuming that the curve corresponds to the transient process of a second order aperiodic link with a transfer function $W_0(s) = \frac{k_0}{(T_1 s+1)(T_2 s+1)}$, transient process can be described by expression $h(t) = k_0 \cdot \left(\frac{-\tau_1}{\tau_1-\tau_2} \cdot e^{\frac{-t}{T_1}} + \frac{-\tau_2}{\tau_1-\tau_2} \cdot e^{\frac{-t}{T_2}} + 1\right) \cdot u$, where $u = 4$ is the input effect, k_0 is the transfer coefficient of the object. From the graph, the transfer coefficient of the object can be taken to be 1.1. Estimation of the unknown constants T_1, T_2, as well as a more accurate value of k_o, providing the best approximation of the calculated curve $h(t)$ and the experimental acceleration curve given numerically, can be obtained with the function $genfit(Vt, Vy, Vs, F)$ of the MathCad 15 package. This function returns the parameter vector T of the function F, which gives the minimum standard deviation of the function $F(t, \ldots, T_n)$ from some function $y(t)$ given by the sets (vectors) of the values of Vy, Vt. The function F must be given in the form of a vector containing the function F itself in the symbolic form, and the expressions for all its derivatives with respect to the parameters T (including k_0, for which we introduce k_0 into the vector T), $u \geq 4 \quad \tau_0 = k_0.F(t, \tau) :=$

$$\begin{bmatrix} \tau_0(\frac{-\tau_1}{\tau_1-\tau_2} e^{\frac{-t}{\tau_1}} + \frac{-\tau_2}{\tau_1-\tau_2} e^{\frac{-t}{\tau_2}} + 1)u \\ (\frac{-\tau_1}{\tau_1-\tau_2} e^{\frac{-t}{\tau_1}} + \frac{-\tau_2}{\tau_1-\tau_2} e^{\frac{-t}{\tau_2}} + 1)u \\ \tau_0[\frac{-t}{\tau_1-\tau_2} e^{\frac{-t}{\tau_1}} + \frac{\tau_1}{(\tau_1-\tau_2)^2} e^{\frac{-t}{\tau_1}} + \frac{-\tau_1}{(\tau_1-\tau_2)\tau_1} e^{\frac{-t}{\tau_1}} + \frac{-\tau_2}{(\tau_1-\tau_2)^2} e^{\frac{-t}{\tau_2}}]u, \\ \tau_0[\frac{-\tau_1}{(\tau_1-\tau_2)^2} e^{\frac{-t}{\tau_1}} + \frac{1}{(\tau_1-\tau_2)^2} e^{\frac{-t}{\tau_2}} + \frac{t}{(\tau_1-\tau_2)\tau_1} e^{\frac{-t}{\tau_2}} + \frac{\tau_2}{(\tau_1-\tau_2)^2} e^{\frac{-t}{\tau_2}}]u, \end{bmatrix}$$

We prescribe the initial increment of the vector parameters $T0, T1, T2$ and consider to the function

$$genfit(Vt, Vv, Vs, F),\ VT := \begin{pmatrix} 1 \\ 1 \\ 4 \end{pmatrix},$$

$TT := genfit(V_t, V_v, VF, F)$. We derive the vector of parameters T as $TT = \begin{pmatrix} 1.245 \\ 89.651 \\ 3.152 \end{pmatrix}$, the vector calculated values of the functions with concrete parameters T as $Y(t) := F(t, TT)_0$ (Table 1).

The transfer function of the object in increments taking into account the delay is $W_0(s) = \frac{k_0}{(T_1 s+1)(T_2 s+1)} e^{-s\tau}$, where $k_0 = 1.245$, $T_1 = 89.65c$, $T_2 = 3.15c$.

Table 1 Table values of experimental data on the acceleration of costs for heat and energy consumption as a function of time t (training day), y is the relation of energy consumption on a student to standard (standard 4.00 kW/month)

t	y	t	y	t	y	t
0.5	0.00	1.5	0.00	2.0	0.00	5.0
25.0	0.90	30.0	1.29	35.0	1.55	40.0
65.0	2.42	70.0	2.60	75.0	2.76	80.0
100.0	3.29	110.0	3.29	120.0	3.47	130.0
170.0	4.15	180.0	4.22	190.0	4.26	

Thus, the transfer function of the two-link adaptive control system under consideration can be represented as $W_0(s) = \frac{1{,}245}{(282{,}3975s^2+92{,}8s+1)}$. The resulting transfer function allows computer experiments to be conducted to study the behavior of the control object (college) when external input parameters change (in our case, heat and power consumption). The adaptive automation system with feedback consists of two series-connected inertial links with one input and one output with a resulting transmission factor equal to 1.245 and time constants $T_1 = 89.65$ and $T_2 = 3.15$. To analyze and synthesize one-dimensional linear (linearized) adaptive systems for automatic control of college, we use the toolkit of the Control System Toolbox SISO (Single Input / Single Output) of MatLab 6.5. The results of computer experiments in the energy management process in the MatLab 6.5 prove the stability of the system for sharp external disturbances. As it is shown above, the dynamics of the management of the two-link adaptive heat and power management system of the college possesses by increased sustainability to spasmodic changes in energy consumption, however this result is achieved by additional financial costs for the purchase of resources [15].

4 Conclusion

The adaptive system for automated management of the process of heat and energy consumption by the college can be optimized through the rational use of heat and electricity. Smoothing of sharp perturbing external factors of heat and power supply of the college allows smoothly and steadily regulates the consumption of these expensive resources within the given limits. Automated adaptive resource saving has a significant impact on the financial management of the educational institution, it allows to conduct modern scientific and technical measures for resource saving to attract additional funds for the adaptive learning process and increase the salaries of teachers.

Acknowledgments We thank Assoc. Prof. Gokhan Chuvalcioglu for valuable discussion during the 6th International IFS and Contemporary Mathematics Conference, 07–10 June 2019, Mersin, TURKEY.

References

1. V. Mityushev, W. Nawalaniec, N. Rylko, *Introduction to Mathematical Modeling and Computer Simulations* (CRC Taylor & Francis, Boca Raton, 2018)
2. G.Yu. Zvezdichev, Problems and contradictions of introduction of adaptive information systems in education. Economics and Management of Innovative Technologies (2018)
3. N. Zaruba, Adaptive approach to educational management: principles of management. Prof. Edu. Russia Abroad **2**, 75–79 (2012)

4. N.P. Guzhov, V.Ya. Olkhovsky, D.A. Pavlyuchenko. *Power Supply Systems* (Publishing house of NSTU, Novosibirsk, 2008)
5. V.I. Livchak. Another argument in favor of increasing the thermal protection of buildings. Energy Saving **6**, 14–16 (2012)
6. Zh.Kh. Zhunussova, S.Sh. Ixanov, K. Dosmagulova, Mathematical model of the effectiveness of adaptive automated control systems in educational organizations, in *Proceedings of the 6th International IFS and CMC*, Turkey, (2019), pp. 50–53
7. D.A. Arzamastsev, A.V. Lipes, A.L. Myzin, *Models and Methods for Optimizing the Development of Energy Systems*. Sverdlovsk (1976), p.233
8. A.A. Voronin, S.P. Mishin. *Optimal Hierarchical Structures* (IPP RAS, Moscow, 2003), p. 210
9. A.O. Blinov, P.P. Pereverzev, N.V. Ugryumova. *The Mechanism of Integration of Methods for Improving the Business Processes of Organizations*. Monograph. Chelyabinsk, (2013), p. 235
10. S. Russell. *Artificial Intelligence: A Modern Approach* (Williams, Moscow, 2006), p.1408
11. Available at: http://robotoved.ru/ai education Russia, Accessed Feb. 2019
12. A.A. Zhdanov. *Autonomous Artificial Intelligence* (BINOM, Laboratory of Knowledge, Moscow, 2013), p. 359
13. S.N. Vasiliev. *Intelligent Control of Dynamic Systems* (Moscow, Fizmatlit, 2000), p. 352
14. A.K. Gultyaev. *MatLAB 5.2. Windows Simulation. A Practical Guide* (SPb, Corona print, 1999), p. 288
15. Zh.Kh. Zhunusova, S.Sh. Ixanov, K.A. Dosmangulova, A mathematical model of the effectiveness of adaptive automated control systems in educational organizations. Almaty Bull. NIA RK **3**, 38-42 (2018)

Part VI
Function Spaces and Applications

A Generalized Hardy Operator on Rearrangement Invariant Spaces

Oscar Blasco and Carolina Espinoza-Villalva

Abstract In this paper we consider a weighted generalization of the Hardy operator acting on a rearrangement invariant function space. We give necessary and sufficient conditions for this linear operator to be bounded on a rearrangement invariant function space in terms of its upper Boyd index and the integrability of the norm of the dilation operator with respect to the considered weight.

Keywords Hardy operator · Rearrangement invariant space · Boyd index

Mathematics Subject Classification (2010) 47B38; 46E30

1 Introduction

In 1925 G.H. Hardy [3, 4] proved the following integral inequality

$$\int_0^\infty (Af(x))^p dx \le \left(\frac{p}{p-1}\right)^p \int_0^\infty f^p(x)dx \tag{1.1}$$

where $1 < p < \infty$, f is a measurable non-negative function defined in $(0, \infty)$ and

$$Af(x) = \frac{1}{x}\int_0^x f(t)dt. \tag{1.2}$$

It was the essential role that the Hardy operator plays in the boundedness of other operators what motivated the study of its several generalizations in different

O. Blasco (✉)
Universidad de Valencia, Valencia, Spain
e-mail: oscar.blasco@uv.es

C. Espinoza-Villalva
Universidad de Sonora, Departamento de Matemáticas, Hermosillo, Mexico

P. Cerejeiras et al. (eds.), *Current Trends in Analysis, its Applications and Computation*, Research Perspectives,
https://doi.org/10.1007/978-3-030-87502-2_32

settings. In particular, it is a cornerstone for the analysis of classical operators such as the Hilbert transform ([2, 11]) or the Hardy-Littlewood maximal operator ([7]) on L^p-spaces and more generally on rearrangement invariant spaces.

Some works concerning different generalizations of (1.2) are given in [8–10, 13]. For instance for $0 < a \leq 1$ the operator P_a given by

$$P_a f(t) = t^{-a} \int_0^t f(s) s^a \frac{ds}{s}, \qquad (0 < t < \infty) \tag{1.3}$$

was shown (see [1, Theorem 5.15]) to be bounded on a r.i. Banach space X of measurable functions defined in $(0, \infty)$ if and only if $\overline{\alpha}_X < a$, where $\overline{\alpha}_X$ stands for the upper Boyd index

$$\overline{\alpha}_X = \lim_{t\to\infty} \frac{\log \|E_{1/t}\|_{\mathcal{B}(X)}}{\log t} = \inf_{t>1} \frac{\log \|E_{1/t}\|_{\mathcal{B}(X)}}{\log t}$$

with $E_t f(s) = f(ts)$ for $t, s > 0$.

A further generalization considered by L. Maligranda in [8] reads as follows: If ψ is measurable and positive in $(0, \infty)$ the operator P_ψ is given by

$$P_\psi f(t) = \psi(t)^{-1} \int_0^t f(s)\psi(s) \frac{ds}{s}, \qquad (0 < t < \infty). \tag{1.4}$$

It was proved (see [8, Theorem 1]) that for a r.i. Banach space X of measurable functions defined in $(0, \infty)$ the condition

$$\int_0^1 \|E_t\|_{\mathcal{B}(X)} M(t, \psi) \frac{dt}{t} < \infty \tag{1.5}$$

implies the boundedness of P_ψ from X into X. Furthermore it was shown ([8, Lemma 2] that condition (1.5) is equivalent to $\overline{\alpha}_X < p_0(\psi)$ where $p_0(\psi) = \sup_{0<s<1} \frac{\log M(s,\psi)}{\log s}$ and

$$M(s, \psi) = \sup_{0<t<\infty} \frac{\psi(ts)}{\psi(t)}, \quad s > 0.$$

Another generalization of (1.2) was considered by J. Xiao in [13]: For a non-negative function ψ defined on the interval $(0, 1)$ and a measurable function f defined on $\mathbb{R}^n$ he introduced

$$A_\psi f(x) = \int_0^1 f(tx)\psi(t) \frac{dt}{t}, \qquad x \in \mathbb{R}^n, \tag{1.6}$$

and he showed that A_ψ defines a bounded operator on $L^p(\mathbb{R}^n)$ for $1 \le p \le \infty$ if and only if

$$\int_0^1 t^{-1-n/p}\psi(t)dt < \infty. \tag{1.7}$$

Throughout this paper we shall consider measurable weights $\omega : (0,\infty) \to (0,\infty)$ and define A_ω as the operator

$$A_\omega f(s) = \int_0^1 f(ts)\omega(t)\frac{dt}{t}, \qquad (0 < s < \infty) \tag{1.8}$$

whenever the integral exists a.e. acting on measurable functions defined on $(0,\infty)$. Such operators are also known as Hausdorff operators, see for instance [5, 6].

Both (1.4) and (1.8) are related by the following inequality

$$P_\omega f \le A_{M(\cdot,\omega)} f, \quad f \ge 0. \tag{1.9}$$

Clearly $M(\cdot,\omega)$ is submultiplicative and, in particular $P_\omega f \le A_\omega f$ for submultiplicative weights ω. We shall analyze A_ω acting on r.i. spaces in this paper. The reader is referred to [12] for the study of optimal domain for this type of operators.

Throughout the paper $X \subset L^0(0,\infty)$ is a *Banach function space* (see [1, Definition 1.3] of measurable functions on $(0,\infty)$, in short $X \in$(BFS).

The associate space of X, denoted by X', consists of all the measurable functions g defined on $(0,\infty)$ such that fg is integrable for all $f \in X$. The associate space X' of X endowed with the norm

$$\|g\|_{X'} = \sup\left\{\int_{\mathcal{R}} |fg| : \|f\|_X \le 1\right\},$$

becomes a Banach function space itself (see [1, Lemma I.2.8]).

The space $X \in$(BFS) is said to be rearrangement invariant (r.i.) if $f \in X$ and g equimeasurable to f implies $g \in X$ and $\|g\|_X = \|f\|_X$.

In particular for a r.i. X defined on $(0,\infty)$ and $f \in X$ one has that $\|f\|_X = \|f^*\|_X$, where $f^*(t) = \inf\{\lambda : |\{x : |f(x)| > \lambda\}| \le t\}$ and, since $((0,\infty), |\cdot|)$ is a resonant space ([1, Theorem 2.7])

$$\|h\|_X = \sup\left\{\int_0^\infty g^*(t)h^*(t)dt : \|g\|_{X'} \le 1\right\}. \tag{1.10}$$

As usual we denote the *dilation operator* by $E_t f(s) = f(ts)$ for $s, t > 0$. It is elementary to see that for $t > 0$ and $f \in X$ one has

$$(E_t f)^* = E_t f^* \tag{1.11}$$

$$\|E_t f\|_X = \sup\{\int_0^\infty f^*(ts)g^*(s)ds : \|g\|_{X'} \leq 1\}. \tag{1.12}$$

Let us recall now that for a submultiplicative weight $\omega(t)$ one can consider the indices

$$\overline{\alpha}(\omega) = \inf_{t>1} \frac{\log \omega(t)}{\log t} = \lim_{t\to\infty} \frac{\log \omega(t)}{\log t}, \text{ and}$$

$$\underline{\alpha}(\omega) = \sup_{0<t<1} \frac{\log \omega(t)}{\log t} = \lim_{t\to 0} \frac{\log \omega(t)}{\log t}$$

which are known to satisfy $-\infty < \underline{\alpha}(\omega) \leq \overline{\alpha}(\omega) < \infty$ (see [8] or [1, Pag 147]).

The aim of this paper is to provide conditions under which A_ω defines a bounded linear operator from a rearrangement invariant Banach function space to itself. We shall prove that for any submultiplicative and locally integrable weight ω such that $\underline{\alpha}(\omega) > 0$ and for any r.i. Banach function space X defined on $(0, \infty)$ the boundedness of A_ω on X is equivalent to either

$$\int_0^1 \|E_t\|_{\mathcal{B}(X)}\omega(t)\frac{dt}{t} < \infty. \tag{1.13}$$

or

$$\overline{\alpha}_X < \underline{\alpha}(\omega). \tag{1.14}$$

The proof of the equivalence between (1.13) and (1.14) follows same ideas as in [8, Lemma 2] and it holds for any submultiplicative and locally integrable weight. Our main contribution consists in showing that the boundedness of A_ω leads for submultiplicative weights to the boundedness of A_{v_ε} for $v_\varepsilon(t) = \omega(t)/t^\varepsilon$ which allows to show the integrability condition (1.13).

2 Some Results on Weights

Definition 2.1 Let $\omega : (0, 1) \to (0, \infty)$ be measurable and let $n \in \mathbb{N}$. Let us define $\omega_0(u) = \omega(u)$ and

$$\omega_n(u) = \int_u^1 \omega_{n-1}(t)\omega(u/t)\frac{dt}{t}, \quad 0 < u < 1.$$

Lemma 2.2 *Let $\omega : (0, 1) \to (0, \infty)$ be a measurable function and $n \in \mathbb{N}$. Then*

$$A_{\omega}^{n+1} f(x) = \int_0^1 f(tx)\omega_n(t)\frac{dt}{t}, \quad n \in \mathbb{N}. \tag{2.1}$$

Proof We will proceed by induction on n. The case $n = 0$ follows from the definition of A_{ω}. If we assume now that the result holds true for n then we can write

$$\begin{aligned} A_{\omega}^{n+1} f(x) &= A_{\omega}(A_{\omega}^{n} f)(x) \\ &= \int_0^1 (A_{\omega}^{n} f)(tx)\omega(t)\frac{dt}{t} \\ &= \int_0^1 \omega(t)\left(\int_0^1 f(stx)\omega_{n-1}(s)\frac{ds}{s}\right)\frac{dt}{t} \\ &= \int_0^1 \omega(t)\left(\int_0^t f(ux)\omega_{n-1}(u/t)\frac{du}{u}\right)\frac{dt}{t} \\ &= \int_0^1 f(ux)\left(\int_u^1 \omega_{n-1}(u/t)\omega(t)\frac{dt}{t}\right)\frac{du}{u} \\ &= \int_0^1 f(ux)\left(\int_u^1 \omega(u/t)\omega_{n-1}(t)\frac{dt}{t}\right)\frac{du}{u} \\ &= \int_0^1 f(ux)\omega_n(u)\frac{du}{u}. \end{aligned}$$

This gives the result. □

Lemma 2.3 *Let $\omega : (0, 1) \to (0, \infty)$ be submultiplicative. Then*

$$\omega(t)\frac{(log\ 1/t)^n}{n!} \le \omega_n(t), \quad n \in \mathbb{N}. \tag{2.2}$$

If $0 < \epsilon < 1$ and we write $\omega_{\epsilon,\infty}(t) = \sum_{n=0}^{\infty} \epsilon^n \omega_n(t)$ then

$$\omega(t)/t^{\epsilon} \le \omega_{\epsilon,\infty}(t), \quad 0 < t < 1. \tag{2.3}$$

Proof We prove (2.2) by induction on n. For $n = 0$ it is obvious. Assume the estimate for $n \in \mathbb{N}$ and observe that

$$\omega_{n+1}(u) \ge \int_u^1 \omega(t)\omega(u/t)\frac{(\log\ 1/t)^n}{n!}\frac{dt}{t} \ge \omega(u)\frac{(\log\ 1/u)^{n+1}}{(n+1)!}.$$

This gives (2.2).

Now (2.3) follows trivially from (2.2) since

$$\omega(t)/t^{\epsilon} = \sum_{n=0}^{\infty} \omega(t) \frac{(\epsilon \log 1/t)^n}{n!} \leq \sum_{n=0}^{\infty} \epsilon^n \omega_n(t) = \omega_{\epsilon,\infty}(t).$$

□

Lemma 2.4 *If A_ω is bounded on X then there exists $\epsilon_0 > 0$ such that $A_{\omega_{\epsilon,\infty}}$ is bounded on X for any $0 < \epsilon \leq \epsilon_0$.*

Proof Choose $\epsilon_0 > 0$ sufficiently small so that $\epsilon_0 \|A_\omega\|_{\mathcal{B}(X)} < 1$. Then for any $0 < \epsilon \leq \epsilon_0$ we have that $I - \epsilon A_\omega$ is bounded on X, invertible and

$$(I - \epsilon A_\omega)^{-1} = \sum_{n=0}^{\infty} \epsilon^n A_\omega^n,$$

where the convergence of the series is in the norm of $\mathcal{B}(X)$. Therefore

$$A_\omega (I - \epsilon A_\omega)^{-1} = \sum_{n=0}^{\infty} \epsilon^n A_\omega^{n+1} \in \mathcal{B}(X).$$

For a non-negative function $f \in X$, the formula given by Lemma 2.3 combined with the Monotone Convergence Theorem gives

$$\begin{aligned} A_\omega (I - \epsilon A_\omega)^{-1} f(x) &= \sum_{n=0}^{\infty} \epsilon^n A_\omega^{n+1} f(x) \\ &= \sum_{n=0}^{\infty} \int_0^1 f(xt) \epsilon^n \omega_n(t) \frac{dt}{t} \\ &= \int_0^1 f(xt) \left(\sum_{n=0}^{\infty} \epsilon^n \omega_n(t) \right) \frac{dt}{t} \\ &= A_{\omega_{\epsilon,\infty}} f(x). \end{aligned}$$

We see the identity holds for an arbitrary $f \in X$ by writing $f = f^+ - f^-$, where f^+ and f^- are the positive and negative parts of f, respectively, and appealing to the linearity of both $A_\omega (I - \epsilon A_\omega)^{-1}$ and the integral. The proof is complete. □

3 The Main Theorem

Proposition 3.1 *Let ω be submultiplicative and locally integrable. The following assertions are equivalent:*

(i) $\int_0^1 \|E_t\|_{\mathcal{B}(\bar{X})}\omega(t)\frac{dt}{t} < \infty$.
(ii) $\overline{\alpha}_X < \underline{\alpha}(\omega)$.

Proof (i) $\Longrightarrow$ (ii) From the definition $t^{\underline{\alpha}(\omega)} \le \omega(t)$ for $0 < t < 1$. The integrability condition implies that $\int_0^1 \|E_t\|_{\mathcal{B}(X)}t^{\underline{\alpha}(\omega)}\frac{dt}{t} < \infty$, and therefore there exists $0 < t_0 < 1$ such that $\|E_{t_0}\|_{\mathcal{B}(X)}t_0^{\underline{\alpha}(\omega)} < 1$. In particular $\|E_{1/s_0}\|_{\mathcal{B}(X)} < s_0^{\underline{\alpha}(\omega)}$ where $s_0 = 1/t_0 > 1$. Hence $\overline{\alpha}_X < \underline{\alpha}(\omega)$.

(ii) $\Longrightarrow$ (i) Let $\epsilon > 0$ such that $\overline{\alpha}_X + \epsilon < \underline{\alpha}(\omega) - \epsilon$. From the definitions there exists $0 < \delta < 1$ so that

$$\|E_t\|_{\mathcal{B}(X)} \le t^{-(\overline{\alpha}_X+\epsilon)} \text{ and } \omega(t) \le t^{\underline{\alpha}(\omega)-\epsilon} \quad \text{for } 0 < t < \delta.$$

Therefore

$$\int_0^\delta \|E_t\|_{\mathcal{B}(X)}\omega(t)\frac{dt}{t} \le \int_0^\delta t^{-(\overline{\alpha}_X+\epsilon)}t^{\underline{\alpha}(\omega)-\epsilon}\frac{dt}{t} < \infty.$$

Since $\int_\delta^1 \omega(t)dt < \infty$ and $\|E_t\|_{\mathcal{B}(X)} \le 1/t \le 1/\delta$ for $\delta < t < 1$ then we obtain (i) and the proof is complete. □

Proposition 3.2 *Let X be a r.i on $(0, \infty)$ and let $\omega : (0, \infty) \to (0, \infty)$ be a measurable weigth such that $\int_0^1 \omega(s)\frac{ds}{s} < \infty$. Assume that A_ω is bounded on X. Then*

$$\sup_{0<t<1} \Omega(t)\|E_t\|_{\mathcal{B}(X)} < \infty \tag{3.1}$$

where $\Omega(t) = \int_0^t \frac{\omega(s)}{s}ds$.

Furthermore $\overline{\alpha}_X \le \overline{\alpha}(\omega)$ for submultiplicative weights ω.

Proof Fix functions $f \in X$ and $g \in X'$ with $\|f\|_X \le 1$ and $\|g\|_{X'} \le 1$.

Define $\Omega(t) = \int_0^t \omega(u)\frac{du}{u}$ for $t \in (0, 1)$. Then, for all $0 < t < 1$ we have

$$\begin{aligned}\int_0^\infty f^*(st)g^*(s)ds &= \frac{1}{\Omega(t)}\int_0^t \omega(u)\left(\int_0^\infty f^*(st)g^*(s)ds\right)\frac{du}{u}\\ &\le \frac{1}{\Omega(t)}\int_0^t \omega(u)\left(\int_0^\infty f^*(su)g^*(s)ds\right)\frac{du}{u}\end{aligned}$$

$$\leq \frac{1}{\Omega(t)} \int_0^\infty g^*(s) \left(\int_0^1 f^*(su)\omega(u) \frac{du}{u} \right) ds$$
$$= \frac{1}{\Omega(t)} \int_0^\infty g^*(s) A_\omega f^*(s) ds.$$

Taking supremum on both sides over all the functions $f \in X$ and $g \in X'$ with $\|f\|_X \leq 1$ and $\|g\|_{X'} \leq 1$, we obtain

$$\|E_t\|_{\mathcal{B}(X)} \leq \frac{1}{\Omega(t)} \|A_\omega\|_{\mathcal{B}(X)} \quad (0 < t < 1).$$

This gives (3.1).

Assuming now that ω is submultiplicative we have for $t, s > 0$

$$\Omega(ts) = \int_0^1 \omega(tsu) \frac{du}{u} \leq \omega(t) \int_0^1 \omega(su) \frac{du}{u} = \omega(t)\Omega(s).$$

Therefore

$$\Omega(1)/\Omega(t) \leq \omega(1/t), 0 < t < 1.$$

Equation (3.1) and the above estimate give $\|E_{1/s}\|_{\mathcal{B}(X)} \leq \frac{\|A_\omega\|_{\mathcal{B}(X)}\omega(s)}{\Omega(1)}$ for $s > 1$ and therefore

$$\frac{\log\left(\|E_{1/s}\|_{\mathcal{B}(X)}\right)}{\log s} \leq \frac{\log(\frac{\|A_\omega\|_{\mathcal{B}(X)}}{\Omega(1)})}{\log s} + \frac{\log(\omega(s))}{\log s}, \quad s > 1.$$

Taking limits as $s \to \infty$ we obtain $\overline{\alpha}_X \leq \overline{\alpha}(\omega)$. □

Theorem 3.3 *Let $\omega : (0, \infty) \to (0, \infty)$ be a submultiplicative function, locally integrable with $\underline{\alpha}(\omega) > 0$ and let X be a r.i. Banach function space on $(0, \infty)$. The following assertions are equivalent:*

(i) $\int_0^1 \|E_t\|_{\mathcal{B}(X)}\omega(t)\frac{dt}{t} < \infty$.
(ii) $\overline{\alpha}_X < \underline{\alpha}(\omega)$.
(iii) *A_ω is bounded on X.*

Proof The equivalence between (i) and (ii) was shown in Proposition 3.1.

(i) $\Longrightarrow$ (iii) From Minkowski's inequality and (1.11) one gets

$$\|A_\omega f\|_X \leq \int_0^1 \|f(t\cdot)\|_X \omega(t) \frac{dt}{t} \leq \|f\|_X \int_0^1 \|E_t\|_{\mathcal{B}(X)}\omega(t)\frac{dt}{t}.$$

(iii) $\Longrightarrow$ (i) Let $v_\epsilon(t) = \omega(t)/t^\epsilon$. It follows from (2.3) that for any $0 < \epsilon < \epsilon_0$ $|A_{v_\epsilon} f(x)| \leq A_{\omega_{\epsilon,\infty}} |f|(x)$, which implies

$$\|A_{v_\epsilon} f\|_X \leq \|A_{\omega_{\epsilon,\infty}} |f|\|_X \leq \|A_{\omega_{\epsilon,\infty}}\|_{\mathcal{B}(X)} \|f\|_X.$$

That is $A_{v_\epsilon} \in \mathcal{B}(X)$.

Let $0 < \epsilon < \underline{\alpha}(\omega)/2$ and select $\delta > 0$ such that $\omega(t) \leq t^{\underline{\alpha}(\omega)-\epsilon}$ for $0 < t < \delta$.

In particular one has that $\int_0^\delta v_\epsilon(t) \frac{dt}{t} \leq \int_0^\delta t^{\underline{\alpha}(\omega)-2\epsilon} \frac{dt}{t} < \infty$. Using now the local integrability of ω we have $\int_0^1 v_\epsilon(t) \frac{dt}{t} < \infty$.

Setting $\Omega_\epsilon(t) = \int_0^t \omega(u) \frac{du}{u^{1+\epsilon}}$, we can apply Proposition 3.2 to obtain that

$$\sup_{0<t<1} \Omega_\epsilon(t) \|E_t\|_{\mathcal{B}(X)} \leq \|A_{\omega_{\epsilon_0,\infty}}\|_{\mathcal{B}(X)}.$$

Hence, since $\omega(s) \geq s^{\underline{\alpha}(\omega)}$ for $0 < s < 1$, given $0 < \epsilon < \min\{\epsilon_0/2, \underline{\alpha}(\omega)/2\}$ and $0 < t < \delta$ we have

$$\Omega_{2\epsilon}(t) \geq \int_0^t u^{\underline{\alpha}(\omega)} u^{-2\epsilon-1} du = \frac{t^{\underline{\alpha}(\omega)-2\epsilon}}{\underline{\alpha}(\omega) - 2\epsilon} \geq \frac{\omega(t) t^{-\epsilon}}{\underline{\alpha}(\omega) - 2\epsilon}.$$

Therefore

$$\int_0^\delta \|E_t\|_{\mathcal{B}(X)} \omega(t) \frac{dt}{t} \leq C \int_0^\delta \frac{\omega(t)}{\Omega_{2\epsilon}(t)} \frac{dt}{t} \leq C(\epsilon) \int_0^1 \frac{dt}{t^{1-\epsilon}} < \infty.$$

Using now that $\int_\delta^1 \|E_t\|_{\mathcal{B}(X)} \omega(t) \frac{dt}{t} < \infty$ (as seen at the end of Proposition 3.1) the proof is complete.

□

Acknowledgments We would like to thank the referee for his/her comments which helped us to improve the paper.

The first author is supported by Ministerio de Ciencia, Innovación y Universidades Project PGC2018-095366-B-I00 and the second author is supported by CONACYT, grant 711238.

References

1. C. Bennett, R. Sharpley, *Interpolation of Operators* (Academic, London, 1988)
2. D.w. Boyd, The Hilbert transform on rearrangement-invariant spaces. Canad. J. Math. **19**, 599–616 (1967)
3. G.H. Hardy, Notes on some points in the integral calculus, LX. An inequality between integrals. Messenger Math. **54**, 150–156 (1925)
4. G. H. Hardy, J.E. Littlewood, G. Polya, *Inequalities* (Cambridge University Press, Cambrigde, 1959)

5. E. Liflyand, Hardy type inequalities in the category of Hausdorff operators, in *Modern Methods in Operator Theory and Harmonic Analysis, Springer Proceedings in Mathematics & Statistics*, vol. 291 (Springer, Cham, 2019), pp. 81–91
6. E. Liflyand, A. Miyachi, Boundedness of multidimensional Hausdorff operators in H^p spaces, $0 < p < 1$. Trans. Amer. Math. Soc. **371**(7), 4793–4814 (2019)
7. G.G. Lorentz, Majorants in spaces of integrable functions. Amer. J. Math. **77**, 484–492 (1955)
8. L. Maligranda, Generalized Hardy inequalities in rearrangement invariant spaces. J. Math. Pures Appl. **59**(4), 405–415 (1980)
9. L. Maligranda, On Hardy's inequality in weighted rearrangement invariant spaces and applications. I, II. Proc. Amer. Math. Soc. **88**(1), 67–74 (1983)
10. S.J. Montgomery-Smith, The Hardy operator and Boyd indices. *Interaction Between Functional Analysis, Harmonic Analysis, and Probability* (Columbia, MO, 1994), 359–364, Lecture Notes in Pure and Applied Mathematics, vol. 175 (Dekker, New York, 1996)
11. T. Shimogaki, Hardy-Littlewood majorants in function spaces. J. Math. Soc. Jpn **17**, 365–373 (1965)
12. J. Soria, P. Tradacete, Optimal rearrangement invariant range for Hardy type operators. Proc. Roy. Soc. Edinburgh Sect. A **146**(4), 865–893 (2016)
13. J. Xiao, L^p and BMO bounds of weighted Hardy–Littlewood averages. J. Math. Anal. Appl. **262**, 660–666 (2001)

Restricted Boundedness of Translation Operators on Variable Lebesgue Spaces

Douadi Drihem

Abstract In this paper, we investigate the inequality

$$\|f(\cdot+h)\|_{p(\cdot)} \leq A\,\|f\|_{p(\cdot)}, \quad h \in \mathbb{R}^n, A > 0$$

under some suitable assumptions on the function f and the variable exponent p.

Keywords Translation operator · Maximal function · Variable exponent

Mathematics Subject Classification (2010) 46E35

1 Introduction

Function spaces with variable exponents have been intensively studied in the recent years by a significant number of authors. The motivation for the increasing interest in such spaces comes not only from theoretical purposes, but also from applications to fluid dynamics [10], image restoration [3] and PDEs with non-standard growth conditions. Some example of these spaces can be mentioned such as: variable Lebesgue space, variable Besov and Triebel-Lizorkin spaces. We only refer to the papers [1, 4, 6–8] and to the monograph [5] for further details and references on recent developments on this field.

The purpose of the present paper is to study the translation operators $\tau_h : f \mapsto f(\cdot + h)$, $h \in \mathbb{R}^n$ in the framework of variable Lebesgue spaces $L^{p(\cdot)}(\mathbb{R}^n)$. Their behavior is well known if p is constant. In general τ_h maps $L^{p(\cdot)}(\mathbb{R}^n)$ to $L^{p(\cdot)}(\mathbb{R}^n)$ for any $\lambda > 0$ if and only if p is constant, see [5, Proposition 3.6.1]. Allowing p to vary from point to point will raise extra difficulties which, in general, are overcome

D. Drihem (✉)
M'sila University, Department of Mathematics, Laboratory of Functional Analysis and Geometry of Spaces, M'sila, Algeria
e-mail: douadidr@yahoo.fr; douadi.drihem@univ-msila.dz

P. Cerejeiras et al. (eds.), *Current Trends in Analysis, its Applications and Computation*, Research Perspectives,
https://doi.org/10.1007/978-3-030-87502-2_33

by imposing some regularity assumptions on this exponent. By these additional assumptions we ensure the boundedness of these operators on variable Lebesgue spaces but with some appropriate assumptions. Before to state the main result, we fix some notation and recall some basics facts on variable Lebesgue spaces. We denote by $B(x, r)$ the open ball in $\mathbb{R}^n$ with center x and radius r. By supp f we denote the support of the function f , i.e., the closure of its non-zero set. By $\mathcal{S}(\mathbb{R}^n)$ we denote the Schwartz space of all complex-valued, infinitely differentiable and rapidly decreasing functions on $\mathbb{R}^n$ and by $\mathcal{S}'(\mathbb{R}^n)$ the dual space of all tempered distributions on $\mathbb{R}^n$. We define the Fourier transform of a function $f \in \mathcal{S}(\mathbb{R}^n)$ by

$$\mathcal{F}(f)(\xi) := (2\pi)^{-n/2} \int_{\mathbb{R}^n} e^{-ix\cdot\xi} f(x)dx, \quad \xi \in \mathbb{R}^n.$$

The variable exponents that we consider are always measurable functions on $\mathbb{R}^n$ with range in $[c, \infty)$ for some $c > 0$. We denote the set of such functions by $\mathcal{P}_0(\mathbb{R}^n)$. The subset of variable exponents with range $[1, \infty)$ is denoted by $\mathcal{P}(\mathbb{R}^n)$. We use the standard notation $p^- := \operatorname*{ess\text{-}inf}_{x\in\mathbb{R}^n} p(x)$ and $p^+ := \operatorname*{ess\text{-}sup}_{x\in\mathbb{R}^n} p(x)$. Everywhere below we shall consider bounded exponents.

The variable exponent Lebesgue space $L^{p(\cdot)}(\mathbb{R}^n)$ is the class of all measurable functions f on $\mathbb{R}^n$ such that for some $\lambda > 0$ the modular $\varrho_{p(\cdot)}(f) := \int_{\mathbb{R}^n} |\frac{f(x)}{\lambda}|^{p(x)} dx$ is finite. This is a quasi-Banach function space equipped with the quasi-norm $\|f\|_{p(\cdot)} := \inf\{\mu > 0 : \varrho_{p(\cdot)}(\frac{f}{\mu}) \leq 1\}$. If $p(x) := p$ is constant, then $L^{p(\cdot)}(\mathbb{R}^n) = L^p(\mathbb{R}^n)$ is the classical Lebesgue space.

An useful property is that $\varrho_{p(\cdot)}(f) \leq 1$ if and only if $\|f\|_{p(\cdot)} \leq 1$ (*unit ball property*), which is clear for constant exponents since the relation between the norm and the modular is obvious in that case. As is known, the following inequalities hold

$$\min\left(\varrho_{p(\cdot)}(f)^{1/p^-}, \varrho_{p(\cdot)}(f)^{1/p^+}\right) \leq \|f\|_{p(\cdot)} \leq \max\left(\varrho_{p(\cdot)}(f)^{1/p^-}, \varrho_{p(\cdot)}(f)^{1/p^+}\right). \tag{1.1}$$

We say that a function $g : \mathbb{R}^n \to \mathbb{R}$ is *locally* log-*Hölder continuous*, if there exists a constant $c_{\log} > 0$ such that $|g(x) - g(y)| \leq \frac{c_{\log}}{\ln(e+1/|x-y|)}$ for all $x, y \in \mathbb{R}^n$. If, for some $g_\infty \in \mathbb{R}$ and $c_{\log} > 0$, there holds $|g(x) - g_\infty| \leq \frac{c_{\log}}{\ln(e+|x|)}$ for all $x \in \mathbb{R}^n$, then we say that g satisfies the log-*Hölder decay condition* (at infinity). Note that every function with log-decay condition is bounded.

The notation $\mathcal{P}^{\log}(\mathbb{R}^n)$ is used for all those exponents $p \in \mathcal{P}(\mathbb{R}^n)$ which satisfy the local log-Hölder continuity condition and the log-Hölder decay condition, where we consider $p_\infty := \lim_{|x|\to\infty} p(x)$. The class $\mathcal{P}_0^{\log}(\mathbb{R}^n)$ is defined analogously. By c we denote generic positive constants, which may have different values at different occurrences. We refer to the recent monograph [5] and the paper [9] for further details, and historical remarks and references on variable exponent spaces.

In this paper we shall show the following result:

Theorem 1.1 *Let* $p \in \mathcal{P}^{\log}(\mathbb{R}^n)$ *with* $1 < p^- \leq p^+ < \infty$, $h \in \mathbb{R}^n$ *and* $k \in \mathbb{N}$. *Then for all* $f \in L^{p(\cdot)}(\mathbb{R}^n)$ *with* $\operatorname{supp} \mathcal{F}f \subset \{\xi \in \mathbb{R}^n : |\xi| \leq 2^{v+1}\}$, $v \in \mathbb{N}_0$, *we have*

$$\|\tau_h f\|_{p(\cdot)} \leq c \, \exp\Big((2 + 2^{vnk} |h|^k) c_{\log}(1/p)\Big) \|f\|_{p(\cdot)},$$

where $c > 0$ *is independent of* h, v *and* k.

We mention that the boundedness of these operators in function spaces play an important role in mathematical analysis. They appear in the localizations of Besov spaces [2], where the author used the boundedness of these operators in Besov spaces which based on the Lebesgue spaces.

2 Auxiliary Results

In this section we present some results which are useful for us. The next lemma often allows us to deal with exponents which are smaller than 1, see [4, Lemma A.6]. Recall that $\eta_{v,m}(x) := 2^{nv}(1 + 2^v |x|)^{-m}$, for any $x \in \mathbb{R}^n$, $v \in \mathbb{Z}$ and $m > 0$. Note that $\eta_{v,m} \in L^1(\mathbb{R}^n)$ when $m > n$ and that $\|\eta_{v,m}\|_1 = c_m$ is independent of v.

Lemma 2.1 *Let* $r > 0$, $v \in \mathbb{N}_0$ *and* $m > n$. *Then there exists* $c = c(r, m, n) > 0$ *such that for all* $g \in \mathcal{S}'(\mathbb{R}^n)$ *with* $\operatorname{supp} \mathcal{F}g \subset \{\xi \in \mathbb{R}^n : |\xi| \leq 2^{v+1}\}$, *we have*

$$|g(x)| \leq c(\eta_{v,m} * |g|^r(x))^{1/r}, \quad x \in \mathbb{R}^n.$$

We will make use of the following statement.

Theorem 2.2 *Let* $p \in \mathcal{P}^{\log}(\mathbb{R}^n)$, $\theta := 2 + \frac{|h|^k}{|Q|^k}$, $k \in \mathbb{N}$, $M := \exp \frac{\theta c_{\log}(1/p)}{p^-}$, *if* $|Q| < \min(|h|, 1)$ *and* $M := 1$, *otherwise. Then for every* $m > 0$ *there exists* $\gamma = \exp(-4m c_{\log}(1/p))$ *such that* $\left(\frac{\gamma}{|Q|} \int_Q |\tau_h f(y)| \, dy\right)^{p(x)}$ *is bounded by*

$$\frac{cM}{|Q|} \int_Q |\tau_h f(y)|^{p(y+h)} \, dy + cB\left((e + |x|)^{-m} + \frac{1}{|Q|} \int_Q (e + |y + h|)^{-m} \, dy\right)$$

for every cube (or ball) $Q \subset \mathbb{R}^n$, *all* $x \in Q$, $h \in \mathbb{R}^n$ *and all* $f \in L^{p(\cdot)}(\mathbb{R}^n) + L^\infty(\mathbb{R}^n)$ *with* $\|f\|_{p(\cdot)} \leq 1$, *where* $B = \min\left(1, |Q|^{\frac{m}{\theta}}\right)$ *and* $c > 0$ *is independent of* h, k *and* x .

Proof Our estimate use partially some decomposition techniques already used in [5, Theorem 4.2.4]. Let $p \in \mathcal{P}^{\log}(\mathbb{R}^n)$ with $1 \le p^- \le p^+ < \infty$ and $p^-_{Q+h} = \operatorname{ess-inf}_{z\in Q} p(z+h)$. Define $q \in \mathcal{P}^{\log}(\mathbb{R}^n \times \mathbb{R}^n \times \mathbb{R}^n)$ by $\frac{1}{q(x,y,h)} = \max\left(\frac{1}{p(x)} - \frac{1}{p(y+h)}, 0\right)$. Then

$$\left(\frac{\gamma}{|Q|}\int_Q |f(y+h)|\,dy\right)^{p(x)} \le \frac{M}{|Q|}\int_Q |f(y+h)|^{p(y+h)}\,dy + \frac{1}{|Q|}\int_Q \gamma^{q(x,y,h)}dy$$

for every cube $Q \subset \mathbb{R}^n$, all $x \in Q$, $h \in \mathbb{R}^n$and all $f \in L^{p(\cdot)}(\mathbb{R}^n)$ with $\|f\|_{p(\cdot)} \le 1$. Indeed, we split $f(y+h)$ into three parts

$$\begin{aligned} f_1(y+h) &= \qquad f(y+h)\chi_{\{y:|f(y+h)|>1\}}(y), \\ f_2(y+h) &= f(y+h)\chi_{\{y:|f(y+h)|\le 1, p(y+h)\le p(x)\}}(y), \\ f_3(y+h) &= f(y+h)\chi_{\{y:|f(y+h)|\le 1, p(y+h)> p(x)\}}(y). \end{aligned}$$

By convexity of $t \mapsto t^p$,

$$\begin{aligned} \left(\frac{\gamma}{|Q|}\int_Q |f(y+h)|\,dy\right)^{p(x)} &\le 3^{p^+-1}\sum_{i=1}^{3}\left(\frac{\gamma}{|Q|}\int_Q |f_i(y+h)|\,dy\right)^{p(x)} \\ &= 3^{p^+-1}\left(I_1 + I_2 + I_3\right). \end{aligned}$$

Estimation of I_1 We divide the estimation in three cases.

Case 1. $p(x) \le p^-_{Q+h}$ By Jensen's inequality,

$$I_1 \le \gamma^{p(x)}\frac{1}{|Q|}\int_Q |f_1(y+h)|^{p(x)}\,dy = I.$$

Since $|\tau_h f_1(y)| > 1$, we have $|\tau_h f_1(y)|^{p(x)} \le |\tau_h f_1(y))|^{p^-_{Q+h}} \le |\tau_h f_1(y))|^{p(y+h)}$ and thus

$$I \le \frac{1}{|Q|}\int_Q |f(y+h)|^{p(y+h)}\,dy.$$

Observe that if $\|f\|_\infty \le 1$, then $f_1(y+h) = 0$ and $I = 0$.

Case 2. $p(x) > p^-_{Q+h} \geq p^-_Q$ Again Jensen's inequality implies that

$$
\begin{aligned}
I_1 &\leq \left(\frac{\gamma}{|Q|}\int_Q |f_1(y+h)|^{p^-_Q}\,dy\right)^{\frac{p(x)}{p^-_Q}} \\
&\leq \left(\frac{\gamma}{|Q|}\int_Q |f(y+h)|^{p(y+h)}\,dy\right)^{\frac{p(x)}{p^-_Q}-1}\left(\frac{\gamma}{|Q|}\int_Q |f(y+h)|^{p(y+h)}\,dy\right) \\
&\leq c\frac{\gamma}{|Q|}\int_Q |f(y+h)|^{p(y+h)}\,dy,
\end{aligned}
$$

by the fact that $\int_Q |f(y+h)|^{p(y+h)}\,dy \leq 1$ and $\left(\frac{1}{|Q|}\right)^{\frac{p(x)}{p^-_Q}-1} \leq c$, which follows from $p \in \mathcal{P}^{\log}(\mathbb{R}^n)$, with $c > 0$ independent of x, h and $|Q|$.

Case 3. $p(x) \geq p^-_Q > p^-_{Q+h}$ We have

$$
\begin{aligned}
I_1 &\leq \left(\frac{\gamma}{|Q|}\int_Q |f_1(y+h)|^{p^-_{Q+h}}\,dy\right)^{\frac{p(x)}{p^-_{Q+h}}} \\
&\leq \left(\frac{\gamma}{|Q|}\int_Q |f(y+h)|^{p(y+h)}\,dy\right)\left(\frac{\gamma}{|Q|}\int_Q |f(y+h)|^{p(y+h)}\,dy\right)^{\frac{p(x)}{p^-_{Q+h}}-1} \\
&\leq \frac{1}{|Q|}\int_Q |f(y+h)|^{p(y+h)}\,dy\left(\frac{1}{|Q|}\right)^{\frac{p(x)}{p^-_{Q+h}}-1}.
\end{aligned}
$$

If $|Q| \geq 1$, then the second term is bounded by 1. Now we suppose that $|Q| < 1$. We use the local log-Hölder condition:

$$
\left(\frac{1}{|Q|}\right)^{\frac{p(x)}{p^-_{Q+h}}-1} = \left(\frac{1}{|Q|}\right)^{\frac{p(x)-p^-_Q}{p^-_{Q+h}}}\left(\frac{1}{|Q|}\right)^{\frac{p^-_Q-p^-_{Q+h}}{p^-_{Q+h}}} \leq c\left(\frac{1}{|Q|}\right)^{\frac{p^-_Q-p^-_{Q+h}}{p^-_{Q+h}}}.
$$

Let $p(x_0) = p^-_Q$ and $p(y_0+h) = p^-_{Q+h}$ with $x_0, y_0 \in Q$. Since $p \in \mathcal{P}^{\log}(\mathbb{R}^n)$, we have

$$
\begin{aligned}
\left(\frac{1}{|Q|}\right)^{\frac{p^-_Q-p^-_{Q+h}}{p^-_{Q+h}}} &= \left(\frac{1}{|Q|}\right)^{\frac{p(x_0)-p(y_0)}{p^-_{Q+h}}}\left(\frac{1}{|Q|}\right)^{\frac{p(y_0)-p(y_0+h)}{p^-_{Q+h}}} \\
&\leq c\left(\frac{1}{|Q|}\right)^{\frac{p(y_0)-p(y_0+h)}{p^-_{Q+h}}}.
\end{aligned}
$$

We see that

$$|p(y_0) - p(y_0 + h)| \le \sum_{i=0}^{N-1} \left| p\big(y_0 + \frac{i}{N}h\big) - p\big(y_0 + \frac{i+1}{N}h\big) \right|,$$

where

$$N := \begin{cases} \left[\frac{|h|^k}{|Q|^k}\right] + 1, & |h| > |Q|, \\ 1, & \text{otherwise.} \end{cases}$$

Therefore,

$$\left(\frac{1}{|Q|}\right)^{\frac{p(y_0)-p(y_0+h)}{p_{Q+h}^-}} \le \prod_{i=0}^{N-1} \left(\frac{1}{|Q|}\right)^{\frac{p(y_0+\frac{i}{N}h)-p(y_0+\frac{i+1}{N}h)}{p_{Q+h}^-}} \le c \exp(N c_{\log}(p)/p^-),$$

where $c > 0$ independent of y_0, h, N and $|Q|$, since

$$\left| p\big(y_0 + \frac{i}{N}h\big) - p\big(y_0 + \frac{i+1}{N}h\big) \right| \le \frac{c_{\log}(1/p)}{\log\left(e + \frac{N}{|h|}\right)} \le \frac{c_{\log}(1/p)}{\log\left(e + \frac{1}{|Q|}\right)}$$

if $|h| > |Q|$.

Estimation of I_2 By Jensen's inequality,

$$I_2 \le \gamma^{p(x)} \frac{1}{|Q|} \int_Q |f_2(y+h)|^{p(x)}\, dy = J.$$

Since $|f_2(y+h)| \le 1$ we have $|f_2(y+h)|^{p(x)} \le |f_2(y+h)|^{p(y+h)}$ and thus

$$J \le \frac{1}{|Q|} \int_Q |f(y+h)|^{p(y+h)}\, dy.$$

Estimation of I_3 Again by Jensen's inequality,

$$\left(\frac{\gamma}{|Q|} \int_Q |f_3(y+h)|\, dy\right)^{p(x)}$$
$$\le \frac{1}{|Q|} \int_Q (|\gamma f(y+h)|)^{p(x)}\, \chi_{\{|f(y+h)|\le 1,\, p(y+h)>p(x)\}}(y) dy.$$

Now, Young's inequality give that the last term is bounded by

$$\frac{1}{|Q|} \int_Q \left(|f(y+h)|^{p(y+h)} + \gamma^{q(x,y,h)}\right) dy.$$

Observe that

$$\frac{1}{q(x,y,h)} = \max\left(\frac{1}{p(x)} - \frac{1}{p(y+h)}, 0\right) \leq \frac{1}{s(x)} + \frac{1}{s(y+h)},$$

where $\frac{1}{s(\cdot)} = \left|\frac{1}{p(\cdot)} - \frac{1}{p_\infty}\right|$. We have

$$\gamma^{q(x,y,h)} = \gamma^{q(x,y,h)/2}\gamma^{q(x,y,h)/2} \leq \gamma^{q(x,y,h)/2}\left(\gamma^{s(x)/4} + \gamma^{s(y+h)/4}\right).$$

We suppose that $|Q| < 1$. Then

$$\begin{aligned}\frac{1}{q(x,y,h)} &\leq \left|\frac{1}{p(x)} - \frac{1}{p(y)}\right| + \left|\frac{1}{p(y)} - \frac{1}{p(y+h)}\right| \\ &\leq \frac{c_{\log}(1/p)}{-\log|Q|} + \sum_{i=0}^{N-1}\left|\frac{1}{p(y+\frac{i}{N}h)} - \frac{1}{p(y+\frac{i+1}{N}h)}\right|.\end{aligned}$$

Therefore,

$$\begin{aligned}\frac{1}{q(x,y,h)} &\leq \frac{c_{\log}(1/p)}{-\log|Q|} + \sum_{i=0}^{N-1}\frac{c_{\log}(1/p)}{\log\left(e+\frac{N}{|h|}\right)} \leq \frac{c_{\log}(1/p)}{-\log|Q|}(1+N) \\ &\leq \frac{c_{\log}(1/p)}{-\log|Q|}\left(2+\frac{|h|^k}{|Q|^k}\right).\end{aligned}$$

Hence, $\gamma^{q(x,y,h)/2} = \gamma^{\frac{q(x,y,h)}{4}}\gamma^{\frac{q(x,y,h)}{4}} \leq |Q|^{\frac{m}{2+\frac{|h|^k}{|Q|^k}}}\gamma^{\frac{q(x,y,h)}{4}}$. If $|Q| \geq 1$, then we use $\gamma^{q(x,y,h)/2} \leq 1$ which follow from $\gamma < 1$. Now by [5, Proposition 4.1.8], we obtain the desired inequality. The proof is complete.

□

3 The Proof of the Main Result

In section we prove our result. Let $p \in \mathcal{P}^{\log}(\mathbb{R}^n)$ and define the translation operator by $(\tau_h f)(\cdot) := f(\cdot + h)$. We recall that the Hardy-Littlewood maximal operator $\mathcal{M}$ is defined on L^1_{loc} by

$$\mathcal{M}f(x) := \sup_{r>0}\frac{1}{|B(x,r)|}\int_{B(x,r)}|f(y)|dy, \quad x \in \mathbb{R}^n.$$

Let

$$M_{B(x,r)}f := \frac{1}{|B(x,r)|}\int_{B(x,r)} |f(y)|\,dy, \quad r > 0, x \in \mathbb{R}^n.$$

Proof of Theorem 1.1 Obviously, we assume that $\|f\|_{p(\cdot)} \neq 0$. Since $L^{p(\cdot)}(\mathbb{R}^n) \subset \mathcal{S}'(\mathbb{R}^n)$, Lemma 2.1 yields $|f| \leq \eta_{v,N} * |f|$, for any $N > n$, $v \in \mathbb{N}_0$. We write

$$\eta_{v,N} * |f|(x+h) = \int_{\mathbb{R}^n} \eta_{v,N}(x-y)\,|f(y+h)|\,dy.$$

We split the integral into two parts, one integral over the set $B(x, 2^{-v})$ and one over its complement. The first part is bounded by $M_{B(x,\,2^{-v})}(\tau_h f)(x)$, and the second one is majorized by $c\sum_{i=0}^{\infty} 2^{(n-N)i} M_{B(x,\,2^{1-v+i})}(\tau_h f)(x)$. Consequently, where $N > n$,

$$\|\tau_h f\|_{p(\cdot)} \leq c\sum_{i=0}^{\infty} 2^{(n-N)i}\left\|M_{B(\cdot,\,2^{1-v+i})}(\tau_h f)\right\|_{p(\cdot)}.$$

We will prove that

$$\left\|\gamma\delta M_{B(\cdot,\,2^{1-v+i})}\Big(\frac{\tau_h f}{\|f\|_{p(\cdot)}}\Big)\right\|_{p(\cdot)} \leq c, \quad i, v \in \mathbb{N}_0,$$

with $c > 0$ independent of i, v and h, $\gamma = \exp\big(-4mc_{\log}(1/p)\big)$ and $\delta = \exp\big(-(2 + 2^{vnk}|h|^k)c_{\log}(1/p)\big)$. Taking into account Theorem 2.2 we have, for any $i \in \mathbb{N}_0$, $m > 0$,

$$\left(\gamma\,\delta 2^{(v-i-1)n}\int_{B(x,\,2^{1-v+i})}\frac{|\tau_h f(y)|}{\|f\|_{p(\cdot)}}dy\right)^{p(x)/p^-} \tag{3.1}$$

we majorized it by, after a simple change of variable,

$$\begin{aligned}&cM_{B(x+h,\,2^{1-v+i})}\Big(|g|^{p(\cdot)/p^-}\Big) + c\,(e+|x|)^{-m}\\&+cM_{B(x+h,\,2^{1-v+i})}\Big((e+|\cdot|)^{-m}\Big),\end{aligned}$$

with $g = \frac{f}{\|f\|_{p(\cdot)}}$. Therefore the expression (3.1) is bounded by

$$c\mathcal{M}\big(|g|^{p(\cdot)/p^-}\big)(x+h) + c\,(e+|x|)^{-m} + c\mathcal{M}(e+|\cdot|^{-m})(x+h). \tag{3.2}$$

Obviously,

$$\varrho_{p(\cdot)}\Big(\gamma\delta M_{B(\cdot,\,2^{1-v+i})}\Big(\frac{\tau_h f}{\|f\|_{p(\cdot)}}\Big)\Big)$$
$$= 3^{p^-}\varrho_{p^-}\Big(\frac{1}{3}\Big(\gamma\delta M_{B(\cdot,\,2^{1-v+i})}\Big(\frac{\tau_h f}{\|f\|_{p(\cdot)}}\Big)\Big)^{p(\cdot)/p^-}\Big).$$

In view of (3.2), the last term can be estimated by

$$c\big\|\mathcal{M}(|g|^{p(\cdot)/p^-})(\cdot+h)\big\|_{p^-}^{p^-}+c\,\big\|(e+|\cdot|)^{-m}\big\|_{p^-}^{p^-}+c\,\big\|\mathcal{M}((e+|\cdot|)^{-m})(\cdot+h)\big\|_{p^-}^{p^-}.$$

First we see that $(e+|\cdot|)^{-m}\in L^{p^-}(\mathbb{R}^n)$ for $m>\frac{n}{p^-}$. Secondly the classical result on the continuity of $\mathcal{M}$ on $L^{p^-}(\mathbb{R}^n)$ implies that

$$\big\|\mathcal{M}(|g|^{p(\cdot)/p^-})(\cdot+h)\big\|_{p^-}^{p^-}=\big\|\mathcal{M}\big(|g|^{p(\cdot)/p^-}\big)\big\|_{p^-}^{p^-}\le c\,\big\||g|^{p(\cdot)/p^-}\big\|_{p^-}^{p^-}$$
$$=c\,\varrho_{p(\cdot)}(g)\le c,$$

and

$$\big\|\mathcal{M}((e+|\cdot|)^{-m})(\cdot+h)\big\|_{p^-}^{p^-}=\big\|\mathcal{M}(e+|\cdot|)^{-m}\big\|_{p^-}^{p^-}\le c\,\,\big\|(e+|\cdot|)^{-m}\big\|_{p^-}^{p^-}\le c,$$

since $m>\frac{n}{p^-}$ (with $c>0$ independent of h). Hence

$$\varrho_{p(\cdot)}\Big(\gamma\delta M_{B(\cdot,\,2^{1-v+i})}\Big(\frac{\tau_h f}{\|f\|_{p(\cdot)}}\Big)\Big)\le C,$$

where $C>0$ independent of i and v. Consequently,

$$\|\tau_h f\|_{p(\cdot)}\le c\exp\big((2+2^{vnk}\,|h|^k)c_{\log}\,(1/p)\big)\,\|f\|_{p(\cdot)}\,.$$

The proof is complete.

Remark 3.1 Using Lemma 2.1 we can extend Theorem 1.1 to the case where $p\in\mathcal{P}_0^{\log}(\mathbb{R}^n)$ with $0<p^-\le p^+<\infty$.

Acknowledgments We thank the referee for carefully reading the paper and for making several useful suggestions and comments, which improved the exposition of the paper substantially.

References

1. A. Almeida, P. Hästö, Besov spaces with variable smoothness and integrability. J. Funct. Anal. **258**,1628–1655 (2010)
2. G. Bourdaud, Localizations des espaces de Besov. Stud. Math. **90**, 153–163 (1988)
3. Y. Chen, S. Levine, R. Rao, Variable exponent, linear growth functionals in image restoration. SIAM J. Appl. Math. **66**(4), 1383–1406 (2006)
4. L. Diening, P. Hästö, S. Roudenko, Function spaces of variable smoothness and integrability. J. Funct. Anal. **256**(6), 1731–1768 (2009)
5. L. Diening, P. Harjulehto, P. Hästö, M. Růžička, *Lebesgue and Sobolev Spaces with Variable Exponents*. Lecture Notes in Mathematics, vol. 2017 (Springer, Berlin 2011)
6. D. Drihem, Atomic decomposition of Besov spaces with variable smoothness and integrability. J. Math. Anal. Appl **389** (1), 15–31 (2012)
7. H. Kempka, J. Vybíral, Spaces of variable smoothness and integrability: characterizations by local means and ball means of differences. J. Fourier. Anal. Appl **18**(4), 852–891 (2012)
8. H. Kempka, J. Vybíral, A note on the spaces of variable integrability and summability of Almeida and Hästö. Proc. Am. Math. Soc. **141**(9), 3207–3212 (2013)
9. O. Kováčik, J. Rákosník, On spaces $L^{p(x)}$ and $W^{1,p(x)}$. Czechoslovak Math. J. **41** (116), 592–618 (1991)
10. M. Růžička, *Electrorheological Fluids: Modeling and Mathematical Theory*. Lecture Notes in Mathematics, vol. 1748 (Springer, Berlin, 2000)

Fourier Convolution Operators with Symbols Equivalent to Zero at Infinity on Banach Function Spaces

C. A. Fernandes, A. Yu. Karlovich, and Yu. I. Karlovich

Abstract We study Fourier convolution operators $W^0(a)$ with symbols equivalent to zero at infinity on a separable Banach function space $X(\mathbb{R})$ such that the Hardy-Littlewood maximal operator is bounded on $X(\mathbb{R})$ and on its associate space $X'(\mathbb{R})$. We show that the limit operators of $W^0(a)$ are all equal to zero.

Keywords Fourier convolution operator · Fourier multiplier · Limit operator · Banach function space · Hardy-Littlewood maximal operator · Equivalence at infinity

Mathematics Subject Classification (2010) Primary 47G10; Secondary 42A45, 46E30

1 Introduction

The set of all Lebesgue measurable complex-valued functions on $\mathbb{R}$ is denoted by $\mathfrak{M}(\mathbb{R})$. Let $\mathfrak{M}^+(\mathbb{R})$ be the subset of functions in $\mathfrak{M}(\mathbb{R})$ whose values lie in $[0, \infty]$. The Lebesgue measure of a measurable set $E \subset \mathbb{R}$ is denoted by $|E|$ and its characteristic function is denoted by χ_E. Following [1, Chap. 1, Definition 1.1], a mapping $\rho : \mathfrak{M}^+(\mathbb{R}) \to [0, \infty]$ is called a Banach function norm if, for all functions

C. A. Fernandes · A. Yu. Karlovich (✉)
Centro de Matemática e Aplicações, Departamento de Matemática, Faculdade de Ciências e Tecnologia, Universidade Nova de Lisboa, Caparica, Portugal
e-mail: caf@fct.unl.pt; oyk@fct.unl.pt

Y. I. Karlovich
Centro de Investigación en Ciencias, Instituto de Investigación en Ciencias Básicas y Aplicadas, Universidad Autónoma del Estado de Morelos, Cuernavaca, Morelos, Mexico
e-mail: karlovich@uaem.mx

P. Cerejeiras et al. (eds.), *Current Trends in Analysis, its Applications and Computation*, Research Perspectives,
https://doi.org/10.1007/978-3-030-87502-2_34

f, g, f_n ($n \in \mathbb{N}$) in $\mathfrak{M}^+(\mathbb{R})$, for all constants $a \geq 0$, and for all measurable subsets E of $\mathbb{R}$, the following properties hold:

(A1) $\rho(f) = 0 \Leftrightarrow f = 0$ a.e., $\quad \rho(af) = a\rho(f), \quad \rho(f+g) \leq \rho(f) + \rho(g)$,

(A2) $0 \leq g \leq f$ a.e. $\Rightarrow \rho(g) \leq \rho(f)$ (the lattice property),

(A3) $0 \leq f_n \uparrow f$ a.e. $\Rightarrow \rho(f_n) \uparrow \rho(f)$ (the Fatou property),

(A4) $|E| < \infty \Rightarrow \rho(\chi_E) < \infty$,

(A5) $|E| < \infty \Rightarrow \int_E f(x)\,dx \leq C_E \rho(f)$

with $C_E \in (0, \infty)$ which may depend on E and ρ but is independent of f. When functions differing only on a set of measure zero are identified, the set $X(\mathbb{R})$ of all functions $f \in \mathfrak{M}(\mathbb{R})$ for which $\rho(|f|) < \infty$ is called a Banach function space. For each $f \in X(\mathbb{R})$, the norm of f is defined by $\|f\|_{X(\mathbb{R})} := \rho(|f|)$. Under the natural linear space operations and under this norm, the set $X(\mathbb{R})$ becomes a Banach space (see [1, Chap. 1, Theorems 1.4 and 1.6]). If ρ is a Banach function norm, its associate norm ρ' is defined on $\mathfrak{M}^+(\mathbb{R})$ by

$$\rho'(g) := \sup\left\{\int_{\mathbb{R}} f(x)g(x)\,dx \;:\; f \in \mathfrak{M}^+(\mathbb{R}),\ \rho(f) \leq 1\right\}, \quad g \in \mathfrak{M}^+(\mathbb{R}).$$

It is a Banach function norm itself [1, Chap. 1, Theorem 2.2]. The Banach function space $X'(\mathbb{R})$ determined by the Banach function norm ρ' is called the associate space (Köthe dual) of $X(\mathbb{R})$. The associate space $X'(\mathbb{R})$ is naturally identified with a subspace of the (Banach) dual space $[X(\mathbb{R})]^*$.

Let $\mathcal{F} : L^2(\mathbb{R}) \to L^2(\mathbb{R})$ denote the Fourier transform

$$(\mathcal{F}f)(x) := \widehat{f}(x) := \int_{\mathbb{R}} f(t)e^{itx}\,dt, \quad x \in \mathbb{R},$$

and let $\mathcal{F}^{-1} : L^2(\mathbb{R}) \to L^2(\mathbb{R})$ be the inverse of $\mathcal{F}$. It is well known that the Fourier convolution operator

$$W^0(a) := \mathcal{F}^{-1} a \mathcal{F}$$

is bounded on the space $L^2(\mathbb{R})$ for every $a \in L^\infty(\mathbb{R})$. Let $X(\mathbb{R})$ be a separable Banach function space. Then by Karlovich and Spitkovsky [9, Lemma 2.12(a)], $L^2(\mathbb{R}) \cap X(\mathbb{R})$ is dense in $X(\mathbb{R})$. A function $a \in L^\infty(\mathbb{R})$ is called a Fourier multiplier on $X(\mathbb{R})$ if the convolution operator $W^0(a)$ maps $L^2(\mathbb{R}) \cap X(\mathbb{R})$ into $X(\mathbb{R})$ and extends to a bounded linear operator on $X(\mathbb{R})$. The function a is called the symbol of the Fourier convolution operator $W^0(a)$. The set $\mathcal{M}_{X(\mathbb{R})}$ of all Fourier multipliers on $X(\mathbb{R})$ is a unital normed algebra under pointwise operations and the norm

$$\|a\|_{\mathcal{M}_{X(\mathbb{R})}} := \left\|W^0(a)\right\|_{\mathcal{B}(X(\mathbb{R}))},$$

where $\mathcal{B}(X(\mathbb{R}))$ denotes the Banach algebra of all bounded linear operators on the space $X(\mathbb{R})$.

Recall that the (non-centered) Hardy-Littlewood maximal operator M of a function $f \in L^1_{\mathrm{loc}}(\mathbb{R})$ is defined by

$$(Mf)(x) := \sup_{J \ni x} \frac{1}{|J|} \int_J |f(y)|\, dy,$$

where the supremum is taken over all finite intervals $J \subset \mathbb{R}$ containing x.

Let $V(\mathbb{R})$ be the Banach algebra of all functions $a : \mathbb{R} \to \mathbb{C}$ with finite total variation

$$V(a) := \sup \sum_{i=1}^{n} |a(t_i) - a(t_{i-1})|,$$

where the supremum is taken over all partitions $-\infty < t_0 < \cdots < t_n < +\infty$ of the real line $\mathbb{R}$ and the norm in $V(\mathbb{R})$ is given by $\|a\|_V := \|a\|_{L^\infty(\mathbb{R})} + V(a)$.

Theorem 1.1 *Let $X(\mathbb{R})$ be a separable Banach function space such that the Hardy-Littlewood maximal operator M is bounded on $X(\mathbb{R})$ and on its associate space $X'(\mathbb{R})$. If $a \in V(\mathbb{R})$, then the convolution operator $W^0(a)$ is bounded on the space $X(\mathbb{R})$ and*

$$\|W^0(a)\|_{\mathcal{B}(X(\mathbb{R}))} \le c_X \|a\|_V \tag{1}$$

where c_X is a positive constant depending only on $X(\mathbb{R})$.

This result follows from [5, Theorem 4.3]. Inequality (1) is usually called the Stechkin type inequality (see also [6, inequality (2.4)]).

Following [3, p. 140], two Fourier multipliers $c, d \in \mathcal{M}_{X(\mathbb{R})}$ are called equivalent at infinity if

$$\lim_{N\to\infty} \left\|\chi_{\mathbb{R}\setminus[-N,N]}(c-d)\right\|_{\mathcal{M}_{X(\mathbb{R})}} = 0.$$

In the latter case we will write $c \overset{\mathcal{M}_{X(\mathbb{R})}}{\sim} d$.

The aim of this paper is to start the study of Fourier convolution operators with symbols equivalent at infinity to well behaved symbols by the method of limit operators in the context of Banach function spaces. We refer to [10] for a general theory of limit operators and to [6–8] for its applications to the study of Fourier convolution operators with piecewise slowly oscillating symbols on

Lebesgue spaces with Muckenhoupt weights, constituting a remarkable example of Banach function spaces.

For a sequence of operators $\{A_n\}_{n\in\mathbb{N}} \subset \mathcal{B}(X(\mathbb{R}))$, let $\operatorname{s-lim}_{n\to\infty} A_n$ denote the strong limit of the sequence, if it exists. For $\lambda, x \in \mathbb{R}$, consider the function $e_\lambda(x) := e^{i\lambda x}$. Let $T \in \mathcal{B}(X(\mathbb{R}))$ and $h = \{h_n\}_{n\in\mathbb{N}} \subset (0,\infty)$ be a sequence satisfying $h_n \to +\infty$ as $n \to \infty$. The strong limit

$$T_h := \operatorname*{s-lim}_{n\to\infty} e_{h_n} T e_{h_n}^{-1} I$$

is called the limit operator of T related to the sequence $h = \{h_n\}_{n\in\mathbb{N}}$, if it exists.

Theorem 1.2 (Main Result) *Let $X(\mathbb{R})$ be a separable Banach function space such that the Hardy-Littlewood maximal operator M is bounded on the space $X(\mathbb{R})$ and on its associate space $X'(\mathbb{R})$. If $a \in \mathcal{M}_{X(\mathbb{R})}$ is such that $a \overset{\mathcal{M}_{X(\mathbb{R})}}{\sim} 0$, then for every sequence $h = \{h_n\}_{n\in\mathbb{N}} \subset (0,\infty)$, satisfying $h_n \to +\infty$ as $n \to \infty$, the limit operator of $W^0(a)$ related to the sequence h is the zero operator.*

As usual, let $C_0^\infty(\mathbb{R})$ denote the set of all infinitely differentiable functions with compact support and let $\mathcal{S}(\mathbb{R})$ be the Schwartz space of rapidly decreasing smooth functions. Finally, denote by $\mathcal{S}_0(\mathbb{R})$ the set of all functions $f \in \mathcal{S}(\mathbb{R})$ such that their Fourier transforms $\mathcal{F}f$ have compact supports.

The paper is organized as follows. In Sect. 2, we discuss approximation by mollifiers in separable Banach function spaces such that M is bounded on $X(\mathbb{R})$. In Sect. 3, we show that under the assumptions of the previous section, the set $\mathcal{S}_0(\mathbb{R})$ is dense in the space $X(\mathbb{R})$. Finally, in Sect. 4, we prove Theorem 1.2, essentially using the density of $\mathcal{S}_0(\mathbb{R})$ in the space $X(\mathbb{R})$.

2 Mollification in Separable Banach Function Spaces

The following auxiliary statement might be of independent interest.

Theorem 2.1 *Let $\varphi \in L^1(\mathbb{R})$ satisfy $\int_{\mathbb{R}} \varphi(x)\,dx = 1$ and*

$$\varphi_\delta(x) := \delta^{-1}\varphi(x/\delta), \quad x \in \mathbb{R}, \quad \delta > 0. \tag{2}$$

Suppose that the radial majorant of φ given by $\Phi(x) := \sup_{|y|\ge|x|} |\varphi(y)|$ belongs to $L^1(\mathbb{R})$. If $X(\mathbb{R})$ is a Banach function space such that the Hardy-Littlewood maximal operator M is bounded on the space $X(\mathbb{R})$, then for all $f \in X(\mathbb{R})$,

$$\sup_{\delta>0} \|f * \varphi_\delta\|_{X(\mathbb{R})} \le L\|f\|_{X(\mathbb{R})}, \tag{3}$$

where $L := \|\Phi\|_{L^1(\mathbb{R})}\|M\|_{\mathcal{B}(X(\mathbb{R}))}$ *and* $\|M\|_{\mathcal{B}(X(\mathbb{R}))}$ *denotes the norm of the sublinear operator* M *on the space* $X(\mathbb{R})$*. If, in addition, the space* $X(\mathbb{R})$ *is separable, then for all* $f \in X(\mathbb{R})$*,*

$$\lim_{\delta \to 0^+} \|f * \varphi_\delta - f\|_{X(\mathbb{R})} = 0. \tag{4}$$

Proof The idea of the proof is borrowed from [11, Theorem 2.4]. By the proof of [2, Lemma 5.7], for every $f \in L^1_{\mathrm{loc}}(\mathbb{R})$,

$$\sup_{\delta>0} |(f * \varphi_\delta)(x)| \le \|\Phi\|_{L^1(\mathbb{R})}(Mf)(x), \quad x \in \mathbb{R}. \tag{5}$$

Inequality (3) follows from inequality (5), the boundedness of the Hardy-Littlewood maximal operator M on the space $X(\mathbb{R})$ and Axiom (A2).

Now assume that the space $X(\mathbb{R})$ is separable. Then by Karlovich and Spitkovsky [9, Lemma 2.12(a)], the set $C_0^\infty(\mathbb{R})$ is dense in the space $X(\mathbb{R})$. Take $f \in X(\mathbb{R})$ and fix $\varepsilon > 0$. Then there exists $g \in C_0^\infty(\mathbb{R})$ such that

$$\|f - g\|_{X(\mathbb{R})} < \frac{\varepsilon}{2(L+1)}. \tag{6}$$

Hence for all $\delta > 0$,

$$\|f * \varphi_\delta - f\|_{X(\mathbb{R})} \le \|(f-g) * \varphi_\delta - (f-g)\|_{X(\mathbb{R})} + \|g * \varphi_\delta - g\|_{X(\mathbb{R})}. \tag{7}$$

Taking into account inequalities (3) and (6), we obtain for all $\delta > 0$,

$$\begin{aligned} \|(f-g) * \varphi_\delta - (f-g)\|_{X(\mathbb{R})} &\le \|(f-g) * \varphi_\delta\|_{X(\mathbb{R})} + \|f-g\|_{X(\mathbb{R})} \\ &\le (L+1)\|f-g\|_{X(\mathbb{R})} < \varepsilon/2. \end{aligned} \tag{8}$$

Let $\{\delta_n\}$ be an arbitrary sequence of positive numbers such that $\delta_n \to 0$ as $n \to \infty$. Since $g \in C_0^\infty(\mathbb{R})$, it follows from [13, Chap. III, Theorem 2(b)] that

$$\lim_{n\to\infty} (g * \varphi_{\delta_n})(x) = g(x) \quad \text{for a.e.} \quad x \in \mathbb{R}. \tag{9}$$

In view of (5), we have for all $n \in \mathbb{N}$,

$$|(g * \varphi_{\delta_n})(x)| \le \|\Phi\|_{L^1(\mathbb{R})}(Mg)(x), \quad x \in \mathbb{R}. \tag{10}$$

Since $g \in C_0^\infty(\mathbb{R}) \subset X(\mathbb{R})$ and the Hardy-Littlewood maximal operator M is bounded on the space $X(\mathbb{R})$, we see that $Mg \in X(\mathbb{R})$. Then Mg has absolutely continuous norm because the Banach function space $X(\mathbb{R})$ is separable (see [1, Chap. 1, Definition 3.1 and Corollary 5.6]). It follows from (9)–(10) and the dominated convergence theorem for Banach function spaces (see [1, Chap. 1, Proposition 3.6]) that

$$\lim_{n\to\infty} \|g * \varphi_{\delta_n} - g\|_{X(\mathbb{R})} = 0.$$

Since the sequence $\{\delta_n\}$ is arbitrary, this means that one can find $\delta_0 > 0$ such that for all $\delta \in (0, \delta_0)$,

$$\|g * \varphi_\delta - g\|_{X(\mathbb{R})} < \varepsilon/2. \tag{11}$$

Combining (7), (8), and (11), we see that for all $\delta \in (0, \delta_0)$ one has

$$\|f * \varphi_\delta - f\|_{X(\mathbb{R})} < \varepsilon,$$

which immediately implies (4). □

3 Density of the Set $\mathcal{S}_0(\mathbb{R})$

Lemma 3.1 *Let $X(\mathbb{R})$ be a Banach function space such that the Hardy-Littlewood maximal operator M is bounded on $X(\mathbb{R})$. Then $\mathcal{S}(\mathbb{R}) \subset X(\mathbb{R})$.*

Proof Suppose that $f \in \mathcal{S}(\mathbb{R})$. Then, in particular,

$$\rho_0(f) := \sup_{x\in\mathbb{R}} |f(x)| < \infty, \quad \rho_1(f) := \sup_{x\in\mathbb{R}} |xf(x)| < \infty.$$

By Grafakos [4, Example 2.1.4],

$$\frac{\chi_{\mathbb{R}\setminus[-1,1]}(x)}{|x|} \le \chi_{\mathbb{R}\setminus[-1,1]}(x)(M\chi_{[-1,1]})(x). \tag{12}$$

Since the function $\chi_{[-1,1]}$ belongs to $X(\mathbb{R})$ by Axiom (A4) and since the operator M is bounded on the space $X(\mathbb{R})$, we have $M\chi_{[-1,1]} \in X(\mathbb{R})$. Let $\psi(x) = |x|$. Then in view of (12) and Axiom (A2), we obtain

$$\begin{aligned}\|f\|_{X(\mathbb{R})} &\le \left\|\chi_{[-1,1]} f\right\|_{X(\mathbb{R})} + \left\|\chi_{\mathbb{R}\setminus[-1,1]}\psi f M\chi_{[-1,1]}\right\|_{X(\mathbb{R})} \\ &\le \rho_0(f)\left\|\chi_{[-1,1]}\right\|_{X(\mathbb{R})} + \rho_1(f)\left\|M\chi_{[-1,1]}\right\|_{X(\mathbb{R})}.\end{aligned}$$

Thus, $f \in X(\mathbb{R})$. □

Theorem 3.2 *Let $X(\mathbb{R})$ be a separable Banach function space such that the Hardy-Littlewood maximal operator M is bounded on $X(\mathbb{R})$. Then the set $\mathcal{S}_0(\mathbb{R})$ is dense in the space $X(\mathbb{R})$.*

Proof Let $f \in X(\mathbb{R})$. Fix $\varepsilon > 0$. By Karlovich and Spitkovsky [9, Lemma 2.12(a)], there exists a function $g \in C_0^\infty(\mathbb{R})$ such that

$$\|f - g\|_{X(\mathbb{R})} < \varepsilon/2. \tag{13}$$

Let

$$\varrho(x) := \begin{cases} e^{1/(x^2-1)} & \text{if } |x| < 1, \\ 0 & \text{if } |x| \geq 1, \end{cases} \qquad \varphi(x) := \frac{(\mathcal{F}^{-1}\varrho)(x)}{\int_{\mathbb{R}} (\mathcal{F}^{-1}\varrho)(y)\,dy}, \qquad x \in \mathbb{R}.$$

As $\varrho \in C_0^\infty(\mathbb{R}) \subset \mathcal{S}(\mathbb{R})$, it follows immediately from [4, Corollary 2.2.15] that $\varphi \in \mathcal{S}_0(\mathbb{R})$. For all $\delta > 0$, we define the family of functions φ_δ by (2). Since $g \in C_0^\infty(\mathbb{R})$ and $\varphi_\delta \in \mathcal{S}(\mathbb{R})$, we infer from [4, Proposition 2.2.11(12)] that

$$[\mathcal{F}(g * \varphi_\delta)](x) = (\mathcal{F}g)(x)(\mathcal{F}\varphi_\delta)(x) = (\mathcal{F}g)(x)(\mathcal{F}\varphi)(\delta x), \qquad x \in \mathbb{R}.$$

As $\mathcal{F}\varphi$ has compact support, we conclude that $\mathcal{F}(g * \varphi_\delta)$ also has compact support. Thus $g * \varphi_\delta \in \mathcal{S}_0(\mathbb{R})$ for every $\delta > 0$. By Lemma 3.1, $g * \varphi_\delta \in X(\mathbb{R})$.

By the definition of the Schwartz class $\mathcal{S}(\mathbb{R})$, there are constants $C_n > 0$ such that

$$|\varphi(x)| \leq C_n(1 + |x|)^{-n}, \qquad x \in \mathbb{R}, \qquad n \in \mathbb{N} \cup \{0\}.$$

Then

$$\Phi(x) = \sup_{|y| \geq |x|} |\varphi(y)| \leq C_n \sup_{|y| \geq |x|} (1 + |y|)^{-n} = C_n(1 + |x|)^{-n}$$

for $x \in \mathbb{R}$ and $n \in \mathbb{N} \cup \{0\}$. This estimate implies that the radial majorant Φ of the function φ is integrable.

Since $\Phi \in L^1(\mathbb{R})$, the space $X(\mathbb{R})$ is separable, and the Hardy-Littlewood maximal operator M is bounded on $X(\mathbb{R})$, it follows from Theorem 2.1 that there is a $\delta > 0$ such that

$$\|g * \varphi_\delta - g\|_{X(\mathbb{R})} < \varepsilon/2. \tag{14}$$

Combining (13) and (14), we see that for every $\varepsilon > 0$ there is a $\delta > 0$ such that $\|f - g * \varphi_\delta\|_{X(\mathbb{R})} < \varepsilon$. Since $g * \varphi_\delta \in \mathcal{S}_0(\mathbb{R})$, the proof is completed. □

4 Proof of Theorem 1.2

Fix a sequence $\{h_n\}_{n\in\mathbb{N}} \subset (0,\infty)$ such that $h_n \to +\infty$ as $n \to \infty$. For every function $f \in \mathcal{S}_0(\mathbb{R})$ there exists a segment $K = [x_1, x_2] \subset \mathbb{R}$ such that $\operatorname{supp} \mathcal{F} f \subset [x_1, x_2]$. Therefore

$$e_{h_n} W^0(a) e_{h_n}^{-1} f = W^0[a(\cdot + h_n)] f = \mathcal{F}^{-1}[a(\cdot + h_n)\chi_K]\mathcal{F} f$$
$$= W^0(a\chi_{K+h_n}) f, \tag{15}$$

where $K + h_n = \{x + h_n : x \in K\}$.

Fix $\varepsilon > 0$. Without loss of generality we may assume that $f \neq 0$. As $a \overset{\mathcal{M}_{X(\mathbb{R})}}{\sim} 0$, there exists $N > 0$ such that

$$\left\| \chi_{\mathbb{R}\setminus[-N,N]} a \right\|_{\mathcal{M}_{X(\mathbb{R})}} < \frac{\varepsilon}{3c_X \|f\|_{X(\mathbb{R})}}, \tag{16}$$

where $c_X > 0$ is the constant from Stechkin's type inequality (1). Since $h_n \to +\infty$ as $n \to \infty$, we conclude that there exists $n_0 \in \mathbb{N}$ such that for all $n > n_0$, one has $K + h_n \subset (N, +\infty) \subset \mathbb{R} \setminus [-N, N]$. Therefore, for $n > n_0$, we have

$$a\chi_{K+h_n} = \chi_{\mathbb{R}\setminus[-N,N]} a \chi_{K+h_n}. \tag{17}$$

By Theorem 1.1, for every $n > n_0$, we have

$$\left\| \chi_{K+h_n} \right\|_{\mathcal{M}_{X(\mathbb{R})}} \le c_X \left\| \chi_{K+h_n} \right\|_V = 3c_X. \tag{18}$$

Combining (15)–(18), we see that for $n > n_0$,

$$\left\| e_{h_n} W^0(a) e_{h_n}^{-1} f \right\|_{X(\mathbb{R})} \le \left\| \chi_{R\setminus[-N,N]} a \chi_{K+h_n} \right\|_{\mathcal{M}_{X(\mathbb{R})}} \|f\|_{X(\mathbb{R})}$$
$$\le \left\| \chi_{R\setminus[-N,N]} a \right\|_{\mathcal{M}_{X(\mathbb{R})}} \left\| \chi_{K+h_n} \right\|_{\mathcal{M}_{X(\mathbb{R})}} \|f\|_{X(\mathbb{R})} < \varepsilon.$$

Hence, for every $f \in \mathcal{S}_0(\mathbb{R})$,

$$\lim_{n\to\infty} \left\| e_{h_n} W^0(a) e_{h_n}^{-1} f \right\|_{X(\mathbb{R})} = 0.$$

Since $\mathcal{S}_0(\mathbb{R})$ is dense in $X(\mathbb{R})$ (see Theorem 3.2), the latter equality immediately implies that

$$\text{s-}\lim_{n\to\infty} e_{h_n} W^0(a) e_{h_n}^{-1} I = 0$$

on the space $X(\mathbb{R})$ in view of [12, Lemma 1.4.1(ii)]. □

Acknowledgments This work was partially supported by the Fundação para a Ciência e a Tecnologia (Portuguese Foundation for Science and Technology) through the projects UID/MAT/00297/2019 (Centro de Matemática e Aplicações). The third author was also supported by the SEP-CONACYT Project A1-S-8793 (México).

References

1. C. Bennett, R. Sharpley, *Interpolation of Operators* (Academic Press, Boston, 1988)
2. D. Cruz-Uribe, A. Fiorenza, *Variable Lebesgue Spaces* (Birkhäuser/Springer, New York, 2013)
3. I. De la Cruz-Rodríguez, Y.I. Karlovich, I. Loreto-Hernández, Commutators of convolution type operators with piecewise quasicontinuous data. Commun. Math. Anal. **17**, 131–150 (2014)
4. L. Grafakos, *Classical Fourier Analysis*, 3rd edn. (Springer, New York, 2014)
5. A.Y. Karlovich, Maximally modulated singular integral operators and their applications to pseudodifferential operators on Banach function spaces. Contemp. Math. **645**, 165–178 (2015)
6. Y.I. Karlovich, I. Loreto Hernández, Algebras of convolution type operators with piecewise slowly oscillating data. I: Local and structural study. Integr. Equ. Oper. Theory **74**, 377–415 (2012)
7. Y.I. Karlovich, I. Loreto Hernández, On convolution type operators with piecewise slowly oscillating data. Oper. Theory Adv. Appl. **228**, 185–207 (2013)
8. Y.I. Karlovich, I. Loreto Hernández, Algebras of convolution type operators with piecewise slowly oscillating data. II: local spectra and fredholmness. Integr. Equ. Oper. Theory **75**, 49–86 (2013)
9. A.Y. Karlovich, I.M. Spitkovsky, The Cauchy singular integral operator on weighted variable Lebesgue spaces. Oper. Theory Adv. Appl. **236**, 275–291 (2014)
10. V. Rabinovich, S. Roch, B. Silbermann, *Limit Operators and Their Applications in Operator Theory* (Birkhäuser, Basel, 2004)
11. H. Rafeiro, S. Samko, Characterization of the range of one-dimensional fractional integration in the space with variable exponent. Oper. Theory Adv. Appl. **181**, 393–416 (2008)
12. S. Roch, P.A. Santos, B. Silbermann, *Non-Commutative Gelfand Theories. A Tool-Kit for Operator Theorists and Numerical Analysts* (Springer, Berlin, 2011)
13. E. Stein, *Singular Integrals and Differentiability Properties of Functions* (Princeton University Press, Princeton, 1970)

Part VII
Generalized Functions and Applications

On Almost Periodicity and Almost Automorphy

Chikh Bouzar

Abstract Some properties shared by almost periodic functions and almost automorphic functions are used to study spaces of smooth functions and distributions as well as algebras of generalized functions. An application to ordinary differential equations is given.

Keywords Almost periodicity · Almost automorphy · Smooth functions · Distributions · Ordinary differential equations · Generalized functions

Mathematics Subject Classification (2010) Primary 42A75-43A60; Secondary 46F05-46F30

1 Introduction

The concepts of Bohr almost periodicity [5] and Bochner almost automorphy [2–4] for functions are nowadays well known and applied in different areas. Although almost automorphy is more general than almost periodicity, the spaces of such functions share some important properties that we use in this work to construct in a unifying abstract setting the spaces of smooth functions and distributions as well as algebras of generalized functions of almost periodic and almost automorphic type studied respectively in [6, 8, 10] and [7]. An application to linear systems of ordinary differential equations is given.

Putting forward the construction, one can obtain this abstract setting not only for distributions and generalized functions as it is done in this paper but also for ultradistributions and generalized ultradistributions.

C. Bouzar (✉)
Laboratory of Mathematical Analysis and Applications, University of Oran 1, Oran, Algeria

P. Cerejeiras et al. (eds.), *Current Trends in Analysis, its Applications and Computation*, Research Perspectives,
https://doi.org/10.1007/978-3-030-87502-2_35

2 Smooth E-Functions

We consider functions defined on the whole space $\mathbb{R}$. Let C_b denote the algebra of continuous and bounded complex-valued functions on $\mathbb{R}$ endowed with the norm $\|.\|_\infty$ of uniform convergence on $\mathbb{R}$. It is well known that $(C_b, \|.\|_\infty)$ is a Banach algebra. Recall the different definitions of an almost periodic function, see [4, 5].

Theorem 2.1 (Definition) *A countinuous function f is called almost periodic, if it satisfies one of the following equivalent propositions :*

1. *(Bohr) For any $\varepsilon > 0$, the set $\{\tau \in \mathbb{R} : \|f(\cdot + \tau) - f(\cdot)\|_\infty < \varepsilon\}$ is relatively dense in $\mathbb{R}$.*
2. *(Approximation) For any $\varepsilon > 0$, there exists a trigonometric polynomial P such that $\|f - P\|_{L^\infty} < \varepsilon$.*
3. *(Bochner) The set $(f(. + h))_{h\in\mathbb{R}}$ is relatively compact in $(C_b, \|.\|_\infty)$.*

S. Bochner introduced and studied a more general class of functions, see [2, 3].

Definition 2.2 A countinuous function f is called almost automorphic, if for any sequence $(s_m)_{m\in\mathbb{N}} \subset \mathbb{R}$, one can extract a subsequence $\left(s_{m_k}\right)_k$ such that

$$g(x) := \lim_{k\to+\infty} f\left(x + s_{m_k}\right) \text{ is well-defined for every } x \in \mathbb{R}$$

and

$$\lim_{k\to+\infty} g\left(x - s_{m_k}\right) = f(x) \text{ for every } x \in \mathbb{R}$$

Remark 2.3 It is known that $C_{ap} \subsetneq C_{aa}$

Notation We denote by E either the space of almost periodic functions C_{ap} or the space of almost automorphic functions C_{aa}

Although $C_{ap} \subsetneq C_{aa}$, these spaces share some important properties that we summarize in the following results.

Theorem 2.4

1. *The space $(E, \|.\|_\infty)$ is a Banach algebra.*
2. *If $f \in E$ and $g \in L^1$, then $f * g \in E$.*
3. *If the derivative f' of $f \in E$ is uniformly continuous on $\mathbb{R}$, then it belongs to E.*
4. *A primitive of $f \in E$ belongs to E if and only if it is bounded.*
5. *If $f \in E$ and F is continuous on $\mathbb{C}$ then $F \circ f \in E$.*
6. *If $f \in E$ and $\lim_{x\longrightarrow+\infty} f(x) = 0$, then $f = 0$.*

Let $p \in [1, +\infty]$ and $\mathcal{D}_{L^p} := \{\varphi \in C^\infty(\mathbb{R}) : \varphi^{(j)} \in L^p(\mathbb{R}), \forall j \in \mathbb{Z}_+\}$, this space endowed with the family of seminorms $|\varphi|_{k,p} := \sum_{j \le k} \left\|\varphi^{(j)}\right\|_{L^p}, k \in \mathbb{Z}_+$, is a Fréchet algebra. Denote $\mathcal{D}_{L^\infty}$ by $\mathcal{B}$ and $\dot{\mathcal{B}}$ the closure in $\mathcal{B}$ of the space $\mathcal{D}$ of smooth functions with compact support.

Definition 2.5 The space of smooth E-functions is denoted and defined by

$$\mathcal{B}_E := \left\{\varphi \in C^\infty(\mathbb{R}) : \varphi^{(j)} \in E, \forall j \in \mathbb{Z}_+\right\}$$

We endow $\mathcal{B}_E$ with the family of seminorms $|\varphi|_k := \sum_{j \le k} \left\|\varphi^{(j)}\right\|_\infty, \ k \in \mathbb{Z}_+$.

Proposition 2.6

1. *$\mathcal{B}_E$ is a Fréchet algebra.*
2. $\mathcal{B}_E * L^1 \subset \mathcal{B}_E$.
3. *Let $f \in \mathcal{B}_E$ and h its primitive, then $h \in \mathcal{B}_E$ if and only if h is bounded.*
4. *If $f \in \mathcal{B}_E$ and F is of class C^∞ on $\mathbb{C}$ then $F \circ f \in \mathcal{B}_E$.*
5. *If $f \in \mathcal{B}_E$ and $\exists k \in \mathbb{Z}_+$ such that $\lim\limits_{x \longrightarrow +\infty} f^{(k)}(x) = 0$, then $f \equiv 0$.*

Proof 1. Using the Leibniz formula we obtain that $\mathcal{B}_E$ is stable under product and as E is a closed subalgebra of the Banach algebra $(\mathcal{C}_b, \|.\|_\infty)$ it holds that $\mathcal{B}_E$ is complete. 2. The proof is obtained from the facts that E is stable under convolution with L^1 and $(f * g)' = f' * g$, where $f \in \mathcal{B}_E$ and $g \in L^1$. 3. If a primitive belongs to $\mathcal{B}_E$ then it is bounded. On the other hand, if a primitive is bounded then it belongs to E consequently it is an element of $\mathcal{B}_E$. 4. By the classical Faà di Bruno formula we have

$$\frac{(F \circ f)^{(j)}}{j!} = \sum_{\substack{l_1+2l_2+\cdots+jl_j=j \\ r=l_1+\cdots+l_j}} \frac{F^{(r)}(f)}{l_1! \cdots l_j!} \prod_{i=1}^{j} \left(\frac{f^{(i)}}{i!}\right)^{l_i}, \forall j \in \mathbb{Z}_+,$$

the properties of the space E give the result. 5. is easy to prove.

Remark 2.7 All the properties of E given by Theorem 2.4 are lifted to $\mathcal{B}_E$.

Proposition 2.8 *We have $\mathcal{B}_E = E \cap \mathcal{B}$.*

Proof It is clear that $\mathcal{B}_E \subset E \cap \mathcal{B}$. As the derivative f' of $f \in \mathcal{B}_E$ is bounded, so f is uniformly continuous and then $f' \in E$, repeating this argument we obtain that $f^{(j)} \in E, \forall j \in \mathbb{N}$, so the reverse inclusion is proved.

Example If $E= \mathcal{C}_{ap}$ then $\mathcal{B}_E = \mathcal{B}_{ap}$ is the space of L. Schwartz [10], if $E= \mathcal{C}_{aa}$ then $\mathcal{B}_E = \mathcal{B}_{aa}$ see [8].

3 *E*-Distributions

The space of L^p–distributions $\mathcal{D}'_{L^p}$, $1 < p \leq +\infty$, is the topological dual of $\mathcal{D}_{L^q}$, where $\frac{1}{p} + \frac{1}{q} = 1$. The topological dual of $\dot{\mathcal{B}}$ is denoted by $\mathcal{D}'_{L^1}$. The space $\mathcal{D}'_{L^\infty}$ is called the space of bounded distributions and denoted by $\mathcal{B}'$, see [10].

Theorem 3.1 (Definition) *The space of E-distributions on* $\mathbb{R}$, *denoted by* $\mathcal{B}'_E$, *is the space of distributions* $T \in \mathcal{B}'$ *satisfying one of the following equivalent propositions :*

1. $T * \varphi \in E$, $\forall \varphi \in \mathcal{D}$.
2. $\exists m \in \mathbb{Z}_+, \exists \left(f_j\right)_{j\leq m} \subset E$ *such that* $T = \sum_{j=0}^{m} f_j^{(j)}$.

Proof $1 \Rightarrow 2$: Let $T \in \mathcal{B}'$, then there exist $m \in \mathbb{Z}_+$, $C > 0$ such that

$$|\langle T, \psi\rangle| \leq C\, |\psi|_{m,1}\,, \forall \psi \in \mathcal{D}_{L^1}.$$

In [10] it is shown that if $k \in \mathbb{N}$ is sufficiently large there exists a fundamental solution h of the linear differential operator $\left(1 - \frac{d^2}{dx^2}\right)^k$ which is of class $\mathcal{C}^{m+2}$ and with integrable derivatives of order $\leq m + 2$, i.e. $h \in \mathcal{D}^{2m+2}_{L^1} := \left\{\varphi \in \mathcal{C}^{2m+2} : \varphi^{(j)} \in L^1, \forall j \leq 2m+2\right\}$, and as $T \in \mathcal{B}'$, it follows that $T * h$ exists. We have $T = \left(1 - \frac{d^2}{dx^2}\right)^k (T * h)$. The space $\mathcal{D}^{2m+2}_{L^1}$ is endowed with the norm $|.|_{2m+2,1}$. It is well known that $\mathcal{D}$ is dense in $\mathcal{D}^{2m+2}_{L^1}$, then T is extended continuously to the space $\mathcal{D}^{2m+2}_{L^1}$. There exists a sequence $(\theta_k)_{k\in\mathbb{N}} \subset \mathcal{D}$ such that $(\theta_k)_k$ converges to h in $\mathcal{D}^{2m+2}_{L^1}$. We have

$$|(T * \theta_k)(x) - (T * h)(x)| \leq C\, |\theta_k - h|_{2m+2,1}$$

which means that the sequence $(T * \theta_k)_{k\in\mathbb{N}}$ converges uniformly to $T * h$ on $\mathbb{R}$. By the properties of the space E we get $g := T * h \in E$.

$2 \Rightarrow 1$: For $\varphi \in \mathcal{D}$, we have $T * \varphi = \sum_{j\leq k} f_j^{(j)} * \varphi = \sum_{j\leq k} f_j * \varphi^{(j)} \in E$ due to the properties of the space E.

The translate $\tau_\omega T$, $\omega \in \mathbb{R}$, of a distribution $T \in \mathcal{D}'$ is defined by $\forall \varphi \in \mathcal{D}$, $\langle \tau_\omega T, \varphi\rangle = \langle T, \tau_{-\omega}\varphi\rangle$, where $\tau_{-\omega}\varphi(x) = \varphi(x - \omega)$, $x \in \mathbb{R}$.

Proposition 3.2

1. *If* $T \in \mathcal{B}'_E$, *then* $T^{(i)} \in \mathcal{B}'_E$, $\forall i \in \mathbb{N}$.
2. $\mathcal{B}'_E \times \mathcal{B}_E \subset \mathcal{B}'_E$.
3. $\mathcal{B}'_E * \mathcal{D}'_{L^1} \subset \mathcal{B}'_E$.

4. *A primitive of an E-distribution is an E-distribution if and only if it is a bounded distribution.*
5. *If* $T \in \mathcal{B}'_E$ *and* $\lim\limits_{\omega \to +\infty} \tau_\omega T = 0$, *then* $T = 0$.

Proof 1. The proof is clear. 2. If $T \in \mathcal{B}'_E$ there exists $\left(f_j\right)_{j \leq k} \subset E$ such that $T = \sum_{j \leq k} f_j^{(j)}$. We have $\varphi T = \sum_{j \leq k} \varphi f_j^{(j)}$, so

$$\varphi T = \sum_{j=0}^{k} \sum_{l=0}^{j} (-1)^l \binom{j}{l} \left(\varphi^{(l)} f_j\right)^{(j-l)},$$

due to the transposed Leibniz formula. Since E is an algebra and $\varphi \in \mathcal{B}_E$, it follows that $\varphi T \in \mathcal{B}'_E$. 3. Let $T \in \mathcal{B}'_E$ and $S \in \mathcal{D}'_{L^1}$, so we have that $T * S \in \mathcal{B}'$. Morever, there exist $(f_l)_{l \leq k} \subset E$ and $\left(g_j\right)_{j \leq m} \subset L^1$ such that $T = \sum_{j \leq k} f_l^{(l)}$ and $S = \sum_{j \leq m} g_j^{(j)}$. Hence

$$(T * S) = \sum_{l=0}^{k} \sum_{j=0}^{m} \left(f_l * g_j\right)^{(l+j)}.$$

By the properties of the space E, and the definition of $\mathcal{B}'_E$, $T * S \in \mathcal{B}'_E$. 4. If $T \in \mathcal{B}'_E$ has a primitive $S \in \mathcal{B}'$, i.e. $S' = T$, then $S * \varphi \in L^\infty$, $\forall \varphi \in \mathcal{D}$, and

$$(S * \varphi)' = S' * \varphi = T * \varphi \in E, \ \forall \varphi \in \mathcal{D},$$

i.e. $(S * \varphi)$ is a bounded primitive of the E- function $(T * \varphi)$. Thus by the properties of the space E we have $S * \varphi \in E, \forall \varphi \in \mathcal{D}$, so $S \in \mathcal{B}'_E$. It is clear that if a primitive belongs to $\mathcal{B}'_E$ then it is a bounded distribution. 5. Let $T \in \mathcal{B}'_E$, , as by hypothesis $\lim_{x \to +\infty} (T * \varphi)(x) = \lim_{x \to +\infty} \left\langle T, \tau_{-x} \check{\varphi} \right\rangle = 0$, it follows, due to the properties of the space E, that $T * \varphi \equiv 0$, $\forall \varphi \in \mathcal{D}$. On the other hand $\langle T, \varphi \rangle = \left(T * \check{\varphi}\right)(0) = 0$, so $T = 0$.

Remark 3.3 All the properties of E and $\mathcal{B}_E$ are lifted to the space $\mathcal{B}'_E$ except the multiplication and composition. These defects will be removed in the context of the algebra of E-generalized functions, see Sect. 5.

The next result shows that $\mathcal{B}_E$ is dense in $\mathcal{B}'_E$.

Proposition 3.4 *Let* $T \in \mathcal{B}'$. *Then* $T \in \mathcal{B}'_E$ *if and only if there exists* $(\varphi_m)_{m \in \mathbb{N}} \subset \mathcal{B}_E$ *such that* $\lim\limits_{m \to +\infty} \varphi_m = T$ *in* $\mathcal{B}'$.

Proof Let $(\varphi_m)_{m \in \mathbb{N}} \subset \mathcal{B}_E$ such that $\lim\limits_{m \to +\infty} = T$ in $\mathcal{B}'$. For any bounded subset $A \subset \mathcal{D}_{L^1}$ we have $\sup_{\psi \in A} |\langle \varphi_m - T, \psi \rangle| \underset{m \to +\infty}{\longrightarrow} 0$. For a fixed $\varphi \in \mathcal{D}$, the set

$A := \left\{ \tau_{-x}\check{\varphi} : x \in \mathbb{R} \right\}$ is bounded in $\mathcal{D}_{L^1}$, so

$$\sup_{x\in\mathbb{R}} \left|(\varphi_m * \varphi)(x) - (T * \varphi)(x)\right| = \sup_{\psi \in A} \left|\langle \varphi_m - T, \psi \rangle\right| \underset{m\to+\infty}{\longrightarrow} 0,$$

i.e. the sequence of functions $(\varphi_m * \varphi)_{m\in\mathbb{N}} \subset E$ converges uniformly to $T * \varphi$, as the space E is complete, it follows that $T * \varphi \in E$, consequenyelly $T \in \mathcal{B}'_E$. Conversely, let $T \in \mathcal{B}'_E$ and take a sequence $(\rho_m)_{m\in\mathbb{N}} \subset \mathcal{D}$ such that $\rho_m \geq 0$, $\operatorname{supp} \rho_m \subset \left[-\frac{1}{m}, \frac{1}{m}\right]$ and $\int_{\mathbb{R}} \rho_m(x)\,dx = 1$. Define $\varphi_m := \rho_m * T \in \mathcal{B}_E$. For any bounded set U of $\mathcal{D}_{L^1}$ we have $\sup_{\varphi\in U} \left|\langle \varphi_m - T, \varphi \rangle\right| \underset{m\to+\infty}{\longrightarrow} 0$. Indeed, since $T \in \mathcal{B}', \exists\, l \in \mathbb{Z}_+, \exists C > 0, \left|\langle \varphi_m - T, \varphi\rangle\right| \leq C \left|\check{\rho}_m * \varphi - \varphi\right|_{l,1}, \forall \varphi \in \mathcal{D}_{L^1}$. On the other hand, by Minkowski's inequality and the mean value theorem we obtain for a $t \in]0, 1[$,

$$\begin{aligned} \left\| \left(\check{\rho}_m * \varphi\right)^{(i)} - \varphi^{(i)} \right\|_{L^1} &\leq \int_{\left[\frac{-1}{m}, \frac{1}{m}\right]} \check{\rho}_m(y) \int_{\mathbb{R}} |y| \left|\varphi^{(i+1)}(x + (t-1)y)\right| dxdy, \\ &\leq \frac{1}{m} \left\| \varphi^{(i+1)} \right\|_{L^1}, \end{aligned}$$

hence

$$\left|\langle \varphi_m - T, \varphi\rangle\right| \leq C \left|\check{\rho}_m * \varphi - \varphi\right|_{l,1} \leq \frac{C}{m} |\varphi|_{l+1,1}, \forall \varphi \in \mathcal{D}_{L^1}.$$

Let U be a bounded set in $\mathcal{D}_{L^1}$, then $\exists C > 0$ such that we obtain

$$\sup_{\varphi\in U} \left|\langle \varphi_m - T, \varphi\rangle\right| \leq \frac{C}{m} \underset{m\to+\infty}{\longrightarrow} 0,$$

which gives the conclusion.

Example If $E = \mathcal{C}_{ap}$ then $\mathcal{B}'_E = \mathcal{B}'_{ap}$ is the space of L. Schwartz in [10], if $E = \mathcal{C}_{aa}$ then $\mathcal{B}'_E = \mathcal{B}'_{aa}$ is the space of [8].

4 Application

The study of linear ordinary differential equations $\sum_{i\leq p} a_i \frac{d^i f}{dx^i} = g$ in the framework of $\mathcal{B}'_E$ can be tackled in the general case of systems of linear ordinary differential equations for an unknown function $U = (U_i)_{1\leq i\leq p}$,

$$U' = AU + S, \tag{4.1}$$

where $A = \left(a_{ij}\right)_{1\leq i,j\leq p}$ is a given square matrix of complex numbers and a given vector of E-distributions $S = (S_i)_{1\leq i\leq p} \in \left(\mathcal{B}'_E\right)^p$.

Theorem 4.1 *Let all S_i, $1 \leq i \leq p$, be E-distributions and assume that the matrix A has no eigenvalues with real part zero. If $U = (U_i)_{1\leq i\leq p} \in \left(\mathcal{B}'\right)^p$ is a solution of the Eq. (4.1) then $U \in \left(\mathcal{B}'_E\right)^p$.*

Proof Consider the Eq. (4.1) and let $\varphi \in \mathcal{D}$, then

$$(U * \varphi)' = A\,(U * \varphi) + (S * \varphi),$$

where $U * \varphi = (U_i * \varphi)_{1\leq i\leq p}$ and $(S * \varphi) = (S_i * \varphi)_{1\leq i\leq p}$, which gives the following system of equations

$$v' = Av + g$$

with $g = S * \varphi \in (E)^p$ and $v = U * \varphi \in (E)^p$, consequently we apply Theorem 8 of [11] in the case of almost periodicity and Theorem 2 of [12] in the case of almost automorphy to obtain the existence of a unique solution $v \in (E)^p$. So $U_i * \varphi \in E, \forall\varphi \in \mathcal{D}, \forall i = 1, \ldots, p,$ i.e. $U \in (\mathcal{B}'_E)^p$.

Corollary 4.2 *If the polynomial $\sum_{i\leq p} a_i\lambda^i$ has no roots with real part zero, then any solution $T \in \mathcal{B}'$ of the inhomogeneous equation $\sum_{i\leq p} a_i \frac{d^iT}{dx^i} = S \in \mathcal{B}'_E$ is an almost automorphic distribution.*

5 *E*-Generalized Functions

In this section we introduce the algebra of E-generalized functions $\mathcal{G}_E$ and we give some of its properties. The proofs of the results of this section are obtained in a straight way.

Let

$$\mathcal{M}_E := \left\{(u_\varepsilon)_\varepsilon \in \left(\mathcal{B}_E\right)^I, \forall k \in \mathbb{Z}_+, \exists m \in \mathbb{Z}_+, |u_\varepsilon|_{k,\infty} = O\left(\varepsilon^{-m}\right), \varepsilon \longrightarrow 0\right\}$$

$$\mathcal{N}_E := \left\{(u_\varepsilon)_\varepsilon \in \left(\mathcal{B}_E\right)^I, \forall k \in \mathbb{Z}_+, \forall m \in \mathbb{Z}_+, |u_\varepsilon|_{k,\infty} = O\left(\varepsilon^{m}\right), \varepsilon \longrightarrow 0\right\}$$

Proposition 5.1

(*i*) *The space $\mathcal{M}_E$ is a subalgebra of $\left(\mathcal{B}_E\right)^I$.*
(*ii*) *The space $\mathcal{N}_E$ is an ideal of $\mathcal{M}_E$.*
(*iii*) $\mathcal{N}_E = \left\{(u_\varepsilon)_\varepsilon \in \mathcal{M}_E : \forall m \in \mathbb{Z}_+, |u_\varepsilon|_{0,\infty} = O\left(\varepsilon^m\right), \varepsilon \longrightarrow 0\right\}$.

Definition 5.2 The algebra of E-generalized functions is the quotient algebra

$$\mathcal{G}_E := \frac{\mathcal{M}_E}{\mathcal{N}_E}$$

If $1 \leq p \leq +\infty$, let $\mathcal{G}_{L^p}$ be the algebra of L^p-generalized functions defined on $\mathbb{R}$, see [1], and denote the algebra $\mathcal{G}_{L^\infty}$ by $\mathcal{G}_{\mathcal{B}}$.

The algebra of tempered generalized functions defined on $\mathbb{C}$ is denoted by $\mathcal{G}_\tau(\mathbb{C})$, see [9].

We give some properties of the algebra of E-generalized functions.

Proposition 5.3

1. *$\mathcal{G}_E$ is a subalgebra of $\mathcal{G}_{\mathcal{B}}$ stable under derivation.*
2. *$\mathcal{G}_E * \mathcal{G}_{L^1} \subset \mathcal{G}_E$.*
3. *If $u = [(u_\varepsilon)_\varepsilon] \in \mathcal{G}_E$ and $F = [(f_\varepsilon)_\varepsilon] \in \mathcal{G}_\tau(\mathbb{C})$, then $F \circ u := [(f_\varepsilon \circ u_\varepsilon)_\varepsilon]$ is a well-defined element of $\mathcal{G}_E$.*

The importance of $\mathcal{G}_E$ is given by the following result. Let $\rho \in \mathcal{S}$ satisfying $\int_{\mathbb{R}} \rho(x)dx = 1$ and set $\rho_\varepsilon(.) := \frac{1}{\varepsilon}\rho\left(\frac{.}{\varepsilon}\right)$, $\varepsilon > 0$.

Proposition 5.4 *The map*

$$\begin{array}{rcl} i_E : \mathcal{B}'_E & \longrightarrow & \mathcal{G}_E \\ T & \longmapsto & (T * \rho_\varepsilon)_\varepsilon + \mathcal{N}_E \end{array}$$

is a linear embedding which commutes with derivatives.

Example If $E = \mathcal{C}_{ap}$ then $\mathcal{G}_E = \mathcal{G}_{ap}$ of [6], and if $E = \mathcal{C}_{aa}$ then $\mathcal{G}_E = \mathcal{G}_{aa}$ of [7].

Remark 5.5 In the same vein as Sect. 4, we can study linear systems of ordinary differential equations in the framework of $\mathcal{G}_E$ as in [6] and [7].

Acknowledgments The author thanks the anonymous referee for suggestions to improve the quality of the text.

References

1. H.A. Biagioni, M. Oberguggenberger, Generalized solutions to the Korteweg-de Vries and the generalized long-wave equations. SIAM J. Math Anal. **23**(4), 923–940 (1992)
2. S. Bochner, Uniform convergence of monotone sequences of functions. Proc. Natl. Acad. Sci. USA **47**, 582–585 (1961)
3. S. Bochner, A new approach to almost periodicity. Proc. Natl. Acad. Sci. USA **48**, 2039–2043 (1962)
4. S. Bochner, Continuous mappings of almost automorphic and almost periodic functions. Proc. Natl. Acad. Sci. USA **52**, 907–910 (1964)

5. H. Bohr, *Almost Periodic Functions* (Chelsea, New York, 1947)
6. C. Bouzar, M.T. Khalladi, Almost periodic generalized functions. Novi Sad J. Math. **41**, 33–42 (2011)
7. C. Bouzar, M.T. Khalladi, F.Z. Tchouar, Almost automorphic generalized functions. Novi Sad J. Math. **45**(1), 207–214 (2015)
8. C. Bouzar, F.Z. Tchouar, Almost automorphic distributions. Mediterr. J. Math. **14**, 151 (2017)
9. M. Nedeljkov, S. Pilipovic, D. Scarpalézos, *Linear Theory of Colombeau Generalized Functions* (Pitman Research Notes in Mathematics, vol. 385 (Longman, Essex, 1998)
10. L. Schwartz, *Théorie des distributions*, 2nd edn. (Hermann, Paris, 1966)
11. S. Zaidman, Notes on abstract differential equations. Ann. Sci. Math. Québec **12**(1), 125–145 (1988)
12. S. Zaidman, Existence of asymptotically almost periodic and of almost automorphic solutions for some classes of abstract differential equations. Ann. Sci. Math. Québec **13**(1), 79–88 (1989)

Note on Vector Valued Hardy Spaces Related to Analytic Functions Having Distributional Boundary Values

Richard D. Carmichael, Stevan Pilipović, and Jasson Vindas

Abstract Analytic functions defined on a tube domain $T^C \subset \mathbb{C}^n$ and taking values in a Banach space X which are known to have X-valued distributional boundary values are shown to be in the Hardy space $H^p(T^C, X)$ if the boundary value is in the vector valued Lebesgue space $L^p(\mathbb{R}^n, X)$, where $1 \le p \le \infty$ and C is a regular open convex cone. Poisson integral transform representations of elements of $H^p(T^C, X)$ are also obtained for certain classes of Banach spaces, including reflexive Banach spaces.

Keywords Hardy spaces · Vector valued analytic functions on tube domains · Vector valued distributional boundary values · Poisson integral transform

Mathematics Subject Classification (2010) Primary 32A35; Secondary 46F20, 32A26

1 Introduction

In [1] Carmichael and Richters proved that if a holomorphic function on a tube domain having as base a regular open convex cone has an L^p function (with $1 \le p \le \infty$) as distributional boundary value, then the holomorphic function should

---

R. D. Carmichael
Department of Mathematics and Statistics, Wake Forest University, Winston-Salem, NC, USA
e-mail: carmicha@wfu.edu

S. Pilipović
Department of Mathematics and Informatics, University of Novi Sad, Novi Sad, Serbia
e-mail: Stevan.Pilipovic@dmi.uns.ac.rs

J. Vindas (✉)
Department of Mathematics: Analysis, Logic and Discrete Mathematics, Ghent University, Gent, Belgium
e-mail: jasson.vindas@UGent.be

P. Cerejeiras et al. (eds.), *Current Trends in Analysis, its Applications and Computation*, Research Perspectives,
https://doi.org/10.1007/978-3-030-87502-2_36

belong to the Hardy space H^p on the tube. The authors have recently obtained a vector valued generalization of this result in [2]. However, we were only able to prove the desired vector valued version in the range $2 \leq p \leq \infty$, and only for Hilbert space valued spaces; see [2, Theorem 4.4].

The aim of this note is to improve the quoted main result from [2] by showing that it holds for any $1 \leq p \leq \infty$ and any Banach space. This will be done in Sect. 3. Further, in Sect. 4, we prove that any element of an X-valued Hardy space is representable as a Poisson integral if X is a dual Banach space satisfying the Radon–Nikodým property. In particular, the latter holds for reflexive Banach spaces.

Distributional boundary value results associated with Hardy spaces have been of importance in particle physics; see [5] for example. The distributional boundary value result of [5] motivated the authors' work in [2] and the current paper.

2 Notation

Let X be a Banach space with norm $\| \cdot \|_X$. Integrals for X-valued functions are interpreted in the Bochner sense [3, 7] and the X-valued Lebesgue spaces $L^p(\mathbb{R}^n, X)$, $p \in (0, \infty]$, are defined in the usual way. The space of X-valued distributions [10] is the space of continuous linear mappings $\mathcal{D}(\mathbb{R}^n) \to X$, denoted as $\mathcal{D}'(\mathbb{R}^n, X)$. In analogy to the scalar valued case, we denote the evaluation of a vector valued distribution $\mathbf{f} \in \mathcal{D}'(\mathbb{R}^n, X)$ at a test function $\varphi \in \mathcal{D}(\mathbb{R}^n)$ as $\langle \mathbf{f}, \varphi \rangle \in X$.

An open convex cone (with vertex at the origin) $C \subset \mathbb{R}^n$ is called *regular* if its closure does not contain any entire straight line. Equivalently, regularity means that the conjugate cone $C^* = \{y \in \mathbb{R}^n : y \cdot x \geq 0, \forall x \in C\}$ has non-empty interior. The tube domain with base C is $T^C = \mathbb{R}^n + iC$. The Cauchy-Szegö kernel of T^C is defined as $K(z) = \int_{C^*} e^{2\pi i z \cdot u} du$ for $z \in T^C$, while its corresponding Poisson kernel is

$$Q(z; u) = \frac{|K(z-u)|^2}{K(2iy)}, \qquad u \in \mathbb{R}^n,\ z = x + iy \in T^C.$$

It should be noted that $Q(z; \cdot) \in L^p(\mathbb{R}^n)$ for any $p \in [1, \infty]$; see [9, 3.7, p. 105].

If C is an open cone and $0 < p \leq \infty$, the X-valued Hardy space consists of those vector valued holomorphic functions $\mathbf{F} : T^C \to X$ such that

$$\sup_{y \in C} \int_{\mathbb{R}^n} \|\mathbf{F}(x + iy))\|_X^p dx < \infty,$$

where the usual modification is made for the case $p = \infty$.

3 Distributional Boundary Values in $L^p(\mathbb{R}^n, X)$

In this section we improve [2, Theorem 4.4, p. 1650]. It is worth pointing out that our method here is much simpler and shorter than the one employed in [2].

Theorem 3.1 *Let X be a Banach spaces, let C be a regular open convex cone, and let* $p \in [1, \infty]$*. Suppose that the vector valued holomorphic function* $\mathbf{F} : T^C \to X$ *has distributional boundary value* $\mathbf{f} \in L^p(\mathbb{R}^n, X)$*, that is,*

$$\lim_{\substack{y \to 0 \\ y \in C}} \int_{\mathbb{R}^n} \mathbf{F}(x + iy)\varphi(x)\mathrm{d}x = \int_{\mathbb{R}^n} \mathbf{f}(x)\varphi(x)\mathrm{d}x \qquad \text{in } X \tag{3.1}$$

holds for each test function $\varphi \in \mathcal{D}(\mathbb{R}^n)$*. Then,* $\mathbf{F} \in H^p(T^C, X)$ *and*

$$\mathbf{F}(z) = \int_{\mathbb{R}^n} \mathbf{f}(u)Q(z; u)\mathrm{d}u, \qquad z \in T^C. \tag{3.2}$$

Proof Define

$$\mathbf{G}(z) = \int_{\mathbb{R}^n} \mathbf{f}(u)Q(z; u)\mathrm{d}u, \qquad z \in T^C.$$

We have shown in [2, Lemma 3.4, p. 1639] that $\mathbf{G} \in H^p(T^C, X)$. Moreover, the quoted lemma also gives that $\mathbf{G}$ has distributional boundary value $\mathbf{f}$. It thus suffices to prove that $\mathbf{F}(z) = \mathbf{G}(z)$, $z \in T^C$, which we verify via the Hahn-Banach theorem and the (scalar valued) edge-of-the-wedge-theorem. Let $\mathbf{w}^* \in X'$. Consider the scalar valued holomorphic function $H_{\mathbf{w}^*}(z) = \langle \mathbf{w}^*, \mathbf{F}(z) - \mathbf{G}(z)\rangle$. It satisfies

$$\lim_{\substack{y \to 0 \\ y \in C}} \int_{\mathbb{R}^n} H_{\mathbf{w}^*}(x+iy)\varphi(x)\mathrm{d}x = \left\langle \mathbf{w}^*, \lim_{\substack{y \to 0 \\ y \in C}} \int_{\mathbb{R}^n} (\mathbf{F}(x + iy) - \mathbf{G}(x + iy))\varphi(x)\mathrm{d}x \right\rangle = 0,$$

for each $\varphi \in \mathcal{D}(\mathbb{R}^n)$. Using [6, Corollary of Theorem B, p. 20], we obtain $\langle \mathbf{w}^*, \mathbf{F}(z) - \mathbf{G}(z)\rangle = 0$ for all $z \in T^C$. Since $\mathbf{w}^* \in X'$ was arbitrary, the Hahn-Banach theorem yields the equality $\mathbf{F}(z) = \mathbf{G}(z)$, $z \in T^C$. This establishes the theorem. □

4 Poisson Integral Representation

We now turn our attention to Banach spaces X where any X-valued holomorphic function $\mathbf{F} \in H^p(T^C, X)$, $1 \le p \le \infty$, admits the Poisson integral representation (3.2) for some $\mathbf{f} \in L^p(\mathbb{R}^n, X)$.

We need to introduce some terminology. A Banach space is said to have the *Radon–Nikodým property* if the Radon–Nikodým theorem holds for vector measures on it; see [3, p. 61] or [7, Chapter 5, p. 102] for precise definitions and background material on Banach spaces with this property. We call X a *dual Banach space* if it is the strong dual of some Banach space.

Theorem 4.1 *Let X be a dual Banach space having the Radon–Nikodým property, let C be a regular open convex cone, and let $p \in [1, \infty]$. If $\mathbf{F} \in H^p(T^C, X)$, then there is $\mathbf{f} \in L^p(\mathbb{R}^n, X)$ such that $\mathbf{F}$ has the Poisson integral representation* (3.2) *and $\mathbf{f}$ is the distributional boundary value of $\mathbf{F}$.*

Proof We first show that $\mathbf{F}$ has distributional boundary value. Let $\varphi \in \mathcal{D}(\mathbb{R}^n)$ and write $\Phi(x, y) = \varphi(x) + i \sum_{j=1}^{n} y_j \partial_j \varphi(x)$ for $x \in \mathbb{R}^n$ and $y = (y_1, \ldots, y_n) \in T^C$. Pick a unit vector $\omega \in C$. Applying the Stokes theorem exactly in the same way as in [4, p. 67], we have, if $y \in C$,

$$\int_{\mathbb{R}^n} \mathbf{F}(x + iy)\varphi(x)\mathrm{d}x = \int_{\mathbb{R}^n} \mathbf{F}(x + iy + i\omega)\Phi(x, \omega)\mathrm{d}x$$
$$+ i \sum_{j=1}^{n} \omega_j \int_0^1 \int_{\mathbb{R}^n} \mathbf{F}(x + it\omega + iy)\partial_j\varphi(x)\mathrm{d}x\mathrm{d}t.$$

Since $\mathbf{F} \in H^p(T^C, X)$, we may take the limit as $y \to 0$ on the right-hand side of the above expression and conclude that $\mathbf{F}$ has distributional boundary value $\mathbf{f} \in \mathcal{D}'(\mathbb{R}^n, X)$, given in fact as

$$\langle \mathbf{f}, \varphi \rangle = \int_{\mathbb{R}^n} \mathbf{F}(x + i\omega)\Phi(x, \omega)\mathrm{d}x + i \sum_{j=1}^{n} \omega_j \int_0^1 \int_{\mathbb{R}^n} \mathbf{F}(x + it\omega)\partial_j\varphi(x)\mathrm{d}x\mathrm{d}t.$$

In view of Theorem 3.1, the representation (3.2) would follow at once if we are able to show that $\mathbf{f} \in L^p(\mathbb{R}^n, X)$. We now focus in showing the latter. The rest of the proof exploits the fact that we can consider a weak* topology on $L^p(\mathbb{R}^n, X)$ due to our assumptions on X. Let Y be a Banach space such that $X = Y'$. For each $y \in C$, write $\mathbf{F}_y(x) = \mathbf{F}(x + iy)$. We split our considerations in two cases.

Case I: $1 < p \leq \infty$ In this case it is well-known (see[1] [3, Theorem 1, Sect. IV.1, p. 98]) that $L^p(\mathbb{R}^n, X)$ is the strong dual of $L^q(\mathbb{R}^n, Y)$ where $1/p + 1/q = 1$. Besides its strong topology, we also provide $L^p(\mathbb{R}^n, X)$ with the weak* topology with respect to this duality. Since the membership $\mathbf{F} \in H^p(T^C, X)$ precisely means that the set $\{\mathbf{F}_y : y \in C\}$ is strongly bounded in $L^p(\mathbb{R}^n, X)$, the Banach–Alaoglu theorem [10] yields the existence of a sequence of points $y_k \in C$ and an X-valued

[1] This result is stated in [3, p. 98] for vector valued L^p-spaces with respect to finite (scalar valued positive) measures, but the proof given there shows that it holds for σ-finite measures, in particular for the Lebesgue measure as we used it here.

function $\mathbf{g} \in L^p(\mathbb{R}^n, X)$ such that $\mathbf{F}_{y_k} \to \mathbf{g}$ as $k \to \infty$, weakly* in $L^p(\mathbb{R}^n, X)$. But, this weak* convergence is stronger than convergence in $\mathcal{D}'(\mathbb{R}^n, X)$, whence $\mathbf{f} = \mathbf{g} \in L^p(\mathbb{R}^n, X)$, as required.

Case II: $p = 1$ Denote by $\mathcal{M}_1(\mathbb{R}^n, X)$ the Banach space of X-valued vector measures with finite variation [7, Chapter 5] (cf. [3]) on the σ-algebra of Borel sets of $\mathbb{R}^n$. We regard $L^1(\mathbb{R}^n, X)$ as a closed subspace of $\mathcal{M}_1(\mathbb{R}^n, X)$. Let $\mathcal{M}(\mathbb{R}^n)$ be the space of (signed) Borel measures on $\mathbb{R}^n$. Denote also by $C_0(\mathbb{R}^n)$ and $C_0(\mathbb{R}^n, X)$ the spaces of continuous and X-valued continuous functions, respectively, vanishing at ∞. Due to the Radon–Nikodým property of X and the fact that $C_0(\mathbb{R}^n)$ has the approximation property (which follows from the fact that it has a Schauder basis [8, Corollary 4.1.4, p. 112]), we have the following natural isomorphisms,

$$\mathcal{M}_1(\mathbb{R}^n, X) \cong \mathcal{M}(\mathbb{R}^n)\hat{\otimes}_\pi X \cong (C_0(\mathbb{R}^n)\hat{\otimes}_\varepsilon Y)',$$

where we have used [7, Theorem 5.22, p. 108] in the first isomorphism and [7, Theorem 5.33, p. 114] in the second one, and we recall that the symbols $\hat{\otimes}_\pi$ and $\hat{\otimes}_\varepsilon$ stand for the projective and injective completed tensor products. Also, reasoning as in [7, Example 3.3, p. 47], one readily verifies that $C_0(\mathbb{R}^n)\hat{\otimes}_\varepsilon Y = C_0(\mathbb{R}^n, Y)$. Summarizing, we may view $\mathcal{M}_1(\mathbb{R}^n, X)$ as the dual of $C_0(\mathbb{R}^n, Y)$. Similarly as in *Case I*, we obtain with the aid of the Banach–Alaoglu theorem that there is an X-valued vector measure $\boldsymbol{\mu}$ such that $\mathbf{f} = \mathrm{d}\boldsymbol{\mu}$. Since X has the Radon–Nikodým property, we must prove that $\boldsymbol{\mu}$ is absolutely continuous with respect to the Lebesgue measure in order to show that $\mathbf{f} \in L^1(\mathbb{R}^n, X)$. By the Hahn-Banach theorem it suffices to show that if $\mathbf{w}^* \in X'$, then the scalar valued measure $\mu_{\mathbf{w}^*} = \langle \mathbf{w}^*, \boldsymbol{\mu} \rangle$ is absolutely continuous with respect to the Lebesgue measure. But note that $\mathrm{d}\mu_{\mathbf{w}^*}$ is the distributional boundary value of $\langle \mathbf{w}^*, \mathbf{F} \rangle$. The function $\langle \mathbf{w}^*, \mathbf{F} \rangle$ clearly belongs to the scalar valued Hardy space $H^1(T^C)$ and the well-known classical result [9, Theorem 5.6, p. 119] says that it has boundary value in $L^1(\mathbb{R}^n)$, so that indeed $\mu_{\mathbf{w}^*} = \langle \mathbf{w}^*, \boldsymbol{\mu} \rangle$ is absolutely continuous with respect to the Lebesgue measure. □

Let us point out that any reflexive Banach space has the Radon–Nikodým property [3, Corollary 4, p. 82], whence we immediately obtain the ensuing corollary.

Corollary 4.2 *Let X be a reflexive Banach space, let $C \subset \mathbb{R}^n$ be a regular open convex cone, and let $p \in [1, \infty]$. Then any $\mathbf{F} \in H^p(T^C, X)$ admits a Poisson integral representation* (3.2) *for some $\mathbf{f} \in L^p(\mathbb{R}^n, X)$. Moreover, the function $\mathbf{f}$ is the distributional boundary value of $\mathbf{F}$.*

Remark 4.3 When $1 \leq p < \infty$, not only is the function $\mathbf{f} \in L^p(\mathbb{R}^n, X)$ from Theorem 4.1 and Corollary 4.2 the distributional boundary value of $\mathbf{F}$, but also $\lim_{y\to 0} \mathbf{F}(\cdot + iy) = \mathbf{f}$ in $L^p(\mathbb{R}^n, X)$ (cf. [2, Lemma 3.4, p. 1639]).

Acknowledgments J. Vindas was supported by Ghent University through the BOF-grants 01J11615 and 01J04017.

References

1. R. Carmichael, S. Richters, Holomorphic functions in tubes which have distributional boundary values and which are H^p functions. SIAM J. Math. Anal. **14**, 596–621 (1983)
2. R.D. Carmichael, S. Pilipović, J. Vindas, Vector valued Hardy spaces related to analytic functions having distributional boundary values. Complex Var. Elliptic Equ. **64**, 1634–1654 (2019)
3. J. Diestel, J. Uhl, *Vector Measures* (American Mathematical Society, Providence, 1977)
4. L. Hörmander, The analysis of linear partial differential operators. I, in *Distribution Theory and Fourier Analysis* (Springer, Berlin, 1990)
5. A. Raina, On the role of Hardy spaces in form factor bounds. Lett. Math. Phys. **2**, 513–519 (1978)
6. W. Rudin, Lectures on the edge-of-the-wedge theorem, in *Conference Board of the Mathematical Sciences Regional Conference Series in Mathematics*, vol. 6 (American Mathematical Society, Providence, 1971)
7. R.A. Ryan, *Introduction to Tensor Products of Banach Spaces* (Springer, London, 2002)
8. Z. Semadeni, *Schauder Bases in Banach Spaces of Continuous Functions*. Lecture Notes in Mathematics, vol. 918 (Springer, Berlin, 1982)
9. E. Stein, G. Weiss, *Introduction to Fourier Analysis on Euclidean Spaces* (Princeton University Press, Princeton, 1971)
10. F. Trèves, *Topological Vector Spaces, Distributions and Kernels* (Academic Press, New York, 1967)

On the Projective Description of Spaces of Ultradifferentiable Functions of Roumieu Type

Andreas Debrouwere, Bojan Prangoski, and Jasson Vindas

Abstract We provide a projective description of the space $\mathcal{E}^{\{\mathfrak{M}\}}(\Omega)$ of ultradifferentiable functions of Roumieu type, where Ω is an arbitrary open set in $\mathbb{R}^d$ and $\mathfrak{M}$ is a weight matrix satisfying the analogue of Komatsu's condition $(M.2)'$. In particular, we obtain in a unified way projective descriptions of ultradifferentiable classes defined via a single weight sequence (Denjoy-Carleman approach) and via a weight function (Braun-Meise-Taylor approach) under considerably weaker assumptions than in earlier versions of these results.

Mathematics Subject Classification (2010) Primary 46E10; Secondary 46F05

Keywords Ultradifferentiable classes of Roumieu type · Projective description

1 Introduction

In his seminal work [8], Komatsu gave an explicit system of seminorms generating the topology of the space $\mathcal{E}^{\{M\}}(\Omega)$ of ultradifferentiable functions of Roumieu type (for short, a projective description of $\mathcal{E}^{\{M\}}(\Omega)$), where Ω is an arbitrary open subset of $\mathbb{R}^d$ and M is a non-quasianalytic weight sequence satisfying the conditions $(M.1)$ and $(M.2)'$ [8, Proposition 3.5]. In [4, Proposition 4.8], the first and the third authors

A. Debrouwere
Department of Mathematics and Data Science, Vrije Universiteit Brussel, Brussels, Belgium
e-mail: andreas.debrouwere@vub.be

J. Vindas (✉)
Department of Mathematics: Analysis Logic and Discrete Mathematics, Ghent University, Gent, Belgium
e-mail: jasson.vindas@UGent.be

B. Prangoski
Faculty of Mechanical Engineering, Ss. Cyril and Methodius University in Skopje, Skopje, Macedonia

P. Cerejeiras et al. (eds.), *Current Trends in Analysis, its Applications and Computation*, Research Perspectives,
https://doi.org/10.1007/978-3-030-87502-2_37

relaxed the non-quasianalyticity assumption on M to $\sup_{p\in\mathbb{N}} pM_p^{-1/p} < \infty$. Similarly, a projective description of the space $\mathcal{E}^{\{\omega\}}(\Omega)$ was implicitly given in [6, Section 3], where ω is a weight function in the sense of Braun et al. [2] that satisfies $\omega(t) = O(t)$. Projective descriptions are indispensable in the study of spaces of vector-valued ultradifferentiable functions of Roumieu type [4, 8], e.g., for achieving completed tensor product representations of such spaces.

The goal of this article is to provide a projective description of spaces of ultradifferentiable functions of Roumieu type defined via a weight matrix [11]. This approach leads to a unified treatment of ultradifferentiable classes defined via a single weight sequence and via a weight function, but also comprises other spaces, e.g., the union of all Gevrey spaces. For the two standard classes we obtain projective descriptions under much weaker assumptions than in the above mentioned works; see Corollaries 3.3 and 3.4.

2 Spaces of Ultradifferentiable Functions of Roumieu Type

Let $M = (M_p)_{p\in\mathbb{N}}$ be a sequence of positive numbers (a *weight sequence*). We consider the following three conditions on M:

$(M.0)$ $M_p \geq ch^p$, $p \in \mathbb{N}$, for some $c, h > 0$;
$(M.1)$ $M_p^2 \leq M_{p-1}M_{p+1}$, $p \in \mathbb{Z}_+$;
$(M.2)'$ $M_{p+1} \leq CH^pM_p$, $p \in \mathbb{N}$, for some $C, H > 0$.

The conditions $(M.1)$ and $(M.2)'$ are denoted by (M_{lc}) and (M_{dc}), respectively, in [11]. We use here the standard notation from [7]. The relation $M \subset N$ between two weight sequences means that there are $C, h > 0$ such that $M_p \leq Ch^pN_p$ for all $p \in \mathbb{N}$. The stronger relation $M \prec N$ means that the latter inequality remains valid for every $h > 0$ and a suitable $C = C_h > 0$. We set $M_\alpha = M_{|\alpha|}$, $\alpha \in \mathbb{N}^d$.

For $h > 0$ and a regular compact set $K \Subset \mathbb{R}^d$ (i.e., $K = \overline{\operatorname{int} K}$) we write $\mathcal{E}^{M,h}(K)$ for the Banach space consisting of all $\varphi \in C^\infty(K)$[1] such that

$$\|\varphi\|_{\mathcal{E}^{M,h}(K)} := \sup_{\alpha\in\mathbb{N}^d} \frac{h^{|\alpha|}\|\partial^\alpha\varphi\|_{L^\infty(K)}}{M_\alpha} < \infty.$$

We set

$$\mathcal{E}^{\{M\}}(K) := \varinjlim_{h\to 0^+} \mathcal{E}^{M,h}(K).$$

[1] We define $C^\infty(K)$ as the space consisting of all $\varphi \in C^\infty(\operatorname{int} K)$ such that $\partial^\alpha\varphi$ extends to a continuous function on K for each $\alpha \in \mathbb{N}^d$.

Given an open set $\Omega \subseteq \mathbb{R}^d$, we define the space of ultradifferentiable functions of Roumieu type (of class $\{M\}$) on Ω as

$$\mathcal{E}^{\{M\}}(\Omega) := \varprojlim_{K \Subset \Omega} \mathcal{E}^{\{M\}}(K).$$

Next, we introduce weight matrices and the associated spaces of ultradifferentiable functions [11]. A *weight matrix* is a sequence $\mathfrak{M} = (M^n)_{n\in\mathbb{N}}$ consisting of weight sequences M^n such that $M^n \leq M^{n+1}$ for all $n \in \mathbb{N}$. We consider the following condition on $\mathfrak{M}$:

$\{\mathfrak{M}.2\}'$ $\quad \forall n \in \mathbb{N}\, \exists m \in \mathbb{N}\, \exists C, H > 0\, \forall p \in \mathbb{N} : M^n_{p+1} \leq CH^p M^m_p.$

The condition $\{\mathfrak{M}.2\}'$ is denoted by $(\mathfrak{M}_{\{\mathrm{dc}\}})$ in [11]. Given a regular compact set $K \Subset \mathbb{R}^d$, we denote

$$\mathcal{E}^{\{\mathfrak{M}\}}(K) := \varinjlim_{n\in\mathbb{N}} \mathcal{E}^{\{M^n\}}(K).$$

Given an open set $\Omega \subseteq \mathbb{R}^d$, we define the space of ultradifferentiable functions of Roumieu type (of class $\{\mathfrak{M}\}$) on Ω as

$$\mathcal{E}^{\{\mathfrak{M}\}}(\Omega) := \varprojlim_{K \Subset \Omega} \mathcal{E}^{\{\mathfrak{M}\}}(K).$$

Finally, we introduce spaces of ultradifferentiable functions defined via a weight function in the sense of Braun et al. [2] and explain how they fit into the weight matrix approach; see [11, Section 5] for more details. By a *weight function* we mean a continuous increasing function $\omega : [0, \infty) \to [0, \infty)$ with $\omega_{|[0,1]} \equiv 0$ and satisfying the following three properties:

(α) $\quad \omega(2t) = O(\omega(t))$;
(γ_0) $\quad \log t = o(\omega(t))$;
(δ) $\quad \phi = \phi_\omega : [0, \infty) \to [0, \infty)$, given by $\phi(x) = \omega(e^x)$, is convex.

The *Young conjugate* ϕ^* of ϕ is defined as

$$\phi^* : [0, \infty) \to [0, \infty), \qquad \phi^*(y) = \sup_{x\geq 0}(xy - \phi(x)).$$

Note that ϕ^* is increasing and convex, $\phi^*(0) = 0$, $(\phi^*)^* = \phi$, $\phi^*(y)/y$ is increasing on $[0, \infty)$ and $\phi^*(y)/y \to \infty$ as $y \to \infty$.

For $\rho > 0$ and a regular compact set $K \Subset \mathbb{R}^d$ we write $\mathcal{E}^{\omega,\rho}(K)$ for the Banach space consisting of all $\varphi \in C^\infty(K)$ such that

$$\|\varphi\|_{\mathcal{E}^{\omega,\rho}(K)} := \sup_{\alpha\in\mathbb{N}^d} \|\partial^\alpha \varphi\|_{L^\infty(K)} \exp\left(-\frac{1}{\rho}\phi^*(\rho|\alpha|)\right) < \infty.$$

Given an open set $\Omega \subseteq \mathbb{R}^d$, we define the space of ultradifferentiable functions of Roumieu type (of class $\{\omega\}$) on Ω as

$$\mathcal{E}^{\{\omega\}}(\Omega) := \varprojlim_{K \Subset \Omega} \varinjlim_{\rho \to \infty} \mathcal{E}^{\omega,\rho}(K).$$

We associate to ω the weight matrix $\mathfrak{M}_\omega = (M^n_\omega)_{n \in \mathbb{N}}$, where the weight sequence M^n_ω is defined as

$$M^n_{\omega,p} := \exp\left(\frac{1}{n}\phi^*(np)\right), \qquad p \in \mathbb{N}.$$

Note that each M^n_ω satisfies $(M.0)$ and $(M.1)$. Furthermore, $\mathfrak{M}_\omega$ satisfies $\{\mathfrak{M}.2\}'$ and $\mathcal{E}^{\{\omega\}}(\Omega) = \mathcal{E}^{\{\mathfrak{M}_\omega\}}(\Omega)$ as locally convex spaces [11, Corollary 5.15].

3 Projective Description of $\mathcal{E}^{\{\mathfrak{M}\}}(\Omega)$

Given a weight matrix $\mathfrak{M}$, we define $V(\mathfrak{M})$ as the set of all weight sequences N such that $M^n \prec N$ for all $n \in \mathbb{N}$. The next theorem is the main result of this article.

Theorem 3.1 *Let $\Omega \subseteq \mathbb{R}^d$ be open and let $\mathfrak{M}$ be a weight matrix satisfying $\{\mathfrak{M}.2\}'$. A function $\varphi \in C^\infty(\Omega)$ belongs to $\mathcal{E}^{\{\mathfrak{M}\}}(\Omega)$ if and only if*

$$\|\varphi\|_{\mathcal{E}^{N,1}(K)} = \sup_{\alpha \in \mathbb{N}^d} \frac{\|\partial^\alpha \varphi\|_{L^\infty(K)}}{N_\alpha} < \infty$$

for all $K \Subset \Omega$ and $N \in V(\mathfrak{M})$. Moreover, the locally convex topology of $\mathcal{E}^{\{\mathfrak{M}\}}(\Omega)$ is generated by the system of seminorms $\{\| \cdot \|_{\mathcal{E}^{N,1}(K)} \mid K \Subset \Omega, N \in V(\mathfrak{M})\}$.

Remark 3.2 Let $\mathfrak{M}$ be a weight matrix satisfying $\{\mathfrak{M}.2\}'$ and suppose that each weight sequence M^n satisfies $(M.0)$ and $(M.1)$. Obviously, every element of $V(\mathfrak{M})$ automatically satisfies $(M.0)$. Define $V^*(\mathfrak{M})$ as the set of all $N \in V(\mathfrak{M})$ for which $(M.1)$ holds. Then, Theorem 3.1 still holds true if we replace $V(\mathfrak{M})$ by $V^*(\mathfrak{M})$. This follows from the fact that for each $N \in V(\mathfrak{M})$ its log-convex minorant N^c belongs to $V^*(\mathfrak{M})$ and satisfies $N^c \leq N$.

Before we prove Theorem 3.1, let us show how it entails the projective description of the spaces $\mathcal{E}^{\{M\}}(\Omega)$ and $\mathcal{E}^{\{\omega\}}(\Omega)$. Following Komatsu [8], we denote by $\mathfrak{R}$ the family of all non-decreasing sequences $r = (r_j)_{j \in \mathbb{N}}$ of positive numbers such that $r_j \to \infty$ as $j \to \infty$. The next result generalizes [8, Proposition 3.5] and [4, Proposition 4.8].

Corollary 3.3 *Let $\Omega \subseteq \mathbb{R}^d$ be open and let M be a weight sequence satisfying $(M.2)'$. A function $\varphi \in C^\infty(\Omega)$ belongs to $\mathcal{E}^{\{M\}}(\Omega)$ if and only if*

$$\|\varphi\|_{\mathcal{E}^{M,r}(K)} := \sup_{\alpha\in\mathbb{N}^d} \frac{\|\partial^\alpha\varphi\|_{L^\infty(K)}}{M_\alpha \prod_{j=0}^{|\alpha|} r_j} < \infty$$

for all $K \Subset \Omega$ and $r \in \mathfrak{R}$. Moreover, the locally convex topology of $\mathcal{E}^{\{M\}}(\Omega)$ is generated by the system of seminorms $\{\|\cdot\|_{\mathcal{E}^{M,r}(K)} \mid K \Subset \Omega, r \in \mathfrak{R}\}$.

Proof This follows from Theorem 3.1 (applied to the constant weight matrix $\mathfrak{M} = (M)_{n\in\mathbb{N}}$) and [8, Lemma 3.4]. □

Given a weight function ω, we define $V(\omega)$ as the set consisting of all weight functions σ such that $\sigma = o(\omega)$.

Corollary 3.4 *Let $\Omega \subseteq \mathbb{R}^d$ be open and let ω be a weight function. A function $\varphi \in C^\infty(\Omega)$ belongs to $\mathcal{E}^{\{\omega\}}(\Omega)$ if and only if*

$$\|\varphi\|_{\mathcal{E}^{\sigma,1}(K)} = \sup_{\alpha\in\mathbb{N}^d} \|\partial^\alpha\varphi\|_{L^\infty(K)} e^{-\phi_\sigma^*(|\alpha|)} < \infty$$

for all $K \Subset \Omega$ and $\sigma \in V(\omega)$. Moreover, the locally convex topology of $\mathcal{E}^{\{\omega\}}(\Omega)$ is generated by the system of seminorms $\{\|\cdot\|_{\mathcal{E}^{\sigma,1}(K)} \mid K \Subset \Omega, \sigma \in V(\omega)\}$.

Proof By Theorem 3.1 and Remark 3.2 (applied to the weight matrix $\mathfrak{M}_\omega$) it suffices to show that

(i) $\forall \sigma \in V(\omega) : M_\sigma^1 \in V(\mathfrak{M}_\omega)$.
(ii) $\forall N \in V^*(\mathfrak{M}_\omega)\, \exists \sigma \in V(\omega) : M_\sigma^1 \subset N$.

The first statement is obvious. We now show the second one. Let $N \in V^*(\mathfrak{M}_\omega)$ be arbitrary. Consider the associated function of N

$$\omega_N(t) = \sup_{p\in\mathbb{N}} \log \frac{t^p N_0}{N_p}, \qquad t \geq 0.$$

Then, $\omega_N = o(\omega)$. By Braun et al. [2, Lemma 1.7 and Remark 1.8], there is a weight function $\sigma \in V(\omega)$ such that $\omega_N = o(\sigma)$. Since $\omega_{M_\sigma^1} \asymp \sigma$ [11, Lemma 5.7], we obtain that

$$\omega_N(t) \leq \omega_{M_\sigma^1}(t) + C, \qquad t \geq 0.$$

Since both N and M_σ^1 satisfy $(M.0)$ and $(M.1)$, the latter inequality yields that $M_\sigma^1 \subset N$ [7, Lemma 3.8]. □

We now turn to the proof of Theorem 3.1. We use the same idea as in Komatsu's proof of [8, Proposition 3.5]. Fix a weight matrix $\mathfrak{M}$ satisfying $\{\mathfrak{M}.2\}'$. Since any open set $\Omega \subseteq \mathbb{R}^d$ admits an exhaustion by compact sets that are finite unions of regular connected compact sets K with smooth boundary[2] (in particular, $\operatorname{int} K$ is a Lipschitz domain), Theorem 3.1 follows from the next result.

Theorem 3.5 *Let $K \Subset \mathbb{R}^d$ be a regular compact set such that* $\operatorname{int} K$ *is a Lipschitz domain and let $\mathfrak{M}$ be a weight matrix satisfying $\{\mathfrak{M}.2\}'$. A function $\varphi \in C^\infty(K)$ belongs to $\mathcal{E}^{\{\mathfrak{M}\}}(K)$ if and only if $\|\varphi\|_{\mathcal{E}^{N,1}(K)} < \infty$ for all $N \in V(\mathfrak{M})$. Moreover, the locally convex topology of $\mathcal{E}^{\{\mathfrak{M}\}}(K)$ is generated by the system of seminorms $\{\| \cdot \|_{\mathcal{E}^{N,1}(K)} \mid N \in V(\mathfrak{M})\}$.*

The rest of this article is devoted to the proof of Theorem 3.5. We start with the following technical lemma (cf. [8, Lemma 3.4]).

Lemma 3.6 *Let $(a_p)_{p\in\mathbb{N}}$ be a sequence of positive numbers.*

(i) $\sup_{p\in\mathbb{N}} \frac{h^p a_p}{M^n_p} < \infty$ *for some $h > 0$ and $n \in \mathbb{N}$ if and only if* $\sup_{p\in\mathbb{N}} \frac{a_p}{N_p} < \infty$ *for all $N \in V(\mathfrak{M})$.*

(ii) $\sup_{p\in\mathbb{N}} a_p N_p < \infty$ *for some $N \in V(\mathfrak{M})$ if and only if* $\sup_{p\in\mathbb{N}} \frac{a_p M^n_p}{h^p} < \infty$ *for all $h > 0$ and $n \in \mathbb{N}$.*

Proof The direct implications are clear. We now show the converse ones.

(i) Suppose that $\sup_{p\in\mathbb{N}} a_p/(n^p M^n_p) = \infty$ for all $n \in \mathbb{N}$. Choose a strictly increasing sequence $(p_n)_{n\in\mathbb{N}}$ of natural numbers with $p_0 = 0$ and

$$\frac{a_{p_n}}{n^{p_n} M^n_{p_n}} \geq n, \qquad n \in \mathbb{Z}_+.$$

Define $N_p = n^p M^n_p$ if $p_n \leq p < p_{n+1}$. Then, $N = (N_p)_{p\in\mathbb{N}}$ belongs to $V(\mathfrak{M})$ but $\sup_{p\in\mathbb{N}} a_p/N_p = \infty$, a contradiction.

(ii) For each $n \in \mathbb{N}$ there is $C_n > 0$ such that

$$n^p M^n_p \leq \frac{C_n}{a_p}, \qquad p \in \mathbb{N}.$$

Define

$$N_p = \sup_{n\in\mathbb{N}} \frac{n^p M^n_p}{C_n}, \qquad p \in \mathbb{N}.$$

Then, $N = (N_p)_{p\in\mathbb{N}}$ belongs to $V(\mathfrak{M})$ and $\sup_{p\in\mathbb{N}} a_p N_p < \infty$. □

[2] This follows from the existence of a positive smooth exhausting function on Ω [9, Proposition 2.28] and Sard's theorem [9, Theorem 6.10].

A set $A \subseteq \mathbb{R}^d$ is said to be *quasiconvex* if there exists $C > 0$ such that any two points $x, y \in A$ can be joined by a curve in A with length at most $C|x - y|$. This notion was introduced by Whitney [13] under the name property (P). The closure of a quasiconvex open set is again quasiconvex [13, Lemma 2]. Moreover, every bounded Lipschitz domain is quasiconvex [3, Section 2.5].

Let $K \Subset \mathbb{R}^d$ be a regular compact set such that $\operatorname{int} K$ is quasiconvex. For $n \in \mathbb{N}$ we denote by $C^n(K)$ the vector space of all $\varphi \in C^n(\operatorname{int} K)$ such that $\partial^\alpha \varphi$ extends to a continuous function on K for each $|\alpha| \leq n$; it is a Banach space when endowed with the norm $\sup_{|\alpha| \leq n} \|\partial^\alpha \varphi\|_{L^\infty(K)}$. By [13, Theorem, p. 485], the space $C^n(K)$ is canonically isomorphic to the Banach space of Whitney jets of order n on K [12]. Let $R > 0$ be such that $K \Subset B(0, R)$. Whitney's extension theorem [14, Theorem I] yields the existence of a continuous linear extension operator $C^n(K) \to C^n(\overline{B}(0, R))$, that is, a continuous linear right inverse of the restriction mapping $C^n(\overline{B}(0, R)) \to C^n(K)$. The latter implies that the inclusion mapping $C^{n+1}(K) \to C^n(K)$ is compact. A standard argument (cf. [7, Proposition 2.2]) therefore gives the following result.

Lemma 3.7 *Let $K \Subset \mathbb{R}^d$ be a regular compact set such that* $\operatorname{int} K$ *is quasiconvex. Then, $\mathcal{E}^{\{\mathfrak{M}\}}(K)$ is a (DFS)-space.*

Next, we show a structural result for the dual of $\mathcal{E}^{\{\mathfrak{M}\}}(K)$; this is the crux of the proof of Theorem 3.5. We need some preparation. Given a Banach space E, a weight sequence $M = (M_p)_{p \in \mathbb{N}}$ and $h > 0$, we define $\Lambda^{M,h}(E)$ as the Banach space consisting of all multi-indexed sequences $e = (e_\alpha)_{\alpha \in \mathbb{N}^d} \in E^{\mathbb{N}^d}$ such that

$$\|e\|_{\Lambda^{M,h}(E)} := \sup_{\alpha \in \mathbb{N}^d} \frac{h^{|\alpha|} \|e_\alpha\|_E}{M_\alpha} < \infty.$$

We define the (LB)-space

$$\Lambda^{\{\mathfrak{M}\}}(E) := \varinjlim_{n \in \mathbb{N}} \varinjlim_{h \to 0^+} \Lambda^{M^n,h}(E)$$

and the Fréchet space

$$\Lambda'^{\{\mathfrak{M}\}}(E) := \varprojlim_{n \in \mathbb{N}} \varprojlim_{h \to 0^+} \Lambda^{1/M^n,1/h}(E).$$

The dual of $\Lambda'^{\{\mathfrak{M}\}}(E)$ may be identified with $\Lambda^{\{\mathfrak{M}\}}(E')$; the dual pairing under this identification is given by

$$\langle e', e \rangle = \sum_{\alpha \in \mathbb{N}^d} \langle e'_\alpha, e_\alpha \rangle, \qquad e' \in \Lambda^{\{\mathfrak{M}\}}(E'), e \in \Lambda'^{\{\mathfrak{M}\}}(E).$$

Proposition 3.8 *Let $K \Subset \mathbb{R}^d$ be a regular compact set such that* $\operatorname{int} K$ *is a Lipschitz domain and let $\mathfrak{M}$ be a weight matrix satisfying $\{\mathfrak{M}.2\}'$. Let B be an equicontinuous subset of $(\mathcal{E}^{\{\mathfrak{M}\}}(K))'$. There exist $N \in V(\mathfrak{M})$ and $C > 0$ such that for each $T \in B$ there is a family $\{F_{\alpha,T} \in L^2(K) \mid \alpha \in \mathbb{N}^d\}$ satisfying*

$$\sup_{\alpha \in \mathbb{N}^d} \|F_{\alpha,T}\|_{L^2(K)} N_\alpha \leq C \tag{3.1}$$

and

$$\langle T, \varphi \rangle = \sum_{\alpha \in \mathbb{N}^d} \int_K F_{\alpha,T}(x) \partial^\alpha \varphi(x) dx, \qquad \varphi \in \mathcal{E}^{\{\mathfrak{M}\}}(K). \tag{3.2}$$

Proof We claim that the continuous linear mapping

$$S : \Lambda'^{\{\mathfrak{M}\}}(L^2(K)) \to (\mathcal{E}^{\{\mathfrak{M}\}}(K))'_\beta, \quad (F_\alpha)_{\alpha \in \mathbb{N}^d} \mapsto \sum_{\alpha \in \mathbb{N}^d} (-1)^{|\alpha|} \partial^\alpha F_\alpha$$

is surjective. Before we prove the claim, let us show how it implies the result. By Lemma 3.6(ii), it suffices to show that for every bounded subset B of $(\mathcal{E}^{\{\mathfrak{M}\}}(K))'_\beta$ (in particular, for every equicontinuous subset B of $(\mathcal{E}^{\{\mathfrak{M}\}}(K))'$) there is a bounded subset A of $\Lambda'^{\{\mathfrak{M}\}}(L^2(K))$ such that $S(A) = B$. Since $(\mathcal{E}^{\{\mathfrak{M}\}}(K))'_\beta$ is a Fréchet-Montel space (Lemma 3.7), this follows from the following general fact [10, Corollary 26.22]: Let $T : E \to F$ be a surjective continuous linear mapping between a Fréchet space E and a Fréchet-Montel space F. Then, for every bounded subset B of F there is a bounded subset A of E such that $T(A) = B$. We now prove the claim. To this end, it suffices to show that the transpose of S is injective and has weak-$*$ closed range. By the remarks preceding this proposition and the fact that $\mathcal{E}^{\{\mathfrak{M}\}}(K)$ is reflexive (Lemma 3.7), we may identify the transpose of S with the mapping

$$S^t : \mathcal{E}^{\{\mathfrak{M}\}}(K) \to \Lambda^{\{\mathfrak{M}\}}(L^2(K)), \; \varphi \to (\partial^\alpha \varphi)_{\alpha \in \mathbb{N}^d}.$$

This mapping is clearly injective. We now show that it has weak-$*$ closed range. Let $(\varphi_j)_j$ be a net in $\mathcal{E}^{\{\mathfrak{M}\}}(K)$ and $F = (F_\alpha)_{\alpha \in \mathbb{N}^d} \in \Lambda^{\{\mathfrak{M}\}}(L^2(K))$ such that $S^t(\varphi_j) \to F$ in the weak-$*$ topology on $\Lambda^{\{\mathfrak{M}\}}(L^2(K))$. In particular, $\partial^\alpha \varphi_j \to F_\alpha$ in $L^1_{\mathrm{loc}}(\operatorname{int} K)$ for all $\alpha \in \mathbb{N}^d$. Consequently, $\partial^\alpha F_0 = F_\alpha \in L^2(\operatorname{int} K)$ (the derivatives should be interpreted in the sense of distributions). By the Sobolev embedding theorem [1, Theorem 4.12 Part II], there is $k \in \mathbb{N}$ such that the continuous embedding $H^k(\operatorname{int} K) \to C(K)$ holds, where $H^k(\operatorname{int} K)$ denotes the Sobolev space of order k. Since $\partial^\alpha F_0 \in H^k(\operatorname{int} K)$ for all $\alpha \in \mathbb{N}^d$, we obtain that $F_0 \in C^\infty(K)$ and

$$\|\partial^\alpha F_0\|_{L^\infty(K)} \leq D \max_{|\beta| \leq k} \|\partial^{\alpha+\beta} F_0\|_{L^2(\operatorname{int} K)} = D \max_{|\beta| \leq k} \|F_{\alpha+\beta}\|_{L^2(\operatorname{int} K)}, \qquad \alpha \in \mathbb{N}^d,$$

for some $D > 0$. Pick $0 < h \leq 1$ and $n \in \mathbb{N}$ such that $F \in \Lambda^{M_n,h}(L^2(K))$. Condition $\{\mathfrak{M}.2\}'$ implies that there are $m \in \mathbb{N}$ and $C, H > 0$ such that $\max_{0\leq j\leq k} M^n_{p+j} \leq CH^p M^m_p$ for all $p \in \mathbb{N}$. Hence,

$$\|F_0\|_{\mathcal{E}^{M_m,h/H}(K)} \leq DCh^{-k}\|F\|_{\Lambda^{M_n,h}(L^2(K))}.$$

This shows that $F_0 \in \mathcal{E}^{\{\mathfrak{M}\}}(K)$ and thus $F = (F_\alpha)_{\alpha\in\mathbb{N}^d} = (\partial^\alpha F_0)_{\alpha\in\mathbb{N}^d} \in \operatorname{Im} S^t$.

□

Remark 3.9 One can also use the dual Mittag-Leffler lemma [7, Lemma 1.4] in the same way as in the proof of [5, Theorem 3.2(ii)] (see also the proof of [7, Proposition 8.6]) to show Proposition 3.8.

Proof of Theorem 3.5 The first statement is a consequence of Lemma 3.6(i). Moreover, it is clear that for each $N \in V(\mathfrak{M})$ the seminorm $\|\cdot\|_{\mathcal{E}^{N,1}(K)}$ is continuous on $\mathcal{E}^{\{\mathfrak{M}\}}(K)$. We now show that for every seminorm q on $\mathcal{E}^{\{\mathfrak{M}\}}(K)$ there are $N \in V(\mathfrak{M})$ and $C > 0$ such that

$$q(\varphi) \leq C\|\varphi\|_{\mathcal{E}^{N,1}(K)}, \qquad \varphi \in \mathcal{E}^{\{\mathfrak{M}\}}(K).$$

Choose an equicontinuous subset B of $\mathcal{E}'^{\{\mathfrak{M}\}}(K)$ such that

$$q(\varphi) = \sup_{T\in B} |\langle T, \varphi\rangle|, \qquad \varphi \in \mathcal{E}^{\{\mathfrak{M}\}}(K).$$

By Proposition 3.8, there exist $N \in V(\mathfrak{M})$ and $C > 0$ such that for each $T \in B$ there is a family $\{F_{\alpha,T} \in L^2(K) | \alpha \in \mathbb{N}^d\}$ satisfying (3.1) and (3.2). Set $L = (N_p/2^p)_{p\in\mathbb{N}} \in V(\mathfrak{M})$. For all $\varphi \in \mathcal{E}^{\{\mathfrak{M}\}}(K)$ it holds that

$$q(\varphi) \leq \sup_{T\in B} \sum_{\alpha\in\mathbb{N}^d} \int_K |F_{\alpha,T}(x)||\partial^\alpha\varphi(x)|dx \leq \sup_{T\in B} \sum_{\alpha\in\mathbb{N}^d} \|F_{\alpha,T}\|_{L^2(K)}\|\partial^\alpha\varphi\|_{L^2(K)}$$

$$\leq C|K|^{1/2} \sum_{\alpha\in\mathbb{N}^d} \frac{\|\partial^\alpha\varphi\|_{L^\infty(K)}}{N_\alpha} \leq 2^d C|K|^{1/2}\|\varphi\|_{\mathcal{E}^{L,1}(K)}.$$

□

Acknowledgments A. Debrouwere was supported by FWO-Vlaanderen through the postdoctoral grant 12T0519N. B. Prangoski was partially supported by the Macedonian Academy of Sciences and Arts through the project "Microlocal analysis and Applications". J. Vindas was supported by Ghent University through the BOF-grants 01J11615 and 01J04017.

References

1. R.A. Adams, J.J.F. Fournier, *Sobolev Spaces* (Elsevier/Academic Press, Amsterdam, 2003)
2. R.W. Braun, R. Meise, B.A. Taylor, Ultradifferentiable functions and Fourier analysis. Results Math. **17**, 206–237 (1990)
3. A. Brudnyi, Y. Brudnyi, *Methods of Geometric Analysis in Extension and Trace Problems*, vol. 1 (Birkhäuser, Basel, 2012)
4. A. Debrouwere, J. Vindas, Solution to the first Cousin problem for vector-valued quasianalytic functions. Ann. Mat. Pura Appl. **196**, 1983–2003 (2017)
5. P. Dimovski, S. Pilipović, B. Prangoski, J. Vindas, Convolution of ultradistributions and ultradistribution spaces associated to translation-invariant Banach spaces. Kyoto J. Math. **56**, 401–440 (2016)
6. T. Heinrich, R. Meise, A support theorem for quasianalytic functionals. Math. Nachr. **280**, 364–387 (2007)
7. H. Komatsu, Ultradistributions I. Structure theorems and a characterization. J. Fac. Sci. Univ. Tokyo IA Math. **20**, 25–105 (1973)
8. H. Komatsu, Ultradistributions III. Vector valued ultradistributions and the theory of kernels. J. Fac. Sci. Univ. Tokyo Sect. IA Math. **29**, 653–717 (1982)
9. J.M. Lee, *Introduction to Smooth Manifolds* (Springer, New York, 2003)
10. R. Meise, D. Vogt, *Introduction to Functional Analysis* (The Clarendon Press/Oxford University Press, New York, 1997)
11. A. Rainer, G. Schindl, Composition in ultradifferentiable classes. Stud. Math. **224**, 97–131 (2014)
12. H. Whitney, Analytic extensions of differentiable functions defined in closed sets. Trans. Am. Math. Soc. **36**, 63–89 (1934)
13. H. Whitney, Functions differentiable on the boundaries of regions. Ann. Math. **35**, 482–485 (1934)
14. H. Whitney, On the extension of differentiable functions. Bull. Am. Math. Soc. **50**, 76–81 (1944)

Singular Solutions to Equations of Fluid Mechanics and Dynamics Near a Hurricane's Eye

Maximilian F. Hasler

Abstract In the first part of this paper, we prove a new theorem concerning non-smooth solutions of nonlinear Navier–Stokes type PDE as they arise in atmospheric and fluid dynamics, but here in arbitrary dimension. In its simplest form, the theorem states that the velocity field must be tangent to the hypersurface on which it has a jump discontinuity, i.e., its singular support. The theorem is proved using Colombeau algebras of generalized functions, providing yet another example of the fruitfulness of this concept for nonlinear problems with singularities, ill posed in distribution theory.
In the second part, we discuss a numerical and analytic study of a two-dimensional model which, in spite of its simplicity, predicts remarkably correctly the "wall of the eye" of a hurricane and allows us to get analytic expressions for the asymptotic behaviour of radial and tangential wind field near this wall. These results are consistent with and confirm the theoretical results of the first part.

Keywords Navier–Stokes equations · Singular solutions · Hurricane's eye

Mathematics Subject Classification (2010) 46F30; Secondary 35A20, 35Q30, 86A10

1 Introduction

During the last decades, much progress has been made concerning the prediction of hurricane tracks, but the dynamics near the hurricane's eye is still not very well understood. This gives an additional "real-world" motivation, complementary to the obvious theoretical interest in the study non-smooth solutions to equations of

M. F. Hasler (✉)
Université des Antilles, Laboratoire MEMIAD, Schoelcher CEDEX, France
e-mail: M.Hasler@memiad.eu

P. Cerejeiras et al. (eds.), *Current Trends in Analysis, its Applications and Computation*, Research Perspectives,
https://doi.org/10.1007/978-3-030-87502-2_38

fluid dynamics derived from Navier–Stokes equations (being part of the famous Millenium problems). We have in mind to focus on the behaviour of the wind field near the eye of a hurricane, where it drops from its maximal value to nearly zero within a very small region of space.

Mathematically speaking, the set of points where the solution makes such jumps is called the singular support of the solution, which is the closure of the points in space-time where it is not smooth. In the case at hand, this singular support is of codimension one, i.e., a hypersurface, border of a domain we will specify later and which will correspond to the hurricane's eye.

Nonlinear differential equations with singularities are genuinely ill defined in the classical theory of distributions which *a priori* does not allow to multiply these generalized functions. Therefore we use the framework of Colombeau algebras of generalized functions to tackle this problem. In this framework, multiplication and differentiation of non-smooth functions is always well defined.

In the second part of this paper, we use convenient reparametrizations in similar differential equations, but limited to two dimensions, in order to determine the wind profile around a hurricane's eye. We will find analytic asymptotic expressions for the speed of the tangent and radial wind components, as was as their derivatives, in the neighborhood of the singular support known as the wall of the eye of the hurricane. We find that the derivatives are indeed unbounded, while the wind field itself has a finite limiting value. This study also sheds new light on the physical problem we consider, and confirms the earlier more theoretical results.

There has been earlier research on singular solutions to Navier–Stokes type equations, in particular by Maslov [8] and his followers Zhikharev [12], Danilov [2] and Dobrokhotov et al. [4, 5, and references therein], who also considered the application to hurricanes. However, all this work focused on a point like vortex-type weak singularity, in contrast to the $(D-1)$-dimensional one considered here. The technical approach used by these authors, based on "asymptotic expansions", is also completely different from ours. On the other hand, the vast literature on propagation of shock waves with, as here, a $(D-1)$-dimensional singular support, is mainly focused on the case where this singular support moves in direction normal to the hypersurface, as in the case of a plane wave or a spherically expanding or collapsing wave. In most other work on singularities in Navier–Stokes equations, authors are rather interested in infinitely large velocity and/or energy density, see recent reviews by Moffatt [9] and Tao [11]. This is not at all the subject of your investigation. Summarizing, we are not aware of any mathematical investigations on singularities of the type we consider here, characterized by the fact that the velocity field is tangent to its $(D-1)$-dimensional singular support, and not directly correlated to the velocity of this moving hypersurface itself.

2 Jump Condition at the Singular Support

2.1 Statement of the Problem

In this part we prove a new theorem for solutions to Navier–Stokes type equations with presence of a jump discontinuity. We first found this result for the case of the two-dimensional model analysed in the second part of the paper, but have generalized it to an arbitrary number of n dimensions and more general terms on the right hand side.

We consider the system of partial differential equations

$$\partial_t u + u \cdot \operatorname{grad} u = A\,(u - u^*)\,, \tag{2.1}$$

where $u : \mathbb{R} \times \mathbb{R}^n \to \mathbb{R}^n$ is an n-dimensional vector field depending on $(t, x) \in \mathbb{R}^{1+n}$, and A is a continuous linear operator without derivatives, or simply a matrix acting on the n-component vector u, while u^* is a given wind field. (In our applications to hurricanes this will be the so-called trade wind.)

We wish to consider a solution of the form

$$u(t, x) = u^*(t, x) + H(S(t, x))\,\hat{u}(t, x)\,, \tag{2.2}$$

where H is the Heaviside step function, equal to 1 for positive arguments and zero else. This means that depending on the sign of $S(t, x)$ the second term is present or not. More precisely, we have $u = u^*$ inside the region $D = \{(t, x) \mid S(t, x) < 0\}$ which represents the hurricane's eye where we have only the weak and roughly constant trade wind, and $u = u^* + \hat{u}$ outside this domain, with potentially very strong winds right next to the eye's wall. To get an idea, the function $S(t, x) = \|x - t\,u^*\| - r_0^2$ would correspond to a ball-shaped "hurricane's eye" of radius r_0 and moving with constant speed u^*. (The choice of the letter S is inspired by Maslov and Zhikharev's "singularity function", see [8, 12].)

In classical distribution theory, this nonlinear differential problem does not make sense, because one cannot consistently multiply distributions. Schwartz' counter-example which has become folklore, considers $H = H^2 = H^3$, which, using Leibniz' rule, yields for the derivative $H' = 2HH' = 3HH'$, whence $HH' = 0$, while the right hand side of the equation is $H' = \delta \neq 0$.

2.2 The Framework of Colombeau Algebras

To formulate, study and solve nonlinear differential equations with irregular solutions, one needs a theory of algebras of generalized function, as are given by the Colombeau type algebras of sequence spaces, in which differentiation and multiplication is always well-defined [1]. More precisely, we consider the spaces $\mathcal{E}(\Omega)$ of smooth functions equipped with seminorms $\mathcal{P}(\Omega) = \{p_{\alpha,K} : f \mapsto \|\partial^\alpha f\|_{K,\infty} ; \alpha \in \mathbb{N}^n, K \Subset \Omega\}$, and the asymptotic scale $M = \{(\varepsilon^m)_{\varepsilon\in\Lambda=(0,1]}; m \in \mathbb{N}\}$. Then the space of "moderate" sequences is $\mathcal{E}_M = \{f \in \mathcal{E}^\Lambda \mid \forall p \in \mathcal{P}\ \exists a \in M : p(a_\varepsilon f_\varepsilon) = O(1)\}$, and the Colombeau-type algebra is the associated Hausdorff space $\mathcal{G}_M = \mathcal{E}_M/\mathcal{N}$ where $\mathcal{N} = \{f \in \mathcal{E}^\Lambda \mid \forall p \in \mathcal{P}\ \forall a \in M : p(f_\varepsilon) = O(a_\varepsilon)\}$ is the closure (intersection of all neighborhoods) of zero for the naturally associated so-called "sharp" topology [3, 10] (for which multiplication is continuous) [7]. The construction is functorial, the spaces are (pre)sheaves of topological algebras, and the point values of generalized functions are generalized numbers $\widetilde{\mathbb{C}}$ which can be infinitely small or large and still mathematically well-defined, as for example $\delta(0)$.

Schwarz distributions $\mathcal{D}'(\Omega)$ and L^1_{loc} are injected into $\mathcal{G}_M(\Omega)$ via $i_\varphi : T \mapsto [\varphi_\varepsilon * T]$, i.e., convolution with a mollifier $\varphi_\varepsilon = \varepsilon^{-n}\varphi(\cdot/\varepsilon)$, where φ has Fourier transform equal to 1 near the origin, as to have $\int_{\mathbb{R}^n} \varphi = 1$ and vanishing higher moments. Consequently, $\boldsymbol{\delta} = i_\varphi\delta = [\varphi_\varepsilon]$ is a Dirac delta function, which is for $n = 1$ the derivative of

$$\mathbf{H}(x) = i_\varphi H(x) = \Big[\int_{-\infty}^{x} \tfrac{1}{\varepsilon}\,\varphi(\tfrac{y}{\epsilon})dy\Big]\,;\quad \mathbf{H}(0) = \tfrac{1}{2}\,. \tag{2.3}$$

The composition of two generalized functions is well defined if the first one is c–bounded [6, Def. 1.2.7], which is the case for smooth functions. Thus, $\mathbf{H} \circ S(x) = [\mathbf{H}_\varepsilon(S(x))] \in \mathcal{G}_M(\mathbb{R}^n)$ is well defined in $\mathcal{G}_M(\Omega)$.

2.3 *Proof of the Jump Condition Theorem*

Writing $D_t = \partial_t + u \cdot \nabla$, we have

$$D_t(u^* + (\mathbf{H} \circ S)\,\hat{u}) = D_t u^* + (\mathbf{H} \circ S)\,D_t\hat{u} + (\delta \circ S)\,(D_t S)\,\hat{u}\,.$$

To prove the result, it is sufficient to take the scalar product of the differential equation with $\hat{u}$. Putting to the right hand side all terms without δ, we have:

$$\hat{u}^2\,(\delta \circ S)\,D_t S = \hat{u} \cdot ((A\,\hat{u} - D_t\hat{u})\,H \circ S - D_t u^*).$$

For $x \in S_0 := \partial D = \{x \mid S(x) = 0\}$, $\delta \circ S(x) = \delta(0)$ is an infinitely large, invertible generalized number, by which we can divide the equation. Using $H(0) = \frac{1}{2}$, we have

$$\forall x \in S_0 : \hat{u}^2\,(\partial_t + (u^* + \tfrac{1}{2}\hat{u}) \cdot \nabla)S = \frac{1}{\delta(0)}\,\hat{u} \cdot (\tfrac{1}{2}(A\hat{u} - D_t\hat{u}) - D_t u^*).$$

All expressions except $1/\delta(0)$ take finite real values in any $x \in S_0$. Therefore, the left hand side must be zero, which we will rewrite as

$$\forall x \in S_0 : \hat{u} \cdot \nabla S = -2\,(\partial_t + u^* \cdot \nabla) S(x).$$

So we have proved the following

Theorem 2.1 *If the PDE* (2.1), $(\partial_t + u \cdot \nabla)u = A(u - u^*)$, *has a solution of the form* $u = u^* + H \circ S\,\hat{u}$, *where H is the Heaviside step function* (2.3), $S \in C^\infty(\mathbb{R}^{n+1})$, *and* $u^*,\ \hat{u} \in C^\infty(\mathbb{R}^{n+1}; \mathbb{R}^n)$, *then* $\hat{u} \cdot \nabla S = -2\,(\partial_t + u^* \cdot \nabla)S$ *on* $S_0 = \{x \mid S(x) = 0\}$.

To get a grasp on the meaning of this jump condition, consider, *e.g.*, $S(x, t) = \|x - tu^*\|^2 - r_0^2$, for which $D = \{x \mid S(x) < 0\}$ is obviously a ball of radius r_0 whose center moves according to $x = t\,u^*$, similar to the eye of a hurricane. In that case, we have $(\partial_t + u^* \cdot \nabla)S = 0$. With this, we can state:

Corollary 2.2 *If the domain* $D = \{x \mid S(x) < 0\}$ *moves with speed* u^*, *i.e.,* $(\partial_t + u^* \cdot \nabla)S = 0$ *on* ∂D, *then the jump* $\hat{u}$ *is tangent (orthogonal to the normal vector* ∇S*) to the boundary of D in each point of this boundary.*

This is (fortunately) in complete agreement with real-world observations and also with the following study of the 2-dimensional case: We will see within a completely different approach that the radial wind component goes to zero when approaching the wall of the hurricane's eye, as predicted by the Corollary.

3 Asymptotics of the Wind Field Near the Eye of a Hurricane

In this part, we consider the two-dimensional differential equation

$$\partial_t(u, v) + (u, v) \cdot \nabla(u, v) = \sigma\,(u, v) + \omega\,(v, -u),$$

which follows from the general Navier–Stokes equations in the limit of a nonviscid fluid and after coordinate transformation to a frame having its origin on the surface of the rotating earth, which induces the Coriolis "force" terms proportional to $\omega = (2/24h)\cos(\mathit{latitude})$.

The parameter σ accounts for friction (negative contribution) but also for energy supply from the hot water surface. Both are proportional to (u, v) only for weak to moderately strong winds: the energy supply can be considered to saturate at some limiting value $\sigma^*\,u^*$, while the friction is known to acquire quadratic terms for larger wind speed. Concerning the energy supply, many more or less complicated piecewise defined heat transfer functions are considered in the literature. We chose a dependence of the form $\sigma\,u = \sigma^*\,u/(1 + |u|/u^*)$ to have the above mentioned limiting behaviour of the energy transfer.

We assume radial symmetry around the center of the hurricane located in $(x, y) = (0, 0)$, i.e., that the wind field (u, v) only depends on $r = x^2 + y^2$, and we reparametrize it using radial and tangential components (a, b), viz: $(u, v) = a(r)(x, y) + b(r)(-y, x)$. This yields the system of equations

$$a' = [(\sigma - a + (\omega + b)b/a]/r, \tag{3.1}$$

$$b' = [(\sigma/a - 2)b - \omega]/r. \tag{3.2}$$

We can integrate these equations numerically, starting at a given point r_i, if we know the initial conditions $a(r_i)$, $b(r_i)$, and of course the parameters σ and ω. We can also make a qualitative study of this system of ODE.

Both approaches lead to the following results: Corresponding to the counter-clockwise rotating winds on the northern hemisphere, the tangent component $b(r)$ has a very small positive value for large r, which increases monotonically as r tends to some finite value R_{eye} at which the derivative diverges, $b'(r) \to -\infty$ as $r \to R_{\text{eye}}$.

The radial component $a(r)$ has a small negative value for large r (airflow towards the hurricane), which decreases to a minimum (large negative value) at $R_{\min} = k\, R_{\text{eye}}$ with some $k \in]2, 9[$, and then tends to zero as $r \to R_{\text{eye}}$, where $a'(r)$ also tends to $-\infty$ as $r \to R_{\text{eye}}$.

When we first noticed this behaviour, we thought that $b(r)$ might be unbounded, which would not be physical but could be explained by the right hand side $\sigma\, u$ which could explain allow an unbounded energy supply. At this point we introduced the more sophisticated form explained earlier, which saturates at a value $\sigma^*\, u^*$. However, a more thorough analysis shows that even with constant σ, we get a finite value b_0 as $r \to R_{\text{eye}}$.

Indeed, we can solve a simplified version of the system which is valid asymptotically, as $r \to R_{\text{eye}}$ and $a \to 0$. In this limit, the equation for b becomes $b'/b \simeq \sigma/(r\, a(r))$, which we can integrate to get $b(r) = b_0 \exp J(r)$ with $J(r) \simeq \int_{R_{\text{eye}}}^{r} \sigma/(s\, a(s))ds$, provided this is well-defined. We find a solution using the ansatz $a(r) \simeq -c\,(r/\,R_{\text{eye}} - 1)^{\alpha}$, which gives

$$b(r) \simeq b_0 \exp\left[-\sigma\,(\tfrac{r}{R_{\text{eye}}} - 1)^{1-\alpha} / \left((1-\alpha)\tfrac{c\,r}{R_{\text{eye}}}\right)\right]$$

and back in (3.1), $\alpha = 1/2$ and $c^2 = (\omega + b_0)b_0$. So we have found that

$$a(r) \simeq -c\sqrt{\frac{r}{R_{\text{eye}}} - 1} \;\text{ with }\; c = \sqrt{(\omega + b_0)\, b_0}$$

and (with $r \simeq R_{\text{eye}}$)

$$b(r) \simeq b_0 \,\exp\left(-2\frac{\sigma}{c}\sqrt{\frac{r}{R_{\text{eye}}} - 1}\right) = b_0 \exp(2\,\sigma\, a(r)/c^2)\,.$$

This concludes the asymptotic analysis of this system near the singularity at $r = R_{\text{eye}}$. We have deduced an analytic form for the tangent and radial wind profile near that point.

Our next goal is the analysis of this simple system is to obtain further analytical results for $r \to R_{\text{eye}}$ and R_{min} in terms of the initial data. But we are also aware that this model has its limits. In particular, we think that an extension to 3D is required to capture more of the dynamics of the real hurricane's eye. The vertical airflows near the eye and between the rainbands are physically important, and taking them into account in our model should improve its potential for predicting realistic values.

4 Summary and Conclusion

The very simple equation $(\partial_t + u \cdot \nabla)u = \sigma\, u + \omega \times u$ predicts qualitatively correctly the existence of the eye of the hurricane and finite limiting speed of wind at the border, in spite of the right hand side possibly violating laws of conservation. We found an analytic expression for the asymptotic form of the solution near the eye's wall.

We also considered solutions of the n-dimensional Euler/Navier–Stokes type equation $(\partial_t + u \cdot \nabla)\, u = A\, u$, in arbitrary dimension, which have a jump $u = u^* + H(S)\, \hat{u}$. We obtained an orthogonality condition involving the jump $\hat{u}$ and the normal vector ∇S on the hypersurface where the jump occurs, i.e., the singular support.

Our theorem was straightforwardly proved in the framework of Colombeau differential algebras, and the result is confirmed by numerical and algebraic analysis of the 2-dimensional case and real-world observations. We plan to continue our work on one hand to try to get explicit non-smooth solution to Euler/Navier–Stokes equations in 2D, and to extend the analytic and numerical study to 3 dimensions.

Let us finally once again insist on two conceptual points: On one hand, the non-linearity in the differential equations which is essential for producing singularities, but which can't be dealt with the linear theory of Schwartz distributions, making algebras of generalized functions an essential tool for further progress in this important field of research. On the other hand, still on the same token, one may admire the simplicity with which we were able to prove our theorem in the framework of Colombeau type algebras. We hope that this additional example of a useful, real-world relevant application will further contribute to popularize this theory.

References

1. J.-F. Colombeau, *Elementary Introduction to New Generalized Functions*. North-Holland Mathematical Studies, vol. 113 (North-Holland, Amsterdam, 1985)
2. V.G. Danilov, V.P. Maslov, V.M. Shelkovich, Algebras of singularities of generalized solutions of strictly hyperbolic systems of quasilinear first-order systems. Teor. Mat. Fiz. **114**(1), 3–55 (1998)
3. A. Delcroix, D. Scarpalezos, Sharp topologies on asymptotic algebras. Monatshefte Math. **129**, 1–14 (2000)
4. S.Y. Dobrokhotov, Hugoniót-Maslov chains for the trajectories of point vortex singularities of the shallow water equations and the Hill equation. Dokl. Akad. Nauk Ross. Akad. Nauk **354**(5), 600–603 (1997). S.Y. Dobrokhotov, Reduction to the Hill equation of the Hugóniot-Maslov chain for the trajectories of solitary vortices of shallow wave equations. Teor. Mat. Fiz. **112**(1), 47–66 (1997). S.Y. Dobrokhotov, Hugoniót-Maslov chains for solitary vortices of the shallow water equations, II. Russ. J. Math. Phys. **6**(3), 282–313 (1999)
5. S.Y. Dobrokhotov, K.V. Pankrashkin, E.S. Semenov, Proof of Maslovs conjecture about the structure of weak point singular solutions of the shallow water equations. Russ. J. Math. Phys. **8**(1), 25–54 (2001). S.Y. Dobrokhotov, K.V. Pankrashkin, E.S. Semenov, On Maslovs conjecture about the structure of weak point singularities of the shallow-water equations. Dokl. Math. **64**(1), 127–130 (2001)
6. M.Grosser, M. Kunzinger, M. Oberguggenberger, R. Steinbauer, *Geometric Theory of Generalized Functions with Applications to General Relativity* (Kluwer Academic, Dordrecht, 2001)
7. M.F. Hasler, Asymptotic extension of topological modules and algebras. Integral Transforms Special Funct. **20**(4), 291–299 (2009)
8. V.P. Maslov, Three algebras corresponding to nonsmooth solutions of systems of quasilinear hyperbolic equations. Uspekhi Matem. Nauk **35**(2), 252–253 (1980)
9. H.K. Moffatt, Singularities in fluid mechanics. Phys. Rev. Fluids **4**(11), 110502 (2019)
10. D. Scarpalezos, Some remarks on functoriality of colombeau's construction: topological and microlocal aspects and applications. Integral Transforms Spec. Funct. **6**, 295–307 (1998); D. Scarpalezos, Colombeau's generalized functions: topological structures; microlocal properties. A simplied point of view, I. Bull. Cl. Sci. Math. Nat. Sci. Math. **25**, 89–114 (2000); D. Scarpalezos, Colombeau's generalized functions: topological structures; microlocal properties. A simplied point of view, II. Publ. Inst. Math. (Beograd) (N.S.) **76**(90), 111–125 (2004)
11. T. Tao, Searching for singularities in the Navier-Stokes equations. Nat. Rev. Phys. **1**, 418–419 (2019)
12. V.N. Zhikharev, On necessary conditions for the existence and uniqueness of the type of solutions with propagating weak point singularity for the 2D hydrodynamic equations (in Russian). VINITI no. B86-8148, Moscow (1986)

Dirac Delta-Function in Optimal Control of Age-Structured Populations

Natali Hritonenko, Nobuyuki Kato, and Yuri Yatsenko

Abstract We demonstrate that use of the Dirac delta-function simplifies qualitative optimization analysis of two contemporary population models. The first problem maximizes the discounted profit from harvesting in the linear age-structured Lotka–McKendrick model. The second problem describes optimal investments in new and old equipment under evolving technology.

Keywords Optimal control · Lotka–McKendrick population model · Harvesting

Mathematics Subject Classification (2010) Primary 92D25; Secondary 46F30

1 Introduction

Generalized functions have long-time links to the optimization theory. L. Pontryagin, I. Ekeland, J.-L. Lions, R. Rockafellar and other great mathematicians spent enormous efforts to prove the existence of solutions to general optimization problems. They highlighted exotic trajectories that arose even in the optimal control of simple ODEs and lead to solutions expressed in terms of measures and generalized functions. The necessity of taking endogenous controls from non-smooth functional spaces with measures was demonstrated in [14, 15] and significant results have been obtained since then.

N. Hritonenko (✉)
Prairie View A&M University, Prairie View, TX, USA
e-mail: nahritonenko@pvamu.edu

N. Kato
Kanazawa University, Kanazawa, Japan
e-mail: nkato@se.kanazawa-u.ac.jp

Y. Yatsenko
Houston Baptist University, Houston, TX, USA
e-mail: yyatsenko@hbu.edu

P. Cerejeiras et al. (eds.), *Current Trends in Analysis, its Applications and Computation*, Research Perspectives,
https://doi.org/10.1007/978-3-030-87502-2_39

The most common reason for getting measure controls and generalized functions is the presence of inequality-constraints on *state variables* in optimal control problems (OCPs) [4]. The state constraints lead to discontinuities in the optimal path of state and co-state variables. Each time when the state trajectory reaches a state constraint, its speed possesses jumps. In worst cases, the optimal state trajectory can touch the state constraint indefinitely many times. It leads to optimal controls and co-state variables (Lagrange multipliers) expressed in terms of measures and functions of bounded variation.

Another more obvious reason for the appearance of measure controls is *discontinuities* or *generalized functions* in the state equations (equality-constraints) or objective function, e.g., [13]. Relevant examples appear in particle mechanics, elasticity theory, nonlinear waves, and other applications [12]. Such problems do not often have solutions in traditional functional classes, and generalized functions and measures should be employed to construct such solutions. Several approaches have been proposed to extend the concept of optimal solutions: vibrational controls, chattering regimes, generalized curves, measure values solutions, relaxed controls [4].

A powerful approach to deal with generalized functions is based on Colombeau algebras. Konjik et al. [12] develop fundamentals of the calculus of variations in Colombeau algebras with generalized functions and demonstrate its applications in geodesics, mechanics, and elastostatics. Their approach is promising for the optimal control, though it needs to be extended to closed sets of control variables. The optimal control was born inside the calculus of variations because of practical needs to consider closed sets of controls. The current extension of the algebras of generalized functions to OCPs cannot consider inequalities-constraints, while most applied optimization problems include them.

Solutions that involve generalized functions and measures appear in the optimal control because of applied nature of the physical, biological, or economic problem under study. This paper demonstrates how the use of a classic generalized function, the Dirac delta-function, simplifies the contemporary analysis of two age-structured population models in different applied areas.

The first discussed problem maximizes the discounted profit from harvesting in the age- structured Lotka–McKendrick population model [1, 3, 7]. Such OCPs often possess bang-bang harvesting controls. We consider a relaxed problem without upper bound on harvesting control and prove that the optimal age-dependent harvesting control is expressed via the Dirac delta-function. This structure is caused by the presence of a state constraint.

The second problem is relevant to new applications of the Lotka-McKendrick model in economics and technology [8, 11]. It describes the optimal distribution of investments into new and old capital under evolving technology. The optimal age- distributions of investment and capital appear to involve the delta-function and its derivative. This OCP does not involve any inequality constraints. It reveals a new phenomenon when generalized functions arise in applied optimization problems with *no constraints and smooth coefficients* just because of the nature of a process under study.

2 Impulse Controls in Optimal Harvesting of Age-Structured Populations

The Lotka–McKendrick model of harvested age-structured population is described as

$$\frac{\partial y(t,a)}{\partial t} + \frac{\partial y(t,a)}{\partial a} = -\mu(t,a)y(t,a) - u(t,a), \tag{2.1}$$

$$y(t,0) = x(t), \quad t \in [0,\infty), \tag{2.2}$$

$$y(0,a) = y_0(a), \quad a \in [0,A), \tag{2.3}$$

where t is time, a is individual age, $A \leq \infty$ is the maximum age, $y(t,a)$ is the population age density, $u(t,a)$ is the harvesting intensity, $\mu(t,a)$ is the given mortality rate, $x(t)$ is the *planting intensity* (the density of zero-age individuals), and $y_0(a)$ is a given population density at time $t = 0$ [1]. The model describes harvested populations in agriculture and aquaculture, when new zero-age individuals are introduced into the population from outside.

Optimal Control Problem (OCP) We consider the problem of maximizing the discounted profit from harvesting over the infinite horizon:

$$\max_{u,x} I = \max_{u,x} \int_0^\infty e^{-rt}\left(z^\alpha(t) - b(t)x(t)\right)dt,\ z(t) = \int_0^A c(t,a)u(t,a)da, \tag{2.4}$$

$$u(t,a) \geq 0, \quad x(t) \geq 0, \quad y(t,a) \geq 0, \tag{2.5}$$

subject to constraints-equalities (2.1)–(2.3). Here, $z(t)$ is the monetary value of harvesting yield, $c(t,a)$ is the unit price of harvested individuals, $b(t)$ is the unit planting cost, $r > 0$ is the discounting rate, and $0 < \alpha < 1$ reflects the concavity, of revenue-product curve. The price $c(t,a)$ increases in a when an individual grows, then slow down, and decreases at the end of its life A.

The majority of research related to optimization in the harvesting model (2.1)–(2.5) involves the constraint-inequality $0 \leq u \leq u_{\max}$, where the upper bound $u_{\max}$ is chosen for mathematical convenience. Optimal harvesting regimes in such PDE-based models often have bang-bang structure [1, 2, 5, 6] but the impulse structure of such regimes has not been suggested. We consider the nonlinear OCP(2.1)–(2.5) without upper constraint $u \leq u_{\max}$ and prove that it possesses an *impulse* optimal harvesting control is expressed via the Dirac delta-function.

We choose u and x as independent controls of the OCP (2.1)–(2.5) and assume $u(t,a)$ to be measurable in t and be a Borel measure with respect to a at a fixed t. Under known u and $x \in L_\infty[0,\infty)$, the dependent state variable $y \in L_\infty([0,\infty) \times [0,A))$ is a weak solution of the boundary problem (2.1)–(2.3).

Hritonenko and Yatsenko [6] analyzed the linear version of OCP (2.1)–(2.5) at $\alpha = 1$ and proved that the optimal harvesting control $u^*(t, a)$ involves the Dirac-function $\delta(a - a^*(t))$ at certain age $a^*(t)$, $0 < a^*(t) < A$, and the optimal state variable $y^*(t, a) = 0$ at $a^*(t) < a < A$. Here we extend this result to the nonlinear problem (2.1)–(2.5) at $0 < \alpha < 1$. Such impulse controls are known in dynamic optimization of homogeneous populations (described by ODEs) but are new in age-dependent populations described by PDEs.

Lemma 2.1 (Necessary Condition for an Extremum) *Let (u^*, x^*) be a solution to the OCP (2.1)–(2.5). Then*

$$I_u(t,a) \leq 0 \text{ at } u^*(t,a) = 0, \quad I_u(t,a) = 0 \text{ at } u^*(t,a) > 0,$$
$$I_x(t) \leq 0 \text{ at } x^*(t) = 0, \quad I_x(t) = 0 \text{ at} x^*(t) > 0, \tag{2.6}$$

where

$$I_u(t,a) = e^{-rt}\left(\alpha c(t,a)x^{\alpha-1}(t) - \lambda(t,a)\right), \tag{2.7}$$

$$I_x(t) = e^{-rt}\left(\lambda(t,0) - \alpha b(t)x^{\alpha-1}(t)\right), \tag{2.8}$$

and the dual variables $\lambda(t, a)$ and $\eta(t, a)$ satisfy the following conditions:

$$\frac{\partial \lambda(t,a)}{\partial t} + \frac{\partial \lambda(t,a)}{\partial a} = (r + \mu(t,a))\lambda(t,a) - \eta(t,a), \tag{2.9}$$

$$\lim_{t\to\infty} e^{-rt}\lambda(t,a) = 0, \ a \in [0, A), \quad \lambda(t, A) = 0, \ t \in [0, \infty), \tag{2.10}$$

$$\eta(t,a) > 0 \text{ at } y^*(t,a) = 0, \quad \eta(t,a) = 0 \text{ at } y^*(t,a) > 0. \tag{2.11}$$

The proof follows the standard method of Lagrange multipliers, which was adjusted to age-structured population models in [5–7]. The condition of complementary slackness (2.11) is an essential part of extremum conditions for the optimal control with state constraints [1, 4, 10]. By (2.11), the dual variable $\eta(t, a) > 0$ when the constraint $y \geq 0$ is active, i.e., $y(t, a) = 0$.

Steady-State Solution in Autonomous Case For clarity, we restrict ourselves to the autonomous (stationary) case of the problem (2.1)–(2.5) when all given functions do not depend on the time t:

$$c(t,a) = c(a), \quad b(t) = b, \quad \mu(t,a) = \mu(a). \tag{2.12}$$

Sustainable development of populations is of increasing importance for related environmental policies. Mathematically, it means finding steady-state solutions of an optimization problem. A steady-state solution of the OCP (2.1)–(2.5) is a trajectory that satisfies the constraints (2.1), (2.2), (2.5) and extremum conditions (2.6)–(2.11), but not necessarily satisfies the initial condition (2.3). At conditions (2.12), we look for a steady-state trajectory

$$x(t) = \widetilde{x} = \text{const}, \quad u(t,a) = \widetilde{u}(a), \quad y(t,a) = \widetilde{y}(a). \tag{2.13}$$

Theorem 2.2 (On a Steady-State Impulse Harvesting) *Let* (2.12) *hold and* $c'(0)/c(0) > r + \mu(0)$, $c'(A)/c(A) < \mu(A)$. *Then the OCP* (2.1)–(2.5) *has the steady-state solution* $(\widetilde{x}, \widetilde{u}, \widetilde{y})$,

$$\widetilde{u}(a) = \widetilde{x}\delta(a - \widetilde{a})e^{-\int_0^{\widetilde{a}} \mu(\xi)d\xi}, \tag{2.14}$$

$$\widetilde{x} = \left(\frac{\alpha}{b}\left(c(\widetilde{a})e^{-\int_0^{\widetilde{a}} \mu(\xi)d\xi}\right)^{\alpha}\right)^{1/(1-\alpha)}, \tag{2.15}$$

$$\widetilde{y}(a) = \begin{cases} \widetilde{x}e^{-\int_0^a \mu(\xi)d\xi}, & 0 < a \le \widetilde{a}, \\ 0, & \widetilde{a} < a \le A, \end{cases} \tag{2.16}$$

$$\widetilde{I} = \frac{1-\alpha}{r}\left(\frac{\alpha}{b}c(\widetilde{a})e^{-\int_0^{\widetilde{a}} \mu(\xi)d\xi}\right)^{\alpha/(1-\alpha)}, \tag{2.17}$$

where the endogenous age $\widetilde{a}$, $0 < \widetilde{a} < A$, *is determined from*

$$\widetilde{a} = \arg\max_{0 \le a \le A}\left[c(a)e^{-\int_0^a \mu(\xi)d\xi}\right]. \tag{2.18}$$

Proof Substituting (2.13) into (2.1) and disregarding the initial condition (2.3), the boundary problem (2.1) and (2.2) and dual PDE equation (2.9) and (2.10) are reduced to the ODEs

$$y'(a) = -\mu(a)y(a) - u(a), \quad y(0) = \widetilde{x}, \tag{2.19}$$

$$\lambda'(a) = (r + \mu(a))\lambda(a) - \eta(a), \quad \lambda(A) = 0. \tag{2.20}$$

The analytic solution of (2.19) is

$$y(a) = xe^{-\int_0^a \mu(\xi)d\xi} - \int_0^a e^{-\int_\tau^a \mu(\xi)d\xi}u(\tau)d\tau. \tag{2.21}$$

Step 1: Let us prove that the optimal state variable $\widetilde{y}(a) = 0$ at $a \in \Delta$, meas$(\Delta) > 0$. By contradiction, assuming that $y(a) > 0$ for $0 \le a \le A$, we obtain $\eta \equiv 0$ by (2.11) and $\lambda \equiv 0$ by (2.10). Then, by Lemma 2.1, $I_u(a) > 0$ and $I_x < 0$, which means instant harvesting of *all* individuals without introducing new ones. Then, by (2.1), the population will extinct over a finite time.

Step 2: If $y(a_e) = 0$ at some instant $0 < a_e < A$, then $y'(a) \le 0$ by (2.19) at $a > a_e$, therefore, $y(a) = 0$ and $u(a) = 0$ for $a_e < a \le A$. Let $y(a) > 0$ for $0 < a < a_e$ and $y(a_e) = 0$. Then, by (2.11), $\eta(a) = 0$ for $0 < a < a_e$ and, by (2.20), $\lambda'(a) = (r + \mu(a))\lambda(a)$ at $0 < a < a_e$.

Step 3: If $c(0) > \lambda(0)\widetilde{x}^{1-\alpha}$, then $I_u(0) > 0$ by Lemma 2.1 and, therefore, all young members of the population should be immediately harvested, which is not realistic. On the other side, if $c(a) > \lambda(a)\widetilde{x}^{1-\alpha}$ on $[0, A]$, then $I_u(a) < 0$ and there is no harvesting. Thus, for the existence of a non-trivial harvesting regime, $c(a) < \lambda(a)\widetilde{x}^{1-\alpha}$ should hold on some initial interval $[0, \widetilde{a})$ and $c(\widetilde{a}) = \lambda(\widetilde{a})\widetilde{x}^{1-\alpha}$ at a certain switching age $0 < \widetilde{a} < A$. Then, by (2.7), $I_u(a) < 0$ and $u^*(a) = 0$ on $[0, \widetilde{a})$. Let us assume that $y(a) > 0$ on a certain interval $[\widetilde{a}, a_e]$. Then, $c(a) > \lambda(a)\widetilde{x}^{1-\alpha}$ and $I_u(a) > 0$ at $\widetilde{a} < a < a_e$, i.e., the optimal harvesting control $\widetilde{u}(a)$ should be maximal on $[\widetilde{a}, a_e]$. It means that it should involve the Dirac delta-function $\delta(a)$:

$$\widetilde{u}(a) = \widetilde{h}\delta(a - \widetilde{a}). \tag{2.22}$$

and the optimal state variable $\widetilde{y}(a)$ should be zero after the age $\widetilde{a}$:

$$\widetilde{y}(a) = \begin{cases} > 0, & 0 < a \le \widetilde{a}, \\ 0, & \widetilde{a} < a \le A. \end{cases} \tag{2.23}$$

Step 4: Let us find the unknown $\widetilde{a}$ and $\widetilde{h}$ in (2.22). Substituting (2.22) into (2.21) and using that $\widetilde{y}(a) = 0$ at $a > \widetilde{a}$ by (2.23), we find $\widetilde{h} = \widetilde{x}e^{-\int_0^{\widetilde{a}} \mu(\xi)d\xi}$ at $a > \widetilde{a}$, which together with (2.22) leads to (2.14) for $\widetilde{u}(a)$. Next, substituting (2.14) into (2.21) and using (2.23), we obtain (2.16) for $\widetilde{y}(a)$. Finally, the substitution of (2.14), (2.12) to (2.4) reduces the OCP (2.1)–(2.5) to

$$\max_{u,x} I = \frac{1}{r} \max_{a,x} \left[x^\alpha \left(c(a)e^{-\int_0^a \mu(\xi)d\xi} \right)^\alpha - bx \right]. \tag{2.24}$$

Taking the derivatives of (2.24) in a and x and setting them to zero, we obtain that the maximum of (2.24) is reached at $\widetilde{a}$ and $\widetilde{x}$ given by (2.18) and (2.15).

□

By Theorem 2.2, there exists a sustainable harvesting regime $(\widetilde{u}, \widetilde{x}, \widetilde{y})$. The optimal harvesting control $\widetilde{u}(a)$ is selective and involves one age $\widetilde{a}$ only. The population density is zero for the individuals older than the harvesting age. The obtained results demonstrate significance of impulse harvesting and have implications for management policies. For instance, they indicate advantages of selective harvesting over clear cutting in forestry.

3 Optimal Investment in Old Capital Under Technological Change and Learning-by-Doing

To demonstrate larger applicability of the age-structured model (2.1)–(2.3), we discuss its use in capital replacement problems of Operations Research and economics presented in [9]. A new feature is that the optimal controls involve the delta-function and its derivative.

We consider a firm that produces a product $y(t)$ and invests $x(t)$ into the new capital assets of vintage t and $u(t,a)$ into the old capital of the age a. The *capital vintage* ν is the time of the asset creation and can be used instead of its age $a = t-\nu$. The evolution of the heterogeneous capital $k(t,a)$ is described by the following linear age-dependent population model [8, 11]:

$$\frac{\partial k(t,a)}{\partial t} + \frac{\partial k(t,a)}{\partial a} = -\mu k(t,a) + u(t,a), \tag{3.1}$$

with the boundary conditions

$$k(t,0) = x(t), \quad t \in [0,\infty), \tag{3.2}$$

$$k(0,a) = k_0(a), \quad a \in [0,\infty), \tag{3.3}$$

where $\mu > 0$ is the capital deterioration rate, $k_0(a)$ is the given age distribution of capital over past vintages at time zero. The linear PDE-based model (3.1)–(3.3) closely resembles (2.1)–(2.3), though has a different interpretation.

We assume that the firm's objective is to maximize its discounted net profit

$$\max_{u,x} \int_0^\infty e^{-rt} \left(y(t) - x(t) - \int_{-\infty}^t u(t,a)da \right) dt, \tag{3.4}$$

$$y(t) = \left(\int_{-\infty}^t e^{\gamma\nu} A(t-\nu) k(t, t-\nu) d\nu \right)^\alpha, \tag{3.5}$$

where $r > 0$ is the discounting rate and the unit price of all capital vintages equals one. The investments $x(t)$ and $u(t,a)$ are control variables, and the unknown state variables $k(t,a)$ and $y(t)$ are determined from (3.1) and (3.5). The parameter $0 < \alpha \le 1$ describes returns to scale. The efficiency of new capital of vintage $\nu = t - a$ exponentially increases with rate $\gamma > 0$ due to the embodied technological change [7]. The factor $A(a)$ in (3.5) describes the effects such as learning-by-doing and spillovers, which depend on the age $a = t - \nu$.

Theorem 3.1 *At increasing bounded learning curve $A(a)$, $A'(0) > \gamma A(0)$, and $\gamma\alpha/(1-\alpha) < r$, the OCP* (3.1)–(3.5) *has the unique steady-state trajectory (k, x, u) with growth rate $g = \gamma\alpha/(1-\alpha) > 0$:*

$$u(t,a) = e^{gt}\hat{u}(a), \quad \hat{u}(a) = e^{(\mu+g)a}\left[f'(a) + \frac{f(a)}{s_{\max} - a}\right]\delta(a - s_{\max}), \tag{3.6}$$

$$k(t,a) = e^{gt} f(a)\delta(a - s_{\max}), \quad x(t) = 0, \quad a \in [0,\infty), \; t \in [0,\infty), \tag{3.7}$$

$$f(a) = (\alpha/(r+\mu))^{\frac{1}{1-\alpha}}\left[e^{-\gamma a}A(a)\right]^{\frac{\alpha}{1-\alpha}}, \tag{3.8}$$

and $s_{\max} = \arg\max\left(e^{-\gamma a}A(a)\right) > 0$ is the optimal age of the only vintage to invest in.

Proof is provided in [9]. It is interesting that the idea of the proof is conceptually similar to elasticity problems of [12], but the regularizing parameter is the substitutability β rather than the modulus of elasticity.

For clarity, this paper focuses on linear PDE-based population models and simple impulse solutions with one delta-function. Similar results can be obtained for a more general nonlinear age- and size-structured population models [6, 7].

Acknowledgments This work was supported by JSPS KAKENHI Grant Number JP17K05273. N. Hritonenko recognizes support of PVAMU FIE Faculty Innovation and Enhancement.

References

1. S. Anita, *Analysis and Control of Age-Dependent Population Dynamics* (Kluwer Academic, Dordrecht, 2000)
2. M. Brokate, Pontryagin's principle for control problems in age-dependent population dynamics. J. Math. Biol. **23**, 75–101 (1985)
3. V. Barbu, M. Iannelli, Optimal control of population dynamics. J. Optim. Theory Appl. **102**, 1–14 (1999)
4. H. Fattorini, *Infinite Dimensional Analysis Optimization and Control Theory* (Cambridge University Press, Cambridge, 1999)
5. N. Hritonenko, Y. Yatsenko, The structure of optimal time- and age-dependent harvesting in the Lotka-McKendrick population model. Math. Biosci. **208**, 48–62 (2007)
6. N. Hritonenko, Y. Yatsenko, Bang-bang, impulse, and sustainable harvesting in age-structured populations. J. Biol. Syst. **20**, 133–153 (2012)
7. N. Hritonenko, Y. Yatsenko, *Mathematical Modeling in Economics, Ecology and the Environment*, 2nd edn. (Springer, New York, 2013)
8. N. Hritonenko, N. Kato, Y. Yatsenko, Optimal control of investments in old and new capital under improving technology. J. Optim. Theory Appl. **172**, 247–266 (2017)
9. N. Hritonenko, Y. Yatsenko, A. Boranbayev, Generalized functions in the qualitative study of heterogeneous populations. Math. Popul. Stud. **26**(3), 146–162 (2019)
10. A.D. Ioffe, V.D. Tikhomirov, *Theory of Extremal Problems* (North-Holland, Amsterdam, 1979)
11. B. Jovanovic, Y. Yatsenko, Investment in vintage capital. *J. Econ. Theory* 147, 551–569 (2012)

12. S. Konjik, M. Kunzinger, M. Oberguggenberger, Foundations of the calculus of variations in generalized function algebras. Acta Appl. Math. **103**, 169–199 (2008)
13. M. Nedeljkov, M. Oberguggenberger, Ordinary differential equations with delta function terms. Publications De l'Institut Mathématique **91**(105), 125–135 (2012)
14. J. Warga, *Optimal Control of Differential and Functional Equations* (Academic Press, Cambridge, 1972)
15. L.C. Young, *Lectures on the Calculus of Variations and Optimal Control Theory* (Chelsea, Providence, 1980)

Generalized Solutions to Equations for Probabilistic Characteristics of Levy Processes

Irina V. Melnikova

Abstract We study relations between stochastic differential equations with inhomogeneities reflecting continuous and discontinuous random perturbations and equations for probabilistic characteristics of processes specified by these stochastic equations. The application of two approaches: based on the Ito formula and on limit relations for process increments, allowed to obtain direct and backward integro-differential equations for various probabilistic characteristics and justify them in distribution spaces.

Keywords Stochastic equation · Ito formula · Generalized transition density · Levy process · Kolmogorov equation

Mathematics Subject Classification (2010) Primary 60G51; Secondary 60J35

1 Introduction

A wide class of processes arising in various fields of natural science, economics and social phenomena, mathematically can be described using differential equations with random perturbations, stochastic differential equations (SDEs). The best-studied class of SDEs is one with random perturbations in the form of Wiener processes. Solutions of such equations (normal diffusion processes), due to continuity of Wiener process trajectories also have continuous trajectories. In addition, normal diffusion processes have the following characteristic property: the variance of the process deviation over time Δt is proportional to Δt. Therefore, modeling within the framework of diffusion-type equations is not suitable for describing processes with jumps and ones with variance proportional to $\Delta t^{\mu}, \mu \neq 1$. The

I. V. Melnikova (✉)
Ural Federal University, Ekaterinburg, Russia
e-mail: Irina.Melnikova@urfu.ru

P. Cerejeiras et al. (eds.), *Current Trends in Analysis, its Applications and Computation*, Research Perspectives,
https://doi.org/10.1007/978-3-030-87502-2_40

behavioral features unusual for normal diffusion processes, can be modeled using Levy processes and more general Levy type processes.

Both in applications and in fundamental science, researchers are often interested not in processes themselves but their characteristics; therefore, the relationship between SDEs and equations of probabilistic characteristics of processes described by SDEs is one of the main directions of stochastic analysis. Most investigated remains the connection for diffusion processes and corresponding partial differential equations for their probabilistic characteristics.

In the paper, we study Levy type stochastic equations and obtain equations for probabilistic characteristics, which in the case are integro-differential (pseudo-differential), in contrast to partial differential equations of parabolic type corresponding to diffusion processes. For this purpose we distinguish two approaches:

- the approach based on the general Ito formula (see, e.g. [1, 2]) allowing to obtain functions of studied processes, which are averaged, and as a result we get the integro-differential equations for probabilistic characteristics.
- the approach allowing to obtain equations for probabilistic characteristics based on the existence of three limits for the random process under study: the limits of the quotient of dividing the local first and second moments by $\Delta t \to 0$ (conditions (3.1)–(3.2)) and the limit (3.3) characterizing the absence of the continuity property of Levy type processes (see, e.g. [3]).

There are deep, not always obvious connections between these approaches, and not all of them, despite many works devoted to the indicated issues, worked out in the desired completeness. In the paper we can see that in the both approaches twice differentiable functions appear in equations for probabilistic characteristics and we show that these functions can be used as test functions. We pay the important attention to substantiation of the resulting direct and backward integro-differential equations, in the general case not having classical solutions, in distribution (generalized functions) spaces.

2 Direct and Backward Equations for Probabilistic Characteristics Based on the Ito Formula

Let a probability space $(\Omega, \mathcal{F}, (\mathcal{F}_t)_{t\geq 0}, P)$ be given. We consider a random process $X = \{X(t), t \geq 0\}$, arising under the influence of continuous and discontinuous random disturbances. In general, this is a Levy type process defined by the stochastic equation

$$X(t) - x = \int_0^t a(X(s-))ds + \int_0^t b(X(s-))dW(s) + \int_0^t \int_{|q|\geq 1} K(X(s-), q)N(ds, dq) + \int_0^t \int_{|q|<1} F(X(s-), q)\widetilde{N}(ds, dq), \quad t \in [0; T]. \tag{2.1}$$

Here $W = \{W(t), t \geq 0\}$ is a standard Wiener process, $N(t, A)$ for any bounded from below set A is a Poisson random measure on $(\mathbb{R}_+ \times (\mathbb{R} \setminus \{0\}), \mathcal{B}(\mathbb{R}_+) \otimes \mathcal{B}(\mathbb{R} \setminus \{0\}))$, and $\widetilde{N}(t, A) := N(t, A) - t\nu(A)$ is a martingale-valued (compensated) Poisson random measure on this space. By the definition, the random variable $N(t, \cdot)(\omega)$ for $\omega \in \Omega$ and $t \geq 0$ is a counting measure on $\mathcal{B}(\mathbb{R} \setminus \{0\})$, and $N = \{N(t, A), t \geq 0\}$ is a Poisson process with intensity $\lambda = \nu(A) := \mathbf{E}[N(1, A)]$. We suppose the following conditions on coefficients supplying existence of a solution to (2.1): functions $a(\cdot)$, $b(\cdot)$, $F(\cdot, q)$ satisfy the Lipschitz and sub-linear growth conditions and $K(\cdot, q)$ is continuous.

Let $f \in C^{1,2}(\mathbb{R}_+, \mathbb{R})$ and $\{X(t), t \geq 0\}$ be a Levy type process defined by (2.1) with $\mathtt{a}(\cdot) = a(X(\cdot))$, $\mathtt{b}(\cdot)$, $\mathtt{F}(\cdot, q)$ square integrable a.s. Then with probability 1 the following equality, the Ito formula for Levy type processes, holds (see, e.g. [2, c. 278]):

$$
\begin{aligned}
f(t, X(t)) - f(0, X(0)) &= \int_0^t f_s'(s, X(s-))ds + \int_0^t \mathtt{a}(s) f_x'(s, X(s-))ds \\
&+ \int_0^t \mathtt{b}(s) f_x'(s, X(s-))dW(s) + \frac{1}{2}\int_0^t \mathtt{b}^2(s) f_{xx}''(s, X(s-))ds \\
&+ \int_0^t \int_{|q|\geq 1} [f(s, X(s-) + \mathtt{K}(s, q)) - f(s, X(s-))]N(ds, dq) \qquad (2.2) \\
&+ \int_0^t \int_{|q|<1} [f(s, X(s-) + \mathtt{F}(s, q)) - f(s, X(s-))]\widetilde{N}(ds, dq) \\
&+ \int_0^t \int_{|q|<1} [f(s, X(s-) + F(s, q)) - f(s, X(s-)) - F(s, q) f_x'(s, X(s-))]\nu(dq)ds.
\end{aligned}
$$

We start by deriving a direct (forward) equation for transition probability $P(\tau, y; t, A)$, the probability of transition from y at time τ to values on A at time t. To this end, using (2.2) first we obtain the equation for the process $\{f(X(t)), t \geq 0\}$, where X is a solution to (2.1) and $f \in C^2(\mathbb{R})$, then we apply the expectation. Using that integrals over W and over compensated Poisson process $\widetilde{N}$ are martingales, hence its expectations are zero, and changing the order of integration in other terms, we obtain

$$
\begin{aligned}
\mathbf{E}[f(X(t))] - f(x) &= \int_0^t \mathbf{E}[a(X(s-)) f'(X(s-)) + \frac{1}{2} b^2(X(s-)) f''(X(s-))]ds \\
&+ \int_0^t \int_{|q|\geq 1} \mathbf{E}[f(X(s-) + K(X(s-), q)) - f(X(s-))]\nu(dq)ds
\end{aligned}
$$

$$+\int_0^t \int_{|q|<1} \mathbf{E}[f(X(s-)+F(X(s-),q)) - f(X(s-))$$
$$-F(X(s-),q)f'(X(s-))]\nu(dq)ds.$$

Further, since Levy type processes possess the property $P(X(s-) = X(s)) = 1$ for any $s > 0$, we obtain the equality

$$\int_{\mathbb{R}} f(y)P(0,x;t,dy) - f(x) = \int_0^t \int_{\mathbb{R}} [a(z)f'(z) + \frac{1}{2}b^2(z)f''(z)]P(0,x;s,dz)ds$$

$$+\int_0^t \int_{|q|\geq 1} \int_{\mathbb{R}} [f(z+K(z,q)) - f(z)]P(0,x;s,dz)\nu(dq)ds$$

$$+\int_0^t \int_{|q|<1} \int_{\mathbb{R}} [f(z+F(z,q)) - f(z) - F(z,q)f'(z)]P(0,x;s,dz)\nu(dq)ds.$$

The right-hand side of the resulting equality are integrals with a variable upper limit, then after differentiating both sides of this equality with respect to parameter t we obtain the direct equation for transition probability:

$$\frac{\partial}{\partial t} \int_{\mathbb{R}} f(y)P(0,x;t,dy) = \int_{\mathbb{R}} [a(y)f'(y) + \frac{1}{2}b^2(y)f''(y)]P(0,x;t,dy)$$

$$+\int_{|q|\geq 1} \int_{\mathbb{R}} [f(y+K(y,q)) - f(y)]P(0,x;t,dy)\nu(dq) \tag{2.3}$$

$$+\int_{|q|<1} \int_{\mathbb{R}} [f(y+F(y,q)) - f(y) - F(y,q)f'(y)]P(0,x;t,dy)\nu(dq).$$

We show that (2.3) is correct in a space of distributions, where functions f will play the role of test functions for functionals determined by the transition probability. To do this, we consider $\Phi = C_c^2(\mathbb{R})$, the linear space of compactly supported twice continuously differentiable functions $f : \mathbb{R} \to \mathbb{R}$ and Φ', the space of linear continuous functionals on Φ. Since integral $\int_{\mathbb{R}} f(y)P(0,x;t,dy)$ exists for any $f \in C_c(\mathbb{R})$ functional $p(0,x;t,\cdot)$ is well defined as follows:

$$\int_{\mathbb{R}} f(y)P(0,x;t,dy) =: \langle f(y), p(0,x;t,y)\rangle, \quad f \in C_c(\mathbb{R}). \tag{2.4}$$

In particular for $f \in \Phi$, we call $p(0,x;t,\cdot)$ the generalized transition probability density of X. If the transition probability has a classical density, then $p(0,x;t,\cdot)$ is a regular generalized function and (2.4) turns into the equality for integrals.

Having defined the functional $p(0, x; t, \cdot)$ on Φ, we pass to the formalization of Eq. (2.3) in Φ' and begin with "differential" terms of the equation. Since there exists $\frac{\partial}{\partial t}\int_{\mathbb{R}} f(y)P(0, x; t, dy)$, for $p(0, x; t, \cdot)$ exists a derivative with respect to t:

$$\frac{\partial}{\partial t}\int_{\mathbb{R}} f(y)P(0, x; t, dy) = \frac{\partial}{\partial t}\langle f(y), p(0, x; t, y)\rangle = \langle f(y), \frac{\partial}{\partial t}p(0, x; t, y)\rangle.$$

Next, we consider integral $\int_{\mathbb{R}}[a(y)f'(y) + \frac{1}{2}b^2(y)f''(y)]P(0, x; t, dy)$. By virtue of conditions on coefficients of (2.1) supplying existence of its solution, functions a, b satisfy the Lipschitz condition. It follows that products af' and $b^2 f''$ define continuous functions with compact supports. Then the integral exists and is equal to

$$\langle f(y), -\frac{\partial}{\partial y}(a(y)p(0, x; t, y)) + \frac{1}{2}\frac{\partial^2}{\partial y^2}\left(b^2(y)p(0, x; t, y)\right)\rangle.$$

Next, we go on to formalize "integral" terms, integrals with respect to q included in the direct equation. Consider integral $\int_{|q|\geq 1}\int_{\mathbb{R}}[f(y + K(y, q)) - f(y)]P(0, x; t, dy)\nu(dq)$. Since, by conditions on coefficients, $K(\cdot, q)$, $|q| \geq 1$, is continuous, then $f(\cdot + K(\cdot, q)) \in C_c(\mathbb{R})$ and the integral is equal to

$$\langle f(y), \int_{|q|\geq 1}(p(0, x; t, y - K(y, q)) - p(0, x; t, y))\,\nu(dq)\rangle.$$

Finally, consider the last term in the right-hand side of (2.3). By virtue of the conditions imposed on F we obtain $f(\cdot + F(\cdot, q))$ and $F(\cdot, q)f'(\cdot)$ belong to $C_c(\mathbb{R})$. Then the term is equal to:

$$\langle f(y), \int_{|q|<1}\Big(p(0, x; t, y - F(y, q)) - p(0, x; t, y) + \frac{\partial}{\partial y}(F(y, q)p(0, x; t, y))\Big)\,\nu(dq)\rangle.$$

Thus, it is shown that if coefficient a, b, K, F of (2.1) satisfy conditions supplying existence of its solution, then the direct equation for the generalized transition probability density of X is correct on functions $f \in \Phi$:

$$\begin{aligned}\langle f(y), \frac{\partial}{\partial t}p(0, x; t, y)\rangle = \langle f(y), &-\frac{\partial}{\partial y}(a(y)p(0, x; t, y)) \\ &+\frac{1}{2}\frac{\partial^2}{\partial y^2}\left(b^2(y)p(0, x; t, y)\right)\rangle \\ + \langle f(y), \int_{|q|\geq 1}(p(0, x; t, y - K(y, q)) &- p(0, x; t, y))\,\nu(dq)\rangle \qquad (2.5)\end{aligned}$$

$$+\langle f(y), \int_{|q|<1} \Big(p(0,x;t,y-F(y,q)) - p(0,x;t,y)$$

$$+ \frac{\partial}{\partial y}(F(y,q)p(0,x;t,y)) \Big) \nu(dq) \rangle.$$

By (2.4), the correctness of the direct equation for the generalized density on Φ leads to the correctness of (2.3) for the transition probability.

Now, briefly, due to the size restriction of the paper, we show that for the important in applications probabilistic characteristic

$$g(t,x) := \mathbf{E}^{t,x}[h(X(T))] = \int_{\mathbb{R}} h(y)P(t,x;T,dy),\ t \in [0;T], \quad h \in C_b(\mathbb{R}) \tag{2.6}$$

the backward equation is correct on Φ under some additional conditions on coefficients of (2.1).

At the first stage, we assume the existence of continuous partial derivatives g'_t, g'_x, g''_{xx} and write the equation for $\{g(t,X(t)), t \in [0;T]\}$ using the Ito formula. By the Markov property of X, we have

$$\mathbf{E}[g(t,X(t))] = \mathbf{E}[\mathbf{E}^{t,X(t)}[h(X(T))]]$$

$$= \int_{\mathbb{R}}\int_{\mathbb{R}} h(y)P(t,x;T,dy)P(0,\xi;t,dx) = \int_{\mathbb{R}} h(y)P(0,\xi;T,dy) = \mathbf{E}[g(0,X(0))].$$

Therefore, the expectation of the right-hand side of the equation written for $\{g(t,X(t))\}$ on the basis of formula (2.2) is zero. Using the Fubini stochastic theorem we can move the expectation under the integral sign, then using the martingale property of W and $\widetilde{N}$ we obtain expectation of integrals over W and $\widetilde{N}$ are equal to zero.

If the evolution of X started at the moment $t \in [0;T]$ from the point $X(t) = x \in \mathbb{R}$, then the resulting equality leads to the backward equation for g:

$$-g'_t(t,x) = a(x)g'_x(t,x) + \frac{1}{2}b^2(x)g''_{xx}(t,x) + \int_{|q|\geq 1}\Big[g\Big(t, x+K(x,q)\Big) - g(t,x)\Big]\nu(dq)$$

$$+ \int_{|q|<1}\Big[g\Big(t,x+F(x,q)\Big) - g(t,x) - F(x,q)g'_x(t,x)\Big]\nu(dq), \quad t \in [0;T]. \tag{2.7}$$

Now we pass to the second stage, the study of the correctness of (2.7) in spaces of generalized functions. On the previous stage (2.7) was obtained under the assumption that there exist continuous partial derivatives g'_t, g'_x, g''_{xx}. The existence of derivatives g'_x, g''_{xx} and corresponding derivatives $p'_x(t,x;T,\cdot)$, $p''_{xx}(t,x;T,\cdot)$, as we can see from formulas (3.1)–(3.2) in the next section, is closely related to existence of derivatives of coefficients. In the general case, even provided that the coefficients of (2.7) are twice continuously differentiable, g_t', g_x', g_{xx}'' may not exist. More accurately: the following conditions guarantee the existence of the derivatives of g: coefficients $a(x), b(x)$, $F(x,q), |q| < 1$, $K(x,q), |q| \geq 1$ are twice continuously differentiable with respect to x and their derivatives satisfy Lipschitz and sub-linear growth conditions, function h is twice continuously differentiable and its derivatives are bounded [4, 5]. Thus, we formalize the backward equation for g with $h \in C_b(\mathbb{R})$ on the space of test functions Φ under the conditions on $a(\cdot), b(\cdot), F(\cdot, q), K(\cdot, q)$ to be twice continuously differentiable on $\mathbb{R}$. Indeed, for $f \in \Phi$ and such a, b, K, F the following equalities are correct:

$$\langle f(x),\, a(x)g_x'(t,x)\rangle = -\langle (a(x)f(x))',\, g(t,x)\rangle,$$

$$\langle f(x),\, b^2(x)g_{xx}''(t,x)\rangle = \langle (b^2(x)f(x))'',\, g(t,x)\rangle,$$

$$\langle f(x),\, F(x,q)g_x'(t,x)\rangle = -\langle (F(x,q)f(x))',\, g(t,x)\rangle,$$

$$\langle f(x),\, g(t,x+F(x,q))\rangle = \langle f(x-F(x,q)),\, g(t,x)\rangle,\ |q| < 1,$$

$$\langle f(x),\, g(t,x+K(x,q))\rangle = \langle f(x-K(x,q)),\, g(t,x)\rangle,\ |q| \geq 1,\ t \in [0;T].$$

Under the indicated conditions on coefficients, these equalities justify the correctness of (2.7) on $f \in \Phi$. It is not difficult to show that under these conditions a backward equation for the transition density can be obtained on Φ as well.

Remark It is important to note that although we justified the direct and backward equations in distribution spaces, the Cauchy problems for them: the problem with an initial condition for the direct equation and with final condition $g(T,x) = h(T)$ for the backward one, are well-posed from the point of view of the theory of ill-posed problems. This is the fundamental difference between the considered finite-dimensional problems and the infinite-dimensional ones, where ill-posedness can arise due to generators that do not generate semigroups of class C_0 [6].

3 The Approach via Limit Relations

This approach goes back to the ideas of A.N. Kolmogorov (see, e.g. [7]) for diffusion processes and is based on three limit values (3.1)–(3.3).

Let $p(t,x;T,y)$ be the transition probability density of a process X and let for any $\varepsilon > 0$ there exist finite limits

$$\lim_{\Delta t \to 0} \frac{1}{\Delta t} \int_{|z-x|<\varepsilon} (z-x)p(t,x;t+\Delta t,z)dz = a(t,x) + O(\varepsilon), \tag{3.1}$$

$$\lim_{\Delta t\to 0}\frac{1}{\Delta t}\int_{|z-x|<\varepsilon}(z-x)^2 p(t,x;t+\Delta t,z)dz = b(t,x)+O(\varepsilon), \quad (3.2)$$

$$\lim_{\Delta t\to 0}\frac{p(t,x;t+\Delta t,z)}{\Delta t} = G(t,x;z), \quad |z-x|>\varepsilon, \quad (3.3)$$

uniform with respect to x, z and t, and with respect to x and t in (3.2). Then for any $f \in C^2(\mathbb{R})$, transition density $p(t,x;T,y)$, $0 \le t \le T < \infty$, satisfies the direct equation [3, pp. 51, 56]:

$$\frac{\partial}{\partial T}\int_{\mathbb{R}} f(y)p(t,x;T,y)dy = \int_{\mathbb{R}}\left[a(t,y)\frac{\partial f(y)}{\partial y}p(t,x;T,y) + \frac{1}{2}b(t,y)\frac{\partial^2}{\partial y^2}p(t,x;T,y)\right]dy$$

$$+\int_{\mathbb{R}} f(y)dy\int_{\mathbb{R}\setminus 0} dz\,[G(T,z;y)p(t,z;T,y) - G(T,y;z)p(t,x;T,y)]$$

and, under the assumption that there exist p'_t, p'_x, p''_{xx}, the backward equation:

$$-p'_t(t,x;T,y) = a(t,x)p'_x(t,x;T,y) + \frac{1}{2}b(t,x)p''_{xx}(t,x;T,y)$$

$$+\int_{\mathbb{R}\setminus 0}(p(t,z;T,y) - p(t,x;T,y))G(t,x;z)dz.$$

Note that in the case of diffusion processes, the third limit is zero, in general case this limit describes the discontinuity of X.

As an example of using the approach, we obtain direct and backward equations for the transition density of $X = \{X(t) = at + bW(t) + cN(t)\}$, where $W = \{W(t),\ t \ge 0\}$ is the standard Wiener process, $N = \{N(t), t \ge 0\}$ is the Poisson process with intensity λ, and a, b, c are constants. We assume that W and N are set independently of each other. Since the density of W is determined by the equality

$$p_W(t,x;T,y) = \frac{1}{\sqrt{2\pi(T-t)}}e^{-\frac{(y-x)^2}{2(T-t)}},$$

density of process $\{at + bW(t)\}$ has the form:

$$p_{a,b,W}(t,x;T,y) = \frac{1}{b\sqrt{2\pi(T-t)}}e^{-\frac{(y-x-a(T-t))^2}{2b^2(T-t)}}.$$

Knowing the law of distribution of a Poisson process, we write out the density of process cN

$$p_{cN}(t,x;T,y) = \sum_{k=0}^{[\frac{y-x}{c}]} \frac{(\lambda(T-t))^k}{k!}\delta(y-x-ck)e^{-\lambda(T-t)}.$$

As a result, we obtain the transition density of X as the convolution of densities of three independent processes:

$$p(t,x;T,y) = \frac{e^{-\lambda(T-t)}}{b\sqrt{2\pi(T-t)}} \sum_{k=0}^{\infty} \frac{(\lambda(T-t))^k}{k!} e^{-\frac{(y-x-ck-a(T-t))^2}{2b^2(T-t)}}.$$

Now we calculate limits (3.1)–(3.2):

$$\lim_{\Delta t\to 0}\frac{1}{\Delta t}\int_{|z-x|<\varepsilon}(z-x)p(t,x;t+\Delta t,z)dz = \left[\begin{array}{ll} a, & \varepsilon \le c, \\ a+c\lambda, & \varepsilon > c \end{array}\right.$$

$$\lim_{\Delta t\to 0}\frac{1}{\Delta t}\int_{|z-x|<\varepsilon}(z-x)^2 p(t,x;t+\Delta t,z)dz = \left[\begin{array}{ll} b^2, & \varepsilon \le c \\ b^2+c^2\lambda, & \varepsilon > c \end{array}\right.$$

and function $G(t,x;z)$:

$$G(t,x;z) = \lim_{\Delta t\to 0}\frac{p(t,x;t+\Delta t,z)}{\Delta t} = \left[\begin{array}{ll} \lambda\delta(z-x-c), & \varepsilon \le c, \\ 0, & \varepsilon > c. \end{array}\right.$$

The found limit values allow us to write the direct and backward equations for the density of X on Φ:

$$\begin{aligned}\langle f(y), p'_T(t,x;T,y)\rangle = \langle f(y), -ap'_y(t,x;T,y) + \frac{b^2}{2}p''_{yy}(t,x;T,y) \\ +\lambda(p(t,x;T,y-c) - p(t,x;T,y))\rangle.\end{aligned}$$

$$\begin{aligned}\langle f(x), -p'_t(t,x;T,y)\rangle = \langle f(x), ap'_x(t,x;T,y) + \frac{b^2}{2}p''_{xx}(t,x;T,y) \\ +\lambda(p(t,x+c;T,y) - p(t,x;T,y))\rangle.\end{aligned}$$

In the considered example, since a and b are constants, we need not additional conditions on coefficients in the backward equation.

Acknowledgment This work was supported by the Russian Academic Excellence Project (agreement no. 02.A03.21.0006 of August 27, 2013).

References

1. D. Applebaum, *Lévy Processes and Stochastic Calculus* (Cambridge University Press, Cambridge, 2004)
2. P.E. Protter, *Stochastic Integration and Differential Equations* (Springer, Berlin, 2005)
3. C. Gardiner, *Stochastic Methods: A Handbook for the Natural and Social Sciences* (Springer, Berlin, 2009)
4. H. Kunita, *Stochastic Flows and Jump-Diffusions* (Springer, Berlin, 2019)
5. H. Kunita, Ito's stochastic calculus: its surprising power for applications. Stoch. Process. Appl. **120**(5), 622–652 (2010)
6. I.V. Melnikova, *Stochastic Cauchy Problems in Infinite Dimensions: Regularized and Generalized Solutions* (CRC Press, Boca Raton, 2016)
7. A.N. Kolmogorov, On analytical methods in the probability theory. Usp. Mat. Nauk. **5**, 5–41 (1938)

Part VIII
Geometric and Regularity Properties of Solutions to Elliptic and Parabolic PDEs

Asymptotic Regularity for a Random Walk over Ellipsoids

Ángel Arroyo

Keywords Dynamic programming principle · Local Hölder estimates · Stochastic games · Coupling of stochastic processes · Ellipsoid process · Equations in nondivergence form

Mathematics Subject Classification (2010) 35B65, 35J15, 60H30, 60J10, 91A50

1 Uniformly Elliptic PDEs in Non-divergence Form

Let $\Omega \subset \mathbb{R}^n$ be a domain with $n \geqslant 2$. Given $0 < \lambda \leqslant \Lambda < \infty$, we consider the partial differential equation in non-divergence form

$$Lu(x) := \mathrm{Tr}\{A(x) \cdot D^2u(x)\}, \tag{1.1}$$

where the coefficients $A(\cdot) = (a_{ij})_{ij}$ are symmetric and uniformly elliptic, that is $\lambda|\xi|^2 \leqslant \xi^\top A(x)\xi \leqslant \Lambda|\xi|^2$ for every $\xi \in \mathbb{R}^n$ and $x \in \Omega$. In 1979, Krylov and Safonov proved in [10] the Harnack inequality for solutions of $Lu = f$ with measurable coefficients, where the function f is continuous and bounded. As a consequence of this, the Hölder regularity for solutions of $Lu = 0$ followed.

One of the most important examples of this equation is the *Laplacian equation* $\Delta u = 0$, which arises when $A(x) \equiv I$. Solutions of the Laplacian equation received the name of *harmonic* and have been widely studied. One of the most important properties of harmonic functions is the so-called *asymptotic mean value property*.

Á. Arroyo (✉)
Departamento de Análisis Matemático y Matemática Aplicada, Universidad Complutense de Madrid, Madrid, Spain
e-mail: ar.arroyo@ucm.es

P. Cerejeiras et al. (eds.), *Current Trends in Analysis, its Applications and Computation*, Research Perspectives,
https://doi.org/10.1007/978-3-030-87502-2_41

Namely, a function $u \in C(\Omega)$ is harmonic in Ω if and only if the asymptotic expansion

$$\fint_{B_\varepsilon(x)} u(\zeta)\, d\zeta = u(x) + o(\varepsilon^2) \qquad (\varepsilon \to 0) \tag{1.2}$$

holds for every $x \in \Omega$. There exists a free-error version of this property, known as the *restricted mean value property* or the *one radius theorem* [9, 17]. It states that a function $u \in C(\overline{\Omega})$ is harmonic in Ω if and only if

$$\fint_{B_{r(x)}(x)} u(\zeta)\, d\zeta = u(x)$$

for each $x \in \Omega$, where $0 < r(x) < \mathrm{dist}(x, \partial\Omega)$. On the other hand, harmonic functions have also appeared in the context of stochastic processes such as the Brownian motion or the random walk.

2 An Asymptotic Mean Value Property

Returning to the general case of a PDE of the form (1.1), it seems a natural question to ask what sort of mean value properties does a solution of $Lu = 0$ satisfy. To answer this question, let us first define $\mathcal{A}(\lambda, \Lambda)$ as the set of all symmetric $n \times n$ real matrices A such that $\lambda|\xi|^2 \leqslant \xi^\top A \xi \leqslant \Lambda|\xi|^2$ for every $\xi \in \mathbb{R}^n$. Since every matrix $A \in \mathcal{A}(\lambda, \Lambda)$ is real, symmetric and positive definite, the *principal square root of* A, which we denote by $\sqrt{A} \in \mathcal{A}(\sqrt{\lambda}, \sqrt{\Lambda})$, is well defined. On the other hand, each matrix $A \in \mathcal{A}(\lambda, \Lambda)$ determines the shape and the orientation of an ellipsoid $E_A \subset \mathbb{R}^n$ centered at the origin and given by the formula

$$E_A := \sqrt{A}\,\mathbb{B} = \{\sqrt{A}\, y \in \mathbb{R}^n \ : \ |y| < 1\},$$

where $\mathbb{B} = B_1(0)$ is the unit ball of $\mathbb{R}^n$.

In what follows, we consider $A : \Omega \to \mathcal{A}(\lambda, \Lambda)$ a matrix-valued function with measurable coefficients and values in $\mathcal{A}(\lambda, \Lambda)$ and we define the family of ellipsoids parametrized by $x \in \Omega$ and given by

$$E_x := E_{A(x)} = \sqrt{A(x)}\,\mathbb{B}. \tag{2.1}$$

In addition, we assume that $A(x)$ has constant determinant, so the ellipsoids E_x have all the same measure. Furthermore, observe that the uniform ellipticity of $A(x)$ implies that the inclusions $B_{\sqrt{\lambda}} \subset E_x \subset B_{\sqrt{\Lambda}}$ hold for every $x \in \Omega$, and thus the *maximum distortion* of E_x is bounded from above by $\sqrt{\Lambda/\lambda}$.

In the next result, we use this link between uniformly elliptic matrices and ellipsoids, (2.1), to provide an asymptotic mean value property for *classical* solutions of $Lu = 0$.

Proposition 2.1 *Let $\phi \in C^2(\Omega)$. Then $L\phi = 0$ in Ω if and only if*

$$\fint_{x+\varepsilon E_x} \phi(\zeta)\, d\zeta = \phi(x) + o(\varepsilon^2) \qquad (\varepsilon \to 0) \tag{2.2}$$

for each $x \in \Omega$.

Proof Let $h \in E_x$. Recalling the second order Taylor's expansion of a twice differentiable function ϕ at $x \in \Omega$ we have

$$\phi(x+\varepsilon h) = \phi(x) + \varepsilon \nabla\phi(x)^\top h + \frac{\varepsilon^2}{2} \operatorname{Tr}\{D^2\phi(x) \cdot hh^\top\} + o(\varepsilon^2) \qquad (\varepsilon \to 0).$$

By the symmetry of the ellipsoid, averaging this expansion over E_x the first order term vanishes,

$$\fint_{E_x} \nabla\phi(x)^\top h\, dh = \nabla\phi(x)^\top \left(\fint_{E_x} h\, dh \right) = 0,$$

while for the second order term, by the linearity of the trace and (2.1), we obtain

$$\begin{aligned} \fint_{E_x} \operatorname{Tr}\{D^2\phi(x) \cdot hh^\top\}\, dh &= \operatorname{Tr}\left\{ D^2\phi(x) \cdot \fint_{E_x} hh^\top\, dh \right\} \\ &= \operatorname{Tr}\left\{ D^2\phi(x) \cdot \sqrt{A(x)} \left(\fint_{\mathbb{B}} ww^\top\, dw \right) \sqrt{A(x)} \right\}. \\ &= \frac{1}{n+2} \operatorname{Tr}\{D^2\phi(x) \cdot A(x)\}, \end{aligned}$$

where the change of variables $h = \sqrt{A(x)}\, w$ has been performed in the second line. Thus, recalling the definition of $L\phi$ in (1.1) and performing the change of variables $\zeta = x + \varepsilon h$ we get

$$\fint_{x+\varepsilon E_x} \phi(\zeta)\, d\zeta = \phi(x) + \frac{\varepsilon^2}{2(n+2)} L\phi(x) + o(\varepsilon^2) \qquad (\varepsilon \to 0),$$

so (2.2) follows after replacing $L\phi(x) = 0$. □

It is worth to remark that the previous proposition is the generalization of the asymptotic mean value characterization of harmonic functions (1.2). Analogously to the harmonic case, we could consider free-error versions of (2.2),

$$\fint_{x+\varepsilon E_x} u(\zeta)\, d\zeta = u(x) \tag{2.3}$$

for each $\varepsilon > 0$. Unfortunately, the connection between solutions of $Lu = 0$ and (2.3) is not as clear as in the harmonic case. However, there exists a relation with stochastic process which is analogous to the link between harmonic functions and random walks.

3 The Ellipsoid Process

In this section we briefly describe the connection between the mean value property (2.3) and a stochastic process related to the random walk. For a more detailed description of the process we refer the reader to [1].

Let $\Omega \subset \mathbb{R}^n$ be a bounded domain. The *ellipsoid process* is a stochastic process describing the location of a particle jumping from one position to another inside a space-dependent ellipsoid contained in Ω. This stochastic process is a generalization of the random walk arising when the ellipsoids are replaced by balls of fixed radius. Let $\{x_0, x_1, x_2, \ldots\} \subset \Omega$ denote the history of the particle, i.e. the sequence describing the successive positions of the particle at each time $j = 0, 1, 2, \ldots$ Given any particle position x_j, the next location of the particle is randomly chosen inside the ellipsoid $x_j + \varepsilon E_{x_j}$ (according to a uniform probability distribution on E_{x_j}). This process stops the first time the particle exits Ω at some $x_\tau \notin \Omega$ and the amount $F(x_\tau)$ is collected, where F is a continuous pay-off function defined outside Ω. More precisely, $F \in C(\Omega_{\sqrt{\Lambda}\,\varepsilon} \setminus \Omega)$, where $\Omega_{\sqrt{\Lambda}\,\varepsilon}$ stands for the $\sqrt{\Lambda}\,\varepsilon$-extension of Ω,

$$\Omega_{\sqrt{\Lambda}\,\varepsilon} := \{x \in \mathbb{R}^n \,:\, \operatorname{dist}(x, \Omega) < \sqrt{\Lambda}\,\varepsilon\}.$$

Let us denote by $u_\varepsilon(x_0)$ the expected pay-off of an ellipsoid process starting from $x_0 \in \Omega$, that is

$$u_\varepsilon(x_0) := \mathbb{E}[F(x_\tau) \,|\, x_0].$$

Then $u_\varepsilon : \Omega_{\sqrt{\Lambda}\,\varepsilon} \to \mathbb{R}$. Moreover, since the ellipsoid process is a Markov chain, computing conditional probabilities using the Markov property we obtain that u_ε satisfies the *dynamic programming principle*

$$u_\varepsilon(x) = \begin{cases} \displaystyle\fint_{x+\varepsilon E_x} u_\varepsilon(\zeta)\, d\zeta & \text{if } x \in \Omega, \\ F(x) & \text{if } x \in \Omega_{\sqrt{\Lambda}\,\varepsilon} \setminus \Omega, \end{cases} \tag{3.1}$$

which inside Ω coincides with (2.3).

4 Asymptotic Regularity: the Coupling Method

We say that a function u_ε is *asymptotically Hölder continuous* if

$$|u_\varepsilon(x) - u_\varepsilon(y)| \leqslant C(|x-y|^\gamma + \varepsilon^\gamma),$$

where the constants $C > 0$ and $\gamma \in (0, 1]$ are independent of ε. In the next theorem we show that, under certain assumptions, the solutions u_ε of the dynamic programming principle (3.1) are asymptotically Hölder continuous.

Theorem 4.1 *Let $\Omega \subset \mathbb{R}^n$ be a domain, $n \geqslant 2$ and $0 < \lambda \leqslant \Lambda < \infty$ such that*

$$1 \leqslant \frac{\Lambda}{\lambda} < \frac{n+1}{n-1}.$$

Suppose that $A : \Omega \to \mathcal{A}(\lambda, \Lambda)$ is a measurable mapping with constant determinant and $B_{2r} \subset \Omega$ for some $r > 0$. If u_ε satisfies

$$u_\varepsilon(x) = \fint_{x+\varepsilon E_x} u_\varepsilon(\zeta)\, d\zeta,$$

then there exists $\gamma = \gamma(n, \lambda, \Lambda) \in (0, 1)$ such that

$$|u_\varepsilon(x) - u_\varepsilon(y)| \leqslant C(|x-y|^\gamma + \varepsilon^\gamma) \tag{4.1}$$

holds for every $x, y \in B_r$ and some constant $C > 0$ depending on n, λ, Λ, r, γ and $\sup_{B_{2r}} |u|$, but independent of ε.

The proof is based on the coupling method for asymptotic regularity developed for different variants of tug-of-war games with noise by Luiro and Parviainen in [11] and extended in several papers, [2, 3, 7, 14]. In turn, this method is inspired by the Ishii-Lions method for viscosity solutions of fully nonlinear elliptic PDEs, [8]. Moreover, for continuous time diffusion processes and the Laplacian, the coupling method for regularity was also used by Cranston in [6] using the techniques developed in [12].

In what follows we describe the general idea of the proof avoiding technical details. The interested readers can find the complete and detailed proof in [1].

Proof Let u_ε be a function satisfying (2.3) and $x, y \in \Omega$ such that $x \neq y$. Since $E_x = \sqrt{A(x)}\,\mathbb{B}$ for every $x \in \Omega$, performing convenient changes of variables we get

$$\begin{aligned} u_\varepsilon(x) - u_\varepsilon(y) &= \fint_{x+\varepsilon E_x} u_\varepsilon(\zeta)\, d\zeta - \fint_{y+\varepsilon E_y} u_\varepsilon(\zeta)\, d\zeta \\ &= \fint_{B_\varepsilon} \left[u_\varepsilon(x + \sqrt{A(x)}\,\zeta) - u_\varepsilon(y + \sqrt{A(y)}\, Q\zeta)\right] d\zeta, \end{aligned}$$

for any orthogonal matrix $Q \in O(n)$. The key idea in the proof of the asymptotic regularity estimate (4.1) is to define a function $\mathcal{U}_\varepsilon : \Omega \times \Omega \to \mathbb{R}$ by $\mathcal{U}_\varepsilon(x, y) = u_\varepsilon(x) - u_\varepsilon(y)$ so that $\mathcal{U}_\varepsilon$ satisfies the following $2n$-dimensional mean value property,

$$\mathcal{U}_\varepsilon(x, y) = \fint_{B_\varepsilon} \mathcal{U}_\varepsilon(x + \sqrt{A(x)}\,\zeta, y + \sqrt{A(y)}\,Q\zeta)\,d\zeta,$$

for every $Q \in O(n)$. In this way, the problem of the regularity of u_ε becomes a question about the absolute size of $\mathcal{U}_\varepsilon(x, y)$. Hence, in order to show that

$$|\mathcal{U}_\varepsilon(x, y)| \leqslant C(|x - y|^\gamma + \varepsilon^\gamma), \tag{4.2}$$

we need to show that the inequality

$$|x - y|^\gamma > \fint_{B_\varepsilon} \left|x - y - (\sqrt{A(x)} - \sqrt{A(y)}\,Q_0)\zeta\right|^\gamma d\zeta \tag{4.3}$$

holds for every $x, y \in \Omega$ with an appropriate choice of $Q_0 \in O(n)$. Let us define the auxiliary function $\varphi : B_\varepsilon \to \mathbb{R}$ given by

$$\varphi(\zeta) = \left|x - y - (\sqrt{A(x)} - \sqrt{A(y)}\,Q_0)\zeta\right|^\gamma.$$

Since $x \neq y$, we can fix a sufficiently small $\varepsilon > 0$ such that $\varphi \in C^2(B_\varepsilon)$. By the mean value property for superharmonic functions, it turns out that if φ is superharmonic in B_ε then the inequality (4.3) holds. Thus we just need to show that $\Delta\varphi(\zeta) \leqslant 0$ for every $\zeta \in B_\varepsilon$. Indeed, computing explicitly the Laplacian of φ we can show that $\Delta\varphi(\zeta) \leqslant 0$ if and only if the inequality

$$\operatorname{Tr}\left\{\begin{pmatrix} -(1-\gamma) & 0 \\ 0 & I_{n-1} \end{pmatrix} \left(A_1 + A_2 - 2\sqrt{A_2}\,Q_0\sqrt{A_1}\right)\right\} \leqslant 0$$

holds for every $A_1, A_2 \in \mathcal{A}(\lambda, \Lambda)$. Therefore, we need to check that there exists a matrix $Q_0 \in O(n)$ for which this trace is non-positive. For this purpose, we use the inequality

$$\max_{Q \in O(n)} \operatorname{Tr}\{MQ\} = \operatorname{Tr}\{\sqrt{MM^\top}\} \geqslant n\,|\det\{M\}|^{1/n},$$

where M is any $n \times n$ real matrix, and we choose Q_0 to be the corresponding maximizer, so we get that

$$\operatorname{Tr}\left\{\begin{pmatrix} -(1-\gamma) & 0 \\ 0 & I_{n-1} \end{pmatrix} \left(A_1 + A_2 - 2\sqrt{A_2}\,Q_0\sqrt{A_1}\right)\right\}$$
$$\leqslant 2\left[(n-1)\Lambda - (1-\gamma)\lambda - n(1-\gamma)^{1/n}\lambda\right]$$

for every $A_1, A_2 \in \mathcal{A}(\lambda, \Lambda)$. Hence, φ is superharmonic in B_ε if and only if

$$\frac{\Lambda}{\lambda} < \frac{n(1-\gamma)^{1/n} + (1-\gamma)}{n-1}.$$

Furthermore, if

$$\frac{\Lambda}{\lambda} < \frac{n+1}{n-1},$$

then we can choose small enough $\gamma > 0$ such that the condition is satisfied and thus (4.3) holds.

Next, in order to prove (4.2), we choose large enough $C > 0$ such that $\mathcal{U}_\varepsilon(x, y) \leqslant C(|x-y|^\gamma + \varepsilon^\gamma)$ for every $x, y \in B_{2r} \setminus B_r$. Our aim is to proceed by contradiction to show that the same inequality also holds for every $x, y \in B_{2r}$. Therefore, let us assume that the inequality does not hold in B_{2r}. Then we define a positive number

$$K := \sup_{x,y \in B_{2r}} \{\mathcal{U}_\varepsilon(x, y) - C|x-y|^\gamma\} > C\varepsilon^\gamma.$$

In particular, this implies that $\mathcal{U}_\varepsilon(x, y) \leqslant K + C|x-y|^\gamma$ for every $x, y \in B_{2r}$. Let $\widetilde{x}, \widetilde{y} \in B_{2r}$ the points where the supremum is attained. Then

$$\begin{aligned} K + C|\widetilde{x} - \widetilde{y}|^\gamma = \mathcal{U}_\varepsilon(\widetilde{x}, \widetilde{y}) &= \fint_{B_\varepsilon} \mathcal{U}_\varepsilon(\widetilde{x} + \sqrt{A(\widetilde{x})}\,\zeta, \widetilde{y} + \sqrt{A(\widetilde{y})}\,Q\zeta)\, d\zeta \\ &\leqslant K + C \fint_{B_\varepsilon} |\widetilde{x} - \widetilde{y} - (\sqrt{A(\widetilde{x})} - \sqrt{A(\widetilde{y})}\,Q)\zeta|^\gamma\, d\zeta. \end{aligned}$$

Hence

$$|\widetilde{x} - \widetilde{y}|^\gamma \leqslant \fint_{B_\varepsilon} |\widetilde{x} - \widetilde{y} - (\sqrt{A(\widetilde{x})} - \sqrt{A(\widetilde{y})}\,Q)\zeta|^\gamma\, d\zeta,$$

which contradicts (4.3) when $Q = Q_0$. Then (4.2) follows and in consequence (4.1) holds. □

5 Limits as $\varepsilon \to 0$

Let $\{u_\varepsilon : \varepsilon > 0\}$ be a collection of functions satisfying (3.1). Recalling Theorem 4.1, since u_ε satisfies the asymptotic Hölder estimate (4.1) for each $\varepsilon > 0$, we can define (passing to a subsequence if necessary) the limit function

$$u_0 = \lim_{\varepsilon \to 0} u_\varepsilon,$$

which turns out to be a viscosity solution to $Lu = 0$ in Ω, i.e.

$$\mathrm{Tr}\{D^2 u_0(x) \cdot A(x)\} = 0$$

holds in the viscosity sense. Therefore, this method provides Hölder estimates for those viscosity solutions of $Lu = 0$ that can be constructed as the limit as $\varepsilon \to 0$ of solutions of (3.1). However, Nadirashvili showed in [13] that there is not necessarily a unique solution of $Lu = 0$, and thus this lack of uniqueness does not allow to extend this estimate for every solution of the PDE. Nevertheless, uniqueness of solutions of $Lu = 0$ holds if one of the following cases is satisfied:

- The dimension is $n = 2$ [4].
- The coefficients $A(\cdot) = (a_{ij})_{ij}$ are continuous in Ω [15].
- The distortion Λ/λ is close enough to 1 (depending on the dimension) [5, 16].

Acknowledgment Many thanks to our TEX-pert for developing this class file.

References

1. Á. Arroyo, M. Parviainen, Asymptotic Hölder regularity for the ellipsoid process. ESAIM Control Optim. Calc. Var. **26**(112), 31 (2020)
2. Á. Arroyo, J. Heino, M. Parviainen, Tug-of-war games with varying probabilities and the normalized $p(x)$-Laplacian. Commun. Pure Appl. Anal. **16**(3), 915–944 (2017)
3. Á. Arroyo, H. Luiro, M. Parviainen, E. Ruosteenoja, Asymptotic Lipschitz regularity for tug-of-war games with varying probabilities. Potential Anal. 53, 565–589 (2020)
4. L. Bers, F. John, M. Schechter, *Partial Differential Equations* (Interscience, New York, 1964). Reprint of the 1964 original. *Lectures in Applied Mathematics* 3A (1979)
5. H.O. Cordes, Zero order a priori estimates for solutions of elliptic differential equations. Proc. Symp. Pure Math. **4**, 157–166 (1961)
6. M. Cranston, Gradient estimates on manifolds using coupling. J. Funct. Anal. **99**(1), 110–124 (1991)
7. J. Han, Local Lipschitz regularity for functions satisfying a time-dependent dynamic programming principle. Commun. Pure Appl. Anal. **19**(5), 2617–2640 (2020)
8. H. Ishii, P.L. Lions, Viscosity solutions of fully nonlinear second-order elliptic partial differential equations. J. Differ. Equ. **83**(1), 26–78 (1990)
9. O.D. Kellogg, Converses of Gauss's theorem on the arithmetic mean. Trans. Am. Math. Soc. **36**(2), 227–242 (1934)
10. N.V. Krylov, M.V. Safonov, An estimate for the probability of a diffusion process hitting a set of positive measure. Dokl. Akad. Nauk SSSR **245**(1), 18–20 (1979)
11. H. Luiro, M. Parviainen, Regularity for nonlinear stochastic games. Ann. Inst. H. Poincaré Anal. Non Linéaire **35**(6), 1435–1456 (2018)
12. T. Lindvall, L.C.G. Rogers, Coupling of multidimensional diffusions by reflection. Ann. Probab. **14**(3), 860–872 (1986)
13. N. Nadirashvili, Nonuniqueness in the martingale problem and the Dirichlet problem for uniformly elliptic operators. Ann. Scuola Norm. Sup. Pisa Cl. Sci. (4) **24**(3), 537–549 (1997)
14. M. Parviainen, E. Ruosteenoja, Local regularity for time-dependent tug-of-war games with varying probabilities. J. Differ. Equ. **261**(2), 1357–1398 (2016)

15. D.W. Stroock, S.R.S. Varadhan, *Multidimensional Diffusion Processes*. Reprint of the 1997 edition. *Classics in Mathematics* (Springer, Berlin, 2006)
16. G. Talenti, Sopra una classe di equazioni ellittiche a coefficienti misurabili. Ann. Mat. Pura Appl. **69**(4), 285–304 (1965)
17. V. Volterra, Alcune osservazioni sopra proprietà atte ad individuare una funzione. Rend. Acadd. d. Lincei Roma. **18**(5), 263–266 (1909)

Short-Time Asymptotics for Game-Theoretic p-Laplacian and Pucci Operators

Diego Berti

Abstract Let Ω be a domain of $\mathbb{R}^N$, $N \geq 2$, with non empty boundary Γ. In these notes, we deal with the solution u of $u_t = F\left(\nabla u, \nabla^2 u\right)$ in $\Omega \times (0, \infty)$, such that u is initially zero in Ω and equals one on Γ for all positive times. Here, F is the *game-theoretic p-Laplacian* Δ_p^G or either one of the *Pucci's extremal operators* $\mathcal{M}^{\pm}$. In the spirit of works by Varadhan and Magnanini-Sakaguchi in the case of the same initial-boundary problem for the *heat equation*, we summarize recent results regarding the connection between the behavior for small times and the geometry of Ω. In particular, we present asymptotic formulas as $t \to 0^+$ for both the values of u and of its q-means on balls touching Γ.

Keywords Game-theoretic p-Laplacian · Pucci operators · Short-time asymptotic analysis · Varadhan formulas · q-Means on balls

Mathematics Subject Classification (2010) Primary 35K55; Secondary 35K20

1 Introduction

We consider the solution $u = u(x, t)$ to the following initial-boundary value problem:

$$u_t - F\left(\nabla u, \nabla^2 u\right) = 0 \quad \text{in } \Omega \times (0, \infty), \tag{1.1}$$

$$u = 1 \quad \text{on } \Gamma \times (0, \infty), \tag{1.2}$$

$$u = 0 \quad \text{on } \Omega \times \{0\}. \tag{1.3}$$

D. Berti (✉)
Department of Engineering Sciences and Methods, Università degli Studi di Modena e Reggio Emilia, Modena, Italy
e-mail: diego.berti@unimore.it

P. Cerejeiras et al. (eds.), *Current Trends in Analysis, its Applications and Computation*, Research Perspectives,
https://doi.org/10.1007/978-3-030-87502-2_42

For a given $N \geq 2$, $\Omega \subset \mathbb{R}^N$ is a domain (possibly unbounded) and Γ is its non empty boundary. In what follows, F shall be the *game-theoretic p-Laplacian* Δ_p^G or either one of the *Pucci's extremal operators* $\mathcal{M}^{\pm}$. We remark that (1.1) shall be satisfied according to the *theory of viscosity solutions* (for which we refer to e.g. [8]) while conditions (1.2)–(1.3) are meant in the classical sense, that is u is supposed to be continuous on $\overline{\Omega} \times [0, \infty)$ away from $\Gamma \times \{0\}$. We also specify that in the case that Ω is unbounded we consider only the bounded solution of (1.1)–(1.3).

With the purpose of understanding how the geometry of Ω affects the short-time diffusion (1.1)–(1.3), we first describe well-known results obtained in the linear case, when $F = \Delta$ is the classical Laplace operator. Varadhan, in its seminal work [16], under the assumption that $\Gamma = \partial\left(\mathbb{R}^N \setminus \overline{\Omega}\right)$, established that

$$\lim_{t \to 0^+} 4t \log\{u(x,t)\} = -d_\Gamma(x)^2 \text{ for any } x \in \overline{\Omega}, \tag{1.4}$$

where with $d_\Gamma(x)$ we mean here the Euclidean distance of a point $x \in \overline{\Omega}$ to Γ, defined as the minimum of $|x - y|$ among all $y \in \Gamma$.

Starting from (1.4), Magnanini and Sakaguchi investigated a more intimate link between the short-time diffusion and the domain, in a series of papers. Here, we only recall some of them: [9–12]. Assume that Ω is of class C^2. Let $x \in \Omega$ be such that there exists an unique $y_x \in \Gamma$ satisfying $|x - y_x| = d_\Gamma(x) = R$. A direct consequence of [10, Theorem 4.2] is the following formula for the *mean value* of $u(\cdot, t)$ on $B_R(x)$:

$$\lim_{t \to 0^+} \left(\frac{R^2}{t}\right)^{\frac{N+1}{4}} \fint_{B_R(x)} u(z,t)\,dz = c_N \left\{\prod_{j=1}^{N-1}\left[1 - R\kappa_j(y_x)\right]\right\}^{-\frac{1}{2}}, \tag{1.5}$$

where $c_N = \frac{2^{\frac{N+1}{2}}}{\sqrt{\pi}} \frac{N}{N+1} \frac{\Gamma(N/2)}{\Gamma(N/4+1/4)}$ and $\kappa_1(\cdot), \ldots, \kappa_{N-1}(\cdot)$ are the principal curvatures of Γ.

In this paper, we shall present extensions of (1.4) and (1.5) to two important one-homogeneous nonlinear operators. The operator Δ_p^G is formally defined by

$$\Delta_p^G u = \frac{1}{p}|\nabla u|^{2-p} \mathrm{div}\left\{|\nabla u|^{p-2}\nabla u\right\},$$

or by

$$\Delta_p^G u = \frac{1}{p}\left\{\Delta u + (p-2)\frac{\langle \nabla^2 u \nabla u, \nabla u\rangle}{|\nabla u|^2}\right\}.$$

We can suppose that $p \in (1, \infty]$, if we specify that for Δ_∞^G we mean the limit of the formal expression as $p \to \infty$. Notice that when $p = 2$, $\Delta_2^G = \Delta/2$. Away from the case $p = 2$, Δ_p^G is not linear and not in divergence form. Moreover, it has (possible) discontinuous coefficients when $\nabla u = 0$. Nevertheless, Δ_p^G is one-homogeneous for any $p \in (1, \infty]$, uniformly elliptic if $p \in (1, \infty)$ and (degenerate) elliptic when $p = \infty$.

As its name suggests, Δ_p^G arises in the context of *game theory*, when one considers the limiting value for vanishing length of steps of certain two-players games (see [13, 14]). The interest on Δ_p^G and on its applications has rapidly grown in recent years. For an overview, we refer to the Ph.D. thesis of the author of these notes [3] and references therein.

For real numbers $0 < \lambda \le \Lambda$, we define $\mathcal{M}^{\pm}$ by

$$\mathcal{M}^-\left(\nabla^2 u\right) = \Lambda \sum_{\lambda_i<0} \lambda_i + \lambda \sum_{\lambda_i>0} \lambda_i \text{ and } \mathcal{M}^+\left(\nabla^2 u\right) = \lambda \sum_{\lambda_i<0} \lambda_i + \Lambda \sum_{\lambda_i>0} \lambda_i,$$

where $\lambda_i = \lambda_i\left(\nabla^2 u\right)$ are the eigenvalues of the Hessian matrix $\nabla^2 u$. Note that, if $\lambda = \Lambda$, then $\mathcal{M}^+ = \mathcal{M}^- = \lambda\Delta$. In the general case ($0 < \lambda \le \Lambda$), $\mathcal{M}^{\pm}$ is still uniformly elliptic while, for $\lambda < \Lambda$, it is fully-nonlinear and obviously not in divergence form. We stress the fact that $\mathcal{M}^{\pm}$, as well as Δ_p^G, is (positively) one-homogeneous and rotation invariant.

Pucci's operators $\mathcal{M}^{\pm}$ (introduced in [15]) arise in the study of stochastic control when the diffusion coefficient is a control variable (see e.g. [2]). Furthermore, they provide natural extremal equations for fully nonlinear operators (see [7, 8]), since they also have the following representation:

$$\mathcal{M}^-\left(\nabla^2 u\right) = \inf_{A\in\mathcal{A}_{\lambda,\Lambda}} \operatorname{tr}\left(A\nabla^2 u\right), \quad \mathcal{M}^+\left(\nabla^2 u\right) = \sup_{A\in\mathcal{A}_{\lambda,\Lambda}} \operatorname{tr}\left(A\nabla^2 u\right)$$

where $\mathcal{A}_{\lambda,\Lambda}$ is the set of $N \times N$ symmetric matrices whose eigenvalues belong to the interval $[\lambda, \Lambda]$.

In Theorems 2.1 and 3.1 of these notes, we present the main results in extending (1.4) and (1.5) when F is either Δ_p^G or $\mathcal{M}^{\pm}$. In Theorem 2.1, as generalization of (1.4), we obtain that

$$\lim_{t\to 0^+} 4t \log\{u(x,t)\} = -\alpha d_\Gamma(x)^2 \text{ for any } x \in \overline{\Omega}.$$

Here, we have introduced the parameter α that depends on which operator is taken into consideration. If we denote with p' the usual conjugate exponent of p, that is $p' = p/(p-1)$ when $p \in (1, \infty)$ and $p' = 1$ when $p = \infty$, we set:

$$\alpha = \begin{cases} p' \text{ when } F = \Delta_p^G \\ \frac{1}{\Lambda} \text{ when } F = \mathcal{M}^+ \\ \frac{1}{\lambda} \text{ when } F = \mathcal{M}^-. \end{cases} \tag{1.6}$$

To obtain the formula, we benefit from the fact that both Δ_p^G and $\mathcal{M}^{\pm}$ take a rather simple form if evaluated on radially symmetric functions. In particular, it is worth emphasizing that Δ_p^G acts linearly on that class. This feature makes possible to employ accurate barriers based on radial solutions of (1.1)–(1.3) obtained when Ω is either the ball or the complement of a ball.

In Theorem 2.2, we present sharp uniform estimates on the rate of convergence of the given Varadhan-type formula, which are instrumental to what comes next. Nevertheless, we point out that Theorem 2.2, to the best of our knowledge, represents a novelty also in the linear case.

Finally, we are able to generalize and extend (1.5) as follows. Let $x \in \Omega$ and $t > 0$. Set $1 \leq q \leq \infty$. The q-mean of $u(\cdot, t)$ on $B_R(x) \subset \Omega$ is the unique real number μ such that

$$\|u(\cdot, t) - \mu\|_{L^q(B_R(x))} \leq \|u(\cdot, t) - \mu'\|_{L^q(B_R(x))} \text{ for } \mu' \in \mathbb{R}. \tag{1.7}$$

In Theorem 3.1, we provide the short-time asymptotics of q-means in the case that $B_R(x)$ satisfies the same assumption for which (1.5) holds. Notice that if $q = 2$ in (1.7), we recover the mean value of $u(\cdot, t)$ on $B_R(x)$.

We end this introduction by noticing that a variant of these techniques (in particular the application of barriers based on radial cases) was used by Berti and Magnanini to obtain similar asymptotic analysis also in the case of a one-parameter family of boundary value elliptic problems, involving again Δ_p^G and $\mathcal{M}^{\pm}$ (see [4, 6]).

2 Varadhan-Type Formulas

The next theorem summarizes [5, Theorem 2.9] and [6, Theorem 1.2 part (i)], which concern the cases $F = \Delta_p^G$ and $F = \mathcal{M}^{\pm}$, respectively.

Theorem 2.1 (Varadhan-Type Formulas) *Let Ω be such that $\Gamma = \partial\left(\mathbb{R}^N \setminus \overline{\Omega}\right)$. Let u be the solution to* (1.1)–(1.3) *when F is Δ_p^G or either one of $\mathcal{M}^{\pm}$.*

Then, we have that

$$\lim_{t \to 0^+} 4t \log\{u(x, t)\} = -\alpha d_\Gamma(x)^2 \ \text{ on } \overline{\Omega}. \tag{2.1}$$

(Here, α is given by (1.6)*.)*

Proof We give a sketch of the proof of (2.1), following [5] and [6]. We emphasize here the common scheme on which the proofs of [5, Theorem 2.9] and [6, Theorem 1.2 part (i)] are based.

The identity (2.1) is plainly true if $x \in \Gamma$, since the left-hand side is zero for each $t > 0$ and $d_\Gamma(x) = 0$. For any $x \in \Omega$, let $y \in \Gamma$ be a point that realizes $|x - y| = d_\Gamma(x)$. Call R this quantity. Since the assumption $\Gamma = \partial\left(\mathbb{R}^N \setminus \overline{\Omega}\right)$ holds,

Fig. 1 The geometrical scheme behind the proof of Theorem 2.1

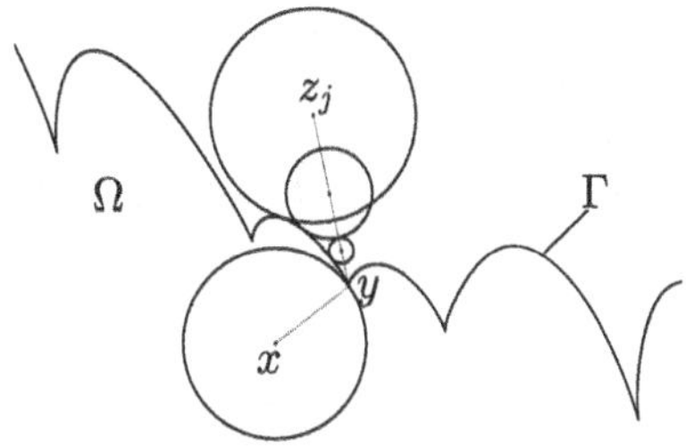

there exists a sequence of points $\{z_j : j \in \mathbb{N}\} \subset \mathbb{R}^N \setminus \overline{\Omega}$ approaching y as $j \to \infty$. (See Fig. 1.) In particular, for each $j \in \mathbb{N}$, it holds that $B_R(x) \subset \Omega \subset \mathbb{R}^N \setminus \overline{B_j}$, where B_j is the ball centered at z_j and with radius $d_\Gamma(z_j)$.

The application of comparison and maximum principles (see [3, Theorems 1.4 and 1.8] and references therein) leads to

$$v_j(x', t) \le u(x', t) \le w(x', t) \text{ for any } (x', t) \in \overline{B_R(x)} \times (0, \infty),$$

where w and v_j are the respective solutions of (1.1)–(1.3) when $\Omega = B_R(x)$ and $\Omega = \mathbb{R}^N \setminus \overline{B_j}$. It follows that in particular at the point x, $v_j(x, t) \le u(x, t) \le w(x, t)$, for any $t > 0$. Thus,

$$\limsup_{t \to 0^+} 4t \log\{w(x, t)\} \ge \limsup_{t \to 0^+} 4t \log\{u(x, t)\} \ge$$
$$\liminf_{t \to 0^+} 4t \log\{u(x, t)\} \ge \liminf_{t \to 0^+} 4t \log\left\{v_j(x, t)\right\}.$$

To conclude we only need to estimate both the left and the right-hand side of the previous chain of inequalities. We claim that

$$\limsup_{t \to 0^+} 4t \log\{w(x, t)\} \le -\alpha d_\Gamma(x)^2 \tag{2.2}$$

and

$$\liminf_{t \to 0^+} 4t \log\left\{v_j(x, t)\right\} \ge -\alpha |z_j - x|^2. \tag{2.3}$$

Passing to the limit, as $j \to \infty$, gives that $|z_j - x|^2 \to |y - x|^2$ and hence the statement is proved, since $|x - y| = d_\Gamma(x)$.

As regards (2.2), we need to consider the solution u^ε to the following auxiliary boundary problem, for $\varepsilon > 0$:

$$\begin{aligned} u^\varepsilon - \varepsilon^2 F\left(\nabla u^\varepsilon, \nabla^2 u^\varepsilon\right) &= 0 \quad \text{in } B_R(x), \\ u^\varepsilon &= 1 \quad \text{on } \partial B_R(x). \end{aligned}$$

By employing a version of the *Laplace transform* in [5] and by applying the comparison principle in [6, Lemma 4.1], it was established the following estimate for w:

$$w(x,t) \leq e^{-t/\varepsilon^2} u^\varepsilon(x),$$

for any $\varepsilon, t > 0$. Since u^ε is radially symmetric (from the fact that F is rotation invariant) then the Dirichlet problem can be rephrased as a problem in the context of ordinary differential equations. This allows us to explicitly compute u^ε, whose expression turns out to be given in terms of so-called *modified Bessel functions*. (See [5, Lemma 2.1] and [6, Lemma 2.2].) Some manipulations, involving the values of ε and t and the explicit asymptotic behavior of u^ε as $\varepsilon \to 0^+$, give (2.2).

To obtain (2.3), we make use of the existence of a global sub-solution Φ of (1.1), which takes the form, for $x' \in \mathbb{R}^N$ and $t > 0$,

$$\Phi(x',t) = t^\gamma e^{-\frac{\alpha|x'|^2}{4t}}.$$

In the case $F = \Delta_p^G$ it is shown in [1, Proposition 2.1] (see also [3, Proposition 2.5]) that if $\gamma = -(N+p-2)/(2p-2)$, then Φ satisfies $\Phi_t - F\left(\nabla\Phi, \nabla^2\Phi\right) \leq 0$, in $\mathbb{R}^N \times (0,\infty)$ and it is bounded away from the origin. The same holds for $\mathcal{M}^\pm$ with a value of γ which depends on λ and Λ (see [6, Lemma 4.3]). Thus, since F is one-homogeneous and translation invariant, we can easily obtain that, for any $t > 0$,

$$v_j(x,t) \geq \delta_j \Phi(x - z_j, t)$$

which ends in (2.3). Here, $\delta_j > 0$ is such that $\delta_j \Phi(x - z_j, t) \leq 1$, for any $t > 0$. □

With some further assumption on the regularity of Ω, we can provide uniform convergence. Let $\omega : (0,\infty) \to (0,\infty)$ be a strictly increasing continuous function such that $\omega(0^+) = 0$. We say that a domain Ω is of class $C^{0,\omega}$, if Γ is locally the graph of a continuous function whose modulus of continuity is controlled by ω. (For detail see [5].)

We then let $\psi_\omega : [0,\infty) \to [0,\infty)$ be the function defined by

$$\psi_\omega(\sigma) = \inf_{s\geq 0} \sqrt{s^2 + [\omega(s) - \sigma]^2} \quad \text{for } \sigma \geq 0.$$

As a consequence of the barriers shortly described in the proof of Theorem 2.1, we obtain the following theorem. The part which concerns Δ_p^G is given in [5, Theorem 2.10] while the part regarding $\mathcal{M}^\pm$ in [6, Theorem 1.2 part (ii)].

Theorem 2.2 (Uniform Estimates) *Let u be the solution to* (1.1)–(1.3) *when F is Δ_p^G or either one of $\mathcal{M}^\pm$.*

Then, uniformly on compact subsets of $\overline{\Omega}$, *we have that*

$$4t \log\{u(x,t)\} + \alpha\, d_\Gamma(x)^2 = O\left(t\ \log \psi_\omega(t)\right) \quad for\ t \to 0^+.$$

In particular, if Ω *is of class* C^2, *then, uniformly on compact subsets of* $\overline{\Omega}$,

$$4t \log\{u(x,t)\} + \alpha\, d_\Gamma(x)^2 = O\left(t\ \log t\right) \quad for\ t \to 0^+. \tag{2.4}$$

3 Asymptotics for q-Means

Assume that Ω is of class C^2 and that $x \in \Omega$ is such that $\overline{B_R(x)} \cap \left(\mathbb{R}^N \setminus \Omega\right) = \{y_x\}$, for some $R > 0$ and $y_x \in \Gamma$ satisfying $\kappa_j(y_x) < \frac{1}{R}$ for $j = 1, \ldots, N-1$. Also, we set

$$\Pi_\Gamma(y_x) = \prod_{j=1}^{N-1} \left[1 - Rk_j(y_x)\right].$$

With the following theorem we present in a unified version the results contained in [5, Theorem 3.5] for $F = \Delta_p^G$ and in [6, Theorem 5.2] when $F = \mathcal{M}^{\pm}$.

Theorem 3.1 (Short-Time Asymptotics for q-Means) *Suppose that* u *is the solution of* (1.1)–(1.3) *when* F *is* Δ_p^G *or either one of* $\mathcal{M}\pm$.

For $1 < q \le \infty$, *let* $\mu_q(x,t)$ *be the* q*-mean of* $u(\cdot,t)$ *on* $B_R(x)$, *defined as in* (1.7).

Then, for $1 < q < \infty$, *it holds that*

$$\lim_{t\to 0^+} \left(\frac{R^2}{t}\right)^{\frac{N+1}{4(q-1)}} \mu_q(x,t) = C_{N,q} \left\{\alpha^{\frac{N+1}{2}} \Pi_\Gamma(y_x)\right\}^{-\frac{1}{2(q-1)}}, \tag{3.1}$$

where

$$C_{N,q} = \left[\frac{N!\ \int_0^\infty \mathrm{Erfc}(\sigma)^{q-1} \sigma^{\frac{N-1}{2}}\, d\sigma}{\Gamma\left(\frac{N+1}{2}\right)^2}\right]^{\frac{1}{q-1}}.$$

Here, Erfc *is the complementary error function, defined for* $\sigma \in \mathbb{R}^N$ *by*

$$\mathrm{Erfc}(\sigma) = \frac{2}{\sqrt{\pi}} \int_\sigma^\infty e^{-s^2}\, ds,$$

and $\Gamma\left(\frac{N+1}{2}\right)$ *is the Euler's gamma function evaluated at* $\frac{N+1}{2}$.

In the case $q = \infty$, we have:

$$\lim_{t \to 0^+} \mu_\infty(x, t) = \frac{1}{2}.$$

Proof Let $v : \overline{\Omega} \times (0, \infty) \to \mathbb{R}$ be such that

$$\mathrm{Erfc}\left(\frac{\sqrt{\alpha}\, v(x', t)}{2\sqrt{t}}\right) = u(x', t) \text{ for any } (x', t) \in \overline{\Omega} \times (0, \infty).$$

Arguing as in [5, Corollary 2.12], after some manipulations based essentially on (2.4) and on an integration by parts, we get that

$$v(\cdot, t) = d_\Gamma + O\left(t \log t\right) \text{ for } t \to 0^+,$$

uniformly on $\overline{B_R(x)} \subset \overline{\Omega}$.

Now, define $\eta : (0, \infty) \to (0, \infty)$ by

$$\eta(t) = \frac{1}{\sqrt{t}} \max\left\{|v(x', t) - d_\Gamma(x')| : x' \in \overline{B_R(x)}\right\} \text{ for } t > 0.$$

For any $(x', t) \in \overline{B_R(x)} \times (0, \infty)$, it holds then that

$$\mathrm{Erfc}\left(\frac{\sqrt{\alpha}\, d_\Gamma(x')}{2\sqrt{t}} + \eta(t)\right) \le u(x', t) \le \mathrm{Erfc}\left(\frac{\sqrt{\alpha}\, d_\Gamma(x')}{2\sqrt{t}} - \eta(t)\right), \tag{3.2}$$

where $\eta(t) = O\left(\sqrt{t} \log t\right)$, as $t \to 0^+$.

With these barriers in mind, we just proceed as in [5, Theorem 3.5] (see also [6, Theorem 5.2]). Since the q-means are monotonic with respect to the pointwise order between functions, formulas (3.1) and the one in the case $q = \infty$ result after computing them for both sides of (3.2). In the case $1 < q < \infty$, the desired asymptotics are a consequence of applications of the co-area formula and the geometrical lemma [10, Lemma 2.1], which generates the term $\Pi_\Gamma(y_x)$. The case $q = \infty$ plainly follows from the fact that the ∞-mean of a function is the arithmetic mean of its supremum and its infimum. □

Acknowledgments The author of these notes was partially supported by the grant of Programa de Exelencia Severo Ochoa (SEV-2015-0554) funded by the Ministerio de Economía y Competitividad and by the Gruppo Nazionale di Analisi Matematica, Probabilità e Applicazioni (GNAMPA) dell'Istituto Nazionale di Alta Matematica (INdAM).

References

1. A. Banerjee, N. Garofalo, Gradient bounds and monotonicity of the energy for some nonlinear singular diffusion equations. Indiana Univ. Math. J. **62**(2), 699–736 (2013)
2. A. Bensoussan, J.-L. Lions, *Applications of Variational Inequalities in Stochastic Control, Studies in Mathematics and Its Applications*, vol. 12 (North-Holland, Amsterdam, 1982). Translated from the French
3. D. Berti, Asymptotic analysis of solutions related to the game-theoretic p-Laplacian. PhD Thesis, Università di Firenze, 2019. https://arxiv.org/abs/1902.10346
4. D. Berti, R. Magnanini, Asymptotics for the resolvent equation associated to the game-theoretic p-Laplacian. Appl. Anal. **98**(10), 1827–1842 (2019)
5. D. Berti, R. Magnanini, Short-time behavior for game-theoretic p-caloric functions. J. Math. Pures Appl. (9) **126**, 249–272 (2019)
6. D. Berti, R. Magnanini, Small diffusione and short-time asymptotics for Pucci operators. Appl. Anal. (2020). https://doi.org/10.1080/00036811.2020.1750602
7. L.A. Caffarelli, X. Cabré, *Fully Nonlinear Elliptic Equations, American Mathematical Society Colloquium Publications*, vol. 43 (American Mathematical Society, Providence, RI, 1995)
8. M.G. Crandall, H. Ishii, P.-L. Lions, User's guide to viscosity solutions of second order partial differential equations. Bull. Am. Math. Soc. (N.S.) **27**(1), 1–67 (1992)
9. R. Magnanini, S. Sakaguchi, Matzoh ball soup: heat conductors with a stationary isothermic surface. Ann. Math. (2) **156**(3), 931–946 (2002)
10. R. Magnanini, S. Sakaguchi, Interaction between degenerate diffusion and shape of domain. Proc. Roy. Soc. Edinb. Sect. A **137**(2), 373–388 (2007)
11. R. Magnanini, S. Sakaguchi, Polygonal heat conductors with a stationary hot spot. J. Anal. Math. **105**, 1–18 (2008)
12. R. Magnanini, S. Sakaguchi, Nonlinear diffusion with a bounded stationary level surface. Ann. Inst. H. Poincaré Anal. Non Linéaire **27**(3), 937–952 (2010)
13. Y. Peres, O. Schramm, S. Sheffield, D.B. Wilson, Tug-of-war and the infinity Laplacian. J. Am. Math. Soc. **22**(1), 167–210 (2009)
14. Y. Peres, S. Sheffield, Tug-of-war with noise: a game-theoretic view of the p-Laplacian. Duke Math. J. **145**(1), 91–120 (2008)
15. C. Pucci, Operatori ellittici estremanti. Ann. Mat. Pura Appl. (4) **72**, 141–170 (1966)
16. S.R.S. Varadhan, On the behavior of the fundamental solution of the heat equation with variable coefficients. Commun. Pure Appl. Math. **20**, 431–455 (1967)

Geometric Properties for a Finsler Bernoulli Exterior Problem

Chiara Bianchini

Abstract The aim of this paper is to prove convexity results for an exterior anisotropic free boundary problem of the Bernoulli type. More precisely we recover the results obtained in Henrot and Shahgholian (Nonlinear Anal 28(5):815–823, 1997) for the exterior problem, in the Finsler setting.

Keywords Class file · Journal

Mathematics Subject Classification (2010) Primary 99Z99; Secondary 00A00

1 Introduction

Let K be a bounded (regular) domain in $\mathbb{R}^N$. The classical Bernoulli problem consists in finding a domain $\Omega \supset K$ and a function $u : \Omega \setminus \overline{K} \to \mathbb{R}$ such that

$$\begin{cases} \Delta u = 0 & \text{in } \Omega \setminus \overline{K}, \\ u = 1 & \text{on } K \\ u = 0, \ |Du| = \tau & \text{on } \partial\Omega, \end{cases} \tag{1.1}$$

where τ is a given positive constant.

Existence of a solution (Ω, u) has been obtained in various ways by several authors, also in the case of some non-linear operators as governing operator (see [12]) and more recently in the more general setting of $\mathcal{A}$-harmonic PDE's (see [1]).

C. Bianchini (✉)
Dipartimento di Matematica e Informatica "U. Dini", Università degli Studi di Firenze, Firenze, Italy
e-mail: chiara.bianchini@unifi.it

P. Cerejeiras et al. (eds.), *Current Trends in Analysis, its Applications and Computation*, Research Perspectives,
https://doi.org/10.1007/978-3-030-87502-2_43

If K is convex we can ask whether this convexity property is inherited by the unknown set Ω. The positive answer has been given firstly by A. Henrot and H. Shahgholian in [11] where the authors proved the following: *let K be a convex set and assume that (Ω, u) is a solution to* (1.1). *Then Ω is also convex and there are no other solutions.* Then such a result has been generalized to more general classes of operators (see [1]).

In this paper we consider the analogous problem of (1.1) in the Finsler setting, that is in the space $\mathbb{R}^N$ endowed by an anisotropic norm H which belongs to the class $\mathcal{H}$:

$$\mathcal{H} = \{H \in C^{2,\alpha}(\mathbb{R}^N \setminus \{O\}),\ H^2 \text{ is uniformly convex}\}. \tag{1.2}$$

The corresponding Finsler Bernoulli problem consists then in finding a couple (u, Ω), where $\Omega \subset \mathbb{R}^N$ contains K and $u : \Omega \to \mathbb{R}$, is such that:

$$\begin{cases} \Delta_H u = 0 & \text{in } \Omega \setminus \overline{K} \\ u = 1 & \text{on } \partial K \\ u = 0,\ H(Du) = \tau & \text{on } \partial\Omega, \end{cases} \tag{1.3}$$

where $\Delta_H u$ is the so called *Finsler Laplacian* associated to the norm H and is defined as

$$\Delta_H u = \operatorname{div}(H(Du)\nabla_\xi H(Du)).$$

In this setting Henrot and Shahgholian's result reads as follows.

Theorem 1.1 *Let H be a norm of $\mathbb{R}^N$ in the class (1.2), and let K be a bounded convex subset of $\mathbb{R}^N$ whose boundary is of class C^2, with $O \in K$. If there exists a solution (Ω, u) to (1.3), then Ω is convex, u has convex super level sets and there is no other solution.*

This result has been proved in [1] by adapting an idea of Lewis [13] and showing that if K is convex then the superlevel sets of u are convex sets (and hence Ω is convex).

In this paper we prove Theorem 1.1 by using a geometric approach. More precisely we first prove the convexity of the set Ω by comparing the capacitary potential of the ring $\mathrm{conv}\left(\Omega \setminus \overline{K}\right)$ with a rearrangement of the capacitary potential of $\Omega \setminus \overline{K}$. Then a monotonicity result is proved, with respect to both data K and τ, which entails the uniqueness result.

2 Notation

Let $N \geq 3$ and $H : \mathbb{R}^N \to \mathbb{R}$ be a norm in $\mathbb{R}^N$, that is a nonnegative positively homogeneous convex function; more explicitly:

- (i) H is convex;
- (ii) $H(\xi) \geq 0$ for $\xi \in \mathbb{R}^N$ and $H(\xi) = 0$ if and only if $\xi = 0$;
- (iii) $H(t\xi) = |t|H(\xi)$ for $\xi \in \mathbb{R}^N$ and $t \in \mathbb{R}$.

We indicate by H_0 the dual norm of H, that is

$$H_0(x) = \sup_{\xi \neq 0} \frac{\langle x; \xi \rangle}{H(\xi)} \quad \text{for } x \in \mathbb{R}^N . \tag{2.1}$$

We denote by B_H and B_{H_0} the unitary balls in the norm H and H_0 respectively; in general, for $r > 0$, we set

$$B_H(r) = \{\xi \in \mathbb{R}^N \;\; H(\xi) < r\} = rB_H , \quad B_{H_0}(r) = \{x \in \mathbb{R}^N \;\; H_0(x) < r\} = rB_{H_0} .$$

We say that a set is a *Wulff shape* with respect to the norm H if it is a ball in the norm H_0.

Given a smooth function u, we will use H_0 to measure the norm of $x \in \mathbb{R}^N$ and H to measure the norm of $Du(x)$ (then H is used on the dual of $\mathbb{R}^N$, that coincides however with $\mathbb{R}^N$).

Notice that by Schneider [14, Corollary 1.7.3], we have that $H_0 \in C^1(\mathbb{R}^N \setminus \{0\})$ if and only B_H is strictly convex. Moreover, we notice that if $H \in C^2(\mathbb{R}^N \setminus \{0\})$ and B_H is uniformly convex (i.e. $H^2 \in C^2_+(\mathbb{R}^N \setminus \{0\})$, see later for the definition), then the same holds for H_0 and B_{H_0}.

The Finsler Laplacian (associated to H) of the function u is given by

$$\Delta_H u = \operatorname{div}\big(H(Du)\nabla_\xi H(Du)\big) = \big(H_{\xi_i}(Du)H_{\xi_j}(Du) + H(Du)H_{\xi_i\xi_j}(Du)\big)\, u_{ij}. \tag{2.2}$$

The Finsler Laplacian have been widely investigated in the literature and goes back to Wulff [15], who considered it to describe the theory of crystals. Many other authors developed related theory in several settings; in particular in the case of overdetermined problems we mention [2–5, 7].

Thanks to the regularity and the homogeneity properties of the norm H, the Finsler Laplacian is a strictly elliptic operator. Moreover several results which are valid in the Euclidean case, hold true in the anisotropic case too. Here we present only few of them.

Proposition 2.1 (Weak Comparison Principle [10]) *Let E be a bounded domain and assume that*

$$-\Delta_H u \leq -\Delta_H v \quad \text{in } E, \qquad \text{and } u \leq v \;\; \text{on } \partial E,$$

then

$$u \leq v \quad a.e.\ in\ E.$$

In particular, the following maximum principle holds.

Proposition 2.2 (Maximum Principle [10]) *If $\Delta_H u = 0$ in E, then*

$$\min_{\partial E} u \leq u(x) \leq \max_{\partial E} u,$$

almost everywhere in E.

For two bounded nested open sets $D_0 \subset D_1$, we define the Finsler capacitary potential of the ring $D_0 \setminus \overline{D}_1$ as the function $u : D_0 \setminus \overline{D}_1 \to \mathbb{R}$ such that

$$\begin{cases} \Delta_H u = 0 & \text{in } D_0 \setminus \overline{D}_1\,, \\ u = 1 & \text{in } \overline{D}_1\,, \\ u = 0 & \text{on } \partial D_0\,. \end{cases} \tag{2.3}$$

In the special case of the annular Wulff ring case $B_{H_0}(r_2) \setminus \overline{B_{H_0}(r_1)}$, the Finsler capacitary potential is the function

$$u_{r_1,r_2}(x) = \frac{H_0^{2-N}(x) - r_2^{2-N}}{r_1^{2-N} - r_2^{2-N}}. \tag{2.4}$$

3 Preliminary Results

We here present some notions and results which will be useful in the proof of the main result. We indicate by $\mathsf{conv}(E)$ the convex hull of the set E. Given a function $u : E \to \mathbb{R}$, we define its quasi-concave envelope as the function u^* such that for every $t \in \mathbb{R}$,

$$\{x \in \mathbb{R}^N \ : u^*(x) < t\} = \mathsf{conv}\Big(\{x \in \mathbb{R}^N \ : \ u(x) < t\}\Big). \tag{3.1}$$

If $\Omega(t)$ denotes the sublevel set of the function u of level t, that is

$$\Omega(t) = \{x \in \mathbb{R}^N \ : \ u(x) < t\},$$

then the function u^* can be equivalently defined as

$$u^*(x) = \sup\{t \in \mathbb{R} \ : \ x \in \mathsf{conv}(\Omega(t))\}$$

and

$$u^*(x) = \max\left\{ \min\{u(x_1),\ldots,u(x_{N+1})\} \ :\ x_i \in \overline{E}, \exists\lambda_1,\ldots\lambda_{N+1} \in [0,1], \sum_{i=1}^{N+1} \lambda_i = 1,\ x = \sum_{i=1}^{N+1} \lambda_i x_i \right\}.$$

Proposition 3.1 ([9]) *Let Ω, K be open bounded convex subsets of $\mathbb{R}^N$, of class C^2, such that $\overline{K} \subseteq \Omega$. Let $u \in C^2(\Omega \setminus \overline{K}) \cap C(\overline{\Omega \setminus K})$ be such that $u \equiv 0$ on $\partial\Omega$ and $u \equiv 1$ on ∂K. If $|Du| > 0$ in $\Omega \setminus \overline{K}$, then for every $x \in \Omega \setminus \overline{K}$ there exist $\lambda_1,\ldots,\lambda_N \in [0,1]$, $\sum_{i=1}^N \lambda_i = 1$ and $x_1,\ldots,x_N \in \Omega \setminus \overline{K}$ such that*

$$\bar{x} = \sum_{i=1}^N \lambda x_i, \qquad \text{and} \qquad u(x_1) = \ldots = u(x_N) = u^*(\bar{x}).$$

Moreover the point x_i belongs to the support hyperplane of the convex set $\{u^* \geq u^*(\bar{x})\}$, for every $i = 1,\ldots,N$. In particular for every i, $Du(x_i)$ is parallel to the normal vector to such hyperplane and

$$\frac{1}{|Du^*(\bar{x})|} = \sum_{i=1}^m \frac{\lambda_i}{|Du(x_i)|}.$$

The previous result has been generalized to a less regular situation in [6].

4 Proof of Theorem 1.1

Proof We compare the capacitary potential u^* of the convex ring $\mathsf{conv}(\Omega) \setminus \overline{K}$ to the quasi-concave envelope w of u.

Notice that, thanks to the Neumann condition on $\partial\Omega$, Ω is a bounded set. Indeed, if there exists an unbounded sequence $\{x_n\} \in \partial\Omega$, the comparison with the Finsler capacitary potential of $\mathbb{R}^N \setminus \overline{K}$ would give a contradiction, thanks to estimates in Theorem 3.3 of [4].

Let t be given by

$$t = \sup\{s > 0 \ :\ s\,\mathsf{conv}(\Omega) \subset \Omega\};$$

it is clear that $0 < t \leq 1$, and $t = 1$ if and only if Ω is convex. We denote by $\bar{x}$ a common point to $\partial\Omega$ and $t\,\mathsf{conv}(\Omega)$.

We consider the rescaling of the capacitary potential u^* given by

$$v(x) = u^*(\frac{x}{t}), \quad \text{for } x \in t\,\mathsf{conv}(\Omega) \setminus \overline{tK}\,.$$

We notice that v is the capacitary potential of $t\Big(\mathsf{conv}(\Omega) \setminus \overline{K}\Big)$ and that

$$Dv(x) = \frac{1}{t} Du^*(\frac{x}{t}),$$

for every $x \in t\big(\mathsf{conv}(\Omega) \setminus \overline{K}\big)$.

By applying the comparison principle to u and v, we obtain $u \geq v$ in $t\,\mathsf{conv}(\Omega) \setminus \overline{K}$ with $u(\bar{x}) = v(\bar{x})$. Hence it holds

$$\tau = H(Du(\bar{x})) \geq H(Dv(\bar{x})) = \frac{1}{t} H(Du^*(\bar{y})), \tag{4.1}$$

where $\bar{y} = \frac{\bar{x}}{t}$ and $H(Du(\bar{x}))$, $H(Du^*(\bar{y}))$ denote the limsup of $H(Du(x))$ and $H(Du^*(y))$ for x, y tending to $\bar{x}$, $\bar{y}$, with $x \in t(\mathsf{conv}(\Omega) \setminus \overline{K})$, $y \in \mathsf{conv}(\Omega) \setminus \overline{K}$, respectively.

Let us now consider the quasi-concave envelope w of the function u (see (3.1) for definition). By Proposition 3.1 there exist $x_1, \ldots, x_N \in \partial\Omega$ such that

$$\bar{y} = \sum_{i=1}^{N} \lambda_i x_i,$$

with $\sum_{i=1}^{N} \lambda_i = 1$ and $\lambda_i \geq 0$, and where

$$\nu_{\mathsf{conv}(\Omega)}(\bar{y}) = \nu_\Omega(x_i) = \nu$$

for any $i = 1, \ldots, N$, and

$$|Dw(y)| = \Big(\sum_{i=1}^{N} \frac{\lambda_i}{|Du(x_i)|}\Big)^{-1}.$$

Hence, since $x_i \in \partial\Omega$, it holds that $H(\nu)|Du(x_i)| = \tau$, which entails that

$$H(Dw(\bar{y})) = H(\nu)|Dw(\bar{y})| = \tau.$$

In the last step we compare w with u^*. Notice that by Bianchini et al. [6] the function w is a subsolution to $-\Delta_H u = 0$ in a viscosity sense, that is

$$\Delta_H w \geq 0 \qquad \text{in } \mathsf{conv}(\Omega) \setminus \overline{K}.$$

Moreover $w(\bar{y}) = u^*(\bar{y}) = 0$. By the viscosity comparison principle (see [8]) we obtain that

$$|Dw(\bar{y})| \leq |Du^*(\bar{y})|,$$

where $Dw(\bar{y})$ is parallel to $Du^*(\bar{y})$ and are both parallel to ν. This guarantees that

$$H(Du^*(\bar{y})) \geq H(Dw(\bar{y})) = \tau.$$

Thanks to (4.1) and the fact that $0 < t \leq 1$ we obtain $t = 1$, that is $\Omega = \mathrm{conv}(\Omega)$.

Hence the function u is the Finsler capacitary potential of the convex ring $\Omega \setminus \overline{K}$ and thanks to [6] it has convex super level sets. Indeed the equation $\Delta_H u = 0$ can be written as

$$\nabla_\xi H(\theta)\, B\, \nabla_\xi H(\theta) + H(\theta)\mathrm{tr}(\nabla_\xi^2 H(\theta) B) = 0,$$

where θ is the direction of the gradient at a generic point x, B is the Hessian matrix of u at x and $\mathrm{tr}(A)$ is the trace of the matrix A.

□

5 Monotonicity and Uniqueness Results

Theorem 5.1 *Let A, B be two convex bounded domains whose boundaries are of class C^2 and such that $O \in A \subseteq B$, let $\tau > 0$. Assume that there exist solutions (u_A, Ω_A), (u_B, Ω_B) to the Bernoulli problem (1.3) related to A and τ and to B and τ, respectively. Then $\Omega_A \subseteq \Omega_B$.*

Proof Assume by contradiction that $\Omega_A \not\subseteq \Omega_B$. Let

$$\mathrm{t} = \sup\{s \in \mathbb{R}^+ \ : \ s\Omega_A \subseteq \Omega_B\};$$

hence $0 < \mathrm{t} < 1$ and there exists $\bar{x} \in \partial\Omega_B \cap \partial(\mathrm{t}\Omega_A)$. For every $x \in \mathrm{t}(\Omega_A \setminus \overline{A})$ we define the function

$$v(x) = u_A(\frac{x}{\mathrm{t}}),$$

which satisfies

$$\begin{cases} \Delta_p^H v = 0 & \text{in } \mathrm{t}(\Omega_A \setminus \overline{A}), \\ v = 1 & \text{on } \partial(\mathrm{t}A), \\ v = 0 & \text{on } \partial(\mathrm{t}\Omega_A) \\ H(Dv) = \dfrac{\tau}{\mathrm{t}} & \text{on } \partial(\mathrm{t}\Omega_A). \end{cases}$$

Let $\mathcal{U}$ be the intersection between an open neighbourhood of $\bar{x}$ and $(\mathrm{t}\Omega_A \cap \Omega_B)$ and denote by Γ the portion of $\partial(\mathrm{t}\Omega_A)$ contained into $\mathcal{U}$. Hence it holds

$$v|_\Gamma = 0 \leq u_B|_\Gamma, \qquad \text{with } u_B(\bar{x}) = v(\bar{x}).$$

Moreover, since Ω_B is tangent to $\mathrm{t}\Omega_A$ at $\bar{x}$ then $Du_B(\bar{x})$ is parallel to $Du_A(\frac{\bar{x}}{\mathrm{t}})$, and we denote by ν their direction. By using the homogeneity of H, we have that

$$\tau = H(Du_B(\bar{x})) = H(\nu)|Du_B(\bar{x})| \geq H(\nu)\frac{|Du_A(\frac{\bar{x}}{\mathrm{t}})|}{\mathrm{t}} = \frac{1}{\mathrm{t}}H(Du_A(\frac{\bar{x}}{\mathrm{t}})) = \frac{\tau}{\mathrm{t}}.$$

Hence we have that $\mathrm{t} \geq 1$, which gives a contradiction and we conclude. □

Theorem 5.2 *Let K be a bounded convex domain whose boundary is of class C^2 and $O \in K$; let $\tau_A \geq \tau_B > 0$. Assume that there exist two solutions (u_A, Ω_A) and (u_B, Ω_B) to the Bernoulli problem (1.3) related to K and τ_A and to K and τ_B, respectively. Then $\Omega_A \subseteq \Omega_B$.*

Proof Let

$$\mathrm{t} = \sup\{s \in \mathbb{R}^+ \ : \ s\Omega_A \subseteq \Omega_B\};$$

there exists $\bar{x} \in \partial\Omega_B \cap \partial(\mathrm{t}\Omega_A)$. For every $x \in \mathrm{t}(\Omega_A \setminus \overline{A})$ we define the function

$$v(x) = u_A(\frac{x}{\mathrm{t}}),$$

which satisfies

$$\begin{cases} \Delta_p^H v = 0 & \text{in } \mathrm{t}(\Omega_A \setminus \overline{A}), \\ v = 1 & \text{on } \partial(\mathrm{t}A), \\ v = 0 & \text{on } \partial(\mathrm{t}\Omega_A) \\ H(Dv) = \dfrac{\tau}{\mathrm{t}} & \text{on } \partial(\mathrm{t}\Omega_A). \end{cases}$$

Let $\mathcal{U}$ be the intersection between an open neighbourhood of $\bar{x}$ and $(\mathrm{t}\Omega_A \cap \Omega_B)$ and denote by Γ the portion of $\partial(\mathrm{t}\Omega_A)$ contained into $\mathcal{U}$. Hence it holds

$$v|_\Gamma = 0 \le u_B|_\Gamma, \qquad \text{with } u_B(\bar{x}) = v(\bar{x})\,.$$

Moreover, since Ω_B is tangent to $\mathrm{t}\Omega_A$ at $\bar{x}$, then $Du_B(\bar{x})$ is parallel to $Dv(\bar{x})$; we denote by ν their direction. Notice that by definition of the function v, also $Du_A(\frac{\bar{x}}{\mathrm{t}})$ has direction ν. By the comparison principle it holds that $|Du_B(\bar{x})| \ge |Dv(\bar{x})|$ and, by using the homogeneity of H and the fact that

$$H(Du_B(\bar{x})) = |Du_B(\bar{x})| H(\nu); \;\; H(Dv(\bar{x})) = |Dv(\bar{x})| H(\nu),$$

we have that

$$\tau_A \ge \tau_B = H(Du_B(\bar{x})) \ge H(Dv(\bar{x})) = \frac{1}{\mathrm{t}} H(Du_A(\frac{\bar{x}}{\mathrm{t}})) = \frac{1}{\mathrm{t}}\tau_A .$$

Hence we have that $\mathrm{t} \ge 1$, which implies that $\Omega_A \subseteq \Omega_B$. □

The monotonicity results in Theorems 5.1 and 5.2 straightforwardly imply the uniqueness of the solution in the following theorem (and hence its proof is omitted).

Theorem 5.3 *Let H be a norm of $\mathbb{R}^N$ in the class (1.2) and let K be a bounded convex subset of $\mathbb{R}^N$ with boundary of class C^2, $\tau > 0$, $O \in K$. If there exists a solution (u, Ω) to (1.3) then it is unique.*

References

1. M. Akman, A. Nanerjee, M.S. Vega Garcia, On a Bernoulli-type overdetermined free boundary problem. https://arxiv.org/pdf/1911.02801.pdf
2. L. Barbu, C. Enache, On a free boundary problem for a class of anisotropic equations. Math. Methods Appl. Sci. **40**(6), 20052012 (2017)
3. L. Barbu, C. Enache, A free boundary problem with multiple boundaries for a general class of anisotropic equations. Appl. Math. Comput. **362**, 124551 (2019)
4. C. Bianchini, G. Ciraolo, Wulff shape characterizations in overdetermined anisotropic elliptic problems. Commun. Partial Differ. Equ. **43**(5), 790–820 (2018)
5. C. Bianchini, G. Ciraolo, P. Salani, An overdetermined problem for the anisotropic capacity. Calc. Var. Partial Differ. Equ. **55**, 84 (2016)
6. C. Bianchini, M. Longinetti, P. Salani, Quasiconcave solutions to elliptic problems in convex rings. Indiana Univ. Math. J. **58**(4), 1565–1589 (2009)
7. A. Cianchi, P. Salani, Overdetermined anisotropic elliptic problems. Math. Ann. **345**, 859–881 (2009)
8. M.G. Crandall, H. Ishii, P.L. Lions, User's guide to viscosity solutions of second order partial differential equations. Bull. Am. Math. Soc. (N.S.) **27**(1), 1–67 (1992)
9. A. Colesanti, P. Salani, Quasi-concave envelope of a function and convexity of level sets of solutions to elliptic equations. Math. Nachr. **258**, 3–15 (2003)
10. V. Ferone, B. Kawohl, Remarks on a Finsler-Laplacian. Proc. Am. Math. Soc. **137**, 247–253 (2009)

11. A. Henrot, H. Shahgholian, Convexity of free boundaries with Bernoulli type boundary condition. Nonlinear Anal. **28**(5), 815–823 (1997)
12. A. Henrot, H. Shahgholian, Existence of classical solutions to a free boundary problem for the p-Laplace operator. II. The interior convex case. Indiana Univ. Math. J. **49**(1), 311–323 (2000)
13. J.L. Lewis, Capacitary functions in convex rings. Arch. Ration. Mech. Anal. **66**(3), 201–224 (1977)
14. R. Schneider, *Convex Bodies: The Brunn-Minkowski Theory* (Cambridge University Press, Cambridge, 1993)
15. G. Wulff, Zur Frage der Geschwindigkeit des Wachstums und der Auflösung der Kristallfläschen. Z. Krist. **34**, 449–530 (1901)

Symmetry Breaking Solutions for a Two-Phase Overdetermined Problem of Serrin-Type

Lorenzo Cavallina and Toshiaki Yachimura

Abstract In this paper, we consider an overdetermined problem of Serrin-type for a two-phase elliptic operator with piecewise constant coefficients. We show the existence of infinitely many branches of nontrivial symmetry breaking solutions which bifurcate from any radially symmetric configuration satisfying some condition on the coefficients.

Keywords Two-phase · Overdetermined problem · Serrin's problem · Transmission condition · Bifurcation · Symmetry breaking

Mathematics Subject Classification (2010) Primary 35N25 035B32; Secondary 35J15 35Q93

1 Introduction and Main Result

In this paper, we consider a bifurcation analysis of a Serrin-type overdetermined problem for an elliptic operator with piecewise constant coefficients. First, let us introduce the problem setting of our overdetermined problem. Let (D, Ω) be a pair of sufficiently smooth bounded domains of $\mathbb{R}^N$ $(N \geq 2)$ such that $\overline{D} \subset \Omega$. Moreover, let n denote the outward unit normal vector of Ω. We consider the

L. Cavallina (✉)
Mathematical Institute, Tohoku University, Aoba-ku, Sendai, Japan
e-mail: cavallina.lorenzo.e6@tohoku.ac.jp

T. Yachimura
Institute for Advanced Study, Kyoto University, Sakyo-ku, Kyoto, Japan
e-mail: yachimura.toshiaki.8n@kyoto-u.ac.jp

P. Cerejeiras et al. (eds.), *Current Trends in Analysis, its Applications and Computation*, Research Perspectives,
https://doi.org/10.1007/978-3-030-87502-2_44

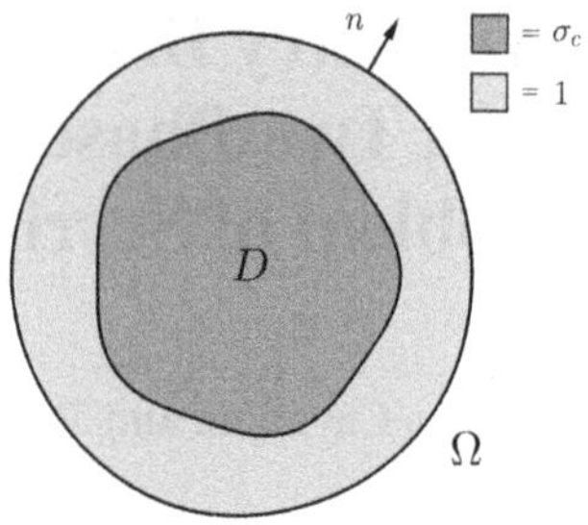

Fig. 1 Problem setting

following two-phase Serrin-type overdetermined problem:

$$\begin{cases} -\operatorname{div}(\sigma \nabla u) = 1 & \text{in } \Omega, \\ u = 0 & \text{on } \partial\Omega, \\ \partial_n u = c & \text{on } \partial\Omega, \end{cases} \tag{1.1}$$

where c is a real constant and $\sigma = \sigma(D, \Omega)$ is the piecewise constant function given by

$$\sigma(x) = \begin{cases} \sigma_c & \text{in } D, \\ 1 & \text{in } \Omega \setminus D, \end{cases}$$

and σ_c is a positive constant such that $\sigma_c \neq 1$ (Fig. 1).

We remark that, if (1.1) is solvable, then the parameter c must be equal to $c(\Omega) = -|\Omega|/|\partial\Omega|$ by integration by parts. In what follows, we will say that a pair of domains (D, Ω) is a solution of problem (1.1) whenever problem (1.1) is solvable for $\sigma = \sigma(D, \Omega)$. Let us define the inner problem and outer problem associated to problem (1.1) (see [5]).

Problem 1 (Inner Problem) For a given domain Ω and a real number $0 < V_0 < |\Omega|$, find a domain $D \subset \overline{D} \subset \Omega$ with volume $|D| = V_0$, such that the pair (D, Ω) is a solution of the overdetermined problem (1.1).

Problem 2 (Outer Problem) For a given domain D and a real number $V_0 > |D|$, find a domain $\Omega \supset \overline{D}$ with volume $|\Omega| = V_0$, such that the pair (D, Ω) is a solution of the overdetermined problem (1.1).

The case where D is empty (one-phase setting) has been studied by many mathematicians in various situations since the pioneering work of Serrin [15], who proved that the overdetermined problem (1.1) without the inclusion D is solvable if and only if the domain Ω is a ball. We refer to [1, 2, 11, 12] and references therein.

However, when D is not empty (two-phase setting), there are a few results for the overdetermined problem (1.1). The paper [4] deals with the inner problem (Problem 1) of the overdetermined problem (1.1), the authors proved the local existence and uniqueness for the inner problem near concentric balls.

The authors, in [5], proved the following local existence and uniqueness results for the outer problem (Problem 2) near concentric balls by perturbation arguments by means of shape derivatives and the implicit function theorem for Banach spaces.

Theorem 1.1 *Let us define*

$$s(k) = \frac{k(N+k-1)-(N+k-2)(k-1)R^{2-N-2k}}{k(N+k-1)+k(k-1)R^{2-N-2k}} \quad \text{for } k = 1, 2, \ldots, \tag{1.2}$$
$$\Sigma = \{s \in (0, \infty) \mid s = s(k) \text{ for some } k = 1, 2, \ldots\}.$$

and let $B_R \subset B_1$ denote concentric balls of radius R and 1 respectively. If $\sigma_c \notin \Sigma$, then for every domain D of class $\mathcal{C}^{2,\alpha}$ sufficiently close to B_R in the $\mathcal{C}^{2,\alpha}$-norm , there exists a domain Ω of class $\mathcal{C}^{2,\alpha}$ sufficiently close to B_1 in the $\mathcal{C}^{2,\alpha}$-norm such that the outer problem (Problem 2) admits a solution for the pair (D, Ω).

Remark 1.2 Notice that, by the definition of $s(k)$ in (1.2), the quantity $s(k)$ is not necessarily positive for all values of N, k and R. Indeed, for fixed N and R, the quantity $s(k)$ tends to -1 as $k \to +\infty$. In particular, this implies that the set Σ is finite.

From Theorem 1.1, problem (1.1) has a solution near concentric balls except for $\sigma_c \in \Sigma$. Our aim in this paper is to examine the case for σ_c near $s(m) \in \Sigma$ in the same situation of Theorem 1.1. In particular, our interest is the shape of the solution of the outer problem near $\sigma_c \in \Sigma$. In what follows, we introduce some notations in order to state the main theorem in this paper precisely (Fig. 2).

Let us take an element $s(m) \in \Sigma$ for some $m \geq 1$ and let X and Y denote the Banach spaces

$$X = \left\{ g \in \mathcal{C}^{2,\alpha}(\partial B_1) \ : \ \int_{\partial B_1} g = 0 \right\}, \quad Y = \left\{ h \in \mathcal{C}^{1,\alpha}(\partial B_1) \ : \ \int_{\partial B_1} h = 0 \right\}$$

endowed with their natural norms. We consider the functional $\Psi : X \times \mathbb{R} \to Y$ defined by

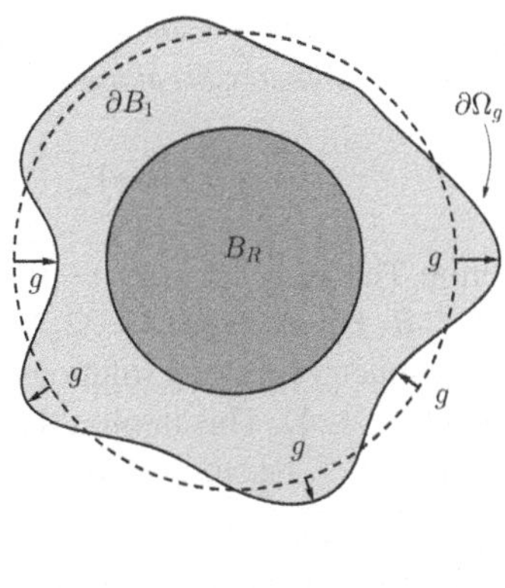

Fig. 2 The geometrical construction used in the definition of $\Psi(g, \lambda)$

$$\Psi(g,\lambda) = \left\{\partial_{n_g} v_g - c_g\right\} J_\tau(g). \tag{1.3}$$

In what follows, we will explain the notation involved in the definition of (1.3). For $g \in X$, let Ω_g be the unique bounded domain whose boundary is defined as

$$\partial\Omega_g = \{x + g(x)n(x) \;:\; x \in \partial B_1\}$$

with outward unit normal vector denoted by n_g. Moreover, let v_g be the solution of the Dirichlet boundary value problem given by the first two equations in (1.1) for $(D,\Omega) = (B_R,\Omega_g)$ and $\sigma_c = s(m) + \lambda$. By the definition (1.3), we notice that $\Psi(0,\lambda) = 0$ for any λ since the pair of concentric balls (B_R,Ω_0) is a solution of overdetermined problem (1.1). By a slight abuse of notation, we will use $\partial_{n_g} v_g$ to denote the function of value

$$\partial_{n_g} v_g\,(x + g(x)n(x)) \quad \text{for } x \in \partial B_1.$$

Finally, $c_g = c(\Omega_g) = -|\Omega_g|/|\partial\Omega_g|$ and $J_\tau(g)$ denotes the tangential Jacobian of the map $x \mapsto x + g(x)n(x)$ from ∂B_1 to $\partial\Omega_g$ (see [9, Definition 5.4.2]).

Now we can present the main result in this paper.

Theorem 1.3 *Let $(D,\Omega) = (B_R,\Omega_g)$ $(0 < R < 1)$. Also we take an element $s(m) \in \Sigma$ for some $m \geq 1$ and suppose that $\sigma_c = s(m) + \lambda$, where $\lambda \in \mathbb{R}$. If we consider the equation*

$$\Psi(g,\lambda) = 0,$$

then $(0,0)$ is a bifurcation point of the equation $\Psi(g,\lambda) = 0$. That is, there exists a smooth function $\varepsilon \mapsto \lambda(\varepsilon) \in \mathbb{R}$ with $\lambda(0) = 0$ such that overdetermined problem (1.1) admits a nontrivial solution of the form $(B_R,\Omega_{g(\varepsilon)})$ for $\sigma_c = s(m) + \lambda(\varepsilon)$ and ε small (notice that, by continuity, $\sigma_c = s(m) + \lambda(\varepsilon) > 0$ if ε is small enough). If $N = 2$, then the symmetry breaking solution $(B_R,\Omega_{g(\varepsilon)})$ satisfies

$$g(\varepsilon) = \varepsilon\cos(m\theta) + o(\varepsilon) \quad \text{in } C^{2,\alpha}(\partial\Omega_0) \quad \text{as } \varepsilon \to 0. \tag{1.4}$$

Moreover, if $N \geq 3$, then there exists a spherical harmonic Y_m of m-th degree, such that the symmetry breaking solution $(B_R,\Omega_{g(\varepsilon)})$ satisfies

$$g(\varepsilon) = \varepsilon Y_m(\theta) + o(\varepsilon) \quad \text{in } C^{2,\alpha}(\partial\Omega_0) \quad \text{as } \varepsilon \to 0. \tag{1.5}$$

From Theorem 1.3, if $D = B_R$, then the outer problem has solutions not only for $\Omega = B_1$ but also for $\Omega = \Omega_g$ given by (1.4) and (1.5). That is, there exist branches of symmetry breaking solutions of the outer problem emanating from the bifurcation points $\sigma_c \in \Sigma$. This implies that the uniqueness of the outer problem does not hold near $\sigma_c \in \Sigma$ because symmetry breaking phenomena occur. Similar results appear in the context of free boundary problems of a circulating flow with surface tension [13] and a model of tumor growth [7, 8].

This paper is organized as follows. In Sect. 2, we prove Theorem 1.3 when $N = 2$. This proof is based on the results obtained in [5] and the Crandall–Rabinowitz theorem. In Sect. 3, we consider high dimensional case $N \geq 3$ and establish Theorem 1.3.

2 Proof of Theorem 1.3 for $N = 2$

In this section, we prove the main Theorem 1.3 when $N = 2$. We obtain the existence of symmetry breaking bifurcation solutions of overdetermined problem (1.1) applying the following version of the Crandall–Rabinowitz theorem.

Theorem 2.1 (Crandall–Rabinowitz Theorem [6]) *Let X, Y be real Banach spaces and $\Psi(x, \lambda)$ be a C^p map ($p \geq 3$) of a neighborhood $(0, \lambda_0)$ in $X \times \mathbb{R}$ into Y. Suppose that*

(i) $\Psi(0, \lambda) = 0$ for all λ in a neighborhood of λ_0.
(ii) There exists $x_0 \in X$ such that Ker $\partial_x \Psi(0, \lambda_0)$ is a one-dimensional space spanned by x_0.
(iii) Im $\partial_x \Psi(0, \lambda_0)$ is a closed subspace of Y which has codimension one.
(iv) $\partial_\lambda \partial_x \Psi(0, \lambda_0)[x_0] \notin$ Im $\partial_x \Psi(0, \lambda_0)$.

Then $(0, \lambda_0)$ is a bifurcation point of the equation $\Psi(x, \lambda) = 0$ in the following sense: In a neighborhood of $(0, \lambda_0)$ the set of solutions of $\Psi(x, \lambda) = 0$ consists of two C^{p-2} smooth curves Γ_1 and Γ_2 which intersect only at the point $(0, \lambda_0)$; Γ_1 is the curve $(0, \lambda)$ and Γ_2 can be parametrized as follows:

$$\Gamma_2 : (x(\varepsilon), \lambda(\varepsilon)), \quad \varepsilon : \textit{small}, \quad (x(0), \lambda(0)) = (0, \lambda_0), \quad x'(0) = x_0.$$

In what follows, we assume that $N = 2$.

Theorem 1.3, $N = 2$ Take an element $s(m) \in \Sigma$ for some $m \geq 1$ and let Ψ be the map defined by (1.3). By definition, notice that $\Psi(g, \lambda) = 0$ if and only if the pair (B_R, Ω_g) solves (1.1) for $\sigma_c = s(m) + \lambda$. By Cavallina [3, Theorem 3.15 (iii)], the map Ψ is Fréchet differentiable infinitely many times in a neighborhood of the origin in X. Moreover, by the explicit formula of its Fréchet derivative $\partial_x \Psi(0, \lambda)$ computed in [5, Theorem 3.5], we know that Ker $\partial_x \Psi(0, \lambda)$ is a two dimensional space, spanned by $\{\cos(m\theta), \sin(m\theta)\}$. As a consequence, we cannot apply the Crandall–Rabinowitz theorem (Theorem 2.1) directly. In order to reduce the kernel to a one dimensional space, we introduce the following spaces of even functions:

$$X^* = \{g \in X \ : \ g(\theta) = g(2\pi - \theta), \ \theta \in [0, 2\pi)\},$$
$$Y^* = \{h \in Y \ : \ h(\theta) = h(2\pi - \theta), \ \theta \in [0, 2\pi)\},$$

where we identified the unit circle $\partial B_1 \subset \mathbb{R}^2$ with the interval $[0, 2\pi)$. Now, we consider the restriction Ψ^* of Ψ on X^*. We claim that Ψ^* is a well-defined mapping

$$\Psi^* : X^* \to Y^*.$$

To show this, notice that $g \in X^*$ implies that the configuration (B_R, Ω_g) is symmetric with respect to the x-axis. Now, by the unique solvability of the Dirichlet boundary value problem given by the first two equations in (1.1), this implies that also v_g shares the same symmetry and, thus, $\Psi^*(g, \lambda) = \Psi(g, \lambda) \in Y^*$ as claimed.

We will now apply Theorem 2.1 to the map Ψ^*. Recall that $\Psi^*(0, \lambda) = 0$ for any λ since the pair of concentric balls (B_R, Ω_0) is a solution of overdetermined problem (1.1). This fact implies that (i) holds true. Let us check condition (ii). In the proof of Theorem 3.6 in [5], we computed the Fréchet derivative $\partial_x \Psi(0, \lambda)$. The case $N = 2$ reads

$$\partial_x \Psi(0, \lambda)[g] = \sum_{k=1}^{\infty} \beta_k(\lambda) \left(\alpha_k^{\text{even}} \cos(k\theta) + \alpha_k^{\text{odd}} \sin(k\theta)\right), \tag{2.1}$$

for

$$g = \sum_{k=1}^{\infty} \left(\alpha_k^{\text{even}} \cos(k\theta) + \alpha_k^{\text{odd}} \sin(k\theta)\right),$$

where

$$\beta_k(\lambda) = \frac{(k+1)(s(m) + \lambda - 1)k + (k + ks(m) + k\lambda)(k-1)R^{-2k}}{2(k + ks(m) + k\lambda)R^{-2k} + 2k(1 - s(m) - \lambda)}. \tag{2.2}$$

Now, a simple computation with (1.2) at hand yields that

$$\beta_m(0) = 0 \quad \text{and} \quad \beta_k(0) \neq 0 \text{ for } k \neq m.$$

Let $x_0 = cos(m\theta)$. Notice that X^* is the subspace of X spanned by $\{\cos(k\theta)\}_{k \geq 1}$. Then, combining (2.1) with the fact that $\beta_m(0) = 0$ by construction, we obtain

$$\text{Ker } \partial_x \Psi^*(0, 0) = \text{span}\{x_0\}.$$

Thus condition (ii) holds true. Moreover, $\partial_x \Psi^*(0, 0)[\cos(k\theta)] = \beta_k(0) \cos(k\theta)$, where $\beta_k(0) \neq 0$ for $k \neq m$. This implies that

$$\text{Im } \partial_x \Psi^*(0, 0) \oplus \text{Ker } \partial_x \Psi^*(0, 0) = Y^*.$$

Therefore, Im $\partial_x \Psi^*(0, 0)$ is codimension one and thus also condition (iii) holds true.

Let us finally check condition (iv). Note that

$$\partial_x \Psi^*(0,\lambda)[x_0] = \beta_m(\lambda)\, x_0.$$

By (2.2), we can easily compute that

$$\partial_\lambda \beta_m(0) = \frac{\{m(m+1) + m(m-1)R^{-2m}\}\{2(m+ms(m))R^{-2m} + 2m(1-s(m))\}}{\{2m(1+s(m))R^{-2m} + 2m(1-s(m))\}^2}, \tag{2.3}$$

where we used the fact that $\beta_m(0) = 0$ by construction. Since $0 < s(m) < 1$ and $m \geq 1$, the right hand side of (2.3) is positive. Thus by (2.1),

$$\partial_\lambda \partial_x \Psi^*(0,0)[x_0] = \partial_\lambda \beta_m(0)\, x_0 \in \operatorname{Ker} \partial_x \Psi^*(0,0) \setminus \{0\}.$$

In other words,

$$\partial_\lambda \partial_x \Psi^*(0,0)[x_0] \notin \operatorname{Im} \partial_x \Psi^*(0,0).$$

Therefore, by the Crandall–Rabinowitz theorem (Theorem 2.1), $(0,0)$ is a bifurcation point of the equation $\Psi(g,\lambda) = 0$ in the sense that there exists a $\mathcal{C}^\infty$ curve $(g(\cdot), \lambda(\cdot))$ from a neighborhood of $0 \in \mathbb{R}$ into $X^* \times \mathbb{R}$, with $(g(0), \lambda(0)) = (0,0)$ and such that, for all small ε, there exists a symmetry breaking solution of overdetermined problem (1.1) for $\sigma_c = s(m) + \lambda(\varepsilon)$, represented by $(B_R, \Omega_{g(\varepsilon)})$, with

$$g(\varepsilon) = \varepsilon \cos(m\theta) + o(\varepsilon) \quad \text{as } \varepsilon \to 0.$$

$\square$

Remark 2.2 Theorem 1.3 ensures the existence of nontrivial solutions of (1.1) of the form (B_R, Ω). In particular, such solutions only partially inherit the symmetry of the core B_R. One might wonder whether nontrivial solutions of the form (D, B_1) exist for some subdomain D other than a ball. Actually this is not the case, since Theorem 5.1 of [14] states that, if $B_1 \setminus D$ is connected and the pair (D, B_1) solves (1.1), then D and B_1 must be concentric balls (Fig. 3).

3 Proof of Theorem 1.3 for $N \geq 3$

The proof of Theorem 1.3 when $N \geq 3$ follows along the same lines as the previous section. Indeed, as in the case $N = 2$, the Crandall–Rabinowitz theorem cannot be applied directly because $\operatorname{Ker} \partial_x \Psi(0,0)$ is not one dimensional. By the N-dimensional analogous of (2.1) (see [5, equation (3.12)]), $\operatorname{Ker} \partial_x \Psi(0,0)$ is the

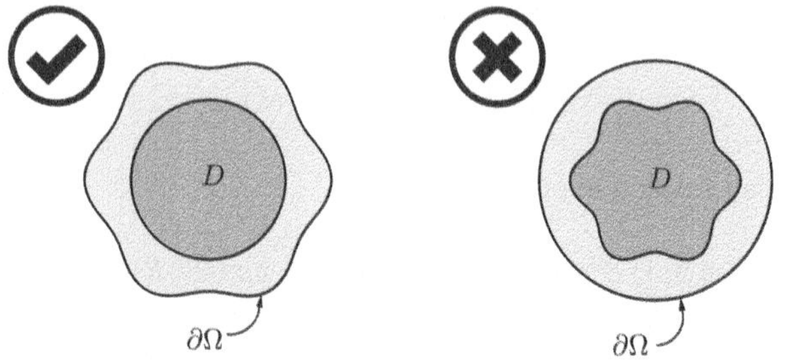

Fig. 3 Left: a symmetry-breaking bifurcating solution of (1.1) given by Theorem 1.3 ($m = 6$, $N = 2$). Right: a symmetry-breaking configuration that cannot be a solution to (1.1) in light of Theorem 5.1 of [14]

subspace of X spanned by the spherical harmonics whose degree is m. In order to reduce the kernel to a one dimensional space, we follow the same ideas as [10] and consider the restriction Ψ^* of Ψ to the space X^* of functions in X that are invariant with respect to some specific group of symmetries $\Gamma \subset O(N)$. Here we recall the definition of Γ-invariance with respect to a subgroup Γ of the orthogonal group $O(N)$. A function $g \in X$ is said to be Γ-invariant if

$$g(\theta) = g(\gamma(\theta)) \quad \text{for all } \theta \in \partial B_1, \ \gamma \in \Gamma.$$

If, for example, we set $\Gamma = \mathrm{Id} \times O(N-1)$, then the space of Γ-invariant spherical harmonics of any given degree $k \in \mathbb{N}$ is a one dimensional space. In particular, if $X^* \subset X$ is the subset of Γ-invariant functions and Ψ^* the restriction of Ψ to X^*, then also $\operatorname{Ker} \partial_x \Psi^*(0,0)$ is a one dimensional space, which can be considered to be spanned by some spherical harmonic $x_0 \in X^*$. The rest of the proof runs just as the one in Sect. 2, by checking conditions (i)–(iv) of the Crandall–Rabinowitz theorem applied to the map Ψ^*.

Remark 3.1 We claim that all nontrivial solutions $(B_R, \Omega_{g(\varepsilon)})$ given by Theorem 1.3 share the same symmetries of the element $x_0 \in X^*$, defined such that $\operatorname{Ker} \partial_x \Psi^*(0,0) = \operatorname{span}\{x_0\}$. To this end, let $\Gamma \subset O(N)$ be a symmetry group such that the function x_0 is Γ-invariant. Now, consider the further restriction Ψ^{**} of Ψ^* to the subspace X^{**} of all Γ-invariant functions in X^*. Notice that, since x_0 is Γ-invariant by hypothesis, then $\operatorname{Ker} \partial_x \Psi^*(0,0) = \operatorname{Ker} \partial_x \Psi^{**}(0,0) = \operatorname{span}\{x_0\}$. Another application of the Crandall–Rabinowitz theorem to Ψ^{**} yields that $g(\varepsilon)$ is also Γ-invariant. The claim follows by the arbitrariness of Γ.

Acknowledgments We would like to sincerely thank the anonymous reviewers for their comments and observations, which helped to improve the overall readability of the paper.

This research was partially supported by the Grant-in-Aid for JSPS Fellows No.18J11430 and No.19J12344.

References

1. C. Bianchini, A. Henrot, P. Salani, An overdetermined problem with non-constant boundary condition. Interfaces Free Bound. **16**(2), 215–241 (2014)
2. B. Brandolini, C. Nitsch, P. Salani, C. Trombetti, On the stability of the Serrin problem. J. Differ. Equ. **245**(6), 1566–1583 (2008)
3. L. Cavallina, Analysis of two-phase shape optimization problems by means of shape derivatives. Doctoral dissertation, Tohoku University, Sendai, 2018. arXiv:1904.10690
4. L. Cavallina, R. Magnanini, S. Sakaguchi, Two-phase heat conductors with a surface of the constant flow property. J. Geom. Anal. **31**, 312–345 (2021). https://doi.org/10.1007/s12220-019-00262-8
5. L. Cavallina, T. Yachimura, On a two-phase Serrin-type problem and its numerical computation. ESAIM Control Optim. Calc. Var. **26**, 65 (2020). https://doi.org/10.1051/cocv/2019048
6. M.G. Crandall, P.H. Rabinowitz, Bifurcation from simple eigenvalue. J. Funct. Anal. **8**, 321–340 (1971)
7. J. Escher, A. Matioc, Bifurcation analysis for a free boundary problem modeling tumor growth. Arch. Math. **97**, 79–90 (2011)
8. A. Friedman, F. Reitich, Symmetry-breaking bifurcation of analytic solutions to free boundary problems: an application to a model of tumor growth. Trans. Am. Math. Soc. **353**, 1587–1634 (2001)
9. A. Henrot, M. Pierre, *Shape Variation and Optimization (a Geometrical Analysis)*. EMS Tracts in Mathematics, vol. 28 (European Mathematical Society (EMS), Zürich, 2018)
10. N. Kamburov, L. Sciaraffia, Nontrivial solutions to Serrin's problem in annular domains. Ann. Inst. H. Poincaré Anal. Non Linéaire **38**(1), 1–22 (2021)
11. R. Magnanini, G. Poggesi, Serrin's problem and Alexandrov's Soap Bubble Theorem: stability via integral identities. Indiana Univ. Math. J. **69**, 1181–1205 (2020). https://doi.org/10.1512/iumj.2020.69.7925
12. C. Nitsch, C. Trombetti, The classical overdetermined Serrin problem. Complex Variables Elliptic Equ. **63**, 7–8, 1107–1122 (2018)
13. H. Okamoto, Bifurcation phenomena in a free boundary problem for a circulating flow with surface tension. Math. Method. Appl. Sci. **6**, 215–233 (1984)
14. S. Sakaguchi, Two-phase heat conductors with a stationary isothermic surface and their related elliptic overdetermined problems. RIMS Kôkyûroku Bessatsu. **B80**, 113–132 (2020)
15. J. Serrin, A symmetry problem in potential theory. Arch. Rat. Mech. Anal. **43**, 304–318 (1971)

Regularity Results for Nonlocal Minimal Surfaces

Eleonora Cinti

Abstract In this note, we present some recent results in the study of nonlocal minimal surfaces. The notion of nonlocal minimal surface was introduced by Caffarelli, Roquejoffre, and Savin, they are boundaries of sets which minimize the nonlocal (or fractional) perimeter. In the last years, much interest has been devoted to the study of their regularity properties. Similarly to the classical local setting, a crucial ingredient in the study of regularity is the classification of minimal cones. In the nonlocal setting, only partial results are available, dealing mainly with the low-dimensional case. We describe the main achievements in the field, focusing in particular on the difference with respect to the classical theory and in the difficulties which arise due to the nonlocal character of the problem.

Keywords Class file · Journal

Mathematics Subject Classification (2010) Primary 99Z99; Secondary 00A00

1 Introduction

We describe some recent results in the study of regularity properties of *nonlocal*, or *fractional, minimal surfaces*. These geometric objects were defined by Caffarelli, Roquejoffre, and Savin in [7], as the boundaries of sets whose characteristic functions minimize a fractional Sobolev norm.

More precisely, in [7] the following notion of *fractional perimeter* was introduced. Let $s \in (0, 1)$. Given E a bounded subset of $\mathbb{R}^n$, the fractional s-perimeter

E. Cinti (✉)
Dipartimento di Matematica, Università di Bologna, Bologna, Italy
e-mail: eleonora.cinti5@unibo.it

P. Cerejeiras et al. (eds.), *Current Trends in Analysis, its Applications and Computation*, Research Perspectives,
https://doi.org/10.1007/978-3-030-87502-2_45

of E is given by

$$\operatorname{Per}_s(E) = c_s \int_E \int_{\mathbb{R}^n \setminus E} \frac{1}{|x-y|^{n+s}}\, dx\, dy = \frac{c_s}{2}[\chi_E]_{W^{s,1}(\mathbb{R}^n)}, \tag{1.1}$$

where χ_E denotes the characteristic function of the set E, $[\,\cdot\,]_{W^{s,1}(\mathbb{R}^n)}$ denotes the seminorm in the fractional Sobolev space $W^{s,1}$, and c_s is a constant depending on s which behaves like $(1-s)$ as $s \uparrow 1$. To be more precise, in [7] the definition of Per_s was given in terms of the squared $W^{s/2,2}$-seminorm of χ_E, but it is easily seen that their definition coincide with the one given above.

Written as in (1.1), one can better appreciate the analogy with the notion of classical perimeter in the sense of De Giorgi, defined as

$$\operatorname{Per}(E) = [\,\chi_E\,]_{BV(\mathbb{R}^n)},$$

where $[\,\cdot\,]_{BV(\mathbb{R}^n)}$ denotes the seminorm in the space BV. In (1.1) we are considering a fractional order derivative of the characteristic function of a set and the two notions are consistent in the sense that $\operatorname{Per}_s \to \operatorname{Per}$ as $s \uparrow 1$ (see e.g. [1, 9, 14]).

Roughly speaking, the s-perimeter captures the interactions between a set E and its complement, these interactions take place in the whole $\mathbb{R}^n$ and are weighted by a kernel with polynomial decay. Due to its nonlocal character, the s-perimeter has several applications, for example in image reconstruction and nonlocal capillarity models, see e.g. [3, 17].

A set E which is a minimizer for the fractional perimeter is called a *fractional (or nonlocal) minimal set*, and its boundary is referred to as a nonlocal minimal surface.

As it happens for the classical notion of area-minimizing surfaces, if the set E is not bounded, in order to give the notion of minimizer for the perimeter functional, one needs to introduce a localized version of perimeter, since the perimeter of an unbounded set E in the whole $\mathbb{R}^n$ could be infinite.

The localized notion of s-perimeter is the following: let Ω be a bounded domain in $\mathbb{R}^n$, we define the fractional s-perimeter of a measurable set $E \subset \mathbb{R}^n$ relative to Ω as

$$\operatorname{Per}_s(E,\Omega) := \int_{E\cap\Omega} \int_{E^c} \frac{1}{|x-y|^{n+s}}\, dx\, dy + \int_{E\setminus\Omega} \int_{\Omega\setminus E} \frac{1}{|x-y|^{n+s}}\, dx\, dy, \tag{1.2}$$

where E^c denotes the complement of E in $\mathbb{R}^n$.

The choice of the set of integration in the definition of the fractional perimeter is the natural one which does not change the variational structure of the functional, once we have fixed the set E outside of Ω. We can now give the definition of minimizer for Per_s in Ω.

Definition 1.1 We say that a set E is a *minimizer* for the s-perimeter in Ω if

$$\mathrm{Per}_s(E,\Omega) \le \mathrm{Per}_s(F,\Omega), \qquad \text{for all } F \text{ such that } E\setminus\Omega = F\setminus\Omega.$$

Moreover, we say that E is a minimizer for the s-perimeter in $\mathbb{R}^n$, if E is a minimizer in a ball B_R, for all radii $R > 0$.

Said in other words, a nonlocal minimal surface in Ω is the boundary of a set E, whose characteristic function minimize the $W^{s,1}$-seminorm, among all sets which coincide with E in the complement of Ω.

In [7] the Euler-Lagrange equation for this functional has been derived: similarly to the classical case, a nonlocal minimal set E must have vanishing *fractional mean curvature* H_s, where H_s is given by the following expression

$$H_s(x) = c_s \lim_{\epsilon\to 0^+} \int_{\mathbb{R}^n\setminus B_\epsilon(x)} \frac{\chi_{\mathbb{R}^n\setminus E}(y) - \chi_E(y)}{|x-y|^{n+s}}\,dy. \tag{1.3}$$

Here c_s denotes again a constant depending on s which behaves like $(1-s)$ as $s \uparrow 1$.

The first example of a surface with zero nonlocal mean curvature is a half-space. Other examples of sets with vanishing nonlocal mean curvature have been studied in the recent contributions [11, 15]. In [15], the nonlocal analogue of catenoids are constructed, but they differ from the standard catenoids since they approach a singular cone at infinity instead of having a logarithmic growth. These surfaces are constructed using perturbative methods, by performing small perturbation along the normal vector to ∂E. Instead in [11] it is proven, just by an easy symmetry argument, that the standard helicoids are surfaces with zero nonlocal mean curvature.

In [7], the study of regularity of nonlocal minimal surfaces has been started. More precisely, Caffarelli, Roquejoffre, and Savin established density estimates, the improvement of flatness for minimizers, a monotonicity formula, a blow-up and a dimension reduction argument. Nevertheless, the regularity theory for minimizers of the fractional perimeter is still widely open. In the following sections we describe the main results and the main open questions in the field.

2 Classification of s-Minimal Cones in Low Dimensions

We start by recalling the following well known results in the regularity theory for classical area-minimizing surfaces.

Every minimal cone in $\mathbb{R}^n$ is a hyperplane, whenever $n < 8$. The condition on the dimension is optimal, indeed in $\mathbb{R}^8$ the Simons cone defined as

$$\mathcal{C} := \{x \in \mathbb{R}^8 \mid x_1^2 + \cdots + x_4^2 = x_5^2 + \cdots + x_8^2\}$$

is a minimizer for the perimeter functional.

The classification of minimal cones is one of the main ingredients in both the classification of entire minimal surfaces (that is surfaces that are minimizer of the perimeter functional in the whole $\mathbb{R}^n$) and in the study of regularity for minimizers of the perimeter in a bounded set Ω. Indeed, the classification of minimal cones leads, on the one hand, to the classification of *any* entire area minimizing surfaces via a blow-down argument. On the other hand the nonexistence of singular minimal cones in space dimension $n \leq 7$ implies, via blow-up and a dimension reduction argument, that any minimal surface is smooth outside a singular set of Hausdorff dimension $n - 8$.

Moreover, again the classification of minimal cones leads to the classification of entire minimal graphs (the so called *Bernstein problem*): If E is a minimizer of the perimeter functional and ∂E is a graph, then E is a half-space, whenever $n < 9$. Note that the critical dimension for a graph to be flat is one more than the one for a general set. The main ingredients in the proof of these results are given by density estimates, perimeter estimates, improvement of flatness for minimizers and a monotonicity formula.

As already mentioned in the Introduction, many of these ingredients in the nonlocal setting were established in [7]. With these tools, Caffarelli, Roquejoffre and Savin could reduce the study of regularity for nonlocal minimal surfaces to the classification of nonlocal minimal cones. More precisely they proved that, if the blow-up, around the origin, of an s-minimal set E is flat, then ∂E is $C^{1,\alpha}$ in a neighborhood of the origin (see [7, Theorem 9.4]). As a consequence of a dimension reduction argument, they proved $C^{1,\alpha}$ regularity outside a singular set of Hausdorff dimension at most $n-2$ (see [7, Theorem 10.4]). The bound $n-2$ on the dimension of the singular set was not optimal due to the fact that in [7] the classification of nonlocal minimal cones was not known, not even in $\mathbb{R}^2$.

Later, in [19] Savin and Valdinoci proved that in $\mathbb{R}^2$ an s-minimal cone is necessarily a half-plane. As a consequence they could improve the bound on the Hausdorff dimension of the singular set from $n - 2$ to $n - 3$ and via a blow-down argument they obtained the classification of any s-minimal surface in $\mathbb{R}^2$.

Moreover, in [2] Barrios, Figalli, and Valdinoci showed that if E is an s-minimal set such that $\partial E \in C^{1,\alpha}$, then ∂E is in fact C^∞ (such a result holds in every dimension). This is a consequence of a more general regularity result for solutions to integro-differential equations via a bootstrap argument. In [16], Figalli and Valdinoci addressed the fractional version of the Bernstein problem and proved that, if there are not s-minimal singular cones in $\mathbb{R}^n$, then the only entire s-minimal graphs in $\mathbb{R}^{n+1}$ are the hyperplanes.[1]

We summarize all these results in the following Theorem.

Theorem 2.1 *The following facts hold:*

1. *Every s-minimal cone in $\mathbb{R}^2$ is a hyperplane [19];*
2. *If E is a minimizer of the s-perimeter in the whole $\mathbb{R}^2$, then E is a half-plane [19];*

[1] Finally, regularity results for nonlocal minimal graphs were recently obtained by Cabré and Cozzi in [4]

3. *If E is a minimizer of the s-perimeter in $\mathbb{R}^n$ and ∂E is a graph, then E is a half-space, whenever $n \leq 3$ [16];*
4. *If E is a minimizer of the s-perimeter, then ∂E is C^∞ outside a singular set Σ of Hausdorff dimension $n-3$ [2, 7, 19].*

In addition, when s is close to 1, Caffarelli and Valdinoci proved that all the regularity results that hold in the classical setting are inherited, by a compactness argument, by s-nonlocal minimal surfaces (see [8, 9]).

Theorem 2.2 (Theorem 5 in [9]) *There exists $\epsilon_0 \in (0, 1)$ such that if $s \geq 1 - \epsilon_0$, then any s-minimal surfaces is C^∞ outside a singular set Σ of Hausdorff dimension $n-8$.*

Finally, in the very recent contribution [5], Cabré, Serra and the author proved flatness for nonlocal s-minimal cones in $\mathbb{R}^3$ for s close to 1. We emphasize that in [5], differently from [9], the proof is not based on a compactness argument and it permits to quantify how much s must be close to 1. This last result holds not only for cones that are minimizers for the s-perimeter, but for the more general class of *stable* cones. Stability here has to be understood in the variational sense, i.e. it corresponds to the fact that the second variation of the s-perimeter is nonnegative (we will comment on the notion of stability in the next section). The following is the main result in [5].

Theorem 2.3 (Theorem 1.2 in [5]) *There exists $s_* \in (0, 1)$ such that for every $s \in (s_*, 1)$ the following statement holds.*

Let $\Sigma \subset \mathbb{R}^3$ be a cone with nonempty boundary of class C^2 away from 0. Assume that Σ is a stable set for the s-perimeter. Then, Σ is a half-space.

The proof of this result uses two crucial ingredients: the fractional Hardy inequality in $\mathbb{R}^2$ (with the precise behavior of its sharp constant as $s \uparrow 1$) and the perimeter estimates for stable sets contained in [12] and that we describe in the next Section.

3 Quantitative Flatness Results and Perimeter Estimates for Stable Sets

We now focus on the two-dimensional result proven by Savin and Valdinoci in [19] (see Theorem 2.1, point *1.*). The proof of this result relies on the following idea: given a minimal cone E in the whole $\mathbb{R}^n$ (i.e. a cone which is a minimizer in B_R for any $R > 0$), one considers perturbations E_R^+ that are small translations, in some direction, of E inside the half ball $B_{R/2}$ (and which coincide with E outside of B_R). A computation shows that the difference between the s-perimeter of E_R^+ and the s-perimeter of E is controlled in the following way:

$$\mathrm{Per}_s(E_R^+, B_R) - \mathrm{Per}_s(E, B_R) \leq C R^{n-2-s}.$$

Hence, when $n = 2$, this difference can be made arbitrarily small as $R \to \infty$. On the other hand, if E was not a half-plane, it could be modified in such a way to decrease its s-perimeter by a small but fixed amount and this leads to a contradiction. It is clear that this argument works only in dimension $n = 2$ (we are using that R^{n-s-2} goes to 0 as $R \to \infty$). We emphasize that the factor R^{n-s} comes from an optimal bound for the s-perimeter of minimizers. Indeed, by a comparison argument one can show that if E is an s-minimal set in B_R, then

$$\mathrm{Per}_s(E, B_R) \le C R^{n-s},$$

and this bound is optimal.

These ideas were recently used in [12] to prove a quantitative version of this two-dimensional flatness result, where *quantitative* has to be understood in the following sense.

Suppose that E is a minimizer for Per_s in a ball B_R for *some* R large enough (and not for *all* R). Is it true that E is "close" to be a half-plane in B_1? Moreover, can we give an estimate on this closeness depending on R? The following result, contained in [12], gives an answer to these questions.

Theorem 3.1 (Theorem 1.3 in [12]) *Let $n = 2$. Let $R \ge 2$ and E be a minimizer for the s-perimeter in the ball $B_R \subset \mathbb{R}^2$.*

Then, there exists a half-plane $\mathfrak{h}$ such that

$$|(E \Delta \mathfrak{h}) \cap B_1| \le C R^{-s/2}. \tag{3.1}$$

Moreover, after a rotation, we have that $E \cap B_1$ is the subgraph of a measurable function $g : (-1, 1) \to (-1, 1)$ with oscillation $\operatorname{osc} g \le C R^{-s/2}$ outside a "bad" set $\mathcal{B} \subset (-1, 1)$ with measure $C R^{-s/2}$.

As mentioned above, the proof of this result is based on the technique developed in [19] which uses perturbations given by small translations of the minimizer E (inside the ball B_R) and introducing quantitative elements which allow to keep track of the dependence on the radius R.

The ideas developed in [12] to prove Theorem 3.1 above have also been used to prove an optimal estimate for the classical perimeter of an s-minimal set E. Of course, such an estimate cannot be deduced just by a comparison argument (indeed, it is a genuine regularity estimate which improves the order of differentiability of χ_E) and needs a more sophisticated argument. More interestingly, this estimate holds true in the more general class of *stable sets*. Here stability has to be understood in the variational sense, that is we require the set to be a minimizer among *small* perturbations, which corresponds, for smooth objects, to the fact that the second variation of the s-perimeter is nonnegative. For the precise notion of stability that we use, we refer to [12, Definition 1.6] and [5, Section 2]. Once one has an estimate for the classical perimeter of E, by a standard interpolation, one can deduce an estimate for its s-perimeter. As already explained, for minimizers the upper bound on the s-perimeter comes easily by comparison, but for stable sets is highly nontrivial.

In order to explain the interest in perimeter estimates for stable objects, we recall some known facts in the classical local setting.

Stable minimal cones (for the classical perimeter) are completely classified: they are hyperplanes in space dimensions $n \leq 7$. In $\mathbb{R}^8$, the Simons cone is an example of stable cone which is singular (i.e., the classification that we have presented in the previous section for classical minimal surfaces holds true for *stable cones*). Once one has a complete classification of stable cones, using a blow-down technique, one could obtain the classification of *any* stable surface in the whole $\mathbb{R}^n$. A crucial tool needed for this argument is an optimal estimate for the perimeter of stable sets. It is well known that any minimizer of the classical perimeter in a ball B_R satisfies the estimate

$$\mathrm{Per}(E, B_R) \leq C R^{n-1}. \tag{3.2}$$

Unfortunately, an estimate like (3.2) is not known to hold for stable sets, unless we are in dimension $n = 3, 4$ and we require some topological assumptions on the set E (see [10, 13, 18]). The difficulty in proving perimeter estimates for stable sets relies on the fact that, when using a comparison argument, we are allowed to consider only competitors which are *small perturbations* of the given set E.

In dimension $n > 3$ the search for a perimeter estimate for stable sets is still completely open. As explained above, having a universal bound for the classical perimeter of embedded minimal surfaces in every dimension $n > 3$ would be a decisive step towards proving the following well-known and long standing conjecture: *The only stable embedded minimal (hyper)surfaces in* $\mathbb{R}^n$ *are hyperplanes as long as the dimension of the ambient space is less than or equal to 7.*

Surprisingly, in the fractional setting, the nonlocal character of the perimeter functional gives somehow more rigidity and allows to obtain the following result (which holds in every dimension):

Theorem 3.2 (Theorem 1.1 in [12]) *Let* $s \in (0, 1)$*,* $R > 0$ *and* E *be a stable set in the ball* $B_{2R} \subset \mathbb{R}^n$ *for the nonlocal* s*-perimeter functional. Then,*

$$\mathrm{Per}(E, B_R) \leq C\, R^{n-1},$$

and

$$\mathrm{Per}_s(E, B_R) \leq C R^{n-s}.$$

As a consequence of Theorem 3.2, in [12] the quantitative flatness result in $\mathbb{R}^2$ was proven to hold also for stable set (and not only for minimizers).

In a similar way to what described for the classical case, once one has a complete classification for s-minimal stable cones, the s-perimeter estimates of Theorem 3.2 would allow to classify any stable s-minimal surface (see Theorem 2.1 in [6]) . In this respect, the difficulties in the nonlocal setting are, in some way, dual to the ones in the local setting: in the first case, we have perimeter estimates in any dimensions but only the classification of stable cones in low dimensions is known; in the second the situation is reversed, since stable cones are completely classified but perimeter estimates are still missing in dimension $n > 3$.

Having in mind this picture, an interesting motivation in the study of nonlocal minimal surfaces is whether nonlocal techniques and nonlocal results could lead to give an answer to some important open questions in the local setting, such as, for example, the complete classification of stable surfaces.

References

1. L. Ambrosio, G. De Philippis, L. Martinazzi, Gamma-convergence of nonlocal perimeter functionals. Manuscripta Math. **134**(3–4), 377–403 (2011)
2. B. Barrios, A. Figalli, E. Valdinoci, Bootstrap regularity for integro-differential operators and its application to nonlocal minimal surfaces. Ann. Sc. Norm. Super. Pisa Cl. Sci. (5) **13**(3), 609–639 (2014)
3. J. Bosch, M. Stoll, A fractional inpainting model based on the vector-valued Cahn-Hilliard equation. SIAM J. Imag. Sci. **8**, 2352–2382 (2015)
4. X. Cabré, M. Cozzi, A gradient estimate for nonlocal minimal graphs. Duke Math. J. **168**(5), 775–848 (2019)
5. X. Cabré, E. Cinti, J. Serra, Stable s-minimal cones in $\mathbb{R}^3$ are flat for $s \sim 1$. J. Reine Angew. Math. **764**, 157–180 (2020)
6. X. Cabré, E. Cinti, J. Serra, Stable solutions to the fractional Allen-Cahn equation in the nonlocal perimeter regime. Submitted. arxivearXiv:2111.06285.pdf
7. L. Caffarelli, J.-M. Roquejoffre, O. Savin, Nonlocal minimal surfaces. Commun. Pure Appl. Math. **63**, 1111–1144 (2010)
8. L. Caffarelli, E. Valdinoci, Uniform estimates and limiting arguments for nonlocal minimal surfaces. Calc. Var. Partial Differ. Equ. **41**(1–2), 203–240 (2011)
9. L. Caffarelli, E. Valdinoci, Regularity properties of nonlocal minimal surfaces via limiting arguments. Adv. Math. **248**, 843–871 (2013)
10. O. Chodosh, C. Li, Stable minimal hypersurfaces in R 4, arXiv:2108.11462v2
11. E. Cinti, J. Dávila, M. del Pino, Solutions of the fractional Allen-Cahn equation which are invariant under screw motion. J. Lond. Math. Soc. **94**, 295–313 (2016)
12. E. Cinti, J. Serra, E. Valdinoci, Quantitative flatness results and BV-estimates for stable nonlocal minimal surfaces. J. Differ. Geom. **112**(3), 447–504 (2019)
13. T.H. Colding, W.P. Minicozzi II, Estimates for parametric elliptic integrands. Int. Math. Res. Not. **6**, 291–297 (2002)
14. J. Dávila, On an open question about functions of bounded variation. Calc. Var. Partial Differ. Equ. **15**(4), 519–527 (2002)
15. J. Dávila, M. del Pino, J. Wei, Nonlocal s-minimal surfaces and Lawson cones. J. Differ. Geom. **109**, 111–175 (2018)
16. A. Figalli, E. Valdinoci, Regularity and Bernstein-type results for nonlocal minimal surfaces. J. Reine Angew. Math. **729**, 263–273 (2017)
17. F. Maggi, E. Valdinoci, Capillarity problems with nonlocal surface tension energies. Commun. Partial Differ. Equ. **42**, 1403–1446 (2017)
18. A.V. Pogorelov, On the stability of minimal surfaces. Sov. Math. Dokl. **24**, 274–276 (1981)
19. O. Savin, E. Valdinoci, Regularity of nonlocal minimal cones in dimension 2. Calc. Var. Partial Differ. Equ. **48**(1–2), 33–39 (2013)

A Short Survey on Overdetermined Elliptic Problems in Unbounded Domains

Pieralberto Sicbaldi

Abstract We present some recent results about overdetermined elliptic problems in unbounded domains.

Keywords Overdetermined elliptic problems · Bifurcation theory · Minimal surfaces · Constant mean curvature surfaces · PDEs · Maximum principle

Mathematics Subject Classification (2010) Primary 35N25; Secondary 53Axx, 35Bxx, 35Nxx

1 Introduction

In these short notes we give an overview on the recent research line about solutions to overdetermined elliptic problems of the following form:

$$\begin{cases} \Delta u + f(u) = 0 & \text{in } \Omega \\ u > 0 & \text{in } \Omega \\ u = 0 & \text{on } \partial\Omega \\ \frac{\partial u}{\partial \nu} = k & \text{on } \partial\Omega\,, \end{cases} \tag{1.1}$$

where Ω is a domain of class C^1 (in order to give a classical sense to the Neumann data), f is a locally Lipschitz function, k is a constant (that must be nonpositive), and ν is the unit normal vector about $\partial\Omega$. The first fact to point out is that if Problem 1.1 has a solution, then the domain Ω is much more regular than C^1: it is in fact of class $C^{2,\alpha}$ for all $\alpha \in (0, 1)$, as shown in [21]. The case of bounded domains

P. Sicbaldi (✉)
Aix Marseille Université - CNRS, Centrale Marseille - I2M, Marseille, France

Departamento de Análisis matemático, Universidad de Granada, Granada, Spain
e-mail: pieralberto@ugr.es

P. Cerejeiras et al. (eds.), *Current Trends in Analysis, its Applications and Computation*, Research Perspectives,
https://doi.org/10.1007/978-3-030-87502-2_46

supporting solutions to Problem 1.1 has been completely solved by J. Serrin in [16] (see also [10]): the ball is the unique such domain and any solution is radial. This result, obtained with the moving plane method, had many applications to Physics and Applied Mathematics (see for example the beautiful survey [18]). The case of unbounded domains Ω is much more difficult. In 1997, H. Berestycki, L. Caffarelli and L. Nirenberg in [3] proposed the following conjecture:

BCN Conjecture If $\mathbb{R}^n \setminus \overline{\Omega}$ is connected, then the existence of a bounded solution to Problem 1.1 implies that Ω is either a ball, a half-space, a generalized cylinder $B^k \times \mathbb{R}^{n-k}$ (B^k is a ball in $\mathbb{R}^k$), or the complement of one of them.

That question was justified by the results of the same authors in [3], and some other results concerning exterior domains, i.e. domains that are the complement of a compact region (see Sect. 4). This conjecture motivated almost all the research activity about overdetermined elliptic problems in unbounded domains.

2 New Nontrivial Solutions

Before 2010, the only known domains supporting solutions to Problem 1.1 were essentially balls, the exterior of balls and the half-space. Solutions in cylinders and in the exterior of cylinders are obviously obtained from the others, adding an empty variable: if $u(x)$ is the solution of Problem 1.1 in a ball B or in the exterior of a ball $\mathbb{R}^n \setminus B$, then $v(x, y) = u(x)$, for $y \in \mathbb{R}^k$ is a solution respectively in the cylinder $B \times \mathbb{R}^k$ or the exterior of it, i.e. in $(\mathbb{R}^n \setminus B) \times \mathbb{R}^k$.

When the function f is given by $f(u) = \lambda\, u$, Problem 1.1 represents the stationary condition for the functional $\Omega \to \lambda_1(\Omega)$ under volume preserving variations of the domain, where $\lambda_1(\Omega)$ is the first eigenvalue of the Dirichlet Laplacian in Ω. This is true also in a Riemannian manifold, replacing the Euclidean Laplacian by the Laplace-Beltrami operator. For this reason, domains where there exists a positive constant λ such that Problem 1.1 has a solution with $f(u) = \lambda u$ are called *extremal domains*. The theory of extremal domains shares strong similarity with the theory of constant mean curvature surfaces. In fact, the result by Serrin recalled before is parallel to the well known result of Alexandrov asserting that round spheres are the only embedded compact constant mean curvature hypersurfaces in the Euclidean space. In [9] F. Pacard and the author proved the existence of extremal domains in a Riemannian manifold which are perturbations of small geodesic balls centered at nondegenerate critical points of the scalar curvature function, paralleling an earlier result of R. Ye which provides constant mean curvature topological spheres close to small geodesic spheres centered at nondegenerate critical points of the scalar curvature function. The analogy with constant mean curvature surfaces has been the key ingredient to investigate the possibility to build new nontrivial extremal domains, and more generally nontrivial solutions to Problem 1.1 for different type of nonlinearities f.

2.1 Bifurcation from a Known Solution

In the Euclidean space cylinders are not the only unbounded constant mean curvature surfaces. In 1841, C. Delaunay discovered the beautiful family of onduloids D_τ in $\mathbb{R}^3$, $\tau \in (0, 1]$, parametrized by

$$X_\tau(\theta, t) = \big(y(t)\cos\theta, y(t)\sin\theta, z(t)\big)$$

for $(\theta, t) \in S^1 \times \mathbb{R}$, where the function y is the smooth solution of $4(y')^2 = 4y^2 - (y^2+\tau)^2$ and z is the solution (up to a constant) of $2z' = y^2+\tau$. They are complete, embedded, non-compact constant mean curvature surfaces invariant under rotation about an axis and periodic in the direction of this axis. This family of surfaces can be seen as a bifurcation of the straight cylinder in the following way. Consider the cylinder $C = \mathbb{S}^1 \times \mathbb{R}/T\,\mathbb{Z}$ in the torus $\mathbb{R}^2 \times \mathbb{R}/T\,\mathbb{Z}$ for some $T > 0$. Consider the set of regular one variable functions v of period 1, with small $C^{2,\alpha}$-norm, and let C_v^T be the perturbation of C done with respect to the function $\tilde{v}(x, T\,t) = v(t)$. If H denotes the mean curvature, it is clear that $H(C_0^T) \equiv 1$ for all $T > 0$, and using the classical bifurcation theory one can prove that at the point $(v, T) = (0, 2\pi)$ the family C_0^T bifurcates in a new branch of perturbed cylinders C_v^T solutions of $H(C_v^T) \equiv 1$. The classification results in the theory of constant mean curvature surfaces allow to conclude that the cylinders C_v^T are elements of the family D_τ.

The same idea has been used in [17] by the author to build new nontrivial extremal domains, which are counterexamples to the BNC conjecture. Consider the cylinder $C = B_1 \times \mathbb{R}/T\,\mathbb{Z}$ in the torus $\mathbb{R}^n \times \mathbb{R}/T\,\mathbb{Z}$ for some $T > 0$, where B_1 is the unit Euclidean ball. It is clear that C is a domain where Problem 1.1 can be solved. Consider the set of regular one variable functions v of period 1, with mean 0 and small $C^{2,\alpha}$-norm, and let C_v^T the perturbation of C obtained perturbing ∂C by the function $\tilde{v}(x, T\,t) = v(t)$. In C_v^T one can consider the first eigenfunction of the Dirichlet Laplacian u_v^T (normalized to have L^2-norm equal to 1). Such a function, in general, does not have constant Neumann data at the boundary. It is then interesting to define the operator F given by

$$F(v, T) = \left.\frac{\partial u_v^T}{\partial \nu_v^T}\right|_{\partial C_v^T} - \fint \left.\frac{\partial u_v^T}{\partial \nu_v^T}\right|_{\partial C_v^T}$$

where ν_v^T is the unit exterior normal vector about ∂C_v^T. In fact, $F(0, T) = 0$ for all $T > 0$, and the existence of a bifurcation point in the branch $\{(0, T)\}$ of solutions of $F(v, T) = 0$ would give the existence of new nontrivial extremal domains (notice that in the new branch the function v cannot be constant, since v has mean 0). A priori, $F(v, T)$ is a function defined on ∂C_v^T, but some easy arguments show that it just depends on the variable t, in which it is periodic of period T and of mean 0. After rescaling we can assume that the period of $F(v, T)$ is 1. So, the operator F can be seen as an operator between some subset of $C_m^{2,\alpha}(\mathbb{R}/\mathbb{Z})$ and $C_m^{1,\alpha}(\mathbb{R}/\mathbb{Z})$, where the subscript m denotes the fact that functions have mean 0. A big difference with respect to the case of constant mean curvature surfaces is that the operator $(v, T) \to H(C_v^T)$ was local, and here the operator F is non-local. The study of the linearized operator prove that the bifurcation theory can be applied to the equation $F(v, T) = 0$ with respect to the branch $\{(0, T)\}_{\{T>0\}}$ at a certain point $(0, T^*)$. This new branch of solutions gives raise to a one-parameter family of unbounded domains Ω_s where Problem 1.1 can be solved for $f(u) = \lambda\, u$. Their boundaries are surfaces invariant under rotation about an axis, periodic in the direction of this axis, and whose mean curvature is not constant. The period T^* depends in fact on the dimension n, and estimations of T^* are given in [15].

A similar method is used in [13] to prove the existence of perturbations of the exterior of a ball in $\mathbb{R}^n$, $n \geq 2$, where one can solve Problem 1.1 for some non-linearities f. Here the new solutions bifurcates from the solution on the exterior of a ball B_{R^*} for a certain radius R^*. This new domains also are counterexamples to the BCN conjecture (in dimension 2 they are the only known counterexamples).

2.2 Perturbation of a Constant-Mean-Curvature Surface

In [5] M. Del Pino, F. Pacard and J. Wei develop a strong method to use directly constant mean curvature surfaces to build new examples of domains where Problem 1.1 can be solved. In their work f is supposed to satisfy the following hypothesis:

$$f(0) = f(1) = 0 \qquad f(s) > 0 \text{ for } s \in (0, 1) \qquad f'(1) < 0$$

(this is the case for example for the Allen-Cahn equation). Under such hypothesis there exists an increasing solution w to the ODE $w'' + f(w) = 0$ in $(0, +\infty)$ with $w(0) = 0$ and $w(+\infty) = 1$, and then the function $u(x_1, x_2, \ldots, x_n) = w(x_1)$ is a bounded solution of Problem 1.1 in the half-space. The main objective of their work is to produce a new counterexample to the BCN conjecture diffeormorphic to a half-space. They obtain it as a perturbation of the epigraph bounded by the Bombieri-De Giorgi-Giusti graph, a very special nontrivial minimal surface, existing only in dimension 9 (and bigger than 9, by an easy generalization), given by the graph of a certain non-constant entire function F:

$$\Gamma = \{(x_1, \ldots, x_9) \mid x_9 = F(x_1, \ldots, x_8)\}\,.$$

This surface, discovered in 1969, gave a counterexample to the Bernstein conjecture (1914) asserting that an entire minimal graph should be a hyperplane. Then, for a sufficiently small $\epsilon > 0$ they are able to solve Problem 1.1 in an epigraph Ω whose boundary lies in a $\mathcal{O}(\epsilon)$-neighborhood of $\epsilon^{-1}\Gamma$. Moreover the solution u obtained is given by $u(x) = w(z) + \mathcal{O}(\epsilon)$, where z is the inner coordinate to $\partial\Omega$. This result produces a counterexample to the BCN conjecture in the form of an epigraph, when f is a nonlinearity of Allen-Cahn type, and it exists in dimension bigger or equal to

9. The strategy to obtain such result is based on perturbative methods, starting from the well known solution w in the half-space, that represents the first approximation of the solution to Problem 1.1 (for this reason the authors need to consider a very large dilation of the Bombieri-De Giorgi Giusti graph).

The same strategy is used by the authors to exhibit two such examples of nontrivial solutions to Problem 1.1, in which the domain Ω is a non-cylindrical domain of revolution in $\mathbb{R}^3$. Both originate from constant mean curvature surfaces: the catenoid, i.e. the minimal surface

$$\Sigma = \{(\cosh z \, \cos\theta, \cosh z \, \sin\theta, z) \ | \ \theta \in \mathbb{S}^1 \, , z \in \mathbb{R}\} ,$$

and the Delaunay surfaces D_τ introduced before. Exactly as for the perturbation of the epigraph of Bombieri-De Giorgi-Giusti, the authors prove that for each ϵ sufficiently small, there exist domains of revolution Ω, which lie within a $\mathcal{O}(\epsilon)$-neighborhood of the region bounded by the surface $\epsilon^{-1}\,\Sigma$ (a large dilation of the catenoid) or the surface $\epsilon^{-1}\,D_\tau$ (a large dilation of a Delaunay surface), where Problem 1.1 is solvable. Here the nonlinearity f is asked to satisfy the same hypothesis as before. In fact, the first approximation of the new solution of Problem 1.1 is again the function w, and such hypothesis on f allow the existence of w.

3 Harmonic Overdetermined Problems

Limits under scaling of sequences of constant mean curvature surfaces produce minimal surfaces. By the strong analogy between constant mean curvature surfaces and extremal domains, it is reasonable to expect that limits under scaling of sequences of extremal domains behave like minimal surfaces. In fact, limits under scaling of sequences of extremal domains are domains where there exists a solution to

$$\begin{cases} \Delta u = 0 \, , \ u > 0 \text{ in } \ \Omega \\ u = 0 \, , \ \frac{\partial u}{\partial \nu} = 1 \ \text{ on } \ \partial\Omega \, . \end{cases} \tag{3.1}$$

We will call it *harmonic overdetermined problem.*

3.1 Catenoidal Domains

One of the most important minimal surface in $\mathbb{R}^3$ (existing in fact in $\mathbb{R}^n$ for $n \geq 3$), is the catenoid Σ defined before. The expected correspondence between minimal surfaces and harmonic overdetermined problems led to the question if it should exist a domain in $\mathbb{R}^n$, $n \geq 2$, that looks like the region inside a catenoid, where

Problem 3.1 can be solved. In the case $n = 2$ such a domain has been found explicitly by F. Hélèin, L. Hauswirth and F. Pacard in [6]. It is

$$\Omega = \left\{(x, y) \in \mathbb{R}^2 \mid |y| < \frac{\pi}{2} + \cosh x\right\}.$$

The function u solving Problem 3.1 in Ω is $u(F(x, y)) = \cosh x \cos y$, where $(x, y) \in S := \mathbb{R} \times \left(-\frac{\pi}{2}, \frac{\pi}{2}\right)$ and

$$F(x, y) = (x + \sinh x \cos y\,,\; y + \cosh x \sin y)\,.$$

In fact, is it easy to prove that F is a conformal diffeomorphism from S to Ω and that u is harmonic and positive in Ω, vanishes and has constant Neumann data at the boundary.

The existence of the same kind of domain in $\mathbb{R}^n$, $n \geq 3$ has been conjectured in the same paper [6], and it has been proved by Y. Liu, K. Wang and J. Wei in [8]: there exists a function $g : [1, +\infty) \to [0, +\infty)$, with the property that

$$\lim_{r \to +\infty} g'(r)\, r^{n-2} \in [0, +\infty)\,,$$

such that the domain

$$\mathbb{R}^n \setminus \left\{(x_1, \ldots, x_n) \in \mathbb{R}^n \;\middle|\; x_n < g(r)\,,\; r := \sqrt{x_1^2 + \cdots + x_{n-1}^2}\right\}$$

supports a solution to Problem 3.1. It is important to remark that in [7] it is proved that this last example of domain does not exist in dimension $n \geq 3$, but the proof of such result is not correct. More precisely, Theorem 8.1 in [7] is wrong, and Conjecture 8.2 in the same paper is disproved by [8] for any $n \geq 3$.

3.2 *Classification of Harmonic Overdetemined Problems in the Plane*

In [20] M. Traizet showed that the correspondence between minimal surfaces and harmonic overdetermined problems is really deep. He establishes a one-to-one correspondence between the following two classes of objects:

1. the set of domains Ω of $\mathbb{R}^2$ such that the solution u to Problem 3.1 satisfies $|\nabla u| < 1$.
2. the set of minimal bigraphs in $\mathbb{R}^3$, i.e. the set of complete, embedded minimal surfaces M in $\mathbb{R}^3$ symmetric with respect to $\{x_3 = 0\}$ and such that $M^+ := M \cap \{x_3 > 0\}$ is a graph over a certain unbounded domain $\hat{\Omega}$ of the plane $\{x_3 = 0\}$.

An extra topological hypothesis is needed: the complement of the domain Ω and the complement of the domain $\hat{\Omega}$ are assumed to be non-thinning at infinity (see Definition 1 in [20]). It is reasonable to think that this hypothesis is non required to establish the correspondence, but this point is still an open question. The two implications in that correspondence can be made explicit. We present here only the way to associate to an element of the first class a minimal bigraph. We start from a solution u of Problem 3.1 in a domain Ω of the plane $\mathbb{R}^2$ considered as the plane $\mathbb{C}$ of the complex variable z. The function $g = 2u_z$ is holomorphic, and we take the holomorphic differential $dh = 2u_z\, dz$ and a point $z_0 \in \partial\Omega$. The Weierstrass representation formula

$$X(z) = (x_1(z), x_2(z), x_3(z)) = \Re \int_{z_0}^{z} \left[\frac{1}{2}(g^{-1} - g)\, dh, \frac{i}{2}(g^{-1} + g)\, dh, dh \right]$$

defines locally a conformal, minimal immersion $X : \Omega \to R^3$. In this case it can be proved

- that X is globally well defined in all Ω;
- that $X(\partial\Omega)$ lies in the horizontal plane $\{x_3 = 0\}$;
- that $M_\Omega^+ = X(\Omega)$ is the graph of a positive function defined in the unbounded domain $\hat{\Omega} \subset \{x_3 = 0\}$ defined by $\hat{X}(\Omega)$, where $\hat{X}(z) = (x_1(z), x_2(z))$, and $\partial\hat{\Omega} = \hat{X}(\partial\Omega)$;
- that M_Ω^+ meets $\{x_3 = 0\}$ orthogonally, and then M_Ω^+ can be completed by symmetry with respect to $\{x_3 = 0\}$ to obtain a minimal bigraph M_Ω.

M_Ω is the minimal bigraph associated to the initial domain Ω. The result in [20] shows that the correspondence $\Omega \to M_\Omega$ is a bijection.

In this correspondence, the vertical minimal catenoid corresponds to the exterior of a disk and the horizontal minimal catenoid corresponds to the Hélèin-Hauswirth-Pacard domain. Using the well developed classification theory for minimal surfaces, from the result of M. Traizet it follows that the only domains in $\mathbb{R}^2$ where there exists a solution to Problem 3.1 and whose boundary is done by a finite number of components are: the half-plane, the exterior of a ball and the Hélèin-Hauswirth-Pacard domain. An other very interesting example of domain (with an infinite number of boundary components) where Problem 3.1 can be solved is the domain that corresponds to the the simply periodic Scherk minimal bigraph. It can be described in this way: for a parameter $\alpha \in \left(0, \frac{\pi}{2}\right)$ take the curve γ given by the implicit equation

$$\cos^2\alpha \cosh\left(\frac{y}{\cos\alpha}\right) = \sin^2\alpha + \cos(2\alpha - x)$$

(it is a closed curve), and consider the periodic unbounded domain Ω_α whose boundary is made by the curve γ and all its translates of a period $T_\alpha = 2\pi(1+\cos\alpha)$ along the x-direction. Again, using the classification theory for minimal surfaces, from the result of M. Traizet it follows that Ω_α is the only simply periodic domain in $\mathbb{R}^2$ where there exists a solution to Problem 3.1 and whose boundary has a finite number of components in the quotient (notice that by the maximum principle, if there exists a solution to Problem 3.1 then Ω cannot be doubly periodic). The result of M. Traizet classifies then all solutions of Problem 3.1 in $\mathbb{R}^2$ for domains with a finite number of boundary components, or periodic with a finite number of boundary components in the quotient. The open problem is to check if there exist non-periodic domains with infinite boundary components supporting solutions to Problem 3.1.

4 Rigidity Results

The main rigidity result on overdetermined elliptic problems is the Serrin's theorem stated in the introduction. Rigidity results for unbounded domains supporting solutions to Problem 1.1 are much more difficult to obtain, and this is now clear if we look at the many nontrivial solutions that have been built in the last years. We recollect here two recent achievements about rigidity of overdetermined elliptic problems in exterior domains and half-space type domains.

4.1 Exterior Domains

An exterior domain is the complement of a bounded region. In the case of exterior domains, the overdetermined elliptic problem that had been considered in the past was a slight modification of Problem 1.1:

$$\begin{cases} \Delta v + g(v) = 0\,, 0 \leq v < a & \text{in } \Omega \\ \frac{\partial v}{\partial \nu} = c\,,\ v = a > 0 & \text{on } \partial\Omega\,. \end{cases} \tag{4.1}$$

In this framework, the main research line has aimed to prove the counterpart of the Serrin's symmetry result, that is to prove that Ω is the complement of a ball. For example under the assumptions that $g(t) \geq 0$ and that $t^{-\frac{n+2}{n-2}} g(t)$ is nonincreasing, A. Aftalion and J. Busca proved in [1] that if Problem 4.1 has a solution then Ω is the complement of a ball. In [11] W. Reichel proved the same symmetry result but under different assumptions: he assumes that $g(t)$ is decreasing for small positive t and that $v \to 0$ at infinity. Finally, Sirakov [19] showed that the condition $v < a$ can be replaced by the assumption $c \geq 0$. The techniques used in these cases are the moving plane from infinity and the Kelvin transform.

With the change of functions $u := a - v$ we have immediately Problem 1.1, but with the extra assumptions $u \leq a$. In this framework, the previous results give us a rigidity result for Problem 1.1 in exterior domains if we are able to prove the extra condition $u \leq a$. This is the case for example when $f \equiv 0$ or when f is such that there exists $L > 0$ with:

$$f(t) > 0 \text{ for } t \in (0, L) \qquad f(t) < 0 \text{ for } t > L \qquad f(L) = 0 \qquad f'(L) < 0\,.$$

If $f(0) = 0$ then we also need to assume that $f'(0) > 0$. Under such assumptions on the nonlinearity f, Ω is the complement of a ball and u is radially symmetric and increasing. We refer to [13] for the details of the proof. We also mention that for the case $f \equiv 0$, i.e. harmonic overdetermined problems, the rigidity results for exterior problems is also proved in [7].

4.2 *Half-Space Type Domains*

A half-space type domain is a domain diffeomorphic to $\mathbb{R}^n_+ = \{(x_1, \ldots, x_n) \mid x_n > 0\}$. The most typical examples are epigraphs.

The example of nontrivial epigraph supporting a solution to Problem 1.1 given by Del Pino, Pacard and Wei (see the previous section) shows that a classification theory for overdetermined elliptic problems in half-space type domains can be very difficult to be understood. Nevertheless, in dimension 2 such classification has been almost completely understood. This fact corresponds to the following result:

Theorem 4.1 (Ros, Ruiz, and Sicbaldi) *Let $\Omega \subset \mathbb{R}^2$ a C^1 domain whose boundary is unbounded and connected (i.e. it is diffeomorphic to a half-space). Assume that there exists a bounded solution u to Problem 1.1 for some (locally) Lipschitz function $f : [0, +\infty) \to \mathbb{R}$ and some non-zero constant k. Then Ω is a half-plane and u is parallel, that is, u depends only on one variable.*

The only two extra hypothesis here are: (1) the solution u is asked to be bounded, (2) the constant k is not 0. The first hypothesis also appears in the BCN conjecture. For the second one, it is not a priori clear that the case $k = 0$ does not occur, and up to now it remains an open question in the classification of solutions to overdetermined problems in $\mathbb{R}^2$. The proof of Theorem 4.1, [12], is divided in several steps, and we present here the ideas of every step. The authors start with a bounded solution u to Problem 1.1 in a half-plane type domain Ω of $\mathbb{R}^2$ (with $k \neq 0$), and by contradiction they suppose that Ω is not a half-space.

Step 1: Boundedness of the Curvature The first step is to show that the curvature of $\partial\Omega$ is bounded. This is obtained by contradiction. Suppose that there exists a sequence of points $p_n \in \partial\Omega$ such that the curvature at p_n tends to infinity. Then, a dilation of the domain Ω near p_n and of the graph of the function u in $\mathbb{R}^3$ near $(p_n, 0)$ allows to define a sequence of domains Ω_n and functions u_n converging to a solution u_∞ of Problem 3.1 in the limit domain Ω_∞, which is a half-space type domain with the property that at least a point of its boundary has non-zero curvature. Making use of the classification results of [20], there exists no limit domain Ω_∞ with such properties, leading to a contradiction. We remark that in this argument some uniform regularity estimates are needed. Via standard regularity for elliptic problems, the boundedness of the curvature of $\partial\Omega$ implies the uniform boundedness of the $C^{2,\alpha}$ norm of the function u in $\overline{\Omega}$.

Step 2: Ω Must Contain a Half-Plane This step is also obtained by contradiction. Assuming that Ω does not contain any half-space, the authors show that the moving plane method can be applied to it to prove that the solution u must be increasing in one variable. The boundedness of the $C^{2,\alpha}$ norm of the function u in $\overline{\Omega}$ (given in the previous step) and some ideas of the famous proof of the De Giorgi's conjecture in dimension 2 allow to prove that the solution u is one-dimensional and Ω is a half-space, leading to a contradiction.

Step 3: Construction of a Parallel Solution as Limit of the Initial One The crucial ingredient in the proof of Theorem 4.1 is the existence of a divergent sequence of points in $p_n \in \partial\Omega$ such that $\partial\Omega$ converges to a straight line near such sequence. In the proof of this geometric property the authors use in a strong way the facts that the curvature of $\partial\Omega$ is bounded and that Ω contains a half-space. In particular, the sequence of functions $u_n(x) = u(x - p_n)$, $x \in \Omega$, converges to a one-dimensional solution u_∞ of Problem 1.1 in a half-space. In other words, this step shows that the existence of the solution u in Ω implies the existence of a one-dimensional solution u_∞ of the same Problem 1.1 in a half-space (with the same f).

Step 4: Comparison of the Two Solutions Using variational methods, the authors are then able to show that the graph of the solution u_∞ stays below the graph of the initial solution u (recall that Ω contains a half-space). The maximum principle gives a direct contradiction.

Acknowledgments The author is supported by Grants: P18-FR-4049 (Superficies y Análisis Geométrico) funded by Junta de Andalucía and PID2020-117868GB-I00 funded by MCIN/AEI/10.13039/501100011033.

References

1. A. Aftalion, J. Busca, Radial symmetry of overdetermined boundary value problems in exterior domains. C. R. Acad. Sci. Paris Sér. I Math. **324**(6), 633–638 (1997)
2. A.D. Alexandrov, Uniqueness theorems for surfaces in the large. Vestnik Leningrad Univ. Math. **11**, 5–17 (1956)
3. H. Berestycki, L. Caffarelli, L. Nirenberg, Monotonicity for elliptic equations in unbounded Lipschitz domains. Commun. Pure Appl. Math. **50**, 1089–1111 (1997)
4. E. Bombieri, E. De Giorgi, E. Giusti, Minimal cones and the Bernstein problem. Invent. Math. **7**, 243–268 (1969)
5. M. Del Pino, F. Pacard, J. Wei, Serrin's overdetermined problem and constant mean curvature surfaces. Duke Math J. **164**, 2643–2722 (2015)
6. F. Hélein, L. Hauswirth, F. Pacard, A note on some overdetermined problems. Pacific J. Math. **250**(2), 319–334 (2011)
7. D. Khavinson, E. Lundberg, R. Teodorescu, An overdetermined problem in potential theory. Pacific J. Math. **265**(1), 85–111 (2012)

8. Y. Liu, K. Wang, J. Wei, On smooth solutions to one phase free boundary problem in $\mathbb{R}^n$. Int. Math. Res. Not. **20**, 15682–15732 (2021)
9. F. Pacard, P. Sicbaldi, Extremal domains for the first eigenvalue of the Laplace-Beltrami operator. Ann. Institut Four. **59**(2), 515–542 (2009)
10. P. Pucci, J. Serrin, *The Maximum Principle.* Progress in Nonlinear Differential Equations and Their Applications (Birkhauser, Basel, 2007)
11. W. Reichel, Radial symmetry by moving planes for semilinear elliptic boundary value problems on exterior domains. Arch. Ration. Mech. Anal. **137**, 381–394 (1997)
12. A. Ros, D. Ruiz, P. Sicbaldi, A rigidity result for overdetermined elliptic problems in the plane. Commun. Pure Appl. Math. **70**, 1223–1252 (2017)
13. A. Ros, D. Ruiz, P. Sicbaldi, Solutions to overdetermined elliptic problems in nontrivial exterior domains. J. Eur. Math. Soc. **22**(1), 253–281 (2020)
14. A. Ros, P. Sicbaldi, Geometry and Topology of some overdetermined elliptic problem. J. Diff. Eq. **255**(5), 951–977 (2013)
15. F. Schlenk, P. Sicbaldi, Bifurcating extremal domains for the first eigenvalue of the Laplacian. Adv. Math. **229**, 602–632 (2012)
16. J. Serrin, A symmetry theorem in potential theory. Arch. Ration. Mech. Anal. **43**, 304–318 (1971)
17. P. Sicbaldi, New extremal domains for the first eigenvalue of the Laplacian in flat tori. Calc. Var. PDE **37**, 329–344 (2010)
18. B. Sirakov, *Overdetermined Elliptic Problems in Physics*. (Kluwer, Alphen aan den Rijn, 2002), p. 573
19. B. Sirakov, Symmetry for exterior elliptic problems and two conjectures in potential theory. Ann. Institut H. Poincaré (C) Nonl. Anal. **18**(2), 135–156 (2001)
20. M. Traizet, Classification of the solutions to an overdetermined elliptic problem in the plane. Geom. Funct. An. **24**(2), 690–720 (2014)
21. L.A. Vogel, Symmetry and regularity for general regions having solutions to certain overdetermined boundary value problems. Atti Sem. Mat. Fis. Univ. Modena **40**(2), 690–720 (1992)

Part IX
Geometries Defined by Differential Forms

Adjoint Representations of Symmetric Groups

Mahir Bilen Can and Miles Jones

Abstract We study the restriction to the symmetric group, $\mathcal{S}_n$ of the adjoint representation of $\mathrm{GL}_n(\mathbb{C})$. We determine the irreducible constituents of the space of symmetric as well as the space of skew-symmetric $n \times n$ matrices as $\mathcal{S}_n$-modules.

1 Introduction

In [2], the first author and Jeff Remmel introduced the notion of a loop-augmented rooted forest; they explained its combinatorial representation theoretic role for the conjugation action of the symmetric group on certain subsets of the partial transformation semigroups. In this note, we present an application of this development in a basic Lie theory context.

Let G be a Lie group, and let $\mathfrak{g}$ denote the Lie algebra of G. The conjugation action $G \times G \to G$, $(g, h) \mapsto ghg^{-1}$, $g, h \in G$ leads to a linear representation of G on its tangent space at the identity element,

$$\begin{aligned} \mathrm{Ad} : G &\to \mathrm{Aut}(\mathfrak{g}) \\ g &\mapsto \mathrm{Ad}_g . \end{aligned} \tag{1}$$

The representation (1), which is known as the *adjoint representation of* G, has a fundamental place in the structure theory of Lie groups. It has a concrete description when G is a closed subgroup of $\mathrm{GL}_n(\mathbb{C})$, the general linear group of $n \times n$ matrices. In this case, $\mathfrak{g}$ is a Lie subalgebra of the Lie algebra of $n \times n$ matrices, and Ad_g ($g \in$

M. B. Can (✉)
Tulane University, New Orleans, LA, USA
e-mail: mcan@tulane.edu

M. Jones
University of California San Diego, La Jolla, CA, USA
e-mail: mej016@ucsd.edu

P. Cerejeiras et al. (eds.), *Current Trends in Analysis, its Applications and Computation*, Research Perspectives,
https://doi.org/10.1007/978-3-030-87502-2_47

G) is defined by $\mathrm{Ad}_g(X) = gXg^{-1}$ for $X \in \mathfrak{g}$. Now let $\mathcal{S}_n$ denote the symmetric group of permutations of $\{1, \ldots, n\}$. We will view $\mathcal{S}_n$ as a subgroup of $\mathrm{GL}_n(\mathbb{C})$ by identifying permutations $\pi \in \mathcal{S}_n$ with invertible $n \times n$ 0/1 matrices with at most one 1 in each row and column. The basic representation theoretic question that we address in the present paper is the following:

> What are the irreducible constituents of the S_n-representation that is obtained from (1) by restriction?

Surprisingly, even though the adjoint representation is at the heart of Lie theory, to the best of our knowledge the answer to our question is missing from the literature, at least, it is not presented in the way that we are answering it. To state our theorem we set up the notation.

It is well-known that the finite dimensional irreducible representations of $\mathcal{S}_n$ are indexed by the integer partitions of n. The Frobenius character map $V \mapsto F_V$ is an assignment of symmetric functions to the finite dimensional representations of $\mathcal{S}_n$. (We will explain this in more detail in the sequel.) In particular, if V is the irreducible representation determined by an integer partition λ, then F_V is a Schur symmetric function, denoted by s_λ; if $V = \bigoplus V_i$ is a decomposition of V into $\mathcal{S}_n$-submodules, then $F_V = \sum F_{V_i}$. Our first main result is as follows.

Theorem 1.1 *Let n be an integer such that $n \geq 2$. The Frobenius character of the adjoint representation of $\mathcal{S}_n$ on $Mat_n(\mathbb{C})$ is given by*

$$F_{Mat_n(\mathbb{C})} = \begin{cases} 2s_2 + 2s_{1,1} & \text{if } n = 2; \\ 2s_3 + 3s_{2,1} + s_{1,1,1} & \text{if } n = 3; \\ 2s_n + 3s_{n-1,1} + s_{n-2,2} + s_{n-2,1,1} & \text{if } n \geq 4. \end{cases} \tag{2}$$

The space of symmetric $n \times n$-matrices, denoted by $\mathrm{Sym}_n(\mathbb{C})$, is closed under the adjoint action of the orthogonal group $\mathrm{O}_n(\mathbb{C}) := \{g \in \mathrm{GL}_n(\mathbb{C}) : gg^\top = id\}$. However, $\mathrm{Sym}_n(\mathbb{C})$ is not closed under the adjoint action of $\mathrm{GL}_n(\mathbb{C})$. Nevertheless, since $\mathcal{S}_n$ is a subgroup of $\mathrm{O}_n(\mathbb{C})$, we see that the representation

$$\mathrm{Ad} : \mathcal{S}_n \to \mathrm{Aut}(\mathrm{Sym}_n(\mathbb{C})) \tag{3}$$

is defined. Moreover, since there is a direct sum decomposition of matrices

$$\mathrm{Mat}_n(\mathbb{C}) = \mathrm{Sym}_n(\mathbb{C}) \oplus \mathrm{Skew}_n(\mathbb{C}),$$

where $\mathrm{Skew}_n(\mathbb{C})$ is the space of $n \times n$ skew-symmetric matrices, a complementary adjoint representation exists:

$$\mathrm{Ad} : \mathcal{S}_n \to \mathrm{Aut}(\mathrm{Skew}_n(\mathbb{C})). \tag{4}$$

Our second main result is about the Frobenius character map image of the corresponding $\mathcal{S}_n$ representations.

Theorem 1.2 *Let n be an integer such that $n \geq 2$. The Frobenius character of the adjoint representation (3) of $\mathcal{S}_n$ on $Sym_n(\mathbb{C})$ is given by*

$$F_{Sym_n(\mathbb{C})} = \begin{cases} 2s_2 + s_{1,1} & \text{if } n = 2; \\ 2s_3 + 3s_{2,1} & \text{if } n = 3; \\ 2s_n + 2s_{n-1,1} + s_{n-2,2} & \text{if } n \geq 4. \end{cases} \tag{5}$$

Corollary 1.1 *Let n be an integer such that $n \geq 2$. The Frobenius character of the adjoint representation (4) of $\mathcal{S}_n$ on $Skew_n(\mathbb{C})$ is given by*

$$F_{Skew_n(\mathbb{C})} = \begin{cases} s_{1,1} & \text{if } n = 2; \\ s_{1,1,1} & \text{if } n = 3; \\ s_{n-1,1} + s_{n-2,1,1} & \text{if } n \geq 4. \end{cases} \tag{6}$$

2 Notation and Background

Let n be a positive integer. We will use the notation $[n]$ and $\overline{[n]}$ to denote the sets $\{1, \ldots, n\}$ and $[n] \cup \{0\}$, respectively. A *partition* of n is a nonincreasing sequence of positive integers $\lambda := (\lambda_1, \ldots, \lambda_l)$ such that $\sum_{i=1}^{l} \lambda_i = n$. In this case, we will write $\lambda \vdash n$. It is well-known that the conjugacy classes in $\mathcal{S}_n$ are parameterized by the partitions of n. Consequently, the irreducible characters of $\mathcal{S}_n$ are in 1-1 correspondence with $\{\lambda \vdash n\}$.

2.1 Symmetric Functions and Plethysm

Let $X := \{x_1, x_2, \ldots\}$ be a set of variables that are algebraically independent over $\mathbb{Q}$. A formal power series $f(X)$ is called a *symmetric function* if f is invariant under every finite permutation of its variables. The set of all homogeneous symmetric functions of degree n has the structure of a $\mathbb{Q}$-vector space, denoted by Λ^n. The direct sum $\bigoplus_{n\geq 0} \Lambda^n$ will be denoted by Λ; it forms a subring of the ring of formal power series with variables in X. Note that $\dim_{\mathbb{Q}} \Lambda^n = |\{\lambda \vdash n\}|$. For our purposes, the following vector space bases for Λ^n will be instrumental: $\{p_\lambda\}_{\lambda\vdash n}$, $\{h_\lambda\}_{\lambda\vdash n}$, and $\{s_\lambda\}_{\lambda\vdash n}$. The first of these bases, called the *power sum basis*, is defined as follows: For $\lambda = (\lambda_1, \ldots, \lambda_l) \vdash n$, p_λ is the product $\prod_i p_{\lambda_i}$, where $p_{\lambda_i} := \sum x_j^{\lambda_i}$ for $i \in \{1, \ldots, l\}$. Likewise, h_λ is the product $\prod_i h_{\lambda_i}$, where h_{λ_i} ($i \in \{1, \ldots, l\}$) is the sum of all monomials of degree λ_i. Finally, the *Schur function basis*, $\{s_\lambda\}_{\lambda\vdash n}$

is defined by setting $s_\lambda := \det(h_{\lambda_i+j-i})_{i,j=1}^{l}$. In particular, for all $k \geq 1$, we have $s_{(k)} = h_k$. From now on, for simplifying the notation, we will denote $s_{(k)}$ by s_k.

The *Frobenius character map* is the map between the class functions on $\mathcal{S}_n$ and Λ^n; it is defined by extending linearly the assignment $\delta_\sigma \mapsto \frac{1}{n!} p_\lambda$, where $\sigma \subset \mathcal{S}_n$ is a conjugacy class of type λ and δ_σ is the indicator function

$$\delta_\sigma(x) = \begin{cases} 1 & \text{if } x \in \sigma, \\ 0 & \text{otherwise.} \end{cases}$$

It turns out that if χ^λ is the irreducible character of $\mathcal{S}_n$ indexed by the partition λ, then $F(\chi^\lambda) = s_\lambda$. In the sequel, we will not distinguish between representations of $\mathcal{S}_n$ and their corresponding characters. In particular, we will often write the *Frobenius character of an orbit* to mean the image under F of the character of the representation of $\mathcal{S}_n$ that is defined by the action on the orbit. If V is an $\mathcal{S}_n$-representation, then we will denote its Frobenius character by F_V. For two irreducible characters χ^μ and χ^λ indexed by partitions λ and μ, the Frobenius character image of the "plethysm" $\chi^\lambda[\chi^\mu]$ is the plethystic substitution $s_\lambda[s_\mu]$ of the corresponding Schur functions. Roughly speaking, the *plethysm of Schur function* s_λ *with* s_μ is the symmetric function obtained from s_λ by substituting the monomials of s_μ for the variables of s_λ. This defines a binary operation on symmetric functions, $[\cdot] : \Lambda \times \Lambda \to \Lambda$ which is uniquely determined by the following three axioms:

P1. For all $m, n \geq 1$, $p_m[p_n] = p_{mn}$.
P2. For all $m \geq 1$, the map $g \mapsto p_m[g]$, $g \in \Lambda$ defines a $\mathbb{Q}$-algebra homomorphism on Λ.
P3. For all $g \in \Lambda$, the map $h \mapsto h[g]$, $h \in \Lambda$ defines a $\mathbb{Q}$-algebra homomorphism on Λ.

For a detailed discussion of the properties of plethysm of symmetric functions, we recommend the article [4].

2.2 Partial Transformations and Loop Augmented Forests

Let us identify $[n]$ with the standard basis for the vector space of $n \times 1$ column matrices with complex entries. We will call a nonzero $n \times n$ 0/1 matrix f with at most one 1 in each of its columns a *partial transformation on* $[n]$. The nomenclature is justified by viewing such a matrix as a function $f : A \to [n]$, where A is a subset of $[n]$. The value of f on $i \in A$ is determined by the matrix multiplication $f \cdot i$. The set of all partial transformations on $[n]$, called the *partial transformation semigroup on* $[n]$, and denoted by $\mathcal{P}_n$, is a semigroup with respect to matrix multiplication. The set of all nilpotent partial transformations will be denoted by $\mathcal{C}_n$. The unit group of $\mathcal{P}_n$ is equal to $\mathcal{S}_n$. Therefore, its conjugation action on $\mathcal{P}_n$ makes sense. Note that $\mathcal{C}_n$ is stable under this conjugation action of $\mathcal{S}_n$.

It will be useful to view $\mathcal{P}_n$ as the set of incidence matrices of labeled directed graphs. For $\tau \in \mathcal{P}_n$, the corresponding graph has a directed edge from the i-th vertex to the j-th vertex if the (i, j)-th entry of τ is 1. Note that the underlying graphs depend only on the $\mathcal{S}_n$-conjugacy classes. A pair (τ, ϕ), where τ is a rooted forest on n vertices and ϕ is a bijective map from $[n]$ onto the vertex set of τ is called a *labeled rooted forest.* The nilpotent partial transformations give the labeled rooted forests. To ease our notation, we will omit writing the corresponding labeling function despite the fact that the action of $\mathcal{S}_n$ does not change the underlying forest but the labeling function only. In particular, when we write $\mathcal{S}_n \cdot \tau$ we actually mean the orbit

$$\mathcal{S}_n \cdot (\tau, \phi) = \{(\tau, \phi') : \phi' = \sigma \cdot \phi, \ \sigma \in \mathcal{S}_n\}. \tag{7}$$

The right hand side of (7) is an $\mathcal{S}_n$-set, hence it defines a representation of $\mathcal{S}_n$. More generally, to any partial transformation τ in $\mathcal{P}_n$, we associate the representation corresponding to the orbit $\mathcal{S}_n \cdot \tau$. We will call the resulting representation the *odun of* τ, and denote it by $o(\tau)$. Since, $o(\tau)$ depends only on the underlying graph, we will write $\mathrm{Stab}_{\mathcal{S}_n}(\tau)$ to denote the stabilizer subgroup of the pair (τ, ϕ). Since $o(\tau)$ is a permutation representation of $\mathcal{S}_n$, we have $o(\tau) = \mathrm{Ind}_{\mathrm{Stab}_{\mathcal{S}_n}(\tau)}^{\mathcal{S}_n} \mathbf{1} = \mathbb{C}[\mathcal{S}_n / \mathrm{Stab}_{\mathcal{S}_n}(\tau)]$. A method for computing the Frobenius character image of $o(\tau)$ is presented in [1].

A *loop-augmented forest* is a rooted forest such that there is at most one loop at each of its roots. In Fig. 1, we depict a loop-augmented forest on 22 vertices and four loops.

Next, we will explain how to interpret loop-augmented forests in terms of partial functions, so, let σ be a loop-augmented forest. Then some of the roots of σ have loops. There is still a partial function for σ, as defined in the previous paragraph for a rooted forest. The loops in this case correspond to the "fixed points" of the associated function. Indeed, for the loop at the i-th vertex we have a 1 at the (i, i)-th entry of the corresponding matrix representation of the partial transformation. Recall that the permutation action of $\mathcal{S}_n$ on the labels translates to the conjugation action on the incidence matrix. By an appropriate relabeling of the vertices, the incidence matrix of a loop-augmented rooted forest can be brought to an upper-triangular form. Clearly, the conjugates of a nilpotent (respectively unipotent) matrix are still nilpotent (respectively unipotent). Let U be an $n \times n$ upper triangular matrix with the Jordan decomposition $U = D + N$, where D (resp. N) is a diagonal (resp. nilpotent) matrix. If σ is from $\mathcal{S}_n$, then $\sigma \cdot U = \sigma U \sigma^{-1} = \sigma D \sigma^{-1} + \sigma N \sigma^{-1}$ shows that the conjugation action on the loop-augmented forests is equivalent

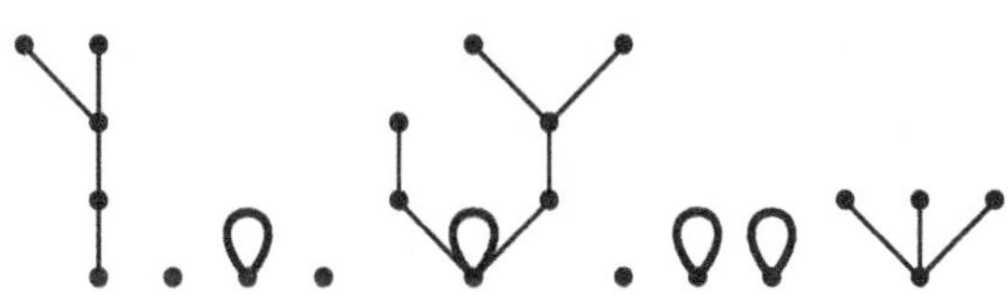

Fig. 1 A loop-augmented forest

to the simultaneous conjugation action on the labeled rooted forests and the $\mathcal{S}_n$ representation on the space of diagonal matrices. The following theorem from [2] describes the character of the odun corresponding to a loop-augmented rooted forest.

Theorem 2.1 *Let f be a partial function representing a loop-augmented forest on n vertices. Then f is similar to a block diagonal matrix of the form $\begin{pmatrix} \sigma & 0 \\ 0 & \tau \end{pmatrix}$ for some $\sigma \in \mathcal{S}_k$ and $\tau \in \mathcal{C}_{n-k}$. Furthermore, if ν, which is a partition of k, is the conjugacy type of σ in $\mathcal{S}_k$ and if the underlying rooted forest of the nilpotent partial transformation τ has λ_1 copies of the rooted tree τ_1, λ_2 copies of the rooted forest τ_2 and so on, then the character of $o(f)$ is given by $\chi_{o(f)} = \chi^{\nu} \cdot \chi_{o(\tau)} = \chi^{\nu} \cdot (\chi^{(\lambda_1)}[\chi_{o(\tau_1)}]) \cdot (\chi^{(\lambda_2)}[\chi_{o(\tau_2)}]) \cdot \dots \cdot (\chi^{(\lambda_r)}[\chi_{o(\tau_r)}])$.*

3 Proof of Theorem 1.1

Let i and j be two integers from $[n]$. We denote by $E_{i,j}$ the $n \times n$ 0/1 matrix with 1 at its (i, j)-th entry and 0's elsewhere. As a vector space, $\mathrm{Mat}_n(\mathbb{C})$ is spanned by $E_{i,j}$'s. In fact, $\{E_{i,j} : i, j \in [n]\}$ constitute a basis,

$$\mathrm{Mat}_n(\mathbb{C}) = \bigoplus_{i,j \in [n]} \mathbb{C}E_{i,j}. \tag{8}$$

Clearly, $E_{i,j}$'s are partial transformation matrices, so, they represent labeled loop-augmented rooted forests. The structures of these forests depend on the indices.

(1) For $i, j \in [n]$, if $i \neq j$, then the labeled loop-augmented rooted forest corresponding to $E_{i,j}$ is as in Fig. 2.
By [1, Corollary 6.2], we see that $F_{o(E_{i,j})} = s_1 \cdot s_1 \cdot s_{n-2}[s_1] = s_1^2 s_{n-2}$. Now we apply the Pieri's formula twice:

$$F_{o(E_{i,j})} = \begin{cases} s_2 + s_{1,1} & \text{if } n = 2; \\ s_3 + 2s_{2,1} + s_{1,1,1} & \text{if } n = 3; \\ s_n + 2s_{n-1,1} + s_{n-2,2} + s_{n-2,1,1} & \text{if } n \geq 4. \end{cases} \tag{9}$$

(2) Let $i, j \in [n]$ be such that $i = j$. In this case, the labeled loop-augmented rooted forest corresponding to $E_{i,i}$ is as in Fig. 3.

Fig. 2 The basis element $E_{i,j}$ as a labeled loop-augmented rooted forest

Fig. 3 The basis element $E_{i,i}$ as a labeled loop-augmented rooted forest

By Theorem 2.1, the Frobenius character of the $\mathcal{S}_n$-module structure on the orbit $\mathcal{S}_n \cdot E_{i,i}$ is given by $F_{o(E_{i,i})} = s_1 \cdot s_{n-1}[s_1] = s_1 s_{n-1}$. By the Pieri's formula, we have

$$F_{o(E_{i,j})} = s_n + s_{n-1,1}. \tag{10}$$

Note that $E_{k,l} \in \mathcal{S}_n \cdot E_{i,j}$ for all $k, l \in [n]$ with $k \neq l$. Indeed, $\mathcal{S}_n$ acts on $E_{i,j}$ by permuting the labels on the vertices. Note also that $E_{k,k} \in \mathcal{S}_n \cdot E_{i,i}$ for all $k \in [n]$. Therefore, in the light of direct sum (8), by combining (9) and (10), we see that the Frobenius character of the adjoint representation of $\mathcal{S}_n$ on $\mathrm{Mat}_n(\mathbb{C})$ is as we claimed in Theorem 1.1. This finishes the proof of our first main result. □

4 Proofs of Theorem 1.2 and Corollary 1.1

For two distinct elements i and j from $[n]$, we set $F_{i,j} := E_{i,j} + E_{j,i}$. A vector space basis for $\mathrm{Sym}_n(\mathbb{C})$ is given by the union

$$\{E_{i,i} : i = 1, \ldots, n\} \cup \{F_{i,j} : i, j \in [n],\ i \neq j\}. \tag{11}$$

Notice that $F_{i,j}$'s are contained in $\mathcal{P}_n$. However, this time, the directed graph corresponding to $F_{i,j}$ is not a forest. See Fig. 4.

Lemma 4.1 *Let i and j be two elements from $[n]$ such that $i \neq j$. The orbit of the adjoint action of $\mathcal{S}_n$ on the matrix $F_{i,j}$ is the same as the permutation action of $\mathcal{S}_n$ on the set of all labelings of the vertices of the directed graph in Fig. 5. In particular, $F_{1,2} \in \mathcal{S}_n \cdot F_{i,j}$.*

Proof The indices i and j are distinct but arbitrary elements from $[n]$, so, it suffices to prove our second claim only. Also, without loss of generality we can assume that $i < j$. By applying the adjoint action of the transposition $(1, j)$ to $F_{i,j}$, we find $\mathrm{Ad}_{(1,j)}(F_{i,j}) = F_{1,j}$. Then we apply the adjoint action of the transposition $(2, j)$

Fig. 4 The basis element $F_{i,j}$ as a directed graph

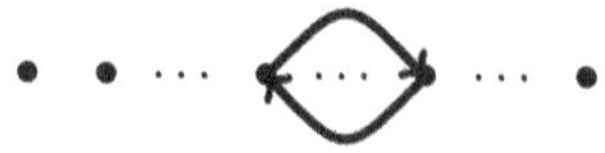

Fig. 5 The arrows are between the i-th and the j-th vertices

to $F_{1,j}$, which gives $\mathrm{Ad}_{(2,j)}(F_{1,j}) = F_{1,2}$. Therefore, we have $\mathrm{Ad}_{(2,j)(1,j)}(F_{i,j}) = F_{1,2}$. This finishes the proof. □

In the notation of [3], any element from (11) is a partial involution. Following our arguments from [2] for the idempotents of $\mathcal{P}_n$, next, we will compute the stabilizer subgroup of the partial involution $F_{i,j}$. Let us mention in passing that, for all i in $[n]$, the matrix $E_{i,i}$ is already an idempotent matrix, so, we know its stabilizer subgroup.

Lemma 4.2 *If i and j are two distinct elements from $[n]$, then the odun of $F_{i,j}$ is given by $o(F_{i,j}) = \bigoplus_{i,j\in[n],\ i\neq j} \mathbb{C}F_{i,j}$. The stabilizer subgroup of $F_{i,j}$ in $\mathcal{S}_n$ is isomorphic to the parabolic subgroup $\mathcal{S}_2 \times \mathcal{S}_{n-2}$.*

Proof As we already mentioned before, the adjoint (conjugation) action of $\mathcal{S}_n$ on $\mathcal{P}_n$ amounts to the permutation action of $\mathcal{S}_n$ on the labels of the associated graphs (Lemma 4.1). Our first claim readily follows from this argument. Now without loss of generality assume that $F_{i,j} = F_{1,2} = \begin{bmatrix} 0 & 1 & 0 \\ 1 & 0 & 0 \\ 0 & 0 & \mathbf{0}_{n-2} \end{bmatrix}$. Clearly, the stabilizer subgroup of $F_{i,j}$ in $\mathcal{S}_n$ consists of the matrices of the form $\begin{bmatrix} \sigma_1 & 0 \\ 0 & \sigma_2 \end{bmatrix}$, where $\sigma_1 \in \mathcal{S}_2$ and $\sigma_2 \in \mathcal{S}_{n-2}$. This finishes the proof of our claim. □

Proof of Theorem 1.2 It follows from Lemma 4.2 that the representation of $\mathcal{S}_n$ on the orbit $o(F_{i,j}) \cong \mathcal{S}_n \cdot F_{1,2}$ is isomorphic to $\mathrm{Ind}_{\mathcal{S}_2\times\mathcal{S}_{n-2}}^{\mathcal{S}_n} \mathbf{1}$. In particular, its Frobenius character image is given by $F_{o(F_{i,j})} = s_2 s_{n-2}$. By using the Pieri's formula, we find that

$$F_{o(F_{i,j})} = \begin{cases} s_2 & \text{if } n = 2; \\ s_3 + s_{2,1} & \text{if } n = 3; \\ s_n + s_{n-1,1} + s_{n-2,2} & \text{if } n \geq 4. \end{cases} \tag{12}$$

The rest of the proof follows from combining (12) with the formula (10). □

Proof of Corollary 1.1 The Frobenius character of $\mathrm{Mat}_n(\mathbb{C})$ is the sum of the Frobenius characters of $\mathrm{Sym}_n(\mathbb{C})$ and $\mathrm{Skew}_n(\mathbb{C})$. The rest of the proof is a consequence of Theorems 1.1 and 1.2. □

Acknowledgments This work was partially supported by a grant from Louisiana Board of Regents. The authors thank John Shareshian and Sheila Sundaram for useful conversations about several other proofs of our results.

References

1. M.B. Can, A representation on the labeled rooted forests. Commun. Algebra **46**(10), 4273–4291 (2018)
2. M.B. Can, J. Remmel, Loop-augmented forests and a variant of Foulkes's conjecture. Algebr. Comb. **1**(5), 573–601 (2018)
3. M.B. Can, Y. Cherniavsky, T. Twelbeck, Bruhat order on partial fixed point free involutions. Electron. J. Combin. **21**(4) (2014). Paper 4.34
4. N.A. Loehr, J. Remmel, A computational and combinatorial exposé of plethystic calculus. J. Algebr. Combin. **33**(2), 163–198 (2011)

Examples of Equivariant Lagrangian Mean Curvature Flow

Jason D. Lotay

Abstract We describe important examples of Lagrangian mean curvature flow in $\mathbb{C}^2$ which are invariant under a circle action. Through these examples, we see compact and non-compact situations, long-time existence, singularities forming via explicit models, and significant objects in Riemannian and symplectic geometry, including the Clifford torus, Whitney sphere and Lawlor necks.

1 Introduction

Lagrangian mean curvature flow has generated major interest from several viewpoints, due to its connections with Riemannian geometry (particularly calibrated geometry), symplectic topology, gauge theory, Calabi–Yau (and, more generally, Kähler–Einstein) manifolds and Mirror Symmetry. In particular, Lagrangian mean curvature flow has the potential to lead to striking applications in diverse areas.

However, there are relatively few cases in which the Lagrangian mean curvature flow is explicitly understood. An exception is the setting of equivariant Lagrangian mean curvature flow in $\mathbb{C}^n$, which was first studied in [4, 6, 7]. In particular, understanding equivariant flows in $\mathbb{C}^2$ was crucial in the ground-breaking result by Neves [7] that *any* embedded Lagrangian in a Calabi–Yau 2-fold is Hamiltonian isotopic to a Lagrangian which develops a finite-time singularity under Lagrangian mean curvature flow.

In the case of $\mathbb{C}^2$ (with its standard symplectic form), which will be the focus of this article, circle-invariant Lagrangian surfaces $L \subseteq \mathbb{C}^2$ are given by curves $\gamma : I \subseteq \mathbb{R} \to \mathbb{C}$ as follows:

$$L = \{(\gamma(s)\cos\phi, \gamma(s)\sin\phi) \in \mathbb{C}^2 : s \in I, \phi \in [0, 2\pi)\}. \tag{1}$$

J. D. Lotay (✉)
Mathematical Institute, University of Oxford, Oxford, UK
e-mail: jason.lotay@maths.ox.ac.uk

P. Cerejeiras et al. (eds.), *Current Trends in Analysis, its Applications and Computation*, Research Perspectives,
https://doi.org/10.1007/978-3-030-87502-2_48

This is the heart of the advantage of studying equivariant Lagrangian mean curvature flow, in that it reduces to a flow of curves in $\mathbb{C}$. (It is worth emphasising that points on γ define circles in the Lagrangian L in (1), except the origin which gives just a point in L.)

Recent exciting progress has been made in obtaining a detailed understanding of equivariant Lagrangian mean curvature flow in $\mathbb{C}^2$ (and, more generally, in $\mathbb{C}^n$), namely in [3, 8–11]. We shall briefly describe the main outcomes of this work in this article, which we organize into three cases that depend on the curve γ in $\mathbb{C}$ defining our equivariant Lagrangian:

- Lagrangian tori T^2, given by embedded closed curves;
- Lagrangian spheres S^2, given by immersed closed curves;
- Lagrangian cylinders $S^1 \times \mathbb{R}$, given by embedded open curves.

(One may also have planes $\mathbb{R}^2$, given by certain open curves, but these have not yet been studied.)

The examples described here provide important results for Lagrangian mean curvature flow, and it certainly motivates future research in the equivariant setting to obtain further progress in our understanding of the flow.

2 Preliminaries

We begin with some fundamental notions we require to describe the examples.

If $z_1 = x_1 + iy_1$, $z_2 = x_2 + iy_2$ are standard coordinates on $\mathbb{C}^2$, we have the standard symplectic form given by

$$\omega = \frac{i}{2}(dz_1 \wedge d\bar{z}_1 + dz_2 \wedge d\bar{z}_2) = dx_1 \wedge dy_1 + dx_2 \wedge dy_2.$$

One may then check that surfaces L as in (1) are Lagrangian, namely $\omega|_L \equiv 0$. Lagrangian mean curvature flow in $\mathbb{C}^2$ (and any Kähler–Einstein manifold) is the mean curvature flow with a Lagrangian initial condition:

$$\frac{\partial L}{\partial t} = H, \tag{2}$$

where H is the mean curvature vector of L.

If we take a circle-invariant Lagrangian L as in (1), then Lagrangian mean curvature flow L_t preserves the circle-invariance, so we may write the flow (2) as a flow of curves γ_t in $\mathbb{C}$:

$$\frac{\partial \gamma}{\partial t} = \kappa - \frac{\gamma^\perp}{|\gamma|^2}, \tag{3}$$

where κ is the curvature vector of γ with respect to the Euclidean metric on $\mathbb{C}$, and $\gamma^\perp$ is the projection of γ to the normal direction to γ.

The key to understanding any geometric flow is to study the formation of singularities. There are two types of singularities in Lagrangian mean curvature flow at a finite time $t = T$, determined by the behaviour of the second fundamental form A_t of the flow L_t, which must blow-up at the singularity.

- Type I: singularities where $\lim_{t\nearrow T}\sup_{L_t}|A_t|^2(T-t) < \infty$.
- Type II: any other singularity at $t = T$.

The simplest examples of Type I singularities are given by *self-shrinkers*: these are solutions to (2) that simply shrink by dilations under the flow.

There are also two main ways of studying these singularities via blow-up.

- Type I blow-up. For a positive sequence $\sigma_i \to \infty$ and $w \in \mathbb{C}^2$, we define

$$L_s^i = \sigma_i(L_{T+\sigma_i^{-2}s} - w) \quad \forall s \in [-\sigma_i^2 T, 0).$$

  The sequence L_s^i subconverges weakly (i.e. as a Brakke flow) as $i \to \infty$, to a limit flow L_s^∞ in $\mathbb{C}^2$ for all $s < 0$, which is called a Type I blow-up (or tangent flow) at (w, T). A Type I blow-up is a self-shrinker and it provides a "first approximation" to the nature of the singularity.
- Type II blow-up. Suppose we have a sequence $(w_i, t_i) \in \mathbb{C}^2 \times (0, T)$ such that $t_i \to T$ and $\sigma_i := A(w_i, t_i) = \sup\{|A_t(z)| : z \in L_t,\ t \le t_i\} > 0$. If

$$L_s^i = \sigma_i(L_{t_i+\sigma_i^{-2}s} - w_i) \quad \forall s \in [-\sigma_i^2 t_i, 0),$$

  then the sequence L_s^i subconverges as $i \to \infty$ and it will define a smooth solution L_s^∞ in $\mathbb{C}^2$ for all $s < 0$, which we call a Type II blow-up. Type II blow-ups give more refined information about the singularity formation.

Suppose L_t is a Type II blow-up. For a sequence $\lambda_i \to \infty$, we let

$$L_s^i := \lambda_i^{-1} L_{\lambda_i^2 s} \qquad \forall s < 0,$$

and define the *blow-down* L_s^∞ as a subsequential limit of the sequence L_s^i. A blow-down of a Type II blow-up is a self-shrinker for Lagrangian mean curvature flow in $\mathbb{C}^2$ and will "approximate" the Type II blow-up. Recently, classification results have been obtained for Type II blow-ups in Lagrangian mean curvature flow in terms of their blow-downs [5].

3 Tori: Embedded Closed Curves

In this section we examine embedded, equivariant, Lagrangian 2-tori L in $\mathbb{C}^2$ which are defined by embedded closed curves γ in $\mathbb{C}$ which do not meet the origin as in Fig. 1.

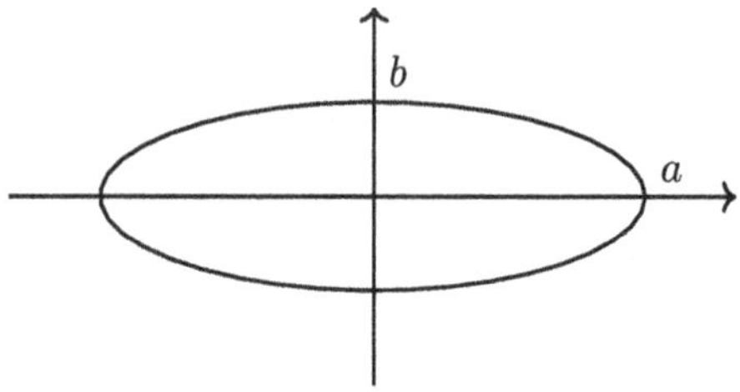

Fig. 1 An ellipse defining an embedded Lagrangian 2-torus T^2

Example 1 (Clifford Torus/Circle) The case when γ is a circle, i.e. when $a = b$ in Fig. 1, corresponds to L being the well-known Clifford torus (which is Lagrangian) in $\mathbb{C}^2$. The Clifford torus is a self-shrinker for Lagrangian mean curvature flow, and this is reflected in the fact that a circle in the plane will self-similarly shrink to a point under the flow (3). Hence, the flow starting at the Clifford torus has a Type I singularity.

Example 2 (Ellipses) Recently, it was shown in [3], that for any Lagrangian torus L defined by an ellipse as in Fig. 1 with $a \neq b$, the Lagrangian mean curvature flow starting at L must develop a finite-time singularity which is *not* modelled on the Clifford torus. Hence, the Clifford torus is *unstable* under Lagrangian mean curvature flow, even under arbitrarily small Hamiltonian perturbations (which correspond to taking variations with ab constant).

Further, it is known by work in [4, 6] that, for $a \gg b$, the Lagrangian mean curvature flow starting at a torus defined by such an ellipse would have to develop a Type II singularity at the origin, whose Type I blow-up is a circle-invariant Lagrangian pair of planes defined by a pair of lines in $\mathbb{C}$ which meet at right angles at the origin. The curve at the singular time will look something like the figure eight curve in Fig. 2, and thus the Lagrangian torus will become a 2-sphere at the singular time. We expect that this same behaviour occurs for any $a \neq b$.

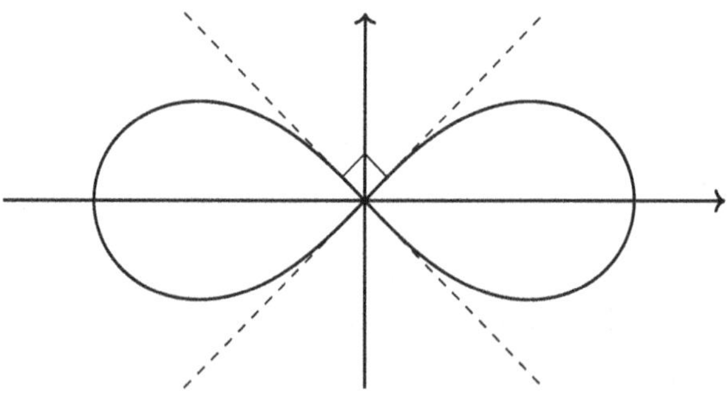

Fig. 2 A figure eight defining an immersed Lagrangian 2-sphere S^2

Example 3 (Star-Shaped Curves) In [4], the authors gave conditions under which a star-shaped curve with respect to the origin will contract to a point under the flow (3) so that the corresponding Lagrangian mean curvature flow has a finite-time Type I singularity. Moreover, the Type I blow-up is a Lagrangian self-shrinker constructed in [1]. However, in $\mathbb{C}^2$, the constraints on the star-shaped curve mean that the Lagrangian tori cannot be embedded.

4 Spheres: Immersed Closed Curves

In this section we consider Lagrangian 2-spheres in $\mathbb{C}^2$ (which must be immersed) and are equivariant, so defined by curves as in Fig. 2.

The curve γ in Fig. 2 and the resulting Lagrangian L defined by γ as in (1) has two distinct properties.

(a) The curve γ is contained in the region defined by the dashed lines, which each make an angle of $\frac{\pi}{4}$ with the horizontal axis, and γ meets any circle centred at the origin in at most 4 points.
(b) The Ricci curvature of the induced metric on L satisfies $\mathrm{Ric} \geq cr^2$ where $c > 0$ is a constant and r is the distance to the origin in $\mathbb{C}^2$.

We shall describe results concerning curves satisfying each of these properties.

Example 4 (Case (a)) For curves γ as in (a), Viana [10] showed that the flow (3) starting at γ has a finite-time singularity at the origin where the flow shrinks to a point. The corresponding Lagrangian mean curvature flow has a Type II singularity at the origin, whose Type I blow-up is a plane with multiplicity two (roughly, two copies of the same plane).

Example 5 (Case (b)) For curves γ as in (b), it was shown in [8] that again the flow (3) starting at γ has a finite-time singularity at 0, where the flow shrinks to a point. Moreover, the Lagrangian mean curvature flow given by γ has a Type II singularity at the origin whose Type II blow-up is the product of a Grim Reaper curve and a real line. The Grim Reaper curve is shown in Fig. 3 and is given by

$$\gamma(t) = \{-\log\cos y + iy \in \mathbb{C} : y \in (-\tfrac{\pi}{2}, \tfrac{\pi}{2})\}. \tag{4}$$

It is a *translator* for curve shortening flow, i.e. it just translates to the right along the flow.

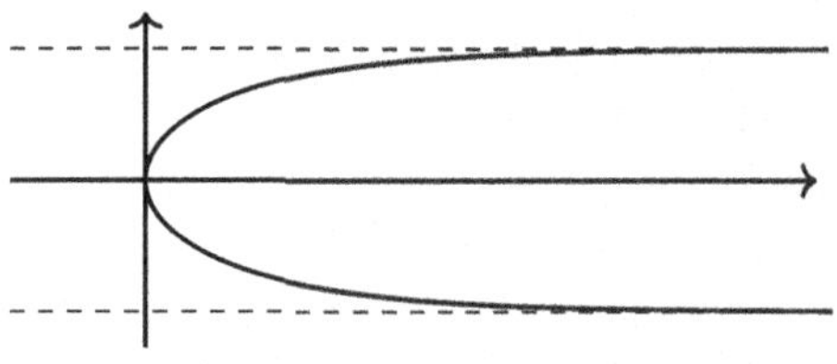

Fig. 3 Grim Reaper curve

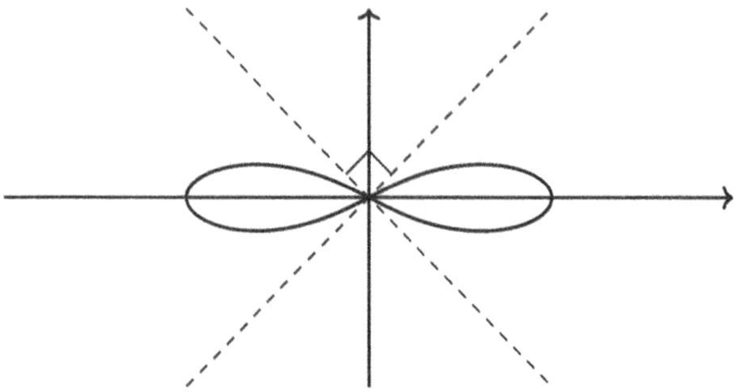

Fig. 4 A collapsing figure eight along the flow (3)

Example 6 (Whitney Sphere/Figure Eight) The figure eight in Fig. 2 defines the well-known Whitney sphere in $\mathbb{C}^2$. The flow (3) starting at this curve cannot be self-similar as there are no Lagrangian self-shrinking spheres [2]. However, by the work in [8, 10] it will still shrink to a point in finite time.

At the finite-time singularity the Lagrangian mean curvature flow will have a Type II singularity whose Type I blow-up is a plane with multiplicity two, and the Type II blow-up is the product of a Grim Reaper curve with a line. Along the flow it deforms as in Fig. 4, "squashing" vertically faster than it does horizontally, and one can imagine the Grim Reaper curve emerging at the extreme left (and right) points of the curve in the limit.

5 Cylinders: Embedded Open Curves

For the final set of examples, we look at embedded non-compact curves asymptotic to straight lines, as in Fig. 5. The behaviour of the flow crucially depends on the angle α in Fig. 5.

Example 7 ($\alpha = \frac{\pi}{2}$: Lawlor Necks) In this case, the curve γ in Fig. 5 is a standard hyperbola asymptotic to a pair of straight lines meeting at right angles. The corresponding Lagrangian L is minimal, i.e. it has $H = 0$, and so is a stationary point for Lagrangian mean curvature flow (2). The Lagrangian L is called a Lawlor neck.

Recently, Su [9] considered a natural class of curves γ asymptotic to a pair of lines with angle $\alpha = \frac{\pi}{2}$ as in Fig. 5, which, in particular, are sandwiched between hyperbolae defining Lawlor necks. Su showed that the Lagrangian mean curvature flow starting at the Lagrangian defined by γ exists for all time and converges to a Lawlor neck.

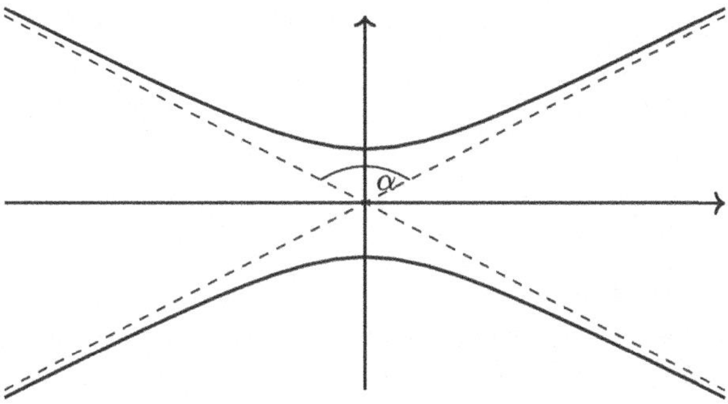

Fig. 5 A non-compact arc defining an asymptotically planar Lagrangian cylinder $S^1 \times \mathbb{R}$

Example 8 ($\alpha \in (0, \frac{\pi}{2})$: Self-expanders) For every $\alpha \in (0, \frac{\pi}{2})$, Anciaux [1] showed that there is a dilation family of curves γ as in Fig. 5 so that the corresponding Lagrangian is a *self-expander*; i.e. it simply expands by dilation under the flow (2).

Su [9] considered certain natural curves γ as in Fig. 5 for $\alpha \in (0, \frac{\pi}{2})$, bounded between two curves defining self-expanders, and showed that (after rescaling) Lagrangian mean curvature flow starting at the Lagrangian defined by γ exists for all time and converges to an Anciaux self-expander.

Example 9 ($\alpha \in (\frac{\pi}{2}, \pi)$: Singularities) When $\alpha \in (\frac{\pi}{2}, \pi)$, it is shown in [7, 11] that for curves γ as in Fig. 5 the flow (3) has a first finite-time singularity at the origin. Moreover, the singularity is Type II and the Type I blow-up is a circle-invariant pair of Lagrangian planes defined by a pair of lines in $\mathbb{C}$ which meet at right angles at 0. A rough picture of what happens at the finite-time singularity is given in Fig. 6. The Lagrangian has become two tranversely intersecting copies of $\mathbb{R}^2$ at the singular time.

Recently, Wood [11] has shown further that the Type I blow-up at the origin is unique, the Type II blow-up is unique (up to scale) and given by the Lawlor neck asymptotic to the Type I blow-up, so that the blow-down of the Type II blow-up is

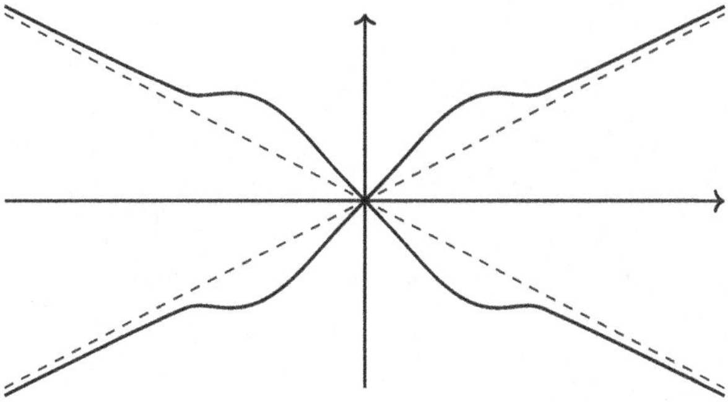

Fig. 6 Singularity formation for an asymptotically planar Lagrangian cylinder

equal to the Type I blow-up. This matches well with the classification results for Type II blow-ups in [5].

6 Outlook

There are many further questions one may ask about general Lagrangian mean curvature flow which may already yield an interesting answer in the equivariant setting, motivated in part by the examples we have discussed.

- When can one relate blow-downs of Type II blow-ups to Type I blow-ups at a finite-time singularity?
- Does the mean curvature necessarily blow-up at a finite-time singularity?
- Can one better understand the role of pseudoholomorphic curves in the formation of singularities in Lagrangian mean curvature flow?

Answering any of these questions would be significant for further study of Lagrangian mean curvature flow.

References

1. H. Anciaux, Construction of Lagrangian self-similar solutions to the mean curvature flow in $\mathbb{C}^n$. Geom. Dedicata **120**, 37–48 (2006)
2. J. Chen, J.M.S. Ma, The space of compact self-shrinking solutions to the Lagrangian mean curvature flow in $\mathbb{C}^2$. J. Reine Angew. Math. **743**, 229–244 (2018)
3. C.G. Evans, J.D. Lotay, F. Schulze, Remarks on the self-shrinking Clifford torus, J. Reine Angew. Math. **2020**, 139–170 (2020)
4. K. Groh, M. Schwarz, K. Smoczyk, K. Zehmisch, Mean curvature flow of monotone Lagrangian submanifolds. Math. Z. **257**(2), 295–327 (2007)
5. B. Lambert, J.D. Lotay, F. Schulze, Ancient solutions in Lagrangian mean curvature flow. Ann. Sc. Norm. Super. Pisa Cl. Sci. **XXII**, 1169–1205 (2021)
6. A. Neves, Singularities of Lagrangian mean curvature flow: monotone case. Math. Res. Lett. **17**(1), 109–126 (2010)
7. A. Neves, Finite time singularities for Lagrangian mean curvature flow. Ann. Math. **177**, 1029–1076 (2013)
8. A. Savas-Halilaj, K. Smoczyk, Lagrangian mean curvature flow of Whitney spheres. Geom. Topol. **23**(2), 1057–1084 (2019)
9. W.-B. Su, Mean curvature flow of asymptotically conical Lagrangian submanifolds. Trans. Am. Math. Soc. **373**, 1211–1242 (2019)
10. C. Viana, A note on the evolution of the Whitney sphere along mean curvature flow, J. Geom. Anal. **31**, 4240–4252 (2021)
11. A. Wood, Singularities of equivariant Lagrangian mean curvature flow. arXiv:1910.06122

Some Useful Facts on Invariant Connections

Gonçalo Oliveira

Abstract This short note gathers a few useful results on connections invariant under a group action. The results here reviewed have proved useful in solving several gauge theoretical equations and may be of relevance to both mathematicians and physicists.

1 Introduction

1.1 Context

The use of Lie group actions in geometry, analysis and physics has long been a fashionable and fruitful tool. For example, there are difficult geometric classification problems which admit interesting very symmetric examples which under such symmetry assumptions can be effectively classified. In many cases, these simplified classifications shed light on the general theory and help to shape the developments which it then takes.

In a related but somewhat transverse direction, consider some geometric partial differential equation (PDE), such as a gauge theoretical equation originating from physics. These are in general difficult to solve, and in a generic geometric setting it is hopeless to expect explicit solutions to exist. However, in very symmetric situations, we may hope to use a Lie group symmetry to reduce these complicated equations to simpler ones depending on less variables, and that may be solved.

This is useful, for instance, when solving gauge theoretical PDEs for connections. When the connections are invariant, such PDEs reduce to lower dimensional equations in the orbit space. This may be a point, as is the case when the group acts transitively in a connected space, and so the original PDEs get reduced to algebraic equations (zero dimensional PDEs). More non-trivial examples happen when the

G. Oliveira (✉)
Universidade Federal Fluminense, Rio de Janeiro, Brazil

P. Cerejeiras et al. (eds.), *Current Trends in Analysis, its Applications and Computation*, Research Perspectives,
https://doi.org/10.1007/978-3-030-87502-2_49

orbit space is 1-dimensional in which case the original problem is reduced to ODEs on it, and so forth. Thomas Parker's beautiful articles [7] and [8] are some of the earliest articles to have effectively used the techniques here reviewed in this line of research.

These notes contain the geometric input required to understand gauge theoretical problems under such symmetry assumptions.

1.2 Summary and Organization

This notes start in Sect. 2 by reviewing the notion of homogeneous principal G-bundles over homogeneous spaces K/H. These are bundles over K/H equipped with a lift of the K-action to their total space and so there is a canonical notion of connections on these which are invariant under the K-action. Wang's theorem, which we review and prove in Sect. 3, classifies these invariant connections.

In Sect. 4, we clarify the relation between two natural definitions of the moduli spaces of invariant connections modulo gauge. Namely, one can consider an action of K on the moduli space of connections up to gauge, denoted $\mathcal{A}/\mathcal{G}$, and define $[\mathcal{A}/\mathcal{G}]^{inv}$ as its fixed locus. Alternatively, we may find more natural/useful to instead consider the space of invariant connections $\mathcal{A}^{inv}$ and quotient it by the action of the invariant gauge transformations $\mathcal{G}^{inv}$, which clearly preserves $\mathcal{A}^{inv}$, yielding $\mathcal{A}^{inv}/\mathcal{G}^{inv}$. We show that there is a well defined surjective map

$$\phi : \mathcal{A}^{inv}/\mathcal{G}^{inv} \to [\mathcal{A}/\mathcal{G}]^{inv},$$

which, for semisimple G, when restricted to the set of irreducible connections is also injevctive.

Several times it is convenient to work with connections on bundles over spaces with a K-action which may not act transitively. Then, we may consider connections whose restriction to each K-orbit is invariant. For these two give a well defined connection on the whole space, they must have some "restricted" behavior at the boundary of the orbit space. There may be several approaches to this, but in Sect. 5 we summarize an approach based on the Eschenburg-Wang analysis for the Cauchy problem for Einstein metrics. This can be effectively employed in the cohomogeneity-1 case and may be adapted to deal with some higher cohomogeneity situations.

2 Principal Bundles with Connections on Homogeneous Spaces

This section introduces the notion of homogeneous bundles over homogeneous spaces following the nomenclature of [3]. A short summary of the content of this section and the next already appear in an Appendix to the author's PhD thesis [6].

Let G be a compact Lie group. Any G-bundle E can be constructed via the associated bundle construction from a principal G-bundle P. So let $\pi : P \to X$ be a principal G-bundle. We shall focus on the case where $H \subset K$ are connected Lie groups, and $X = K/H$ is a homogeneous space.

So K acts transitively on X with isotropy subgroup H, which is normal in K. As usual K acts on the left on X and G on the right on P.

Definition 2.1 A lifting of the K-action to P is an homomorphism $\rho : K \to \mathrm{Diff}(P)$, such that:

- ρ covers the K action on the base, i.e.

$$\forall\, k \in K, \quad \pi \circ \rho(k) = k \circ \pi,$$

  where in the right hand side $k \in \mathrm{Diff}(X)$ is the diffeomorphism defined by k.
- ρ is a bundle map, i.e. commutes with right G action on P.

Definition 2.2 A principal G-bundle $\pi : P \to X$ as above is called K-homogeneous if it is equipped is a lifting ρ of the K-action. Two K-homogeneous principal G-bundles $(\pi_i : P_i \to X, \rho_i)$, for $i = 1, 2$, as above are said to be isomorphic if there is a bundle isomorphism $\phi : P_1 \to P_2$ which intertwines the two K-actions, i.e.

$$\phi \circ \rho_1 = \rho_2 \circ \phi.$$

Remark 2.3 Let H be the isotropy subgroup at $x \in X$. The restriction of the lifts $\rho|_H$ preserve the fibre at x. Since the G action is injective and transitive along the fibres, we obtain an homomorphism $\lambda : H \to G$. Notice that this depends on the choice of a point in the fibre above x, which we shall think as fixed from now on. We shall call λ the **isotropy homomorphism.**

Moreover, the isotropy isomorphism can be used to recover the principal G-bundle P, via the associated bundle construction

$$P = K \times_{H,\lambda} G.$$

These bundles are clearly reducible, since we can embed $i : K \hookrightarrow P$ as a subbundle, via the map $k \mapsto i(k) = [k, 1]$, where $1 \in G$ is the identity.
Let F be a differentiable manifold with a G action $\eta : G \to Diff(F)$, construct the associated bundle $E = P \times_{G,\eta} F$ with fibre F. We can naturally define a K-action on E, via the lifting ρ to P. We shall also denote this action by ρ. Then there is an isomorphism of homogeneous bundles

$$P \times_{G,\eta} G \cong K \times_{H,\eta\circ\lambda} F. \tag{2.1}$$

3 Invariant Sections and Connections

This section defines invariant sections and connections, culminating with the proof of Wang's theorem classifying invariant connections.

Definition 3.1 $s_E \in \Gamma(E)$ is said to be an **invariant section** under the K action on E if for all $k \in K$

$$\rho(k) \circ s_E = s_E \circ k.$$

Proposition 3.2 *There is a one to one correspondence between K-invariant sections of E and points in F fixed by the action $\eta \circ \lambda$ of H in F.*

Proof Since E and $E' = K \times_{H,\eta\circ\lambda} F$ are isomorphic as homogeneous bundles we can use the correspondence between sections of E' and H-equivariant maps $f : K \to F$. Then, K-invariant sections must correspond to constant maps. The H-equivariance condition on the map imposes that its image is a point fixed by $\eta \circ \lambda(H)$. □

Remark 3.3 In the case where F is a vector space and $\eta \circ \lambda : H \to GL(F)$ is a representation we can split F into irreducible components $F = \bigoplus_i F_i$. Then invariant sections of $E = K \times_{H,\eta\circ\lambda} F$ are given by vectors in F_0 the component corresponding to the trivial representation in which $\eta \circ \lambda$ acts with eigenvalue 1.

We shall now restrict to the case of the above remark, i.e. F be a vector space and η a representation. The goal now is to define an invariant connection in a workable and equivalent way. We shall now describe connections on associated bundles of the form $E = P \times_\eta F$. Notice that this contains the case of connections on principal bundles as P can be seen as such an associated bundle via $P = P \times_{\mathrm{id}_G} G$. A connection on $E = P \times_\eta F$ is given by a 1 form in P with values in $End(F)$, say $B \in \Omega^1(P, End(F))$, such that

1. For all $g \in G$,

$$R_g^* B = Ad_{\eta(g^{-1})} B.$$

2. For all $X \in \mathfrak{g}$,

$$B\left(\frac{d}{dt}\Big|_{t=0} pe^{tX}\right) = d\eta(X).$$

However, as seen in (2.1), these connections can also be described as $A \in \Omega^1(K, End(F))$ such that

1. For all $h \in H$,

$$R_h^* A = Ad_{\eta(\lambda(h)^{-1})} A.$$

2. For all $Y \in \mathfrak{h}$,

$$A\left(\frac{d}{dt}\Big|_{t=0} pe^{tY}\right) = d(\eta \circ \lambda)(Y).$$

In fact, in view of the embedding $i : K \hookrightarrow P$ one has $A = i^*B$. So we shall stick to this later description of connections in any homogeneous principal and vector bundles.

Definition 3.4 A connection $A \in \Omega^1(K, End(F))$ on $K \times_{H,\eta\circ\lambda} F$ is said to be invariant if the connection 1 form A is left invariant, i.e.

$$L_k^* A = A,$$

for all $k \in K$.

In the case of homogeneous spaces Wang's theorem classifies all invariant connections on a homogeneous principal bundle. [3] volume II., theorem 11.5.

Theorem 3.5 (Wang's Theorem [10]) *Let $P = K \times_{H,\lambda} G$ be a principal homogeneous G-bundle. Then K-invariant connections A on P are in one to one correspondence with linear maps $\Lambda : \mathfrak{k} \to \mathfrak{g}$, such that*

1. *$\Lambda \circ ad(h) = ad(\lambda(h)) \circ \Lambda$, for all $h \in H$, and*
2. *$\Lambda|_{\mathfrak{h}} = d\lambda$.*

The curvature of the connection can be written in terms of Λ as

$$F(X, Y) = [\Lambda(X), \Lambda(Y)] - \Lambda\,[X, Y],$$

for $X, Y \in \mathfrak{k}$ extended as left invariant vector fields.

Proof Given $A \in \Omega^1(K, \mathfrak{g})$ we can restrict it to the identity to get a map $\Lambda : \mathfrak{k} \to \mathfrak{g}$, properties 1. and 2. in the theorem are obvious consequences of the conditions A need to satisfy to be a connection. Conversely, given a map $\Lambda : \mathfrak{k} \to \mathfrak{g}$ we can extend it as a left invariant 1-form. This will be a connection due to the properties 1. and 2. in the theorem.

To compute the curvature of an invariant connection $A \in \Omega^1(K, \mathfrak{g}))$ on left invariant vector fields X, Y.

$$\begin{aligned} F(X, Y) &= dA(X, Y) + \frac{1}{2}[A \wedge A](X, Y) \\ &= X \cdot (A(Y)) - A \cdot (A(X)) - A([X, Y]) + [A(X), A(Y)] \\ &= [A(X), A(Y)] - A([X, Y]), \end{aligned}$$

using that $A(X), A(Y)$ are constant. □

From now on we must consider Homogeneous spaces $X = K/H$, where H is connected and we can find an ad-$\mathfrak{h}$ invariant complement to $\mathfrak{h}$, i.e.

$$\mathfrak{k} = \mathfrak{m} \oplus \mathfrak{h},$$

with $[\mathfrak{m}, \mathfrak{h}] \subset \mathfrak{m}$, this shall be called a reductive decomposition. Fixing a choice of $\mathfrak{m}$, Λ is totally determined from its values in $\mathfrak{m}$, since $\Lambda = \lambda_*|_{\mathfrak{h}} + \Lambda|_{\mathfrak{m}}$. In such a situation, with a lift of the action to the principal bundle, invariant connections always exist and we define the **canonical invariant connection** to be given by $\Lambda|_{\mathfrak{m}} = 0$, i.e

$$A = d\lambda|_{\mathfrak{h}}$$

Remark 3.6 Notice that according to our definition of canonical invariant connection, it is only unique once a specific complement $\mathfrak{m}$ have been chosen.

Example 3.7 A rather special case is when the bundle P is the H-bundle $K \to K/H$. Pick a reductive decomposition $\mathfrak{k} = \mathfrak{m} \oplus \mathfrak{h}$ and let θ be the Maurer Cartan form. This can be projected onto each $\mathfrak{m}$ or $\mathfrak{h}$, denote this last one by $\theta|_{\mathfrak{h}}$. The action is the left multiplication of H onto K. Hence the isotropy homomorphism λ_k at $k \in K$ satisfying $hk = k\lambda(h)$ for $h \in H$ is given by $\lambda_k(h) = k^{-1}hk$. If we pick k to be the identity, then the canonical invariant connection A is defined by taking the left invariant extension of the projection $p_{\mathfrak{g}}$ into $\mathfrak{h}$. So for Y a left invariant vector field, we have

$$A(Y) = p_{\mathfrak{h}}\theta(Y) = p_{\mathfrak{h}}(Y).$$

In this way we can view the connection as a 1 form in K with values in $\mathfrak{h}$. Since it obviously agrees with the Maurer Cartan form along the fibres, to show this really is a connection we just need to verify the equivariance condition. Let $Y \in T_k K$ and compute

$$\begin{aligned}(R_h^* A)_k (Y) &= A_{kh}(dR_h Y) = p_{\mathfrak{h}}(dL_{(kh)^{-1}} dR_h Y) \\ &= p_{\mathfrak{h}}(Ad_{h^{-1}} dL_{k^{-1}} Y) \\ &= Ad_{h^{-1}} A_k(Y).\end{aligned}$$

Moreover, this canonical invariant invariant connection on $K \to K/H$ has as horizontal space the left translations of $\mathfrak{m}$.

Let $P = K \times_{H,\lambda} G$ be a principal G bundle over K/H. Notice that $\mathfrak{k}$ and $\mathfrak{g}$ are two representations of H, respectively given by $Ad : H \to GL(\mathfrak{k})$ and $Ad \circ \lambda$. Then, Wang's theorem 3.5 can be phrased in the following way.

Theorem 3.8 (Wang, Version 2) *There is a one to one correspondence between K-invariant connections on P and morphisms of H-representations $\Lambda : \mathfrak{k} \to \mathfrak{g}$ such that $\Lambda_{\mathfrak{h}} = d\lambda$.*

The following example gives a way of constructing a special class of invariant connections.

Example 3.9 Suppose K is simply connected and let $\Lambda : \mathfrak{k} \to \mathfrak{g}$ be a morphism of Lie Algebras. Then for each Lie group G with Lie algebra $\mathfrak{g}$, we can uniquely integrate Λ to a morphism $\tilde{\Lambda} : K \to G$ of Lie groups. Define

$$\lambda : H \to G,$$

through the composition $H \hookrightarrow K \to G$ via $\tilde{\Lambda}$. Then, the map Λ obviously define a morphism of H representations satisfying $\Lambda|_{\mathfrak{h}} = d\lambda$ and hence gives rise to an invariant connection.

4 Moduli Space of Invariant Connections Modulo Gauge

Let us consider a reductive homogeneous space of the form K/H for some compact K and and a homogeneous principal G-bundle $P_\lambda = K \times_{(\lambda, H)} G$. The lift of the K action to P_λ will be denoted by $\rho(k)[k', g] = [kk', g]$ and it acts on a connection form $A \in \Omega^1(K, \mathfrak{g})$ by $\rho(k)A = L_k^* A$.

Definition 4.1 A gauge transformation on P_λ is a $g \in \Omega^0(K, G)$ such that

$$g(kh) = \lambda(h)g(k)\lambda(h)^{-1}$$

for all $k \in K$ and $h \in H$. A gauge transformation acts on a connection via

$$g \cdot A = Ad(g^{-1})A.$$

We can define the moduli space of invariant connections in two different ways. The first and more standard one is by taking $\mathcal{A}^{inv}/\mathcal{G}^{inv}$, where $\mathcal{A}^{inv}$ are those connections A such that $\rho(k)A = A$ for any $k \in K$ and $\mathcal{G}^{inv}$ the gauge transformations g such that $L_k^* g = g$ for all $k \in K$, whose action can be seen to preserve $\mathcal{A}^{inv}$. We now take one second definition which requires some preparation. First we consider the full moduli space of connections up to gauge $\mathcal{A}/\mathcal{G}$. Then we define an action of K on $\mathcal{A}/\mathcal{G}$ by letting $k \in K$ act on a gauge equivalence class by

$$\rho(k)[A] = [\rho(k)A],$$

where A is some element in the class $[A]$.

Lemma 4.2 *This action is well defined!*

Proof Let $A_1, A_2 \in [A]$ we need to prove that for $k \in K$, $\rho(k)A_1$ and $\rho(k)A_2$ are gauge equivalent.

As A_2 is gauge equivalent to A_1 there is g such that $g \cdot A_2 = A_1$ and so

$$\rho(k)A_2 = \rho(k)g{\cdot}A_1 = L_k^* Ad(g^{-1})A_1 = Ad(g^{-1} \circ L_k)L_k^* A_1 = (g^{-1} \circ L_k){\cdot}\rho(k)A_1,$$

which is gauge equivalent to $\rho(k)A_1$ as $g^{-1} \circ L_k$ is a gauge transformation. To check that recall that for $h \in H \subset K$, $R_h^*(g^{-1} \circ L_k) = R_h^* L_k^* g^{-1} = L_k^* R_h^* g^{-1} = L_k^* Ad(\lambda(h^{-1}))g^{-1} = Ad(\lambda(h^{-1}))(g^{-1} \circ L_k)$ as R_h and L_k commute and so $g^{-1} \circ L_k$ satisfies the correct equivariance condition for a gauge transformation. □

We can then define the moduli space of invariant connections in one other way, namely

$$[\mathcal{A}/\mathcal{G}]^{inv} = \{[A] \in \mathcal{A}/\mathcal{G} \mid \rho(k)[A] = [A], \text{ for all } k \in K\}.$$

The key point is that

Proposition 4.3 *The map*

$$\begin{aligned} \phi : \mathcal{A}^{inv}/\mathcal{G}^{inv} &\to [\mathcal{A}/\mathcal{G}]^{inv} \\ A &\mapsto [A], \end{aligned} \tag{4.1}$$

is surjective.

Proof Let $[A] \in [\mathcal{A}/\mathcal{G}]^{inv}$, we need to check that there is an invariant connection $\overline{A}$ in the gauge equivalence class of $[A]$. To do this we fix $A \in [A]$, and define our candidate to be

$$\overline{A} = \int_K \rho(k)A \, d\mu_K,$$

where $d\mu_K$ denotes the Haar measure on K, which recall is compact. This $\overline{A}$ is obviously invariant and so in the image of ϕ, so we only have to check that it is indeed gauge equivalent to A. Notice that as $[A] \in [\mathcal{A}/\mathcal{G}]^{inv}$, for any element $k \in K$ there is a gauge transformation g_k such that $\rho(k)A = g_k \cdot A$. In fact, we can see that this defines a group homomorphism $K \to \mathcal{G}$, and

$$\overline{A} = \int_K \rho(k)A \, d\mu_K = \left(\int_K g_k d\mu_K \right) \cdot A,$$

which shows that $\overline{A}$ is gauge equivalent to A. □

The previous map is not injective in general, we shall now investigate when such happens.

Lemma 4.4 *Let G be semisimple and A_1, A_2 be different connections in $\mathcal{A}^{inv}/\mathcal{G}^{inv}$ such that $\phi(A_1) = \phi(A_2)$, then A_1 and A_2 must be reducible.*

Proof If $\phi(A_1) = \phi(A_2)$, then $A_1, A_2 \in \mathcal{A}^{inv}$ must be gauge equivalent invariant connections, i.e. $[A_1] = [A_2]$ and so $A_1 = g \cdot A_2$ for some gauge transformation $g \notin \mathcal{G}^{inv}$. As A_1, A_2 are invariant we conclude that for all $k \in K$,

$$A_1 = L_k^* A_1 = L_k^*(g \cdot A_2) = (g \circ L_k) \cdot A_2.$$

Taking k to be of the form $\exp(tX)$ for $X \in \mathfrak{k}$ we conclude that

$$d_{A_2}(dg(X)) = 0,$$

for all $X \in \mathfrak{k}$. Therefore, either $dg(X) = 0$ which cannot happen as $g \notin \mathcal{G}^{inv}$, or $dg(X) \neq 0$ and spans a nonvanishing vector sub-bundle of $\ker(d_{A_2}) \subset \Gamma(X, \mathfrak{g}_P)$, which must therefore be nonzero. As G is semisimple A_2 must be reducible and so must A_1. □

Corollary 4.5 *The map ϕ from Eq.* (4.1) *is surjective and restricted to the irreducible connections also injective (for semisimple G).*

5 Extending Invariant Tensors Over Singular Orbits

Consider a space X with an action of a Lie group K and a G-bundle over X equipped with a lift of the K to its total space. Recall from Definition 3.4 this implies it covers the K-action on X and commutes with the right G action on the total space. A connection on such a bundle is called invariant if its restriction to any K-orbit is invariant as in Definition 3.4. Not all of these give rise to globally well defined smooth connections on the whole X as they must satisfy some smoothness conditions when the K-orbits are different from the generic ones, as happens for instance at the boundary of the orbit space.

In the case when K acts on X with cohomogeneity one, meaning that the generic orbit of the K-action has codimension one, a possible approach is to use the Eschenburg and Wang's method in [2], which we review in this section. This method was effectively used in [4], and the reader may view that as an example of application. We should also point out that there are other ways to proceed such as employing the techniques of [9], or in case the connections satisfy a PDE using a suitable removable singularity result such as in [1] and [5].

Let K act with cohomogeneity-1 on X and H be the principal isotropy, i.e. the isotropy of the generic orbits. A codimension k singular orbit Q is then of the form $Q = K/H_-$ with $H_-/H \cong S^{k-1}$. Notice that if $V = \mathbb{R}^k$, then H_- acts transitively on the spheres in V, with isotropy H. The normal bundle of Q is then

$$E = K \times_{H_-} V,$$

with projection $\pi_E : E \to Q$. Moreover, a neigbourhood of the zero section is K-equivariantly diffeomorphic to a neighborhood of the singular orbit Q in X. Moreover, we can suppose E is equipped with a Riemannian metric which we may assume to be K invariant if this is compact.

Let $s \in \Omega^0(E, T)$ be a K-invariant tensor, i.e. a K-invariant section of some K-homogeneous vector bundle T of tensors over E. In what follows let V' denote the tensor representation with which T is associated, i.e. $T = K \times_H V'$. As s is K-invariant and K acts on E with cohomogeneity-1, s is completely determined by its values along the line

$$\ell_{v_0} := \{tv_0, \ t \in \mathbb{R}^+\},$$

where $v_0 \in E$ is some vector in the fiber over a point $p \in Q$. Moreover, as s in K invariant and the isotropy of any point tv_0, for $t \neq 0$ is H, the values

$$s_t = s(tv) \in (V')^{H_-},$$

must be H-invariant (where H acts via $H \subset H_-$).

Eschenburg and Wang explain how to reverse this, i.e. they explain the conditions on a family of H-invariant vectors $s_t \in (V')^{H_-}$, so that the function

$$s : V\backslash\{0\} \to V', \ \ s(h_-(t, 0, \ldots, 0)) = h_- \cdot s_t, \tag{5.1}$$

for $t \in \mathbb{R}^+$ and $h_- \in H_-$, extends over the origin. The answer is most easily understood by considering the space W of H_--equivariant maps $S^{k-1} \to V'$. Note that by construction $s|_{S^{k-1}} \in W$ (where s is as above). Moreover, the evaluation map

$$ev : W \to (V')^{H_-}, \ \ ev(s) = s(1, 0, \ldots, 0)$$

is an isomorphism of vector spaces. In what follows, we shall denote by W_m the space of those maps which are the restriction to S^{k-1} of degree-m H_--equivariant maps $V \to V'$.

Proposition 5.1 (Eschenburg and Wang [2]) *Let $s_t : \mathbb{R}_0^+ \to V'$ be smooth with Taylor expansion $\sum_i s_i t^i$. Then, the map s in 5.1 extends smoothly over the origin if an only if $s_i \in W_i$.*

The following comment is crucial in easing most of the applications of this proposition.

Remark 5.2 Notice that $W = \sum_{i=0}^{+\infty} W_m$ and is finite dimensional and that multiplying an element in W by $|\cdot|^2$ increases its degree by 2. Hence we can find a set of generator of minimal degree.

Acknowledgments I would like to thank an anonymous referee for carefully reading these notes and making valuable suggestions towards improving the exposition. The author is supported by Fundação Serrapilheira 1812-27395, by CNPq grants 428959/2018-0 and 307475/2018-2, and by FAPERJ through the grant Jovem Cientista do Nosso Estado E-26/202.793/2019.

References

1. A. Clarke, G. Oliveira, Spin(7)-instantons from evolution equations (2019). Preprint arXiv:1903.05526
2. J.-H. Eschenburg, M.Y. Wang., The initial value problem for cohomogeneity one Einstein metrics. J. Geometr. Anal. **10**(1), 109–137 (2000)
3. S. Kobayashi, K. Nomizu, *Foundations of Differential Geometry (2 vols.)* (Wiley, New York, 1996)
4. J.D. Lotay, G. Oliveira, $SU(2)^2$-invariant G_2-instantons. Math. Ann. **371**(1–2), 961–1011 (2018)
5. G. Oliveira, Calabi–Yau monopoles for the Stenzel metric. Commun. Math. Phys. **341**(2), 699–728 (2016)
6. G. Oliveira, Monopoles in higher dimensions. Imperial College London PhD thesis (2014)
7. T. H. Parker, Non-minimal Yang-Mills fields and dynamics. Invent. Math. **107**(1), 397–420 (1992)
8. T. H. Parker, *A Morse Theory for Equivariant Yang-Mills*. Duke Math. J. **66**(2), 337–356 (1992)
9. L. Verdianu, W. Ziller, Smoothness conditions in cohomogeneity one manifolds (2018). Preprint arXiv:1804.04680
10. H.C. Wang, On invariant connections over a principal fibre bundle. Nagoya Math. J. **13**, 1–19 (1958)

Almost Symplectic Structures on $Spin(7)$-Manifolds

Sema Salur and Eyup Yalcinkaya

Abstract A non-degenerate differential 2-form on an even dimensional manifold M^{2n} is called an almost-symplectic structure. A necessary condition for the existence of an almost-symplectic structure is that all odd-dimensional Stiefel-Whitney classes of M should vanish. In this paper, we prove that all odd-dimensional Stiefel-Whitney classes of a smooth, closed, connected, orientable 8-manifold with spin structure vanish. We also study the almost-symplectic structures on certain classes of $Spin(7)$-manifolds.

Keywords Holonomy · Almost symplectic structure · $Spin(7)$-manifold

Mathematics Subject Classification (2010) Primary 53D15; Secondary 53C29

1 Introduction

In this paper we study geometric structures on $Spin(7)$ manifolds. A $Spin(7)$-manifold is an 8-dimensional Riemannian manifold with the holonomy group inside the exceptional Lie group $Spin(7)$. Manifolds with special holonomy are spaces whose infinitesimal symmetries allow them to play a crucial role in M-theory compactifications. They represent the tiny curled up dimensions hiding at every point of spacetime. Examples of manifolds with special holonomy are 6-dimensional Calabi-Yau manifolds, 7-dimensional G_2 manifolds and 8-dimensional $Spin(7)$ manifolds. Despite extensive research on Calabi-Yau manifolds, the geometric properties of G_2 and $Spin(7)$ manifolds are not well understood. In this paper we initiate a

S. Salur (✉)
Department of Mathematics, University of Rochester, Rochester, NY, USA
e-mail: salur@math.rochester.edu

E. Yalcinkaya
The Scientific and Technological Research Council of Turkey, Ankara, Turkey
e-mail: eyup.yalcinkaya@tubitak.gov.tr

P. Cerejeiras et al. (eds.), *Current Trends in Analysis, its Applications and Computation*, Research Perspectives,
https://doi.org/10.1007/978-3-030-87502-2_50

program to study almost symplectic structures on Riemannian 8-manifolds with spin structure.

In particular we prove

Theorem *All the odd-dimensional Stiefel-Whitney classes of a smooth, closed, connected, orientable 8-manifold with spin structure vanish.*

Note that a manifold M with $Spin(7)$-structure is orientable and spin. The theorem above implies that the obstructions for the existence of almost symplectic (and hence almost complex) structures on a manifold with full $Spin(7)$ holonomy vanish as well. There are inclusions between the groups

$$SU(2) \longrightarrow SU(3) \longrightarrow G_2 \longrightarrow Spin(7),$$

and

$$SU(2) \times SU(2) \longrightarrow Sp(2) \longrightarrow SU(4) \longrightarrow Spin(7).$$

These are the only connected Lie subgroups of $Spin(7)$ which can be holonomy groups of Riemannian metrics on 8-manifolds. Hence the theorem above also holds for 8-manifolds with reduced holonomy groups.

2 $Spin(7)$-Structures

In this section we review the basics of $Spin(7)$ geometry. More on the subject can be found in [4, 6, 8] and [13].

Let $(x^1, \ldots, x^8)$ be coordinates on $\mathbb{R}^8$. The standard Cayley 4-form on $\mathbb{R}^8$ can be written as

$$\begin{aligned}\Phi_0 = dx^{1234} + dx^{1256} + dx^{1278} + dx^{1357} - dx^{1368} - dx^{1458} - dx^{1467}\\ - dx^{2358} - dx^{2367} - dx^{2457} + dx^{2468} + dx^{3456} + dx^{3478} + dx^{5678}\end{aligned}$$

where $dx^{ijkl} = dx^i \wedge dx^j \wedge dx^k \wedge dx^l$.

The subgroup of $GL(8, \mathbb{R})$ that preserves Φ_0 is the group $Spin(7)$. It is a 21-dimensional compact, connected and simply-connected Lie group which preserves the orientation on $\mathbb{R}^8$ and the Euclidean metric g_0.

Definition 2.1 A differential 4-form Φ on an oriented 8-manifold M is called admissible if it can be identified with Φ_0 through an oriented isomorphism between T_pM and $\mathbb{R}^8$ for each point $p \in M$.

Definition 2.2 Let $\mathcal{A}(M)$ denotes the space of admissible 4-forms on M. A $Spin(7)$-structure on an 8-dimensional manifold M is an admissible 4-form $\Phi \in \mathcal{A}(M)$. If M admits such structure, (M, Φ) is called a manifold with $Spin(7)$-structure.

Each 8-manifold with a $Spin(7)$-structure Φ is canonically equipped with a metric g. Hence, we can think of a $Spin(7)$-structure on M as a pair (Φ, g) such that for all $p \in M$ there is an isomorphism between T_pM and $\mathbb{R}^8$ which identifies (Φ_p, g_p) with (Φ_0, g_0).

Existence of a $Spin(7)$-structure on an 8-dimensional manifold M is equivalent to a reduction of the structure group of the tangent bundle of M from $SO(8)$ to its subgroup $Spin(7)$. The following result gives the necessary and sufficient conditions so that the 8-manifold admits $Spin(7)$-structure.

Theorem 2.3 ([4, 6]) *Let M be a differentiable 8-manifold. M admits a $Spin(7)$-structure if and only if $w_1(M) = w_2(M) = 0$ and for appropriate choice of orientation on M we have that*

$$p_1(M)^2 - 4p_2(M) \pm 8\chi(M) = 0.$$

Furthermore, if $\nabla\Phi = 0$, where ∇ is the Riemannian connection of g, then $\mathrm{Hol}(M) \subseteq Spin(7)$, and M is called a $Spin(7)$-manifold. All $Spin(7)$ manifolds are Ricci flat.

Let (M, g, Φ) be a $Spin(7)$ manifold. The action of $Spin(7)$ on the tangent space gives an action of $Spin(7)$ on the spaces of differential forms, $\Lambda^k(M)$, and so the exterior algebra splits orthogonally into components, where Λ^k_l corresponds to an irreducible representation of $Spin(7)$ of dimension l:

$$\Lambda^1(M) = \Lambda^1_8, \quad \Lambda^2(M) = \Lambda^2_7 \oplus \Lambda^2_{21}, \quad \Lambda^3(M) = \Lambda^3_8 \oplus \Lambda^3_{48},$$

$$\Lambda^4(M) = \Lambda^4_+(M) \oplus \Lambda^4_-(M), \quad \Lambda^4_+(M) = \Lambda^4_1 \oplus \Lambda^4_7 \oplus \Lambda^4_{27}, \quad \Lambda^4_- = \Lambda^4_{35}$$

$$\Lambda^5(M) = \Lambda^5_8 \oplus \Lambda^5_{48} \quad \Lambda^6(M) = \Lambda^6_7 \oplus \Lambda^6_{21}, \quad \Lambda^7(M) = \Lambda^7_8;$$

where $\Lambda^4_\pm(M)$ are the $\pm$-eigenspaces of $*$ on $\Lambda^4(M)$ and

$$\Lambda^2_7 = \{\alpha \in \Lambda^2(M) | *(\alpha \wedge \Phi) = 3\alpha\}, \quad \Lambda^2_{21} = \{\alpha \in \Lambda^2(M) | *(\alpha \wedge \Phi) = -\alpha\},$$

$$\Lambda^3_8 = \{*(\beta \wedge \Phi) | \beta \in \Lambda^1(M)\}, \quad \Lambda^3_{48} = \{\gamma \in \Lambda^3(M) | \gamma \wedge \Phi = 0\},$$

$$\Lambda^4_1 = \{f\Phi | f \in \mathcal{F}(M)\}$$

The Hodge star $*$ gives an isometry between Λ^k_l and Λ^{8-k}_l.

3 Almost Symplectic Structures and $Spin(7)$-Structures

In this section we show that all the odd-dimensional Stiefel-Whitney classes on a closed, connected orientable 8-manifold with spin structure vanish.

An almost symplectic manifold M is a n-dimensional manifold ($n = 2m$) with a non degenerate 2-form ω. If in addition, ω is closed then M is called a symplectic manifold. An almost symplectic structure defines an $Sp(m, \mathbb{R})$ structure. A necessary and sufficient condition for the existence of an almost symplectic structure on M is the reduction of the structure group of the tangent bundle to the unitary group $U(m)$. It is therefore necessary that all odd-dimensional Stiefel-Whitney classes of M to vanish [9].

For any manifold M and integer $k \geq 0$, one can construct a graded linear map $Sq^k : H^*(M, \mathbb{Z}_2) \to H^*(M, \mathbb{Z}_2)$ of degree k. This is called the kth Steenrod square. One can define Stiefel-Whitney classes using both Steenrod squares and the Thom isomorphism.

There is also a unique class $\nu_k \in H^k(M, \mathbb{Z}_2)$ such that for any $x \in H^{n-k}(M, \mathbb{Z}_2)$, $Sq^k(x) = \nu_k \cup x$. We call this class ν_k, the kth Wu class.

Now suppose M is a smooth, closed, connected n-dimensional manifold. Wu's theorem states that the total Stiefel-Whitney class of the tangent bundle of M, denoted by w, Steenrod squares and Wu classes are all related by the equation $w = Sq(\nu)$, for more on the subject see [11]. This gives the following formula:

$$w_k = \sum_{i+j=k} Sq^i(\nu_j)$$

One can also compute the action of the Steenrod squares on the Stiefel-Whitney classes. This is called the Wu formula:

$$Sq^i(w_j) = \sum_{t=0}^{i} \binom{j+t-i-1}{t} w_{i-t} w_{j+t}$$

for $0 \leq i \leq j$. Thus we obtain

$$w_1 = Sq^0(\nu_1) = \nu_1,$$

$$w_2 = Sq^0(\nu_2) + Sq^1(\nu_1) = \nu_2 + \nu_1 \cup \nu_1,$$

$$w_3 = Sq^0(\nu_3) + Sq^1(\nu_2) = \nu_3 + Sq^1(\nu_2) = \nu_3 + Sq^1(w_2) + Sq^1(w_1 \cup w_1),$$

$$w_4 = Sq^0(\nu_4) + Sq^1(\nu_3) + Sq^2(\nu_2) = \nu_4 + Sq^1(\nu_3) + \nu_2 \cup \nu_2$$

$$w_5 = Sq^0(\nu_5) + Sq^1(\nu_4) + Sq^2(\nu_3) = \nu_5 + Sq^1(\nu_4) + Sq^2(w_1 \cup w_2)$$

And one can write the corresponding Wu classes as polynomials in the Stiefel-Whitney classes as follows: For simplicity, we replace the cup product symbol by multiplication sign.

$$\nu_1 = w_1,$$

$$\nu_2 = w_2 + w_1^2,$$

$$\nu_3 = w_1 w_2,$$

$$\nu_4 = w_4 + w_3 w_1 + w_2^2 + w_1^4,$$

$$\nu_5 = w_4 w_1 + w_3 w_1^2 + w_2^2 w_1 + w_2 w_1^3,$$

In a spin manifold, $w_1 = w_2 = 0$ which imply $v_1 = v_2 = 0$ which then gives $w_3 = v_3$. One can also see that $w_3 = 0$ as follows: Note that by definition of Wu classes, $v_3 \cup x = Sq^3(x)$ for all $x \in H^{(n-3)}(M, \mathbb{Z}_2)$. Then one can see that Sq^3 is a linear combination of $Sq^1 \circ Sq^2$ and $Sq^2 \circ Sq^1$ and $Sq^1 \circ Sq^1 \circ Sq^1$ so that we get

$$v_3 \cup x = (aSq^1Sq^2 + bSq^2Sq^1 + cSq^1Sq^1Sq^1)(x) = Sq^1(y) + Sq^2(z)$$

for some y, z. This term is equal to $v_1 \cup y + v_2 \cup z = 0$. As $v_3 \cup x = 0$ for all x, Poincare duality then gives $v_3 = 0$ and hence $w_3 = 0$.

The Wu relations also imply that $w_4 = \nu_4$. Since $w_1 = 0$ (as M is orientable) this gives us $w_5 = Sq^1(w_4)$. Equivalently, w_5 is the image of w_4 under the Bockstein map induced by

$$0 \longrightarrow \mathbb{Z}_2 \longrightarrow \mathbb{Z}_4 \longrightarrow \mathbb{Z}_2 \longrightarrow 0$$

This implies that w_5 is the mod-2 reduction of the integral Stiefel-Whitney class W_5, which is the element of $H^5(M, \mathbb{Z})$, that is the image of w_4 under the Bockstein map induced by

$$0 \longrightarrow \mathbb{Z} \longrightarrow \mathbb{Z} \longrightarrow \mathbb{Z}_2 \longrightarrow 0$$

Note also that v_4 is by definition the Poincare dual to the $\mathbb{Z}_2$ linear map

$$Sq^4 : H^4(M, \mathbb{Z}_2) \to H^8(M, \mathbb{Z}_2) = \mathbb{Z}_2$$

which implies

$$w_4.x = v_4.x = x.x$$

for any element x of $H^4(M, \mathbb{Z}_2)$. In other words, w_4 just represents the mod-2 intersection form.

One can then use the Hirzebruch-Hopf theorem [7] to show that w_4 has an integer lift and therefore w_4 is in the image of $H^4(M, \mathbb{Z}) \longrightarrow H^4(M, \mathbb{Z}_2)$ and so $W_5 = 0$.

The commutative diagram of short exact sequences

$$\begin{array}{ccccccccc} 0 & \longrightarrow & \mathbb{Z} & \longrightarrow & \mathbb{Z} & \longrightarrow & \mathbb{Z}_2 & \longrightarrow & 0 \\ & & \downarrow & & \downarrow & & \downarrow & & \\ 0 & \longrightarrow & \mathbb{Z}_2 & \longrightarrow & \mathbb{Z}_4 & \longrightarrow & \mathbb{Z}_2 & \longrightarrow & 0 \end{array}$$

induces a commutative diagram of the corresponding long exact sequences and hence implies that $w_5 = 0$.

And finally, in order to show that $w_7 = 0$ we use a result of W. Massey. In [10, Thm II] it was shown that if M^{2m} is orientable (where $2m \equiv 0 \pmod 4$), then $w_{2m-1} = 0$. Hence for an 8-manifold we obtain $w_7 = 0$. One can read more about the proof in [10, Section 5].

This completes the proof of the main theorem:

Theorem 3.1 All the odd-dimensional Stiefel-Whitney classes of a smooth, closed, connected, orientable 8-manifold with spin structure vanish.

4 A Motivating Example

Next, we discuss a special class of $Spin(7)$ manifolds that admits an almost complex structure and show how it is related to the $Spin(7)$-structure.

Let (M, Φ) be a $Spin(7)$ manifold (or more generally manifold with $Spin(7)$-structure) admitting a non-vanishing 2-plane field $\Lambda = \{u, v\} \in TM$. In [12], E. Thomas shows that the Euler characteristic $\chi(M) = 0$ and the signature $\sigma(M) \equiv 0 \pmod 4$ provides the necessary and sufficient conditions for the existence of a 2-plane field on an 8-manifold. Now define, $[u, v]^{\perp} = \{w \in TM | < u, w >=< v, w >= 0\}$. One can show that $[u, v]^{\perp}$ carries a non-degenerate 2-form $\omega_{u,v}$ which is compatible with the almost complex structure $J_{u,v} : [u, v]^{\perp} \to [u, v]^{\perp}$ and given by

$$\omega_{u,v}(w, z) = \Phi(w, u, v, z) \text{ and } J_{u,v}(z) = u \times v \times z.$$

Definition 4.1 Let (M, Φ) be a $Spin(7)$ manifold. Then $J_{u,v}(z) = u \times v \times z$ is the triple cross product defined by the identity:

$$< J_{u,v}(z), w >= \Phi(u, v, z, w). \tag{4.1}$$

Theorem 4.2 *Let (M, Φ) be a $Spin(7)$ manifold with a non-vanishing oriented 2-plane field. Then, $J_{u,v}(z) = u \times v \times z$ defines an almost complex structure on M compatible with the $Spin(7)$ structure.*

Proof Let $\{u, v\} \in TM$ be two vectors generating the non-vanishing oriented 2-plane field. $J(z)$ is well defined since by Equation (1), $< J(z), w >= \Phi(u, v, z, w)$.

Next, we show $J^2(z) = -id$. This can be done using the properties of the $Spin(7)$-structure on M. Let $z_i, z_j \in TM$,
Then

$$\begin{aligned} < u \times v \times (u \times v \times z_i), z_j > &= \Phi(u, v, (u \times v \times z_i), z_j) \\ &= -\Phi(u, v, z_j, (u \times v \times z_i)) \\ &= - < u \times v \times z_j, u \times v \times z_i > \\ &= -\delta_{ij} \end{aligned}$$

The last equality holds since the map J is orthogonal. Note that the map J only depends on the oriented 2-plane $\Lambda = \{u, v\}$. □

5 Interesting Questions

One major problem in the field of manifolds with special holonomy is a lack of an existence theorem that gives necessary and sufficient conditions for a 7-dimensional manifold to admit a G_2 metric. In an earlier paper, [3], Arikan, Cho and Salur proposed a program to study the relations between (almost) contact structures and G_2 structures. The main goal is to understand the topological obstructions for the existence of a G_2 metric on a Riemannian 7-manifold with spin structure. In that paper, they proved the following theorem:

Theorem 5.1 Every 7-manifold with a spin structure admits an almost contact structure.

Since every 7-manifold with spin structure admits a G_2 structure this implies:

Corollary 5.2 Every manifold with G_2-structure admits an almost contact structure.

As one might expect, a promising direction for future investigation is to obtain similar results for almost complex (and hence almost symplectic) 8-manifolds with $Spin(7)$ structures. Understanding almost complex structures on a Spin(7) manifold might help us to understand the properties of the Spin(7) metric. We plan to investigate these relations in a future paper.

Also in the papers, [1, 2] and [5], it is shown that the rich geometric structures of a G_2 manifold N with 2-plane fields provide complex and symplectic structures to certain 6-dimensional subbundles of $T(N)$. Using the 2-plane fields, one can introduce a mathematical definition of "mirror symmetry" for Calabi-Yau and G_2 manifolds. More specifically, one can assign a G_2 manifold (N, φ, Λ), with the calibration 3-form φ and an oriented 2-plane field Λ, a pair of parametrized tangent bundle valued 2 and 3-forms of N. These forms can then be used to define different complex and symplectic structures on certain 6-dimensional subbundles of $T(N)$. When these bundles are integrated they give mirror CY manifolds. This is one way of explaining duality between the symplectic and complex structures on the CY 3-folds inside of a G_2 manifold. Similarly, one can construct these structures and define mirror dual Calabi Yau manifolds inside a $Spin(7)$ manifold which admits an almost complex structure. These topics will be also studied in a future paper.

Acknowledgments The authors would like to thank Claude Le Brun and Jonathan Pakianathan for their helpful advice on various technical issues discussed in this paper. The first named author is partially supported by NSF grant DMS–1711178.

References

1. S. Akbulut, S. Salur, *Mirror Duality via G_2 and $Spin(7)$ Manifolds*, Special Volume For the Arithmetic & Geometry Around Quantization, ed. by Y. Manin, M. Marcolli. Prog. in Mathematics Series (Birkhauser, Basel, 2009)
2. S. Akbulut,S. Salur, Harvey Lawson manifolds and dualities. Real and complex submanifolds, Daejeon, Korea, in *Springer Proceedings in Math and Stat.* (2015), pp. 505–514
3. M.F. Arikan, H. Cho, S. Salur, Existence of compatible contact structures on G_2 manifolds. Asian J. Math. **17**(2), 321–334 (2013)
4. M. Fernandez, A Classification of Riemannian Manifolds with Structure Group $Spin(7)$. Annali di Mat. Pura ed App. 143, 101–122 (1986)
5. R. Glover, S. Salur, Lagrangian-type submanifolds of G_2 and $Spin(7)$ manifolds and their deformations. Differ. Geom. Appl. **65**, 241–255 (2019)
6. F.R. Harvey, H.B. Lawson, Calibrated geometries. Acta. Math. **148**, 47–157 (1982)
7. F. Hirzebruch, H. Hopf, Felder von Flachenelementen in 4-dimensionalen Mannigfaltigkeiten. Math. Ann. **136**, 156–172 (1958)
8. D. Joyce, *Riemannian Holonomy Groups and Calibrated Geometry*, Oxford Graduate Texts in Mathematics, vol. 12 (OUP, Oxford, 2007)
9. P. Liberman, Sur les structures presque complexe et autres structures infinitesimales regulieres. Bull. Soc. Math. France **83**, 195–224 (1955)
10. W. Massey, On the Stiefel-Whitney classes of a manifold. Am. J. Math. **82**(1), 92–102 (1960)
11. J. Milnor, J. Stasheff, *Characteristic Classes*, vol. 76. Annals of Mathematics Studies (Princeton University. Press, Princeton, 1974)
12. E. Thomas, Vector fields on manifolds. Bull. Am. Math. Soc. **75**, 643–683 (1969)
13. E. Yalcinkaya, İ. Ünal, On symplectic 8-manifolds admitting Spin(7)-structure. Turkish J. Math. **44**, 1792–1801 (2020)

On Sub-Riemannian and Riemannian Spaces Associated to a Lorentzian Manifold

Roman Sverdlov and Dimiter Vassilev

Abstract We present a certain construction of a sub-Riemannian and Riemannian spaces naturally associated to a Lorentzian manifold. Some additional structures and relations between geometric properties of the corresponding spaces will be explored. The emphasis will be on keeping the text as self-sufficient as possible while linking various well developed fields.

1 Some Notations and Conventions

Throughout the text we let $d \geq 1$ be a positive integer, which equals the space dimension of the Lorentzian manifold M. Latin indices will vary from 1 to d, while Greek indices will vary from 0 to d, and we will assume a summation on repeated indices. We will consider a Lorentzian metric g of signature $(1, d)$ on M. In local coordinates, $g = (g_{\alpha\beta})$. We shall raise and lower indices using the metric and its inverse $g^{-1} = (g^{\alpha\beta})$. In particular, for a tangent vector $y = y^\alpha \frac{\partial}{\partial x^\alpha} \in T_x M$ we have $y_\alpha = g_{\alpha\beta} y^\beta$ and $y^\alpha = g^{\alpha\beta} y_\beta$ and its length is given by

$$|y|^2 \overset{def}{=} g_x(y, y) = g_{\alpha\beta}(x)\, y^\alpha y^\beta = y^\alpha y_\alpha \in \mathbb{R}.$$

The simplest example is the case of Minkowski space where $M = \mathbb{R}^{1,d}$ and

$$|y|^2 = (y^0)^2 - (y^1)^2 - \cdots - (y^d)^2.$$

In the general case, given a point and a normal coordinate system centered at it the metric takes the above form at the given point.

R. Sverdlov · D. Vassilev (✉)
University of New Mexico, Albuquerque, NM, USA
e-mail: rsverdlov@unm.edu; vassilev@unm.edu

P. Cerejeiras et al. (eds.), *Current Trends in Analysis, its Applications and Computation*, Research Perspectives,
https://doi.org/10.1007/978-3-030-87502-2_51

As usual, the Lorentzian metric induces a decomposition of the tangent vectors at each point, which with our agreement of the signature of the metric means that a tangent vector $v \in TM$ is timelike if $|v| > 0$, lightlike if $|v| = 0$, and spacelike if $|v| < 0$.

We shall assume throughout that the considered spacetime is non-compact and time oriented, i.e., there exists a continuous timelike vector field. The future cone at every point is the part of the timelike double cone that contains the fixed timelike global vector field.

The authors would like to thank the referee for the careful reading and comments of the referee.

2 Causal Set Theory

The causal order of space-time has a long history. One of the more recent developments lead to a candidate of a framework on which a theory of quantum gravity can be based, see [7, 11] and references therein. Causal set theory provides a way of discretization that avoids preferred frame, [3], while preserving Lorentz invariance as a fundamental property. Both of these underlining notions are present in our constructions as well. Let (M, g) be a noncompact time-oriented Lorentzian manifold. We define a point p to be in the past of a point q, $p \prec q$, if there is a smooth, future-directed timelike curve from p to q. The future $I^+(p)$ and past $I^-(p)$ of p are defined, respectively, by

$$I^+(p) = \{q \mid p \prec q\} \quad \text{and} \quad I^-(p) = \{q \mid q \prec p\}.$$

(M, g) is called future (past) distinguishing if $I^+(p) = I^+(q)$ $(I^-(p) = I^-(q))$ implies $p = q$.

To a certain extent the Lorentzian geometry of space-time can be recovered from its causal order.

Theorem 1 ([8, 9]) *Assume M is both future and past distinguishing. Then, the causal structure determines the metric up to a conformal factor.*

The modern version of Causal Set Theory began with the paper of Bombelli, Lee, Meyer and Sorkin [4].

Definition 1 A causal set is a partially ordered set $(C, \prec)$ where $\prec$ is (i) acyclic; (ii) transitive and (iii) locally finite, i.e, for all $x, y \in C$ the set $A(x, y) = \{z \mid x \prec z \prec y\}$ is finite.

The absence of cycles in the above definition can be replaced with irreflexivity of the partial order. The general idea is that a causal set replaces the continuum manifold, while the latter is regarded as an approximation of the causal set.

A causal set $(C, \prec)$ with elements given through an injection in a spacetime (M, g) and order induced from the causal structure of the spacetime is said to be an embedding. The approximation is of density ρ_c if there exists an order preserving injection $\Phi : C \to M$ such that $\Phi(C)$ is uniformly distributed with density ρ_c. Here, every spacetime region of volume V contains approximately $\rho_c V$ elements of C. This approach leads to several problems, including symmetry breaking.

The adopted approach to handle this uniformity issue has been to introduce the concept of *sprinkling*, where we begin with a spacetime (M, g) and then randomly "sprinkle" elements onto M via a Poisson process. Thus, the probability of finding n elements in a spacetime region of volume V is $P_V(n) = \frac{(\rho_c V)^n}{n!} e^{-\rho_c V}$. The set of events within a proper time τ_0 in the future of a point $p = 0$ is the region between the light cone and the hyperboloid $|y|^2 = (y^0)^2 - (y^1)^2 - \cdots - (y^d)^2 \approx \tau_0^2$. This is a non-compact region and almost surely contains infinite number of points q directly in the future of p, i.e, $p \prec q$ and there is no r such that $p \prec r \prec q$. For a causal set that is approximated, for example, by Minkowski spacetime, every element therefore has an infinite number of nearest neighbours in its future and past light-cones. This "non-locality" complicates the definitions of discrete version of continuum quantities, including D'Alembertian, leading to non-convergent infinite sums.

The sub-Riemannian and Riemannian spaces introduced in the subsequent sections arose in our goal to remove the non-localities by modeling and discretizing a future directed timelike sector of the tangent bundle to Lorentzian space as a (sub-)Riemannian manifold.

3 Sub-Riemannian Space

In this section we associate a sub-Riemannian space to the considered spacetime. For more details on the relevant definitions and results in sub-Riemannian geometry we refer to [1, 2], and [10]. The space we shall define is a sub-bundle of the tangent bundle. The latter and the cotangent bundle have appeared in theories of a maximal proper acceleration seeking geometric formulation of quantum mechanics whose early ideas can be found in [5, 6].

Definition 2 Let M be a time-oriented Lorentzian manifold. Define the *phase space manifold* $\mathcal{M}$ to be the elements of TM consisting of all points whose tangent component is a timelike future-oriented vector.

Definition 3 Let $\pi : TM \to M$ be the natural projection. For a point $\xi \in \mathcal{M}$ we shall say that (x, y) is a *normal coordinate system centered* at ξ if x is a normal coordinate system centered at $\pi(\xi) \in M$ and y is the fiber coordinate.

Thus, locally, using normal coordinate systems centered at the corresponding point ξ, we have

$$\mathcal{M} = \{\xi \in TM \mid y^0 > 0, \ |y|^2 > 0\}.$$

Definition 4 Let (x, y) be a normal coordinate system at the point $\xi \in \mathcal{M}$. Consider the following forms on $\mathcal{M}$ defined near ξ,

$$\theta^k = y^k dx^0 - y^0 dx^k.$$

The *horizontal space at* ξ is the joint kernel of the forms θ^k at ξ,

$$\mathcal{H}_\xi = \cap_{k=1}^d Ker\, \theta^k|_\xi \subset T_\xi \mathcal{M}.$$

Let us observe that we obtain a (well defined) sub-bundle $\mathcal{H}$ of the tangent bundle of the phase space $\mathcal{M}$ due to the following proposition, which also exhibits the Lorentz invariance of the horizontal space.

Proposition 1 *We have* $span\{\theta^1, \ldots, \theta^d\} = span\{\eta^0, \eta^1, \ldots, \eta^d\}$, *where*

$$\eta^\alpha \overset{def}{=} dx^\alpha - \frac{y^\alpha}{|y|^2} y_\beta dx^\beta.$$

Furthermore, the horizontal space $\mathcal{H}$ is Lorentz invariant under Lorentz transformations (of the spacetime coordinates).

Proof The first claim follows from the following easily verifiable identities

$$\eta^\alpha(\frac{\partial}{\partial y^\beta}) = 0,\ \eta^\alpha(y^\beta \frac{\partial}{\partial x^\beta}) = y^\alpha - \frac{y^\alpha y_\beta y^\beta}{|y|^2} = 0,\ \text{and}\ \theta^k = y^k \eta^0 - y^0 \eta^k.$$

For the proof of the second part let x and $\tilde{x}$ are normal coordinates centered at the same point p of M, hence $x = \Lambda \tilde{x}$ where Λ is a Lorentzian matrix. Using vector notation we have at p the identities $dx = \Lambda d\tilde{x}$, $y = \Lambda \tilde{y}$. The invariance follows from $\eta = dx - \frac{\langle y, dx \rangle}{|y|^2} y$, where we used the notation $\langle y, dx \rangle = y_\alpha dx^\alpha$. □

In fact, we are in the realm of sub-Riemannian geometry since the horizontal vector fields and their commutators span the whole tangent space, i.e., the horizontal space is *bracket generating* (completely non-holonomic).

Proposition 2 *For $\xi \in \mathcal{M}$ and (x, y) a normal coordinate system of $\mathcal{M}$ centered at ξ we have that*

$$\mathcal{H}_\xi = span\,\{V = y^\beta \frac{\partial}{\partial x^\beta}, \frac{\partial}{\partial y^\alpha}\}.$$

The horizontal space $\mathcal{H}$ is a rank $d+2$ bracket generating sub-bundle (distribution) of $T\mathcal{M}$ which satisfies Hörmander's condition of step one.

Proof It is obvious that the right-hand side involves $d+2$ linearly independent vectors that are annihilated by the 1-forms defining the horizontal space. Furthermore, since we have

$$[V, \frac{\partial}{\partial y^\alpha}] = \frac{\partial}{\partial x^\alpha} \quad \text{and} \quad \mathcal{H}^{[1]} \stackrel{def}{=} \mathcal{H} + [\mathcal{H}, \mathcal{H}] = T\mathcal{M},$$

the horizontal bundle $\mathcal{H}$ together with its commutator span the tangent space. □

Physically, the vector V should be thought as a vector defining the future direction. It can be used to define a causal structure on the sub-Riemannian and Riemannian spaces we define.

By Chow-Rashevskii' theorem, see [1, 2] and [10], the bracket generating condition is sufficient for any two points of $\mathcal{M}$ to be connected by a horizontal curve, i.e., a curve whose velocity lies in the horizontal direction. We note explicitly that in our case $\mathcal{H}$ is not strong bracket generating due to $\mathcal{H}_\xi + [\frac{\partial}{\partial y^\alpha}, \mathcal{H}]_\xi \neq T_\xi\mathcal{M}$. Recall, [1] and [10], that *strong bracket generating* or *fat distribution* means either of the following equivalent conditions, where $\xi \in \mathcal{M}$ and $w,\ w' \in \mathcal{H}_\xi$, $w \neq 0$, with horizontal extensions W and W': (i) $\mathcal{H} + [W, \mathcal{H}]_\xi = T_\xi\mathcal{M}$; (ii) the curvature (Levi) form $\mathcal{L} : \mathcal{H} \times \mathcal{H} \to T\mathcal{M}/\mathcal{H}$,

$$\mathcal{L}_z(w, w') = [W, W']_\xi \ mod\ \mathcal{H}_\xi$$

defines a surjective map $\mathcal{L}(w, .)$, i.e., the *dual curvature is symplectic*. In general, the strong bracket generating property (which does not hold here) excludes the existence of abnormal (sub-Riemannian) geodesics, see Sect. 3.1.1.

3.1 Sub-Riemannian Metrics

Let f and h be smooth positive functions on $\mathcal{M}$. For $0 < b < 1$, $\xi = (x, y) \in \mathcal{M}$ and $y = y^\alpha \frac{\partial}{\partial x^\alpha}$, recalling that $y_\alpha = g_{\alpha\beta} y^\beta$, define

$$G = G_b(\xi) \stackrel{def}{=} f(|y|)\frac{y_\alpha y_\beta}{|y|^2}\, dx^\alpha \otimes dx^\beta \ + \ h(|y|)\left(\frac{y_\alpha y_\beta}{|y|^2} \ - \ b\, g_{\alpha\beta}\right) dy^\alpha \otimes dy^\beta.$$

Theorem 2 *The above formula for G defines a positive definite and Lorentz invariant symmetric tensor on $\mathcal{H}$, i.e., G is invariant under the transformations*

$$(x, y) \mapsto (\Lambda x, \Lambda y),$$

where Λ is a Lorentz transformation with respect to the given Lorentzian metric g on M.

Proof The Lorentz invariance is obvious. We sketch the proof of the positivity. For $W = u^\alpha \frac{\partial}{\partial y^\alpha} + aV \in \mathcal{H}$, where V was defined in Proposition 2, we have

$$G(W, W) = a^2 f(|y|)\, |y|^2 \; + \; h(|y|)\, \frac{g(y, u)^2 - b|u|^2|y|^2}{|y|^2},$$

where $u = u^\alpha \frac{\partial}{\partial x^\alpha}$. Consider the two cases (recall $|y|^2 > 0$), depending on u being timelike or non-timelike. If u is timelike we have the reverse Cauchy-Schwarz inequality $(y^\alpha u_\alpha)^2 \geq |y|^2|u|^2$, hence

$$G(W, W) \geq a^2 f(|y|)\, |y|^2 \; + \; h(|y|)\,(1 - b)|u|^2.$$

In the second case, where $|u|^2 \leq 0$, we use that $b|u|^2|y|^2 \leq 0$. □

When $M = \mathbb{R}^{1,d}$ is a Lorentz spacetime, then G is also invariant under translations in x.

3.1.1 The Sub-Riemannian Distance and Geodesics

For the sake of giving some context and the possible difficulties we may encounter we recall some known results.

As usual, we define the Carnot-Carathédory (sub-Riemannian) distance using the arclength of "admissible" curves. For sufficiently smooth (e.g. locally rectifiable) $\gamma : [0, 1] \to \mathcal{M}$ which is horizontal, $z'(\tau) \in \mathcal{H}_{\gamma(\tau)}$, we define the Carnot-Carathédory (sub-Riemannian) length of γ by the formula

$$l(\gamma) = \int_0^1 \sqrt{G_b(z', z')} d\tau.$$

For $z_1, z_2 \in \mathcal{M}$ the Carnot-Carathédory (sub-Riemannian) distance(abbr. CC-distance) between the two points is

$$d_{CC}(z_0, z_1) = \inf\{l(\gamma) \mid \gamma(0) = z_0, \gamma(1) = z_1, \gamma'(\tau) \in \mathcal{H}_{\gamma(\tau)}\}.$$

As well known, the CC-distance defines a topology equivalent to the manifold topology.A (horizontal) curve is called a *minimizing CC-geodesic* if it achieves the CC-distance between its endpoints. It is called a CC-geodesic if it is locally a minimizing CC-geodesic.

Let $W_0, \ldots, W_{d+2}$ be a local orthonormal frame for the horizontal distribution $\mathcal{H}$. For $\theta \in T^*_\xi\mathcal{M}$ define the Hamiltonian function

$$H(\theta) = \frac{1}{2}\left(\langle\theta, W_0(\xi)\rangle^2 + \cdots + \langle\theta, W_{d+2}(\xi)\rangle^2\right).$$

H is a fiber-quadratic positive semi-definite form on $T^*\mathcal{M}$ of rank $d+3 = \dim \mathcal{H}$. Recall that $T^*\mathcal{M}$ has the canonical symplectic form.

The projection to $\mathcal{M}$ of an integral curve for the Hamiltonian vector field with Hamiltonian H is a CC-geodesic.

In the Riemannian case this characterizes the geodesics. In the sub-Riemannian case there could be CC-geodesics which are not the projections of integral curves for the Hamiltonian vector field of H. Let $\mathcal{H}^\perp \subset T^*\mathcal{M}$ be the annihilator of the distribution $\mathcal{H}$. Let ω be the restriction to $\mathcal{H}^\perp$ of the canonical symplectic form of $T^*\mathcal{M} \subset T(TM)$. A *characteristic curve* for $\mathcal{H}^\perp$ is an absolutely continuous nowhere vanishing curve $\eta : [0, 1] \to \mathcal{H}^\perp$ whose derivative lies in the kernel of ω whenever it exists, $\omega(\theta'(t), \Theta) = 0$ for all $\Theta \in T_{\theta(t)}\mathcal{H}^\perp$. An admissible curve γ on $\mathcal{M}$ is *singular* if and only if it is the projection of a characteristic curve, which depends only on the distribution, not on the sub-Riemannian metric. We note that there is another (equivalent) definition of singular curves as the critical points of the end-point map. It is also known that if ω is "symplectic", i.e., has trivial kernel then there are no characteristics ("strong bracket generating case"). However, every CC-geodesic is a singular curve or a normal geodesics.

A sub-Riemannian metric space is complete if and only if the closed metric balls (or all sufficiently small balls) are compact. In this case, there exists a minimizing CC-geodesics between any two given point. This is the case when the sub-Riemannian metric G is the restriction of a complete Riemannian metric on $\mathcal{M}$.

4 The Riemannian Space

Below we shall use the notation set at the beginning of Sect. 3.1. The following proposition is an easy corollary of the previous constructions, see in particular Theorem 2.

Theorem 3 *For* $a > 0$, $\hat{G}_{a,b} \overset{def}{=} G_b - af\, g_{\alpha\beta}\eta^\alpha \otimes \eta^\beta$ *defines a "Lorentz invariant" Riemannian metric on* $\mathcal{M}$. *Explicitly, dropping a and b in the notation,*

$$\hat{G} = f \cdot \left(\frac{(1+a) y_\alpha y_\beta}{|y|^2} - a g_{\alpha\beta} \right) dx^\alpha \otimes dx^\beta + h \cdot \left(\frac{y_\alpha y_\beta}{|y|^2} - b g_{\alpha\beta} \right) dy^\alpha \otimes dy^\beta .$$

The Riemannian metrics $\hat{G}_{a,b}$ are a Riemannian approximation of the sub-Riemannian metric G_b, which in the limit $a \to \infty$ converge to the sub-Riemannian space.

5 Some Remarks on the Minkowski Space Case

Assume that g is the Minkowski metric on $\mathbb{R}^{1,d}$. We note that, the horizontal space $\mathcal{H}$ is invariant under Lorentz transformations and in the case of Minkowski space also under translations in the space time variable x. The exhibited Lorentz invariances imply

Proposition 3 *Let $\gamma(\tau) = (x(\tau), y(\tau))$ be a smooth phase space curve, Λ a Lorentz transformation of $\mathbb{R}^{1,d}$, and $\gamma_\Lambda(\tau) \stackrel{def}{=} (\Lambda x(\tau), \Lambda y(\tau))$. If γ is a Riemannian geodesic (for $\hat{G}$) then γ_Λ is also a geodesic. The same is true in the setting of CC-geodesics.*

An interesting case comes from letting $f = k = const, h = l/|y|^2, l = const$. With these assumptions, the Riemannian metric $\hat{G}$ is complete, hence, the sub-Riemannian metric is complete as well. Furthermore, using vector notation and letting $\langle y, x'\rangle = y_\alpha dx^\alpha/d\tau$ etc., with prime denoting derivative with respect to the parameter τ, the geodesic equations of the Riemannian metric $\hat{G}$ are

$$x'' = (a+1)\left(\frac{a-1}{a}\frac{\langle y, x'\rangle\langle y, y'\rangle}{|y|^2} - \langle x', y'\rangle\right)\frac{y}{|y|^2} + \frac{1+a}{a}\frac{\langle y, x'\rangle}{|y|^2}y'$$

$$y'' = \left(\frac{k(a+1)}{la}\langle y, x'\rangle^2 - |y'|^2\right)\frac{y}{|y|^2} - \frac{k(a+1)}{la}\langle y, x'\rangle x' + \frac{2}{|y|}|y|'y'.$$

Proposition 4 *If $\gamma(\tau) = (x(\tau), y(\tau))$ is a geodesic of $(\mathcal{M}, \hat{G})$ such that $\gamma(0) = (0, v)$ and $\gamma'(0) = (u, w)$ with v, v' and w parallel vectors in $\mathbb{R}^{1,d}$, then the same condition holds throughout the definition of γ*

As a corollary, we have that every timelike line in Minkowski space is locally the projection of some geodesic of $(\mathcal{M}, \hat{G})$.

References

1. A. Agrachev, D. Barilari, U. Boscain, *A Comprehensive Introduction to Sub-Riemannian Geometry*. Cambridge Studies in Advanced Mathematics (Cambridge University Press, Cambridge, 2019)
2. A. Bellaïche, *The Tangent Space in Sub-Riemannian Geometry. Sub-Riemannian Geometry*. Progress in Mathematics, vol. 144 (Birkhäuser, Basel, 1996), pp. 1–78
3. L. Bombelli, J. Henson, R.D. Sorkin, Discreteness without symmetry breaking: a theorem. Modern Phys. Lett. A **24**(32), 2579–2587 (2009)
4. L. Bombelli, J. Lee, D. Meyer, R.D. Sorkin, Space-time as a causal set. Phys. Rev. Lett. **59**(5), 521–524 (1987)
5. E.R. Caianiello, Geometry from quantum mechanics. Nuovo Cimento B (11) **59**(2), 350–366 (1980)
6. E.R. Caianiello, Quantization as geometry in phase space. Rend. Sem. Mat. Fis. Milano **53**(1983), 245–271 (1986)

7. F. Dowker, *Causal Sets and the Deep Structure of Spacetime*. 100 Years of Relativity, pp. 445–464. https://doi.org/10.1142/9789812700988_0016
8. S.W. Hawking, A.R. King, P.J. McCarthy, A new topology for curved space-time which incorporates the causal, differential, and conformal structures. J. Math. Phys. **17**(2), 174–181 (1976)
9. D.B. Malament, The class of continuous timelike curves determines the topology of spacetime. J. Math. Phys. **18**(7), 1399–1404 (1977)
10. R. Montgomery, *A Tour of Subriemannian Geometries, Their Geodesics and Applications*. Mathematical Surveys and Monographs, vol. 91 (American Mathematical Society, Providence, RI, 2002)
11. S. Surya, The causal set approach to quantum gravity. Living Rev. Relativ. **22**, 5 (2019)

The Orbit Space and Basic Forms of a Proper Lie Groupoid

Jordan Watts

Abstract A classical result in differential geometry states that for a free and proper Lie group action, the quotient map to the orbit space induces an isomorphism between the de Rham complex of differential forms on the orbit space and the basic differential forms on the original manifold. In this paper, this result is generalized to the case of a proper Lie groupoid, in which the orbit space is equipped with the quotient diffeological structure. As an application of this, we obtain a de Rham theorem for the de Rham complex on the orbit space.

Keywords Diffeology · Lie groupoid · de Rham complex · Basic forms · Linearization

Mathematics Subject Classification (2010) 58H05,22A22

1 Introduction

Given a proper Lie group action on a smooth manifold, the Slice Theorem of Koszul and Palais [9, 12] states that any orbit of the action has an invariant tubular neighbourhood equivariantly diffeomorphic to a linear model about the orbit. From this it follows that in the case of a free and proper action, the quotient map induces an isomorphism between the complex of differential forms on the orbit space and the complex of basic differential forms on the manifold.

Using the theory of diffeology, which generalises the theory of smooth manifolds (see Definition 2.1), this result was extended by the author to include any compact Lie group action (not necessarily free) [14, Chapter 3], and further to Lie group actions for which the identity component of the Lie group acts properly by Karshon-Watts [7]. The purpose of this note is to extend the result to proper Lie groupoids;

J. Watts (✉)
Department of Mathematics, Central Michigan University, Mount Pleasant, MI, USA
e-mail: jordan.watts@cmich.edu

P. Cerejeiras et al. (eds.), *Current Trends in Analysis, its Applications and Computation*, Research Perspectives,
https://doi.org/10.1007/978-3-030-87502-2_52

namely, to prove Theorem 1.1 below. Denote by **Bi** the bicategory of Lie groupoids with (right principal) bibundles as arrows and isomorphisms of bibundles as 2-arrows (see [10, 11] for definitions and more details). Given a Lie groupoid $G = (G_1 \rightrightarrows G_0)$, we define a differential form α on G_0 to be basic if $s^*\alpha = t^*\alpha$, where s and t are the source and target maps.

Theorem 1.1 *Given a proper Lie groupoid $G = (G_1 \rightrightarrows G_0)$, the quotient map $G_0 \to G_0/G_1$ induces an isomorphism between the diffeological de Rham complex of the orbit space G_0/G_1 and the complex of basic forms on G_0; moreover, this isomorphism is natural on the full sub-bicategory of proper Lie groupoids in* **Bi**.

The natural isomorphism in Theorem 1.1 suggests that functors should be involved. Indeed, we show that there is a 2-functor Ψ from **Bi** to the category of diffeological spaces, sending a Lie groupoid G to its orbit space G_0/G_1, a bibundle $P: G \to H$ to a smooth map $\Psi_P: G_0/G_1 \to H_0/H_1$, and a 2-arrow between bibundles to a trivial 2-arrow. If Ω^*_{basic} is the (contravariant) functor sending a Lie groupoid to its complex of basic forms, and Ω^* is the functor sending a diffeological space to its de Rham complex, then Theorem 1.1 states that when restricting to proper groupoids, there is a natural isomorphism from $\Omega^* \circ \Psi$ to Ω^*_{basic} given by the pullback map induced by the quotient map from the base of a groupoid to its orbit space.

The proof of Theorem 1.1 relies on the corresponding result for compact Lie group actions (see Theorem 3.7), and a linearisation theorem for proper Lie groupoids (see Theorem 4.1). This linearisation theorem has been developed throughout a series of works, which include authors such as Zung, Weinstein, Crainic-Struchiner, and del Hoyo-Fernandes [2, 3, 17–19]; for our purposes we adopt the language of del Hoyo-Fernandes.

We obtain some immediate consequences of Theorem 1.1. It follows from the Slice Theorem for a proper Lie group action on a manifold that the cohomology of the basic forms is isomorphic to the singular cohomology on the orbit space. This result was extended to proper Lie groupoids by Pflaum-Posthuma-Tang [13, Section 8]. Thus together with Theorem 1.1, we obtain a de Rham theorem for the orbit space of a proper Lie groupoid which is *intrinsic*, in the sense that it only depends on the diffeology (*i.e.* smooth structure) of the orbit space, and not on the original Lie groupoid.

Corollary 1.2 *Given a proper Lie groupoid $G = (G_1 \rightrightarrows G_0)$, the de Rham cohomology of the orbit space G_0/G_1 is isomorphic to the singular cohomology of G_0/G_1.*

Another application is a reinterpretation of Corollary 3.6 and Theorem 1.1 in terms of exactness in the following sequence:

$$0 \longrightarrow \Omega^*(G_0/G_1) \xrightarrow{\pi_G^*} \Omega^*(G_0) \xrightarrow{s^*-t^*} \Omega^*(G_1) \tag{1.1}$$

Corollary 1.3 *For any Lie groupoid $G = (G_0 \rightrightarrows G_1)$, the sequence 1.1 is exact at $\Omega^*(G_0/G_1)$. If G is proper, then the sequence is also exact at $\Omega^*(G_0)$.*

At this point it is natural for the reader to ask whether the properness condition on a Lie groupoid in Theorem 1.1 can be relaxed. At this time, the author is not aware of an example in which the isomorphism does *not* hold. Indeed, even when the quotient has trivial topology but non-trivial diffeology, such as the 1-dimensional irrational torus, Theorem 1.1 still holds; see [7, Example 5.11].

This paper is broken down as follows: Section 2 reviews the theory of diffeology. Section 3 reviews and develops needed results on basic differential forms on Lie groupoids from a categorical perspective. Section 4 reviews linearizations, and proves Theorem 1.1.

We end this introduction with a brief survey of related ideas in the literature. For instance, the functor Ψ between Lie groupoids and diffeological spaces appears in the literature already, sometimes in disguise. It follows from [6] that the restriction of Ψ to effective orbifolds, viewed as effective proper étale Lie groupoids, is essentially injective on objects. There is a similar functor to (Sikorski) differential spaces that factors through Ψ; see [15, Theorem B] for details. The point-of-view of stacks over manifolds is taken in [16], in which a functor sending a stack to its underlying diffeological coarse moduli space is constructed; as differentiable stacks are represented by Lie groupoids, this is just a stacky manifestation of the functor Ψ above extended to all stacks. Finally, a more detailed study of when and how a diffeologically smooth map between orbit spaces is in the image of the functor Ψ appears in [8].

The author would like to thank Rui Loja Fernandes, Eugene Lerman, and Ioan Mărcuţ for many illuminating discussions about Lie groupoids, and Yael Karshon for her comments and encouragement to publish this note.

2 Background

It is assumed that the reader is familiar with Lie groupoids and (right principal) bibundles between them; see [10, 11] for an exposition on these. For the purposes of this paper, all Lie groupoids $G = (G_1 \rightrightarrows G_0)$ are assumed to be finite dimensional, paracompact, and Hausdorff. Groupoid homomorphisms $F\colon G \to H$ are denoted using $F_0\colon G_0 \to H_0$ for the map between bases, and $F_1\colon G_1 \to H_1$ for the map between arrow spaces. A more detailed exposition of diffeological spaces appears in [5], but we give a brief review of them now.

Definition 2.1 (Diffeology) Let X be a set. A **parametrisation** of X is a map of sets $p\colon U \to X$ where U is an open subset of some Euclidean space (no fixed dimension). A **diffeology** $\mathcal{D}$ on X is a family of parametrisations satisfying:

1. **(Covering)** $\mathcal{D}$ contains all constant maps into X.
2. **(Locality)** Let $p\colon U \to X$ be a parametrisation, $\{U_\alpha\}$ an open cover of U, and $\{p_\alpha\colon U_\alpha \to X\} \subseteq \mathcal{D}$ such that $p|_{U_\alpha} = p_\alpha$ for each α. Then $p \in \mathcal{D}$.

3. **(Smooth Compatibility)** For any $(p\colon U \to X) \in \mathcal{D}$ and any smooth parametrisation $f\colon V \to U$ of U, the composition $p \circ f$ is in $\mathcal{D}$.

A set X equipped with a diffeology $\mathcal{D}$ is called a **diffeological space**, and is denoted by $(X, \mathcal{D})$. When the diffeology is understood, we will drop the symbol $\mathcal{D}$. The parametrisations $p \in \mathcal{D}$ are called **plots**. Given two diffeological spaces $(X, \mathcal{D}_X)$ and $(Y, \mathcal{D}_Y)$, a map $\varphi\colon X \to Y$ is **(diffeologically) smooth** if $\varphi \circ p \in \mathcal{D}_Y$ for any plot $p \in \mathcal{D}_X$.

Diffeological spaces with smooth maps between them form a category **Diffeol**, which contains the category of smooth manifolds as a full subcategory. Indeed, the **standard diffeology** on M is the set of all smooth parametrisations $f\colon U \to M$, where "smooth" is taken in the usual sense.

The usefulness of **Diffeol** is illustrated by the fact that it is complete and co-complete [1, 5]. For instance, given a diffeological space X and an equivalence relation $\sim$ on X with quotient map $\pi : X \to X/\sim$, the quotient comes equipped with the **quotient diffeology**. This is the family of all parametrisations $p\colon U \to X/\sim$ satisfying the condition that each point $u \in U$ has an open neighbourhood V such that $p|_V$ factors through π. As another example, given a subset $Z \subseteq X$, the **subset diffeology** on Z consists of all plots of X whose image lies in X. Finally, given another diffeological space Y, the **product diffeology** on $X \times Y$ is the collection of all parametrisations (p_1, p_2) such that p_1 is a plot of X and p_2 is a plot of Y.

Example 2.2 Let G be a Lie groupoid and fix $x \in G_0$. The **orbit** $\mathcal{O}$ of G through x is the set

$$\mathcal{O} = \{y \in G_0 \mid \exists g \in G_1 \text{ so that } s(g) = x \text{ and } t(g) = y\}.$$

The subset diffeology on $\mathcal{O}$ gives it the structure of an immersed submanifold of G_0 (see [2, Section 1.2]). The **orbit space** G_0/G_1 of G is the quotient of G_0 by the equivalence relation $\sim$ given by: $x \sim y$ if x and y are in the same orbit. G_0/G_1 comes equipped with the quotient diffeology induced by the standard manifold diffeology on G_0; denote the quotient map by $\pi_G\colon G_0 \to G_0/G_1$.

The fibred product $G_0 \times_\pi G_0$ is the arrow space of the **relation groupoid** of $\sim$

$$\{(x_1, x_2) \in G_0 \times G_0 \mid \exists g \in G_1 \text{ such that } s(g) = x_1 \text{ and } t(g) = x_2\}$$

whose source and target are restrictions of the first and second projection maps, respectively, to G_0. The arrow space comes equipped with the subset diffeology induced by $G_0 \times G_0$, with respect to which the map $(s, t)\colon G_1 \to G_0 \times_\pi G_0$ is a smooth surjection. ◇

Definition 2.3 (Differential Forms) A **differential** k**-form** α on a diffeological space X is an assignment to each plot $p\colon U \to X$ a differential form $\alpha_p \in \Omega^k(U)$, satisfying $\alpha_{p \circ f} = f^*\alpha_p$ for any smooth parametrisation $f\colon V \to U$ of U. Denote the collection of k-forms on X by $\Omega^k(X)$. The **exterior derivative** $d\colon \Omega^k(X) \to \Omega^{k+1}(X)$ is defined plot-wise: $(d\alpha)_p = d(\alpha_p)$.

The 0-forms of a diffeological space are exactly the smooth functions $f\colon X \to \mathbb{R}$; the collection of all differential forms on X forms a de Rham complex $(\Omega^*(X), d)$, although we will abbreviate the notation to just $\Omega^*(X)$; and a smooth map $\varphi\colon X \to Y$ induces a **pullback map** $f^*\colon \alpha \mapsto f^*\alpha$, which is a map of complexes. In particular, we have a contravariant functor Ω sending a diffeological space X to the complex $\Omega^*(X)$ and a smooth map to its corresponding pullback map. Finally, on a manifold equipped with the standard diffeology, the diffeological de Rham complex is exactly the standard one.

3 Basic Differential Forms and Bibundles

In this section, we collect a number of results regarding basic forms of a Lie groupoid. After defining them and relating them to the classical notion for a Lie group action (Lemma 3.3), we prove that pullbacks of forms from an orbit space are basic (Corollary 3.6). We also show that there is a functor between **Bi** and **Diffeol** (Theorem 3.8), as well as that basic forms "pullback" by bibundles to basic forms (Proposition 3.9). These results will be needed in the sequel.

Definition 3.1 (Basic Differential Forms) Given a Lie groupoid G, a differential form α on G_0 is **basic** if $s^*\alpha = t^*\alpha$. Together with the standard differential, these form a complex, denoted by $\Omega^*_{\text{basic}}(G)$.

Remark 3.2 Pflaum-Posthuma-Tang define a form $\alpha \in \Omega^k(M)$ to be basic with respect to a Lie groupoid G if $\rho(X)\lrcorner\alpha = 0$ for all smooth sections X of the associated Lie algebroid to G with anchor map ρ, and if α is G-invariant [13, Definition 8.1]. Regarding the second condition, the first condition implies that α descends to a smooth section $\widetilde{\alpha}$ of the kth wedge power of the normal bundle to an orbit, for each orbit in G_0; the second condition now requires that $([s_*v]-[t_*v])\lrcorner\widetilde{\alpha}$ vanishes for each $v \in TG_1$ (abusing notation). It is not difficult to show that this definition is equivalent to Definition 3.1 using the standard identification of the Lie algebroid with the pullback by the unit map of the subbundle $\bigcup_{x\in M} T(t^{-1}(x)) \subset TG_1$; we do not need this result, however, in this paper, and so we provide no further detail.

Recall that for a Lie group K and a K-manifold M, a differential form on M is basic if it is K-invariant and horizontal, the latter term meaning that it vanishes on vectors tangent to the K-orbits.

Lemma 3.3 (Equivalence of Notions of Basicness) *Let K be a Lie group and M a K-manifold. A differential form α on M is basic with respect to the action if and only if it is basic with respect to the action groupoid $K \ltimes M := (K \times M \rightrightarrows M)$.*[1]

Proof Fix $(k, x) \in K \times M$, and a vector $v \in T_{(k,x)}(K \times M)$. Denote by $\mathfrak{k}$ the Lie algebra of K. Via a left trivialisation of TK we identify $T(K \times M)$ with $K \times \mathfrak{k} \times TM$. Under this identification, there is some $\xi \in \mathfrak{k}$ and $u \in T_xM$ such that $v = (k, \xi, u)$. It follows that $s_*v = u$ and $t_*v = \xi_M|_{k\cdot x} + k_*u$, where ξ_M is the vector field on M induced by $\xi \in \mathfrak{k}$.

Given $v_1, \ldots, v_\ell \in T_{(k,x)}(K \times M)$, setting $v_i = (k, \xi_i, u_i)$ as above, for any ℓ-form α on M:

$$s^*\alpha(v_1, \ldots, v_l) = \alpha(u_1, \ldots, u_l), \tag{3.1}$$

$$t^*\alpha(v_1, \ldots, v_l) = \alpha\big((\xi_1)_M|_{k\cdot x} + k_*u_1, \ldots, (\xi_l)_M|_{k\cdot x} + k_*u_l\big). \tag{3.2}$$

If follows that if α is K-invariant and horizontal, then $s^*\alpha = t^*\alpha$.

Conversely, suppose that $s^*\alpha = t^*\alpha$. Fix $u_1, \ldots, u_l \in T_xM$. Since s is a surjective submersion, there exist for each $i = 1, \ldots, l$ vectors $v_i = (k, \xi_i, u_i) \in K \times \mathfrak{k} \times TM$ such that $s_*v_i = u_i$. Without loss of generality, we may take $\xi_i = 0$ for each i. By Eqs. (3.1) and (3.2), it follows that α is K-invariant.

Now suppose that u_1 is tangent to the K-orbit through x. There exists $\zeta \in \mathfrak{k}$ such that $u_1 = \zeta_M|_x$. Let $\xi_1 = -\mathrm{Ad}_k(\zeta)$. Then

$$t_*v_1 = -(\mathrm{Ad}_k(\zeta))_M|_{k\cdot x} + k_*u_1 = 0.$$

By (3.1) and (3.2) we conclude that α is horizontal. The proof is complete. □

The following is a useful tool when checking whether a differential form on the base of a Lie groupoid is the pullback of a form on its orbit space by the quotient map. A proof appears in [5, Article 6.38].

Proposition 3.4 (Pullbacks from a Quotient) *Let X be a diffeological space equipped with an equivalence relation $\sim$, with quotient $Y = X/\sim$ and quotient map $\pi: X \to Y$. For any k and $\alpha \in \Omega^k(X)$, there exists $\beta \in \Omega^k(Y)$ such that $\pi^*\beta = \alpha$ if and only if $p_1^*\alpha = p_2^*\alpha$ for any plots $p_1, p_2 : U \to X$ such that $\pi \circ p_1 = \pi \circ p_2$.*

Remark 3.5 Proposition 3.4 can be restated in terms of the relation groupoid as follows. Let $\mathrm{pr}_i : X \times_\pi X \to X$ be the ith standard projection map. For any k and $\alpha \in \Omega^k(X)$, there exists $\beta \in \Omega^k(Y)$ such that $\pi^*\beta = \alpha$ if and only if $\mathrm{pr}_1^*\alpha = \mathrm{pr}_2^*\alpha$.

We now can prove part of Theorem 1.1.

[1] The statement of this lemma was communicated to the author by Eugene Lerman; however, the proof is the author's.

Corollary 3.6 (Pullbacks from the Quotient Are Basic) *Let G be a Lie groupoid. The pullback map $\pi_G^*\colon \Omega^*(G_0/G_1) \to \Omega^*(G_0)$ is an injection with image in $\Omega^*_{\mathrm{basic}}(G)$.*

Proof Fix $\beta \in \Omega^k(G_0/G_1)$ and let $\alpha = \pi_G^*\beta$. By Proposition 3.4 and Remark 3.5, $\mathrm{pr}_1^*\alpha = \mathrm{pr}_2^*\alpha$. Thus $(s,t)^*(\mathrm{pr}_1^*\alpha - \mathrm{pr}_2^*\alpha) = 0$ (see Example 2.2), and so $s^*\alpha = t^*\alpha$. Injectivity follows from the definition of the quotient diffeology on G_0/G_1. □

In the case of a compact Lie group action, π_G^* from Corollary 3.6 becomes an isomorphism. This follows from Lemma 3.3 and [14, Theorem 3.20] (see also [7]), stated below.

Theorem 3.7 (Group Action Case) *Let K be a compact Lie group and M a K-manifold, with quotient map $\pi\colon M \to M/K$. The pullback map π^* is an isomorphism between the de Rham complexes of differential forms on M/K and basic differential forms on M.*

Sending a Lie groupoid to its diffeological orbit space constitutes a functor.

Theorem 3.8 (The Functor Ψ) *There is a functor $\Psi\colon$ **Bi** $\to$ **Diffeol** sending a Lie groupoid G to its orbit space G_0/G_1, a bibundle $P\colon G \to H$ to a unique smooth map $\Psi_P\colon G_0/G_1 \to H_0/H_1$ such that $\Psi_P \circ \pi_G \circ a_L = \pi_H \circ a_R$, and a 2-arrow in* **Bi** *to a trivial 2-arrow in* **Diffeol**.

Proof Let $P\colon G \to H$ a bibundle between Lie groupoids; denote the anchor maps to G and H by $a_L^P\colon P \to G_0$ (or just a_L if P is understood) and $a_R^P\colon P \to H_0$, respectively. Fix $x \in G_0$. Define $\Psi_P\colon G_0/G_1 \to H_0/H_1$ by $\Psi_P([x]) = \pi_H \circ a_R \circ \sigma(x)$ where σ is a local section of a_L about x. To show Ψ_P is well-defined, let y be in the same orbit as x, and σ' a local section of a_L about y. It suffices to show there exists $h \in H_1$ so that $a_R(\sigma'(y)) = s(h)$ and $a_R(\sigma(x)) = t(h)$. Let $g \in G_1$ so that $s(g) = x$ and $t(g) = y$. Then $a_L(g\cdot\sigma(x)) = y$. Since a_L is a principal H-bundle, there exists a unique h such that $(g\cdot\sigma(x))\cdot h = \sigma'(y)$. Well-definedness of Ψ_P now follows from the G-invariance of a_R. The identity $\Psi_P \circ \pi_G \circ a_L = \pi_H \circ a_R$ follows from the definition of Ψ_P, and implies uniqueness.

To show that Ψ_P is smooth, through local sections of a_L and local lifts of a plot $p\colon U \to G_0/G_1$ (which exist by definition of the quotient diffeology), one obtains that $\Psi_P \circ p$ is locally a plot, and hence globally a plot by the Locality Axiom.

That a 2-arrow $\alpha\colon P \Rightarrow Q$ is sent to a trivial 2-arrow follows from the uniqueness of Ψ_P and Ψ_Q and the $(G\text{-}H)$-equivariance of α.

Given a third Lie groupoid $K = (K_1 \rightrightarrows K_0)$ and bibundle $Q\colon H \to K$, the composition $Q \circ P$ is the quotient by the diagonal H-action $(P \times_{H_0} Q)/H$. To show $\Psi_{Q\circ P} = \Psi_Q \circ \Psi_P$, it suffices to show

$$\Psi_Q \circ \Psi_P \circ \pi_G \circ a_L^P(p) = \pi_K \circ a_R^Q(q) \quad \forall (p,q) \in P \times_{H_0} Q.$$

This follows from $\pi_H \circ a_R^P(p) = \pi_H \circ a_L^Q(q)$, which in turn follows from the definition of $P \times_{H_0} Q$. Thus $\Psi(P) := \Psi_P$ respects composition, and all other identities are straightforward to check. □

We now show that bibundles induce maps between complexes of basic forms, similar to what smooth maps between manifolds do with the corresponding de Rham complexes. Versions of this result appear in the literature; for instance, [4, 13].

Proposition 3.9 (Pullbacks of Basic Forms by Bibundles) *Let $P\colon G \to H$ be a bibundle between Lie groupoids with anchor maps $a_L\colon P \to G_0$ and $a_R\colon P \to H_0$. For any $\beta \in \Omega^k_{basic}(H)$ there exists a unique form in $\Omega^k_{basic}(G)$, denoted $P^*\beta$, such that $a_L^*(P^*\beta) = a_R^*\beta$. In fact, Ω^k_{basic} is a functor on* **Bi** *sending P to a homomorphism of complexes $P^*\colon \Omega^k_{\text{basic}}(H) \to \Omega^k_{\text{basic}}(G)$, and a 2-arrow in* **Bi** *to a trivial 2-arrow. In particular, a Morita equivalence induces an isomorphism between complexes of basic forms.*

Proof Fix $\beta \in \Omega^k_{\text{basic}}(H_0)$. The action groupoid $P \rtimes H$ has source the action map act_H and target pr_1; with respect to these we have

$$\text{act}_H^* \, a_R^*\beta = \text{pr}_2^* s^*\beta = \text{pr}_2^* t^*\beta = \text{pr}_1^* a_R^*\beta.$$

Thus $a_R^*\beta$ is basic with respect to $P \rtimes H$.

Since P is a principal H-bundle over G_0, $P \rtimes H$ is isomorphic as a groupoid to $P \times_{G_0} P \rightrightarrows P$. But this is the relation groupoid for the action groupoid $G \ltimes P$, and so it follows that $a_R^*\beta$ is basic with respect to $G \ltimes P$ as well. By Remark 3.5 there exists a form α on G_0 such that $a_L^*\alpha = a_R^*\beta$; α is unique since a_L is a surjective submersion. Similarly, $\text{pr}_1 : G_1 \times_{G_0} P \to G_1$ is a surjective submersion, and so the result follows from

$$\text{pr}_1^* s^*\alpha = \text{pr}_2^* a_L^*\alpha = \text{pr}_2^* a_R^*\beta = \text{act}_G^* \, a_R^*\beta = \text{act}_G^* \, a_L^*\alpha = \text{pr}_1^* t^*\alpha.$$

It follows from uniqueness and the definitions that $(Q \circ P)^* = P^*Q^*$ for composable bibundles P and Q, and that isomorphic bibundles yield the same pullback map between complexes. The remaining statements are straightforward to check. □

4 Linearisations and Proof of Theorem 1.1

In this section we review linearisations in the context of Lie groupoids, and prove Theorem 1.1. We follow the notation and terminology of [3]; more details can be found in [2, Subsection 1.2].

Fix a Lie groupoid G. A submanifold $S \subseteq G_0$ is **saturated** if it is a union of orbits of G. The restriction of G to S is denoted by $G_S \rightrightarrows S$. The normal bundle $\nu(S)$ of S in G_0 along with the normal bundle $\nu(G_S)$ of G_S in G_1 form a Lie groupoid $\nu(G_S) \rightrightarrows \nu(S)$ whose structure maps are induced by the differentials of those of G. The standard bundle projections form a homomorphism $(\nu(G_S) \rightrightarrows \nu(S)) \to (G_S \rightrightarrows S)$ that provides a local linear model of G about S.

A **groupoid neighbourhood** of $G_S \rightrightarrows S$ is a subgroupoid $U = (U_1 \rightrightarrows U_0)$ of G in which U_0 is an open neighbourhood of S and U_1 an open neighbourhood of G_S; it is **full** if $U_1 = s^{-1}(U_0) \cap t^{-1}(U_0)$. We say that G is **linearisable** at a saturated submanifold S if there exist full groupoid neighbourhoods U of $G_S \rightrightarrows S$ in G and V of $G_S \rightrightarrows S$ in $\nu(G_S) \rightrightarrows \nu(S)$, and a Lie groupoid isomorphism $\Lambda\colon V \to U$ whose restriction to G_S is the identity. A celebrated result is the following:

Theorem 4.1 ([2], Corollary 5.13 of [3]) *Proper Lie groupoids are linearisable at each of their orbits.*

We now show given an orbit $\mathcal{O}$ of a proper Lie groupoid G that the condition that the de Rham complex of the orbit space of $\nu(G_{\mathcal{O}}) \rightrightarrows \nu(\mathcal{O})$ is isomorphic via pullback of the quotient map to basic forms of $\nu(G_{\mathcal{O}}) \rightrightarrows \nu(\mathcal{O})$ is a local one.

Lemma 4.2 (Locality) *Let G be a proper Lie groupoid, fix $x \in G_0$, and let $\mathcal{O}$ be the orbit of x. If V is a full groupoid neighbourhood of $\mathcal{O}$ in $N := (\nu(G_{\mathcal{O}}) \rightrightarrows \nu(\mathcal{O}))$ then $\pi_V^*\colon \Omega^*(V_0/V_1) \to \Omega^*_{\text{basic}}(V)$ is an isomorphism of complexes.*

Proof Let V^N be the saturation of V in N; *i.e.* $(V^N)_0$ is the union of orbits of points in V_0 and $(V^N)_1 = \nu(G_{\mathcal{O}})_{(V^N)_0}$. By [2, Example 3.2] V and V^N are Morita equivalent. Let G_x be the stabiliser of x in G. By [2, Example 3.3], there is a linear action of G_x on the normal space $\nu(\mathcal{O})_x$ to $\mathcal{O}$ at x whose action groupoid $G_x \ltimes \nu(\mathcal{O})_x$ is Morita equivalent to N; this Morita equivalence restricts to one between the restriction W of $G_x \ltimes \nu(\mathcal{O})_x$ to $W_0 := \nu(\mathcal{O})_x \cap (V^N)_0$ and V^N. By Proposition 3.9 and Theorem 3.8, it suffices to prove $\pi_W^*\colon \Omega^*(W_0/W_1) \to \Omega^*_{\text{basic}}(W)$ is an isomorphism. But as G_x is a compact Lie group, this follows from Theorem 3.7 using G_x-invariant partitions of unity on W_0. □

Proof *(of Theorem 1.1)* By Corollary 3.6, it is sufficient to show that the image of π_G^* is $\Omega^*_{\text{basic}}(G)$. Fix $\alpha \in \Omega^k_{\text{basic}}(G)$. By Proposition 3.4, it suffices to show $p_1^*\alpha = p_2^*\alpha$ for any two plots $p_1, p_2\colon U \to G_0$ such that $\pi_G \circ p_1 = \pi_G \circ p_2$; fix two such plots. It further suffices to show this equality near each point $u \in U$; fix such a point. Let $\mathcal{O}$ be the orbit of $p_1(u)$ (and hence $p_2(u)$ as well). By Theorem 4.1, G is linearisable; let $V = (V_1 \rightrightarrows V_0)$ and $W = (W_1 \rightrightarrows W_0)$ be full groupoid neighbourhoods of $G_{\mathcal{O}} \rightrightarrows \mathcal{O}$ in $\nu(G_{\mathcal{O}}) \rightrightarrows \nu(\mathcal{O})$ and $G_1 \rightrightarrows G_0$, respectively, and let $\Lambda\colon V \to W$ be an isomorphism that fixes $G_{\mathcal{O}} \rightrightarrows \mathcal{O}$. Set $B := p_1^{-1}(W_0) \cap p_2^{-1}(W_0)$ and let $i\colon W \to G$ be the inclusion morphism. The pullback $\Lambda_0^* i_0^* \alpha$ is basic with respect to V.

By Lemma 4.2, there exists $\beta \in \Omega^k(V_0/V_1)$ such that $\pi_V^*\beta = \Lambda_0^* i_0^*\alpha$. By Theorem 3.8, Λ descends to a diffeomorphism $\Psi_\Lambda \colon V_0/V_1 \to W_0/W_1$ such that $\Psi_\Lambda \circ \pi_V = \pi_W \circ \Lambda_0$, from which it follows that $\pi_W^*(\Psi_\Lambda^{-1})^*\beta = i_0^*\alpha$. Setting $q_n = p_n|_B$ for $n = 1, 2$,

$$p_n|_B^*\alpha = q_n^* i_0^*\alpha = q_n^*\pi_W^*(\Psi_\Lambda^{-1})^*\beta.$$

The proof now is reduced to showing that $\pi_W \circ q_1 = \pi_W \circ q_2$.

By Theorem 3.8 the inclusion i descends to a smooth map $j \colon W_0/W_1 \to G_0/G_1$ such that for $n = 1, 2$,

$$j \circ \pi_W \circ q_n = \pi_G \circ i_0 \circ q_n = \pi_G \circ p_n|_B.$$

Thus $j \circ \pi_W \circ q_1 = j \circ \pi_W \circ q_2$, and the proof is now reduced to showing that j is injective.

Fix $x_1, x_2 \in W$ such that $j(\pi_W(x_1)) = j(\pi_W(x_2))$. There exists $g \in G_1$ such that $s(g) = x_1$ and $t(g) = x_2$. Since W is full, $g \in W_1$ as well, and so $\pi_W(x_1) = \pi_W(x_2)$. This shows that j is injective.

Naturality of the isomorphism of complexes follows from Proposition 3.9 and Theorem 3.8. This completes the proof. □

References

1. J.C. Baez, A.E. Hoffnung, Convenient categories of smooth spaces. Trans. Am. Math. Soc. **363**, 5789–5825 (2011)
2. M. Crainic, I. Struchiner, On the linearization theorem for proper Lie groupoids. Ann. Sci. École Norm. Sup (4), **46**, 5 (2013)
3. M.L. del Hoyo, R.L. Fernandes, Riemannian metrics on Lie groupoids. J. Reine Angew. Math. **735**, 143–173 (2018)
4. B. Hoffman, R. Sjamaar, Stacky Hamiltonian actions and symplectic reduction. Int. Math. Res. Not. IMRN **20**, 15209–15300 (2021)
5. P. Iglesias-Zemmour, *Diffeology*. Math. Surveys and Monographs (Amer. Math. Soc., 2012)
6. P. Iglesias-Zemmour, Y. Karshon, M. Zadka, Orbifolds as diffeologies. Trans. Am. Math. Soc. **362**(6), 2811–2831 (2010)
7. Y. Karshon, J. Watts, Basic forms and orbit spaces: A diffeological approach. SIGMA Symmetry Integrability Geom. Methods Appl. **12**(026), 19 p. (2016)
8. Y. Karshon, M. Zoghi, An effective orbifold groupoid is determined up to Morita equivalence by its underlying diffeological orbifold (preprint)
9. J.-L. Koszul, Sur certains groupes de transformations de Lie. *Géométrie Différentielle, Colloques Internationaux du Centre National Recherche Scientifique (Strasbourg, 1953)* (Centre National Recherche Scientifique, Paris, 1953), pp. 137–141
10. E. Lerman, Orbifolds as stacks? Enseign. Math. (2) **56**(3-4), 315–363 (2010)
11. I. Moerdijk, J. Mrčun, Lie groupoids, sheaves and cohomology, *Poisson geometry, deformation quantisation and group representations, 145–272, London Math. Soc. Lecture Note Ser., 323* (Cambridge University Press, Cambridge, 2005)
12. R. Palais, On the existence of slices for actions of non-compact Lie groups. Ann. Math. **73**, 295–323 (1961)

13. M.J. Pflaum, H. Posthuma, X. Tang, Geometry of orbit spaces of proper Lie groupoids. J. Reine Angew. Math. **694**, 49–84 (2014)
14. J. Watts, *Diffeologies, Differential Spaces, and Symplectic Geometry*. Ph.D. Thesis. University of Toronto, Canada (2012). http://arxiv.org/abs/1208.3634
15. J. Watts, The differential structure of an orbifold. Rocky Mountain J. Math. **47**, 289–327 (2017)
16. J. Watts, S. Wolbert, Diffeological coarse moduli spaces of stacks over manifolds (preprint). http://arxiv.org/abs/1406.1392
17. A. Weinstein, Linearization problems for Lie algebroids and Lie groupoids. Lett. Math. Phys., **52**, 93–102 (2000)
18. A. Weinstein, Linearization of regular proper groupoids. J. Inst. Math. Jussieu **1**(3), 493–511 (2002)
19. N.T. Zung, Proper groupoids and momentum maps: linearization, affinity and convexity. Ann. Sci. École Norm. Sup. (4) **39**(5), 841–869 (2006)

Two-Frame Fields on Simply Connected Spin 7-Manifolds

Emily Windes and Üstün Yıldırım

Abstract In this paper, we show that simply connected, closed, spin, 7-manifolds with a fixed non-vanishing vector field admit a second linearly independent vector field. As a corollary, this shows that every almost contact structure on a simply connected, closed, (almost) G_2-manifold is B-compatible with the G_2 structure in the sense of (Firat Arikan et al., Asian J. Math. **17**(2), 321–334 (2013)).

Keywords 7-manifolds · Spin · G_2 · Contact structure · Postnikov resolution · Frame fields

Mathematics Subject Classification (2010) Primary 57R25; Secondary 55S45

1 Introduction

One major direction of research in the field of G_2 geometry is the problem of finding the right analogue to the Calabi conjecture in Kähler geometry. One key idea in the proof of the Calabi conjecture is the idea of "triality" where one exploits the relationship between the Kähler form, the complex structure, and the metric to find distinguished metrics on Kähler manifolds. The obvious analogue of the Kähler form in the G_2 case is the three form φ while the analogue of the complex structure might be the cross product. However, this G_2 triality is too rigid to yield any Calabi conjecture-like statements due to the fact that the G_2-structure φ determines both the metric and the cross product structure on a G_2-manifold all by itself. Thus it makes sense to consider other structures on G_2-manifolds in the search for a more flexible triality. It is a fact that G_2-manifolds are always contact manifolds making the contact structure a good candidate in this sense.

E. Windes (✉) · Ü. Yıldırım
University of Rochester, Department of Mathematics, Rochester, NY, USA
e-mail: ewindes@ur.rochester.edu

P. Cerejeiras et al. (eds.), *Current Trends in Analysis, its Applications and Computation*, Research Perspectives,
https://doi.org/10.1007/978-3-030-87502-2_53

In [1] Arikan, Cho, and Salur defined two ways in which a (co-oriented) almost-contact structure on a G_2-manifold can be compatible with the G_2-structure φ. They called them A- and B-compatible. In their paper, they mainly investigated the A-compatible condition. However, they also showed that A-compatible almost contact structures cannot exist on closed manifolds. In this paper, we reduce the problem of the existence of B-compatible contact structures to the problem of extending a given one-frame field to a two-frame field. More precisely, we prove:

Theorem 1.1 *Given a one-frame field X on a closed, simply connected, spin, 7-manifold M, M admits a one-frame field Y that is linearly independent from X.*

Next, we define B-compatibility and explain the reduction to Theorem 1.1.

Definition 1.2 ([1]) A (co-oriented) almost contact structure ξ on a G_2-manifold is said to be B-compatible with the G_2-structure φ if there exist global, non vanishing vector fields X and Y on M such that $\alpha = \iota_X \iota_Y \varphi$ is a contact form for ξ.

Let ξ be a contact structure on a G_2-manifold (M, φ) with contact one form α. Suppose that there exists a global, non vanishing vector field which is linearly independent from $\alpha^\sharp$ (the metric dual of α). Then there exists a unit vector field Y which is everywhere perpendicular to $\alpha^\sharp$. Also assume that $\alpha^\sharp$ is a unit vector field (by normalizing it). Next, choose $X = Y \times \alpha^\sharp$. Then we have

$$\iota_X \iota_Y \varphi = ((Y \times \alpha^\sharp) \times Y)^\flat = (-(Y \cdot \alpha^\sharp)Y + (Y \cdot Y)\alpha^\sharp)^\flat = (\alpha^\sharp)^\flat = \alpha$$

Therefore an almost contact structure is always B-compatible if there exists a non vanishing, global vector field on M which is linearly independent from $\alpha^\sharp$. The question of whether a vector field has such a linearly independent partner is a homotopy lifting problem.

In [6], Emery Thomas showed that all 7-manifolds admit two-frame fields (X, Y). The main difference in our theorem is we show for simply connected, spin, 7-manifolds, one can choose X to be any non-vanishing vector field. In fact, our theorem does not necessarily apply to non simply connected manifolds. For example if $M = S^1 \times S^6$, one can take X to be the vector field generated by the obvious S^1 action. Then, existence of a linearly independent vector field Y would imply that S^6 has a non trivial vector field by restriction.

It is well known that topologically G_2 structures are equivalent to spin structures on a 7-manifold. Therefore, we have:

Corollary 1.3 *Every almost contact structure on a simply connected, 7-manifold with a G_2-structure is B-compatible with the G_2-structure.*

2 The Lifting Problem

In this section, M will be a closed, 7-dimensional, spin manifold and all vector fields will be non vanishing. More details about topics in this section can be found in [2, 7].

A manifold admits a spin structure if and only if the second Stiefel-Whitney class of its tangent bundle, w_2, vanishes. Now suppose that there exists a vector field u on M. In this case, the structure group of the tangent bundle reduces to $Spin(6)$. Thus we have a map ξ from M to $BSpin(6)$, the classifying space of $Spin(6)$. Therefore the question of whether or not there is a second vector field linearly independent from u is the same question as whether or not the map ξ lifts to $BSpin(5)$. Notice that there is a fibration $S^5 \to BSpin(5) \xrightarrow{\pi} BSpin(6)$ where the map π is induced by the inclusion $Spin(5) \to Spin(6)$. Thus the problem of proving Theorem 1 can be summed up by the following diagram:

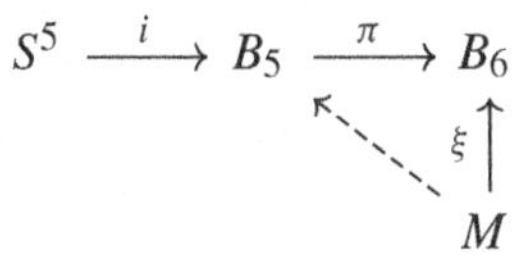

where $B_n = BSpin(n)$. To solve this problem, we consider a Postnikov resolution of the top row which gives us the following diagram:

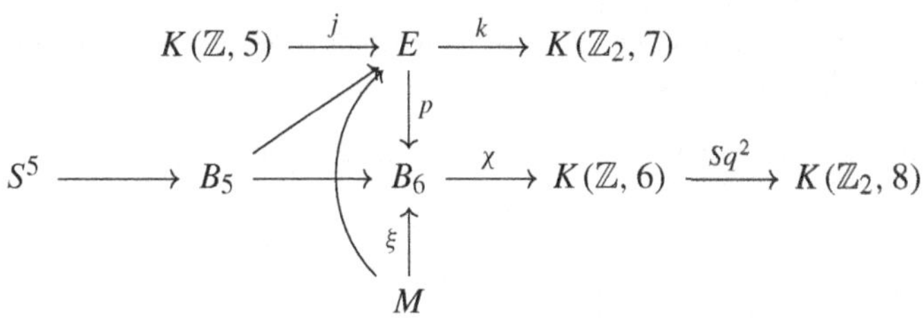

(2.1)

It is convenient to give maps suggestive names based on the representability theorem for singular cohomology. For example, χ is the map which pulls back the fundamental class ι_6 of $K(\mathbb{Z}, 6)$ to the euler class, χ, of B_6. Similarly, Sq^2 is the map which represents the class $Sq^2(\lfloor \iota_6 \rfloor_2) \in H^8(K(\mathbb{Z}, 6); \mathbb{Z}_2)$ (where $\lfloor \cdot \rfloor_2$ is mod 2 reduction). Notice that there are only two nontrivial levels of our Postnikov tower since M is a 7-manifold and therefore has trivial cohomology in dimensions 8 and above. The maps χ and k are referred to as obstructions because they are the obstructions to lifting ξ to B_5 in the following sense: If $\chi \circ \xi \sim *$ (homotopically trivial) then ξ lifts to a map, say, ξ', from M to E. Similarly if $k \circ \xi' \sim *$ then ξ' lifts to one level higher in the Postnikov tower but since the tower ends at k this is enough to say that ξ lifts all the way to B_5. For details of this approach, we refer the reader to [7]. A more elementary description of Postnikov towers can be found in [3].

In this paper, we assume that the manifold is closed and simply connected and therefore, $H^6(M, \mathbb{Z}) = 0$. Hence, any map $M \to K(\mathbb{Z}, 6)$ is homotopically trivial. In other words, the first obstruction vanishes. Thus, it is enough to show that the second obstruction vanishes as well. To get a hold on the secondary obstruction, we will need to introduce several new tools.

3 Secondary Compositions and Secondary Operations

Almost all the constructions here will arise from a situation where one has a null-homotopic composition of maps $A \xrightarrow{\alpha} B \xrightarrow{\beta} C$, where for our purposes A, B, and C will always be CW complexes. A pair of maps α, β along with a null homotopy H from $*$ to $\beta \circ \alpha$ will often be referred to as *a sequence with homotopy* and denoted by (α, β, H). Note that such a contracting homotopy H also defines a map from A to PC, the pathspace of C. More details can be found in [2]. For the most part, we follow his conventions with very minor variations. Here are the relevant definitions:

Definition 3.1 The *direction reversal map* of the unit interval is given by $\tau : I \to I$, $\tau(t) = 1 - t$ and the notation $H_\tau(t, x)$ means $H(\tau(t), x)$.

Definition 3.2 The *adjoint* of a map $f : I \times X \to Y$ is $f^\natural : X \to Y^I$ given by $f^\natural(x)(t) = f(x, t)$. The adjoint of a map $f : X \to Y^I$ is $g^\natural : I \times X \to Y$ $g^\natural(t, x) = g(x)(t)$. The analogous definitions will apply to the reduced cone T_0X and the reduced suspension ΣX.

Definition 3.3 Given a map $f : X \to Y$ we construct the fiber square:

$$\begin{array}{ccc} W_f & \longrightarrow & PY \\ \downarrow & & \downarrow \\ X & \xrightarrow{f} & Y \end{array}$$

called the *homotopy fiber* of f, where $W_f = \{(x, \gamma) \in X \times PY | f(x) = \gamma(1)\}$

Definition 3.4 Given a sequence with homotopy (α, β, H) a *lifting* of α is the map $\bar{\alpha} : A \to W_\beta$ obtained by the naturality of pullbacks from the following data:

$$\begin{array}{ccccc} A & = & A & = & A \\ \downarrow{\scriptstyle \alpha} & & \downarrow{\scriptstyle \beta\alpha} & & \downarrow{\scriptstyle H^\natural} \\ B & \xrightarrow{\beta} & C & \longleftarrow & P_0C \end{array}$$

More precisely, we form a pullback from the top row of the diagram which gives us A again. Then we also form a pullback from the bottom row of the diagram. This gives us the space W_β. Since we have maps from the top row to the bottom row, you obtain the map $\bar{\alpha} : A \to W_\beta$

Definition 3.5 On the other hand, a *colifting* of β is the map $\tilde{\beta} : W_\alpha \to \Omega C$ obtained by the naturality of pullbacks from the following data:

$$\begin{array}{ccccc} A & \xrightarrow{\alpha} & B & \longleftarrow & P_0 B \\ \downarrow{\scriptstyle H_\tau^\natural} & & \downarrow{\scriptstyle \beta} & & \downarrow{\scriptstyle P_0\beta} \\ P_1 C & \longrightarrow & C & \longleftarrow & P_0 C \end{array}$$

A composition of the form $\bar{\alpha} \circ \tilde{\beta}$ is called *a secondary composition*. Let $C_0 \xrightarrow{\theta} C_1 \xrightarrow{\varphi} C_2$ be a pair of maps and C_2 a simply connected H-space.

Definition 3.6 Given a space X the *source* $S_\theta(X)$ is the set of homotopy classes of maps $\epsilon : X \to C_0$ such that the composition $\theta \circ \epsilon$ is null-homotopic. The *target* $T_{\Omega\varphi}(X) = [X, \Omega C_2]/im(\Omega\varphi_\sharp)$ where $\Omega\varphi_\sharp : [X, \Omega C_1] \to [X, \Omega C_2]$ is given by $\Omega\varphi_\sharp(g) = \Omega\varphi \circ g$. We will denote the image of $g : X \to \Omega C_2$ in the target by $[[g]]$. Both S_θ and $T_{\Omega\varphi}$ are functors. A *secondary cohomology operation* is a natural transformation $\Theta : S_\theta(-) \to T_{\Omega\varphi}(-)$ of these functors. Furthermore, $im(\Omega\varphi_\sharp)$ is called the *indeterminancy* of Θ.

One can show that, given a sequence with homotopy, (φ, θ, H), and a representative $\epsilon : X \to C_0$ of an element in $S_\theta(X)$, the value of $[[\tilde{\varphi} \circ \bar{\epsilon}]]$ in $T_{\Omega\varphi}(X)$ does not depend on the choice of lifting, $\bar{\epsilon}$. Therefore, these secondary compositions are invariants of the homotopy class of (φ, θ, H). Thus there is a well-defined, natural transformation $\Theta : S_\theta(-) \to T_{\Omega\varphi}(-)$ given by the formula $\Theta(\epsilon) = [[\tilde{\varphi} \circ \bar{\epsilon}]]$ for each homotopy class of (φ, θ, H). In this case, Θ is said to be *associated* to (φ, θ, H) and the homotopy H is referred to as a *tethering*.

Our strategy for approaching Theorem 1.1 will be to express the obstruction to lifting ξ in terms of the image of a secondary operation Θ. Then, we will discover that by an appropriate choice of the sequence with homotopy, this obstruction will be zero. In general, calculating the value of secondary operations depends on knowing the cohomology rings of all the spaces involved as well as being able to arrange the diagrams in a favorable way. Now that we have the basic setup, we will collect the information needed to solve the problem below:

Fact 1 The Euler class mod 2, $\lfloor \chi \rfloor_2 = w_n$ where w_n is the Stiefel-Whitney class and ξ has degree n. Similarly, if c_i is a Chern class, $\lfloor c_i \rfloor_2 = w_{2i}$.

Wu Formulas The Stiefel-Whitney classes have the following relationships over the Steenrod algebra: $Sq^i(w_j) = \Sigma_{t=0}^{i} \binom{j+t-i-1}{t} w_{i-t} w_{j+1}$.

Adem Relations For all $i, j > 0$ such that $i < 2j$, and with mod 2 binomial coefficients, $Sq^i Sq^j = \Sigma_{k=0}^{\lfloor k/2 \rfloor} \binom{j-k-1}{i-2k} Sq^{i+j-k} Sq^k$.

Definition 3.7 A sequence of numbers $I = \{i_1, i_1, \ldots, i_k\}$ is called *admissible* if $i_j \geq 2i_{j+1}$. The *excess* of I, $e(I)$, is the measure of how close a sequence is to being admissible. That is, $e(I) = (i_1 - 2i_2) + (i_2 - 2i_3) + \cdots + (i_{k-1} - 2i_k) + i_k$.

Fact 2 Let ι_q be the fundamental class of the Eilenberg-MacLane space $K(\mathbb{Z}, q)$. Then the cohomology ring $H^*(K(\mathbb{Z}, q), \mathbb{Z}_2)$ is the polynomial ring with generators $Sq^I(\iota_q)$ where I runs through admissible sequences of excess $e(I) < q$ and where i_k, the last entry in I is not 1. The requirement that $i_k > 1$ comes from another important fact: $Sq^1(\iota_q) = 0$. For a recent treatment of this fact see [5].

Fact 3 There is an exceptional isomorphism between $Spin(6)$ and $SU(4)$. Therefore the cohomology ring of B_6 with integer coefficients is $\mathbb{Z}[c_2, c_3, c_4]$.

Fact 4 All the Stiefel-Whitney classes of an orientable, 7-dimensional, compact, spin manifold, M, are zero. It is also a fact that the first nonzero Stiefel-Whitney class must be w_i where i is a power of two. Therefore, since the manifold is orientable and spin, $w_1 = w_2 = w_3 = 0$. Also, from [4], we know that $w_7 = w_6 = w_5 = 0$. So the only Stiefel-Whitney class which might be non-zero is w_4. We can use the properties of Wu classes to show that w_4 is also zero. Let v_k be the Wu classes. The facts already stated imply that $v_1 = v_2 = v_3 = 0$. Let v be the total Wu class, w the total Stiefel-Whitney class, and Sq the total Steenrod square. Then $Sq(v) = w$. Also from [4] we know that $v_k = 0$ if $k > \frac{1}{2}n$ which, in our case, means $v_k = 0$ for $k = 4, 5, 6$, and 7. Thus $w = Sq(1) = 1$ and we see that all the Stiefel-Whitney classes of M are zero.

In general, the Wu formulas give us the following:

$$Sq^2 w_6 = \binom{6-2-1}{0} w_2 w_6 + \binom{6+1-2-1}{1} w_1 w_7 + \binom{6+2-2-1}{2} w_0 w_8.$$

In the context we are interested in, the only non-zero Stiefel Whitney classes are the mod 2 reductions of the generators of $H^*(BSU(4))$ which are w_4, w_6, and w_8 so $Sq^2 w_6 = 0 \Rightarrow Sq^2 \circ \chi$ is homotopically trivial. These are exactly the maps occurring in diagram (2.1). Therefore, we may fix a nullhomotopy and consider the secondary cohomology operation associated with this homotopy. Furthermore, the secondary compositions representing elements in the target of this operation are exactly the ones representing the obstruction to lifting ξ. From [2, p. 61], we know that up to homotopy, there is only one colifting and effectively only one homotopy that we can choose. Therefore there is a unique secondary operation associated to the composition $Sq^2 \circ \chi$. Thus we may choose the map k above to be the colifting $\widetilde{Sq^2}$. Figuring out whether or not we can choose the composition $k \circ \xi'$ to be trivial now amounts to figuring out whether 0 is contained in the target of this secondary composition or not. We analyze the composition by factoring it through the lower line in the diagram below:

$$
\begin{array}{ccccccccc}
 & & & & M & & & & \\
 & & & \overset{\xi'}{\swarrow} & \downarrow \xi & & & & \\
K(\mathbb{Z},5) & \longrightarrow & E & \xrightarrow{p} & B_6 & \xrightarrow{\chi} & K(\mathbb{Z},6) & \xrightarrow{Sq^2} & K(\mathbb{Z}_2,8) \\
\downarrow & & \downarrow i & & \downarrow w_4 & & \downarrow & \begin{pmatrix} Sq^2 \\ Sq^3 \end{pmatrix} & \| \\
K(\mathbb{Z}_2,5)\times K(\mathbb{Z}_2,4) & \longrightarrow & W & \xrightarrow{j} & K(\mathbb{Z}_2,4) & \xrightarrow{(Sq^2 Sq^1)} & K(\mathbb{Z}_2,6)\times K(\mathbb{Z}_2,5) & \longrightarrow & K(\mathbb{Z}_2,8) \\
 & & \downarrow \tilde{\varphi} & & & & & & \\
 & & K(\mathbb{Z}_2,7) & & & & & &
\end{array}
\tag{3.1}
$$

For simplicity, we denote $(Sq^2 Sq^1)$ by θ and $\begin{pmatrix} Sq^2 \\ Sq^3 \end{pmatrix}$ by φ. We have that $k = \widetilde{Sq^2} = \tilde{\varphi} \circ i$. Next we will see why this diagram helps us. First of all, we will see how it is constructed. Since B_6 has the Stiefel-Whitney class w_4, we can think about starting with that map. Next, note that we can map $K(\mathbb{Z}_2, 4)$ to $K(\mathbb{Z}_2, 6) \times K(\mathbb{Z}_2, 5)$ by applying Sq^2, Sq^1. Our motivation for doing that is the Adem relation $Sq^2 Sq^2 = Sq^3 Sq^1$. This, combined with the fact that $Sq^2 Sq^3$ maps $K(\mathbb{Z}_2, 6) \times K(\mathbb{Z}_2, 5)$ to $K(\mathbb{Z}_2, 8)$ ensures that the second row is still a nullhomotopy and ends at the same space as the first row. It remains to fill in the middle map so that the right half of the diagram commutes. We can achieve this by letting the middle map from $K(\mathbb{Z}, 6)$ to $K(\mathbb{Z}_2, 6) \times K(\mathbb{Z}_2, 5)$ be reduction mod 2 on the first factor and zero on the second factor. Then, since $Sq^2 w_4 = w_6 = \lfloor \chi \rfloor_2$ and $Sq^1 w_4 = Sq^1 \lfloor c_2 \rfloor_2 = 0$, as Sq^1 is the Bockstein homomorphism. Thus the second square from the right commutes. The square on the far right commutes trivially. Next, we form W as the pullback bundle of the map (Sq^2, Sq^1) so its fiber is $\Omega(K(\mathbb{Z}_2, 6) \times K(\mathbb{Z}_2, 5))$. Commutativity of the rest of the diagram now follows easily.

To finish up the argument, consider the map $\bar{\epsilon} = i \circ \xi'$. Note that since $w_4(M) = 0$ by Fact 4, $j \circ \bar{\epsilon}$ is also zero so we know that $\bar{\epsilon}$ factors through a map f from M to $K(\mathbb{Z}_2, 5) \times K(\mathbb{Z}_2, 4)$. Finally, from [4], we know that Sq^2 and Sq^1 as maps from $H^5(M) \to H^7(M)$ and from $H^6(M) \to H^7(M)$ are zero. Therefore, again by commutativity, $0 = \begin{pmatrix} Sq^2 \\ Sq^3 \end{pmatrix} \circ f = \tilde{\varphi} \circ \bar{\epsilon} = k \circ \xi'$ and Theorem 1.1 is proved.

4 Conclusion

Theorem 1.1 can be strengthened in a few different ways which we plan to address in a future paper. First of all, Thomas' result about two-frame fields applies to any closed $(4s + 3)$-manifold. So one might hope that for any closed $(4s + 3)$-manifold one can choose the first vector field in the two-frame field freely. Here, we have three conditions; simply connected, spin, 7 dimensional. We will investigate which of these properties can be dropped so that Theorem 1.1 still applies.

Acknowledgments We thank Professor John Harper for the many hours he spent teaching us the contents of his book and discussing our geometry problems. We are truly lucky to have his support and expertise in topology.

References

1. M. Firat Arikan, H. Cho, S. Salur, et al., Existence of compatible contact structures on G_2-manifolds. Asian J. Math. **17**(2), 321–334 (2013)
2. J.R. Harper, *Secondary Cohomology Operations* (vol. 49) (American Mathematical Soc., 2002)
3. A. Hatcher, *Algebraic Topology* (Cambridge University Press, 2002)
4. W.S. Massey, On the stiefel-whitney classes of a manifold. Am. J. Math. **82**(1), 92–102 (1960)
5. R.E. Mosher, M.C. Tangora, *Cohomology Operations and Applications in Homotopy Theory* (Courier Corporation, 2008)
6. E. Thomas, Postnikov invariants and higher order cohomology operations. Ann. Math., 184–217 (1967)
7. E. Thomas, *Seminar on Fiber Spaces: Lectures Delivered in 1964 in Berkeley and 1965 in Zürich* (vol. 13) (Springer, 2006)

Part X
Partial Differential Equations on Curved Spacetimes

Local Existence of Solutions to the Euler–Poisson System, Including Densities Without Compact Support

Uwe Brauer and Lavi Karp

Abstract Local existence and uniqueness for a class of solutions for the Euler Poisson system is shown, whose properties can be described as follows. Their density ρ either falls off at infinity or has compact support. Their mass and the energy functional is finite and they also include the static spherical solutions for $\gamma = \frac{6}{5}$. The result is achieved by using weighted Sobolev spaces of fractional order and a new non-linear estimate that allows to estimate the physical density by the regularised non linear matter variable.

Keywords Euler–Poisson systems · Hyperbolic symmetric systems · Energy estimates · Makino variable · Weighted fractional Sobolev spaces

Mathematics Subject Classification (2010) Primary 35Q75; Secondary 35Q76,35J61, 35Q31

1 Introduction

We consider the Euler–Poisson system

$$\partial_t \rho + v^a \partial_a \rho + \rho \partial_a v^a = 0 \tag{1.1}$$

U. B. gratefully acknowledges support from Grant MTM2016-75465 by MINECO, Spain and UCM-GR17-920894.

U. Brauer (✉)
Departamento Matemática Aplicada, Universidad Complutense Madrid, Madrid, Spain
e-mail: karp@braude.ac.il

L. Karp
Department of Mathematics, ORT Braude College, Karmiel, Israel
e-mail: oub@mat.ucm.es

P. Cerejeiras et al. (eds.), *Current Trends in Analysis, its Applications and Computation*, Research Perspectives,
https://doi.org/10.1007/978-3-030-87502-2_54

$$\rho\left(\partial_t v^a + v^b\partial_b v^a\right) + \partial^a p = -\rho\partial^a\phi \tag{1.2}$$

$$\Delta\phi = 4\pi G\rho \tag{1.3}$$

where G denotes the gravitational constant. Using suitable physical units we can set $G = 1$. Here we have used the summation convention, for example, $v^k\partial_k := \sum_{k=1}^3 v^k\partial_k$, a convention we will use in the rest of the paper wherever it seems appropriate to us. Moreover $\partial^a\phi := \delta^{ab}\partial_b\phi$, and we will wherever it is convenient denote $\partial^a\phi$ by $\nabla\phi$. In this paper, we consider the barotropic equation of state

$$p = K\rho^\gamma \quad 1 < \gamma,\ 0 < K, \tag{1.4}$$

and we study this system with initial data for the density which either has compact support or falls off at infinity in an appropriate way. It is well known that the usual symmetrization of the Euler equations is badly behaved in such cases. The coefficients of the system degenerate or become unbounded when ρ approaches zero. It was observed by Makino [11] that this difficulty can be to some extent circumvented by using a new matter variable w in place of the density. For this reason, we introduce the quantity

$$w = \frac{2\sqrt{K\gamma}}{\gamma-1}\rho^{\frac{\gamma-1}{2}}, \tag{1.5}$$

which allows treating the situation where $\rho = 0$. Replacing the density ρ by the Makino variable w, the system (1.1)–(1.3) coupled with the equation of state (1.4) takes the following form:

$$\partial_t w + v^a\partial_a w + \frac{\gamma-1}{2}w\partial_a v^a = 0 \tag{1.6}$$

$$\partial_t v^a + v^b\partial_b v^a + \frac{\gamma-1}{2}w\partial^a w = -\partial^a\phi \tag{1.7}$$

$$\Delta\phi = 4\pi\rho \tag{1.8}$$

which we will sometimes denote as the Euler–Poisson–Makino system. The Euler–Poisson system consists of a hyperbolic system of evolution equations and the elliptic Poisson equation.

2 Main Results

We obtain local existence and uniqueness for classical solutions of the Euler–Poisson–Makino system (1.6)–(1.8) for densities without compact support but with a polynomial decay at infinity, and with the equation of state (1.4). The class of solutions we obtain has finite mass, a finite energy functional, moreover, they contain the static spherically symmetric solutions for the adiabatic constant $\gamma = \frac{6}{5}$ (see Sect. 2.1). These solutions are continuously differentiable and they are also classical solutions of the Euler–Poisson system (1.1)–(1.3).

The Euler–Poisson–Makino system is considered in the weighted Sobolev spaces of fractional order $H_{s,\delta}$. So we first define these spaces.

Let $\{\psi_j\}_{j=0}^{\infty}$ dyadic partition of unity in $\mathbb{R}^3$, that is, $\psi_j \in C_0^{\infty}(\mathbb{R}^3)$, $\psi_j(x) \geq 0$, $\mathrm{supp}(\psi_j) \subset \{x : 2^{j-2} \leq |x| \leq 2^{j+1}\}$, $\psi_j(x) = 1$ on $\{x : 2^{j-1} \leq |x| \leq 2^j\}$ for $j = 1, 2, \ldots$, $\mathrm{supp}(\psi_0) \subset \{x : |x| \leq 2\}$, $\psi_0(x) = 1$ on $\{x : |x| \leq 1\}$ and

$$|\partial^{\alpha}\psi_j(x)| \leq C_{\alpha}2^{-|\alpha|j}, \tag{2.1}$$

where the constant C_{α} does not depend on j. We denote by H^s the Sobolev spaces with the norm given by

$$\|u\|_{H^s}^2 = \int (1 + |\xi|^2)^s |\hat{u}(\xi)|^2 d\xi,$$

where $\hat{u}$ is the Fourier transform of u. The scaling by a positive number ϵ is denoted by $f_{\varepsilon}(x) = f(\varepsilon x)$.

Definition 2.1 (Weighted Fractional Sobolev Spaces) Let $s, \delta \in \mathbb{R}$, the weighted Sobolev space $H_{s,\delta}$ is the set of all tempered distributions such that the norm

$$\left(\|u\|_{H_{s,\delta}}\right)^2 = \sum_{j=0}^{\infty} 2^{(\frac{3}{2}+\delta)2j} \|(\psi_j u)_{(2^j)}\|_{H^s}^2 \tag{2.2}$$

is finite.

The largest integer less than or equal to s is denoted by $[s]$. In this setting our main result is the following.

Theorem 2.2 (Local Existence and Uniqueness of Classical Solutions to the Euler–Poisson–Makino System) *Let* $1 < \gamma < \frac{5}{3}$, $-\frac{3}{2} + \frac{2}{\left[\frac{2}{\gamma-1}\right]-1} \leq \delta < -\frac{1}{2}$, $\frac{5}{2} < s$ *if* $\frac{2}{\gamma-1}$ *is an integer and* $\frac{5}{2} < s < \frac{5}{2} + \frac{2}{\gamma-1} - \left[\frac{2}{\gamma-1}\right]$ *otherwise. Suppose* $(w_0, v_0^a) \in H_{s,\delta}$ *and* $w_0 \geq 0$*, then there exists a positive* T *which depends on the* $H_{s,\delta}$*-norm of the initial data and there exists and a unique solution* (w, v^a) *of the Euler–Poisson–Makino system (1.6)–(1.8) such that*

$$(w, v^a) \in C\left([0, T], H_{s,\delta}\right) \cap C^1\left([0, T], H_{s-1,\delta+1}\right)$$

and $0 \leq w(t, \cdot)$ *in* $[0, T]$.

The proof of this theorem is outlined in Sect. 3, while the complete version can be found in [1]. It has been common in recent years to use the term well-posedness as a synonym for existence and uniqueness, however well-posedness, as originally introduced by Hadamard [7], includes the continuity of the flow map with respect to the initial data. This, however, has been proven recently by the authors [2].

In any case, our result as stated here has a series of noteworthy corollaries which we list below:

2.1 *Properties of the Solutions*

We start with static solutions of the Euler–Poisson system. Those solutions must be spherical symmetric (see for example [9]) and they can be obtained by solving the Lane–Emden equation [5]. The linear stability has been an open problem for a long time, so it is interesting to see whether a class of solutions can be constructed which include static solutions. To the best of our knowledge, this has not been achieved for solutions with a finite radius. For $\gamma = \frac{6}{5}$ there is one parameter family (parameterized by the central density) of solutions which have *finite mass* but *infinite radius*, and it is given by

$$\rho(t,x) = \rho(|x|) = a^{\frac{5}{2}} \left(a^2 + |x|^2\right)^{-\frac{5}{2}} \sim |x|^{-5}, \tag{2.3}$$

where a is a positive constant see [5]. The corresponding solutions in the Makino variable are given by

$$w(x,t) = a^{\frac{1}{4}} \left(a^2 + |x|^2\right)^{-\frac{1}{4}} \sim |x|^{-\frac{1}{2}}. \tag{2.4}$$

Such static solutions are included in the class of solutions whose existence is guaranteed by Theorem 2.2, as it is stated in the following corollary.

Corollary 2.3 (The Static Solutions of the Euler–Poisson System) *Let* $\gamma = \frac{6}{5}$, $-\frac{23}{18} < \delta < -1$ *and* $\frac{5}{2} < s$. *Then there exists a positive* T *and a unique solution* (w, v^a) *to the Euler–Poisson–Makino system (1.6)–(1.8) such that*

$$(w, v^a) \in C\left([0,T], H_{s,\delta}\right) \cap C^1\left([0,T], H_{s-1,\delta+1}\right),$$

and for which the initial data include the static solution $w_0(x) = \left(a^2 + |x|^2\right)^{-\frac{1}{4}}$.

Proof The proof is straightforward. As discussed above for $\gamma = \frac{6}{5}$, ρ is given by equation (2.3), while w is given by Eq. (2.4). Note that $(a^2 + |x|^2)^{-\frac{1}{4}} \in H_{s,\delta}$ if $\delta < -1$. On the other hand, the lower bound for δ in Theorem 2.2 for $\gamma = \frac{6}{5}$ gives us $-\frac{3}{2} + \frac{2}{9} = -\frac{23}{18} < -1$. □

Note that the local existence and uniqueness result is obtained in terms of the Makino variable. Nevertheless, setting $\rho(t,x) = c_{K,\gamma} w^{\frac{2}{\gamma-1}}(t,x)$, $c_{K,\gamma} = \left(\frac{2\sqrt{K\gamma}}{\gamma-1}\right)^{\frac{-2}{\gamma-1}}$, we also get a classical solution to the Euler–Poisson system (1.1)–(1.3).

Corollary 2.4 (Local Solutions of the Original Euler–Poisson System) *Let* $1 < \gamma < \frac{5}{3}$, $-\frac{3}{2} + \frac{3}{\left[\frac{2}{\gamma-1}\right]} \leq \delta < -\frac{1}{2}$, $\frac{5}{2} < s$ *if* $\frac{2}{\gamma-1}$ *is an integer and* $\frac{5}{2} < s < \frac{5}{2} + \frac{2}{\gamma-1} - \left[\frac{2}{\gamma-1}\right]$ *otherwise. Suppose* $(\rho_0^{\frac{2}{\gamma-1}}, v_0^a) \in H_{s,\delta}$. *Then there exists a positive* T *and a unique* C^1*–solution* (ρ, v^a) *to the Euler–Poisson system (1.1)–(1.3) with the equation of state (1.4) such that*

$$\left(\rho(t,\cdot), v^a(t,\cdot)\right) \in L^\infty([0,T], H_{s,\delta}).$$

Please note that the initial data in Corollary 2.4 are given by the Makino variable w and not by the physical quantity ρ. It is an open problem to solve the Euler–Poisson system entirely in terms of ρ for situations in which ρ could be zero.

There exists a wide range of publications concerning the non-linear stability of stationary solutions of the Euler-Poisson system relying on the method of energy functionals, see for example Rein [8, 14]. Having this context in mind we turn to the question of finite mass and finite energy functional.

Corollary 2.5 (Finite Mass and Finite Energy Functional) *The solutions obtained by Theorem 2.2 have the properties that,*

1. $\rho(t,\cdot) \in L^1(\mathbb{R}^3)$, *that is, they have finite mass.*
2. *The energy functional*

$$E = E(\rho, v^a) := \int \left(\frac{1}{2}\rho|v^a|^2 + \frac{K\rho^\gamma}{\gamma-1}\right) dx - \frac{1}{2} \iint \frac{\rho(t,x)\rho(t,y)}{|x-y|} dx dy \tag{2.5}$$

is well defined for those solutions.

2.2 The Advantages of the $H_{s,\delta}$ Spaces

In this section, we discuss the consequences of our main result, Theorem 2.2 and possible applications, and compare them with previous results obtained by other authors.

- We recall that the Euler–Poisson system (1.1)–(1.3) degenerates when the density approaches to zero and the only known method to solve an initial value problem in this context is to regularize the Euler equations by introducing the Makino variable (1.5). All the previous local existence results [3, 6, 11], including the present paper, have used this technique. It should, however, be noted that Bezard uses the ordinary Sobolev spaces H^s, and therefore his claim that his solutions include static spherical solutions if $\gamma = \frac{6}{5}$ is simply not correct, since the initial data of the corresponding Makino variable do not belong to H^s.

Thus in order to include the spherically symmetric static solutions of the Lane–Emden equation for $\gamma = \frac{6}{5}$ in our class of solutions, it is necessary to express it in terms of the Makino variable w. But from (2.4) we see that this function does not belong to the Sobolev H^s space.

- To overcome the difficulty with the Makino variable Gamblin uses uniformly locally Sobolev spaces H^s_{ul} spaces which were introduced by Kato. However, as it was pointed out by Majda [10, Thm 2.1, p. 5 0], for first-order symmetric hyperbolic systems with a given initial data $u_0 \in H^s_{ul}$, $\frac{3}{2} + 1 < s$ the corresponding solutions belong only to $C\left([0,T]; H^s_{loc}\right) \cap C^1\left([0,T]; H^{s-1}_{loc}\right) \cap L^\infty\left([0,T]; H^s_{ul}\right)$. Furthermore, continuity in the H^s_{ul} norm causes a loss of regularity [6, Theorem 2.4]. We prove well-posedness in the $H_{s,\delta}$ spaces, Theorem A.4, and circumvent these weaknesses of the uniformly locally Sobolev spaces.
- Another benefit of the $H_{s,\delta}$ spaces concerns the treatment of the Poisson equations. The Laplacian is a Fredholm operator in those spaces [4, 12], and for certain values of δ is an isomorphism. Thus with the aid of the nonlinear power estimate, Proposition A.3, we are able to treat both the hyperbolic and the elliptic part in the same type of Sobolev spaces. On the contrary, the H^s_{ul} spaces are not suited for the Poisson equation. To circumvent this difficulty Gamblin demanded that the initial density ρ_0 belong to $W^{1,p}$, $1 \leq p < 3$. Therefore he has two types of initial data, namely, $\rho_0 \in W^{1,p}$ and the Makino variable $\rho_0^{\frac{\gamma-1}{2}} \in H^s_{ul}$. However, his initial data for the velocity v_0^a belongs to H^s_{ul}. Under these initial conditions, Gamblin proved that for $\frac{7}{2} < s < \frac{2}{\gamma-1}$ the solutions are:

$$(\rho, v^a) \in \cap_{i=1,2} C^i\left([0,T^*]; H^{s'-i}_{ul}\right), \quad s' < s, \quad \rho \in L^\infty\left([0,T]; W^{1,p} \cap H^{s_\epsilon}_{ul}\right),$$

where $s_\epsilon = \min\{\frac{2}{\gamma-1} - \epsilon, s\}$ if $\frac{2}{\gamma-1} \notin \mathbb{N}$ and $s_\epsilon = s$ otherwise. Thus the density belongs to $W^{1,p}$ and falls off at infinity, while the velocity is in H^s_{ul} and therefore does not tend to zero. Such a class of solutions, even if it contains spherically symmetric static solutions, does not model isolated bodies in an appropriate way.
- The uniform Sobolev spaces H^s_{ul}, that Gamblin used in order to include the static solutions for $\gamma = \frac{6}{5}$ are not suited for the Einstein–Euler system in an asymptotically flat setting. Recall that in these functional spaces the Einstein constraint equations cannot be solved, while they can be solved using the $H_{s,\delta}$ spaces. The last question is important if one considers the Euler–Poisson system as the Newtonian limit of the Einstein–Euler system.

Oliynyk [13] proved the Newtonian limit in an asymptotically flat setting. He showed that solutions of the Einstein-Euler system converge to solutions of the Euler–Poisson system, under the restriction that the density has compact support. In order to generalize his result to the case where the density only falls off in an appropriate way one needs a functional setting that is suited for both systems. While the weighted fractional Sobolev spaces are known to be appropriate, there is no existence result known for the Einstein equations (plus matter fields) in an asymptotically flat situation using the functional setting of H^s_{ul} spaces.

3 Structure of the Proof

The most obvious way to solve system (1.6)–(1.8) would be to apply some sort of iteration procedure or a fixed–point argument directly to that system. But since the system is coupled to an elliptic equation, it seemed more convenient and transparent to split up the proof in several parts. Firstly we prove local existence and well-posedness for a general symmetric hyperbolic system ($A^0 \neq \mathrm{Id}$) in the weighted Sobolev spaces.

Since the density falls off but could become zero, we will need the established tool of regularizing the system, by introducing a new matter variable, the Makino variable (1.5). In this setting, the power $w^{\frac{2}{\gamma-1}}$ must be estimated in the weighted fractional norm. The estimates of the power in the H^s spaces under certain restrictions on the power and s are known (see e.g. [15]). An essential ingredient of our proof is a nonlinear power estimate in the weighted fractional Sobolev spaces that preserves the regularity and improves the fall off at infinity (Proposition A.3). It enables us to apply the known estimates for the Poisson equation (1.8) in these spaces. We then prove the existence of solutions to the Euler–Poisson–Makino system by using a fixed–point argument. In any case, either for the fixed–point or for the direct iteration we are faced with the well-known fact that we have to use a *higher* and a *lower* norm. We show boundness in the higher norm and contraction in the lower. Under this circumstances, the existence of a fixed point in the higher norm is well known.

Appendix A Useful Propositions

The following Proposition was proved by Kateb in the H^s spaces.

Proposition A.1 *Let $u \in H_{s,\delta} \cap L^\infty$, $1 < \beta$, $0 < s < \beta + \frac{1}{2}$, and $\delta \in \mathbb{R}$, then*

$$\| |u|^\beta \|_{H_{s,\delta}} \leq C(\|u\|_{L^\infty}) \|u\|_{H_{s,\delta}}. \tag{A.1}$$

Proposition A.2 (Sobolev Embedding)

(i) If $\frac{3}{2} < s$ *and* $\beta \le \delta + \frac{3}{2}$, *then* $\|u\|_{L^\infty_\beta} \le C\|u\|_{H_{s,\delta}}$.
(ii) Let m *be a nonnegative integer,* $m + \frac{3}{2} < s$, *and* $\beta \le \delta + \frac{3}{2}$, *then*

$$\|u\|_{C^m_\beta} \le C\|u\|_{H_{s,\delta}}.$$

Proposition A.3 (Nonlinear Estimate of Power of Functions) *Suppose that* $w \in H_{s,\delta}$, $0 \le w$ *and* β *is a real number greater or equal* 2. *Then*

1. *If* β *is an integer,* $\frac{3}{2} < s$ *and* $\frac{2}{\beta-1} - \frac{3}{2} \le \delta$, *then*

$$\|w^\beta\|_{H_{s-1,\delta+2}} \le C_n \left(\|w\|_{H_{s,\delta}}\right)^\beta. \tag{A.2}$$

2. *If* $\beta \notin \mathbb{N}$, $\frac{5}{2} < s < \beta - [\beta] + \frac{5}{2}$ *and* $\frac{2}{[\beta]-1} - \frac{3}{2} \le \delta$, *then*

$$\|w^\beta\|_{H_{s-1,\delta+2}} \le C_n \left(\|w\|_{H_{s,\delta}}\right)^{[\beta]}. \tag{A.3}$$

Theorem A.4 (Well Posedness of First Order Hyperbolic Symmetric Systems in $H_{s,\delta}$**)** *Let* $\frac{5}{2} < s$, $-\frac{3}{2} \le \delta$, $U_0 \in H_{s,\delta}$, *and* $F(t,\cdot) \in C([0,T^0], H_{s,\delta})$ *for some positive* T^0. *Then there exists a positive* $T \le T^0$ *and a unique solution* U *to the system*

$$\begin{cases} \partial_t U + A^a(U)\partial_a U + B(U)U = F(t,x) \\ U(0,x) = U_0(x) \end{cases}, \tag{A.4}$$

such that

$$U \in C([0,T], H_{s,\delta}) \cap C^1([0,T], H_{s-1,\delta+1}).$$

References

1. U. Brauer, L. Karp, Local existence of solutions to the Euler–Poisson system, including densities without compact support. J. Differ. Equ. **264**(2), 755–785 (2018). https://doi.org/10.1016/j.jde.2017.09.024
2. U. Brauer, L. Karp, Continuity of the flow map for symmetric hyperbolic systems and its application to the Euler–Poisson system, accepted for publication. J. d'Analyse Math. (2019)
3. M. Bézard, Existence locale de solutions pour les équations d'Euler-Poisson. Jpn. J. Indust. Appl. Math. **10**(3), 431–450 (1993). http://dx.doi.org/10.1007/BF03167283
4. M. Cantor, Spaces of functions with asymptotic conditions on $\mathbb{R}^n$. Ind. Univ. Math. J. **24**(9), 897–902 (1975)
5. S. Chandrasekhar, *An Introduction to the Study of Stellar Structure* (Dover Publications, Inc., 1939)

6. P. Gamblin, Solution régulière à temps petit pour l'équation d'Euler-Poisson. Comm. Partial Differ. Equ. **18**(5-6), 731–745 (1993).
http://dx.doi.org/10.1080/03605309308820948
7. J. Hadamard, Sur les problèmes aux dérivées partielles et leur signification physique. Princeton Univ. Bull. 49–52 (1902)
8. J. Jang, Nonlinear instability in gravitational Euler-Poisson systems for $\gamma = \frac{6}{5}$. Arch. Ration. Mech. Anal. **188**(2), 265–307 (2008).
http://dx.doi.org/10.1007/s00205-007-0086-0
9. L. Lichtenstein, Über einige Eigenschaften der Gleichgewichtsfiguren rotierender Flüssigkeiten, deren Teilchen einander nach dem Newtonsch'schen Gesetz anziehen., Mathematische Zeitschrift **28**, 635–640 (1928)
10. A. Majda, *Compressible Fluid Flow and Systems of Conservation Laws in Several Space Variables* (Springer, New York, 1984)
11. T. Makino, On a local existence theorem for the evolution equation of Gaseous stars, in *Patterns and Waves*, eds. by T. Nishida, M. Mimura, H. Fujii (North-Holland, Amsterdam, 1986)
12. R.M. McOwen, The behavior of weighted Sobolev spaces. Commun. Pure Appl. Math. **32**, 783–795 (1979)
13. T.A. Oliynyk, The Newtonian limit for perfect fluids. Comm. Math. Phys. **276**(1), 131–188 (2007).
http://dx.doi.org/10.1007/s00220-007-0334-z
14. G. Rein, Non-linear stability of gaseous stars. Arch. Ration. Mech. Anal. **168**(2), 115–130 (2003).
http://dx.doi.org/10.1007/s00205-003-0260-y
15. T. Runst, W. Sickel, *Sobolev Spaces of Fractional Order, Nemytskij Operators, and Nonlinear Partial Differential Equations* (Walter de Gruyter, 1996)

Isometric Flows of G_2-structures

Sergey Grigorian

Abstract We survey recent progress in the study of flows of isometric G_2-structures on seven-dimensional manifolds, that is, flows that preserve the metric, while modifying the G_2-structure. In particular, heat flows of isometric G_2-structures have been recently studied from several different perspectives, in particular in terms of 3-forms, octonions, vector fields, and geometric structures. We will give an overview of each approach, the results obtained, and compare the different perspectives.

Keywords G_2-structure · Geometric flows

Mathematics Subject Classification (2010) 53C10, 53C29, 58E30, 58E15

1 Introduction

One of the most challenging problems in differential geometry is the question of existence conditions for torsion-free G_2-structures on smooth seven-dimensional manifolds. Such G_2-structures are precisely the ones that correspond to metrics with holonomy contained in G_2. One approach that has been pioneered by Robert Bryant [4] is to considered heat-like flows of G_2-structures with the hope that under certain conditions they may converge to a torsion-free G_2-structure. A difficulty that is encountered in such an approach is that in general, deformations of a G_2-structure also affect the corresponding metric, and so any heat equation for the G_2-structure becomes nonlinear. This is not unlike the situation for the Ricci flow,

This work was supported by the National Science Foundation [DMS-1811754].

S. Grigorian (✉)
University of Texas Rio Grande Valley, Edinburg, TX, USA
e-mail: sergey.grigorian@utrgv.edu

P. Cerejeiras et al. (eds.), *Current Trends in Analysis, its Applications and Computation*, Research Perspectives,
https://doi.org/10.1007/978-3-030-87502-2_55

where the underlying geometry changes along the flow, however in the G_2 case, we have two separate but closely related objects, the G_2-structure and the metric, both of which vary along the flow. Given a Riemannian metric on a 7-manifold that admits G_2-structures, there is a family of G_2-structures that correspond to it, so a possible approach could be to separate as much as possible the deformations of the metric from the deformations of G_2-structures that preserve the metric. Indeed, as was shown by Karigiannis [13], given a decomposition of 3-forms according to representations of G_2, the deformations of the G_2-structure 3-form that preserve the metric are precisely the ones that lie in the seven-dimensional representation Λ^3_7. Bryant's original Laplacian flow of closed G_2-structures has no component in Λ^3_7 [4], and as such is transverse to directions that preserve the metric. This allowed for more tractable analytic properties. In contrast, a similar flow for co-closed G_2-structures that was proposed in [15] does have a component in Λ^3_7, which, as shown in [9], causes non-parabolicity of the flow. This suggests that the freedom of G_2-structures to move in directions that preserve the metric is some kind of degeneracy and thus suitable gauge-fixing conditions within the metric class are needed to address it.

These considerations show that it is necessary to have a clearer picture of G_2-structures within a fixed metric class. In [4], Bryant observed that such G_2-structures are parametrized by sections of an $\mathbb{R}P^7$-bundle, or more concretely, by pairs (a, α) where a is a real-valued function and α is a vector field such that $a^2 + |\alpha|^2 = 1$, and $\pm(a, \alpha)$ define the same G_2-structure. If φ is a fixed G_2-structure, then any other G_2-structure $\sigma_{(a,\alpha)}(\varphi)$ within the same metric class is given by:

$$\sigma_{(a,\alpha)}(\varphi) = \left(a^2 - |\alpha|^2\right)\varphi - 2a\alpha\lrcorner\psi + 2\alpha \wedge (\alpha\lrcorner\varphi), \tag{1}$$

where $\psi = *\varphi$.

Given that the group G_2 may be defined as the automorphism group of the octonions, a G_2-structure defines an octonion structure on the manifold, and in [10], this observation was used to interpret the above pair (a, α) as a unit octonion V, and then (1) is just the 3-form that corresponds to a modified octonion product defined by V. Thus, a flow of isometric G_2-structures can be interpreted as a flow of the unit octonion section V. In particular, a natural heat flow of isometric G_2-structures was introduced in [10]. Given an octonionic covariant derivative D, constructed from the Levi-Civita connection and the torsion of the initial G_2-structure φ, the heat flow of isometric G_2-structures is then the semilinear, parabolic equation

$$\frac{\partial V}{\partial t} = \Delta_D V + |DV|^2 V \tag{2}$$

with some initial condition $V(0) = V_0$ and where $\Delta_D = -D^* D$ is the Laplacian operator corresponding to D. This was obtained as the negative gradient flow of an energy functional with respect to D. The critical points of the flow (2) correspond to G_2-structures for which the torsion tensor is divergence-free, i.e. satisfies $\operatorname{div} T = 0$, where divergence is taken with respect to the Levi-Civita connection. This is significant for several reasons. The divergence of torsion is precisely the term that causes the non-parabolicity of the Laplacian flow of co-closed G_2-structures from [15] as mentioned above, and $\operatorname{div} T = 0$ for closed G_2-structures. Thus, closed G_2-structures are automatically critical points of (2). Secondly, T has been interpreted in [10] as an imaginary octonion-valued 1-form, which is added to the Levi-Civita connection to obtain the octonionic covariant derivative D, hence the condition $\operatorname{div} T = 0$ is precisely analogous to the Coulomb gauge condition in gauge theory. This analogy makes this condition a reasonable candidate for a gauge-fixing condition within a fixed metric class.

Soon after the introduction of the flow (2) in [10], it was further studied from different perspectives by several authors: Bagaglini in [1]; Dwivedi, Gianniotis, and Karigiannis in [8]; the author in [11]; Loubeau and Sá Earp in [17].

Equivalently to the flow of octonions (2), one can consider directly the evolution of the 3-form φ via the equation

$$\frac{\partial \varphi}{\partial t} = 2\,(\operatorname{div} T) \lrcorner \psi \tag{3}$$

where T is the torsion tensor that corresponds to the G_2-structure 3-form at time t. This is the way the flow was formulated in [1] and in [8] (although here we are following [10, 11] and added a factor of 2 in (3). In [17], a more general approach is taken and a harmonic heat flow of geometric structures is considered. In the case of G_2-structures, it is shown to reduce to (3). In this survey we will review the above approaches to the flow of isometric G_2-structures and outline the key analytic results.

2 Isometric G_2-structures

A G_2-structure on a 7-manifold is defined by a smooth *positive* 3-form φ [3, 12]. This is a nowhere-vanishing 3-form that defines a Riemannian metric g_{φ}, such that for any vectors u and v, the following holds

$$g_{\varphi}(u, v)\,\mathrm{vol}_{\varphi} = \frac{1}{6}\,(u \lrcorner \varphi) \wedge (v \lrcorner \varphi) \wedge \varphi. \tag{4}$$

At any point, the stabilizer of g_{φ} (along with orientation) is $SO(7)$, whereas the stabilizer of φ is $G_2 \subset SO(7)$. This shows that at a point, positive 3-forms that correspond to the same metric, i.e., are *isometric*, are parametrized by $SO(7)/G_2 \cong \mathbb{RP}^7 \cong S^7/\mathbb{Z}_2$. Therefore, on a Riemannian manifold, metric-compatible G_2-structures are parametrized by sections of an $\mathbb{RP}^7$-bundle, or alternatively, by sections of an S^7-bundle, with antipodal points identified. This is precisely the parametrization given by (1).

Alternatively, a G_2-structure in a fixed metric class can be interpreted as a reduction of the principal $SO(7)$-bundle P of orthonormal frames to a principal G_2-subbundle, and hence each such reduction corresponds to a section σ of an $SO(7)/G_2$-bundle N and equivalently, an $SO(7)$-equivariant map $s : P \longrightarrow SO(7)/G_2 \cong S^7/\mathbb{Z}_2$. This is the picture used in [17].

We may also use the G_2-structure φ and the metric to define the *octonion bundle* $\mathbb{O}M \cong \Lambda^0 \oplus TM$ on M as a rank 8 real vector bundle equipped with an octonion product of sections given by

$$A \circ_\varphi B = \left(ab - g(\alpha, \beta), a\beta + b\alpha + \alpha \times_\varphi \beta\right) \tag{5}$$

for any sections $A = (a, \alpha)$ and $B = (b, \beta)$. We set the metric $g = g_\varphi$, since we are fixing the metric, even though the G_2-structure may change. Here we define $\times_\varphi$ by $g\left(\alpha \times_\varphi \beta, \gamma\right) = \varphi(\alpha, \beta, \gamma)$ and given $A \in \Gamma(\mathbb{O}M)$, we write $A = (\operatorname{Re} A, \operatorname{Im} A)$. The metric on TM is extended to $\mathbb{O}M$ to give the octonion inner product $\langle A, B\rangle = ab + g(\alpha, \beta)$, which is Hermitian with respect to the octonion product. In the formula (1), the pair (a, α) can now be interpreted as a unit octonion section.

The *intrinsic torsion* of a G_2-structure is defined by $\nabla\varphi$, where ∇ is the Levi-Civita connection for the metric g that is defined by φ. Following [14], we have

$$\nabla_a \varphi_{bcd} = 2T_a{}^e \psi_{ebcd} \text{ and } \nabla_a \psi_{bcde} = -8T_{a[b}\varphi_{cde]} \tag{6}$$

where T_{ab} is the *full torsion tensor*, note that an additional factor of 2 is for convenience, and $\psi = *\varphi$ is the 4-form that is the Hodge dual of φ with respect to the metric g. The G_2-structure is known as *torsion-free* if $T = 0$, and in that case ∇ has holonomy contained in G_2. Conversely, if ∇ has holonomy contained in G_2, then there exists a torsion-free G_2-structure within the metric class. Let $V = (a, \alpha)$ be a unit octonion section, then define $\sigma_V(\varphi) = \sigma_{(a,\alpha)}(\varphi)$, as in (1). It has been shown in [10] that the torsion of the G_2-structure $\varphi_V = \sigma_V(\varphi)$ is given by

$$T^{(V)} = VTV^{-1} - (\nabla V)V^{-1} \tag{7}$$

where T is the torsion of φ, interpreted as a 1-form with values in the bundle of imaginary octonions $\operatorname{Im}\mathbb{O}M$. If we now define an octonion covariant derivative D on sections of $\mathbb{O}M$ via

$$DV = \nabla V - VT, \tag{8}$$

the expression (7) simply becomes

$$T^{(V)} = -(DV)V^{-1}. \tag{9}$$

As shown in [10], the derivative D has other nice properties—it is metric-compatible, and satisfies a partial product rule with respect to octonion product on

$\mathbb{O}M$, that is, $D(UV) = (\nabla U)V + U(DV)$. Now given (9), the divergence of $T^{(V)}$ can be expressed as

$$\operatorname{div} T^{(V)} = -(\Delta_D V)V^{-1} - |DV|^2. \tag{10}$$

3 Energy Functional

Given that the torsion varies across G_2-structures within the same metric class, an obvious question is how to pick a representative of the class with the "best" torsion. A reasonable way to try and characterize the best torsion is to look for critical points of a functional. Therefore, given the set $\mathcal{F}_g$ of all G_2-structures that are compatible with a given metric g, and assuming M is compact, define the functional $\mathcal{E} : \mathcal{F}_g \longrightarrow \mathbb{R}$ by

$$\mathcal{E}(\varphi) = \int_M \left|T^{(\varphi)}\right|^2 \mathrm{vol}, \tag{11}$$

where $T^{(\varphi)}$ is the torsion of a G_2-structure φ. This is the functional used by Dwivedi, Gianniotis, and Karigiannis in [8].

As we have seen in the previous section, given a G_2-structure φ, any other G_2-structure within the same metric class is given by $\sigma_V(\varphi)$ for a unit octonion section V. Therefore, the functional (11) is equivalent to the functional $\mathcal{E}_{\mathbb{O}} : \Gamma(S\mathbb{O}M) \longrightarrow \mathbb{R}$ given by

$$\mathcal{E}_{\mathbb{O}}(V) = \int_M \left|T^{(V)}\right|^2 \mathrm{vol} = \int_M |DV|^2 \mathrm{vol} \tag{12}$$

where we have also applied (9). Hence, in fact, the functional $\mathcal{E}_\varphi$ is equivalent to an energy functional with respect to the derivative D. This is the functional used in [10, 11].

On the other hand, following the approach in [17], recall that a principal H-subbundle of a principal G-bundle P may be characterized by an equivariant map $s : P \longrightarrow G/H$, or equivalently, as a section σ of the associated bundle $N = P \times_G (G/H) \cong P/H$. Assuming that G is semi-simple, so that it admits a bi-invariant metric, we may define a metric η on N, together with the corresponding Levi-Civita connection ∇^η. Moreover, given a metric g on the base manifold, we may induce a metric on $T^*M \otimes \sigma^*TN$, which is compatible with the splitting $TN = \mathcal{V}N \oplus \mathcal{H}N$ induced by ∇^η. Using this metric, we may then define an energy functional $\mathcal{E}_\Gamma : \Gamma(N) \longrightarrow \mathbb{R}$ on sections of N:

$$\mathcal{E}_\Gamma(\sigma) = \int_M |d\sigma|^2 \mathrm{vol}. \tag{13}$$

Alternatively, suppose that moreover G is compact, so that P is compact. Then, let us define an energy functional on G-equivariant maps $s : P \longrightarrow G/H$:

$$\mathcal{E}_G(s) = \int_P |ds|^2 \operatorname{vol}_P \tag{14}$$

where an induced metric on $T^*P \otimes s^*T(G/H)$ is used. It is then shown in [17], that for any section $\sigma \in \Gamma(N)$ and its corresponding G-equivariant map $s \in C_G^\infty(P, G/H)$, $\mathcal{E}_G(s) = c_1\mathcal{E}_\Gamma(\sigma) + c_2$ where c_1 and c_2 are uniform constants.

Consider the orthogonal splitting $d\sigma = d^{\mathcal{V}}\sigma + d^{\mathcal{H}}\sigma$ into horizontal and vertical parts. Since the horizontal component of the metric is given by π^*g, where $\pi : N \longrightarrow M$ is the bundle projection map, we find that for any $X \in TM$,

$$\left|d^{\mathcal{H}}\sigma(X)\right|^2 = (\pi^*g)(d\sigma(X), d\sigma(X)) = g((\pi\circ\sigma)_* X, (\pi\circ\sigma)_* X) = g(X, X).$$

Thus, the horizontal part of $d\sigma$ contributes only a constant term to (13), and it is thus sufficient to consider just the vertical component

$$\mathcal{E}_\Gamma^{\mathcal{V}}(\sigma) = \int_M \left|d^{\mathcal{V}}\sigma\right|^2 \operatorname{vol}. \tag{15}$$

In the G_2 case, Loubeau and Sá Earp show in [17] that this functional is equivalent to (11).

Theorem 3.1 ([17]) *If M is seven-dimensional, P is the $SO(7)$-principal bundle of oriented orthonormal frames, and N is an associated $SO(7)/G_2$-bundle over M, then $\left|d^{\mathcal{V}}\sigma\right|^2 = \frac{8}{3}\left|T^{(\sigma)}\right|^2$ where $T^{(\sigma)}$ is the torsion tensor of the G_2-structure defined by the section σ.*

4 Gradient Flow

Given the functionals defined in the previous section, we may consider critical points and negative gradient flows of the functionals. This is summarized below.

Space	Functional	Critical points	Negative gradient flow
$\mathcal{F}_g$	$\mathcal{E}(\varphi)$	$\operatorname{div} T^{(\varphi)} = 0$	$\frac{\partial\varphi_t}{\partial t} = 2\operatorname{div} T^{(\varphi_t)} \lrcorner \psi_t$
$\Gamma(S\mathbb{O}M)$	$\mathcal{E}_{\mathbb{O}}(V)$	$\Delta_D V + \|DV\|^2 V = 0$	$\frac{\partial V_t}{\partial t} = \Delta_D V_t + \|DV_t\|^2 V_t$
$\Gamma(N)$	$\mathcal{E}_\Gamma(\sigma)$	$\tau^{\mathcal{V}}(\sigma) = 0$	$\frac{\partial\sigma_t}{\partial t} = \tau^{\mathcal{V}}(\sigma_t)$
$C_G^\infty(P, G/H)$	$\mathcal{E}_G(s)$	$\tau^{\mathcal{H}}(s) = 0$	$\frac{\partial s_t}{\partial t} = \tau^{\mathcal{H}}(s_t)$

where $\tau^{\mathcal{V}}(\sigma) := \mathrm{Tr}_g\left(\nabla^{\eta} d^{\mathcal{V}} \sigma\right)$ is the *vertical tension field* of the functional $\mathcal{E}_{\Gamma}(\sigma)$ and $\tau^{\mathcal{H}}(s) := \mathrm{Tr}_g^{\mathcal{H}}(\nabla^{\eta} ds)$ is the *horizontal tension field* of the functional $\mathcal{E}_G(s)$. It is proved in [20, Theorem 1] that $\sigma \in \Gamma(N)$ is a harmonic *section*, i.e. a critical point of the functional (13), if and only if the corresponding G-equivariant map $s \in C_G^{\infty}(P, G/H)$ is a *horizontally harmonic map*, that is $\tau^{\mathcal{H}}(s) = 0$. In the expression for $\tau^{\mathcal{H}}(s)$, the trace is just over the horizontal distribution in TP. It should be emphasized that the reason that the critical points of $\mathcal{E}_G$ are not exactly harmonic maps is that we are varying over only the equivariant maps, rather than arbitrary maps. On the other hand, Wood does prove in [20, Theorem 3], that if G/H is a normal G-homogeneous manifold and the metric on P is constructed from any compatible metric on G, then σ is a harmonic section if and only s is a harmonic map, that is, $\tau(s) := \mathrm{Tr}_g(\nabla^{\eta} ds) = 0$. Crucially, these conditions are satisfied for $G = SO(7)$, $H = G_2$, and P the orthonormal frame bundle on M. Moreover, as shown in [17], given these conditions, a family $\sigma_t \in \Gamma(N)$ satisfies the harmonic section flow $\frac{\partial \sigma_t}{\partial t} = \tau^{\mathcal{V}}(\sigma_t)$ if and only if there is a corresponding family $s_t \in C_G^{\infty}(P, G/H)$ that satisfies the harmonic map flow $\frac{\partial s_t}{\partial t} = \tau(s_t)$. Also, Wood has shown in [19] that equivariance is preserved along the harmonic map flow, so that if the initial condition is equivariant, then the flow will continue to be equivariant. This shows a close relationship between harmonic map theory and the theory of harmonic sections, and hence the flow (3) of isometric G_2-structures.

On the other hand, one must be careful when applying harmonic map results. In particular, the energy $\mathcal{E}_G(s)$ contains a topological term that can never be arbitrarily small, and thus standard small initial energy long-time existence results [5] for harmonic maps cannot be applied. Similarly, while a constant map is always harmonic, an equivariant map $s : P \longrightarrow G/H$ can never be constant (if $H \neq G$). Thus existence of non-trivial harmonic equivariant maps and hence harmonic sections is not guaranteed, as expected.

Some results from the theory of harmonic maps do carry over, at least in the G_2-case. It was shown in [8, 11] that almost monotonicity and ε-regularity results similar to the harmonic map heat flow [5, 6, 18] hold for the flow (3).

Let $p_{x_0,t_0}(x,t)$ be the backward heat kernel on M, that is, the solution of the backward heat equation for $0 \leq t \leq t_0$ that converges to a delta function at $(x,t) = (x_0, t_0)$. Then, given a time-dependent octonion section V_t or equivalently, a 3-form $\varphi_t = \sigma_{V(t)}(\varphi)$ for some fixed G_2-structure φ, define the $\mathcal{F}$-functional [11]

$$\mathcal{F}(x_0, t_0, t) = (t_0 - t) \int_M \left| T^{(V_t)}(x) \right|^2 p_{x_0,t_0}(x,t) \,\mathrm{vol}(x), \tag{16}$$

where $T^{(V_t)} = -(DV_t)V_t^{-1}$ is the torsion of the G_2-structure φ_t. In [8], the analogous quantity is denoted by $\Theta_{(x_0,t_0)}(\varphi(t))$. It is then shown in both [8, Theorem 5.3] and [11, Proof of Corollary 7.2] that $\mathcal{F}$ satisfies an *almost* monotonicity formula along the flow (2). Suppose V_t is a solution of the flow (2) for $0 \leq t < t_0$ with initial energy $\mathcal{E}(0) = \mathcal{E}_0$. Then, there exists a constant $C > 0$, that only depends on the background geometry, such that for any t and τ satisfying $t_0 - 1 \leq \tau \leq t < t_0$, $\mathcal{F}$

satisfies the following relation

$$\mathcal{F}(x_0, t_0, t) \leq C\mathcal{F}(x_0, t_0, \tau) + C(t-\tau)\left(\mathcal{E}_0 + \mathcal{E}_0^{\frac{1}{2}}\right). \tag{17}$$

In [8], the last term in (17) was $C(t-\tau)(\mathcal{E}_0 + 1)$, which of course follows from (17) for a different constant C. In both [8] and [11] similar versions of an ε-regularity result is proven for $\mathcal{F}$. We'll state it as in [11].

Theorem 4.1 ([8, Theorem 5.7] and [11, Theorem 7.1]) *Given $\mathcal{E}_0$, there exist $\varepsilon > 0$ and $\beta > 0$, both depending on M and β also depending on $\mathcal{E}_0$, such that if V is a solution of the flow (2) on $M \times [0, t_0)$ with energy bounded by $\mathcal{E}_0$, and if*

$$\mathcal{F}(x_0, t_0, t) \leq \varepsilon \tag{18}$$

for $t \in [t_0 - \beta, t_0)$, then V extends smoothly to $U_{x_0} \times [0, t_0]$ for some neighborhood U_{x_0} of x_0 with $|DV| = \left|T^{(V)}\right|$ bounded uniformly.

Then, Theorem 4.1 was used in [8, 11] to show long-time existence of the isometric heat flow and convergence to a G_2-structure with $\operatorname{div} T = 0$ given sufficiently small initial pointwise torsion.

Given a G_2-structure 3-form φ, in [8] a concept of *entropy* was defined:

$$\lambda(\varphi, \sigma) = \max_{(x,t)\in M\times(0,\sigma)} \left\{ t \int_M \left|T^{(\varphi)}(y)\right|^2 p_{(x,t)}(y, 0)\operatorname{vol}(y) \right\}. \tag{19}$$

This mirrors similar entropy concepts defined for the mean curvature flow, Yang-Mills flow, and the harmonic map heat flow, in [7, 16], and [2], respectively. The quantity $\lambda(\varphi, \sigma)$ is shown in [8] to be invariant under the scaling $(\varphi, \sigma) \mapsto (c^3\varphi, c^2\sigma)$. While the same quantity could be defined for an octonion section V, if considered as a function of V, λ would lose the scaling property for V. So in this case, using the 3-form has an advantage. Overall, one of the key results in [8] is long term existence and convergence of the flow (3) given sufficiently small entropy.

Theorem 4.2 ([8, Theorem 5.15]) *Let φ_0 be a G_2-structure on a compact 7-manifold M. For any $\delta, \sigma > 0$, there exists $\varepsilon > 0$, such that if $\lambda(\varphi_0, \sigma) < \varepsilon$, then the flow (3) with initial condition $\varphi(0) = \varphi_0$ exists for all time and converges smoothly to a G_2-structure φ_∞ that satisfies $\operatorname{div} T^{(\varphi_\infty)} = 0$ and $\left|T^{(\varphi_\infty)}\right| < \delta$.*

Although good progress has been made on properties of the flows (2) and (3), many questions still remain. For example, is it possible to prove long-time existence given small initial energy, rather than entropy or pointwise torsion? If we combine the equivariant harmonic map approach with the octonion approach, then everything could be reformulated in terms of equivariant maps from the orthonormal frame bundle P to S^7 equipped with the octonion product. It is likely that the additional algebraic structure could help achieve stronger results.

References

1. L. Bagaglini, The energy functional of G_2-structures compatible with a background metric. J. Geom. Anal. (2019). https://doi.org/10.1007/s12220-019-00264-6
2. J. Boling, C. Kelleher, J. Streets, Entropy, stability and harmonic map flow. Trans. Am. Math. Soc. **369**(8), 5769–5808 (2017). https://doi.org/10.1090/tran/6949
3. R.L. Bryant, Metrics with exceptional holonomy. Ann. Math. (2) **126**(3), 525–576 (1987). https://doi.org/10.2307/1971360
4. R.L. Bryant, Some remarks on G_2-structures, in *Proceedings of Gökova Geometry-Topology Conference 2005 (GGT), Gökova* (2006), pp. 75–109. math/0305124
5. Y.M. Chen, W.Y. Ding, Blow-up and global existence for heat flows of harmonic maps. Invent. Math. **99**(3), 567–578 (1990). https://doi.org/10.1007/BF01234431
6. Y.M. Chen, M. Struwe, Existence and partial regularity results for the heat flow for harmonic maps. Math. Z. **201**(1), 83–103 (1989). https://doi.org/10.1007/BF01161997
7. T.H. Colding, W.P. Minicozzi II, Generic mean curvature flow I: generic singularities. Ann. Math. (2) **175**(2), 755–833 (2012). https://doi.org/10.4007/annals.2012.175.2.7
8. S. Dwivedi, P. Gianniotis, S. Karigiannis, A gradient flow of isometric G_2 structures. J. Geom. Anal. (2019). 1904.10068. https://doi.org/10.1007/s12220-019-00327-8
9. S. Grigorian, Short-time behaviour of a modified Laplacian coflow of G2-structures. Adv. Math. **248**, 378–415 (2013). 1209.4347. https://doi.org/10.1016/j.aim.2013.08.013
10. S. Grigorian, G_2-structures and octonion bundles. Adv. Math. **308**, 142–207 (2017). 1510.04226. https://doi.org/10.1016/j.aim.2016.12.003
11. S. Grigorian, Estimates and monotonicity for a heat flow of isometric G_2-structures. Calc. Var. Partial Differ. Equ. **58**(5), Art. 175, 37 (2019). https://doi.org/10.1007/s00526-019-1630-0
12. N.J. Hitchin, The geometry of three-forms in six dimensions. J. Differ. Geom. **55**(3), 547–576 (2000). math/0010054. http://projecteuclid.org/euclid.jdg/1090341263
13. S. Karigiannis, Deformations of G_2 and $Spin(7)$ structures on manifolds. Can. J. Math. **57**, 1012 (2005). math/0301218. https://doi.org/10.4153/CJM-2005-039-x
14. S. Karigiannis, Flows of G_2-structures, I. Q. J. Math. **60**(4), 487–522 (2009). math/0702077. https://doi.org/10.1093/qmath/han020
15. S. Karigiannis, B. McKay, M.-P. Tsui, Soliton solutions for the Laplacian coflow of some G_2-structures with symmetry. Differ. Geom. Appl. **30**(4), 318–333 (2012). 1108.2192. https://doi.org/10.1016/j.difgeo.2012.05.003
16. C. Kelleher, J. Streets, Entropy, stability, and Yang-Mills flow. Commun. Contemp. Math. **18**(2), 1550032, 51 (2016). https://doi.org/10.1142/S0219199715500327
17. E. Loubeau, H.N. Sá Earp, Harmonic flow of geometric structures (2019). 1907.06072
18. M. Struwe, On the evolution of harmonic maps in higher dimensions. J. Differ. Geom. **28**(3), 485–502 (1988). http://projecteuclid.org/euclid.jdg/1214442475
19. C.M. Wood, An existence theorem for harmonic sections. Manuscripta Math. **68**(1), 69–75 (1990). https://doi.org/10.1007/BF02568751
20. C.M. Wood, Harmonic sections and equivariant harmonic maps. Manuscripta Math. **94**(1), 1–13 (1997). https://doi.org/10.1007/BF02677834

Remarks on the Navier-Stokes Equations in Homogeneous and Isotropic Spacetimes

Makoto Nakamura

Abstract The Navier-Stokes equations are derived in homogeneous and isotropic spacetimes. The effects of the spatial expansion and contraction are considered.

Keywords Navier-Stokes equations · Cosmological principle

Mathematics Subject Classification (2010) Primary 35Q30; Secondary 35Q75

1 Introduction

We recall the Navier-Stokes equations in the Minkowski spacetime. Let us consider the incompressible viscous fluid with the density ρ, the rate of the viscosity μ and the pressure p. The Navier-Stokes equations are given by the form

$$\begin{cases} \operatorname{div} u = 0, \\ \partial_0 u^k + \sum_{j=1}^{n} u^j \partial_j u^k + \frac{1}{\rho}\partial_k p - \frac{\mu}{\rho}\Delta u^k = 0 \end{cases} \tag{1.1}$$

for $1 \le k \le n$ on $[0,T) \times \mathbb{R}^n$ for $T > 0$, where $x = (x^0, x^1, \cdots, x^n) \in \mathbb{R}^{1+n}$, $u = u(x) = (u^1(x), \cdots, u^n(x))$ denotes the velocity vector, $\operatorname{div} u := \sum_{1\le j\le n} \partial_j u^j$, Δ is the Laplacian defined by $\Delta := \sum_{j=1}^{n} \partial_j^2$. By the Helmholtz projection defined by

$$(Hf)^k := f^k - \partial_k \frac{1}{\Delta} \operatorname{div} f \tag{1.2}$$

M. Nakamura (✉)
Faculty of Science, Yamagata University, Yamagata, Japan
e-mail: nakamura@sci.kj.yamagata-u.ac.jp

P. Cerejeiras et al. (eds.), *Current Trends in Analysis, its Applications and Computation*, Research Perspectives,
https://doi.org/10.1007/978-3-030-87502-2_56

for $1 \le k \le n$ and any function $f = (f^1, \cdots, f^n)$, Eq. (1.1) are rewritten as

$$\begin{cases} \operatorname{div} u = 0, \\ \partial_0 u^k + (Hf)^k - \frac{\mu}{\rho} \Delta u^k = 0 \end{cases} \tag{1.3}$$

for $1 \le k \le n$, where we have put $f^k := \sum_{j=1}^n u^j \partial_j u^k$.

We consider the extension of the Navier-Stokes equations in the Minkowski spacetime to the equations in homogeneous and isotropic spacetimes. Under the assumption of the cosmological principle, namely, that the space is homogeneous and isotropic, the solution of the Einstein equations with the spatial curvature 0 is given by the Friedmann-Lemaître-Robertson-Walker metric (the FLRW metric) defined by

$$-c^2(d\tau)^2 = g_{\alpha\beta} dx^\alpha dx^\beta = -c^2(dx^0)^2 + a(x^0)^2 \sum_{j=1}^n (dx^j)^2, \tag{1.4}$$

where τ denotes the proper time, and $a(\cdot)$ is the scale-function of the space defined by

$$a(x^0) := \begin{cases} a(0)\left(1 + \frac{n(1+\sigma)\partial_0 a(0) x^0}{2a(0)}\right)^{2/n(1+\sigma)} & \text{if } \sigma \ne -1, \\ a(0) \exp\left(\frac{\partial_0 a(0) x^0}{a(0)}\right) & \text{if } \sigma = -1. \end{cases} \tag{1.5}$$

(see [1] for the references of the Einstein equations and the FLRW metric). Put

$$T_0 := -\frac{2a(0)}{n(1+\sigma)\partial_0 a(0)}. \tag{1.6}$$

To describe the generalization of the Navier-Stokes equations (1.3) in homogeneous and isotropic spacetimes, we use the following convention. The Greek letters $\alpha, \beta, \gamma, \cdots$ run from 0 to n, and the Latin letters $j, k, \ell, \cdots$ run from 1 to n. We use the Einstein rule for the sum of indices of tensors, for example, $T^\alpha{}_\alpha := \sum_{\alpha=0}^n T^\alpha{}_\alpha$ and $T^i{}_i := \sum_{i=1}^n T^i{}_i$. We denote the speed of light by $c > 0$, the metric by $g_{\alpha\beta} dx^\alpha dx^\beta$ for $x = (x^0, \cdots, x^n) \in \mathbb{R}^{1+n}$. We denote by $(g_{\alpha\beta})$ the matrix whose components are given by $\{g_{\alpha\beta}\}_{0 \le \alpha, \beta \le n}$. Put $g := \det(g_{\alpha\beta})$. Let $(g^{\alpha\beta})$ be the inverse matrix of $(g_{\alpha\beta})$. The change of upper and lower indices is done by $g_{\alpha\beta}$ and $g^{\alpha\beta}$, for example, $T^\alpha{}_\beta := g^{\alpha\gamma} T_{\gamma\beta}$ for any tensor $T_{\alpha\beta}$.

To derive the Navier-Stokes equations, we use a stress-energy tensors $T^{\alpha\beta}$ (see Sect. 2, below). The conservation law of $T^{\alpha\beta}$ is expressed by $\nabla_\alpha T^{\alpha\beta} = 0$. We consider the nonrelativistic limit of this conservation law. Namely,

$$\lim_{c \to \infty} \nabla_\alpha T^{\alpha\beta} = 0. \tag{1.7}$$

We put

$$\partial^k := a^{-2}\partial_k \text{ and } f^k := u^j\partial_j u^k$$

for $1 \le k \le n$. We note $\partial_j\partial^j = a^{-2}\Delta$ and $\partial^k/\partial_\ell\partial^\ell = \partial_k/\Delta$. Equation (1.7) yield the Navier-Stokes equations

$$\begin{cases} \partial_j u^j + \frac{n\partial_0 a^2}{2a^2} = 0, \\ \partial_0 u^k + (Hf)^k + \frac{\partial_0 a^2}{a^2}u^k - \frac{\mu}{\rho}\partial_j\partial^j u^k = 0, \end{cases} \tag{1.8}$$

where the outline of the derivation is described in Sect. 2.

For the Cauchy problem of (1.8), we have the following results.

Proposition 1.1 *The scale-function in* (1.8) *must satisfy*

$$a(t) = a_0\left(1 + \frac{2a_1 t}{a_0}\right)^{1/2}. \tag{1.9}$$

Namely, σ must satisfy $\sigma = 4/n - 1$ in (1.5).

When we consider the problem in the Minkowski spacetime with $a_0 = 1$, namely $a(\cdot) = 1$, the Cauchy problem (1.8) is rewritten as

$$\begin{cases} \partial_j U^j = 0 & \text{on } [0, T) \times \mathbb{R}^n, \\ \partial_t U^k + f(U)^k + \frac{1}{\rho}\partial^k\tilde{p} - \frac{\mu}{\rho}\Delta U^k = 0 & \text{on } [0, T) \times \mathbb{R}^n, \\ U(0, \cdot) = U_0(\cdot) \end{cases} \tag{1.10}$$

for $1 \le k \le n$, where $U = (U^1, \cdots, U^n)$, $U_0 = (U_0^1, \cdots, U_0^n)$ and $\tilde{p}$ is the pressure for U. The following lemma shows one relation between the solutions u of (1.8) and U of (1.10).

Proposition 1.2 *Put*

$$\phi(x) := \frac{1}{2}a_0 a_1 \sum_{j=1}^{n}(x^j)^2 \tag{1.11}$$

for $x \in \mathbb{R}^n$. If u is the solution of (1.8), *then the function U defined by*

$$U^j(t, x) = a(t)u^j\left(t, \frac{x}{a(t)}\right) + \frac{1}{a(t)}(\partial_j\phi)\left(\frac{x}{a(t)}\right) \tag{1.12}$$

for $1 \le j \le n$ is the solution of (1.10). *Conversely, the function u defined by* (1.12) *for the solution U of* (1.10) *is the solution of* (1.8).

Proposition 1.2 is a useful tool to consider the problem (1.8) since the transformation reduces it to the well-known problem (1.10) in the Minkowski spacetime. For example, we show an application of Proposition 1.2 to the Cauchy problem of (1.8). For any fixed r_0 with $1 \le r_0 < \infty$, we put

$$\alpha(r) := \frac{n}{2}\left(\frac{1}{r_0} - \frac{1}{r}\right)$$

for $1 \le r < \infty$. For $T > 0$ and r_1 with $1 \le r_1 < \infty$, we set function spaces

$$\begin{aligned} X_0 &:= \left\{U_0 \in \left(L^{r_0}(\mathbb{R}^n)\right)^n;\ \partial_j U_0^j = 0\right\}, \\ X(T) &:= \{U;\ t^{\alpha(r)} U \in BC((0,T), H\left(L^r(\mathbb{R}^n)\right)^n\right), \\ &\qquad t^{1/2+\alpha(r_1)} \nabla U \in BC((0,T), \left(H\left(L^{r_1}(\mathbb{R}^n)\right)^n\right)^n)\}, \end{aligned}$$

where $U := (U^1, \cdots, U^n)$ and $U_0 := (U_0^1, \cdots, U_0^n)$, with the norms

$$\begin{aligned} \|U\|_A &:= \sup_{0<t<T} t^{\alpha(r)} \|U(t)\|_r, \\ \|\nabla U\|_{A_1} &:= \sup_{0<t<T} t^{1/2+\alpha(r_1)} \|\nabla U(t)\|_{r_1}, \\ \|U\|_X &:= \max\left\{\|U\|_A,\ \|\nabla U\|_{A_1}\right\} \end{aligned}$$

and the metric

$$d_X(U,V) := \max\left\{\|U-V\|_A,\ \|\nabla(U-V)\|_{A_1}\right\}.$$

For the function ϕ defined by (1.11), we set function spaces

$$\begin{aligned} Y_0 &:= \left\{a_0^{-1} U_0(a_0 \cdot) - a_0^{-2} \nabla\phi(\cdot);\ U_0 \in X_0\right\}, \\ Y(T) &:= \left\{a(t)^{-1} U(t, a(t)\cdot) - a(t)^{-2} \nabla\phi(\cdot);\ U \in X(T)\right\}. \end{aligned}$$

We put

$$T_1 := \begin{cases} \infty & a_1 \ge 0, \\ T_0 & a_1 < 0, \end{cases}$$

where T_0 is defined by (1.6). We say that the solution u of (1.8) is global if $u \in Y(T_1)$ since the space does not exist after T_0 when $a_1 < 0$.

We have the following results.

Theorem 1.3 *Let $n \geq 2$. Let ρ and μ be constants with $\mu/\rho > 0$. Let $a_0 > 0$, $a_1 \in \mathbb{R}$. Let us consider the Cauchy problem* (1.8). *Then the following results hold.*

1. *(Local and global solutions) For sufficiently small number $\varepsilon > 0$, put r_0, r_1 as $1/r_0 = 1/r_1 = 1/n - \varepsilon$, and let r be a number with $0 \leq 1/r \leq 1/n - \varepsilon$. Then for any $u_0 \in Y_0$, there exists a unique solution $u \in Y(T)$ of* (1.8)*, where T with $0 < T < T_1$ can be taken of the form*

$$T = \left(\frac{C}{\|U_0\|_{(L^{r_0}(\mathbb{R}^n))^n}} \right)^{2/n\varepsilon} \tag{1.13}$$

for a constant $C > 0$ which is independent of u_0, where

$$U_0(x) := a_0 u_0 \left(\frac{x}{a_0} \right) + \frac{1}{a_0} \nabla \phi \left(\frac{x}{a_0} \right) \tag{1.14}$$

and $\nabla \phi := (\partial_1, \cdots, \partial_n)\phi$. Moreover, if $a_1 < 0$ and U_0 is small such that

$$\|U_0\|_{(L^{r_0}(\mathbb{R}^n))^n} \leq C T_0^{-n\varepsilon/2}$$

holds, then T can be taken as $T = T_1$. Namely, u is a unique global solution in $Y(T_1)$.

2. *(Global solutions) Let $r_0 = r_1 = n$, and let r be a number with $0 < 1/r < 1/n$. If $\|U_0\|_{(L^{r_0}(^n))^n}$ is sufficiently small, then there exists a unique global solution $u \in Y(T_1)$ of* (1.8).

In Theorem 1.3, we have shown the results in Lebesgue spaces corresponding to the results in [2]. Compared with the results in [2], we obtain the small global solution in (1) in Theorem 1.3 when the space is contracting, namely, $a_1 < 0$.

2 Derivation of the Equations

In this section, we show the outline to derive the equations in (1.8). We show some results on the stress-tensor $P^{\alpha\beta}$ with Lamé's elastic constants λ and μ. We put $\partial^\alpha := g^{\alpha\beta}\partial_\beta$. In the case of (1.4), we have

$$\partial^0 = -\frac{1}{c^2}\partial_0, \quad \partial^j = \frac{1}{a^2}\partial_j.$$

When we take the nonrelativistic limit ($c \to \infty$), we naturally regard ∂^0 as the zero-operator ($\partial^0 = 0$). Let p be a function which denotes the pressure. For $0 \leq \alpha \leq n$, let v^α be a contravariant tensor which satisfies $\lim_{c\to\infty} \partial_j v^0 = 0$ for $1 \leq j \leq n$.

Let $P^{\alpha\beta}$ be a stress-tensor defined by

$$P^{\alpha\beta} := -pg^{\alpha\beta} + \lambda g^{\alpha\beta}\nabla_\gamma v^\gamma + \mu(\nabla^\alpha v^\beta + \nabla^\beta v^\alpha) \tag{2.1}$$

for $0 \le \alpha, \beta \le n$. Then we have the following results step by step.

$$(1)\quad \lim_{c\to\infty} v^\alpha \nabla_\alpha v^\beta = v^0 \partial_0 v^\beta + v^j \partial_j v^\beta + 2v^0 \Gamma^\beta{}_{0j} v^j,$$

$$(2)\quad \nabla_\gamma v^\gamma = \frac{\partial_0 a^n}{a^n} v^0 + \partial_\gamma v^\gamma,$$

$$(3)\quad \lim_{c\to\infty} \nabla^\beta \nabla_\gamma v^\gamma = \partial^\beta \partial_j v^j,$$

$$(4)\quad \lim_{c\to\infty} \nabla_\alpha \nabla^\alpha v^\beta = \partial_j \partial^j v^\beta,$$

$$(5)\quad \lim_{c\to\infty} \nabla_\alpha \nabla^\beta v^\alpha = \partial^\beta \partial_j v^j,$$

$$(6)\quad \lim_{c\to\infty} \nabla_\alpha (pg^{\alpha\beta}) = \partial^\beta p,$$

$$(7)\quad \lim_{c\to\infty} \nabla_\alpha P^{\alpha\beta} = -\partial^\beta p + \mu \partial_j \partial^j v^\beta + (\mu + \lambda)\partial^\beta \partial_j v^j$$

for $0 \le \beta \le n$, where we regard ∂^0 as the zero-operator in the right hand side of each equation.

By the metric (1.4), any velocity-tensor u^α must satisfy the equation

$$-c^2 = -c^2(u^0)^2 + a(x^0)^2 \sum_{j=1}^n (u^j)^2.$$

So that, u^0 satisfies $\lim_{c\to\infty} u^0 = \pm 1$. We assume

$$\lim_{c\to\infty} u^0 = 1 \ \text{ and } \ \lim_{c\to\infty} \partial_\alpha u^0 = 0 \tag{2.2}$$

for $0 \le \alpha \le n$, which means that the local time tends to the proper time in the nonrelativistic limit. The Navier-Stokes equations in homogeneous and isotropic spacetimes are given as follows.

Now, let us consider the spacetime which has the metric (1.4). Let u^α denote a velocity-tensor. Assume (2.2). Let $P^{\alpha\beta}$ be the stress-tensor with $v^\alpha := u^\alpha$. Let ρ be a function which denotes the density of mass. Let $T^{\alpha\beta}$ be a stress-energy tensor defined by

$$T^{\alpha\beta} := \left(\rho + \frac{p}{c^2}\right) u^\alpha u^\beta - P^{\alpha\beta} \tag{2.3}$$

for $0 \le \alpha, \beta \le n$. Then we have the following results step by step.

1. We have

$$\lim_{c\to\infty} \nabla_\alpha \left(\left(\rho + \frac{p}{c^2}\right) u^\alpha u^\beta\right) = I^\beta$$

for $0 \le \beta \le n$, where we have put

$$I^\beta := \partial_0 \rho u^\beta + \partial_j \rho u^j u^\beta + \rho\left(\left(\frac{\partial_0 a^n}{a^n} + \partial_j u^j\right) u^\beta + \partial_0 u^\beta + u^j \partial_j u^\beta + 2\Gamma^\beta{}_{0j} u^j\right) \tag{2.4}$$

and we regard $u^0 = 1$ in the right hand side.

2. We have

$$\lim_{c\to\infty} \nabla_\alpha T^{\alpha\beta} = I^\beta + \partial^\beta p - \mu \partial_j \partial^j u^\beta - (\mu + \lambda)\partial^\beta \partial_j u^j$$

for $0 \le \beta \le n$, where we regard $u^0 = 1$ and $\partial^0 = 0$ in the right hand side. Let us consider the nonrelativistic limits of the conservation $\nabla_\alpha T^{\alpha\beta} = 0$ for $0 \le \beta \le n$. The equations

$$\lim_{c\to\infty} \nabla_\alpha T^{\alpha\beta} = 0 \tag{2.5}$$

for $0 \le \beta \le n$ are equivalent to

$$\partial_0 \rho + \partial_j \rho\, u^j + \rho\left(\partial_j u^j + \frac{\partial_0 a^n}{a^n}\right) = 0 \tag{2.6}$$

and

$$\rho\left(\partial_0 u^k + u^j \partial_j u^k + \frac{\partial_0 a^2}{a^2} u^k\right) + \partial^k p - \mu \partial_j \partial^j u^k - (\mu + \lambda)\partial^k \partial_j u^j = 0 \tag{2.7}$$

for $1 \le k \le n$.

3. When $\rho(\neq 0)$ is a constant, Eq. (2.5) are equivalent to

$$\partial_j u^j + \frac{n \partial_0 a^2}{2a^2} = 0 \tag{2.8}$$

and

$$\partial_0 u^k + f^k + \frac{\partial_0 a^2}{a^2} u^k + \frac{1}{\rho} \partial^k p - \frac{\mu}{\rho} \partial_j \partial^j u^k = 0 \tag{2.9}$$

for $1 \leq k \leq n$. Equation (2.9) is rewritten as

$$\partial_0 u^k + (Hf)^k + \frac{\partial_0 a^2}{a^2} u^k - \frac{\mu}{\rho} \partial_j \partial^j u^k = 0 \tag{2.10}$$

for $1 \leq k \leq n$, where $(Hf)^k$ is defined by (1.2).

Therefore, we have obtained the required equations.

Acknowledgments The author is very thankful to Professor Anahit Galstyan and Professor Karen Yagdjian for valuable comments during the conference.

References

1. S. Carroll, *Spacetime and Geometry. An Introduction to General Relativity* (Addison Wesley, San Francisco, CA, 2004), xiv+513 pp.
2. T. Kato, Strong L^p-solutions of the Navier-Stokes equation in $\mathbb{R}^m$ with applications to weak solutions. Math. Z. **187**, 471–480 (1984)

Part XI
Partial Differential Equations with Nonstandard Growth

Ultra-Parabolic Kolmogorov-Type Equation with Multiple Impulsive Sources

Ivan Kuznetsov and Sergey Sazhenkov

Abstract Existence and uniqueness of entropy solutions of the Cauchy–Dirichlet problem for the non-autonomous ultra-parabolic equation with partial diffusivity and multiple impulsive sources is established. The limiting passage from the equation incorporating a single distributed source to the multi-impulsive equation is fulfilled, as the distributed source collapses to a parameterized multi-atomic Dirac delta measure.

Keywords Ultra-parabolic equation · Entropy solution · Impulsive source

Mathematics Subject Classification (2010) 35D30, 35K70, 35R12

1 Introduction

The focus of this article is centered on the well-posedness topics for the initial-boundary value problem for the quasi-linear ultra-parabolic equation with partial diffusivity and multiple impulsive source terms. Such equations are commonly called *Kolmogorov-type equations*. Besides, they are also called *impulsive equations* due to the presence of impulsive source terms. Impulsive sources involve parameterized Dirac delta measures in their construction. From the physical viewpoint, they reflect phenomena of instantaneous loading, i.e., drastic change of mass, energy, impulse, etc. at a moment. The theory of impulsive partial differential equations is rather new and far from complete. In the present article, we develop the results earlier obtained in [3–5] for quasi-linear equations with autonomous coefficients

I. Kuznetsov · S. Sazhenkov (✉)
Lavrentyev Institute of Hydrodynamics, Novosibirsk, Russia

Novosibirsk State University, Novosibirsk, Russia
e-mail: kuznetsov_i@hydro.nsc.ru

P. Cerejeiras et al. (eds.), *Current Trends in Analysis, its Applications and Computation*, Research Perspectives,
https://doi.org/10.1007/978-3-030-87502-2_57

and single impulsive source onto the cases with non-autonomous coefficients and multiple impulsive sources, which is the novelty of this research.

2 Problem Π_0: The Basic Formulation

Let Ω be a bounded domain of spatial variables $\boldsymbol{x} \in \mathbb{R}^d$ with a smooth boundary $\partial\Omega$ ($\partial\Omega \in C^2$). Let $t \in [0, T]$ and $s \in [0, S]$ be two independent time-like variables. Here T and S are given positive constants. Let $\tau_1, \tau_2, \dots, \tau_N \in (0, T)$ be fixed time moments, labeled so that $0 < \tau_1 < \dots < \tau_N < T$. Denote

$$G_{T,S} := \Omega \times (0, T) \times (0, S), \quad \Xi^1 := \overline{\Omega} \times [0, S], \quad \Xi^2 := \overline{\Omega} \times [0, T],$$
$$\Gamma_l := \partial\Omega \times [0, T] \times [0, S], \quad \Gamma_0^1 := \overline{\Omega} \times \{t = 0\} \times [0, S],$$
$$\Gamma_T^1 := \overline{\Omega} \times \{t = T\} \times [0, S], \quad \Gamma_{\tau_j}^1 := \overline{\Omega} \times \{t = \tau_j\} \times [0, S] \ (j = 1, \dots, N),$$
$$\Gamma_0^2 := \overline{\Omega} \times [0, T] \times \{s = 0\}, \quad \Gamma_S^2 := \overline{\Omega} \times [0, T] \times \{s = S\}.$$

In this article, the focus of our study is centered on the following problem.

Problem Π_0 It is necessary to find a function $u\colon G_{T,S} \mapsto \mathbb{R}$ satisfying the quasi-linear ultra-parabolic equation

$$\partial_t u + \partial_s a(u) + \operatorname{div}_x \boldsymbol{\varphi}(\boldsymbol{x}, t, s, u) = \operatorname{div}_x(\mathbb{A}(\boldsymbol{x}, t, s, u)\nabla_x u) + \sum_{j=1}^{N} \beta_j(\boldsymbol{x}, s, u)\delta_{(t=\tau_j-0)}, \quad (\boldsymbol{x}, t, s) \in G_{T,S}, \tag{2.1a}$$

the initial condition with respect to time-like variable t

$$u|_{t=0} = u_0^{(1)}(\boldsymbol{x}, s), \quad (\boldsymbol{x}, s) \in \Xi^1, \tag{2.1b}$$

the initial and final conditions with respect to time-like variable s

$$u|_{s=0} \approx u_0^{(2)}(\boldsymbol{x}, t), \quad u|_{s=S} \approx u_S^{(2)}(\boldsymbol{x}, t), \quad (\boldsymbol{x}, t) \in \Xi^2, \tag{2.1c}$$

and the homogeneous boundary condition

$$u|_{\Gamma_l} = 0. \tag{2.1d}$$

In the formulation of Problem Π_0, we suppose that the initial and final data $u_0^{(1)}$, $u_0^{(2)}$, $u_S^{(2)}$, the nonlinear convective fluxes a and $\boldsymbol{\varphi} = (\varphi_1, \dots, \varphi_d)$, the diffusion matrix $\mathbb{A}$ and the impulsive perturbations β_j are given and satisfy the conditions stated further. In (2.1a) and further, by $\delta_{(t=\tau-0)}$ we denote the Dirac delta measure centered 'on the left-hand side' of a point $t=\tau$ in $\mathbb{R}$, i.e.,

$$\langle \delta_{(\cdot=\tau-0)}, \phi \rangle = \phi(\tau-0) \tag{2.2}$$

for any integrable in a neighborhood of $\{t=\tau\} \subset \mathbb{R}$ function ϕ having the trace at the point $t=\tau$ from the left: $\phi(\tau-0) = \lim\limits_{t\to\tau-0} \phi(t)$.

In (2.1c) the relation sign $\approx$ means that $u_0^{(2)}$ and $u_S^{(2)}$ may be unattained by a solution u on some parts of the sets Γ_0^2 and Γ_S^2, respectively. The fact whether $\approx$ becomes equality ($=$), or not, is figured out *a posteriori*, i.e., after a solution of equation (2.1a) is constructed somehow.

Conditions on $u_0^{(1)}\&u_0^{(2)}\&u_S^{(2)}$ The initial and final data meet the regularity requirements $u_0^{(1)} \in C^{2+\alpha}(\Xi^1)$, $u_0^{(2)}, u_S^{(2)} \in C^{2+\alpha}(\Xi^2)$ ($\alpha \in (0,1)$) and the following consistency conditions: $u_0^{(1)}$ vanishes in a neighbourhood of $\partial\,\Xi^1$ and $u_0^{(2)}$ and $u_S^{(2)}$ vanish in a neighbourhood of $\partial\,\Xi^2$.

Conditions on $a\&\boldsymbol{\varphi}\&\mathbb{A}\&\boldsymbol{\beta}$

(i) Matrix $\mathbb{A} = (a_{ij}(\boldsymbol{x},t,s,\lambda))_{i,j=1,\ldots,d}$ is symmetric and uniformly positive-definite, i.e., $a_{ij}(\boldsymbol{x},t,s,\lambda) = a_{ji}(\boldsymbol{x},t,s,\lambda)$ and there exists a positive constant a_* independent of $\boldsymbol{x}$, t, s and λ such that $\sum\limits_{i,j=1}^{d} a_{ij}(\boldsymbol{x},t,s,\lambda)\xi_i\xi_j \geq a_*|\boldsymbol{\xi}|^2 \quad \forall\boldsymbol{\xi} \in \mathbb{R}^d,\ \forall(\boldsymbol{x},t,s,\lambda) \in G_{T,S}\times\mathbb{R}_\lambda$.
Components a_{ij} meet the regularity requirement $a_{ij} \in C^2_{\text{loc}}(G_{T,S}\times\mathbb{R}_\lambda)$.

(ii) Convective fluxes a and $\boldsymbol{\varphi}$ and impulsive perturbations β_j ($j=1,\ldots,N$) meet the regularity requirements $a \in C^2_{\text{loc}}(\mathbb{R})$, $a(0)=0$, $\boldsymbol{\varphi} \in C^2_{\text{loc}}(G_{T,S}\times\mathbb{R}_\lambda)^d$, $\boldsymbol{\varphi}(\boldsymbol{x},t,s,0) = (0,\ldots,0)$, and $\beta_j \in C^1_0(\Xi^1\times\mathbb{R}_\lambda)$.

(iii) Flux a satisfies the following *genuine nonlinearity condition*: the Lebesgue measure of the set $\{\lambda\in\mathbb{R}\colon\ \xi_1 + a'(\lambda)\xi_2 = 0\}$ is zero for every fixed pair $(\xi_1,\xi_2)\in\mathbb{R}^2$ such that $\xi_1^2+\xi_2^2=1$, i.e., for every fixed $(\xi_1,\xi_2)\in\mathbb{S}^1$.

(iv) Flux $\boldsymbol{\varphi}$ satisfies the following *growth condition*: there exist constants $b_1, b_2 > 0$ such that for all $(\boldsymbol{x},t,s)\in G_{T,S}$ and $\lambda\in\mathbb{R}$ the inequality $-\operatorname{div}_x\boldsymbol{\varphi}(\boldsymbol{x},t,s,\lambda)\lambda \leq b_1\lambda^2 + b_2$ takes place.

Remark 2.1 Note that, with $u\big|_{s=0} = u_0^{(2)}$ and $u\big|_{s=S} = u_S^{(2)}$ on Ξ^2 on the place of (2.1c), Problem Π_0 becomes ill-posed. Indeed, since flux a is nonlinear and, in general, non-monotonous, it may be impossible to equate a solution u of Problem Π_0 to $u_0^{(2)}$ and $u_S^{(2)}$ on the **entire** sets Γ_0^2 and Γ_S^2. Therefore, we permit that a possible weak solution of Problem Π_0 may deviate from $u_0^{(2)}$ and $u_S^{(2)}$ on Γ_0^2 and Γ_S^2, respectively. Like in [4, 5], we set up a more loose non-classical condition (2.1c) following the original ideas presented in [1, 7].

Remark 2.2 In the formulation of Problem Π_0 notice that, in the sense of distributions, Eq. (2.1a) can be equivalently written with the help of identity (2.2) as the system consisting of the equation

$$\partial_t u + \partial_s a(u) + \mathrm{div}_x \boldsymbol{\varphi}(\boldsymbol{x}, t, s, u) = \mathrm{div}_x(\mathbb{A}(\boldsymbol{x}, t, s, u)\nabla_x u) \quad \text{in } G_{T,S} \backslash (\cup_{j=1}^N \Gamma_{\tau_j}^1)$$

and the impulsive conditions ($j = 1, \ldots, N$)

$$u(\boldsymbol{x}, \tau_j + 0, s) = u(\boldsymbol{x}, \tau_j - 0, s) + \beta_j(\boldsymbol{x}, s, u(\boldsymbol{x}, \tau_j - 0, s)), \quad (\boldsymbol{x}, s) \in \Xi^1. \tag{2.3}$$

Our first aim in this article is to establish the existence and uniqueness results for Problem Π_0. To this end, we revisit and extend our research originally carried out in [2, 4, 5]. This way we introduce and study the suitable concept of *entropy solutions* to Problem Π_0. This concept arises from the strictly parabolic regularization of Problem Π_0.

3 Parabolic Regularization of Problem Π_0

We construct an entropy solution of Problem Π_0 as a limit of the family of weak solutions u_ε of the following strictly parabolic model, as $\varepsilon \to 0+$.

Problem Π_ε Under Conditions on $u_0^{(1)}\&u_0^{(2)}\&u_S^{(2)}$ and on $a\&\boldsymbol{\varphi}\&\mathbb{A}\&\boldsymbol{\beta}$, it is necessary to find u_ε: $G_{T,S} \mapsto \mathbb{R}$ satisfying the strictly parabolic equation

$$\partial_t u_\varepsilon + \partial_s a(u_\varepsilon) + \mathrm{div}_x \boldsymbol{\varphi}(\boldsymbol{x}, t, s, u_\varepsilon) = \mathrm{div}_x(\mathbb{A}(\boldsymbol{x}, t, s, u_\varepsilon)\nabla_x u_\varepsilon) + \varepsilon \partial_{ss}^2 u_\varepsilon \tag{3.1}$$

in $G_{T,S} \backslash (\bigcup_{j=1}^N \Gamma_{\tau_j}^1)$, the impulsive conditions (2.3) for $j = 1, \ldots, N$, the initial condition (2.1b), the homogeneous boundary condition (2.1d) and the (exact!) initial and final conditions

$$u_\varepsilon|_{s=0} = u_0^{(2)}(\boldsymbol{x}, t), \quad u_\varepsilon|_{s=S} = u_S^{(2)}(\boldsymbol{x}, t), \quad (\boldsymbol{x}, t) \in \Xi^2. \tag{3.2}$$

Here $\varepsilon \in (0, 1]$ is an arbitrarily fixed small parameter.

The notion of weak solution to Problem Π_ε is quite standard in the theory of parabolic problems. We notice that, in order to find a weak solution of Problem Π_ε, it is necessary and sufficient to fulfill the following $N + 1$ steps. On the first step, in the subdomain $\Omega \times (0, \tau_1) \times (0, S) \subset G_{T,S}$, we find a weak solution u_ε of the system consisting of Eq. (3.1) and conditions (2.1b), (2.1d) and (3.2). On the j-th step ($j = 2, \ldots, N + 1$), in the subdomain $\Omega \times (\tau_{j-1}, \tau_j) \times (0, S) \subset G_{T,S}$, we find the weak solution u_ε of the system consisting of Eq. (3.1) and conditions (2.1d), (3.2) and

$$u_\varepsilon(\boldsymbol{x}, t, s)|_{t=\tau_{j-1}} = u_\varepsilon(\boldsymbol{x}, \tau_{j-1} - 0, s) + \beta_{j-1}(\boldsymbol{x}, s, u(\boldsymbol{x}, \tau_{j-1} - 0, s)), \ (\boldsymbol{x}, s) \in \Xi^1. \tag{3.3}$$

In the right-hand side of (3.3), $u_\varepsilon(\boldsymbol{x}, \tau_{j-1} - 0, s)$ is the trace on $\{t = \tau_{j-1} - 0\}$ of the weak solution u_ε obtained on the $(j-1)$-th step.

Thus, Problem Π_ε is, in fact, the set of $N + 1$ initial boundary value problems that should be solved successively. Each of these problems is well-posed in the class of weak solutions due to the well-known theory of quasilinear parabolic equations of the second order [6]. Aggregating the results of the $N + 1$ steps we conclude that the following assertion holds true.

Proposition 3.1 *Under Conditions on $u_0^{(1)}$& $u_0^{(2)}$&$u_S^{(2)}$ and on a&$\boldsymbol{\varphi}$&$\mathbb{A}$&$\boldsymbol{\beta}$, for any fixed $\varepsilon \in (0, 1]$ there exists the unique weak solution $u_\varepsilon \in L^\infty(G_{T,S}) \cap L^2(0, T; W_2^1(\Xi^1))$ of Problem Π_ε.*

Moreover, the energy estimate $\|\nabla_x u_\varepsilon\|^2_{L^2(G_{T,S})} + \varepsilon\|\partial_s u_\varepsilon\|^2_{L^2(G_{T,S})} \le M_0$ and the maximum principle $\|u_\varepsilon\|_{L^\infty(G_{T,S})} \le M_1$ hold, where constants M_0 and M_1 do not depend on ε.

4 Entropy Solutions of Problem Π_0

The first main result of this article arises as $\varepsilon \to 0+$ in the formulation of the Problem Π_ε. Similarly to [2, Theorem 1] and [5, Theorem 3.1 (assertion 2)], we establish the following theorem.

Theorem 4.1

(1) ***(Convergence result.)*** *There exist a subsequence $\varepsilon_k \underset{k\to\infty}{\longrightarrow} 0+$ and a limiting function $u \in L^\infty(G_{T,S}) \cap L^2((0, T) \times (0, S); \overset{o}{W}{}_2^1(\Omega))$ such that the sequence of weak solutions u_{ε_k} of Problems Π_{ε_k} converges to u strongly in $L^1(G_{T,S})$, as $k \to \infty$. In other words, the family of weak solutions $\{u_\varepsilon\}_{\varepsilon\in(0,1]}$ of Problem Π_ε is relatively compact in $L^1(G_{T,S})$, as $\varepsilon \to 0+$.*

(2) ***(Existence of entropy solutions.)*** *Function $u = \lim_{k\to\infty} u_{\varepsilon_k}$ serves as an entropy solution of Problem Π_0 in the sense of Definition 4.2 formulated below.*

(3) ***(Uniqueness of entropy solutions.)*** *Let, in addition to Conditions on a&$\boldsymbol{\varphi}$&$\mathbb{A}$&$\boldsymbol{\beta}$, the convective flux $\boldsymbol{\varphi}$ and the diffusion matrix $\mathbb{A}$ be autonomous, i.e., be independent of $\boldsymbol{x}$, t and s explicitly. Then the entropy solution of Problem Π_0 is unique.*

Definition 4.2 Function $u \in L^\infty(G_{T,S}) \cap L^2((0,T)\times(0,S); \overset{o}{W}{}_2^1(\Omega))$ is called ***an entropy solution*** to Problem Π_0, if it satisfies the entropy inequality

$$\begin{aligned}&\partial_t \eta(u) + \partial_s q_a(u) + \mathrm{div}_x \boldsymbol{q}_\varphi(\boldsymbol{x},t,s,u) - \mathrm{Div}_x \boldsymbol{q}_\varphi(\boldsymbol{x},t,s,u)\\ &\qquad + (\mathrm{Div}_x \boldsymbol{\varphi}(\boldsymbol{x},t,s,u))\eta'(u) - \mathrm{div}_x(\mathbb{A}(\boldsymbol{x},t,s,u)\nabla_x \eta(u)) \le\\ &\qquad - \eta''(u)|\mathbb{A}^{1/2}(\boldsymbol{x},t,s,u)\nabla_x u|^2 \quad \text{in } G_{T,S}\backslash(\cup_{j=1}^N \Gamma^1_{\tau_j}),\end{aligned} \tag{4.1a}$$

the maximum principle $\|u\|_{L^\infty(G_{T,S})} \le M_1$, the initial condition (2.1b) (in the trace sense) and the impulsive conditions (2.3), possesses the strong traces $u_0^{\mathrm{tr},(2)}$ and $u_S^{\mathrm{tr},(2)}$ on the planes Γ_0^2 and Γ_S^2, respectively, and satisfies the following two entropy boundary conditions:

$$\begin{aligned}&q_a(u_0^{\mathrm{tr},(2)}(\boldsymbol{x},t)) - q_a(u_0^{(2)}(\boldsymbol{x},t))\\ &\qquad - \eta'(u_0^{(2)}(\boldsymbol{x},t))(a(u_0^{\mathrm{tr},(2)}(\boldsymbol{x},t)) - a(u_0^{(2)}(\boldsymbol{x},t))) \le 0, \quad (\boldsymbol{x},t) \in \Xi^2,\end{aligned} \tag{4.1b}$$

$$\begin{aligned}&q_a(u_S^{\mathrm{tr},(2)}(\boldsymbol{x},t)) - q_a(u_S^{(2)}(\boldsymbol{x},t))\\ &\qquad - \eta'(u_S^{(2)}(\boldsymbol{x},t))(a(u_S^{\mathrm{tr},(2)}(\boldsymbol{x},t)) - a(u_S^{(2)}(\boldsymbol{x},t))) \ge 0, \quad (\boldsymbol{x},t) \in \Xi^2.\end{aligned} \tag{4.1c}$$

In (4.1a), (4.1b) and (4.1c), $\eta \in C^2(\mathbb{R})$ is an arbitrary convex test-function: $\eta''(z) \ge 0\ \forall z \in \mathbb{R}$, and $(\eta, q_a, \boldsymbol{q}_\varphi)$ is ***the convex entropy—entropy flux triple*** such that $\boldsymbol{q}_\varphi(\boldsymbol{x},t,s,z) = \int_0^z \eta'(\tilde z)\partial_{\tilde z}\boldsymbol{\varphi}(\boldsymbol{x},t,s,\tilde z)\,d\tilde z$ and $q_a(z) = \int_0^z a'(\tilde z)\eta'(\tilde z)\,d\tilde z$, $\forall z \in \mathbb{R}$.

In (4.1a) the differential operator Div_x is defined by the formula

$$\mathrm{Div}_x \boldsymbol{\psi}(\boldsymbol{x},t,s,u) = (\mathrm{div}_x \boldsymbol{\psi}(\boldsymbol{x},t,s,z))|_{z=u(\boldsymbol{x},t,s)} \quad \forall \boldsymbol{\psi} \in C^1(G_{T,S}\times \mathbb{R}_z)^d.$$

In particular, operators Div_x and div_x relate through the identity

$$\mathrm{div}_x \boldsymbol{\psi}(\boldsymbol{x},t,s,u) = \mathrm{Div}_x \boldsymbol{\psi}(\boldsymbol{x},t,s,u) + \partial_u \boldsymbol{\psi}(\boldsymbol{x},t,s,u)\cdot \nabla_x u.$$

Entropy inequality (4.1a) is understood in the sense of distributions. Entropy boundary conditions (4.1b) and (4.1c) are understood a.e. in Ξ^2, and impulsive conditions (2.3) are understood a.e. in Ξ^1.

5 Problem Π_0 as a Singular Limit Model Arising due to the Collapse of a Distributed Source

In this section let us consider the following Cauchy–Dirichlet problem for Eq. (2.1a) incorporating the distributed source term instead of the multiple impulsive ones.

Problem Π_γ It is necessary to find a function u: $G_{T,S} \mapsto \mathbb{R}$ satisfying the quasi-linear ultra-parabolic equation

$$\partial_t u + \partial_s a(u) + \mathrm{div}_x \boldsymbol{\varphi}(\boldsymbol{x}, t, s, u) = \mathrm{div}_x(\mathbb{A}(\boldsymbol{x}, t, s, u)\nabla_x u) + Z_\gamma(\boldsymbol{x}, t, s, u), \quad (\boldsymbol{x}, t, s) \in G_{T,S},$$

the initial and final conditions (2.1b) and (2.1c) and the homogeneous boundary condition (2.1d).

We suppose that the initial and final data are given and satisfy Conditions on $u_0^{(1)}\&u_0^{(2)}\&u_S^{(2)}$; and the diffusion matrix $\mathbb{A}$, convective fluxes a and $\boldsymbol{\varphi}$ and the source term Z_γ are given and satisfy the following requirements.

Conditions on $a\&\boldsymbol{\varphi}\&\mathbb{A}\&Z_\gamma$

- **(i)** Matrix $\mathbb{A} = (a_{ij})_{i,j=1,\ldots,d}$ and fluxes a and $\boldsymbol{\varphi}$ satisfy all requirements in items (i)–(iii) of Conditions on $a\&\boldsymbol{\varphi}\&\mathbb{A}\&\boldsymbol{\beta}$.
- **(ii)** Z_γ belongs to the space $L^\infty(0, T; C_{\mathrm{loc}}(\Xi^1 \times \mathbb{R}))$.
- **(iii)** Flux $\boldsymbol{\varphi}$ and source Z_γ satisfy the following *growth condition*: there exist constants $b_\gamma^{(1)}$, $b_\gamma^{(2)} > 0$ such that for all $(\boldsymbol{x}, t, s) \in G_{T,S}$ and $\lambda \in \mathbb{R}$ the inequality $(Z_\gamma(\boldsymbol{x}, t, s) - \mathrm{div}_x \boldsymbol{\varphi}(\boldsymbol{x}, t, s, \lambda))\lambda \le b_\gamma^{(1)}\lambda^2 + b_\gamma^{(2)}$ takes place.

We remark that label $\boldsymbol{\gamma}$ is dumb so far.

For Problem Π_γ the notion of entropy solutions is introduced quite analogously to [5, Definition 3.4] with several necessary natural modifications. The following results on well-posedness of Problem Π_γ take place.

Proposition 5.1

(1) ***(Existence of entropy solutions.)*** *Under Conditions on $u_0^{(1)}\&u_0^{(2)}\&u_S^{(2)}$ and on $a\&\boldsymbol{\varphi}\&\mathbb{A}\&Z_\gamma$ there exists at least one entropy solution to Problem Π_γ.*

(2) ***(Uniqueness of entropy solutions.)*** *Let, in addition to Conditions on $a\&\boldsymbol{\varphi}\&\mathbb{A}\&Z_\gamma$, the convective flux $\boldsymbol{\varphi}$ and the diffusion matrix $\mathbb{A}$ be autonomous, i.e., be independent of $\boldsymbol{x}$, t and s explicitly. Then the entropy solution of Problem Π_γ is unique.*

Proof In order to justify the proposition, it is sufficient to keep track of the proof of Theorem 3.1 in [5] and fulfill necessary natural changes in it. □

Now suppose that the source term has the specific form

$$Z_{\gamma}(x, t, s, \lambda) = \sum_{j=1}^{N} K_{\gamma_j}(t; \tau_j)\beta_j(x, s, \lambda), \tag{5.1}$$

where $\gamma = (\gamma_1, \gamma_2, \ldots, \gamma_N)$ is multi-index, γ_j ($j = 1, 2, \ldots, N$) are small positive parameters, β_j are given smooth functions satisfying the requirements of item (ii) of Conditions on $a\&\varphi\&\mathbb{A}\&\beta$, τ_j are given fixed moments of time labeled so that $0 < \tau_1 < \ldots < \tau_N < T$, and K_{γ_j} are given by the formula

$$K_{\gamma_j}(t; \tau_j) = \mathbf{1}_{(t \le \tau_j)} \frac{2}{\gamma_j} \omega\left(\frac{t - \tau_j}{\gamma_j}\right) \quad (\gamma_j > 0), \tag{5.2a}$$

where $\omega\colon \mathbb{R} \mapsto \mathbb{R}^+$ is a standard regularizing kernel having the properties

$$\omega \in C_0^{\infty}(\mathbb{R}), \quad \omega(-t) = \omega(t)\ \forall t \in \mathbb{R}, \quad \operatorname{supp}\omega \subset [-1, 1], \quad \int_{\mathbb{R}} \omega(t)\,dt = 1. \tag{5.2b}$$

For each $j = 1, \ldots, N$, function $K_{\gamma_j} = K_{\gamma_j}(t; \tau_j)$ is a weak* approximation of the Dirac delta-function $\delta_{(t=\tau_j-0)}$. Indeed, from (5.2) we easily deduce that

$$K_{\gamma_j}(\cdot; \tau_j) \underset{\gamma_j \to 0+}{\longrightarrow} \delta_{(t=\tau_j-0)} \text{ weakly}^\star \text{ in } \mathcal{M}(\mathbb{R}), \tag{5.3}$$

where $\mathcal{M}(\mathbb{R})$ is the space of Radon measures on $\mathbb{R}$: $\mathcal{M}(\mathbb{R}) = (C_0(\mathbb{R}))^*$.

Remark 5.2 Analogously to [5, Remark 1.5], we have that the function Z_γ of the form (5.1) meets the requirements of items (ii) and (iii) of Conditions on $a\&\varphi\&\mathbb{A}\&Z_\gamma$ for all small fixed $\gamma_j > 0$. This implies that Proposition 5.1 holds true for such source terms Z_γ.

Let us denote by $u_\gamma = u_\gamma(x, t, s)$ the entropy solution of Problem Π_γ incorporating the source term of the form (5.1). In view of the limiting relation (5.3), there arises a question whether the family $\{u_\gamma\}$ converges to an entropy solution u_* of Problem Π_0. The answer to this question is positive. It constitutes the second main result of this article and reads as follows.

Theorem 5.3 (Convergence Result) *Let the source term Z_γ in Problem Π_γ be given by* (5.1), *where functions K_{γ_j} satisfy demands* (5.2) *and functions β_j satisfy the requirements of item (ii) of Conditions on $a\&\varphi\&\mathbb{A}\&\beta$. Let the convective fluxes a and φ and the diffusion matrix $\mathbb{A}$ satisfy Conditions on $a\&\varphi\&\mathbb{A}\&\beta_j$ and the* *additional demands*

$$\max_{\lambda\in\mathbb{R}} |a'(\lambda)| \le M_2, \qquad \max_{G_{T,S}\times\mathbb{R}_\lambda} |\partial_\lambda \boldsymbol{\varphi}(\boldsymbol{x},t,s,\lambda)| \le M_2,$$

$$\max_{G_{T,S}\times\mathbb{R}_\lambda} |\partial_\lambda \mathbb{A}(\boldsymbol{x},t,s,\lambda)| \le M_2, \quad M_2 = \text{const} < +\infty.$$

Then there exist a subsequence $\{u_{\boldsymbol{\gamma}^{(k)}}\}_{k\in\mathbb{N}}$ *(*$\boldsymbol{\gamma}^{(k)} \underset{k\to\infty}{\longrightarrow} (0,\dots,0)$*) of entropy solutions to Problem* $\Pi_{\boldsymbol{\gamma}^{(k)}}$ *and a limiting function* $u_* \in L^\infty(G_{T,S}) \cap L^2((0,T)\times(0,S); \overset{o}{W}{}_2^1(\Omega))$ *such that*

$$u_{\boldsymbol{\gamma}^{(k)}} \underset{k\to\infty}{\longrightarrow} u_* \text{ strongly in } L^1(G_{T,S}) \\ \text{and weakly in } L^2((0,T)\times(0,S); \overset{o}{W}{}_2^1(\Omega)). \tag{5.4}$$

Furthermore, u_* *is an entropy solution of Problem* Π_0 *in the sense of Definition 4.2.*

Addition to Theorem 5.3 Let hypotheses of Theorem 5.3 hold and, additionally, $\boldsymbol{\varphi}$ and $\mathbb{A}$ be autonomous, i.e., be independent of $\boldsymbol{x}$, t and s explicitly. Then the limiting relations (5.4) hold true for **the whole** family $\{u_{\boldsymbol{\gamma}}\}$ as $\gamma_j \to 0+$ $(j = 1,\dots,N)$, i.e., for any subsequence $\boldsymbol{\gamma}^{(k)} \underset{k\to\infty}{\longrightarrow} (0,\dots,0)$.

Proof In order to establish Theorem 5.3, it is sufficient to keep track of the proof of assertion 1 in Theorem 9.1 in [5] and fulfill necessary natural changes in this proof. Addition to Theorem 5.3 holds true thanks to the uniqueness result in Theorem 4.1. □

Acknowledgments The work was supported by the Ministry of Science and Higher Education of the Russian Federation (project no. III.22.4.2) and by the Russian Foundation for Basic Research (grant no. 18-01-00649). The authors are very grateful to Professor Stanislav N. Antontsev (CMAFCIO, Universidade de Lisboa, Portugal) for fruitful discussions and to the supervisors of the session 'Partial Differential Equations with Nonstandard Growth' at the 12th International ISAAC Congress held in Aveiro in 2019, Professor Hermenegildo Borges de Oliveira (University of Algarve, Faro, Portugal) and Professor Sergey I. Shmarev (University of Oviedo, Spain) for kind invitation to take part in the session and for fruitful discussions.

References

1. C. Bardos, A.Y. Leroux, Y.C. Nedelec, First order quasilinear equations with boundary conditions. Commun. Partial Differ. Equ. **4**(9), 1017–1034 (1979)
2. I.V. Kuznetsov, Genuinely nonlinear forward-backward ultra-parabolic equations. Sib. Elect. Math. Rep. **14**, 710–731 (2017)
3. I. Kuznetsov, Kinetic and entropy solutions of quasilinear impulsive hyperbolic equations. Math. Model. Nat. Phenom. **13**(2), 1–7 (2018)

4. I.V. Kuznetsov, S.A. Sazhenkov, Genuinely nonlinear impulsive ultra-parabolic equations and convective heat transfer on a shock wave front. IOP Conf. Ser. Earth Environ. Sci. **193**(012037), 1–7 (2018)
5. I.V. Kuznetsov, S.A. Sazhenkov, Singular limits of the quasi-linear Kolmogorov-type equation with a source term, 9 Jul 2019, pp. 1–59. arXiv:1907.04250v1 [math.AP]
6. O.A. Ladyženskaja, V.A. Solonnikov, N.N. Ural′ceva, *Linear and Quasi-Linear Equations of Parabolic Type* (AMS, Providence, RI, 1968)
7. F. Otto, Initial-boundary value problem for a scalar conservation law. C. R. Acad. Sci. Paris Ser. I Math. **322**, 729–734 (1996)

Part XII
Quaternionic and Clifford Analysis

On a Sylvester Equation over a Division Ring

Vladimir Bolotnikov

Abstract The matrix Sylvester equation $AX - BX = C$ is considered over a division ring. In the case where A and B are Jordan blocks, the solvability criterion is given along with the description of all solutions.

1 Introduction

Given complex matrices $A \in \mathbb{C}^{n\times n}$, $B \in \mathbb{C}^{m\times m}$ and $C \in \mathbb{C}^{n\times m}$, the Sylvester equation

$$AX - XB = C \tag{1.1}$$

can be equivalently written in the form $G\mathbf{x} = \mathbf{c}$ where $G = I_m \otimes A - B^\top \otimes I_n$ and $\mathbf{x}$, $\mathbf{c}$ are the columns constructed from the entries of $X = [x_{ij}]$ and $C = [c_{ij}]$ by

$$\mathbf{x} = \mathrm{Col}_{1\le i\le n}\left(\mathrm{Col}_{1\le i\le m} x_{ij}\right), \quad \mathbf{c} = \mathrm{Col}_{1\le i\le n}\left(\mathrm{Col}_{1\le i\le m} c_{ij}\right).$$

In this way, the consistency criterion and the description of all solutions of the Sylvester equation (1.1) reduce to the standard questions concerning the equation $G\mathbf{x} = \mathbf{c}$. In particular, (1.1) has a unique solution X (corresponding to the column $\mathbf{x} = G^{-1}\mathbf{c}$) if and only if G is invertible or, equivalently, if and only if the spectrums of A and B are disjoint. For more explicit formulas for solutions of (1.1) (in terms of A, B and C rather than their entries) can be found in the survey [6].

In the case Eq. (1.1) is considered over a *division ring* $\mathbb{F}$, even its scalar version is non-trivial. It has been settled in [3] and [4] (see also [2], [5, Section 6]). Here, we consider the matrix equation (1.1) with A and B being algebraic Jordan blocks. The regular (determinate) case is presented in Sect. 3 along with the formula for

V. Bolotnikov (✉)
Department of Mathematics, William & Mary, Williamsburg, VA, USA
e-mail: vladi@math.wm.edu

P. Cerejeiras et al. (eds.), *Current Trends in Analysis, its Applications and Computation*, Research Perspectives,
https://doi.org/10.1007/978-3-030-87502-2_58

the unique solution. The singular case is discussed in Sect. 4, where we present the solvability criterion and the parametrization of all solutions in case the equation is solvable.

2 Background

Given a division ring $\mathbb{F}$ with the center $Z_{\mathbb{F}}$, we denote by $\mathbb{F}[z]$ the ring of polynomials in one formal variable z which commutes with coefficients from $\mathbb{F}$. The center of $\mathbb{F}[z]$ is formed by polynomials with coefficients in $Z_{\mathbb{F}}$ (central polynomials); in other words, $Z_{\mathbb{F}[z]} = Z_{\mathbb{F}}[z]$.

Left and right evaluations of an $f \in \mathbb{F}[z]$ at $\alpha \in \mathbb{F}$ can be defined as the remainders of f when divided by $\boldsymbol{\rho}_\alpha(z) = z - \alpha$ on the left and on the right, respectively. As is easily verified, for any $\alpha \in \mathbb{F}$ and $f \in \mathbb{F}[z]$,

$$f(z) = f^{\mathfrak{e}_\ell}(\alpha) + \boldsymbol{\rho}_\alpha \cdot (L_\alpha f) = f^{\mathfrak{e}_r}(\alpha) + (R_\alpha f)(z) \cdot \boldsymbol{\rho}_\alpha \qquad (\boldsymbol{\rho}_\alpha(z) := z - \alpha), \tag{2.1}$$

where $f^{\mathfrak{e}_\ell}(\alpha)$ and $f^{\mathfrak{e}_r}(\alpha)$ are left and right evaluations of f at α:

$$f^{\mathfrak{e}_\ell}(\alpha) = \sum_{j=0}^{m} \alpha^j f_j \quad \text{and} \quad f^{\mathfrak{e}_r}(\alpha) = \sum_{j=0}^{m} f_j \alpha^j \quad \text{if} \quad f(z) = \sum_{j=0}^{m} z^j f_j, \tag{2.2}$$

and where $L_\alpha f$ and $R_\alpha f$ are the polynomials given by

$$(L_\alpha f)(z) = \sum_{i+j=0}^{m-1} \alpha^i f_{i+j+1} z^j, \quad (R_\alpha f)(z) = \sum_{i+j=0}^{m-1} z^j f_{i+j+1} \alpha^i. \tag{2.3}$$

For $g \in Z_{\mathbb{F}}[z]$, we simply write $g(\alpha)$, as in this case, $g^{\mathfrak{e}_\ell}(\alpha) = g^{\mathfrak{e}_r}(\alpha)$ for any $\alpha \in \mathbb{F}$. Also, we see from (2.3) that $L_\alpha g = R_\alpha g$ for any $g \in Z_{\mathbb{F}}[z]$ and $\alpha \in \mathbb{F}$.

Remark 2.1 The objects introduced in (2.2), (2.3) are related as follows:

$$\alpha \cdot (L_\alpha f)^{\mathfrak{e}_r}(\beta) - (L_\alpha f)^{\mathfrak{e}_r}(\beta) \cdot \beta = f^{\mathfrak{e}_\ell}(\alpha) - f^{\mathfrak{e}_r}(\beta), \tag{2.4}$$

for any $\alpha, \beta \in \mathbb{F}$ and $f \in \mathbb{F}[z]$.

Indeed, evaluating the first equality in (2.1) at $z = \beta$ on the right gives

$$f^{\mathfrak{e}_r}(\beta) = f^{\mathfrak{e}_\ell}(\alpha) + (L_\alpha f)^{\mathfrak{e}_r}(\beta) \cdot \beta - \alpha \cdot (L_\alpha f)^{\mathfrak{e}_r}(\beta)$$

which is equivalent to (2.4). To keep notation compact, we will use notation

$$f \cdot X := f(z) \otimes X, \quad (fX)^{\mathfrak{e}_r}(\beta) := \left[(f x_{i,j})^{\mathfrak{e}_r}(\beta)\right] \tag{2.5}$$

for $f \in \mathbb{F}[z]$, $X = [x_{i,j}] \in \mathbb{F}^{n\times m}$ and $\beta \in \mathbb{F}$. Also, we will write f' for the formal derivative of $f \in \mathbb{F}[z]$. Upon differentiating equalities (2.1) we get

$$f' = L_\alpha f + \rho_\alpha \cdot (L_\alpha f)' = R_\alpha f + (R_\alpha f)' \cdot \rho_\alpha. \tag{2.6}$$

Given $\alpha \in \mathbb{F}$, we let $[\alpha] = \{h\alpha h^{-1} : h \in \mathbb{F}\backslash\{0\}\}$ to denote its similarity (conjugacy) class. An element $\alpha \in \mathbb{F}$ is called *algebraic* (over $Z_{\mathbb{F}}$) if it is annihilated by some central polynomial. In this case, all such polynomials form a two-sided ideal of $\mathbb{F}[z]$ whose generator (the monic central polynomial of the least degree that vanishes at α) $\boldsymbol{\mu}_\alpha$ is called the *minimal central polynomial* of α. Observe that

$$\boldsymbol{\mu}_\alpha = \rho_\alpha \cdot (L_\alpha \boldsymbol{\mu}_\alpha) = (L_\alpha \boldsymbol{\mu}_\alpha) \cdot \rho_\alpha, \quad \boldsymbol{\mu}'_\alpha(\alpha) \neq 0. \tag{2.7}$$

Indeed, the equalities follow from (2.1) since $\boldsymbol{\mu}_\alpha(\alpha) = 0$ and $\boldsymbol{\mu}_\alpha \in Z_{\mathbb{F}}[z]$, while the last relation holds since $\boldsymbol{\mu}'_\alpha \in Z_{\mathbb{F}}[z]$ and $\deg \boldsymbol{\mu}'_\alpha < \deg \boldsymbol{\mu}_\alpha$. We also recall that $\boldsymbol{\mu}_\alpha(\beta) = 0$ if and only if $\beta \in [\alpha]$ and $\boldsymbol{\mu}_\alpha = \boldsymbol{\mu}_\beta$ for all $\beta \in [\alpha]$.

3 The Regular Case

We extend evaluation functionals (2.2) to square matrices using the standard polynomial calculus. Then we have the following result.

Lemma 3.1 *Given $A \in \mathbb{F}^{n\times n}$ and $B \in \mathbb{F}^{m\times m}$, let us assume that there is a central polynomial $p(z) = z^\kappa + p_{\kappa-1}z^{\kappa-1} + \ldots + p_0$ ($p_j \in Z_{\mathbb{F}}$) such that $p(A) = 0$ and $p(B)$ is invertible. Then for any $C \in \mathbb{F}^{n\times m}$ Eq.* (1.1) *has a unique solution given by the formula*

$$X = -\sum_{i=1}^{\kappa} p_j \sum_{i=0}^{j-1} A^i C B^{j-i-1} \cdot p(B)^{-1}. \tag{3.1}$$

Proof Since $p \in Z_{\mathbb{F}}[z]$ and $p(A) = 0$, we have for any $X \in \mathbb{F}^{n\times k}$,

$$\begin{aligned}\sum_{j=1}^{\kappa} p_j \sum_{i=0}^{j-1} A^i(AX - XB)B^{j-i-1} &= \sum_{j=0}^{\kappa} p_j(A^j X - XB^j)\\ &= p(A)X - Xp(B) = -Xp(B).\end{aligned} \tag{3.2}$$

If X satisfies (1.1), we replace $AX - XB$ on the left side by C and see that X (on the right side) is uniquely defined from (3.2) by the formula (3.1). To verify that X

of the form (3.1) satisfies (1.1), we use the computation as above but with C instead of X:

$$\sum_{j=1}^{\kappa} p_j \sum_{i=0}^{j-1} (A^{i+1} C B^{j-i-1} - A^i C B^{j-i}) = -Cp(B).$$

If X is defined as in (3.1), the expression on the left side of the latter equality can be written as $AXp(B) - Xp(B)B$. Since the matrices B and $p(B)$ commute, we therefore, have

$$AXp(B) - XBp(B) = -Cp(B),$$

which is equivalent to (1.1), since $p(B)$ is invertible. □

We next consider the particular case where A and B are Jordan blocks.

Corollary 3.2 *Let $\alpha \in \mathbb{F}$ be algebraic. Then the Sylvester equation*

$$AX - XB = C, \quad A = \begin{bmatrix} \alpha & 0 & \dots & 0 \\ 1 & \alpha & \dots & 0 \\ \vdots & \ddots & \ddots & \vdots \\ 0 & \dots & 1 & \alpha \end{bmatrix}, \quad B = \begin{bmatrix} \beta & 1 & \dots & 0 \\ \vdots & \ddots & \ddots & \vdots \\ 0 & \dots & \beta & 1 \\ 0 & \dots & 0 & \beta \end{bmatrix} \tag{3.3}$$

has a unique solution for any $C \in \mathbb{F}^{n\times m}$ if and only if $\beta \notin [\alpha]$.

Proof If $\mu_\alpha \in Z_{\mathbb{F}}[z]$ is the minimal polynomial of α, then for A and B as in (3.3), we have $\mu_\alpha^n(A) = 0$, while $\mu_\alpha^n(B)$ is an upper triangular matrix with $\mu_\alpha^n(\beta)$ on the main diagonal. If $\beta \notin [\alpha]$, then $\mu_\alpha(\beta) \neq 0$ and therefore, $\mu_\alpha^n(B)$ is invertible. Hence, the "if" part of the statement follows from Lemma 3.1 with $p = \mu_\alpha^n$. For the "only if" part, see Remark 4.6 below. □

4 The Singular Case

We now consider Eq. (3.3) with algebraic $\alpha \sim \beta$. Our goal is to establish necessary and sufficient conditions for the equation to have a solution (see part (1) in Theorem 4.2 below) and to describe all solutions if these conditions are met. Since the equation is linear, it suffices to construct a particular solution (see part (2) in Theorem 4.2) and to describe all solutions to the homogeneous equation $AX - XB = 0$ (see Theorem 4.4). We start with several preliminary results.

Lemma 4.1 *Let $\alpha \in \mathbb{F}$ be algebraic with the minimal polynomial μ_α.*

1. If the equality $\alpha y - y\beta = \gamma$ holds for some $\beta, y, \gamma \in \mathbb{F}$, then

$$(L_\alpha \mu_\alpha y)^{\mathfrak{e}_r}(\beta) = y\mu'_\alpha(\beta) + ((L_\alpha \mu_\alpha \gamma)')^{\mathfrak{e}_r}(\beta), \tag{4.1}$$

2. *If* $u = ((L_\alpha \boldsymbol{\mu}_\alpha)' y)^{\mathfrak{e}_r}(\beta)$, *then*

$$\alpha u - u\beta = (L_\alpha \boldsymbol{\mu}_\alpha y)^{\mathfrak{e}_r}(\beta) - y\boldsymbol{\mu}_\alpha'(\beta). \tag{4.2}$$

3. *If* $v = (L_\alpha \boldsymbol{\mu}_\alpha y)^{\mathfrak{e}_r}(\beta)$ *and* $\beta \sim \alpha$, *then* $\alpha v - v\beta = 0$.

Proof It follows from (2.2) that for all $f, g \in \mathbb{F}[z]$ and $\beta \in \mathbb{F}$,

$$(fg)^{\mathfrak{e}_r}(\alpha) = \sum f_k g^{\mathfrak{e}_r}(\beta)\beta^k = \left(f \cdot g^{\mathfrak{e}_r}(\beta)\right)^{\mathfrak{e}_r}(\beta). \tag{4.3}$$

To prove part (1), we apply (4.3) to $f = (L_\alpha \boldsymbol{\mu}_\alpha)'$ and $g = \boldsymbol{\rho}_\alpha y$; taking into account that $(\boldsymbol{\rho}_\alpha y)^{\mathfrak{e}_r}(\beta) = y\beta - \alpha y = -\gamma$ (by the assumption) we get

$$((L_\alpha \boldsymbol{\mu}_\alpha)' \boldsymbol{\rho}_\alpha y)^{\mathfrak{e}_r}(\beta) = -((L_\alpha \boldsymbol{\mu}_\alpha)' \gamma)^{\mathfrak{e}_r}(\beta). \tag{4.4}$$

On the other hand, differentiating the second representation in (2.7) gives

$$\boldsymbol{\mu}_\alpha' = L_\alpha \boldsymbol{\mu}_\alpha + (L_\alpha \boldsymbol{\mu}_\alpha)' \boldsymbol{\rho}_\alpha.$$

Multiplying this identity by y on the right and then evaluating the resulting identity at β on the right we get

$$y\boldsymbol{\mu}_\alpha'(\beta) = (\boldsymbol{\mu}_\alpha' y)^{\mathfrak{e}_r}(\beta) = (L_\alpha \boldsymbol{\mu}_\alpha y)^{\mathfrak{e}_r}(\beta) + ((L_\alpha \boldsymbol{\mu}_\alpha)' \boldsymbol{\rho}_\alpha y)^{\mathfrak{e}_r}(\beta),$$

which, due to (4.4), is equivalent to (4.1). To prove part (2) we invoke part (1) according to which the expression on the right side of (4.2) equals $((L_\alpha \boldsymbol{\mu}_\alpha)' \gamma)^{\mathfrak{e}_r}(\beta)$ with $\gamma = \alpha y - y\beta$. We transform it using (4.3) and the definition of u as follows:

$$\begin{aligned}
((L_\alpha \boldsymbol{\mu}_\alpha)'(\alpha y - y\beta))^{\mathfrak{e}_r}(\beta) &= -((L_\alpha \boldsymbol{\mu}_\alpha)' \cdot (\boldsymbol{\rho}_\alpha y)^{\mathfrak{e}_r}(\beta))^{\mathfrak{e}_r}(\beta) \\
&= -((L_\alpha \boldsymbol{\mu}_\alpha)' \cdot \boldsymbol{\rho}_\alpha y)^{\mathfrak{e}_r}(\beta) \\
&= -(\boldsymbol{\rho}_\alpha \cdot (L_\alpha \boldsymbol{\mu}_\alpha)' y)^{\mathfrak{e}_r}(\beta) \\
&= -(\boldsymbol{\rho}_\alpha \cdot ((L_\alpha \boldsymbol{\mu}_\alpha)' y)^{\mathfrak{e}_r}(\beta))^{\mathfrak{e}_r}(\beta) \\
&= -(\boldsymbol{\rho}_\alpha u)^{\mathfrak{e}_r}(\beta) = \alpha u - u\beta,
\end{aligned}$$

which completes the proof of part (2). The last part follows by letting $f = \boldsymbol{\mu}_\alpha y$ in (2.4), due to equalities $\boldsymbol{\mu}_\alpha(\alpha) = \boldsymbol{\mu}_\alpha(\beta) = 0$. □

For the rest of the section, we will assume (without much loss of generality) that $n \geq m$. Making use of the entries $c_{i,j}$ of the matrix $C = [c_{i,j}]$ on the right side of the Sylvester equation (3.3), we next introduce the elements $\Gamma_{k,j} \in \mathbb{F}$ for all

$$(k, j) \quad \text{such that} \quad 1 \leq k < n, \;\; 1 \leq j < m, \;\; k + j \leq n, \tag{4.5}$$

by the double recursion

$$\Gamma_{k,j} = (L_\alpha \boldsymbol{\mu}_\alpha \cdot (\Gamma_{k+1,j-1} + c_{k+1,j}))^{\mathfrak{e}_r}(\beta) \cdot (\boldsymbol{\mu}'_\alpha(\beta))^{-1}$$
$$- ((L_\alpha \boldsymbol{\mu}_\alpha)' \cdot (\Gamma_{k,j-1} - \Gamma_{k-1,j} + c_{k,j}))^{\mathfrak{e}_r}(\beta) \cdot (\boldsymbol{\mu}'_\alpha(\beta))^{-1} \tag{4.6}$$

with the initial conditions

$$\Gamma_{0,j} = \Gamma_{k,0} = 0 \quad \text{for all} \quad k, j \geq 1. \tag{4.7}$$

Theorem 4.2 *Let* $\Gamma_{k,j}$ *be defined as in* (4.6), (4.7). *If Eq.* (3.3) *has a solution* $X = [x_{i,j}]$, *then necessarily,* $x_{k,j} = \Gamma_{k,j}$ *for all* (k, j) *as in* (4.5), *and*

$$(L_\alpha \boldsymbol{\mu}_\alpha \cdot (\Gamma_{1,j-1} + c_{1,j}))^{\mathfrak{e}_r}(\beta) = 0 \quad for \quad j = 1, \ldots, m. \tag{4.8}$$

Conversely, if conditions (4.8) *are satisfied, then*

$$\Delta_{k,j} := \alpha\Gamma_{k,j} - \Gamma_{k,j}\beta + \Gamma_{k-1,j} - \Gamma_{k,j-1} - c_{k,j} = 0 \tag{4.9}$$

for all (k, j) *as in* (4.5).

Proof Let us introduce the Jordan block $F_k = J_k(0) = \left[\delta_{i,j+1}\right]_{i,j=1}^{k}$, where $\delta_{i,j}$ is the Kronecker symbol, so that A and B in (3.3) can be written as

$$A = \alpha I_n + F_n \quad \text{and} \quad B = \beta I_m + F_m^\top. \tag{4.10}$$

Making use of (4.10), we can write (3.3) in the "polynomial" form as

$$-(\boldsymbol{\rho}_\alpha \cdot I_n - F_n)X + X(\boldsymbol{\rho}_\beta \cdot I_m - F_m^\top) = C, \tag{4.11}$$

and multiply both parts of the latter equality by the two-diagonal matrix

$$G = (L_\alpha \boldsymbol{\mu}_\alpha) \cdot I_n - (L_\alpha \boldsymbol{\mu}_\alpha)' \cdot F_n. \tag{4.12}$$

on the left. Due to equalities (2.7) and (2.6) (with $f = \boldsymbol{\mu}_\alpha$), we get

$$(-\boldsymbol{\mu}_\alpha \cdot I_n + \boldsymbol{\mu}'_\alpha \cdot F_n - (L_\alpha \boldsymbol{\mu}_\alpha)' \cdot F_n^2)X + GX(\boldsymbol{\rho}_\beta \cdot I_m - F_m^\top) = GC. \tag{4.13}$$

Recall that $\boldsymbol{\mu}_\alpha(\beta) = 0$ (since $\beta \sim \alpha$) and $\boldsymbol{\mu}_\alpha, \boldsymbol{\mu}'_\alpha \in Z_{\mathbb{F}}[z]$. Therefore we have

$$(\boldsymbol{\mu}_\alpha X)^{\mathfrak{e}_r}(\beta) = X\boldsymbol{\mu}_\alpha(\beta) = 0 \quad \text{and} \quad (\boldsymbol{\mu}'_\alpha X)^{\mathfrak{e}_r}(\beta) = X\boldsymbol{\mu}'_\alpha(\beta).$$

Taking this into account along with equality $\boldsymbol{\rho}_\beta(\beta) = 0$, we evaluate both parts in (4.13) at β on the right:

$$F_n X \boldsymbol{\mu}'_\alpha(\beta) - ((L_\alpha \boldsymbol{\mu}_\alpha)' F_n^2 X)^{\mathfrak{e}_r}(\beta) - (G X F_m^\top)^{\mathfrak{e}_r}(\beta) = (GC)^{\mathfrak{e}_r}(\beta).$$

Substituting (4.12) into the latter equality gives

$$\begin{aligned} &F_n X \boldsymbol{\mu}'_\alpha(\beta) - ((L_\alpha \boldsymbol{\mu}_\alpha) X F_m^\top)^{\mathfrak{e}_r}(\beta)) - ((L_\alpha \boldsymbol{\mu}_\alpha)'(F_n^2 X - F_n X F_m^\top))^{\mathfrak{e}_r}(\beta) \\ &= ((L_\alpha \boldsymbol{\mu}_\alpha) C)^{\mathfrak{e}_r}(\beta) - ((L_\alpha \boldsymbol{\mu}_\alpha)' F_n C)^{\mathfrak{e}_r}(\beta), \end{aligned}$$

which can be written, as the evaluation functionals are additive, as

$$\begin{aligned} F_n X \boldsymbol{\mu}'_\alpha(\beta) &= ((L_\alpha \boldsymbol{\mu}_\alpha) \cdot (X F_m^\top + C))^{\mathfrak{e}_r}(\beta)) \\ &\quad - ((L_\alpha \boldsymbol{\mu}_\alpha)' \cdot (F_n X F_m^\top - F_n^2 X + F_n C))^{\mathfrak{e}_r}(\beta). \end{aligned} \tag{4.14}$$

Writing the $n \times m$ matrices in (4.14) explicitly as

$$F_n X = \begin{bmatrix} 0 & \dots & 0 \\ x_{11} & \dots & x_{1m} \\ \vdots & & \vdots \\ x_{n-1,1} & \dots & x_{n-1,m} \end{bmatrix}, \quad X F_m^\top + C = \begin{bmatrix} c_{11} & x_{11}+c_{12} & \dots & x_{1,m-1}+c_{1,m} \\ \vdots & \vdots & & \vdots \\ c_{n1} & x_{n1}+c_{n,2} & \dots & x_{n,m-1}+c_{n,m} \end{bmatrix},$$

$$\begin{aligned} &F_n X F_m^\top - F_n^2 X + F_n C \\ &= \begin{bmatrix} 0 & 0 & \dots & 0 \\ c_{11} & x_{11}+c_{12} & \dots & x_{1,m-1}+c_{1,m} \\ c_{21}-x_{11} & x_{21}-x_{12}+c_{22} & \dots & x_{2,m-1}-x_{1,m}+c_{2,m} \\ \vdots & \vdots & & \vdots \\ c_{n-1,1}-x_{n-2,1} & x_{n-1,1}-x_{n-2,2}+c_{n-1,2} & \dots & x_{n-1,m-1}-x_{n-2,m}+c_{n-1,m} \end{bmatrix} \end{aligned}$$

we see that upon letting

$$x_{-1,j} = x_{0,j} = x_{k,0} = 0 \quad \text{for all} \quad k,\, j \geq 1, \tag{4.15}$$

the entries of the latter matrices can be written more uniformly as

$$\begin{aligned} &[F_n X]_{k,j} = x_{k-1,j}, \quad \left[X F_m^\top + C\right]_{k,j} = x_{k,j-1} + c_{kj}, \\ &\left[F_n X F_m^\top - F_n^2 X + F_n C\right]_{k,j} = x_{k-1,j-1} - x_{k-2,j} + c_{k-1,j} \end{aligned}$$

for $k = 1, \dots, n$ and $j = 1, \dots, m$. Plugging the latter formulas into (4.14) and recalling (2.5), we write the matrix equality (4.14) entry-wise as

$$\begin{aligned} x_{k,j} \boldsymbol{\mu}'_\alpha(\beta) &= (L_\alpha \boldsymbol{\mu}_\alpha \cdot (x_{k+1,j-1} + c_{k+1,j}))^{\mathfrak{e}_r}(\beta) \\ &\quad - ((L_\alpha \boldsymbol{\mu}_\alpha)' \cdot (x_{k,j-1} - x_{k-1,j} + c_{k,j}))^{\mathfrak{e}_r}(\beta) \end{aligned} \tag{4.16}$$

for $0 \le k < n$ and $1 \le j \le m$. Due to (4.15), letting $k = 0$ in (4.16) gives

$$(L_\alpha \boldsymbol{\mu}_\alpha \cdot (x_{1,j-1} + c_{1,j}))^{\mathfrak{e}_r}(\beta) = 0 \quad \text{for} \quad j = 1, \dots, m. \tag{4.17}$$

For $k \ge 1$ and $j = 1, \dots, m$, relations (4.16) along with the initial conditions $x_{0,j} = x_{k,0} = 0$ in (4.15) recursively determine $x_{k,j}$ for all indices (k, j) as in (4.5). Since the recursion formula (4.16) and the initial conditions (4.7) for $\Gamma_{k,j}$ are the same as the ones for $x_{k,j}$, we conclude that $x_{k,j} = \Gamma_{k,j}$ for all (k, j) as in (4.5). In particular, $x_{1,j} = \Gamma_{1,j}$ for all $j = 1, \dots, m$ and therefore, conditions (4.8) hold, due to (4.17).

For the converse statement, we first verify that the formula

$$\Delta_{k,j} = (L_\alpha \boldsymbol{\mu}_\alpha \cdot (\Gamma_{k-1,j} - \Gamma_{k,j-1} - c_{k,j}))^{\mathfrak{e}_r}(\beta) \cdot (\boldsymbol{\mu}'_\alpha(\beta))^{-1}, \tag{4.18}$$

holds for all indices (k, j) as in (4.5). To this end, we first multiply both sides in (4.6) by α on the left, then by β on the right and then take the difference of the resulting equalities. By invoking part (2) in Lemma 4.1 with $y = \Gamma_{k,j-1} - \Gamma_{k-1,j} + c_{k,j}$ and part (3) with $y = \Gamma_{k-1,j} + c_{k+1,j}$ and recalling that β commutes with $\boldsymbol{\mu}'_\alpha(\beta)$ (since $\boldsymbol{\mu}'_\alpha \in Z_{\mathbb{F}}[z]$), we arrive at

$$\begin{aligned}\alpha\Gamma_{k,j} - \Gamma_{k,j}\beta = &-(L_\alpha \boldsymbol{\mu}_\alpha \cdot (\Gamma_{k,j-1} - \Gamma_{k-1,j} + c_{k,j}))^{\mathfrak{e}_r}(\beta) \cdot (\boldsymbol{\mu}'_\alpha(\beta))^{-1} \\ &+ \Gamma_{k,j-1} - \Gamma_{k-1,j} + c_{k,j},\end{aligned}$$

which being substituted into the right side of (4.9) implies (4.18).

Let us now assume that equalities (4.8) hold. Letting $k = 1$ in (4.18) and making use of initial conditions (4.7), we get

$$\Delta_{1,j} = -(L_\alpha \boldsymbol{\mu}_\alpha \cdot (\Gamma_{1,j-1} + c_{1,j}))^{\mathfrak{e}_r}(\beta) \cdot (\boldsymbol{\mu}'_\alpha(\beta))^{-1} = 0 \quad (1 \le j \le m) \tag{4.19}$$

where the rightmost equality in (4.19) are justified by (4.8). To proceed by induction, let us assume that $\Delta_{k,j} = 0$ for some (k, j) such that $j \le m$ and $k + j < n$. Writing (4.9) equivalently as

$$\alpha\Gamma_{k,j} - \Gamma_{k,j}\beta = \Gamma_{k,j-1} - \Gamma_{k-1,j} + c_{k,j},$$

we apply part (1) in Lemma 4.1 to $y = \Gamma_{k,j}$ and $\gamma = \Gamma_{k,j-1} - \Gamma_{k-1,j} + c_{k,j}$ arriving at

$$\begin{aligned}(L_\alpha \boldsymbol{\mu}_\alpha \Gamma_{k,j})^{\mathfrak{e}_r}(\beta) &= \Gamma_{k,j}\boldsymbol{\mu}'_\alpha(\beta) + ((L_\alpha \boldsymbol{\mu}_\alpha)' \cdot (\Gamma_{k,j-1} - \Gamma_{k-1,j} + c_{k,j}))^{\mathfrak{e}_r}(\beta) \\ &= (L_\alpha \boldsymbol{\mu}_\alpha \cdot (\Gamma_{k+1,j-1} + c_{k+1,j}))^{\mathfrak{e}_r}(\beta),\end{aligned}$$

where the second equality follows from (4.6). Thus,

$$(L_\alpha \mu_\alpha \cdot (\Gamma_{k,j} - \Gamma_{k+1,j-1} - c_{k+1,j}))^{\mathfrak{e}_r}(\beta) = 0,$$

which together with formula (4.18) (with $k+1$ instead of k) implies

$$\Delta_{k+1,j} = (L_\alpha \mu_\alpha \cdot (\Gamma_{k,j} - \Gamma_{k+1,j-1} - c_{k+1,j}))^{\mathfrak{e}_r}(\beta) \cdot (\mu'_\alpha(\beta))^{-1} = 0.$$

This proves the implication $\Delta_{k,j} = 0 \quad \Rightarrow \quad \Delta_{k+1,j} = 0$ which together with equalities (4.19) as the induction basis completes the proof of (4.9) by the induction argument. □

Remark 4.3 Let us observe that the element $\Gamma_{k,j}$ produced by the recursion (4.6), (4.7) within the index range (4.5), is *independent* of the elements $c_{i,\ell}$ with $\ell > j$ or $i + \ell > k + j + 1$. In other words, the closed formula for $\Gamma_{k,j}$ may contain only the entries $c_{i,\ell}$ from the j leftmost columns and above the $(k+j)$-th counter-diagonal of C.

Theorem 4.4 *Given a matrix $C = [c_{k,j}] \in \mathbb{F}^{n\times m}$ $(n \geq m)$ and algebraic elements $\alpha \sim \beta$, let $\Gamma_{k,j}$ be defined by* (4.6), (4.7) *for all (k, j) as in* (4.5).

1. *Equation* (3.3) *has a solution if and only if equalities* (4.8) *hold.*
2. *If this is the case, a particular solution X to Eq.* (3.3) *can be constructed as follows: pick any $c_{n+i,j}$ $(i, j = 1, \ldots, m)$ and apply the recursion* (4.6), (4.7) *to the extended set $\{c_{i,j}\}$ to produce $\Gamma_{k,j}$ for all*

$$(k, j) \quad \textit{such that} \quad 1 \leq k < n+m, \quad 1 \leq j < m, \quad k+j \leq n+m. \tag{4.20}$$

Then the matrix $X = \left[\Gamma_{k,j}\right] \in \mathbb{F}^{n\times m}$ solves Eq. (3.3).

Proof The necessity of conditions (4.8) was confirmed in Theorem 4.2. Conversely, assuming (4.8) to be in force, let us use the entries $c_{i,j}$ of the matrix C along with the arbitrarily chosen elements $c_{n+i,j}$ $(i, j = 1, \ldots, m)$ to produce via the recursion (4.6), (4.7) the elements $\Gamma_{k,j}$ for all (k, j) as in (4.20). By Remark 4.3, these elements are independent of $c_{n+i,j}$ for $i + j > m + 1$, whereas the elements $\Gamma_{1,1}, \ldots \Gamma_{1,m-1}$ in conditions (4.8) (and hence, the conditions themselves) do not depend on the chosen $c_{n+i,j}$ at all.

We now apply the converse statement in Theorem 4.2 to the extended set $\{\Gamma_{k,j} : 1 \leq j < m, \quad k+j \leq n+m\}$ by just keeping m the same and replacing n by $n+m$. Since the requisite conditions (4.8) are satisfied, we conclude that equalities (4.9) hold for all (k, j) as in (4.20). In particular, they hold for all $k = 1, \ldots, n$ and $j = 1, \ldots, m$. Thus,

$$\alpha\Gamma_{k,j} + \Gamma_{k-1,j} - \Gamma_{k,j}\beta - \Gamma_{k,j-1} = c_{k,j} \quad \text{for} \quad 1 \leq k \leq n; \ 1 \leq j \leq m,$$

which means that the matrix $X = \left[\Gamma_{k,j}\right] \in \mathbb{F}^{n\times m}$ satisfies (3.3). □

We next consider the homogeneous Sylvester equation

$$AX = XB, \qquad A = \begin{bmatrix} \alpha & 0 & \dots & 0 \\ 1 & \alpha & \cdots & 0 \\ \vdots & \ddots & \ddots & \vdots \\ 0 & \dots & 1 & \alpha \end{bmatrix}, \qquad B = \begin{bmatrix} \beta & 1 & \dots & 0 \\ \vdots & \ddots & \ddots & \vdots \\ 0 & \dots & \beta & 1 \\ 0 & \dots & 0 & \beta \end{bmatrix}. \tag{4.21}$$

Theorem 4.5 *Given $n \geq m \geq 1$ and algebraic elements $\alpha \sim \beta \in \mathbb{F}$, a matrix $X \in \mathbb{F}^{n\times m}$ solves* (4.21) *if and only if it is of the Hankel form*

$$X = \left[\gamma_{i+j-1}\right]: \quad \gamma_\ell = 0 \ \ (\ell \leq n-1) \ \ \textit{and} \ \ \alpha\gamma_\ell = \gamma_\ell\beta \ \ (\ell \geq n). \tag{4.22}$$

Proof The fact that any matrix X as in (4.22) satisfies (4.21) is verified directly by comparing the corresponding entries in matrices AX and XB.

Conversely, let us assume that X satisfies (4.21) (i.e., (3.3) with $C = 0$). Then it follows by (the proof of) Theorem 4.2 that, under the conventions (4.15), the entries of X also satisfy equalities (4.16) with $c_{k,j} = c_{k,j+1} = 0$:

$$x_{k,j}\boldsymbol{\mu}'_\alpha(\beta) = (L_\alpha\boldsymbol{\mu}_\alpha x_{k+1,j-1})^{\mathfrak{e}_r}(\beta) - ((L_\alpha\boldsymbol{\mu}_\alpha)'\cdot(x_{k,j-1}-x_{k-1,j}))^{\mathfrak{e}_r}(\beta) \tag{4.23}$$

for $1 \leq j \leq m$ and $1 \leq k \leq n-1$. Again by Theorem 4.2, all the entries above the n-th counter-diagonal (i.e., $x_{k,j}$ with $k + j \leq n$) in any solution X to (4.21) are determined uniquely. Since the zero matrix is a solution of (4.21), it now follows that

$$x_{k,j} = 0 \quad \text{for all} \quad 0 \leq j \leq m; \;\; k+j \leq n. \tag{4.24}$$

Taking (4.24) into account, we equate the corresponding entries (on and below the n-th counter-diagonal) in the matrix equality $AX = BX$ to get

$$\alpha x_{k,j} + x_{k-1,j} = x_{k,j}\beta + x_{k,j-1} \quad (1 \leq j \leq m; \;\; n+1 \leq k+j \leq n+m). \tag{4.25}$$

Assuming that for some $\ell \leq n$, all entries $x_{i,j}$ with $i + j = \ell$ are equal to each other, which can be written equivalently as

$$x_{k-1,j} = x_{k,j-1} = \gamma_{\ell-1}, \quad \text{whenever} \quad k+j = \ell+1. \tag{4.26}$$

we will show that the entries in the next counter-diagonal of X are also equal to each other, i.e.,

$$x_{k,j} = x_{k+1,j-1} = \gamma_\ell, \quad \text{whenever} \quad k+j = \ell+1. \tag{4.27}$$

Indeed, on account of (4.26), we have from (4.23) and (4.25)

$$x_{k,j}\boldsymbol{\mu}'_\alpha(\beta) = (L_\alpha\boldsymbol{\mu}_\alpha x_{k+1,j-1})^{\mathfrak{e}_r}(\beta) \quad \text{and} \quad \alpha x_{k,j} = x_{k,j}\beta, \tag{4.28}$$

whenever $k + j = \ell + 1$. In particular, $\alpha x_{k+1,j-1} = x_{k+1,j-1}\beta$ and therefore, part (1) in Lemma 4.1 applies to $\gamma = 0$ and $y = x_{k+1,j-1}$ and leads us to

$$(L_\alpha \boldsymbol{\mu}_\alpha x_{k+1,j-1})^{\mathfrak{e}_r}(\beta) = x_{k+1,j-1} \cdot \boldsymbol{\mu}'_\alpha(\beta).$$

Combining the latter equality with the first equality in (4.28) and cancelling $\boldsymbol{\mu}'_\alpha(\beta) \neq 0$ we arrive at (4.27). It now follows from the second relation in (4.28) that $\alpha\gamma_\ell = \gamma_\ell\beta$. Since the entries on the $(n-1)$-th counter-diagonal of X are equal (to zero, by (4.24)), we may use the induction argument to justify the Hankel structure (4.22) of X. □

Throughout this section, we assumed that $n \geq m$. The case $n \leq m$ can be handled similarly: multiplying both sides in (4.11) by $\widetilde{G} = (R_\beta\boldsymbol{\mu}_\alpha) \cdot I_m - (R_\beta\boldsymbol{\mu}_\alpha)' \cdot F_m^\top$ on the right and then evaluating the resulting identity at α on the left, eventually lead us to n conditions (similar to those in (4.8)) given in terms of left evaluations at α (rather than right evaluations at β) that turn out to be necessary and sufficient for the *singular* equation (3.3) to have a solution. As for Theorem 4.5, it still holds true for $n \leq m$ once we replace n by m in (4.22). Combining both cases we conclude that $X \in \mathbb{F}^{n\times m}$ satisfies (4.21) if and only it is of Hankel structure with all non-zero entries in it last $\min(m, n)$ counter-diagonals and intertwining α and β.

Remark 4.6 By Theorems 4.4 and 4.5, whenever algebraic α and β are similar, Eq. (3.3) either has infinitely many solutions or has no solution. This justifies the "only if" part in Corollary 3.2.

Being specified to the case $n = m = 1$, Corollary 3.2 and Theorem 4.4 recover the results in [3, 4] concerning the scalar Sylvester equation

$$\alpha x - x\beta = \gamma, \qquad \alpha, \beta, \gamma \in \mathbb{F} \tag{4.29}$$

with an algebraic α. More directly, writing (4.29) in the form $\boldsymbol{\rho}_\alpha x - x\boldsymbol{\rho}_\beta = -\gamma$ (see (4.11)), multiplying both sides by $L_\alpha\boldsymbol{\mu}_\alpha$ on the left and evaluating the resulting identity at β on the right, we get

$$x\boldsymbol{\mu}_\alpha(\beta) = -(L_\alpha\boldsymbol{\mu}_\alpha\gamma)^{\mathfrak{e}_r}(\beta), \tag{4.30}$$

which in case $\beta \not\sim \alpha$, implies the formula $x = -(L_\alpha\boldsymbol{\mu}_\alpha\gamma)^{\mathfrak{e}_r}(\beta) \cdot (\boldsymbol{\mu}_\alpha(\beta))^{-1}$ (more compact than (3.1)) for the unique solution to Eq. (4.29).

If $\alpha \sim \beta$ (i.e., $\boldsymbol{\mu}_\alpha(\beta) = 0$) then (4.30) implies $\big(L_\alpha\boldsymbol{\mu}_\alpha\gamma\big)^{\mathfrak{e}_r}(\beta) = 0$. The latter condition is identical to that in (4.8) (with $j = 1$ and $\gamma = c_{11}$) and hence, is necessary and sufficient for Eq. (4.29) to have a solution. The formula $x = -((L_\alpha\boldsymbol{\mu}_\alpha)'\gamma)^{\mathfrak{e}_r}(\beta)$ for a particular solution follows from (4.16) (for $k = j = 1$ and with $c_{11} = \gamma$ and $c_{12} = 0$).

The case where α and β are both transcendental is less clear at the moment. Even with $\alpha \not\sim \beta$, Eq. (4.29) may have no solutions; see [1]. We are not aware of solvability or uniqueness criteria for this case.

Acknowledgment The project was partially supported by Simons Foundation grant 524539.

References

1. R.F. Berghout, On the equation $ax - xb = c$. J. Austral. Math. Soc. **8**, 383–384 (1968)
2. P.M. Cohn, The range of derivations on a skew field and the equation $ax - xb = c$. J. Indian Math. Soc. **37**, 61–69 (1973)
3. N. Jacobson, The equation $x' \equiv xd - dx = b$. Bull. Am. Math. Soc. **50**, 902–905 (1944)
4. R.E. Johnson, On the equation $\chi\alpha = \gamma\chi + \beta$ over an algebraic division ring. Bull. Am. Math. Soc. **50**, 202–207 (1944)
5. T.Y. Lam, A. Leroy, Wedderburn polynomials over division rings. I. J. Pure Appl. Algebra **186**(1), 43–76 (2004)
6. P. Lancaster, Explicit solutions of linear matrix equations. SIAM Rev. **12**, 544–566 (1970)

Hypercomplex Iterated Function Systems

Peter Massopust

Abstract We introduce the novel concept of hypercomplex iterated function system (IFS) on the complete metrizable space $(\mathbb{A}_{n+1}^k, d)$ and define its hypercomplex attractor. Systems of hypercomplex function systems arising from hypercomplex IFSs and their backward trajectories are also introduced and it is shown that the attractors of such backward trajectories possess different local (fractal) shapes.

1 Preliminaries and Notation

This paper intends to merge two areas of mathematics: Clifford Algebra and Analysis and Fractal Geometry. The former has a long successful history of extending concepts from classical analysis and function theory to a noncommutative division algebra setting and the latter has developed into an area of bourgeoning research over the last few decades. Attractors of so-called iterated function systems (IFSs) are fractal sets in the sense given first by B. Mandelbrot [1]. Fractals generated by IFSs have the following significant approximation property, called the Collage Theorem[2]: Every nonempty compact subset of a complete metric space can be approximated arbitrarily close (in the sense of the Hausdorff metric) by the attractor of an IFS, that is, by a fractal set. (For the precise mathematical statement, we refer to [2].) This property has important and fundamental implications in image compression and image analysis; cf. for instance, [3]. Moreover, it has initiated the construction of fractal functions and fractal surfaces and their application to approximation and interpolation theory. A collection of related results can be found in, e.g., [4]. This monograph also provides the fundamental result by D. Hardin [5], namely, that every continuous compactly supported and refinable function, i.e., every wavelet, is a piecewise fractal function.

P. Massopust (✉)
Technical University of Munich, Center of Mathematics, Garching by Munich, Germany
e-mail: massopust@ma.tum.de

P. Cerejeiras et al. (eds.), *Current Trends in Analysis, its Applications and Computation*, Research Perspectives,
https://doi.org/10.1007/978-3-030-87502-2_59

Here, we initiate the inclusion of fractal techniques into the Clifford setting. The non-commutativity generates more intricate fractal patterns. The structure of this paper is as follows. After a short and terse introduction to Clifford algebras and IFSs, we define hypercomplex IFSs and their attractors. The final section extends these concepts to systems of function systems as defined in [6].

1.1 A Brief Introduction to Clifford Algebras

In this section, we give a terse introduction to the concept of Clifford algebra, mainly, to set notation and terminology, and refer the reader to, for instance, [7–9]. To this end, denote by $\{e_1, \ldots, e_n\}$ the canonical basis of the Euclidean vector space $\mathbb{R}^n$. The real Clifford algebra $C\ell(n)$ generated by $\mathbb{R}^n$ is defined by the multiplication rules $e_i e_j + e_j e_i = -2\delta_{ij}$, $i, j \in \{1, \ldots, n\} =: \mathbb{N}_n$, where δ_{ij} is the Kronecker symbol. The dimension of $C\ell(n)$ regarded as a real vector space is 2^n.

An element $x \in C\ell(n)$ can be represented in the form $x = \sum\limits_A x_A e_A$ with $x_A \in \mathbb{R}$ and $\{e_A : A \subseteq \mathbb{N}_n\}$, where $e_A := e_{i_1} e_{i_2} \cdots e_{i_m}$, $1 \le i_1 < \cdots < i_m \le n$, and $e_\emptyset =: e_0 := 1$. A conjugation on Clifford numbers is defined by $\overline{x} := \sum\limits_A x_A \overline{e}_A$ where $\overline{e}_A := \overline{e}_{i_m} \cdots \overline{e}_{i_1}$ with $\overline{e}_i := -e_i$ for $i \in \mathbb{N}_n$, and $\overline{e}_0 := e_0 = 1$. The Clifford norm of the Clifford number $x = \sum\limits_A x_A e_A$ is $|x| := \sqrt{\sum\limits_A |x_A|^2}$.

An important subspace of $C\ell(n)$ is the space of hypercomplex numbers or paravectors. These are Clifford numbers of the form $x = x_0 + \sum\limits_{i=1}^n x_i e_i$. The subspace of hypercomplex numbers is denoted by $\mathbb{A}_{n+1} := \operatorname{span}_{\mathbb{R}}\{e_0, e_1, \ldots, e_n\} = \mathbb{R} \oplus \mathbb{R}^n$. Given a Clifford number $x \in C\ell(n)$, we assign to x its hypercomplex or paravector part by means of the mapping $\pi : C\ell(n) \to \mathbb{A}_{n+1}, x \mapsto x_0 + \sum\limits_{i=1}^n x_i e_i$.

Note that each hypercomplex number x can be identified with an element $(x_0, x_1, \ldots, x_n) =: (x_0, \boldsymbol{x}) \in \mathbb{R} \times \mathbb{R}^n$. For many applications in Clifford theory, one therefore identifies $\mathbb{A}_{n+1}$ with $\mathbb{R}^{n+1}$.

The scalar part, Sc, and vector part, Vec, of a hypercomplex number $\mathbb{A}_{n+1} \ni x = x_0 + \sum\limits_{i=1}^n x_i e_i$ is given by x_0 and $\boldsymbol{x} = \sum\limits_{i=1}^n x_i e_i$, respectively.

The conjugate $\overline{x}$ of the hypercomplex number $x = s + \boldsymbol{x}$ is the hypercomplex number $\overline{x} = s - \boldsymbol{x}$. The Clifford norm of $x \in \mathbb{A}_{n+1}$ is given by $|x| = \sqrt{x\overline{x}} = \sqrt{s^2 + |\boldsymbol{x}|^2} = \sqrt{s^2 + \sum\limits_{i=1}^n x_i^2}$.

We denote by $M_k(\mathbb{A}_{n+1})$ the right module of $k \times k$-matrices over $\mathbb{A}_{n+1}$. Every element $H = (H_{ij})$ of $M_k(\mathbb{A}_{n+1})$ induces a right linear transformation $L : \mathbb{A}_{n+1}^k \to C\ell(n)^k$ via $L(x) = Hx$ defined by $L(x)_i = \sum_{j=1}^{k} H_{ij}x_j$, $H_{ij} \in \mathbb{A}_{n+1}$. To obtain an endomorphism $\mathcal{L} : \mathbb{A}_{n+1}^k \to \mathbb{A}_{n+1}^k$, we set $\mathcal{L}(x)_i := \pi(L(x)_i)$, $i = 1, \ldots, k$. In this case, we write $\mathcal{L} = \pi \circ L$. For example, if $n := 3$ (the case of real quaternions) $L : \mathbb{A}_4^k \to \mathbb{A}_4^k$ and thus $\mathcal{L} = L$.

A function $f : \mathbb{A}_{n+1} \to \mathbb{A}_{n+1}$ is called a hypercomplex or paravector-valued function. Any such function is of the form $f(x_0 + \boldsymbol{x}) = f_0(x_0, |\boldsymbol{x}|) + \omega(\boldsymbol{x}) f_1(x_0, |\boldsymbol{x}|)$, where $f_0, f_1 : \mathbb{R} \times \mathbb{R}^n \to \mathbb{R}$ and $\omega(\boldsymbol{x}) := \frac{\boldsymbol{x}}{|\boldsymbol{x}|} \in \mathbb{S}^n$ with $\mathbb{S}^n$ denoting the unit sphere in $\mathbb{R}^n$. For some properties of hypercomplex functions, see, for instance[10].

1.2 Iterated Function Systems and Their Attractors

Let (X, d) be a complete metrizable space with metric d. For a map $f : X \to X$, we define the Lipschitz constant associated with f by

$$\mathrm{Lip}(f) = \sup_{x,y \in X, x \neq y} \frac{d\big(f(x), f(y)\big)}{d(x, y)}.$$

A map f is said to be *Lipschitz* if $\mathrm{Lip}(f) < +\infty$ and a *contraction* if $\mathrm{Lip}(f) < 1$.

Definition 1.1 Let (X, d) be a complete metrizable space and let $\mathcal{F}$ be a finite set of contractions on X. Then the pair $(X, \mathcal{F})$ is called an iterated function system (IFS) on X.

With the finite set of contractions $\mathcal{F}$ on X, one associates a set-valued operator, again denoted by $\mathcal{F}$, acting on the hyperspace $\mathscr{H}(X)$ of nonempty compact subsets of X endowed with the Hausdorff-Pompeiu metric $d_{\mathscr{H}}$ by

$$\mathcal{F}(E) := \bigcup_{f \in \mathcal{F}} f(E), \qquad E \in \mathscr{H}(X).$$

The Hausdorff-Pompeiu metric $d_{\mathscr{H}}$ is defined by

$$d_{\mathscr{H}}(S_1, S_2) := \max\{d(S_1, S_2), d(S_2, S_1)\},$$

where $d(S_1, S_2) := \sup_{x \in S_1} d(x, S_2) := \sup_{x \in S_1} \inf_{y \in S_2} d(x, y)$.

The completeness of (X, d) implies that the set-valued operator $\mathcal{F}$ is contractive on the complete metrizable space $(\mathscr{H}(X), d_{\mathscr{H}})$ with Lipschitz constant $\operatorname{Lip}\mathcal{F} = \max\{\operatorname{Lip}(f) : f \in \mathcal{F}\}$. Therefore, by the Banach Fixed Point Theorem, $\mathcal{F}$ has a unique fixed point in $\mathscr{H}(X)$. This fixed point if called the *attractor* of or the *fractal (set)* generated by the IFS $(X, \mathcal{F})$. The attractor or fractal F satisfies the self-referential equation

$$F = \mathcal{F}(F) = \bigcup_{f \in \mathcal{F}} f(F), \tag{1.1}$$

i.e., F is made up of a finite number of images of itself. Equation (1.1) reflects the fractal nature of F showing that it is as an object of immense geometric complexity.

The proof of the Banach Fixed Point Theorem also shows that the fractal F can be iteratively obtained via the following procedure: Choose an arbitrary $F_0 \in \mathscr{H}(X)$ and set

$$F_n := \mathcal{F}(F_{n-1}), \qquad n \in \mathbb{N}. \tag{1.2}$$

Then $F = \lim\limits_{n\to\infty} F_n$, where the limit is taken in the Hausdorff-Pompeiu metric $d_{\mathscr{H}}$.

In Fig. 1, two examples of fractal sets are displayed. For more details about IFSs and fractals and their properties, we refer the interested reader to the large literature on these topics and list only two references [2, 4] pertaining to the present exhibition.

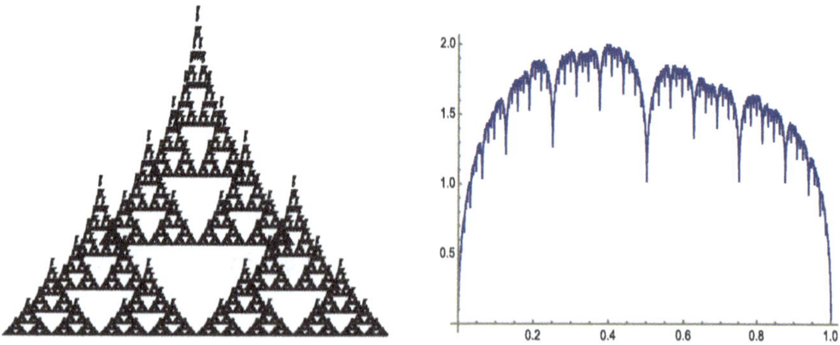

Fig. 1 Left: A fractal set in $X := [0, 1]^2$ generated by the maps $f_1(x, y) := (\frac{1}{2}x, \frac{1}{2}y)$, $f_2(x, y) := (\frac{1}{2}(x+1), \frac{1}{2}y)$, and $f_3(x, y) := (\frac{1}{2}(x+\frac{1}{2}), \frac{3}{4}y+\frac{\sqrt{3}}{4})$. Right: The graph of a *fractal function* in $X := [0, 1]\times[0, 3]$ generated by the maps $f_1(x, y) := (\frac{1}{2}x, x+\frac{3}{4}y)$ and $f_2(x, y) := (\frac{1}{2}(x+1), x^2+\frac{3}{4}y)$

2 Hypercomplex IFSs

Let $k \in \mathbb{N}$ and consider the set $\mathbb{A}_{n+1}^k := \bigtimes_{i=1}^{k} \mathbb{A}_{n+1}$. We represent elements $\xi \in \mathbb{A}_{n+1}^k$ as column vectors. The hypercomplex conjugate * of $\xi \in \mathbb{A}_{n+1}^k$ is defined by

$$\begin{pmatrix} \xi_1 \\ \vdots \\ \xi_k \end{pmatrix}^* := \begin{pmatrix} \overline{\xi_1} & \cdots & \overline{\xi_k} \end{pmatrix}.$$

Similarly, for any matrix $H = (H_{ij})$ over $\mathbb{A}_{n+1}$, we define $(H_{ij})^* := (\overline{H_{ji}})$. The norm of $\xi \in \mathbb{A}_{n+1}^k$ is defined to be

$$\|\xi\| := \sqrt{\xi^* \xi} = \sqrt{\sum_{i=1}^{k} |\xi_i|^2}. \tag{2.1}$$

Then $(\mathbb{A}_{n+1}^k, d)$ is a complete metrizable space with metric $d(\xi, \eta) := \|\xi - \eta\|$.

We define a norm on $M_k(\mathbb{A}_{n+1})$ as follows. For $H \in M_k(\mathbb{A}_{n+1})$, set

$$|||H||| := \sup \left\{ \frac{\|H\xi\|}{\|\xi\|} : 0 \neq \xi \in \mathbb{A}_{n+1}^k \right\}.$$

Definition 2.1 Let $k \in \mathbb{N}$ and let $\mathcal{F} := \{f_1, \ldots, f_m\}$ be a finite collection of contractive hypercomplex functions on $\mathbb{A}_{n+1}^k$. Then the pair $(\mathbb{A}_{n+1}^k, \mathcal{F})$ is called a *hypercomplex IFS* (on $\mathbb{A}_{n+1}^k$).

Definition 2.2 Let $k \in \mathbb{N}$. An element $F \in \mathscr{H}(\mathbb{A}_{n+1}^k)$ is called a *hypercomplex attractor* of or the *hypercomplex fractal (set)* generated by the hypercomplex IFS $(\mathbb{A}_{n+1}^k, \mathcal{F})$ if F satisfies the self-referential equation

$$F = \mathcal{F}(F) = \bigcup_{i=1}^{m} f_i(F). \tag{2.2}$$

Note that by the Banach Fixed Point Theorem F as defined above is unique.

Remark 2.3 Although the *point sets* $\mathbb{A}_{n+1}^k$ and $(\mathbb{R}^{n+1})^k$ are isomorphic under an obvious bijection, the difference in the *non-commutative algebraic structure* yields a broader class of attractors. (See the examples below.)

As an example of a hypercomplex IFS, we consider right linear maps $L_i : \mathbb{A}_{n+1}^k \to C\ell(n)^k$ and define right affine maps by

$$A_i(\xi) := L_i(\xi) + b_i = H_i \xi + b_i,$$

with $H_i \in M_k(\mathbb{A}_{n+1})$ and $b_i \in \mathbb{A}^k_{n+1}$, $i \in \mathbb{N}_m$. The right affine maps A_i generate hypercomplex functions f_i via

$$f_i(\xi) := \pi(L_i(\xi)) + b_i. \tag{2.3}$$

It follows immediately from the definition of contraction applied to $\mathbb{A}^k_{n+1}$, that each f_i is contractive provided that $h := \max\{|||H_i||| : i \in \mathbb{N}_m\} < 1$. The unique fractal generated by the hypercomplex IFS $(\mathbb{A}^k_{n+1}, \mathcal{F})$ is then given by the nonempty compact subset $F \subset \mathbb{A}^k_{n+1}$ satisfying

$$F = \bigcup_{i \in \mathbb{N}_m} \pi(L_i(F)) + b_i.$$

Example Let $n := 3$ and $k := 2$. Then $\mathbb{A}_4$ can be identified with the noncommutative associative division algebra $\mathbb{H}$ of quaternions with $e_0 = 1$, $e_i e_j + e_j e_i = -2\delta_{ij}$, and $e_i e_j = \varepsilon_{ijk} e_k$, $i, j, k = 1, 2, 3$. Here, ε_{ijk} denotes the Levi-Cività symbol. We note that $\mathbb{H} \cdot \mathbb{H} = \mathbb{H}$ and therefore $\pi(L_i) = L_i$.

On $\mathbb{H}^2$ we consider the three affine mappings

$$f_i(\xi, \eta) := \begin{pmatrix} q & 0 \\ 0 & q \end{pmatrix} \begin{pmatrix} \xi \\ \eta \end{pmatrix} + b_i,$$

with $b_1 := 0$, $b_2 := \begin{pmatrix} 1-q \\ 0 \end{pmatrix}$, $b_3 := \begin{pmatrix} \frac{1}{2}(1-q) \\ 1-q \end{pmatrix}$, and $q := 0.3e_0 - 0.1e_1 + 0.4e_2 - 0.2e_3$. Note that $|q| = \sqrt{0.3} < 1$. Hence, the quaternionic IFS $(\mathbb{H}^2, \mathcal{F})$ with $\mathcal{F} = \{f_1, f_2, f_3\}$, possesses a unique attractor $F \in \mathscr{H}(\mathbb{H}^2)$.

In Fig. 2, some projections of the attractor F onto subspaces of $\mathbb{H}^2$ are displayed.

3 Systems of Hypercomplex Function Systems

Employing the setting first introduced in [6], then generalized in [11], and finally applied to non-stationary IFSs in [12], we will replace the single set-valued map $\mathcal{F}$ in a hypercomplex IFS by a sequence of function systems consisting of different families $\mathcal{F}$ in order to define an iterative process $\{F_n\}_{n\in\mathbb{N}}$ with initial $F_0 \in \mathscr{H}(\mathbb{A}^k_{n+1})$ as in (1.2).

To this end, consider the complete metrizable space $(\mathbb{A}^k_{n+1}, d)$ and let $\{\mathcal{T}_\ell\}_{\ell\in\mathbb{N}}$ be a sequence of transformations on $\mathbb{A}^k_{n+1}$, i.e., $\mathcal{T}_\ell : \mathbb{A}^k_{n+1} \to \mathbb{A}^k_{n+1}$.

Definition 3.1 A subset $\mathscr{I}$ of $\mathbb{A}^k_{n+1}$ is called a hypercomplex invariant set of the sequence $\{\mathcal{T}_\ell\}_{\ell\in\mathbb{N}}$ if

$$\forall \ell \in \mathbb{N}\, \forall x \in \mathscr{I} : \mathcal{T}_\ell(x) \in \mathscr{I}.$$

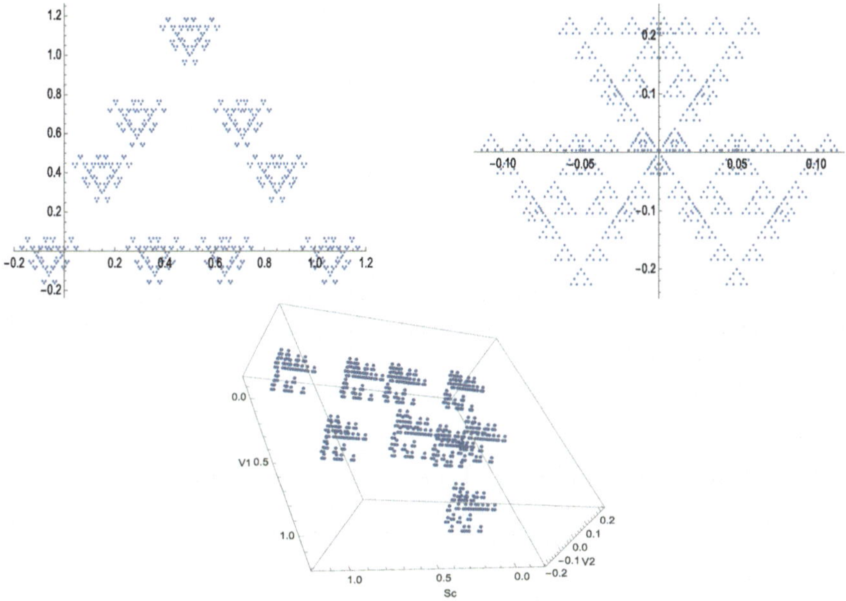

Fig. 2 Left: The projection of F onto the $e_0 - e_1$-plane. Middle: The projection of F onto the $e_1 - e_3$-plane. Right: The projection of F onto the hyperplane spanned by $\{e_0, e_1, e_2\}$

A criterion for obtaining a hypercomplex invariant domain for a sequence $\{\mathcal{T}_\ell\}_{\ell\in\mathbb{N}}$ of transformations on a complete metrizable space is the following which first appeared in [6]. We will state the result for our setting.

Proposition 3.2 *Let $\{\mathcal{T}_\ell\}_{\ell\in\mathbb{N}}$ be a sequence of transformations on the complete metrizable space $(\mathbb{A}_{n+1}^k, d)$. Assume that there exists an $x_0 \in \mathbb{A}_{n+1}^k$ such that for all $x \in \mathbb{A}_{n+1}^k$*

$$d(\mathcal{T}_\ell(x), x_0) \leq s\, d(x, x_0) + M, \tag{3.1}$$

for some $s \in [0, 1)$ and $M > 0$. Then any ball $B_r(x_0) := \{x \in \mathbb{A}_{n+1}^k : |x - x_0| < r\}$ with radius $r > M/(1-s)$ is a hypercomplex invariant set for $\{\mathcal{T}_\ell\}_{\ell\in\mathbb{N}}$.

Proof The proof of this statement follows directly from [6, Lemma 3.7] and [6, Remark 3.8]. □

In case the transformations $\mathcal{T}_\ell$ are maps of the form (2.3) then condition (3.1) is satisfied with $x_0 = 0$, $M := \sup_{\ell\in\mathbb{N}} \|b_\ell\| < \infty$, and any H_ℓ with $\sup_{\ell\in\mathbb{N}} |||H_\ell||| =: s < 1$. For, if we choose $x_0 := 0$,

$$\begin{aligned} d(\mathcal{T}_\ell(x), 0) &= \|\mathcal{T}_\ell(x)\| = \|H_\ell x + b_\ell\| \leq \|H_\ell x\| + \|b_\ell\| \\ &\leq |||H_\ell||| \cdot \|x\| + \|b_\ell\|. \end{aligned}$$

Hence, every ball centered at the origin of $\mathbb{A}_{n+1}^k$ of radius greater than $M/(1-s)$ is a hypercomplex invariant set for $\{\mathcal{T}_\ell\}_{\ell\in\mathbb{N}}$.

Now suppose that $\{\mathcal{F}_\ell\}_{\ell\in\mathbb{N}}$ is a sequence of set-valued maps $\mathcal{F}_\ell : \mathscr{H}(\mathbb{A}_{n+1}^k) \to \mathscr{H}(\mathbb{A}_{n+1}^k)$ defined by

$$\mathcal{F}_\ell(F_0) := \bigcup_{i=1}^{n_\ell} f_{i,\ell}(F_0), \quad F_0 \in \mathscr{H}(\mathbb{A}_{n+1}^k), \tag{3.2}$$

where $\mathcal{F}_\ell = \{f_{i,\ell} : i \in \mathbb{N}_{n_\ell}\}$ is a family of hypercomplex contractions constituting a hypercomplex IFS on the complete metrizable space $(\mathbb{A}_{n+1}^k, d)$. Setting $s_{i,\ell} := \text{Lip}(f_{i,\ell})$, we obtain that $\text{Lip}(\mathcal{F}_\ell) = \max\{s_{i,\ell} : i \in \mathbb{N}_{n_\ell}\} < 1$.

The following definition is taken from [6, Section 4] using our setting with $X = \mathbb{A}_{n+1}^k$.

Definition 3.3 Let $F_0 \in \mathscr{H}(\mathbb{A}_{n+1}^k)$. For any $\ell \in \mathbb{N}$, the sequence

$$\Psi_\ell(F_0) := \mathcal{F}_1 \circ \mathcal{F}_2 \circ \cdots \circ \mathcal{F}_\ell(F_0) \tag{3.3}$$

is called the backward trajectory of F_0.

Backward trajectories converge under rather mild conditions and their limits generate new types of fractal sets. For more details, we refer the interested reader to [6, 11, 12].

The next theorem summarizes the convergence result for backward trajectories.

Theorem 3.4 *Let $\{\mathcal{F}_\ell\}_{\ell\in\mathbb{N}}$ be a family of set-valued maps of the form* (3.2) *whose elements are collections $\mathcal{F}_\ell = \{f_{i,\ell} : i \in \mathbb{N}_{n_\ell}\}$ of hypercomplex contractions constituting hypercomplex IFSs on the complete metrizable space $(\mathbb{A}_{n+1}^k, d)$. Suppose that*

(i) *there exists a nonempty closed hypercomplex invariant set $\mathscr{I} \subseteq \mathbb{A}_{n+1}^k$ for $\{f_{i,\ell}\}$, $i \in \mathbb{N}_{n_\ell}$, $\ell \in \mathbb{N}$,*
(ii) *and*

$$\sum_{\ell=1}^{\infty} \prod_{j=1}^{\ell} \text{Lip}(\mathcal{F}_j) < \infty. \tag{3.4}$$

Then the backward trajectories $\{\Psi_\ell(F_0)\}$ converge for any initial $F_0 \subseteq \mathscr{I}$ to a unique hypercomplex attractor $F \subseteq \mathscr{I}$.

Proof The proof can be found in [6]. Note that, as the proof only involves point sets and moduli of numbers, it immediately applies to the hypercomplex setting. □

Example Here, we take $n := 3$ and $k := 1$ and write $\mathbb{H}$ for $\mathbb{A}_4^1$. Define a family $\{\mathcal{F}_\ell\}_{\ell \in \mathbb{N}}$ of set-valued epimorphism on $\mathscr{H}$ whose members are as follows:

$$\mathcal{F}_\ell := \begin{cases} \mathcal{F}_1, & 10(j-1) < \ell \leq 10j - 5, \\ \mathcal{F}_2, & 10j - 5 < \ell \leq 10j, \end{cases} \qquad j \in \mathbb{N},$$

where

$$\mathcal{F}_1 := \{q_1 x, q_1 x + \mathbb{1} - q_1\},$$
$$\mathcal{F}_2 := \{0.7\hat{q}_2 x \hat{q}_2 + 0.1, 0.7\hat{q}_2 x \hat{q}_2 - 0.1\},$$

with $q_1 := 0.75e_0$, $\mathbb{1} := e_0 + e_1 + e_2 + e_3$, and $\hat{q}_2 := \sqrt{\frac{10}{3}}(0.3e_0 - 0.1e_1 + 0.4e_2 - 0.2e_3)$.

Figure 3 displays the projection of the attractor F of the backward trajectory $\{\Psi_\ell\}_{\ell \in \mathbb{N}}$ onto the $e_0 - e_1$-plane at two different levels.

This example shows that hypercomplex attractors generated by backwards trajectories exhibit more flexibility in their shapes and that a proper choice of IFSs reveals different local behavior. This is due to the fact that in the sequence

$$\mathcal{F}_1 \circ \mathcal{F}_2 \circ \cdots \mathcal{F}_{\ell-1} \circ \mathcal{F}_\ell(F_0), \qquad F_0 \in \mathscr{H}(\mathbb{A}_{n+1}^k),$$

the global shape of the hypercomplex attractor is determined by the initial maps $\mathcal{F}_1 \circ \mathcal{F}_2 \circ \cdots$ whereas the local shape is determined by the terminal maps $\mathcal{F}_{\ell-1} \circ \mathcal{F}_\ell \circ \cdots$. In addition, the non-commutative algebraic structure of $\mathbb{H}$ yields a broader class of attractors than in the case $\mathbb{R}^4$.

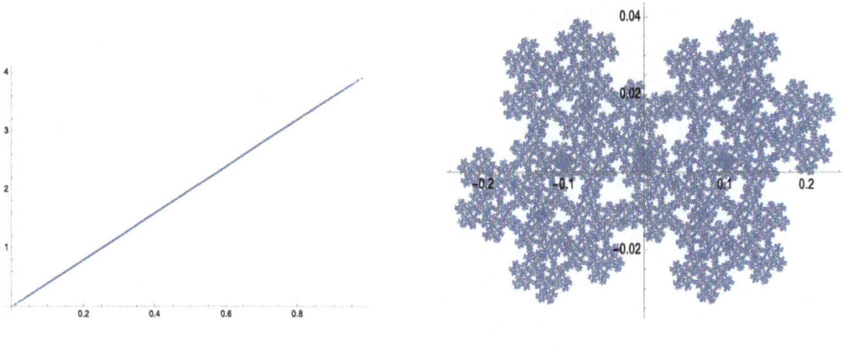

Fig. 3 The projection of the attractor F of the backward trajectory $\{\Psi\}_{k \in \mathbb{N}}$ onto the $e_0 - e_1$-plane at two different levels. The attractor is smooth at one level (left) and fractal at another (right)

References

1. B. Mandelbrot, *The Fractal Geometry of Nature* (W.H.Freeman, New York, 1977)
2. M.F. Barnsley, *Fractals Everywhere* (Dover, Boston, 2012)
3. M.F. Barnsley, L.P. Hurd, *Fractal Image Compression* (AK Peters, Wellesley, 1993)
4. P. Massopust, *Fractal Functions, Fractal Surfaces, and Wavelets*, 2nd edn. (Academic, San Diego, 2016)
5. D. Hardin, Wavelets are piecewise fractal functions, in *Fractals in Multimedia*, ed. by M.F. Barnsley, D. Saupe, E.R. Vrscay (Springer, New York, 2012), pp.121–135
6. D. Levin, N. Dyn, P. Viswanathan, Non-stationary versions of fixed point theory, with applications to fractals and subdivision. J. Fixed Point Theory Appl. **21**(1), 1–26 (2019)
7. R. Abłamowicz, G. Sobczyk, *Lectures on Clifford (Geometric) Algebras and Applications* (Birkhäuser, New York, 2004)
8. R. Delanghe, F. Sommen, V. Souček, *Clifford Algebra and Spinor-Valued Functions. A Function Theory for the Dirac Operator* (Springer, Dordrecht, 1992)
9. I.L. Kantor, A.S. Solodovnik, *Hypercomplex Numbers. An Elementary Introduction to Algebras* (Springer, New York, 1989)
10. W. Sprössig, On Operators and Elementary Functions in Clifford Analysis. Z. Anal. Anwendung **19**(2), 349–366 (1999)
11. N. Dyn, D. Levin, P. Massopust, Attractors of trees of maps and of sequences of maps between spaces with applications to subdivision. J. Fixed Point Theory Appl. **22**(1), 1–24 (2020)
12. P. Massopust, Non-stationary fractal interpolation. Mathematics **7**, 666 (2019). https://doi.org/10.3390/math7080666

Part XIII
Recent Progress in Evolution Equations

Compactness of Localization Operators on Modulation Spaces of ω-Tempered Distributions

Chiara Boiti and Antonino De Martino

Abstract We give sufficient conditions for compactness of localization operators on modulation spaces $\mathbf{M}^{p,q}_{m_\lambda}(\mathbb{R}^d)$ of ω-tempered distributions whose short-time Fourier transform is in the weighted mixed space $L^{p,q}_{m_\lambda}$ for $m_\lambda(x) = e^{\lambda\omega(x)}$.

In this paper we study some properties of localization operators, which are pseudo-differential operators of time-frequency analysis suitable for applications to the reconstruction of signals, because they allow to recover a filtered version of the original signal. To introduce the problem, let us recall the *translation* and *modulation* operators

$$T_x f(y) = f(y-x), \quad M_\xi f(y) = e^{iy\cdot\xi} f(y), \qquad x, y \in \mathbb{R}^d,$$

and, for a window function $\psi \in L^2(\mathbb{R}^d)$, the *short-time Fourier transform* (briefly STFT) of a function $f \in L^2(\mathbb{R}^d)$

$$V_\psi f(z) = \langle f, M_\xi T_x \psi \rangle = \int_{\mathbb{R}^d} f(y)\overline{\psi(y-x)}e^{-iy\cdot\xi}\, dy, \qquad z = (x,\xi) \in \mathbb{R}^{2d}.$$

With respect to the inversion formula for the STFT (see [13, Cor. 3.2.3])

$$f = \frac{1}{(2\pi)^d \langle \gamma, \psi \rangle} \int_{\mathbb{R}^{2d}} V_\psi f(x,\xi) M_\xi T_x \gamma\, dx d\xi,$$

C. Boiti (✉)
Dipartimento di Matematica e Informatica, Università di Ferrara, Ferrara, Italy
e-mail: chiara.boiti@unife.it

A. De Martino
Dipartimento di Matematica, Politecnico di Milano, Milano, Italy
e-mail: antonino.demartino@polimi.it

P. Cerejeiras et al. (eds.), *Current Trends in Analysis, its Applications and Computation*, Research Perspectives,
https://doi.org/10.1007/978-3-030-87502-2_60

which gives a reconstruction of the signal f, the localization operator, as defined in (0.2), modifies $V_\psi f(x,\xi)$ by multiplying it by a suitable $a(x,\xi)$ before reconstructing the signal, so that a filtered version of the original signal f is recovered.

Another important operator in time-frequency analysis that we shall need in the following is the *cross-Wigner transform* defined, for $f, g \in L^2(\mathbb{R}^d)$, by

$$\mathrm{Wig}(f,g)(x,\xi) = \int_{\mathbb{R}^d} f(x+\frac{t}{2})\overline{g(x-\frac{t}{2})}e^{-i\xi\cdot t}\,dt \qquad x,\xi \in \mathbb{R}^d.$$

The *Wigner transform* of f is then defined by $\mathrm{Wig}\, f := \mathrm{Wig}(f,f)$.

The above Fourier integral operators, with standard generalizations to more general spaces of functions or distributions, have been largely investigated in time-frequency analysis. In particular, results about boundedness or compactness related to the subject of this paper can be found, for instance, in [1, 7, 10–12, 16, 17].

Inspired by Cordero and Gröchenig [7] and Fernández and Galbis [10], our aim in this paper is to study boundedness of localization operators on modulation spaces in the setting of ω-tempered distributions, for a weight functions ω defined as below:

Definition 0.1 A *non-quasianalytic subadditive weight function* is a continuous increasing function $\omega : [0,+\infty) \to [0,+\infty)$ satisfying the following properties:

(α) $\omega(t_1+t_2) \leq \omega(t_1)+\omega(t_2), \qquad \forall t_1, t_2 \geq 0;$
(β) $\int_1^{+\infty} \frac{\omega(t)}{t^2}\,dt < +\infty;$
(γ) $\exists A \in \mathbb{R}, B > 0$ s.t $\omega(t) \geq A + B\log(1+t), \qquad \forall t \geq 0;$
(δ) $\varphi_\omega(t) := \omega(e^t)$ is convex.

We then consider $\omega(\xi) := \omega(|\xi|)$ for $\xi \in \mathbb{C}^d$.

Definition 0.2 The space $\mathcal{S}_\omega(\mathbb{R}^d)$ is defined as the set of all $u \in L^1(\mathbb{R}^d)$ such that $u, \hat{u} \in C^\infty(\mathbb{R}^d)$ and

(i) $\forall \lambda > 0, \alpha \in \mathbb{N}_0^d$: $\sup_{x\in\mathbb{R}^d} e^{\lambda\omega(x)}|D^\alpha u(x)| < +\infty,$
(ii) $\forall \lambda > 0, \alpha \in \mathbb{N}_0^d$: $\sup_{\xi\in\mathbb{R}^d} e^{\lambda\omega(\xi)}|D^\alpha \hat{u}(\xi)| < +\infty,$

where $\mathbb{N}_0 := \mathbb{N} \cup \{0\}$.

Note that for $\omega(t) = \log(1+t)$ we obtain the classical Schwartz class $\mathcal{S}(\mathbb{R}^d)$, while in general $\mathcal{S}_\omega(\mathbb{R}^d) \subseteq \mathcal{S}(\mathbb{R}^d)$. For more details about the spaces $\mathcal{S}_\omega(\mathbb{R}^d)$ we refer to [3–6]. In particular, we can define on $\mathcal{S}_\omega(\mathbb{R}^d)$ different equivalent systems of seminorms that make $\mathcal{S}_\omega(\mathbb{R}^d)$ a Fréchet nuclear space. It is also an algebra under multiplication and convolution.

The corresponding strong dual space is denoted by $\mathcal{S}'_\omega(\mathbb{R}^d)$ and its elements are called *ω-tempered distributions*. Moreover, $\mathcal{S}'(\mathbb{R}^d) \subseteq \mathcal{S}'_\omega(\mathbb{R}^d)$ and the Fourier transform, the short-time Fourier transform and the Wigner transform are continuous from $\mathcal{S}_\omega(\mathbb{R}^d)$ to $\mathcal{S}_\omega(\mathbb{R}^d)$ and from $\mathcal{S}'_\omega(\mathbb{R}^d)$ to $\mathcal{S}'_\omega(\mathbb{R}^d)$.

The "right" function spaces in time-frequency analysis to work with the STFT are the so-called *modulation spaces*, introduced by H. Feichtinger in [9]. In this context, we consider the weight $m_\lambda(z) := e^{\lambda\omega(z)}$, for $\lambda \in \mathbb{R}$, and define $L^{p,q}_{m_\lambda}(\mathbb{R}^{2d})$ as the space of measurable functions f on $\mathbb{R}^{2d}$ such that

$$\|f\|_{L^{p,q}_{m_\lambda}} := \int_{\mathbb{R}^d}\left(\int_{\mathbb{R}^d}|f(x,\xi)|^p m_\lambda(x,\xi)^p\,dx\right)^{\frac{q}{p}}d\xi\Bigg)^{\frac{1}{q}} < +\infty,$$

for $1 \le p, q < +\infty$, with standard changes if p (or q) is $+\infty$. We define then, for $1 \le p, q \le +\infty$, the modulation space

$$\mathbf{M}^{p,q}_{m_\lambda}(\mathbb{R}^d) := \{f \in \mathcal{S}'_\omega(\mathbb{R}^d) : V_\varphi f \in L^{p,q}_{m_\lambda}(\mathbb{R}^{2d})\},$$

which is independent of the window function $\varphi \in \mathcal{S}_\omega(\mathbb{R}^d) \setminus \{0\}$ and is a Banach space with norm $\|f\|_{\mathbf{M}^{p,q}_{m_\lambda}} := \|V_\varphi f\|_{L^{p,q}_{m_\lambda}}$ (see [4]). Moreover, for $1 \le p, q < +\infty$, the space $\mathcal{S}_\omega(\mathbb{R}^d)$ is a dense subspace of $\mathbf{M}^{p,q}_{m_\lambda}$ by Boiti et al. [4, Prop. 3.9]. We shall denote $\mathbf{M}^p_{m_\lambda}(\mathbb{R}^d) = \mathbf{M}^{p,p}_{m_\lambda}(\mathbb{R}^d)$ and $\mathbf{M}^{p,q}(\mathbb{R}^d) = \mathbf{M}^{p,q}_{m_0}(\mathbb{R}^d)$.

As in [13, Thm. 12.2.2] if $p_1 \le p_2$, $q_1 \le q_2$, and $\lambda \le \mu$ then $\mathbf{M}^{p_1,q_1}_{m_\mu} \subseteq \mathbf{M}^{p_2,q_2}_{m_\lambda}$ with continuous inclusion (see [8, Lemma 2.3.16]). Set

$$m_{\lambda,1}(x) := m_\lambda(x,0), \quad m_{\lambda,2}(\xi) := m_\lambda(0,\xi),$$
$$v_\lambda(z) = e^{|\lambda|\omega(z)}, \quad v_{\lambda,1}(x) := v_\lambda(x,0), \quad v_{\lambda,2}(\xi) := v_\lambda(0,\xi),$$

and prove the following generalization of [7, Prop. 2.4]:

Proposition 0.3 *Let* $1 \le p, q, r, t, t' \le +\infty$ *such that* $\frac{1}{p} + \frac{1}{q} - 1 = \frac{1}{r}$ *and* $\frac{1}{t} + \frac{1}{t'} = 1$. *Then, for all* $\lambda, \mu \in \mathbb{R}$ *and* $1 \le s \le +\infty$,

$$\mathbf{M}^{p,st}_{m_{\lambda,1}\otimes m_{\mu,2}}(\mathbb{R}^d) * \mathbf{M}^{q,st'}_{v_{\lambda,1}\otimes v_{\lambda,2}m_{-\mu,2}}(\mathbb{R}^d) \hookrightarrow \mathbf{M}^{r,s}_{m_\lambda}(\mathbb{R}^d)$$

and

$$\|f * g\|_{\mathbf{M}^{r,s}_{m_\lambda}} \le \|f\|_{\mathbf{M}^{p,st}_{m_{\lambda,1}\otimes m_{\mu,2}}} \|g\|_{\mathbf{M}^{q,st'}_{v_{\lambda,1}\otimes v_{\lambda,2}m_{-\mu,2}}}. \tag{0.1}$$

Proof For the Gaussian function $g_0(x) = e^{-\pi|x|^2} \in \mathcal{S}_\omega(\mathbb{R}^d)$ consider on $\mathbf{M}^{r,s}_{m_\lambda}$ the modulation norm with respect to the window function $g(x) := g_0 * g_0(x) = 2^{-d/2}e^{-\frac{\pi}{2}|x|^2} \in \mathcal{S}_\omega(\mathbb{R}^d)$. Since $m_\lambda(x,\xi) \le m_\lambda(x,0)v_\lambda(0,\xi)$ and $\overline{g_0(-x)} = g_0(x)$, by Gröchenig [13, Lemma 3.1.1], Young and Hölder inequalities:

$$\|f * h\|_{\mathbf{M}^{r,s}_{m_\lambda}} = \|V_g(f*h)\|_{L^{r,s}_{m_\lambda}} = \left(\int_{\mathbb{R}^d}\left(\int_{\mathbb{R}^d}|V_g(f*h)|^r m^r_\lambda(x,\xi)\,dx\right)^{\frac{s}{r}}d\xi\right)^{\frac{1}{s}}$$

$$\le \left(\int_{\mathbb{R}^d}\left(\int_{\mathbb{R}^d}|(f * M_\xi g_0) * (h * M_\xi g_0)(x)|^r m_\lambda(x,0)^r\,dx\right)^{\frac{s}{r}} v^s_\lambda(0,\xi)\,d\xi\right)^{\frac{1}{s}}$$

$$= \left(\int_{\mathbb{R}^d} \|(f * M_\xi g_0) * (h * M_\xi g_0)\|^s_{L^r_{m_{\lambda,1}}} v^s_\lambda(0,\xi)\, d\xi \right)^{\frac{1}{s}}$$

$$\leq \left(\int_{\mathbb{R}^d} \|f * M_\xi g_0\|^s_{L^p_{m_{\lambda,1}}} \|h * M_\xi g_0\|^s_{L^q_{v_{\lambda,1}}} v^s_\lambda(0,\xi)\, d\xi \right)^{\frac{1}{s}}$$

$$= \left(\int_{\mathbb{R}^d} \|V_{g_0} f\|^s_{L^p_{m_{\lambda,1}}} m^s_\mu(0,\xi) \|V_{g_0} h\|^s_{L^q_{v_{\lambda,1}}} m^s_{-\mu}(0,\xi) v^s_\lambda(0,\xi)\, d\xi \right)^{\frac{1}{s}}$$

$$\leq \|f\|_{\mathbf{M}^{p,st}_{m_{\lambda,1}\otimes m_{\mu,2}}} \|h\|_{\mathbf{M}^{q,st'}_{v_{\lambda,1}\otimes v_{\lambda,2} m_{-\mu,2}}}.$$

□

Given two window functions $\psi, \gamma \in \mathcal{S}_\omega(\mathbb{R}^d) \setminus \{0\}$ and a symbol $a \in \mathcal{S}'_\omega(\mathbb{R}^{2d})$, the corresponding *localization operator* $L^a_{\psi,\gamma}$ is defined, for $f \in \mathcal{S}_\omega(\mathbb{R}^d)$, by

$$L^a_{\psi,\gamma} f = V^*_\gamma(a \cdot V_\psi f) = \int_{\mathbb{R}^{2d}} a(x,\xi) V_\psi f(x,\xi) M_\xi T_x \gamma \, dx d\xi, \tag{0.2}$$

where V^*_γ is the adjoint of V_γ. As in [2, Lemma 2.4] we have that $L^a_{\psi,\gamma}$ is a Weyl operator L^{a^w} with symbol $a^w = a * \mathrm{Wig}(\gamma,\psi)$:

$$L^{a^w} f := \frac{1}{(2\pi)^d} \int_{\mathbb{R}^{2d}} \hat{a}^w(\xi,u) e^{-i\xi\cdot u} T_{-u} M_\xi f \, du d\xi. \tag{0.3}$$

Moreover, if $f, g \in \mathcal{S}_\omega(\mathbb{R}^d)$ then by definition of adjoint operator we can write

$$\langle L^a_{\psi,\gamma} f, g\rangle = \langle a \cdot V_\psi f, V_\gamma g\rangle = \langle a, \overline{V_\psi f} V_\gamma g\rangle,$$

and, similarly as in [13, Thm. 14.5.2] (see also [8, Teo. 2.3.21]), we have, for $a^w \in \mathbf{M}^{\infty,1}_{m_\mu}(\mathbb{R}^{2d})$ with $\mu \geq 0$,

$$\|L^{a^w} f\|_{\mathbf{M}^{p,q}_{m_\lambda}} = \|L^a_{\psi,\gamma} f\|_{\mathbf{M}^{p,q}_{m_\lambda}} \leq \|a^w\|_{\mathbf{M}^{\infty,1}_{m_\mu}} \|f\|_{\mathbf{M}^{p,q}_{m_\lambda}}, \tag{0.4}$$

for all $f \in \mathbf{M}^{p,q}_{m_\lambda}$ and $\lambda \in \mathbb{R}$.

Theorem 0.4 *Let $\psi, \gamma \in \mathcal{S}_\omega(\mathbb{R}^d) \setminus \{0\}$ and $a \in \mathbf{M}^\infty_{m_\lambda}(\mathbb{R}^{2d})$ for some $\lambda \geq 0$. Then $L^a_{\psi,\gamma}$ is bounded from $\mathbf{M}^{p,q}_{m_\lambda}(\mathbb{R}^d)$ to $\mathbf{M}^{p,q}_{m_\lambda}(\mathbb{R}^d)$, for $1 \leq p, q < +\infty$, and*

$$\|L^a_{\psi,\gamma}\|_{op} \leq \|a\|_{\mathbf{M}^\infty_{m_{-\lambda,2}}} \|\psi\|_{\mathbf{M}^1_{v_\lambda}} \|\gamma\|_{\mathbf{M}^p_{m_\lambda}}.$$

Proof By definition $V_\psi : \mathbf{M}^{p,q}_{m_\lambda} \to L^{p,q}_{m_\lambda}(\mathbb{R}^{2d})$ and, by Boiti et al. [4, Prop. 3.7], $V^*_\gamma : L^{p,q}_{m_\lambda}(\mathbb{R}^{2d}) \to \mathbf{M}^{p,q}_{m_\lambda}(\mathbb{R}^d)$. Let $f \in \mathbf{M}^{p,q}_{m_\lambda}(\mathbb{R}^d)$. To prove that $L^a_{\psi,\gamma} f = V^*_\gamma(a \cdot V_\psi f) \in \mathbf{M}^{p,q}_{m_\lambda}$, it is then enough to show that $a \cdot V_\psi f \in L^{p,q}_{m_\lambda}(\mathbb{R}^{2d})$. By the inversion formula [4, Prop. 3.7], given two window functions $\Phi, \Psi \in \mathcal{S}_\omega(\mathbb{R}^{2d})$ with $\langle \Phi, \Psi \rangle \neq 0$, we have, for $z = (z_1, z_2) \in \mathbb{R}^{2d} \times \mathbb{R}^{2d}$,

$$
\begin{aligned}
&\left(\int_{\mathbb{R}^d} \left(\int_{\mathbb{R}^d} |a(x,\xi)|^p |V_\psi f(x,\xi)|^p e^{p\lambda\omega(x,\xi)}\, dx \right)^{\frac{q}{p}} d\xi \right)^{\frac{1}{q}} \\
&\leq \frac{1}{(2\pi)^d} \frac{1}{|\langle \Phi, \Psi \rangle|} \left(\int_{\mathbb{R}^d} \left(\int_{\mathbb{R}^d} \left(\int_{\mathbb{R}^{4d}} |V_\Psi a(z)|^p |M_{z_2} T_{z_1} \Phi(x,\xi)|^p dz \right) \right.\right. \\
&\quad \left.\left. \cdot |V_\psi f(x,\xi)|^p e^{p\lambda\omega(x,\xi)} dx \right)^{\frac{q}{p}} d\xi \right)^{\frac{1}{q}} \\
&\leq \frac{1}{(2\pi)^d} \frac{1}{|\langle \Phi, \Psi \rangle|} \left(\int_{\mathbb{R}^d} \left(\int_{\mathbb{R}^d} \left(\int_{\mathbb{R}^{4d}} \left(|V_\Psi a(z)| e^{\lambda\omega(z)} \right)^p |M_{z_2} T_{z_1} \Phi(x,\xi)|^p dz \right) \right.\right. \\
&\quad \left.\left. \cdot |V_\psi f(x,\xi)|^p e^{p\lambda\omega(x,\xi)} dx \right)^{\frac{q}{p}} d\xi \right)^{\frac{1}{q}} \\
&\leq C \|V_\Psi a\|_{L^\infty_{m_\lambda}} \cdot \|V_\psi f\|_{L^{p,q}_{m_\lambda}} = C \|a\|_{\mathbf{M}^\infty_{m_\lambda}} \cdot \|f\|_{\mathbf{M}^{p,q}_{m_\lambda}},
\end{aligned}
$$

for some $C > 0$. Therefore $a \cdot V_\psi f \in L^{p,q}_{m_\lambda}(\mathbb{R}^{2d})$ and $L^a_{\psi,\gamma} f \in \mathbf{M}^{p,q}_{m_\lambda}(\mathbb{R}^d)$.

To prove that $L^a_{\psi.\gamma}$ is bounded, consider $g \in \mathcal{S}_\omega(\mathbb{R}^d)$ and set $\Psi = \mathrm{Wig}(g, g) \in \mathcal{S}_\omega(\mathbb{R}^{2d})$. For $\xi = (\xi_1, \xi_2) \in \mathbb{R}^{2d}$, we set $\tilde{\xi} = (\xi_2, -\xi_1)$. By Cordero and Gröchenig [7, Lemma 2.2]

$$
\begin{aligned}
\|\mathrm{Wig}(\gamma, \psi)\|_{\mathbf{M}^{1,p}_{m_{\lambda,2}}} &= \|V_\Psi \mathrm{Wig}(\gamma, \psi)\|_{L^{1,p}_{m_{\lambda,2}}} \\
&= \left(\int_{\mathbb{R}^{2d}} \left(\int_{\mathbb{R}^{2d}} \left| V_g \psi\left(z + \frac{\tilde{\xi}}{2}\right) V_g \gamma\left(z - \frac{\tilde{\xi}}{2}\right) \right| dz \right)^p m^p_{\lambda,2}(\xi)\, d\xi \right)^{\frac{1}{p}}.
\end{aligned}
$$

By the change of variables $z + \frac{\tilde{\xi}}{2} = \tilde{z}$ and [4, formula (3.12)] we obtain (cf. also [7, Prop. 2.5]):

$$
\begin{aligned}
\|\mathrm{Wig}(\gamma, \psi)\|_{\mathbf{M}^{1,p}_{m_{\lambda,2}}} &= \left(\int_{\mathbb{R}^{2d}} \left(\int_{\mathbb{R}^{2d}} |V_g \psi(\tilde{z})||V_g \gamma(\tilde{z} - \tilde{\xi})|\, d\tilde{z} \right)^p m^p_{\lambda,2}(\xi)\, d\xi \right)^{\frac{1}{p}}. \\
&= \left(\int_{\mathbb{R}^{2d}} (|V_g \psi(\tilde{z})| * |V_g \gamma(-\tilde{z})|)^p(\tilde{\xi})\, m^p_{\lambda,2}(\tilde{\xi})\, d\tilde{\xi} \right)^{\frac{1}{p}} \\
&\leq \|V_g \psi\|_{L^1_{v_\lambda}} \|V_g \gamma\|_{L^p_{m_\lambda}} = \|\psi\|_{\mathbf{M}^1_{v_\lambda}} \|\gamma\|_{\mathbf{M}^p_{m_\lambda}}. \qquad (0.5)
\end{aligned}
$$

Therefore $\mathrm{Wig}(\gamma,\psi) \in \mathbf{M}^1_{m_{\lambda,2}}(\mathbb{R}^{2d})$ and hence, from Proposition 0.3 (with $p = t = r = +\infty$, $q = s = t' = 1$, $\lambda = 0$ and $\mu = -\lambda$), we have that $\mathbf{M}^\infty_{m_{-\lambda,2}} * \mathbf{M}^1_{m_{\lambda,2}} \subseteq \mathbf{M}^{\infty,1}$, so that $a^w = a * \mathrm{Wig}(\gamma,\psi) \in \mathbf{M}^{\infty,1}$ and by (0.4) with $\mu = 0$

$$\|L^a_{\psi,\gamma}\|_{op} \leq \|a^w\|_{\mathbf{M}^{\infty,1}}.$$

From (0.1) and (0.5) we finally have

$$\begin{aligned}\|L^a_{\psi,\gamma}\|_{op} &\leq \|a * \mathrm{Wig}(\gamma,\psi)\|_{\mathbf{M}^{\infty,1}} \leq \|a\|_{\mathbf{M}^\infty_{m_{-\lambda,2}}} \|\mathrm{Wig}(\gamma,\psi)\|_{\mathbf{M}^1_{m_{\lambda,2}}} \\ &\leq \|a\|_{\mathbf{M}^\infty_{m_{-\lambda,2}}} \|\psi\|_{\mathbf{M}^1_{v_\lambda}} \|\gamma\|_{\mathbf{M}^p_{m_\lambda}}.\end{aligned}$$

□

A boundedness result analogous to that of Theorem 0.4 is proved, with different techniques, in [16] under further restrictions on the symbol $a(x,\xi)$ and without estimates on the norm of $L^a_{\psi,\gamma}$.

Set now

$$\mathbf{M}^{0,1}_{m_\lambda}(\mathbb{R}^d) = \{f \in \mathbf{M}^{\infty,1}_{m_\lambda}(\mathbb{R}^d) : \lim_{|x|\to\infty} \|V_g f(x,.)\|_{L^1_{m_\lambda}} e^{\lambda\omega(x)} = 0\}$$

and prove the following compactness result (cf. also [1, Prop. 2.3] and [12, Thm. 3.22]):

Theorem 0.5 *If $a^w \in \mathbf{M}^{0,1}_{m_\lambda}(\mathbb{R}^{2d})$ for some $\lambda \geq 0$, then L^{a^w} is a compact mapping of $\mathbf{M}^{p,q}_{m_\lambda}(\mathbb{R}^d)$ into itself, for $1 \leq p, q < +\infty$.*

Proof The operator L^{a^w} maps $\mathbf{M}^{p,q}_{m_\lambda}(\mathbb{R}^d)$ into itself by (0.4). To prove that L^{a^w} is compact we first assume $a^w \in \mathcal{S}_\omega(\mathbb{R}^{2d})$. From (0.3)

$$\begin{aligned}L^{a^w} f(y) &= \frac{1}{(2\pi)^d}\int_{\mathbb{R}^{2d}} \hat{a}^w(\xi,u) e^{-i\xi\cdot u} e^{i\xi\cdot(y+u)} f(y+u)\,du\,d\xi \\ &= \frac{1}{(2\pi)^d}\int_{\mathbb{R}^{2d}} \hat{a}^w(\xi,x-y) e^{i\xi\cdot y} f(x)\,dx\,d\xi \\ &= \int_{\mathbb{R}^d} k(x,y) f(x)\,dx, \qquad (0.6)\end{aligned}$$

with kernel $k(x,y) = \frac{1}{(2\pi)^d}\int_{\mathbb{R}^d} \hat{a}^w(\xi,x-y)e^{i\xi\cdot y}d\xi$. Note that $k(x,y) \in \mathcal{S}_\omega(\mathbb{R}^{2d})$ because it is the inverse Fourier transform (with respect to the first variable) of the translation (with respect to the second variable) of $\hat{a}^w \in \mathcal{S}_\omega(\mathbb{R}^{2d})$.

Now, let $\phi \in \mathcal{S}_\omega(\mathbb{R}^d)$ and $\alpha_0, \beta_0 > 0$ such that $\{\phi_{jl}\}_{j,l\in\mathbb{Z}^d} = \{M_{\beta_0 l}T_{\alpha_0 j}\phi\}_{j,l\in\mathbb{Z}^d}$ is a tight Gabor frame for $L^2(\mathbb{R}^d)$ (see [13, Def. 5.1.1] for the definition). Then $\{\Phi_{jlmn}\}_{j,l,m,n\in\mathbb{Z}^d} = \{\phi_{jl}(x)\phi_{mn}(y)\}_{j,l,m,n\in\mathbb{Z}^d}$ is a tight Gabor frame for $L^2(\mathbb{R}^{2d})$. Since $k \in \mathcal{S}_\omega(\mathbb{R}^{2d})$ we have that $\langle k, \Phi_{jlmn}\rangle = V_\phi k(\alpha_0 j, \alpha_0 m, \beta_0 l, \beta_0 n) \in \ell^1$ and (see [4, Lemma 3.15])

$$k = \sum_{j,l,m,n\in\mathbb{Z}^d} \langle k, \Phi_{jlmn}\rangle \Phi_{jlmn}.$$

Therefore from (0.6)

$$L^{a^w} f = \sum_{j,l,m,n\in\mathbb{Z}^d} \langle k, \Phi_{jlmn}\rangle \langle \phi_{jl}, f\rangle \phi_{mn},$$

with $\langle k, \Phi_{jlmn}\rangle \in \ell^1$, $(\phi_{jl})_{j,l\in\mathbb{Z}^d}$ equicontinuous in $\mathbf{M}^{p',q'}_{m_{-\lambda}} = (\mathbf{M}^{p,q}_{m_\lambda})^*$ and $(\phi_{mn})_{m,n\in\mathbb{Z}^d}$ bounded in $\bigcup_{n\in\mathbb{N}} n\{f \in \mathbf{M}^{p,q}_{m_\lambda} : \|f\|_{\mathbf{M}^{p,q}_{m_\lambda}} < 1\}$, so that L^{a^w} is a nuclear operator from $\mathbf{M}^{p,q}_{m_\lambda}$ to $\mathbf{M}^{p,q}_{m_\lambda}$ (see [15, §17.3]). From [15, §17.3, Cor. 4] we thus have that L^{a^w} is compact.

Let us finally consider the general case $a \in \mathbf{M}^{0,1}_{m_\lambda}(\mathbb{R}^{2d})$. By Boiti et al. [4, Prop. 3.9] there exist $a_n \in \mathcal{S}_\omega(\mathbb{R}^{2d})$ converging to a in $\mathbf{M}^{\infty,1}_{m_\lambda}$ and hence, by (0.4)

$$\|L^{a^w} - L^{a_n^w}\|_{\mathbf{M}^{p,q}_{m_\lambda}\to\mathbf{M}^{p,q}_{m_\lambda}} \leq \|a - a_n\|_{\mathbf{M}^{\infty,1}_{m_\lambda}} \to 0.$$

Since the set of compact operators is closed we have that L^{a^w} is compact on $\mathbf{M}^{p,q}_{m_\lambda}(\mathbb{R}^d)$. □

We have the following generalization of [10, Lemma 3.4] and [11, Prop. 5.2]:

Lemma 0.6 *Let $g_0 \in \mathcal{S}_\omega(\mathbb{R}^d)$ and $a \in \mathbf{M}^{\infty}_{m_\lambda}(\mathbb{R}^d)$, with $\lambda \geq 0$, such that*

$$\lim_{|x|\to+\infty} \sup_{|\xi|\leq R} |V_{g_0} a(x,\xi)| e^{\lambda\omega(x,\xi)} = 0, \qquad \forall R > 0. \tag{0.7}$$

Then $a * H \in \mathbf{M}^{0,1}_{m_\lambda}(\mathbb{R}^d)$ for any $H \in \mathcal{S}_\omega(\mathbb{R}^d)$.

Proof The case $\lambda = 0$ has been proved in [10, Lemma 3.4]. Let $\lambda > 0$. Since $g_0 \in \mathcal{S}_\omega(\mathbb{R}^d)$ and $H \in \mathcal{S}_\omega(\mathbb{R}^d)$, by Gröchenig and Zimmermann [14, Thm. 2.7] we have that $V_{g_0}H \in \mathcal{S}_\omega(\mathbb{R}^{2d})$ and hence, for a fixed $\ell > 0$ (to be chosen later depending on λ), there exists $c_\lambda > 0$ such that

$$|V_{g_0}H(x,\xi)| \leq c_\lambda e^{-3\ell\lambda\omega(x)} e^{-3\ell\lambda\omega(\xi)}, \qquad \forall x, \xi \in \mathbb{R}^d.$$

Now, as in the proof of Proposition 0.3, for $g = g_0 * g_0$, we have that $|V_g(a * H)(\cdot, \xi)| = |V_{g_0}a(\cdot, \xi) * V_{g_0}H(\cdot, \xi)|$. Since ω is increasing and subadditive we have

$$
\begin{aligned}
&|V_g(a * H)(x, \xi)| \leq \int_{\mathbb{R}^d} |V_{g_0}a(x-y, \xi)||V_{g_0}H(y, \xi)|dy \\
&\leq c_\lambda e^{-3\ell\lambda\omega(\xi)} \int_{\mathbb{R}^d} |V_{g_0}a(x-y, \xi)|e^{-3\ell\lambda\omega(y)}dy \\
&= c_\lambda e^{-3\ell\lambda\omega(\xi)} \int_{\mathbb{R}^d} |V_{g_0}a(x-y, \xi)|e^{-3\ell\lambda\omega(y)}\, e^{\lambda\omega(x-y,\xi)}e^{-\lambda\omega(x-y,\xi)}dy \\
&\leq c_\lambda e^{-3\ell\lambda\omega(\xi)}e^{-\lambda\omega(x)} \int_{\mathbb{R}^d} |V_{g_0}a(x-y, \xi)|e^{\lambda\omega(x-y,\xi)}e^{-(3\ell-1)\lambda\omega(y)}dy.
\end{aligned}
$$

Since $a \in \mathbf{M}^\infty_{m_\lambda}(\mathbb{R}^d)$ we have that

$$
\begin{aligned}
&e^{\lambda\omega(x)+2\ell\lambda\omega(\xi)}|V_g(a * H)(x, \xi)| \\
&\leq \ c_\lambda e^{-\ell\lambda\omega(\xi)} \int_{\mathbb{R}^d} |V_g a(x-y, \xi)|e^{\lambda\omega(x-y,\xi)}e^{-(3\ell-1)\lambda\omega(y)}dy \qquad (0.8) \\
&\leq \ c_\lambda e^{-\ell\lambda\omega(\xi)}\|a\|_{\mathbf{M}^\infty_{m_\lambda}} \int_{\mathbb{R}^d} e^{-(3\ell-1)\lambda\omega(y)}dy < +\infty, \qquad (0.9)
\end{aligned}
$$

if $\ell > \frac{1}{3} + \frac{d}{3B\lambda}$, where B is the constant of condition (γ) in Definition 0.1. Since $\lim_{|\xi|\to+\infty}\omega(\xi) = +\infty$, from (0.9) we have that for all $\varepsilon > 0$ there exists $R_1 > 0$ such that

$$
e^{\lambda\omega(x)+2\ell\lambda\omega(\xi)}|V_g(a * H)(x, \xi)| < \varepsilon, \quad \forall x, \xi \in \mathbb{R}^d, \quad |\xi| \geq R_1. \qquad (0.10)
$$

We now choose $\delta > 0$ small enough so that

$$
\delta\left(1 + c_\lambda \int_{\mathbb{R}^d} e^{-(3\ell-1)\lambda\omega(y)}\right)dy \leq \varepsilon. \qquad (0.11)
$$

From the hypothesis (0.7) we can choose $R_2 > 0$ sufficiently large so that

$$
\sup_{|\xi|\leq R_1} |V_{g_0}a(x, \xi)|e^{\lambda\omega(x,\xi)} < \delta, \quad |x| \geq R_2, \qquad (0.12)
$$

$$
\int_{|y|>R_2} e^{-(3\ell-1)\lambda\omega(y)}\,dy < \frac{\delta}{c_\lambda e^{-\ell\lambda\omega(\xi)}\|a\|_{\mathbf{M}^\infty_{m_\lambda}}}, \qquad |\xi| \leq R_1. \qquad (0.13)
$$

Therefore for $|x| \geq 2R_2$, $|y| \leq R_2$ (so that $|x-y| \geq R_2$) and $|\xi| \leq R_1$, by (0.8), (0.9), (0.13), (0.12) and (0.11):

$$\begin{aligned} & e^{\lambda\omega(x)+2\ell\lambda\omega(\xi)}|V_g(a*H)(x,\xi)| \\ \leq\ & c_\lambda e^{-\ell\lambda\omega(\xi)}\|a\|_{\mathbf{M}^\infty_{m_\lambda}} \int_{|y|>R_2} e^{-(3\ell-1)\lambda\omega(y)}dy \\ & +c_\lambda e^{-\ell\lambda\omega(\xi)} \int_{|y|\leq R_2} |V_{g_0}a(x-y,\xi)|e^{\lambda\omega(x-y,\xi)}e^{-(3\ell-1)\lambda\omega(y)}dy \\ <\ & \delta + c_\lambda\delta \int_{\mathbb{R}^d} e^{-(3\ell-1)\lambda\omega(y)}dy \leq \varepsilon. \end{aligned}$$

The above estimate, together with (0.10), gives

$$e^{\lambda\omega(x)} \int_{\mathbb{R}^d} |V_g(a*H)(x,\xi)|e^{\lambda\omega(\xi)}d\xi \leq \varepsilon \int_{\mathbb{R}^d} e^{-(2\ell-1)\lambda\omega(\xi)}d\xi, \qquad |x| \geq 2R_2.$$

Choosing now $\ell > \frac{1}{2} + \frac{d}{2B\lambda} > \frac{1}{3} + \frac{d}{3B\lambda}$ so that $e^{-(2\ell-1)\lambda\omega(\xi)} \in L^1(\mathbb{R}^d)$, we finally obtain

$$\lim_{|x|\to\infty} e^{\lambda\omega(x)}\|V_g(a*H)(x,.)\|_{L^1_{m_\lambda}} = 0.$$

□

Theorem 0.7 *Let $\psi,\gamma \in \mathcal{S}_\omega(\mathbb{R}^d)$, $g_0 \in \mathcal{S}_\omega(\mathbb{R}^{2d})$ and $a \in \mathbf{M}^\infty_{m_\lambda}(\mathbb{R}^{2d})$ satisfying* (0.7), *for some $\lambda \geq 0$. Then $L^a_{\psi,\gamma} : \mathbf{M}^{p,q}_{m_\lambda}(\mathbb{R}^d) \to \mathbf{M}^{p,q}_{m_\lambda}(\mathbb{R}^d)$ is compact, for $1 \leq p,q < +\infty$.*

Proof Set $H := W(\gamma,\psi) \in \mathcal{S}_\omega(\mathbb{R}^{2d})$. Since $a \in \mathbf{M}^\infty_{m_\lambda}(\mathbb{R}^{2d})$, by Lemma 0.6 we have that $a^w = a*H \in \mathbf{M}^{0,1}_{m_\lambda}(\mathbb{R}^{2d})$ and hence $L^a_{\psi,\gamma} = L^{a^w}$ is compact by Theorem 0.5. □

Acknowledgments The authors are grateful to Proff. C. Fernández, A. Galbis and D. Jornet for helpful discussions. The first author is member of the GNAMPA-INdAM.

References

1. A. Bényi, K. Gröchenig, C. Heil, K. Okoudjou, Modulation spaces and a class of bounded multilinear pseudodifferential operators. J. Oper. Theory **54**(2), 387–399 (2005)
2. P. Boggiato, E. Cordero, K. Gröchenig Generalized anti-wick operators with symbols in distributional Sobolev space. Integr. Equ. Oper. Theory **48**, 424–442 (2004)
3. C. Boiti, D. Jornet, A. Oliaro, A regularity of partial differential operators in ultradifferentiable spaces and Wigner type transforms. J. Math. Anal. Appl. **446**, 920–944 (2017)

4. C. Boiti, D. Jornet, A. Oliaro, The Gabor wave front set in spaces of ultradifferentiable functions. Monatsh. Math. **188**, 199–246 (2018)
5. C. Boiti, D. Jornet, A. Oliaro, Real Paley-Wiener Theorems in spaces of ultradifferentiable functions. J. Funct. Anal. **278**(4), 108348, 1–45 (2020)
6. C. Boiti, D. Jornet, A. Oliaro, G. Schindl, Nuclearity of rapidly decreasing ultradifferentiable functions and time-frequency analysis Collect. Math. **72**(2), 423–442 (2021)
7. E. Cordero, K. Gröchenig, Time-frequency analysis of localization operators. J. Funct. Anal. **205**(1), 107–131 (2005)
8. A. De Martino, Trasformata di Gabor ed applicazioni allo studio della regolarità di operatori pseudo-differenziali. Master Thesis, University of Ferrara, 2019
9. H.G. Feichtinger, Modulation spaces on locally compact abelian groups. Technical report, University of Vienna, 1983
10. C. Fernández, A. Galbis, Compactness of time-frequency localization operators on $L^2(\mathbb{R}^d)$. J. Funct. Anal. **233**, 335–350 (2006)
11. C. Fernández, A. Galbis, Annihilating sets for the short time Fourier transform. Adv. Math. **224**(5), 1904–1926 (2010)
12. C. Fernández, A. Galbis, E. Primo, Compactness of Fourier integral operators on weighted modulation spaces. Trans. Am. Math. Soc. **372**(1), 733–753 (2019)
13. K. Gröchenig, *Foundations of Time-Frequency Analysis* (Birkhäuser, Boston, 2001)
14. K. Gröchenig, G. Zimmermann, Spaces of test functions via the STFT. J. Funct. Spaces Appl. **2**(1), 25–53 (2004)
15. H. Jarchow, *Locally Convex Spaces* (Vieweg Teubner Verlag, Stuttgart, 2014)
16. E. Primo, Boundedness and compactness of operators related to time-frequency analysis. Ph-D thesis, Universitat de València, 2018
17. J. Toft, The Bargman transform on modulation and Gelfand-Shilov spaces, with applications to Toeplitz and pseudo-differential operators. J. Pseudo-Differ. Oper. Appl. **3**(2), 145–227 (2012)

Lifespan Estimates for a Special Quasilinear Time-Dependent Damped Wave Equation

Giovanni Girardi and Sandra Lucente

Abstract In this paper we consider a quasilinear Cauchy problem for the scale invariant damped wave equation

$$v_{tt} - \Delta v + \frac{\mu}{(1+t)} v_t + \frac{\mu}{2}\left(\frac{\mu}{2} - 1\right) \frac{v}{(1+t)^2} = \left| \frac{\mu}{2(1+t)} v + v_t \right|^p,$$

with $\mu \geq 0$, $v = v(t, x)$ and $x \in \mathbb{R}^n$. The particular structure of the nonlinear term, guarantees a blow up result and a lifespan estimate, assuming radial initial data having slow decay. In particular the range of admissible exponents p depends on μ, n and the rate of the initial data decay.

Keywords Scale invariant damped wave · Slow decay · Lifespan

Mathematics Subject Classification (2010) Primary 35B33; Secondary 35L70

1 Introduction

Let us consider the scale invariant damped wave equation

$$v_{tt}(t, x) - \Delta v(t, x) + \frac{\mu}{(1+t)} v_t(t, x) + \frac{\mu}{2}\left(\frac{\mu}{2} - 1\right) \frac{v(t, x)}{(1+t)^2} = G, \tag{1.1}$$

G. Girardi
Department of Mathematics, University of Bari, Bari, Italy
e-mail: giovanni.girardi@uniba.it

S. Lucente (✉)
Department of Physics, University of Bari, Bari, Italy
e-mail: sandra.lucente@uniba.it

P. Cerejeiras et al. (eds.), *Current Trends in Analysis, its Applications and Computation*, Research Perspectives,
https://doi.org/10.1007/978-3-030-87502-2_61

where $x \in \mathbb{R}^n$, $t \geq 0$, $\mu \geq 0$. We want to choose a special nonlinear term G, depending on v_t such that blow-up or global existence results can be established by means of a natural transformation that acts on the linear part of the equation.

In [1] the scale invariant damped wave operator is rewritten as

$$u_{tt} - \Delta u + \frac{\mu}{1+t}u_t + \frac{\mu(\mu-2)}{4(1+t)^2}u = \partial_{(\mu),t}\partial_{(\mu),t}u - \Delta u$$

where

$$\partial_{(\mu),t} := \partial_t + \frac{\mu}{2(1+t)}$$

is a covariant time-derivative. It follows that the meaningful nonlinear term is

$$G := G(t, v, v_t) := \left|\left(\partial_t + \frac{\mu}{2(1+t)}\right)v\right|^p \quad \text{with } p > 1. \tag{1.2}$$

The aim of this paper is to prove a blow up result for the solution of (1.1), with G as in (1.2), under a slow decay condition on radial initial data. The case $G = G(v) = |v|^p$ is considered in [2].

We will prove the following result.

Theorem 1.1 *Let $n \geq 2$ and $\varepsilon > 0$. Consider the Cauchy problem*

$$\begin{cases} v_{tt} - \Delta v + \frac{\mu}{(1+t)}v_t + \frac{\mu}{2}\left(\frac{\mu}{2}-1\right)\frac{v}{(1+t)^2} = \left|\frac{\mu}{2(1+t)}v + v_t\right|^p, \\ v(0,x) = 0, \\ v_t(0,x) = \varepsilon g(x), \end{cases} \tag{1.3}$$

with g a radial smooth function such that

$$g(|x|) \geq \frac{M}{(1+|x|)^{h+1}}, \quad \text{with} \quad -1 < h < \frac{1}{p-1} - 1 - \frac{\mu}{2}, \tag{1.4}$$

for some $M > 0$. If $1 < p < 1 + 2/\mu$, then the maximal solution $v : [0, T(\varepsilon)) \times \mathbb{R}^n \to \mathbb{R}$ of (1.3) blows up in finite time. Moreover, the lifespan satisfies

$$T(\varepsilon) \leq C\varepsilon^{-\frac{2(p-1)}{2-(\mu+2h+2)(p-1)}},$$

with $C > 0$, independent of ε.

Remark 1.2 Let us recall that $p_F(d) = 1 + 2/d$ is the Fujita type exponent that plays a crucial role in the semilinear damped wave equations, i.e. $G = |u|^p$. For summarizing the long literature on this subject see [3]. By using this notation we can say that our result requires

$$1 < p < \min\{p_F(\mu), p_F(2 + 2h + \mu)\}.$$

This means a candidate to a critical curve appears. It is still necessary to prove the global existence for p greater than this value.

Remark 1.3 We remark that the range of admissible h in (1.4) is not empty due to our assumption $1 < p < p_F(\mu)$.

Remark 1.4 We underline that no assumption on the support of the initial data has been required in this theorem.

Remark 1.5 For $\mu = 0$ our result corresponds to Theorem 1.2 in [4].

2 Rewriting the Main Theorem

Let

$$u(t, x) = (1 + t)^{\frac{\mu}{2}} v(t, x).$$

We see that

$$\partial_{(\mu),t} v = (1 + t)^{-\frac{\mu}{2}} u_t .$$

This means that (1.3) reduces to the following nonlinear wave equation with polynomial potential:

$$\begin{cases} u_{tt} - \Delta u = (1 + t)^{-\frac{\mu}{2}(p-1)}|u_t|^p, & t \geq 0, \quad x \in \mathbb{R}^n, \\ u(0, x) = 0, \\ u_t(0, x) = \varepsilon g(x), & x \in \mathbb{R}^n. \end{cases}$$

In particular, we consider a radial initial datum, i.e. $g(x) = g(|x|)$ and hence radial solutions. Thus, we can rewrite our problem in the following form

$$\begin{cases} u_{tt} - u_{rr} - \frac{n-1}{r} u_r = (1 + t)^{-\frac{\mu}{2}(p-1)}|u_t|^p, & (t, r) \in [0, \infty)^2, \\ u(0, r) = 0, \\ u_t(0, r) = \varepsilon g(r). \end{cases} \tag{2.1}$$

Theorem 1.1 will be a consequence of the following one, proved in the next section.

Theorem 2.1 *Let $n \geq 2$ and $1 < p < p_F(\mu)$. Assume that there exists $M > 0$ such that for any $r \in [0, \infty)$ it holds*

$$g(r) \geq \frac{M}{(1+r)^{h+1}}, \quad \text{with} \quad -1 < h < \frac{1}{p-1} - 1 - \frac{\mu}{2}. \tag{2.2}$$

Given $\varepsilon > 0$, the lifespan $T(\varepsilon)$ of classical solutions to (2.1) *satisfies*

$$T(\varepsilon) \leq C\varepsilon^{-\frac{2(p-1)}{2-(\mu+2h+2)(p-1)}}, \tag{2.3}$$

with $C > 0$, independent of ε.

3 Proof of Theorem 2.1

In order to prove that if $u = u(t, x)$ is a solution to (2.1), with g satisfying (2.2), then u blows up in finite time even for small ε, we will use the crucial Lemma 2.5 and Lemma 2.6 in [4], that for our nonlinear terms gives the following.

Lemma 3.1 *Let $n \geq 2$ and $m = [n/2]$. Let $u^0 = u^0(r, t)$ be a solution to the free linear Cauchy problem associated to* (2.1), *that is*

$$\begin{cases} u^0_{tt} - u^0_{rr} - \frac{n-1}{r} u^0_r = 0, & (t, r) \in [0, \infty)^2, \\ u^0(0, r) = 0, \\ u^0_t(0, r) = \varepsilon g(r). \end{cases}$$

If $u = u(t, r)$ is a solution to (2.1), *then there exists a constant $\delta_m > 0$ such that*

$$u(t, r) \geq \varepsilon u^0(t, r) + \frac{1}{8r^m} \int_0^t d\tau \int_{r-t+\tau}^{r+t+\tau} \lambda^m (1+\tau)^{-\frac{\mu}{2}(p-1)} |u_t(\tau, \lambda)|^p d\lambda,$$

$$u^0(t, r) \geq \frac{1}{8r^m} \int_{r-t}^{r+t} \lambda^m g(\lambda) d\lambda,$$

for any

$$r - t > \frac{2t}{\delta_m}.$$

Let us fix $\delta > 0$ a suitable small constant; for any dimension $n \geq 2$ and $m = [n/2]$ we consider δ_m introduced in Lemma 3.1 and we define the set Σ_δ such that

$$\Sigma_\delta = \left\{ (t, r) \in (0, \infty)^2 : r - t \geq \max \left\{ \frac{2}{\delta_m} t, \delta \right\} \right\}. \tag{3.1}$$

Then, by Lemma 3.1 we get for each $(t, r) \in \Sigma_\delta$,

$$
\begin{aligned}
u(t,r) \geq \varepsilon u^0(t,r) &\geq \frac{\varepsilon}{8r^m} \int_{r-t}^{r+t} \lambda^m g(\lambda) d\lambda \\
&\geq \frac{M\varepsilon}{8r^m} \int_{r-t}^{r+t} \lambda^m (1+\lambda)^{-(h+1)} d\lambda \,.
\end{aligned}
$$

Then, (3.1) implies that

$$
\begin{aligned}
u(t,r) &\geq \frac{M\varepsilon}{8r^m} \left(\frac{1+\delta}{\delta}\right)^{-(h+1)} \int_{r-t}^{r+t} \lambda^{m-(h+1)} d\lambda \\
&\geq \frac{M\varepsilon}{8r^m} \left(\frac{1+\delta}{\delta}\right)^{-(h+1)} (r+t)^{-(h+1)} \int_{r-t}^{r+t} \lambda^m d\lambda. \\
&\geq \frac{M\varepsilon}{8r^m} \left(\frac{1+\delta}{\delta}\right)^{-(h+1)} \frac{2t(r-t)^m}{(r+t)^{(h+1)}}.
\end{aligned}
$$

Hence, since $(t, r) \in \Sigma_\delta$ we obtain

$$
u(t,r) \geq \frac{C_0 t^{m+1}}{r^m (r+t)^{(h+1)}} \text{ for } (t,r) \in \Sigma_\delta, \tag{3.2}
$$

where we set

$$
C_0 := \varepsilon \frac{2^{m-2} M}{\delta_m^m} \left(\frac{\delta}{1+\delta}\right)^{h+1} > 0. \tag{3.3}
$$

Let us suppose now that there exist a, b, and C, positive constants such that $u(t, r)$ satisfies the following estimate

$$
u(t,r) \geq \frac{C t^a}{r^m (r+t)^b} \text{ for } (t,r) \in \Sigma_\delta. \tag{3.4}
$$

In particular, from (3.2) and (3.3), the estimate (3.4) is true with $a = m + 1$, $b = h + 1$, $C = C_0$. By Lemma 3.1 we have

$$
u(t,r) \geq \frac{1}{8r^m} \int_0^t d\tau \int_{r-t+\tau}^{r+t-\tau} \frac{\lambda^m}{(1+\tau)^{\frac{\mu}{2}(p-1)}} |u_t(\tau,\lambda)|^p d\lambda \,.
$$

Exchanging the order of the integrals, we get

$$u(t,r) \geq \frac{1}{8r^m}\int_{r-t}^{r} d\lambda \int_0^{\lambda-(r-t)} \frac{\lambda^m}{(1+\tau)^{\frac{\mu}{2}(p-1)}}|u_t(\tau,\lambda)|^p d\tau$$
$$+\frac{1}{8r^m}\int_{r}^{r+t} d\lambda \int_0^{r+t-\lambda} \frac{\lambda^m}{(1+\tau)^{\frac{\mu}{2}(p-1)}}|u_t(\tau,\lambda)|^p d\tau. \tag{3.5}$$

We neglect the first term. By applying the Hölder inequality, since we are assuming $u(0,\lambda)=0$, we find

$$\begin{aligned}|u(r+t-\lambda,\lambda)|^p &= \left|\int_0^{r+t-\lambda} u_t(\tau,\lambda)\,d\tau\right|^p \\ &\leq \left(\int_0^{r+t-\lambda}(1+\tau)^{\frac{\mu}{2}}\,d\tau\right)^{p-1}\int_0^{r+t-\lambda}\frac{|u_t(\tau,\lambda)|^p}{(1+\tau)^{\frac{\mu}{2}(p-1)}}\,d\tau \\ &\leq (1+r+t-\lambda)^{(\frac{\mu}{2}+1)(p-1)}\int_0^{r+t-\lambda}\frac{|u_t(\tau,\lambda)|^p}{(1+\tau)^{\frac{\mu}{2}(p-1)}}\,d\tau.\end{aligned}$$

This means that

$$\int_0^{r+t-\lambda}\frac{|u_t(\tau,\lambda)|^p}{(1+\tau)^{\frac{\mu}{2}(p-1)}}\,d\tau \geq (1+r+t-\lambda)^{-(\frac{\mu}{2}+1)(p-1)}|u(r+t-\lambda,\lambda)|^p \tag{3.6}$$

Hence, by (3.5) and (3.6) for any $(t,r)\in\Sigma_\delta$ we get

$$u(t,r) \geq \frac{1}{8r^m}\int_r^{r+t}\lambda^m(1+r+t-\lambda)^{-(\frac{\mu}{2}+1)(p-1)}|u(r+t-\lambda,\lambda)|^p\,d\lambda. \tag{3.7}$$

Being $\lambda \geq r$, then $(r+t-\lambda,\lambda)\in\Sigma_\delta$. By (3.7), applying estimate (3.4) we find

$$u(t,r) \geq \frac{C^p}{8r^m(r+t)^{bp}}\int_r^{r+t}\lambda^{m(1-p)}(1+r+t-\lambda)^{-(\frac{\mu}{2}+1)(p-1)}(r+t-\lambda)^{ap}\,d\lambda.$$

While searching a finite lifespan of a solution, it is not restrictive to assume $t>1$. On the other hand $(t,r)\in\Sigma_\delta$, hence $\lambda \geq r \geq t+\delta > 1$. We deduce that

$$\begin{aligned}u(t,r) &\geq \frac{C^p}{8r^m(r+t)^{bp+(m+\frac{\mu}{2}+1)(p-1)}}\int_r^{r+t}(r+t-\lambda)^{ap}\,d\lambda \\ &= \frac{C^p t^{pa+1}}{8(pa+1)r^m(r+t)^{bp+(m+\frac{\mu}{2}+1)(p-1)}}.\end{aligned}$$

Hence, we conclude

$$u(t,r) \geq \frac{C^* t^{a^*}}{r^m (r+t)^{b^*}} \text{ for } (t,r) \in \Sigma_\delta \tag{3.8}$$

where

$$a^* = pa+1, \qquad b^* = pb + \Big(m + \frac{\mu}{2} + 1\Big)(p-1), \qquad C^* = \frac{C^p}{8(pa+1)}.$$

Let us define the sequences $\{a_k\}$, $\{b_k\}$, $\{C_k\}$

$$\begin{aligned} a_{k+1} &= pa_k + 1, \qquad a_1 = m+1, \\ b_{k+1} &= pb_k + \Big(m + \frac{\mu}{2} + 1\Big)(p-1), \quad b_1 = h+1, \\ C_{k+1} &= \frac{C_k^p}{8(pa_k+1)} = \frac{C_k^p}{8a_{k+1}}, \quad C_1 = C_0, \end{aligned}$$

where $k \in \mathbb{N}$, $k \geq 1$ and C_0 given by (3.3). Hence, we have

$$a_{k+1} = p^k\Big(m+1+\frac{1}{p-1}\Big) - \frac{1}{p-1}, \tag{3.9}$$

$$b_{k+1} = p^k\Big(h+m+\frac{\mu}{2}+2\Big) - \Big(m+\frac{\mu}{2}+1\Big), \tag{3.10}$$

$$C_{k+1} \geq H\frac{C_k^p}{p^k} \tag{3.11}$$

for some constant $H = H(p,\mu,m) > 0$ independent of k. Thanks to inequality (3.11) we get

$$C_{k+1} \geq \exp\Big(p^k\left(\log(C_0) - S_p(k)\right)\Big), \tag{3.12}$$

$$S_p(k) = \Sigma_{j=0}^k d_j, \quad d_j = \frac{j\log(p) - \log H}{p^j}, \quad d_0 = 0. \tag{3.13}$$

We remark that for j sufficiently large d_j is positive, and moreover $d_{j+1}/d_j \to 1/p$ as $j \to \infty$; thus, the ratio criterion on series with positive terms allows to conclude that there exists $S_{p,H}$ positive such that the sequence $S_p(k) \to S_{p,H}$ and for large k it holds $S_p(k) \leq S_{p,H}$. It follows that

$$C_{k+1} \geq \exp(p^k(\log C_0 - S_{p,H})). \tag{3.14}$$

Thus, we proceeded as for deducing (3.8) from (3.4). By using estimates (3.9), (3.10) and (3.14), for any $k \in \mathbb{N}$ we get

$$u(t,r) \geq \frac{(r+t)^{m+\frac{\mu}{2}+1}}{r^m t^{\frac{1}{p-1}}} \exp(p^k J(t,r)), \tag{3.15}$$

where we set

$$J(t,r) := \log(C_0) - S_{p,H} + \Big(m+1+\frac{1}{p-1}\Big)\log(t) - \Big(h+m+\frac{\mu}{2}+2\Big)\log(r+t).$$

Suppose that there exists $(t_0, r_0) \in \Sigma_\delta$ such that $J(t_0, r_0) > 0$; then, by (3.15) we would conclude that the solution blows up in finite time, in fact

$$u(t_0, r_0) \to \infty \text{ for } k \to \infty.$$

By our definition of $J(t,r)$, we find $J(t_0, r_0) > 0$ if and only if

$$\Big(\frac{1}{p-1} - h - 1 - \frac{\mu}{2}\Big)\log(t_0) > \log\Big(\frac{e^{S_{p,H}}}{C_0}\Big(2+\frac{r_0-t_0}{t_0}\Big)^{h+m+\frac{\mu}{2}+2}\Big).$$

In particular, for $\Big(t_0, t_0 + \max\{\frac{2t_0}{\delta_m}, \delta\}\Big) \in \Sigma_\delta$ it is enough to prove that

$$\Big(\frac{1}{p-1} - h - 1 - \frac{\mu}{2}\Big)\log(t_0) > \log\Big(\frac{e^{S_{p,H}}}{C_0}\Big(2+\frac{2}{\delta_m}\Big)^{h+m+\frac{\mu}{2}+2}\Big).$$

Therefore, such point $(t_0, r_0) \in \Sigma_\delta$ exists for any $h > 0$ such that

$$h < \frac{1}{p-1} - 1 - \frac{\mu}{2},$$

i.e., our decay condition (2.2) is satisfied. Under this latter assumption on h the coefficient in the left side is positive and by using (3.3) we find that $J(t_0, r_0) > 0$ once we have

$$t_0 > C\varepsilon^{-\left(\frac{1}{p-1}-h-1-\frac{\mu}{2}\right)^{-1}}, \tag{3.16}$$

where

$$C = \Big(\frac{e^{S_{p,H}}\delta_m^m}{2^{m-2}M}\Big(\frac{1+\delta}{\delta}\Big)^{h+1}\Big(2+\frac{2}{\delta_m}\Big)^{1+h+m}\Big)^{\frac{1}{\frac{1}{p-1}-h-1-\frac{\mu}{2}}},$$

which is positive. As a consequence of (3.16) we get the lifespan estimate (2.3) and we conclude the proof of Theorem 2.1.

Acknowledgments The Authors thank Prof. M. Reissig and the ISAAC board for organizing 12th ISAAC Congress and M. D'Abbicco, M. Ebert for the interesting works in the session "Recent progress in evolution equations".

References

1. V. Georgiev, S. Lucente, *Quasilinear wave equation with decaying time-potential* (submitted)
2. F. Chiarello, G. Girardi, S. Lucente, Fujita modified exponent for scale invariant damped semilinear wave equations. J. Evol. Eqs. **21**, 2735–2748 (2021)
3. A. Palmieri, M. Reissig, A competition between Fujita and Strauss type exponents for blow-up of semi-linear wave equations with scale-invariant damping and mass. J. Differ. Eq. **266**, 1176–1220 (2019)
4. H. Takamura, Blow-up for semilinear wave equations with slowly decaying data in high dimensions. Differ. Integ. Eq. **8**, 647–661 (1995)

On a Criterion for Log-Convex Decay in Non-selfadjoint Dynamics

Jon Johnsen

Abstract The short-time and global behaviour are studied for autonomous linear evolution equations defined by generators of uniformly bounded holomorphic semigroups in a Hilbert space. A general criterion for log-convexity in time of the norm of the solution is treated. Strict decrease and differentiability at the initial time results, with a derivative controlled by the lower bound of the negative generator, which is proved strictly accretive with equal numerical and spectral abscissas.

Keywords Log-convex decay · Non-selfadjoint · Hyponormal · Strictly accretive operators · Short-time behaviour

1 Introduction

The subjects here are the *global* and the *short-time* behaviour of the solutions to the Cauchy problem of an autonomous linear evolution equation, with data $u_0 \neq 0$,

$$\partial_t u + Au = 0 \quad \text{for } t > 0, \qquad u(0) = u_0 \quad \text{in } H. \tag{1.1}$$

In case the generator $-A$ is non-selfadjoint, this is particularly interesting. "Non-self-adjoint operators is an old, sophisticated and highly developed subject" to quote the recent treatise of Sjöstrand [17]; also the exposition of Helffer [5, Ch. 13] on their pseudo-spectral theory could be mentioned; or [18].

Logarithmically convex decay of the solutions was seemingly first studied in the paper [8]. This is given a more concise exposition here, with additional examples.

The main purpose below, however, is to improve the results in [8] by adding in Section 2 a much sharper necessary condition on A for the log-convex decay, leading to the improved Theorem 7 below.

J. Johnsen (✉)
Department of Mathematical Sciences, Aarhus University, Aarhus, Denmark
e-mail: jojoh@math.au.dk

P. Cerejeiras et al. (eds.), *Current Trends in Analysis, its Applications and Computation*, Research Perspectives,
https://doi.org/10.1007/978-3-030-87502-2_62

It is assumed that A is an accretive operator with domain $D(A)$ in a complex Hilbert space H, with norm $|\cdot|$ and inner product $(\cdot\,|\,\cdot)$, and that $-A$ generates a uniformly bounded, holomorphic C_0-semigroup e^{-zA} for z in an open sector having the form $\Sigma_\delta = \{\, z \in \mathbb{C} \mid -\delta < \arg z < \delta \,\}$. Focus is here on the "height" function

$$h(t) = |e^{-tA} u_0|. \tag{1.2}$$

This was shown in [8] to be a *log-convex* function, that is, for $0 \le r \le s \le t < \infty$

$$\left|e^{-sA} u_0\right| \le \left|e^{-rA} u_0\right|^{1-\frac{s-r}{t-r}} \left|e^{-tA} u_0\right|^{\frac{s-r}{t-r}}, \tag{1.3}$$

if and only if the possibly non-normal generator $-A$ has the special property that for every $x \in D(A^2)$,

$$2\big(\operatorname{Re}(Ax \,|\, x)\big)^2 \le \operatorname{Re}(A^2 x \,|\, x)|x|^2 + |Ax|^2|x|^2. \tag{1.4}$$

The present paper and [8] grew out of the author's joint work [1, 2] on the inverse heat equation and its well-posedness under the Dirichlet condition. But the main parts also apply to solutions of the similar Neumann problem studied in [9–11].

To elucidate the importance of (1.3), hence of (1.4), two remarks are made.

1° The log-convexity in (1.3) implies that the solutions u of (1.1) have important global properties in common with those of the heat equation (the case $A = -\Delta$ in $H = L_2(\Omega)$ for a bounded domain $\Omega \subset \mathbb{R}^n$). Namely, for (1.1) it holds, cf. (1.2),

(i) $h(t)$ is strictly positive,
(ii) $h(t)$ is strictly decreasing,
(iii) $h(t)$ is strictly convex.

Here the strict decrease and strict convexity combine to a noteworthy and precise dynamical property. For example, even if A has eigenvalues in $\mathbb{C} \setminus \mathbb{R}$, they do not give rise to oscillations in the *size* of the solution $e^{-tA}u_0$—this is ruled out by strict convexity, which thus can be seen as a stiffness in the decay of $h(t)$.

In addition, (1.1) also shares the short-time behaviour with the heat equation, for in terms of the numerical range $\nu(A) = \big\{\, (Ax \,|\, x) \;\big|\; x \in D(A),\ |x| = 1 \,\big\}$ and its lower bound $m(A) = \inf \operatorname{Re} \nu(A)$, the onset of decay of the height $h \in C^\infty(\,]0, \infty[\,) \cap C([0, \infty[\,)$ is constrained by the properties:

(iv) $h(t)$ is right differentiable at $t = 0$, with
(v) $h'(0) \le -m(A) < 0$ for $|u_0| = 1$, though
(vi) $h'(0) = -\operatorname{Re}(Au_0 \,|\, u_0)$ whenever $u_0 \in D(A)$, $|u_0| = 1$.

For the considered A, (iv)–(vi) follow from log-convexity; cf. the below Theorem 3.4.

More generally, one could try to work with the A that merely have strictly convex height functions, but this class is not easy to characterise. One may therefore view (1.4) as a very large class of (possibly non-normal) generators having the described dynamical properties in common with the selfadjoint cases.

2° Secondly, the operators satisfying (1.4) may be seen to comprise the A that are selfadjoint, $A^* = A$, or normal, $A^*A = AA^*$. But as observed in [1], one only needs the following two half-way houses,

$$D(A) \subset D(A^*), \qquad |Ax| \geq |A^*x| \text{ for every } x \in D(A). \tag{1.5}$$

This property is *hyponormality* for unbounded operators, as studied by Janas [7]. Clearly A is normal if and only if both A, A^* are hyponormal, so this operator class is quite general. As symmetric operators have a full inclusion $A \subset A^*$, they are also encompassed by the hyponormal class. But there is more:

Example Truly hyponormal operators are easily exemplified: for the advection-diffusion operators $A^{\pm}u = -u'' \pm u'$ in $L^2(\alpha, \beta)$, for $\alpha < \beta$ in $\mathbb{R}$, it is classical that the minimal realisation $A^{\pm}_{\text{min}}$ has the domain $D(A^{\pm}_{\text{min}}) = H^2_0(\alpha, \beta)$ because of the ellipticity (cf. [4, Thm. 6.24]). The maximal realisation has $D(A^{\pm}_{\text{max}}) = H^2(\alpha, \beta)$, for when $f = -u'' \pm u'$ holds for $u, f \in L^2$, then $-u' \pm u \in L^2$ as primitives of f, so $u' \in L^2$; hence $u'' \in L^2$. Via the formal adjoints $A^{\mp}$ this gives (cf. [4, Lem. 4.3])

$$D(A^{\pm}_{\text{min}}) \subsetneq D(A^{\mp}_{\text{max}}) = D((A^{\pm}_{\text{min}})^*). \tag{1.6}$$

Partial integration for $u \in H^2(\alpha, \beta)$ yields $\| -u'' \pm u'\|^2 = \|u''\|^2 + \|u'\|^2 \mp (|u'(\beta)|^2 - |u'(\alpha)|^2)$, where the last two terms vanish for $u \in D(A^{\pm}_{\text{min}})$, so that $\|A^{\pm}_{\text{min}}u\| = \|(A^{\pm}_{\text{min}})^*u\|$. Hence the $A^{\pm}_{\text{min}}$ are nonnormal, but nonetheless hyponormal.

That every hyponormal operator A in H necessarily satisfies the log-convexity condition (1.4) is recalled from [8] for the reader's convenience: the inclusion $D(A) \subset D(A^*)$ gives at once for $x \in D(A^2)$ that

$$2(\text{Re}(Ax \mid x))^2 \leq \frac{1}{2}|(A + A^*)x|^2|x|^2 \leq \big(|Ax|^2 + \text{Re}(A^2x \mid x)\big)|x|^2, \tag{1.7}$$

for in the last step the inequality in (1.5) gives, because $D(A^2) \subset D(A) \subset D(A^*)$, that

$$|(A + A^*)x|^2 = |Ax|^2 + |A^*x|^2 + 2\,\text{Re}(Ax \mid A^*x) \leq 2|Ax|^2 + 2\,\text{Re}(A^2x \mid x). \tag{1.8}$$

It is noteworthy, though, that whilst hyponormality expresses a certain interrelationship between A and its adjoint, criterion (1.4) instead involves A and its square A^2. In addition it was exemplified in [8] that (1.4) is unfulfilled for certain explicitly given $A \in \mathbb{B}(H)$, even for some symmetric $n \times n$-matrices, $n \geq 2$.

Moreover, the mixed Dirichlet–Neumann and Dirichlet–Robin realisations A^{+}_{DN} and A^{-}_{DR}, respectively, are variational and elliptic, so they generate holomorphic semigroups in $L^2(\alpha, \beta)$. But none of them are hyponormal, cf. Example 3 below. This delicate situation around the $A^{\pm}$ should motivate the present analysis of the generators that have log-convex decay. It is envisaged that (1.4) can give interesting examples when A is a suitable realisation of a partial differential operator.

In the above discussion of log-convexity of $h(t)$, its importance for the dynamics of (1.1) was explained in (i)–(vi) via the more general strict convexity. So it is natural to pose the question: does log-convexity have advantages in itself? At least it gives rise to the (perhaps new) proof technique used in the next section.

2 A New Necessary Condition for Log-Convex Decay

The reader is assumed familiar with semigroup theory, for which [3, 15] could be references; the simpler Hilbert space case is exposed e.g. in [4, Ch. 14].

It is recalled that there is a bijection between the C_0-semigroups e^{-tA} in $\mathbb{B}(H)$ that are uniformly bounded, i.e. $\|e^{-tA}\| \leq M$ for $t \geq 0$, and holomorphic in $\Sigma_\delta \subset \mathbb{C}$ for $\delta \in]0, \frac{\pi}{2}[$, and the densely defined, closed operators A in H satisfying a resolvent estimate $|\lambda| \|(A + \lambda I)^{-1}\| \leq C$ for all $\lambda \in \{0\} \cup \Sigma_{\delta+\pi/2}$.

It is classical that, since $\sigma(A) \subset \{ z \in \mathbb{C} \mid \operatorname{Re} z \geq \varepsilon \}$ for some $\varepsilon > 0$, there is a bound $\|e^{-tA}\| \leq M_\eta e^{-t\eta}$ for $t \geq 0$, $0 < \eta < \varepsilon$. This yields the crude decay estimate

$$h(t) \leq M_\eta e^{-t\eta} |u_0|. \tag{2.1}$$

Here η is restricted by $0 \leq \eta < \underline{\sigma}(A)$, using the spectral abscissa of A,

$$\underline{\sigma}(A) = \inf \operatorname{Re} \sigma(A). \tag{2.2}$$

The below analyses all rely on the recent result that such semigroups consist of *injections*, which, mentioned for precision, holds without the uniform boundedness:

Lemma 2.1 ([9] and [10]) *If $-A$ generates a holomorphic semigroup e^{-zA} in $\mathbb{B}(X)$ for some complex Banach space X, and e^{-zA} is holomorphic in the open sector $\Sigma_\delta \subset \mathbb{C}$ given by $|\arg z| < \delta$ for some $\delta > 0$, then e^{-zA} is injective on X for each such z.*

The injectivity is clearly equivalent to the geometric property that two solutions $e^{-tA}v$ and $e^{-tA}w$ to the differential equation $u' + Au = 0$ cannot have any points of confluence in X for $t > 0$ when $v \neq w$. One obvious consequence of this is its backward uniqueness: $u(T) = 0$ implies $u(t) = 0$ for $0 \leq t \leq T$.

Lemma 2.1 is also important because it allows a calculation of $h'(t)$, $h''(t)$, using differential calculus in Banach spaces as exposed in e.g. [6, Ch. 1] or [13]. This uses that $u(t) = e^{-tA}u_0 \neq 0$ for all $t > 0$ when $u_0 \neq 0$, cf. Lemma 2.1, whence $h(t) > 0$:

As the inner product on H, despite its sesquilinearity, is differentiable on the induced real vector space $H_{\mathbb{R}}$ with derivative $(\cdot \mid y) + (x \mid \cdot)$ at $(x, y) \in H_{\mathbb{R}} \oplus H_{\mathbb{R}}$, which applies to the composite map between open sets $\mathbb{R}_+ \to (H_{\mathbb{R}} \setminus \{0\}) \oplus (H_{\mathbb{R}} \setminus \{0\}) \to \mathbb{R}_+ \to \mathbb{R}_+$ given by $t \mapsto \sqrt{(u(t) \mid u(t))}$, the Chain Rule for real Banach spaces gives

$$h'(t) = \frac{(u' \mid u) + (u \mid u')}{2\sqrt{(u \mid u)}} = -\frac{\operatorname{Re}(Au \mid u)}{|u|}; \tag{2.3}$$

$$h''(t) = \frac{(A^2u \mid u) + 2(Au \mid Au) + (u \mid A^2u)}{2|u|} - \frac{(\operatorname{Re}(Au \mid u))^2}{|u|^3}. \tag{2.4}$$

The second line follows from the first, since $u'' = (e^{-tA}u_0)'' = A^2 e^{-tA}u_0 = A^2u$.

When A satisfies (1.4), the short-time behaviour at $t = 0$ is via the information on $h'(0)$ in (iv)–(vi) specifically controlled by $\nu(A)$, and not by its spectrum $\sigma(A)$. Moreover, the proofs in [8] also gave that $h'(0) = \inf h' < 0$, which when combined with (vi) shows that A is a bit better than accretive ($m(A) \geq 0$) because its numerical range is contained in the *open* right half-plane, $\nu(A) \subset \{ z \in \mathbb{C} \mid \operatorname{Re} z > 0 \}$. It seems useful to call A a *positively accretive* operator, when it has this property (milder than strict accretivity [12]), and it was shown in [8] that (1.4) implies this.

But there is a significantly sharper necessary condition, which is given already now because of the novelty. Its proof exploits the log-convexity directly:

Proposition 2.2 *If the generator A has log-convex height functions $h(t)$ on $[0, \infty[$ for every $u_0 \neq 0$ and the one-sided derivative $h'(0)$ exists and fulfils that $h'(0) = -\operatorname{Re}(Au_0 \mid u_0)$ when $u_0 \in D(A)$ satisfies $|u_0| = 1$, then A is strictly accretive and*

$$m(A) = \underline{\sigma}(A) > 0. \tag{2.5}$$

Proof The log-convexity means that the continuous function $\log h(t)$ is convex on $[0, \infty[$, so it is classical that its graph lies entirely above each of its half-tangents. Applying this at $t = 0$ for $u_0 \in D(A)$, $|u_0| = 1$, and invoking (2.1), one finds that

$$\log h(0) + t\frac{h'(0)}{h(0)} \leq \log h(t) \leq \log M_\eta - t\eta \qquad \text{for } t > 0. \tag{2.6}$$

Indeed, $h(t)$ extends to $t < 0$ in a C^1-fashion along its (half-)tangent at $t = 0$, after which the Chain Rule applies to $\log h(t)$. (Differentiability of $h(t)$ holds for $t > 0$ by (2.3), for $t \leq 0$ by construction.)

Now, the above inequalities being valid for all $t > 0$, the graphs of the two first order polynomials cannot intersect, so their slopes fulfil $h'(0) \leq -\eta$ (as $h(0) = |u_0| = 1$). Hence $-h'(0) \geq \underline{\sigma}(A)$, as the spectral abscissa is the supremum of the possible η; cf. (2.1) ff. The assumption on $h'(0)$ in the statement now gives that for any $u_0 \in D(A)$ having $|u_0| = 1$,

$$\mathrm{Re}(Au_0 \mid u_0) \geq \underline{\sigma}(A). \tag{2.7}$$

This entails the inequality $m(A) \geq \underline{\sigma}(A)$, hence strict accretivity since $\underline{\sigma}(A) > 0$.

However, the strict inequality $m(A) > \underline{\sigma}(A)$ is impossible, for it would imply that $\overline{\nu(A)}$ is contained in the closed half-plane $\Pi_{m(A)} = \{ z \mid \mathrm{Re}\, z \geq m(A) \}$ and that $\mathbb{C} \setminus \Pi_{m(A)} = \{ z \mid \mathrm{Re}\, z < m(A) \}$ contains some $\lambda \in \sigma(A)$ as well as $\mathbb{R}_-$ in the resolvent set $\rho(A)$; but then $\sigma(A)$ and $\rho(A)$ intersect the same connectedness component of $\mathbb{C} \setminus \overline{\nu(A)}$, contradicting [15, Thm. 1.3.9]. Hence $m(A) = \underline{\sigma}(A)$ as claimed. ∎

3 Main Results

For the reader's sake, some basics are recalled: a positive function $f : \mathbb{R} \to [0, \infty[$ is *log-convex* if $\log f(t)$ is convex, or more precisely, for all $r \leq t$ in $\mathbb{R}$ and for $0 < \theta < 1$,

$$f((1-\theta)r + \theta t) \leq f(r)^{1-\theta} f(t)^{\theta}. \tag{3.1}$$

Note, though, that t^{θ} and $t^{1-\theta}$ do not require their continuous extensions to $t = 0$ when we take $f = h$ below, for since e^{-tA} is holomorphic, $h(t) > 0$ or equivalently $e^{-tA}u_0 \neq 0$ holds for $t \geq 0$ by Lemma 2.1.

For the intermediate point $s = (1-\theta)r + \theta t$ an exercise yields $\theta = (s-r)/(t-r)$, so log-convexity therefore means that, for $0 \leq r < s < t$,

$$f(s) \leq f(r)^{1-\frac{s-r}{t-r}} f(t)^{\frac{s-r}{t-r}}. \tag{3.2}$$

This leads to (1.3) for the semigroup. There A is just a positive scalar if $\dim H = 1$, so (1.3) is then an identity. For $\dim H > 1$, the possible validity of (1.3) is by no means obvious to discuss for the operator function e^{-tA} in $\mathbb{B}(H)$.

Log-convexity is stronger than strict convexity for non-constant functions:

Lemma 3.1 *If $f : I \to [0, \infty[$ is log-convex on an interval or halfline $I \subset \mathbb{R}$, then f is convex—and if f is not constant in any subinterval, then f is* strictly *convex on I.*

Proof Convexity on I follows from Young's inequality for the dual exponents $1/\theta$ and $1/(1-\theta)$:

$$f((1-\theta)r + \theta t) \leq f(r)^{1-\theta} f(t)^{\theta} \leq (1-\theta) f(r) + \theta f(t). \tag{3.3}$$

In case $f(r) \neq f(t)$, the last inequality will be strict, as equality holds in Young's inequality if and only if the numerators are identical (cf. [14, p. 14]). This yields strict convexity in this case.

If there is a common value $C = f(r) = f(t)$ for some $r < t$ in I, there is by assumption a $u \in]r, t[$ so that $f(u) \neq f(r)$, and because of the convexity of f this entails that $f(u) < f(r) = f(t)$: when $r < s \leq u$ one may write $s = (1-\theta)r + \theta u$ and $s = (1-\omega)r + \omega t$ for suitable $\theta, \omega \in]0, 1[$, so clearly

$$\begin{aligned} f(s) &\leq (1-\theta)f(r) + \theta f(u) \\ &< (1-\theta)f(r) + \theta f(t) = C = (1-\omega)f(r) + \omega f(t); \end{aligned} \tag{3.4}$$

similarly for $u \leq s < t$; so f is strictly convex. ∎

As examples it is noted that whilst e^t is log-convex, $f(t) = e^t - 1$ is not log-convex as $(\log f)'' < 0$. However, when $f\colon I \to]0, \infty[$ is log-convex, so is the stretched function defined for $a < b$ in I as

$$f_{a,b}(t) = \begin{cases} f(t) & \text{for } t < a, \\ f(a) & \text{for } a \leq t < b, \\ f(t-b) & \text{for } b \leq t. \end{cases} \tag{3.5}$$

This follows from the geometrically obvious fact that the convexity of $\log f$ survives the stretching. Since $f_{a,b}$ clearly is not strictly convex, the last assumption of Lemma 3.1 is necessary. Moreover, a small exercise yields, cf. [8],

Lemma 3.2 *If $f\colon [0, \infty[\to \mathbb{R}_+$ is convex and $f(t) \to 0$ for $t \to \infty$, then f is* strictly *monotone decreasing.*

By now it is obvious that if a height function $h(t)$ is log-convex on $[0, \infty[$ for some $u_0 \neq 0$, it fulfils the first assumption in Lemma 3.2 by the convexity statement in Lemma 3.1, and the second assumption holds because of (2.1). Therefore such $h(t)$ is necessarily strictly decreasing on $[0, \infty[$—hence non-constant in any subinterval, and by Lemma 3.1 therefore strictly convex.

That $h(t) > 0$ allows an analysis of its log-convexity using a characterisation of the log-convex C^2-functions as the set of solutions to a differential inequality:

Lemma 3.3 *If $f \in C([0, \infty[, \mathbb{R}_+)$ is C^2 for $t > 0$, the following are equivalent:*

(I) *$f'(t)^2 \leq f(t)f''(t)$ holds whenever $0 < t < \infty$.*
(II) *$f(t)$ is log-convex on the open halfline $]0, \infty[$, cf. (3.2).*

In the affirmative case $f(t)$ is log-convex also on the closed halfline $[0, \infty[$.

Proof By the assumptions $F(t) = \log f(t)$ is defined for $t \geq 0$ and C^2 for $t > 0$ and

$$F''(t) = \left(\frac{f'(t)}{f(t)}\right)' = \frac{f''(t)f(t) - f'(t)^2}{f(t)^2}. \tag{3.6}$$

Hence (I) is equivalent to $F''(t) \geq 0$ for $t > 0$, which is the criterion for the C^2-function F to be convex for $t > 0$; which is a paraphase of the condition (II) for log-convexity of the positive function $f(t)$ for $t > 0$.

Letting $r \to 0^+$ for fixed $s < t$, the continuity of $f(r)$ and of, say $\exp(\frac{t-s}{t-r} \log f(r))$, yields that (3.2) is valid for $0 = r < s < t$. So f is log-convex on $[0, \infty[$. ■

The formulation of the lemma was inspired by the discussion of convexity notions in [14]. Whilst f in C^2 is convex if and only if $f'' \geq 0$, this positivity is clearly fulfilled if f satisfies (I), as $f(t) > 0$ is assumed—but the positivity then holds in a qualified way, equivalent to log-convexity, since (I) $\iff$ (II).

The differential inequality in (I) of Lemma 3.3 is straightforwardly seen to amount to the following for $h(t)$, cf. (2.3)–(2.4),

$$2(\mathrm{Re}(Au \mid u))^2 \leq \big(\mathrm{Re}(A^2u \mid u) + |Au|^2\big)|u|^2. \tag{3.7}$$

Clearly this is fulfilled for every $t > 0$ when A fulfils (1.4) above, for $u(t) = e^{-tA}u_0$ belongs to the subspace $D(A^n) \subset D(A^2)$ for every $n \geq 2$, and all $u_0 \in H$, when the semigroup is holomorphic. Moreover, the continuity of $h(t)$ and of its derivatives h', h'' given above show that $h \in C^2$ for $t > 0$. So according to Lemma 3.3, condition (1.4) implies that $h(t) = |e^{-tA}u_0|$ is log-convex on the closed half-line $[0, \infty[$.

Conversely, when the height function $h(t)$ is log-convex for each $u_0 \neq 0$, then the generator $-A$ fulfils (1.4). Indeed, h then fulfils (I) above by the log-convexity, hence (3.7) holds. Especially it is seen by insertion of an arbitrary $u_0 \in D(A^2)$ in (3.7) and commutation of A and A^2 with the semigroup that

$$2(\mathrm{Re}(e^{-tA}Au_0 \mid e^{-tA}u_0))^2 \leq \Big(\mathrm{Re}(e^{-tA}A^2u_0 \mid e^{-tA}u_0) + |e^{-tA}Au_0|^2\Big)|e^{-tA}u_0|^2. \tag{3.8}$$

By passing to the limit for $t \to 0^+$ it follows by continuity that (1.4) holds for $x = u_0$.

Consequently (1.4) characterises the generators $-A$ of uniformly bounded, analytic semigroups having log-convex height functions for all non-trivial initial data.

The above discussion now allows the following sharpening of [8, Thm. 2.5]:

Theorem 3.4 *When $-A$ denotes a generator of a uniformly bounded, holomorphic C_0-semigroup e^{-tA} in a complex Hilbert space H, then the following are equivalent:*

(I) $2\big(\mathrm{Re}(Ax \mid x)\big)^2 \leq \mathrm{Re}(A^2x \mid x)|x|^2 + |Ax|^2|x|^2$ *for every* $x \in D(A^2)$.

(II) $h(t) = |e^{-tA}u_0|$ *is log-convex for every* $u_0 \neq 0$*, that is,*

$$\big|e^{-sA}u_0\big| \leq \big|e^{-rA}u_0\big|^{\frac{t-s}{t-r}} \big|e^{-tA}u_0\big|^{\frac{s-r}{t-r}}. \tag{3.9}$$

whenever $0 \leq r < s < t$,

In the affirmative case, $h(t)$ is for $u_0 \neq 0$ strictly positive, strictly decreasing and strictly convex on the closed halfline $[0, \infty[$ and moreover differentiable from the right at $t = 0$, with a derivative in $[-\infty, 0[$, which for $|u_0| = 1$ satisfies

$$h'(0) = \inf_{t>0} h'(t) \leq -m(A) < 0; \tag{3.10}$$

and if $u_0 \in D(A)$ with $|u_0| = 1$, then $h \in C^1([0, \infty[\,, \mathbb{R}) \bigcap C^\infty(\mathbb{R}_+, \mathbb{R})$ and

$$h'(0) = -\operatorname{Re}(Au_0 \mid u_0). \tag{3.11}$$

Furthermore $\underline{\sigma}(A) = m(A) > 0$ holds, in particular such A are strictly accretive.

Proof That (I) $\Longleftrightarrow$ (II) was seen in the considerations after Lemma 3.3. The strict positivity was derived after Lemma 2.1, strict decrease and strict convexity after Lemma 3.2.

Convexity of h entails $h''(t) \geq 0$ for $t > 0$, so $h'(t)$ is increasing on $\mathbb{R}_+$ and $\lim_{t \to 0^+} h'(t) = \inf_{t>0} h'$ exists in $[-\infty, 0[$, as $h' < 0$. By the Mean Value Theorem, some $t' \in \,]0, t[$ fulfils

$$(h(t) - h(0))/t = h'(t') < 0. \tag{3.12}$$

Therefore $h(t)$ is (extended) differentiable from the right at $t = 0$, with $h'(0) = \inf h'$. Since the strong continuity and strict decrease of h gives $|e^{-tA}u_0| \nearrow 1$ for $t \to 0^+$, an application of (2.3) yields

$$h'(0) = \inf h' \leq \limsup_{t \to 0^+} h'(t) \leq \limsup_{t \to 0^+} (-m(A)|e^{-tA}u_0|) \leq -m(A). \tag{3.13}$$

In case $u_0 \in D(A)$ and $|u_0| = 1$, one can exploit that $h'(0) = \lim_{t \to 0^+} h'(t)$ by commuting A with e^{-tA} in (2.3), which in the limit gives, because of the strong continuity at $t = 0$ and the continuity of inner products,

$$h'(0) = \lim_{t \to 0^+} -\operatorname{Re}(e^{-tA}Au_0 \mid e^{-tA}u_0) = -\operatorname{Re}(Au_0 \mid u_0). \tag{3.14}$$

In addition, it is seen that $h'(0)$ is a real number for $u_0 \in D(A)$, so $h \in C^1([0, \infty[\,, \mathbb{R})$ for such u_0. For general $u_0 \in H$ it follows from the Chain Rule that $h \in C^\infty(\mathbb{R}_+, \mathbb{R})$.

Finally, the last line of the statement results from Proposition 2.2. ∎

The conclusions of the theorem apply in particular to every hyponormal generator $-A$, cf. the account in (1.7) that such A always satisfy the criterion (1.4).

It is instructive to review condition (1.4) in case the generator A is variational. That is, for some Hilbert space $V \subset H$ algebraically, topologically and densely and some sesquilinear form $a\colon V \times V \to \mathbb{C}$, which is V-bounded and V-elliptic in the sense that (with $\|\cdot\|$ denoting the norm in V) for some $C_0 > 0$

$$\operatorname{Re} a(u,u) \geq C_0 \|u\|^2 \qquad \text{for all } u \in V, \tag{3.15}$$

it holds for A that $(Au \mid v) = a(u,v)$ for all $u \in D(A)$ and $v \in V$. Lax–Milgram's lemma on the properties of A is exposed in [4, Ch. 12] and [5, Ch. 3]. It is classical that $-A$ generates a holomorphic semigroup e^{-tA} in $\mathbb{B}(H)$; an explicit proof is e.g. given in [1, Lem. 4].

For such A, the log-convexity criterion (1.4) can be stated for V-elliptic variational A as a comparison of sesquilinear forms, cf. [8],

$$\big(\operatorname{Re} a(u,u)\big)^2 \leq \operatorname{Re}\big(a_{\mathrm{Re}}(Au,u)\big)(u \mid u) \qquad \text{for } u \in D(A^2). \tag{3.16}$$

Example To see that variational operators need not be hyponormal, one may take $H = L_2(\alpha,\beta)$, with norm $\|f\|_0 = (\int_\alpha^\beta |f(x)|^2\,dx)^{1/2}$, for reals $\alpha < \beta$ and let $V = \{ v \in H^1(\alpha,\beta) \mid u(\alpha) = 0 \}$ be a subspace of the first Sobolev space with norm given by $\|f\|_1^2 = \int_\alpha^\beta (|f(x)|^2 + |f'(x)|^2)\,dx$ and the sequilinear forms

$$a(u,v) = \int_\alpha^\beta u'(x)\overline{v'(x)} + u'(x)\overline{v(x)}\,dx. \tag{3.17}$$

This is clearly V-bounded, and also V-elliptic: partial integration gives $\operatorname{Re} a(u,u) = \|u'\|_0^2 + \frac{1}{2}|u(\beta)|^2$, and $\operatorname{Re} a(u,u) \geq C_0\|u\|_1^2$ follows for all $u \in V$ and e.g. $C_0 = \min(\frac{1}{2}, (\beta-\alpha)^{-2})$ by ignoring the last term and using Poincaré's inequality (its standard proof, e.g. [4, Thm. 4.29], applies to V).

The induced A_{DN}^+ acts in the distribution space $\mathscr{D}'(\alpha,\beta)$ of Schwartz [16] as $A_{\mathrm{DN}}^+ u = -u'' + u'$, which is the advection-diffusion operator with a mixed Dirichlet and Neumann condition,

$$D(A_{\mathrm{DN}}^+) = \big\{ u \in H^2(\alpha,\beta) \bigm| u(\alpha) = 0,\ u'(\beta) = 0 \big\}. \tag{3.18}$$

(The Dirichlet realisation of $u' - u''$ has been studied at length; cf. [18, Ch. 12].)

As $(A_{\mathrm{DN}}^+)^*$ is induced by $\overline{a(v,u)}$, one finds similarly $(A_{\mathrm{DN}}^+)^* u = -u'' - u' = A_{\mathrm{DR}}^- u$ with the domain characterised by a mixed Dirichlet and Robin condition,

$$D((A_{\mathrm{DN}}^+)^*) = D(A_{\mathrm{DR}}^-) = \big\{ u \in H^2(\alpha,\beta) \bigm| u(\alpha) = 0,\ u'(\beta) + u(\beta) = 0 \big\}. \tag{3.19}$$

As both $D(A^+_{\mathrm{DN}})$ and $D((A^+_{\mathrm{DN}})^*)$ contain functions outside their intersection, (1.5) shows that neither A^+_{DN} nor $(A^+_{\mathrm{DN}})^* = A^-_{\mathrm{DR}}$ is hyponormal. This is part of the motivation for the study of condition (1.4).

References

1. A.-E. Christensen, J. Johnsen, Final value problems for parabolic differential equations and their well-posedness. Axioms **7**, Article No. 31, 1–36 (2018)
2. A.-E. Christensen, J. Johnsen, On parabolic final value problems and well-posedness. C. R. Acad. Sci. Paris, Ser. I **356**, 301–305 (2018)
3. K.-J. Engel, R. Nagel, *One-Parameter Semigroups for Linear Evolution Equations*. Graduate Texts in Mathematics, vol. 194 (Springer, New York, 2000)
4. G. Grubb, *Distributions and Operators*. Graduate Texts in Mathematics, vol. 252 (Springer, New York, 2009)
5. B. Helffer, *Spectral Theory and Its Applications*. Cambridge Studies in Advanced Mathematics, vol. 139 (Cambridge University Press, Cambridge, 2013)
6. L. Hörmander, *The Analysis of Linear Partial Differential Operators*. Grundlehren der Mathematischen Wissenschaften (Springer, Berlin, 1983)
7. J. Janas, On unbounded hyponormal operators. III. Studia Math. **112**(1), 75–82 (1994)
8. J. Johnsen, Characterization of log-convex decay in non-selfadjoint dynamics. Electron. Res. Announc. Math. Sci. **25**, 72–86 (2018)
9. J. Johnsen, A class of well-posed parabolic final value problems, Advances in Microlocal and Time-Frequency Analysis (eds. Boggiato, P. et al.). Birkhäuser, 2020. Appl. Num. Harm. Ana. **99**, 259–280 (2020)
10. J. Johnsen, Well-posed final value problems and Duhamel's formula for coercive Lax–Milgram operators. Elec. Res. Arch. **27**, 20–36 (2019)
11. J. Johnsen, Isomorphic well-posedness of final value problems for the heat equation with the homogeneous Neumann condition. Integr. Equ. Oper. Theory **92**, 43 (2020)
12. T. Kato, *Perturbation Theory for Linear Operators*. Classics in Mathematics (Springer, Berlin, 1995). Reprint of the 1980 edition
13. S. Lang, *Differential Manifolds* (Addison Wesley, Boston, 1972)
14. C.P. Niculescu, L.-E. Persson, *Convex Functions and Their Applications. A Contemporary Approach*. CMS Books in Mathematics/Ouvrages de Mathématiques de la SMC, vol. 23 (Springer, New York, 2006)
15. A. Pazy, *Semigroups of Linear Operators and Applications to Partial Differential Equations*. Applied Mathematical Sciences, vol. 44 (Springer, New York, 1983)
16. L. Schwartz, *Théorie des distributions*, Revised and enlarged ed. (Hermann, Paris, 1966)
17. J. Sjöstrand, *Non-self-Adjoint Differential Operators, Spectral Asymptotics and Random Perturbations*. Pseudodifferential Operators: Theory and Applications, vol. 14 (Birkhäuser, Basel, 2019)
18. L.N. Trefethen, M. Embree, *Spectra an Pseudospectra. The Behaviour of Nonnormal Matrices and Operators* (Princeton University Press, Princeton, 2005)

A Uniform Resolvent Estimate for a Helmholtz Equation with Some Large Perturbations in An Exterior Domain

Hideo Nakazawa

Abstract We derive a uniform resolvent estimate for a stationary dissipative wave equation without smallness conditions. Existing results required a smallness condition for the coefficient of the dissipation. This paper removes the assumption of the smallness. Our proof is based on an energy estimate for stationary problems.

Keywords Uniform resolvent estimate · Large perturbation · Helmholtz equation · Energy-dependent potential · Dissipative wave equation

Mathematics Subject Classification (2010) 35J05, 47A40, 81Q12

1 Introduction and Result

Several results on uniform resolvent estimates for Helmholtz or stationary Schrödinger equations have been obtained so far by many authors since Kato-Yajima [1] and Watanabe [12]. In particular, some uniform resolvent estimates have been proved in Mochizuki [4] and Mochizuki–Nakazawa [6] for a stationary Schrödinger equation in a magnetic field in an exterior domain under some smallness conditions. In this paper, we see that even without smallness condition for the perturbation term, a uniform resolvent estimate for the stationary problem of the wave equation with an energy-dependent potential as perturbation term can be derived. The key point of the proof is the use of a suitable inequality derived in Nakazawa [10], which is an energy type estimate of solutions for stationary problems.

H. Nakazawa (✉)
Department of Mathematics, Faculty of Medicine, Nippon Medical School, Musashino City, Tokyo, Japan
e-mail: nakazawa-hideo@nms.ac.jp

P. Cerejeiras et al. (eds.), *Current Trends in Analysis, its Applications and Computation*, Research Perspectives,
https://doi.org/10.1007/978-3-030-87502-2_63

Let $\Omega \subseteq \mathbb{R}^N$ ($N \geq 3$) be an exterior domain outside the bounded star-shaped obstacle $\mathcal{O}$ with smooth boundary

$$\Omega = \mathbb{R}^N \setminus \mathcal{O}, \quad \mathcal{O} \subset \mathbb{R}^N$$

($\mathcal{O}$ may be the empty set). We consider the following Helmholtz equations with energy-dependent potentials

$$\begin{cases} \left(-\Delta - i\kappa b(x) - \kappa^2\right) u(x) = f(x), \ (\kappa \in \mathbb{C}), & x \in \Omega, \\ u(x) = 0, \quad x \in \partial\Omega, \end{cases} \tag{1.1}$$

where we assume that the function $b(x)$ is a non-negative smooth function, $f(x)$ belongs to a subset of $L^2(\Omega)$, and $\kappa \in \mathbb{C}$. This equation is the stationary problem of the following dissipative wave equations

$$\begin{cases} w_{tt} - \Delta w + b(x)w_t = f(x)e^{-i\kappa t}, & (x,t) \in \Omega \times (0,\infty), \\ w(x,0) = u(x), \quad w_t(x,0) = -i\kappa u(x), & x \in \Omega, \\ w(x,t) = 0, \quad x \in \partial\Omega \end{cases} \tag{1.2}$$

obtained by setting $w(x,t) = e^{-i\kappa t}u(x)$. In the study of the asymptotic behavior of solutions and the scattering theory for (1.2), stationary methods in scattering theory often play an important role [2, 3, 5–7, 9]. In particular, it is pointed out in [8] that the uniform resolvent estimate in complex upper-half plane for (1.1) is useful for proving the principle of limiting amplitude for (1.2). In this process, we need to assume a smallness condition for the function $b(x)$. However, in [2], the principle of limiting amplitude is obtained without such a smallness condition. The authors of [2] studied the initial value problem in $\mathbb{R}^3$ by using the explicit formula of the fundamental solution of the Helmholtz equation in $\mathbb{R}^3$.

This paper aims to remove the smallness conditions for $b(x)$ to derive the uniform resolvent estimate for (1.1) in the complex upper-half plane for any $N \geq 3$. In the followings, $r = |x|$, and the space $C^k(X)$ denotes the set of all k times continuously differentiable functions defined on a set X. We put

$$u_r = \frac{x}{r} \cdot \nabla u, \quad D_r^{\pm} = u_r + \frac{N-1}{2r}u \mp i\kappa u \ (\pm\Im\kappa \geq 0). \tag{1.3}$$

The result is given as followings

Theorem 1.1 *Assume that the space dimension N satisfies $N \geq 3$, and the function $b(x) \in C(\Omega)$ satisfies*

$$0 \leq b(x) \leq b_1(1+r)^{-\frac{3}{2}-b_2}, \qquad x \in \Omega \tag{1.4}$$

for some positive constants b_1 and b_2. Let $u \in C^2(\Omega)$ a solution of (1.1) with $\Im\kappa \geq 0$. Then it holds that

$$|\kappa|^2 \int_\Omega (1+r)^{-1-\delta}|u(x)|^2dx + \int_\Omega \{1+(\Im\kappa)r\}\left|D_r^+u(x)\right|^2 dx$$

$$+\int_\Omega \frac{|u(x)|^2}{r^2}dx \leq C\int_\Omega (1+r)^{1+\max\{1,\delta\}}|f(x)|^2dx$$

for some $\delta > 0$, where $C > 0$ is independent of κ.

Remark 1.2 A similar proof is available for the usual stationary Schrödinger equation. The details will be published elsewhere.

Remark 1.3 In [2], the function $b(x)$ satisfies

$$0 \leq b(x) \leq b_0 r^{-3-\delta}$$

for some positive (non-small) b_0 and δ and large $r \gg 1$. Our result cover $b_2 \leq \frac{3}{2}+\delta$.

At the end of this section, our paper is organized as follows. In Sect. 2, some inequalities are derived. These play a fundamental role in the proof of Theorem 1.1. In Sect. 3, Theorem 1.1 is proved under a smallness assumption on b_1. In Sect. 4, another inequality is derived, which works for proving Theorem 1.1 without smallness. The mechanism is explained in the final section.

2 Some Inequalities

In this section, let $u = u(x)$ be a solution of (1.1).

Lemma 2.1 *Assume κ satisfies $\Im\kappa \geq 0$. Let two functions $\varphi = \varphi(r)$ and $\psi = \psi(r)$ are both non-negative smooth functions satisfying*

$$\varphi(\cdot) \in L^1\left([0,\infty);\mathbb{R}_+\right)$$

and

$$\psi_r \leq \frac{\psi}{r}.$$

Put

$$g(r) = \int_r^\infty \varphi(s)ds.$$

For any small positive ε_1 *and* ε_2*, it holds that*

$$|\kappa|^2 \int_\Omega \left\{(1-\varepsilon_1)\varphi + \frac{b\psi}{2}\right\} |u(x)|^2 dx + \int_\Omega \left(\varphi - \frac{b\psi}{2}\right) \left|u_r + \frac{N-1}{2r}u\right|^2 dx$$
$$+ \int_\Omega \left\{\frac{\psi_r}{2} + \Im(\kappa)\psi + \left(\frac{b\psi}{2} - \varphi\right) - \varepsilon_2\right\} |D_r^+ u|^2 dx + \int_\Omega W_N |u|^2 dx \quad (2.1)$$
$$\leq \frac{2}{\varepsilon_1}\int_\Omega g^2 \varphi^{-1} |f|^2 dx + \frac{1}{4\varepsilon_2}\int_\Omega \psi^2 |f|^2 dx, \qquad (\Im\kappa \geq 0),$$

where

$$W_N = -\frac{(N-1)(N-3)}{8}\left(\frac{\psi}{r^2}\right)_r + \Im(\kappa)\frac{(N-1)(N-3)}{4r^2}\psi.$$

Proof We shall state only the outline of the proof of Lemma 2.1. See [9] for details.

A first identity is obtained from ($B_R = \{x \in \Omega \mid |x| < R\}$, $R > 0$)

$$\int_0^\infty \varphi(r) \times \Re\left\{\int_{B_R} (\text{eq.}(1.1)) \times \left(\overline{i\kappa u}\right) dx\right\} dr, \tag{2.2}$$

where $\overline{i\kappa u}$ denotes the complex conjugate of $i\kappa u$. Put $v = e^\rho u$ and $h = e^\rho f$ to lead to a second inequality, where

$$\rho = i\kappa r + \frac{N-1}{2r}\log r.$$

Then v solves

$$-\Delta v + 2\rho_r v_r + \tilde{b} v = h, \tag{2.3}$$

where

$$\tilde{b} = -i\kappa b + \frac{(N-1)(N-3)}{4r^2}.$$

Then we can derive

$$\Re\left\{\int_\Omega (\text{eq.}(2.3)) \times \overline{v_r}\psi dx\right\}. \tag{2.4}$$

Combining these two identities (2.2) and (2.4) with a Young type inequality, we obtain the desired results. □

Next, let us explain some Hardy type inequalities related to the operator $D_r^{\pm}$ in (1.3).

Lemma 2.2 ([7, 10, 11]) Let $\Omega = \mathbb{R}^N$ or $\Omega = \mathbb{R}^N \setminus \mathcal{O}$ (The assumption that the obstacle is star-shaped is unnecessary). Assume that $N \geq 1$, $v \in C_0^{\infty}(\Omega)$, $(0 \leq)\phi(r) \in C^1([0, \infty))$, $h(r) \in C^1([0, \infty))$. Then it holds that

$$\int_{\Omega} \phi |D_r^{\pm} v|^2 dx \geq \pm\Im(k) \int_{\Omega} 2\phi h |v|^2 dx + \int_{\Omega} \Phi_{\phi,h} |v|^2 dx, \quad (\pm\Im k \geq 0), \quad (2.5)$$

where

$$\Phi_{\phi,h}(r) = -\left\{(\phi h)_r + \phi h^2\right\}.$$

Remark 2.3 This inequality was first derived in [4], where it is not given in this form.

Taking

$$h(r) = \frac{1}{2r}, \qquad \phi(r) = 1$$

in (2.5), we have easily obtain

Corollary 2.4 *Let Ω be under the same hypotheses of Lemma 2.2. Let $v \in C_0^{\infty}(\Omega)$. It holds that*

$$\int_{\Omega} |D_r^{\pm} v(x)|^2 dx \geq \int_{\Omega} \frac{|v(x)|^2}{4r^2} dx + \{\pm\Im(\kappa)\} \int_{\Omega} \frac{|v(x)|^2}{r} dx \ (\pm\Im\kappa \geq 0).$$

Remark 2.5 This inequality is used as an aid to derive the uniform estimate

$$\int_{\Omega} \frac{|u(x)|^2}{r^2} dx \leq C \int_{\Omega} (1+r)^{1+\max\{1,\,\delta\}} |f(x)|^2 dx.$$

3 Proof of Theorem 1.1 Under the Smallness Condition

We choose

$$\varphi(r) = (1+r)^{-1-\delta}, \qquad \psi(r) = \psi_0 r + \psi_1 - \psi_2 (1+r)^{-\delta_1},$$

where $\delta > 0$, $\delta_1 \in (0, 1)$, $\psi_0 \geq \psi_1 > \psi_2 > 0$. Consider the case

$$(i) \ \varphi - \frac{b\psi}{2} \geq 0.$$

This essentially corresponds to the case

$$0 \leq b(x) \leq b_0(1+r)^{-2-\delta} \text{ with small } b_0 > 0.$$

In this case, the second term in the left-hand side of (2.1) in Lemma 2.1 is non-negative and the third term of (2.1) may be negative in general. However, for $\delta_1 \geq \frac{2}{\psi_1}$, paying attention to the smallness of the function $b(x)$, we obtain the following estimate for the coefficient function of $|D_r^+ u|^2$ in the third term in the left-hand-side of (2.1) in Lemma 2.1:

$$\left\{ \frac{\psi_r}{2} + \Im(\kappa)\psi + \left(\frac{b\psi}{2} - \varphi \right) - \varepsilon_2 \right\} \geq C\{1 + \Im(\kappa)\}$$

for some constant $C > 0$ independent of κ. Being $W_N \geq \frac{C}{r^2}$ for another $C > 0$, we conclude the desired result. □

4 Another Inequality

This section describes another inequality related to operator $D_r^{\pm}$ in (1.3). This is useful for estimating terms that can be negative if the function $b(x)$ is large.

Proposition 4.1 ([10]) *Let $\Omega = \mathbb{R}^N$ or $\Omega = \mathbb{R}^N \setminus \mathcal{O}$ (the assumption that the obstacle is star-shaped is unnecessary). Assume that $N \geq 1$, $v \in C^1(\Omega)$, and two functions $m = m(x)$ and $n = n(x)$ satisfy $m + n \geq 0$, $m + n \not\equiv 0$. Then it holds that*

$$|k|^2 \int_\Omega \frac{mn}{m+n} |v|^2 dx \leq \int_\Omega m \left| D_r^{\pm} v \right|^2 dx + \int_\Omega n \left| v_r + \frac{N-1}{2r} v \right|^2 dx.$$

Proof The following two relations

$$2\Re\left(\overline{\mp i\kappa v} \cdot v_r\right) = \left| D_r^{\pm} v \right|^2 - \left| v_r + \frac{N-1}{2r} v \right|^2 - |\kappa|^2 |v|^2 - (\pm \Im(\kappa)) \frac{N-1}{r} |v|^2,$$

$$\overline{D_r^{\pm} v} \left(v_r + \frac{N-1}{2r} v \right) = \left| v_r + \frac{N-1}{2r} v \right|^2 + \overline{\mp i\kappa v} v_r + \left(\overline{\mp i\kappa}\right) \frac{N-1}{2r} |v|^2$$

allow the following calculation:

$$
\begin{aligned}
0 &\le \left| \frac{m}{\sqrt{m+n}} D_r^{\pm} v + \frac{n}{\sqrt{m+n}} \left(v_r + \frac{N-1}{2r} v \right) \right|^2 \\
&= \frac{m^2}{m+n} \left| D_r^{\pm} v \right|^2 + \frac{n^2}{m+n} \left| v_r + \frac{N-1}{2r} v \right|^2 \\
&\quad + 2\Re \left\{ \frac{mn}{m+n} \overline{D_r^{\pm} v} \left(v_r + \frac{N-1}{2r} v \right) \right\} \\
&= \frac{m^2}{m+n} |D_r^{\pm} v|^2 + \frac{n^2}{m+n} \left| v_r + \frac{N-1}{2r} v \right|^2 + \frac{2mn}{m+n} \left| v_r + \frac{N-1}{2r} v \right|^2 \\
&\qquad + \frac{mn}{m+n} 2\Re\left(\overline{\mp i\kappa v} v_r \right) + 2\Re\left(\overline{\mp i\kappa} \right) \frac{N-1}{2r} |v|^2 \frac{mn}{m+n} \\
&= m \left| D_r^{\pm} v \right|^2 + n \left| v_r + \frac{N-1}{2r} v \right|^2 - \frac{mn}{m+n} |\kappa|^2 |v|^2 .
\end{aligned}
$$

Integration over Ω gives the desired inequality. □

5 Proof of Theorem 1.1 Without the Smallness Condition

Let us treat the case

$$
(ii)\ \varphi - \frac{b\psi}{2} \le 0
$$

with φ and ψ given in Sect. 3. Choose the two functions m and n in Proposition 4.1 for a sufficiently small positive ε_3 as follows:

$$
m = (1-\varepsilon_3)\frac{\psi_r}{2} + \left(\frac{b\psi}{2} - \varphi \right), \qquad n = \varphi - \frac{b\psi}{2}.
$$

Then we find that $m + n = (1-\varepsilon_3)\dfrac{\psi_r}{2} \ge 0$ and $m + n \not\equiv 0$ hold. Therefore by Lemma 2.1 and Proposition 4.1, being $\displaystyle\int W_N |u|^2 dx \ge 0$, for any $\varepsilon_4 \in (0, 1)$, we have

$$
\begin{aligned}
&|\kappa|^2 \int_\Omega \{\varepsilon_4 \varphi + G(M)\} |u|^2 dx + \int_\Omega \left\{ \varepsilon_3 \frac{\psi_r}{2} - \varepsilon_2 + \Im(\kappa)\psi \right\} |D_r^{+} u|^2 dx \\
&\le C(\varepsilon_1, \varepsilon_2) \int_\Omega \left(\frac{g^2}{\varphi} + \psi^2 \right) |f|^2 dx
\end{aligned}
$$

for some positive $C(\varepsilon_1, \varepsilon_2)$, where

$$G(M) = (1-\varepsilon_1-\varepsilon_4)\varphi + M + \frac{mn}{m+n}, \qquad M \equiv \frac{b\psi}{2}.$$

Thus, if $G(M) \geq 0$ and $\varepsilon_3\frac{\psi_r}{2} - \varepsilon_2 \geq {}^{\exists}C > 0$, we obtain Theorem 1.1.

Now, we find that

$$G(M) \geq 0 \Leftrightarrow M^2 - 2\varphi M + \varphi^2 - \frac{(1-\varepsilon_3)(2-\varepsilon_1-\varepsilon_4)\psi_r\varphi}{2} \leq 0.$$

That is equivalent to

$$\varphi \leq M \leq \varphi + \sqrt{D'}, \tag{5.1}$$

where

$$D' = \frac{(1-\varepsilon_3)(2-\varepsilon_1-\varepsilon_4)\psi_r\varphi}{2} \geq 0.$$

Next, the assumption (1.4) in Theorem 1.1 implies (ii) once $\delta \geq b_2 - \frac{1}{2}$. Let us prove the second inequality in (5.1)

$$M \leq \varphi + \sqrt{D'} \Leftrightarrow b \leq \frac{2}{\psi}\left(\varphi + \sqrt{\frac{(1-\varepsilon_3)(2-\varepsilon_1-\varepsilon_4)\psi_r\varphi}{2}}\right).$$

If we choose ε_j $(j = 1, 3, 4)$ as $0 < \varepsilon_1 + \varepsilon_4 < 1$, choose ψ_0 as

$$0 < \psi_0 \leq \frac{2(1-\varepsilon_3)(2-\varepsilon_1-\varepsilon_4)}{b_1^2},$$

and choose δ and δ_1 as $0 < \delta_1 \leq \delta \leq 2b_2$, we then have

$$b \leq \frac{2}{\psi}\sqrt{\frac{(1-\varepsilon_3)(2-\varepsilon_1-\varepsilon_4)\psi_r\varphi}{2}} \Leftrightarrow M \leq \sqrt{D'} \Rightarrow M \leq \varphi + \sqrt{D'}$$

by (1.4) to obtain (5.1) and consequently (ii). Finally choose ε_2 as

$$0 < \varepsilon_2 \leq \frac{\varepsilon_3\psi_0}{3},$$

we then have

$$\varepsilon_3 \frac{\psi_r}{2} - \varepsilon_2 \geq \frac{\varepsilon_2}{2} > 0.$$

The proof of Theorem 1.1 is now completed by combining the inequality obtained in the above discussion with the one in Corollary 2.4 (see Remark 2.5). □

Acknowledgments The author would like to thank the anonymous referees for carefully reading our manuscript and for giving such constructive comments which substantially helped to improve the quality of the paper. This work was supported by JSPS KAKENHI (Grant-in-Aid for Scientific Research (C)), Grant Number JP 16K05243.

References

1. T. Kato, K. Yajima, Some examples of smooth operators and the associated smoothing effect. Rev. Math. Phys. **1**(4), 481–496 (1989)
2. S. Mizohata, K. Mochizuki, On the principle of limiting amplitude for dissipative wave equations. J. Math. Kyoto Univ. **6**, 109–127 (1966)
3. K. Mochizuki, Scattering theory for wave equations with dissipative terms. Publ. Res. Inst. Math. Sci. **12**(2), 383–390 (1976/1977)
4. K. Mochizuki, Uniform resolvent estimates for magnetic Schrödinger operators and smoothing effects for related evolution equations. Publ. Res. Inst. Math. Sci. **46**(4), 741–754 (2010)
5. K. Mochizuki, H. Nakazawa, Energy decay and asymptotic behavior of solutions to the wave equations with linear dissipation. Publ. Res. Inst. Math. Sci. **32**(3), 401–414 (1996)
6. K. Mochizuki, H. Nakazawa, Uniform resolvent estimates for magnetic Schrödinger operators in a 2D exterior domain and their applications to related evolution equations. Publ. Res. Inst. Math. Sci. **51**(2), 319–336 (2015)
7. K. Mochizuki, H. Nakazawa, Uniform resolvent estimates for Helmholtz equations in 2D exterior domain and their applications, (Kyoto, 2015). Surikaisekikenkyusho Kokyuroku **1962**, 68–76 (2015).
8. K. Mochizuki, H. Nakazawa, Uniform resolvent estimates for stationary dissipative wave equations in an exterior domain and their application to the principle of limiting amplitude, in *New Trends in Analysis and Interdisciplinary Applications*. Trends and Perspectives in Applied Mathematics (Birkhäuser/Springer, Cham, 2017), pp. 521–527
9. H. Nakazawa, The principle of limiting absorption for the non-selfadjoint Schrödinger operator with energy dependent potential. Tokyo J. Math. **23**(2), 519–536 (2000)
10. H. Nakazawa, Some remarks on wave equations with dissipations (Japanese), in *Seminar Notes of Mathematical Sciences*, vol. 14 (Ibaraki University, Ibaraki, 2013), pp. 64–71
11. H. Nakazawa, Uniform resolvent estimates for Schrödinger equations in an exterior domain in $\mathbb{R}^2$ and their applications to scattering problems. Bull. Liberal Arts Sci. Nippon Medical School **42**, 1–12 (2013)
12. K. Watanabe, Smooth perturbations of the selfadjoint operator $|\Delta|^{-\alpha/2}$. Tokyo J. Math. **14**(1), 239–250 (1991)

Part XIV
Wavelet Theory and Its Related Topics

A Survey on the Time-Frequency Analysis on the Half Real Line

Yun-Zhang Li

Abstract During past more than 30 years, the time-frequency analysis on $L^2(\mathbb{R})$ such as wavelet and Gabor analysis has been extensively studied, but the time-frequency analysis on $L^2(\mathbb{R}_+)$ has not, where $L^2(\mathbb{R}_+)$ is the space of square integrable functions on the half real line $\mathbb{R}_+$. It is because $\mathbb{R}$ is a group under addition, while $\mathbb{R}_+$ is not. This leads to $L^2(\mathbb{R}_+)$ admitting no nontrivial shift-invariant system. The present article gives a survey on the time-frequency analysis on $L^2(\mathbb{R}_+)$. It only focuses main ideas instead of concrete results. Readers may refer to listed references for details.

Keywords Frame · Time-frequency · Wavelet analysis · Gabor analysis · Dilation-and-modulation system

Mathematics Subject Classification (2010) Primary 42C40; Secondary 42C15

1 Introduction

A central part of harmonic analysis deals with functions on groups and ways to decompose such functions in terms of either series representations or integral representations of certain "basic functions". Structured basic functions are important in mathematics and engineering. Wavelet and Gabor frames are such basic functions representing square integrable functions on $\mathbb{R}$ which is a locally compact abelian group under addition and the usual topology. The study of frames on general locally compact abelian groups has appeared in several publications including [2, 3, 5, 6, 12, 15, 16, 18]. And the study of wavelet and Gabor frames for $L^2(\mathbb{R})$

Y.-Z. Li (✉)
College of Mathematics, Faculty of Science, Beijing University of Technology, Beijing, P. R. China
e-mail: yzlee@bjut.edu.cn

P. Cerejeiras et al. (eds.), *Current Trends in Analysis, its Applications and Computation*, Research Perspectives,
https://doi.org/10.1007/978-3-030-87502-2_64

has seen great achievements during past more than 30 years, but the study of the time-frequency analysis for $L^2(\mathbb{R}_+)$ has not.

This article gives a survey of the study of structured frames for $L^2(\mathbb{R}_+)$ with $\mathbb{R}_+ = [0, \infty)$ or $(0, \infty)$. In practice, the time variable cannot be negative. The space $L^2(\mathbb{R}_+)$ models the causal signal space, but it is not closed under the Fourier transform since the Fourier transform of a compactly supported nonzero function in $L^2(\mathbb{R}_+)$ lies outside this space. Also it admits no nontrivial shift-invariant system since $\mathbb{R}_+$ is not a group under addition. This leads to the fact that $L^2(\mathbb{R}_+)$ cannot admit wavelet and Gabor frames. Therefore, the time-frequency analysis for $L^2(\mathbb{R}_+)$ is essentially different from that for $L^2(\mathbb{R})$.

The rest of this article is organized as follows. Section 2 focuses on Walsh series-based wavelet analysis in $L^2(\mathbb{R}_+)$, and Sect. 3 on a kind of dilation-and-modulation systems in $L^2(\mathbb{R}_+)$. Herein, we only state main ideas instead of details. Readers may refer to listed references for details.

2 Walsh Series-Based Wavelet Analysis in $L^2(\mathbb{R}_+)$ with $\mathbb{R}_+ = [0, \infty)$

Note that $L^2(\mathbb{R}_+)$ can be considered as a subspace of $L^2(\mathbb{R})$ consisting of all functions in $L^2(\mathbb{R})$ vanishing outside $\mathbb{R}_+$. There have been many papers on "wavelet" frames in $L^2(\mathbb{R}_+)$. Also due to $\mathbb{R}_+$ being not a group under addition, such frames do not have complete affine structure, and include boundary and interior ones. And boundary ones are related to inhomogeneous refinement equations. For details, we refer to [4, 7, 17, 19, 29] and references therein. Walsh series-based wavelet analysis for $L^2(\mathbb{R}_+)$ is similar to that for $L^2(\mathbb{R})$, and is based on a new addition operation on $\mathbb{R}_+$ which makes $\mathbb{R}_+$ a group under it. Let us recall the ideas on Walsh series-based wavelet analysis from [1, 8–11, 24, 27, 28, 32] and references therein.

Let p be a integer greater than 1. We denote by $\mathbb{Z}_+$ and $\mathbb{N}$ the set of nonnegative integers and the set of positive integers respectively, and by $\mathbb{N}_p$ the set of $\{0, 1, \cdots, p-1\}$. Define the addition $\oplus$ and subtraction $\ominus$ on $\mathbb{N}_p$ by

$$x \oplus y = (x + y)(\text{mod } p) = \begin{cases} x + y, & x + y < p; \\ x + y - p, & x + y \geq p, \end{cases}$$

and

$$x \ominus y = (x - y)(\text{mod } p) = \begin{cases} x - y, & x \geq y; \\ x - y + p, & x < y \end{cases}$$

for $x, y \in \mathbb{N}_p$, respectively. Given $x \in \mathbb{R}_+$, we denote by $[x]$ its integer part, and by $\{x\}$ its fraction part. Then x has following unique representation:

$$x = \sum_{j=1}^{\infty} x_{-j} p^{j-1} + \sum_{j=1}^{\infty} x_j p^{-j} = [x] + \{x\}, \tag{2.1}$$

where $x_j, x_{-j} \in \mathbb{N}_p$ for $j \in \mathbb{N}$, and $x_{-j} = [p^{j-1}x](\mathrm{mod}\, p)$, $x_j = [p^j x](\mathrm{mod}\, p)$. For $y, \omega \in \mathbb{R}_+$, we define y_j, y_{-j} and ω_j, ω_{-j} similarly. Define the addition $\oplus$ and subtraction $\ominus$ on $\mathbb{R}_+$ by

$$x \oplus y = \sum_{j=1}^{\infty} \left(x_{-j} \oplus y_{-j}\right) p^{j-1} + \sum_{j=1}^{\infty} \left(x_j \oplus y_j\right) p^{-j} \tag{2.2}$$

and

$$x \ominus y = \sum_{j=1}^{\infty} \left(x_{-j} \ominus y_{-j}\right) p^{j-1} + \sum_{j=1}^{\infty} \left(x_j \ominus y_j\right) p^{-j} \tag{2.3}$$

for $x, y \in \mathbb{R}_+$, respectively. It is easy to check that $\mathbb{R}_+$ and $\mathbb{Z}_+$ are groups under "$\oplus$", and that the inverse operation of "$\oplus$" is "$\ominus$". Define the *dilation operator* D_λ and the *translation operator* T_x on $L^2(\mathbb{R}_+)$ with $\lambda > 1$ and $x \in \mathbb{R}_+$ by

$$D_\lambda f(\cdot) = \lambda^{\frac{1}{2}} f(\lambda \cdot) \text{ and } T_x f(\cdot) = f(\cdot \ominus x) \text{ for } f \in L^2(\mathbb{R}_+), \tag{2.4}$$

respectively, and write $f_{j,k} = D_{p^j} T_k f$ for $j \in \mathbb{Z}$ and $k \in \mathbb{Z}_+$. Readers refer to [1, 8–11, 27, 28, 32] for basic properties of $\oplus$, D_λ and T_x. Now let us define the affine systems in $L^2(\mathbb{R}_+)$. Given a finite subset Ψ of $L^2(\mathbb{R}_+)$, the *affine system* $X(\Psi)$ generated by Ψ is defined by

$$X(\Psi) = \left\{\psi_{j,k} : j \in \mathbb{Z},\, k \in \mathbb{Z}_+,\, \psi \in \Psi\right\}. \tag{2.5}$$

Similarly to the cases in $L^2(\mathbb{R})$, $X(\Psi)$ is called a *Walsh p-adic wavelet basis* (*Walsh p-adic wavelet frame*) for $L^2(\mathbb{R}_+)$ if it is an orthonormal basis (a frame) for $L^2(\mathbb{R}_+)$. Walsh p-adic wavelet Riesz basis and Walsh p-adic wavelet Parseval frame can be defined similarly.

It is well known that wavelet bases and wavelet frames for $L^2(\mathbb{R})$ may be characterized in Fourier transform domain, and that the (generalized) multiresolution analysis ((G)MRA for simplicity) is an important tool to construct wavelets (wavelet frames). For Walsh p-adic wavelet bases (frames) in $L^2(\mathbb{R}_+)$, a theory parallel to wavelet bases (frames) in $L^2(\mathbb{R})$ holds. It is based on the Walsh p-adic Fourier transform defined on $L^2(\mathbb{R}_+)$.

Define

$$\chi(x, \omega) = e^{\frac{2\pi i}{p}\sum_{j=1}^{\infty}(x_j\omega_{-j}+x_{-j}\omega_j)} \quad \text{for } x, \omega \in \mathbb{R}_+. \tag{2.6}$$

Then the system $\{\chi(k, \cdot)\}_{k\in\mathbb{Z}_+}$ is an orthonormal basis for $L^2[0, 1]$. For $z \in l^2(\mathbb{Z}_+)$, its *Walsh p-adic Fourier series* is defined by

$$\tilde{z}(\cdot) = \sum_{k=0}^{\infty} z(k)\overline{\chi(k, \cdot)}. \tag{2.7}$$

The *Walsh p-adic Fourier transform* on $L^1(\mathbb{R}_+)$ and $L^2(\mathbb{R}_+)$ are defined by

$$\tilde{f}(\cdot) = \int_{\mathbb{R}_+} f(x)\overline{\chi(x, \cdot)}dx \text{ for } f \in L^1(\mathbb{R}_+) \tag{2.8}$$

and

$$\tilde{f}(\cdot) = \lim_{a\to+\infty}\int_0^a f(x)\overline{\chi(x, \cdot)}dx \text{ for } f \in L^2(\mathbb{R}_+), \tag{2.9}$$

respectively, where the limit is in $L^2(\mathbb{R}_+)$ norm. Similarly to the usual Fourier transform, (2.7) and (2.9) define unitary operators from $l^2(\mathbb{Z}_+)$ onto $L^2[0, 1]$ and from $L^2(\mathbb{R}_+)$ onto itself, respectively. We refer to [13, 25] for basics of the Walsh p-adic Fourier series (transform).

It is well known that MRAs and GMRAs are determined by refinable functions, and that a refinable function is determined by its symbol. In $L^2(\mathbb{R}_+)$, Walsh p-refinable functions may be defined similarly. A function $\varphi \in L^2(\mathbb{R}_+)$ is said to be *Walsh p-refinable* if

$$\varphi(\cdot) = p\sum_{k\in\mathbb{Z}_+}\alpha(k)\varphi(p\cdot\ominus k), \tag{2.10}$$

equivalently,

$$\tilde{\varphi}(\cdot) = \tilde{\alpha}(p^{-1}\cdot)\tilde{\varphi}(p^{-1}\cdot) \tag{2.11}$$

for some sequence $\alpha = \{\alpha(k)\}_{k\in\mathbb{Z}_+} \in l^2(\mathbb{Z}_+)$, where $\tilde{\alpha}$ is called the symbol of φ. Given φ satisfying (2.10), define the sequence $\{V_j\}_{j\in\mathbb{Z}}$ of closed subspaces of $L^2(\mathbb{R}_+)$ by

$$V_j = \overline{\text{span}}\left\{\varphi_{j,k} : j \in \mathbb{Z}, k \in \mathbb{Z}_+\right\} \text{ for } j \in \mathbb{Z}. \tag{2.12}$$

Then $\{V_j\}_{j\in\mathbb{Z}}$ is a GMRA for $L^2(\mathbb{R}_+)$, and is an MRA for $L^2(\mathbb{R}_+)$ if we make suitable restriction on φ. Using the (G)MRA structure, we may construct Walsh p-adic wavelet bases (frames), and derive the corresponding extension principles. Readers refer to [1, 8–11, 24, 27, 28, 32] and references therein for details.

3 $\mathcal{MD}$-Frame Theory in $L^2(\mathbb{R}_+)$ with $\mathbb{R}_+ = (0, \infty)$

Affine systems in Sect. 2 are based on the fact that $[0, \infty)$ is a group under "$\oplus$". Observe that $\mathbb{R}$ is a group under addition, and that a Gabor system in $L^2(\mathbb{R})$ has the form

$$\{e^{2\pi imb\cdot}\psi(\cdot - na) : m, n \in \mathbb{Z}\} \tag{3.1}$$

which is the shift-and-modulation version of fixed function $\psi \in L^2(\mathbb{R})$ with modulation factors being periodic with respect to addition. So it is strongly dependent on the additive group structure of $\mathbb{R}$. On the other hand, $\mathbb{R}_+ = (0, \infty)$ is a group under multiplication although it is not a group under addition. This motivates us to consider multiplication-based time-frequency analysis in $L^2(\mathbb{R}_+)$. In this section, we focus on a class of dilation-and-modulation system frames in $L^2(\mathbb{R}_+)$ based on multiplication.

In what follows, unless specified, a and b are two constants greater than 1. A measurable function h defined on $\mathbb{R}_+$ is said to be *b-dilation periodic* if $h(b\cdot) = h(\cdot)$ on $\mathbb{R}_+$. Define the sequence $\{\Lambda_m\}_{m\in\mathbb{Z}}$ of b-dilation periodic functions by

$$\Lambda_m(\cdot) = \frac{1}{\sqrt{b-1}} e^{\frac{2\pi im\cdot}{b-1}} \text{ on } [1, b) \text{ for each } m \in \mathbb{Z}. \tag{3.2}$$

For a nonempty finite subset Ψ of $L^2(\mathbb{R}_+)$, we define the *dilation-and-modulation system* ($\mathcal{MD}$-system) in $L^2(\mathbb{R}_+)$ generated by Ψ as

$$\mathcal{MD}(\Psi, a, b) = \left\{ \Lambda_m D_{a^j}\psi : m, j \in \mathbb{Z}, \psi \in \Psi \right\}, \tag{3.3}$$

where $D_{a^j}\psi(\cdot) = a^{\frac{j}{2}}\psi(a^j\cdot)$. In particular, we write $\mathcal{MD}(\Psi, a, b) = \mathcal{MD}(\psi, a, b)$ if Ψ is a singleton $\{\psi\}$. Our goal is to study the frame properties of $\mathcal{MD}$-systems of the form (3.3). Before proceeding, the following two facts should be clarified:

- The factor Λ_m in (3.3) is essentially different from the factor $e^{2\pi imb\cdot}$ in Gabor system (3.1).

  Given a constant $c > 1$,

$$\frac{e^{2\pi imbc\cdot}}{e^{2\pi imb\cdot}} = e^{2\pi imb(c-1)\cdot}.$$

It cannot be a constant function if $m \neq 0$. Thus $e^{2\pi imb\cdot}$ with $m \neq 0$ cannot be c-dilation periodic for each $c > 1$. However, Λ_m in (3.3) is b-dilation periodic.
- A $\mathcal{MD}$-frame for $L^2(\mathbb{R}_+)$ cannot be derived from an affine system in $H^2(\mathbb{R})$ via Fourier transform.

Wavelet frames for the Hardy space $H^2(\mathbb{R})$ were studied in [14], [26], [30], where

$$H^2(\mathbb{R}) = \{f \in L^2(\mathbb{R}) : \hat{f}(\cdot) = 0 \text{ a.e. on } (-\infty, 0)\},$$

the *Fourier transform* of a function $f \in L^1(\mathbb{R}) \cap L^2(\mathbb{R})$ is defined by

$$\hat{f}(\xi) = \int_{\mathbb{R}} f(x) e^{-2\pi i x\xi} dx$$

and extended to $L^2(\mathbb{R})$ by the Plancherel theorem. Obviously, $L^2(\mathbb{R}_+)$ is the Fourier transform of $H^2(\mathbb{R})$. But a $\mathcal{MD}$-system of the form (3.3) is essentially different from the Fourier transform version

$$\{a^{\frac{j}{2}} e^{-2\pi i a^j bm\cdot} \hat{\psi}(a^j \cdot) : m, j \in \mathbb{Z}\} \tag{3.4}$$

of an arbitrary affine system $\{a^{\frac{j}{2}} \psi(a^j \cdot - bm) : m, j \in \mathbb{Z}\}$ in $H^2(\mathbb{R})$. It is because $e^{-2\pi i a^j bm\cdot}$ in (3.4) is $\frac{1}{a^j b}\mathbb{Z}$-periodic with respect to addition, and the period varies with j, while Λ_m in (3.3) is b-dilation periodic, and it is unrelated to j.

It is well known that Fourier transform plays a significant role in time-frequency analysis for $L^2(\mathbb{R})$. But it cannot be used in time-frequency analysis for $L^2(\mathbb{R}_+)$ since $L^2(\mathbb{R}_+)$ is not closed under Fourier transform. For doing time-frequency analysis for $L^2(\mathbb{R}_+)$, we introduce Θ_β transform with $\beta > 1$ based on the dilation operation.

Definition 3.1 Given a constant $\beta > 1$, define $\Theta_\beta : L^2(\mathbb{R}_+) \to L^2_{loc}(\mathbb{R}_+ \times \mathbb{R})$ by

$$\Theta_\beta f(x, \xi) = \sum_{l \in \mathbb{Z}} \beta^{\frac{l}{2}} f(\beta^l x) e^{-2\pi i l\xi} \tag{3.5}$$

for $f \in L^2(\mathbb{R}_+)$ and a.e. $(x, \xi) \in \mathbb{R}_+ \times \mathbb{R}$.

Remark 3.2

(i) Θ_β is well-defined since

$$\int_{\beta^j}^{\beta^{j+1}} \sum_{l \in \mathbb{Z}} \beta^l |f(\beta^l x)|^2 dx = \|f\|^2_{L^2(\mathbb{R}_+)} < \infty \text{ for } j \in \mathbb{Z} \text{ and } f \in L^2(\mathbb{R}_+).$$

(ii) By [22, Lemma 2.3], Θ_β has the quasi-periodicity:

$$\Theta_\beta f(\beta^j x,\, \xi + m) = \beta^{-\frac{j}{2}} e^{2\pi i j \xi} \Theta_\beta f(x,\, \xi) \text{ for } f \in L^2(\mathbb{R}_+),\ m,$$
$$j \in \mathbb{Z} \text{ and a.e. } (x,\, \xi) \in \mathbb{R}_+ \times \mathbb{R},$$

and Θ_β is a unitary operator from $L^2(\mathbb{R}_+)$ onto $L^2([1,\, \beta) \times [0,\, 1))$.

The works on $\mathcal{MD}$-system frames may be summarized as follows. Readers refer to [20–23, 31] for details.

Case 1 $a = b$.

In this case, write $\mathcal{MD}(\Psi,\, a,\, b) = \mathcal{MD}(\Psi,\, a)$ for simplicity. Frames of the form $\mathcal{MD}(\Psi,\, a)$ and dual frame pairs of the form $(\mathcal{MD}(\Psi,\, a),\, \mathcal{MD}(\Phi,\, a))$ were characterized using the Θ_a-transform (take $\beta = a$ in Θ_β); given a frame $\mathcal{MD}(\Psi,\, a)$, a Θ_a-transform expression of Φ was obtained such that $(\mathcal{MD}(\Psi,\, a),\, \mathcal{MD}(\Phi,\, a))$ is a pair of dual frames; frames of the form $\mathcal{MD}(\Psi,\, a)$ with Ψ being a set of characteristic functions were studied. On the other hand, frame properties of vector-valued $\mathcal{MD}$-systems were considered.

Case 2 $\log_b a = \frac{p}{q}$ with p and q being coprime positive integers.

In this case, $b^p = a^q$. Write $\beta = b^p$. Completeness and frame characterization of $\mathcal{MD}(\psi,\, a,\, b)$ with $\psi \in L^2(\mathbb{R}_+)$ were presented in terms of Θ_β-transform-based matrix-valued functions. A density result was also obtained, which reads as

Theorem 3.3 *For $a,\, b > 1$ with $\log_b a$ being a rational number, the following are equivalent:*

(i) $\log_b a \le 1$.
(ii) There exists $\psi \in L^2(\mathbb{R}_+)$ such that $\mathcal{MD}(\psi,\, a,\, b)$ is complete in $L^2(\mathbb{R}_+)$.
(iii) There exists $\psi \in L^2(\mathbb{R}_+)$ such that $\mathcal{MD}(\psi,\, a,\, b)$ is a frame for $L^2(\mathbb{R}_+)$.

Acknowledgement This work was supported by National Natural Science Foundation of China (Grant No. 11971043). The author thank the referees for their valuable suggestions.

References

1. S. Albeverio, S. Evdokimov, M. Skopina, p-adic multiresolution analysis and wavelet frames. J. Fourier Anal. Appl. **16**, 693–714 (2010)
2. M. Bownik, K. Ross, The structure of translation-invariant spaces on locally compact abelian groups. J. Fourier Anal. Appl. **21**, 849–884 (2015)
3. C. Cabrelli, V. Paternostro, Shift-invariant spaces on LCA groups. J. Funct. Anal. **258**, 2034–2059 (2010)
4. Z. Chen, C.A. Micchelli, Y. Xu, A construction of interpolating wavelets on invariant sets. Math. Comp. **6**, 1569–1587 (1999)
5. O. Christensen, S.S. Goh, Fourier-like frames on locally compact abelian groups. J. Approx. Theory **192**, 82–101 (2015)

6. S. Dahlke, A note on generalized Weyl-Heisenberg frames. Appl. Math. Lett. **7**, 79–82 (1994).
7. W. Dahmen, B. Han, R.Q. Jia, A. Kunoth, Biorthogonal multiwavelets on the interval: cubic Hermite splines. Constr. Approx. **16**, 221–259 (2000)
8. S. Evdokimov, M. Skopina, On orthogonal p-adic wavelet bases. J. Math. Anal. Appl. **424**, 952–965 (2015)
9. Yu.A. Farkov, *Orthogonal p-Wavelets on* $\mathbb{R}_+$, *in: Wavelets and Splines* (St. Petersburg University Press, St. Petersburg, 2005), pp. 4–26
10. Yu.A. Farkov, On wavelets related to the Walsh series. J. Approx. Theory **161**, 259–279 (2009)
11. Yu.A. Farkov, V.Y. Protasov, Dyadic wavelets and refinable functions on a half-line. Mat. Sb. **197**, 129–160 (2006). *English transl. Sb. Math.* **197**, 1529–1558 (2006)
12. H.G. Feichtinger, K. Gröchenig, Banach spaces related to integrable group representations and their atomic decompositions, I.. J. Funct. Anal. **86**, 307–340 (1989)
13. B.I. Golubov, A.V. Efimov, V.A. Skvortsov, *Walsh Series Transforms* (Nauka, Moscow, 1987); (English transl. Kluwer, Dordrecht, 1991)
14. E. Hernández, G. Weiss, *A First Course on Wavelets* (CRC Press, Boca Raton, 1996)
15. M.S. Jakobsen, J. Lemvig, Reproducing formulas for generalized translation invariant systems on locally compact abelian groups. Trans. Amer. Math. Soc. **368**, 8447–8480 (2016)
16. M.S. Jakobsen, J. Lemvig, Co-compact Gabor systems on locally compact abelian groups. J. Fourier Anal. Appl. **22**, 36–70 (2016)
17. R.Q. Jia, Q. Jiang, Z. Shen, Convergence of cascade algorithms associated with nonhomogeneous refinement equations. Proc. Amer. Math. Soc. **129**, 415–427 (2001)
18. G. Kutyniok, D. Labate, The theory of reproducing systems on locally compact abelian groups. Colloq. Math. **106**, 197–220 (2006)
19. S. Li, Convergence of cascade algorithms in Sobolev spaces associated with inhomogeneous refinement equations. J. Approx. Theory **104**, 153–163 (2000)
20. Y.-Z. Li, Y.-H. Wang, The density theorem of a class of dilation-and-modulation systems on the half real line. Results Math. **74**, Art. 190, 19pp. (2019)
21. Y.-Z. Li, Y.-H. Wang, The dilation-and-modulation frame sets on the half real line. Acta Math. Sinica (Chin. Ser.) **63**, 1–16 (2020)
22. Y.-Z. Li, W. Zhang, Multi-window dilation-and-modulation frames on the half real line. Sci. China Math. **63**, 2423–2438 (2020)
23. Y.-Z. Li, W. Zhang, Dilation-and-modulation systems on the half real line. J. Inequal. Appl. **2016**, Paper No. 186, 11 (2016)
24. Meenakshi, P. Manchanda, A.H. Siddiqi, Wavelets associated with nonuniform multiresolution analysis on positive half-line. Int. J. Wavelets Multiresolut. Inf. Process. **10**, 1250018, 27 (2012)
25. F. Schipp, W.R. Wade, P. Simon, *Walsh Series: An Introduction to Dyadic Harmonic Analysis* (Adam Hilger, Bristol/New York, 1990)
26. K. Seip, Regular sets of sampling and interpolation for weighted Bergman spaces. Proc. Amer. Math. Soc. **117**, 213–220 (1993)
27. F.A. Shah, Construction of wavelet packets on p -adic field. Int. J. Wavelets Multiresolut. Inf. Process. **7**, 553–565 (2009)
28. F.A. Shah, On some properties of p-wavelet packets via the Walsh-Fourier transform. J. Nonlinear Anal. Optim. **3**, 185–193 (2012)
29. Q. Sun, Homogeneous and nonhomogeneous refinable distributions in $F^{q,\gamma}$, in *Wavelet Analysis and Applications*, Guangzhou (1999), pp. 235–244; AMS/IP Studies in Advanced Mathematics, vol. 25 (American Mathematical Society, Providence, 2002)
30. H. Volkmer, Frames of wavelets in Hardy space. Anal. **15**, 405–421 (1995)
31. Y.-H. Wang, Y.-Z. Li, A class of vector-valued dilation-and-modulation frames on the half real line. Math. Methods Appl. Sci. **41**, 3900–3912 (2018)
32. Y. Zhang, Y.-Z. Li, The intersection and union of dilates of singly generated Walsh p-adic shift-invariant spaces. Acta Math. Sinica (Chin. Ser.) **62**, 1–12 (2019)

Some Notes on the Inequalities for Parseval Generalized Continuous Frames

Dengfeng Li and Lingge Liu

Abstract In this short article, we establish some new inequalities for Parseval generalized continuous frames and give another proof of existing two inequalities for Parseval generalized continuous frames.

Keywords Frame · Generalized continuous frame · Parseval generalized continuous frame

Mathematics Subject Classification (2010) Primary 42C15

1 Introduction

Hilbert spaces are the natural framework for mathematical version of many areas of physics, quantum mechanics and image analysis. Often the main problem is to decompose an arbitray element in terms of simpler basic elements. Then what are these simpler basic elements and how efficient are they? The natural choice for these basic elements is an orthonormal basis of the Hilbert space of the problem. But orthonormal bases are difficult to work with because they decompose every element in a unique way. This lack of flexibility was not welcomed by practitioners in the field. Thus a more flexible alternative was formally introduced by Duffin and Schaeffer [1] in 1952 for studying some deep problems in nonharmonic Fourier series. This alternative is called a frame (discrete frame). Discrete frames, which are redundant sets of elements that provide robust, stable and usually non-unique representations of every element in underlying Hilbert space, have been a focus of study in the last three decades in applications where redundancy plays a useful

D. Li (✉)
School of Mathematics and Physics, Wuhan Textile University, Wuhan, PR China

L. Liu
School of Mathematics and Statistics, Henan University, Kaifeng, PR China

P. Cerejeiras et al. (eds.), *Current Trends in Analysis, its Applications and Computation*, Research Perspectives,
https://doi.org/10.1007/978-3-030-87502-2_65

and vital role, e.g., signal and image processing [2], neural network [3], digital communication [4, 5] and so on.

With the in-depth study of discrete frame theory, discrete frames have been extended to various more general forms, such as continuous frame, generalized frame and others. In particular, the concept of generalized continuous frames on the basis of continuous frame and generalized frame was proposed in [6]. Let $\mathcal{H}$ and $\mathcal{V}$ be complex Hilbert spaces, (Ω, μ) be a measure space with positive radon measure μ, $\{\mathcal{V}_\omega : \omega \in \Omega\}$ be a sequence of closed subspaces of $\mathcal{V}$ and $\mathcal{B}(\mathcal{H}, \mathcal{V}_\omega)$ be the collection of all bounded linear operators from $\mathcal{H}$ into $\mathcal{V}_\omega$. The definition of a generalized continuous frame is as follows.

Definition 1.1 A sequence $\mathcal{F} \equiv \{\Lambda_\omega \in \mathcal{B}(\mathcal{H}, \mathcal{V}_\omega) : \omega \in \Omega\}$ is called a generalized continuous frame for $\mathcal{H}$ with respect to $\{\mathcal{V}_\omega : \omega \in \Omega\}$, if
(1) $\mathcal{F}$ is weakly-measurable, i.e., for any $f \in \mathcal{H}$, $\omega \to \Lambda_\omega(f)$ is a measurable function on Ω;
(2) there exist two positive constants A and B such that

$$A\|f\|^2 \leq \int_\Omega \|\Lambda_\omega f\|^2 d\mu(\omega) \leq B\|f\|^2, \quad \forall f \in \mathcal{H}, \tag{1.1}$$

where A and B are called the lower and the upper generalized continuous frame bounds, respectively. If only the right-hand inequality of (1.1) holds, then we call $\{\Lambda_\omega : \omega \in \Omega\}$ a generalized continuous Bessel sequence for $\mathcal{H}$ with respect to $\{\mathcal{V}_\omega : \omega \in \Omega\}$. We call $\{\Lambda_\omega : \omega \in \Omega\}$ an A-tight generalized continuous frame if $A = B$. Moreover, if $A = B = 1$, $\{\Lambda_\omega : \omega \in \Omega\}$ is called a Parseval generalized continuous frame.

Remark 1.2 If $\mathcal{V}_\omega = \mathbb{C}$ for all $\omega \in \Omega$, then Λ_ω is a bounded linear functional from $\mathcal{H}$ to $\mathbb{C}$, thereby, by the Riesz representation theorem, there exists $g_\omega \in \mathcal{H}$ such that $\Lambda_\omega(f) = \langle f, g_\omega \rangle$ for all $f \in \mathcal{H}$. Therefore, in the case, the generalized continuous frame is just the continuous frame [7–9]. If Ω is a countable set and μ is a counting measure, then the generalized continuous frame is just the generalized frame in [10].

In the study of efficient algorithms for signal reconstruction, some equalities and inequalities for Parseval discrete frames are found in [11, 12]. Afterwards, the results related to these equalities and inequalities are obtained in [13, 14], and the equalities and inequalities related to generalized frames are also established [15, 16]. In particular, the following equalities and inequalities of generalized continuous frames are presented in [17].

If $\{\Lambda_\omega : \omega \in \Omega\}$ is a Parseval generalized continuous frame for $\mathcal{H}$ with respect to $\{\mathcal{V}_\omega : \omega \in \Omega\}$, then for all $f \in \mathcal{H}$, and $\Omega_1 \subset \Omega$ and $\Omega_1^c = \Omega \setminus \Omega_1$,

$$\int_{\Omega_1} \|\Lambda_\omega f\|^2 d\mu(\omega) + \left\| \int_{\Omega_1^c} \Lambda_\omega^* \Lambda_\omega f d\mu(\omega) \right\|^2 = \int_{\Omega_1^c} \|\Lambda_\omega f\|^2 d\mu(\omega) + \left\| \int_{\Omega_1} \Lambda_\omega^* \Lambda_\omega f d\mu(\omega) \right\|^2, \tag{1.2}$$

and

$$\frac{3}{4}\|f\|^2 \le \int_{\Omega_1} \|\Lambda_\omega f\|^2 d\mu(\omega) + \left\| \int_{\Omega_1^c} \Lambda_\omega^* \Lambda_\omega f d\mu(\omega) \right\|^2 \le \|f\|^2, \tag{1.3}$$

where Λ_ω^* is the adjoint operator of Λ_ω. In the discrete case, (1.2) and (1.3) are in the following form:

$$\sum_{k\in\mathbb{J}} |\langle x, x_k\rangle|^2 + \left\| \sum_{k\in\mathbb{J}^c} \langle x, x_k\rangle x_k \right\|^2 = \sum_{k\in\mathbb{J}^c} |\langle x, x_k\rangle|^2 + \left\| \sum_{k\in\mathbb{J}} \langle x, x_k\rangle x_k \right\|^2,$$

and

$$\frac{3}{4}\|x\|^2 \le \sum_{k\in\mathbb{J}} |\langle x, x_k\rangle|^2 + \left\| \sum_{k\in\mathbb{J}^c} \langle x, x_k\rangle x_k \right\|^2 \le \|x\|^2,$$

respectively, where $\{x_k\}_{k\in\mathbb{N}}$ is a Parseval frame for $\mathcal{H}$, that is, $\|x\|^2 = \sum_{k\in\mathbb{N}} |\langle x, x_k\rangle|^2$ for all $x \in \mathcal{H}$, where $\mathbb{N}$ is a countable set, and $\mathbb{J} \subset \mathbb{N}$, $\mathbb{J}^c = \mathbb{N} \setminus \mathbb{J}$.

The objective of this short article is twofold. One is to establish new inequalities for Parseval generalized continuous frames based on the equality (1.2), the other is to give another proof of inequalities (1.3).

The organization of this article is as follows. Section 2 lists the auxiliary lemmas used in the proof of main results, and the main results and their proofs will be given in the last section.

2 Two Auxiliary Lemmas

In order to prove main results, we need two auxiliary conclusions which are listed in this section.

Lemma 2.1 ([18]) *Suppose that T is a bounded linear operator on $\mathcal{H}$. If T is positive, then T is self-adjoint and there exists a unique bounded self-adjoint operator V such that $T = V^2$. Moreover, if T is invertible, then V is also invertible.*

Lemma 2.2 ([19]) *Suppose that S and T are bounded self-adjoint positive operators. If $ST = TS$, then ST is positive.*

In addition, the following operators S_{Ω_1} and $S_{\Omega_1^c}$ are also needed in the proof of main results. Let $\{\Lambda_\omega : \omega \in \Omega\}$ is a generalized continuous frame for $\mathcal{H}$ with respect to $\{\mathcal{V}_\omega : \omega \in \Omega\}$. For any $\Omega_1 \subset \Omega$, set $\Omega_1^c = \Omega \backslash \Omega_1$. If the operators S_{Ω_1} and $S_{\Omega_1^c}$ are defined by

$$S_{\Omega_1} f = \int_{\Omega_1} \Lambda_\omega^* \Lambda_\omega f d\mu(\omega), \quad S_{\Omega_1^c} f = \int_{\Omega_1^c} \Lambda_\omega^* \Lambda_\omega f d\mu(\omega),$$

then it is easy to verify that S_{Ω_1} and $S_{\Omega_1^c}$ are well defined, bounded linear and positive.

3 Main Results and their Proofs

It is obvious to see that for all $f \in \mathcal{H}$, (1.2) can be written as

$$\int_{\Omega_1} \|\Lambda_\omega f\|^2 d\mu(\omega) - \left\|\int_{\Omega_1} \Lambda_\omega^* \Lambda_\omega f d\mu(\omega)\right\|^2 = \int_{\Omega_1^c} \|\Lambda_\omega f\|^2 d\mu(\omega) - \left\|\int_{\Omega_1^c} \Lambda_\omega^* \Lambda_\omega f d\mu(\omega)\right\|^2. \tag{3.1}$$

It follows from the definitions of S_{Ω_1} and $S_{\Omega_1^c}$ that for all $f \in \mathcal{H}$, (3.1) can be rewritten as

$$\begin{aligned} \int_{\Omega_1} \|\Lambda_\omega f\|^2 d\mu(\omega) - \left\|\int_{\Omega_1} \Lambda_\omega^* \Lambda_\omega f d\mu(\omega)\right\|^2 &= \langle S_{\Omega_1} f, f\rangle - \langle S_{\Omega_1} f, S_{\Omega_1} f\rangle \\ &= \langle (S_{\Omega_1} - S_{\Omega_1}^2) f, f\rangle. \end{aligned} \tag{3.2}$$

Now, main results and their proofs are stated as follows.

Theorem 3.1 *Suppose that $\{\Lambda_\omega : \omega \in \Omega\}$ is a Parseval generalized continuous frame for $\mathcal{H}$ with respect to $\{\mathcal{V}_\omega : \omega \in \Omega\}$. Then for all $f \in \mathcal{H}$, we have*

$$0 \le \int_{\Omega_1} \|\Lambda_\omega f\|^2 d\mu(\omega) - \left\|\int_{\Omega_1} \Lambda_\omega^* \Lambda_\omega f d\mu(\omega)\right\|^2 \le \frac{1}{4}\|f\|^2. \tag{3.3}$$

Proof Since S_{Ω_1} and $S_{\Omega_1^c}$ are bounded positive operators, by Lemma 2.1, S_{Ω_1} and $S_{\Omega_1^c}$ are self-adjoint, thereby $S_{\Omega_1} - \frac{1}{2}I$ are also self-adjoint, where I is the identity operator on $\mathcal{H}$. Employing (3.2) yields that

$$\begin{aligned}\int_{\Omega_1} \|\Lambda_\omega f\|^2 d\mu(\omega) - \left\| \int_{\Omega_1} \Lambda_\omega^* \Lambda_\omega f d\mu(\omega) \right\|^2 &= \langle (S_{\Omega_1} - S_{\Omega_1}^2) f, f \rangle \\ &= \left\langle \left(\frac{I}{4} - \left(S_{\Omega_1} - \frac{I}{2} \right)^2 \right) f, f \right\rangle \\ &= \frac{1}{4}\|f\|^2 - \left\| \left(S_{\Omega_1} - \frac{I}{2} \right) f \right\|^2 \\ &\leq \frac{1}{4}\|f\|^2.\end{aligned}$$

Note that $\{\Lambda_\omega : \omega \in \Omega\}$ is a Parseval generalized continuous frame, thus $S_{\Omega_1} + S_{\Omega_1^c} = I$. It derives from $S_{\Omega_1} S_{\Omega_1^c} = S_{\Omega_1^c} S_{\Omega_1}$ and Lemma 2.2 that $S_{\Omega_1} S_{\Omega_1^c}$ is a positive operator. Therefore by (3.2),

$$\begin{aligned}\int_{\Omega_1} \|\Lambda_\omega f\|^2 d\mu(\omega) - \left\| \int_{\Omega_1} \Lambda_\omega^* \Lambda_\omega f d\mu(\omega) \right\|^2 &= \langle S_{\Omega_1}(I - S_{\Omega_1}) f, f \rangle \\ &= \langle S_{\Omega_1} S_{\Omega_1^c} f, f \rangle \geq 0.\end{aligned}$$

So (3.3) holds. The proof is completed. □

Corollary 3.2 *Suppose that $\{\Lambda_\omega : \omega \in \Omega\}$ is a Parseval generalized continuous frame for $\mathcal{H}$ with respect to $\{\mathcal{V}_\omega : \omega \in \Omega\}$. Then for all $f \in \mathcal{H}$, we have*

$$\frac{1}{2}\|f\|^2 \leq \left\| \int_{\Omega_1} \Lambda_\omega^* \Lambda_\omega f d\mu(\omega) \right\|^2 + \left\| \int_{\Omega_1^c} \Lambda_\omega^* \Lambda_\omega f d\mu(\omega) \right\|^2 \leq \|f\|^2. \tag{3.4}$$

Proof Employing both (3.1) and (3.3), for all $f \in \mathcal{H}$, we have

$$0 \leq \int_{\Omega_1^c} \|\Lambda_\omega f\|^2 d\mu(\omega) - \left\| \int_{\Omega_1^c} \Lambda_\omega^* \Lambda_\omega f d\mu(\omega) \right\|^2 \leq \frac{1}{4}\|f\|^2. \tag{3.5}$$

Combining (3.3) with (3.5), we obtain

$$
\begin{aligned}
0 \leq & \int_{\Omega_1^c} \|\Lambda_\omega f\|^2 d\mu(\omega) + \int_{\Omega_1} \|\Lambda_\omega f\|^2 d\mu(\omega) \\
& - \left(\left\| \int_{\Omega_1^c} \Lambda_\omega^* \Lambda_\omega f d\mu(\omega) \right\|^2 + \left\| \int_{\Omega_1} \Lambda_\omega^* \Lambda_\omega f d\mu(\omega) \right\|^2 \right) \\
\leq & \frac{1}{2} \|f\|^2. \qquad (3.6)
\end{aligned}
$$

Since $\{\Lambda_\omega : \omega \in \Omega\}$ is a Parseval generalized continuous frame, we have

$$
f \in \mathcal{H}, \int_{\Omega_1^c} \|\Lambda_\omega f\|^2 d\mu(\omega) + \int_{\Omega_1} \|\Lambda_\omega f\|^2 d\mu(\omega) = \|f\|^2,
$$

this and (3.6) imply that (3.4) holds. The proof is ended. □

The inequalities (1.3) have been obtained in [17]. Here we give another proof of these inequalities. For completeness, we describe them in the form of a theorem as follows.

Theorem 3.3 *Suppose that $\{\Lambda_\omega : \omega \in \Omega\}$ is a Parseval generalized continuous frame for $\mathcal{H}$ with respect to $\{\mathcal{V}_\omega : \omega \in \Omega\}$. Then for all $f \in \mathcal{H}$, (1.3) holds, namely*

$$
\frac{3}{4} \|f\|^2 \leq \int_{\Omega_1} \|\Lambda_\omega f\|^2 d\mu(\omega) + \left\| \int_{\Omega_1^c} \Lambda_\omega^* \Lambda_\omega f d\mu(\omega) \right\|^2 \leq \|f\|^2.
$$

Proof Since $S_{\Omega_1} S_{\Omega_1^c}$ is a positive operator, it follows from $S_{\Omega_1} S_{\Omega_1^c} = S_{\Omega_1^c} - S^2_{\Omega_1^c}$ that $\langle S^2_{\Omega_1^c} f, f \rangle \leq \langle S_{\Omega_1^c} f, f \rangle$. So

$$
\left\| \int_{\Omega_1^c} \Lambda_\omega^* \Lambda_\omega f d\mu(\omega) \right\|^2 \leq \int_{\Omega_1^c} \|\Lambda_\omega f\|^2 d\mu(\omega).
$$

Because $\{\Lambda_\omega : \omega \in \Omega\}$ is a Parseval generalized continuous frame, we get

$$
\begin{aligned}
& \int_{\Omega_1} \|\Lambda_\omega f\|^2 d\mu(\omega) + \left\| \int_{\Omega_1^c} \Lambda_\omega^* \Lambda_\omega f d\mu(\omega) \right\|^2 \leq \int_{\Omega_1} \|\Lambda_\omega f\|^2 d\mu(\omega) \\
& + \int_{\Omega_1^c} \|\Lambda_\omega f\|^2 d\mu(\omega) = \|f\|^2.
\end{aligned}
$$

On the other hand, for all $f \in \mathcal{H}$, we have

$$\begin{aligned}\int_{\Omega_1} \|\Lambda_\omega f\|^2 d\mu(\omega) + \left\| \int_{\Omega_1^c} \Lambda_\omega^* \Lambda_\omega f d\mu(\omega) \right\|^2 &= \langle S_{\Omega_1} f, f\rangle + \langle S_{\Omega_1^c} f, S_{\Omega_1^c} f\rangle \\ &= \langle (I - S_{\Omega_1^c} + S_{\Omega_1^c}^2) f, f\rangle \\ &= \frac{3}{4}\|f\|^2 + \left\| \left(S_{\Omega_1^c} - \frac{1}{2} I \right) f \right\|^2 \\ &\geq \frac{3}{4}\|f\|^2.\end{aligned}$$

So (1.3) is true. This completes the proof. □

Remark 3.4 It is easy to see that the inequality (3.4) can be also proved with (3.3) and the inequality in Theorem 3.3.

Acknowledgments The authors would like to express their gratitude to the referee for his (or her) valuable comments and suggestions.

This work is supported by National Natural Science Foundation of China (No.61471410).

References

1. R.J. Duffin, A.C. Schaeffer, A class of nonharmonic Fourier series. Trans. Amer. Math. Soc. **72**, 341–366 (1952)
2. E.J. Candés, D.L. Donoho, New tight frames of curvelets and optimal representations of objects with piecewise $\mathcal{C}^2$ singularities. Comm. Pure Appl. Math. **56**, 216–266 (2004)
3. E.J. Candés, Harmonic analysis and neural networks. Appl. Comput. Harmon. Anal. **6**, 197–218 (1999)
4. T. Strohmer, R. W. Heath Jr, Grassmannian frames with applications to coding and communication. Appl. Comput. Harmon. Anal. **14**, 257–275 (2003)
5. R.W. Heath, A.J. Paulraj, Linear dispersion codes for MIMO systems based on frame theory. IEEE Trans. Signal Process. **50**, 2429–2441 (2002)
6. M.R. Abdollahpour, M.H. Faroughi, Generalized continuous frames in Hilbert spaces. Southeast Asian Bull. Math. **32**, 1–19 (2008)
7. S.T. Ali, J.P. Antoine, J.P. Gazeau, Continuous frames in Hilbert spaces. Ann. Phys. **222**, 1–37 (1993)
8. J.P. Gabardo, D. Han, Frames associated with measurable spaces. Adv. Comput. Math. **18**, 127–147 (2003)
9. A. Rahimi, A. Najati, Y.N. Dehghan, Continuous frames in Hilbert spaces. Methods Funct. Anal. Topology **12**, 170–182 (2006)
10. W.C. Sun, Generalized frames and g-Riesz bases. J. Math. Anal. Appl. **322**, 437–452 (2006)
11. R. Balan, P.G. Casazza, D. Edidin, G. Kutyniok, Decompositions of frames and a new frame identity. Proc. SPIE. **5914**, 379–388 (2005)
12. R. Balan, P.G. Casazza, D. Edidin, G. Kutyniok, A new identity for Parseval frames. Proc. Amer. Math. Soc. **135**, 1007–1015 (2007)

13. P. Gävruta, On some identities and inequalities for frames in Hilbert spaces. J. Math. Anal. Appl. **321**, 469–478 (2006)
14. X.G. Zhu, G.C. Wu, A note on some equalities for frames in Hilbert spaces. Appl. Math. Lett. **23**, 788–790 (2010)
15. D.F. Li, W.C. Sun, Some equalities and inequalities for generalized frames. Chin. J. Contemp. Math. **29**, 301–308 (2008)
16. J.Z. Li, Y.C. Zhu, Some equalities and inequalities for g-Bessel sequences in Hilbert spaces. Appl. Math. Lett. **25**, 1601–1607 (2012)
17. X.C. Xiao, X.M. Zeng, Some equalities and inequalities of G-continuous frames. Sci. China Math. **53**, 2621–2632 (2010)
18. O. Christensen, *An Introduction to Frames and Riesz Bases* (Birkhäuser, Boston, 2003)
19. K. Erwin, *Introductory Functional Analysis with Applications* (Wiley, New York, 1978)

Centering Projection Methods for Wavelet Feasibility Problems

Neil D. Dizon, Jeffrey A. Hogan, and Scott B. Lindstrom

Abstract We revisit the feasibility approach to the construction of compactly supported smooth orthogonal wavelets on the line. We highlight its flexibility and illustrate how symmetry and cardinality properties are easily embedded in the design criteria. We solve the resulting wavelet feasibility problems using recently introduced centering methods, and we compare performance. Solutions admit real-valued compactly supported smooth orthogonal scaling functions and wavelets with near symmetry and near cardinality properties.

Keywords Wavelets · Feasibility problem · Douglas–Rachford · Centering · Circumcentering

Mathematics Subject Classification (2020) 90C26, 47H10, 65K10, 65T60

1 Wavelet Construction as a Feasibility Problem

Wavelets are traditionally constructed through *multiresolution analysis* (MRA) which was introduced by Mallat [17] and Meyer [18]. Following MRA, Daubechies derived the first known examples of compactly supported smooth wavelets with orthonormal shifts [7, 8]. While these wavelets have been demonstrably useful in many signal processing applications, symmetry and cardinality properties are also often desired. It is known that symmetry is incompatible with real-valuedness, orthogonality, smoothness and compact support [8, Theorem 8.1.4]. In the same way, the cardinality property cannot be imposed together with all

N. D. Dizon (✉) · J. A. Hogan
The University of Newcastle, Newcastle, NSW, Australia
e-mail: neilkristofer.dizon@uon.edu.au; jeff.hogan@newcastle.edu.au

S. B. Lindstrom
The Hong Kong Polytechnic University, Hung Hom, Hong Kong
e-mail: scott.b.lindstrom@polyu.edu.hk

P. Cerejeiras et al. (eds.), *Current Trends in Analysis, its Applications and Computation*, Research Perspectives,
https://doi.org/10.1007/978-3-030-87502-2_66

of compact support, continuity, and orthogonal shifts [20]. Recognizing these theoretical obstructions, we relax perfect symmetry or cardinality and impose only near symmetry or near cardinality. A construction technique that readily accounts for these design criteria and that easily extends to higher dimensions is preferable.

Wavelet construction has been recently formulated as a feasibility problem originally aimed at generating compactly supported smooth wavelets with orthonormal shifts [12–14]. This approach handily accounts for other design criteria and allows for construction of non-tensorial wavelets in higher dimensions.

Outline and Contributions In the remainder of this section, we recall one reformulation of wavelet construction as a feasibility problem. In Sect. 2, we recall the two centering methods we will compare: a generically proper variant of circumcentering reflections method (CRM) [1, 3–5] and a new method due to Lindstrom [15]. Section 3 contains our principal contribution: an experimental comparison of 2-stage global-then-local search methods, first introduced in [10], that combine the Douglas–Rachford method together with centering methods. This is the first such comparison for a feasibility problem, and also the first for a nonconvex problem. The results shed light on the algorithms more generally, while offering a path forward for wavelet feasibility problems specifically.

MRA Conditions and Wavelet Properties The traditional approach to the construction of wavelet orthonormal bases is based on MRA. For a more detailed discussion of the concepts that follow, refer to [8, 9, 12–14]. Henceforth, $\hat{f}$ denotes the Fourier transform of a function $f \in L^2(\mathbb{R}, \mathbb{C})$, $\overline{A}$ is the conjugate of A and denotes elementwise conjugation when A is a matrix, $A[j, k]$ is the (j, k)-entry of a matrix A, $\|A\|$ is the Frobenius norm of the matrix A, and $\mathrm{cl}(S)$ is the closure of a set S.

Definition 1.1 A *multiresolution analysis* for $L^2(\mathbb{R}, \mathbb{C})$ consists of a sequence of closed subspaces $\{V_j\}_{j\in\mathbb{Z}}$ of $L^2(\mathbb{R}, \mathbb{C})$ and a *scaling function* $\varphi \in V_0$ such that the following conditions hold:

(i) the spaces V_j are nested, i.e., $V_j \subset V_{j+1}$ for all $j \in \mathbb{Z}$,
(ii) $\mathrm{cl}\left(\bigcup_{j\in\mathbb{Z}} V_j\right) = L^2(\mathbb{R}, \mathbb{C})$ and $\bigcap_{j\in\mathbb{Z}} V_j = \{0\}$,
(iii) $f(\cdot) \in V_0$ if and only if $f(\cdot - k) \in V_0$ for all $k \in \mathbb{Z}$,
(iv) $f(\cdot) \in V_j$ if and only if $f(2(\cdot)) \in V_{j+1}$ for all $j \in \mathbb{Z}$, and
(v) $\{\varphi(\cdot - k)\}_{k\in\mathbb{Z}}$ forms an orthonormal basis for V_0.

If φ arises from an MRA, then we may write $\frac{1}{2}\varphi\left(\frac{x}{2}\right) = \sum_{k\in\mathbb{Z}} h_k\varphi(x - k)$ with $\{h_k\}_{k\in\mathbb{Z}} \in \ell^2(\mathbb{Z})$. Taking the Fourier transforms of both sides of this equation, we obtain the *scaling equation* in the Fourier domain given by $\hat{\varphi}(2\xi) = H(\xi)\hat{\varphi}(\xi)$ where $H(\xi) = \sum_k h_k e^{-2\pi ik\xi}$ is the *scaling filter*. Moreover, we can find a *wavelet function* $\psi \in V_1 \setminus V_0$ satisfying $\frac{1}{2}\psi\left(\frac{x}{2}\right) = \sum_{k\in\mathbb{Z}} g_k\varphi(x - k)$ where $\{g_k\}_{k\in\mathbb{Z}} \in \ell^2(\mathbb{Z})$. Taking the Fourier transforms of both sides of this equation, one obtains $\hat{\psi}(2\xi) = G(\xi)\hat{\varphi}(\xi)$ where $G(\xi) = \sum_k g_k e^{-2\pi ik\xi}$ is the *wavelet filter*. If φ has orthonormal shifts and $\left\{\psi_{j,k} := 2^{-j/2}\psi(2^{-j}x - k)\right\}_{j,k\in\mathbb{Z}}$ forms an orthonormal basis for $L^2(\mathbb{R}, \mathbb{C})$, then the wavelet matrix

$$U(\xi) := \begin{bmatrix} H(\xi) & G(\xi) \\ H\left(\xi + \frac{1}{2}\right) & G\left(\xi + \frac{1}{2}\right) \end{bmatrix} \quad (1.1)$$

is unitary for almost every $\xi \in \mathbb{R}$ and $H(0) = 1$. This definition introduces a *consistency condition* that $U(\xi + \frac{1}{2}) = JU(\xi)$ where J is the row-swap matrix.

The effectiveness of a wavelet orthonormal basis $\{\psi_{j,k}\}_{j,k\in\mathbb{Z}}$ lies in its efficient analysis and synthesis of signals. To allow for speedy and accurate computation of the wavelet coefficients, we desire compact support. The scaling function and wavelet are compactly supported on the interval $[0, M - 1]$ if and only if we can write $H(\xi) = \sum_{k=0}^{M-1} h_k e^{2\pi i k\xi}$ and $G(\xi) = \sum_{k=0}^{M-1} g_k e^{2\pi i k\xi}$ as trigonometric polynomials of degree $M - 1$ [12, 14]. Thus, we are able to write the wavelet matrix in the form

$$U(\xi) = \sum_{k=0}^{M-1} A_k e^{2\pi i k\xi} \quad \text{where} \quad A_k = \begin{bmatrix} h_k & g_k \\ (-1)^k h_k & (-1)^k g_k \end{bmatrix}.$$

Additionally, if ψ has D *continuous and bounded derivatives*, then this allows for better approximation using relatively fewer wavelet coefficients. Consequently, H, G and U satisfy

$$\left.\frac{d^k H(\xi)}{d\xi^k}\right|_{\xi=\frac{1}{2}} = 0 \iff \left.\frac{d^k G(\xi)}{d\xi^k}\right|_{\xi=0} = 0 \iff \left(\left.\frac{d^k U(\xi)}{d\xi^k}\right|_{\xi=0}\right)[1, 2] = 0$$

for all $k \in \{0, 1, \ldots, D\}$, where the differentiation of the matrix is interpreted element-wise [8, 12, 14].

Furthermore, *symmetry* is another design criterion that we want φ and ψ to possess. It is known that symmetric filters applied to image processing can deal better with boundaries than asymmetric ones. A scaling function φ is symmetric about $x = P \in (0, M - 1)$ if and only if $H(\xi) = e^{4\pi i P\xi} H(-\xi)$. If $K = \operatorname{diag}(-1, 1) \in \mathbb{C}^{2\times 2}$, then the symmetry condition can be written in terms of the wavelet matrix as $U(\xi) = e^{4\pi i P\xi} KU(\xi)K$ [9]. Note that when the scaling function is symmetric, the associated wavelet is either symmetric or anti-symmetric depending on the length of support. For conciseness, we simply say that the wavelet is symmetric.

On the other hand, *cardinality* is also often sought in certain applications. A scaling function φ is *cardinal* at $P \in \mathbb{Z}$ if $\varphi(k) = \delta_{kP}$ for all $k \in \mathbb{Z}$, where δ is the Kronecker delta. A cardinal φ admits a reconstruction formula for recovery of any function in V_0 from its integer samples. A necessary condition for φ to be cardinal at $P \in \mathbb{Z}$ is $H(\xi) + (-1)^P H(\xi + \frac{1}{2}) = e^{2\pi i P\xi}$ [9]. Note that cardinality is desired only for the scaling function. For brevity in describing our wavelets, any mention of cardinal wavelet means that the associated scaling function is cardinal.

If we further want to guarantee that φ and ψ are *real-valued*, we impose the condition that $H(\xi) = \overline{H(-\xi)}$ and $G(\xi) = \overline{G(-\xi)}$ which is equivalent to $U(\xi) = \overline{U(-\xi)}$ [9, 12].

At this point we see that wavelet construction may be reduced to generating a matrix $U(\xi)$ satisfying the above conditions.

Discretization by Uniform Sampling Since a trigonometric polynomial of degree $M-1$ is determined by M points, we discretize $U(\xi)$ by a uniform sampling at M points in $\{\frac{j}{M}\}_{j=0}^{M-1} \subseteq [0, 1)$. By denoting each sample point by $U_j = U(\frac{j}{M})$, we form an *ensemble* $\mathcal{U} := (U_0, U_1, \ldots, U_{M-1}) \in (\mathbb{C}^{2\times 2})^M$. The coefficient matrices A_k are computed from an ensemble through an invertible M-point discrete Fourier transform $\mathcal{F}_M\colon (\mathbb{C}^{2\times 2})^M \to (\mathbb{C}^{2\times 2})^M : \mathcal{U} \mapsto \mathcal{A} := (A_0, \ldots, A_{M-1})$ where

$$A_k = (\mathcal{F}_M \mathcal{U})_k = \frac{1}{M} \sum_{j=0}^{M-1} U_j e^{-2\pi i jk/M}, \text{ for } k \in \{0, 1, \ldots, M-1\}. \tag{1.2}$$

The discretized version of the consistency condition requires $U_{j+\frac{M}{2}} = JU_j$ for every $j \in \{0, 1, \ldots, M-1\}$. For $U(\xi)$ to be unitary almost everywhere, we need to enforce $U(\xi)$ to be unitary at $2M$ samples. Given the sample points in $\mathcal{U}$, the other set of M samples may be computed to form another ensemble using $\widetilde{\mathcal{U}} := \mathcal{F}_M^{-1} \chi_M \mathcal{F}_M(\mathcal{U})$, where $(\chi_M)_j = e^{\pi i j/M}$ for $j = \{0, 1, \ldots, M-1\}$. Moreover, the regularity condition is imposed by forcing $(\sum_{j=0}^{M-1} j^\ell A_j)[1, 2] = 0$ for all $\ell \in \{0, 1, \ldots, D\}$ where

$$\sum_{j=0}^{M-1} j^\ell A_j = \frac{1}{M} \sum_{k=0}^{M-1} \alpha_{\ell k} U_k \text{ and } \alpha_{\ell k} = \frac{1}{M} \sum_{j=0}^{M-1} j^\ell e^{-2\pi i kj/M}.$$

For the symmetry condition, we require $U_j = e^{4\pi i Pj/M} K U_{M-j} K$ for all $j \in \{1, \ldots, \frac{M}{2}\}$. Cardinality is imposed by forcing $U_j[1, 1] + (-1)^P U_{j+\frac{M}{2}}[1, 1] = e^{2\pi i Pj/M}$, and the real-valuedness condition requires $U_j = \overline{U_{M-j}}$ for $j \in \{1, \ldots, \frac{M}{2}\}$.

The Wavelet Feasibility Problem The *feasibility problem* is to find a point in the intersection of a finite number of constraint sets. To reformulate wavelet construction as a feasibility problem, we treat the wavelet *properties* as *constraints* imposed on the discrete version of the wavelet matrix $U(\xi)$. We denote the collection of ensembles in $(\mathbb{C}^{2\times 2})^M$ that satisfies the consistency condition by $(\mathbb{C}^{2\times 2})^M_J$, and the collection of all 2-by-2 unitary matrices by $\mathbb{U}(2)$. For an even integer $M \geq 4$ and

$D = \frac{M-2}{2}$ (unless otherwise specified), we define $B_1, B_2, B_3, B_4, B_5^{(S)}, B_5^{(C)} \subseteq (\mathbb{C}^{2\times 2})_J^M$ as follows.

$$B_1 := \left\{\mathcal{U} : U_0 = \begin{pmatrix} 1 & 0 \\ 0 & z \end{pmatrix}, \ |z| = 1, \ U_j \in \mathbb{U}(2), \ j \in \{0, 1, \dots, \tfrac{M}{2}\}\right\},$$

$$B_2 := \{\mathcal{U} : (\mathcal{F}_M \chi_M (\mathcal{F}_M)^{-1}(\mathcal{U}))_j \in \mathbb{U}(2), \ j \in \{0, 1, \dots, \tfrac{M}{2}\}\},$$

$$B_3 := \left\{\mathcal{U} : \left(\textstyle\sum_{j=0}^{M-1} \alpha_{\ell k} U_k\right)[1, 2] = 0, \ 0 \le \ell \le D\right\},$$

$$B_4 := \{\mathcal{U} : U_j = \overline{U_{M-j}}, \ j \in \{1, 2, \dots, \tfrac{M}{2}\}\},$$

$$B_5^{(S)} := \left\{\mathcal{U} : \|U_j - e^{2\pi i P j/M} K U_{M-j} K\| < \gamma, \ j \in \{1, 2, \dots, M/2\}\right\}$$

$$B_5^{(C)} := \left\{\mathcal{U} : \left|U_j[1, 1] + (-1)^P U_{j+\frac{M}{2}}[1, 1] - e^{2\pi i P j/M}\right| < \gamma, \ j \in \{1, 2, \dots, \tfrac{M}{2}\}\right\}.$$

Note that B_1 and B_2 are nonconvex constraint sets that correspond to the unitarity condition at $2M$ sample points. The subspaces B_3 and B_4 are constraint sets for regularity and real-valuedness, respectively. Moreover, $B_5^{(S)}$ and $B_5^{(C)}$ are convex sets that promote near symmetry and near cardinality properties, respectively. Notice the introduction of a small positive number γ in the definition of $B_5^{(S)}$ and $B_5^{(C)}$ to get around the theoretical obstructions for obtaining perfect symmetry and cardinality [9]. In summary, we have the following feasibility problems.

Problem 1.2 (Nearly Symmetric Wavelets) *The feasibility problem for constructing compactly supported real-valued smooth nearly symmetric orthogonal wavelets is to find an ensemble* $\mathcal{U} \in \bigcap_{k=1}^4 B_k \cap B_5^{(S)} \subseteq (\mathbb{C}^{2\times 2})_J^M$.

Problem 1.3 (Nearly Cardinal Wavelets) *The feasibility problem for constructing compactly supported real-valued smooth nearly cardinal orthogonal wavelets is to find an ensemble* $\mathcal{U} \in \bigcap_{k=1}^4 B_k \cap B_5^{(C)} \subseteq (\mathbb{C}^{2\times 2})_J^M$.

2 Centering Methods for Feasibility Problems

The original works that solved wavelet feasibility problems for compactly supported smooth orthogonal wavelets employed the *Douglas–Rachford (DR) algorithm* [11, 16] to solve *Pierra's product space reformulation* [19] of the feasibility problem. The method demonstrated surprising robustness in this context [6, 12–14]. Convergence plots frequently feature the tell-tale characteristics of local spiraling during convergence; such features are described in [15]. The spiraling is associated with longer runs for numerical implementations [12] and presents an opportunity to accelerate convergence [15].

In this section, we recall the DR operator, the generalized circumcentered reflections method operator (GCRM) [4, 10] and the new centering operator L_T introduced by Lindstrom in [15]. We expect the two centering methods to accelerate convergence to feasible solutions.

Let $\mathcal{H}$ be a real Hilbert space with induced norm $\|\cdot\|$. For a closed subset C of $\mathcal{H}$, we define the operator $P_C : \mathcal{H} \to C$ by $P_C(x) = \operatorname{argmin}_{z \in C} \|z - x\|$; it is a selector for the closest point projection for C. Its associated *reflector* is defined as $R_V := 2P_V - Id$ where Id is the identity map. Given three points $x, y, z \in \mathcal{H}$, we denote $C(x, y, z)$ to be their *circumcenter*, which is equidistant to the given points and lies on the affine subspace they define. The circumcenter exists whenever x, y, z are not simultaneously distinct and colinear; for more on existence and formulae for computation, see [1, 2].

Definition 2.1 Let V and W be nonempty subsets of $\mathcal{H}$.

1. The *DR operator* for V and W is defined as $T(x) := x - P_V(x) + P_W R_V(x)$.
2. The *circumcentering reflections method* operator is defined as $CRM(x) := C(x, R_V(x), R_W R_V(x))$. For history and properties, see [3–5].
3. The *GCRM operator* is defined as

$$C_{V,W}(x) := \begin{cases} T(x) & \text{if } x, R_V x, R_W R_V x \text{ are colinear;} \\ CRM(x) & \text{otherwise.} \end{cases}$$

4. The centering operator L_T from [15] is defined as:

$$L_T(x) := \begin{cases} C\left(x, 2T(x) - x, \pi_T(x)\right) & \text{if } x, 2T(x) - x, \pi_T(x) \text{ are not colinear;} \\ T^2(x) & \text{otherwise.} \end{cases}$$

where $\pi_T(x) = 2\left(T^2(x) - T(x)\right) + 2P_{\operatorname{span}(T^2(x) - T(x))}(T(x) - x) + x$.

Lindstrom discovered that for some prototypical feasibility problems for which Lyapunov functions are known, CRM returns the minimizer of a quadratic surrogate for the local Lyapunov function [15]. Lindstrom showed that L_T's lack of dependence on subproblems (in our case, reflections) allows it to recapture this property in settings where CRM may not, such as for the primal-dual implementation of ADMM/Douglas–Rachford for basis pursuit. In our setting, this possible improvement in stability carries the computational cost that one application of L_T requires two applications of the pair of projections P_V and P_W, instead of just one pair for CRM.

For numerical implementations, we set up a 2-stage DR-GCRM and a 2-stage DR-L_T. In stage 1, we exploit the greater global robustness of DR to find local basins of attraction to feasible points, and thereafter, in stage 2, we apply centering methods to obviate local spiraling thereto. It has already been shown experimentally that this approach consistently outperforms a full run of DR in the context of solving wavelet feasibility problems [10]. In the next section, we use GCRM and L_T as the local methods of 2-stage global-then-local search algorithms, in order to solve Problems 1.2 and 1.3.

3 Numerical Results

We use a product space technique similar to those employed in [10, 12–14] to convert our many-set feasibility problems into 2-set problems amenable to solution by the methods described above.

Problem 1.2: The constraints for obtaining nearly symmetric wavelets are

$$V := B_1 \times B_2 \times (B_3 \cap B_4) \times B_5^{(S)} \subseteq \left((\mathbb{C}^{2\times 2})_J^M \right)^4,$$

$$W := \left\{ (\mathcal{U}_j)_{j=1}^4 \in \left((\mathbb{C}^{2\times 2})_J^M \right)^4 : \mathcal{U}_1 = \mathcal{U}_2 = \mathcal{U}_3 = \mathcal{U}_4 \right\}.$$

Problem 1.3: The constraints for obtaining nearly cardinal wavelets are

$$V := B_1 \times B_2 \times (B_3 \cap B_4) \times B_5^{(C)} \subseteq \left((\mathbb{C}^{2\times 2})_J^M \right)^4,$$

$$W := \left\{ (\mathcal{U}_j)_{j=1}^4 \in \left((\mathbb{C}^{2\times 2})_J^M \right)^4 : \mathcal{U}_1 = \mathcal{U}_2 = \mathcal{U}_3 = \mathcal{U}_4 \right\}.$$

The projection of a 4-tuple of ensembles onto W is obtained by averaging the 4 ensembles. Notice that the set V and its projection P_V are different for the two problems, though this should create no confusion because we will only discuss one problem at a time. We have $P_V = P_{B_1} \times P_{B_2} \times (P_{B_3} P_{B_4}) \times P_{B_5^{(\eta)}}$, where η is C or S respectively for the two problems. Because $P_{B_3}(B_4) \subset B_4$ and B_4 is a subspace, the identity $P_{B_3} P_{B_4} = P_{B_3 \cap B_4}$ admits the *constraint-reduction reformulation* we have used; see [6].

In what follows, we solve Problems 1.2 and 1.3 with $M = 6, D = 1$ and $\gamma = 0.5$. We compare the performance of 2-stage DR-GCRM with 2-stage DR-L_T. We initialize at 100 random ensembles that satisfy the consistency condition. Throughout, we let $\{x_n\}_{n\in\mathbb{N}}$ be the sequence of iterates generated by the projection algorithm under consideration. We fix a tolerance $\varepsilon := 10^{-9}$ and use the stopping criterion $\varepsilon_n := \|P_V P_W(x_n) - P_W(x_n)\| < \varepsilon$, whereupon $P_W(x_n)$ is a feasible point. In implementing a 2-stage method, we first run DR until the gap distance ε_n reaches a 10^{-2} threshold; thereafter we switch to applying GCRM or L_T. We declare a particular run to have *solved* the feasibility problem whenever it attains the threshold of ε within 20,000 iterations. We provide statistics on the *number of iterations* needed, which is our main performance measure. We do not report the number of iterates required for ε_n to obtain the threshold 10^{-2}, because it is the same for both 2-stage algorithms. We only report the number of iterates needed thereafter. Examples of scaling functions derived from solutions of the wavelet feasibility problems are shown in Fig. 1.

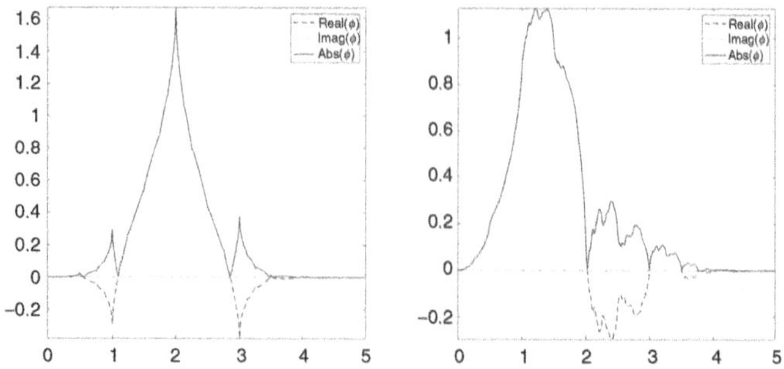

(a) Nearly symmetric about $x = 2$ (b) Nearly cardinal at $x = 1$

Fig. 1 Scaling functions plotted by employing the *cascade algorithm* on filters generated from the feasible ensembles solved in Problems 1.2 and 1.3, respectively

Table 1 Performance during stage 2 of a 2-stage search

				When solved by all				
	Algorithm	Cases solved	Solved by all	Wins	Q1	Mean	Q3	Median
Problem 1	DR	51	51	0	194	211	215	201
	GCRM	51	51	13	36	28	40	38
	L_T	51	51	38	29	36	39	33
Problem 2	DR	96	79	0	176	182	186	185
	GCRM	79	79	22	31	33	35	33
	L_T	96	79	57	28	32	33	31

Table 1 summarizes the numerical results. L_T solved every problem DR solved. For Problem 1.3, GCRM was less stable than L_T, which is consistent with what one might expect, given that L_T is constructed to retain the property of minimizing a surrogate Lyapunov function in situations where GCRM's dependence on subproblems may cause instability [15]. Interestingly, for Problem 1.2, GCRM also solved every problem DR solved. When both algorithms converged, L_T and GCRM performed quite similarly, which is what one would expect if both methods are constructing, from their respective sampling points, relatively similar quadratic surrogates for the underlying Lyapunov function. However, one should remember that computing a single centering step of L_T requires computing twice the number of projection substeps that are needed by a single step of GCRM.

4 Conclusion

We have shown how the symmetry and cardinality constraints are readily accounted for in the feasibility approach to wavelet construction. Numerical results also shed light on local behaviour of L_T and GCRM. We speculate that both are viable heuristics that may be applied to deal with wavelet feasibility problems for higher dimensional constructions, and we suggest this as the next step of research.

References

1. H.H. Bauschke, H. Ouyang, X. Wang, On circumcenter mappings induced by nonexpansive operators (2018). Preprint arXiv:1811.11420
2. H.H. Bauschke, H. Ouyang, X. Wang, On circumcenters of finite sets in Hilbert spaces (2018). Preprint arXiv:1807.02093
3. R. Behling, J.Y. Bello-Cruz, L.R. Santos, On the linear convergence of the circumcentered-reflection method. Oper. Res. Lett. **46**(2), 159–162 (2018)
4. R. Behling, J.Y.B. Cruz, L.R. Santos, Circumcentering the Douglas–Rachford method. Num. Algor. **78**(3), 759–776 (2018)
5. R. Behling, J.Y. Bello-Cruz, L.R. Santos, On the circumcentered-reflection method for the convex feasibility problem (2019). Preprint arXiv:2001.01773
6. M. Dao, N. Dizon, J. Hogan, M. Tam, Constraint reduction reformulations for projection algorithms with applications to wavelet construction. J. Optim. Theory Appl. **190**, 201–233 (2021)
7. I. Daubechies, Orthonormal bases of compactly supported wavelets. Commun. Pur. Appl. Math. **41**, 909–996 (1988)
8. I. Daubechies, *Ten Lectures on Wavelets* (Society for Industrial and Applied Mathematics, Philadelphia, 1992)
9. N.D. Dizon, J.A. Hogan, J.D. Lakey, Optimization in the construction of nearly cardinal and nearly symmetric wavelets, in *13th International Conference on Sampling Theory and Applications (SampTA)* (IEEE, Piscataway, 2019), pp. 1–4
10. N.D. Dizon, J.A. Hogan, S.B. Lindstrom, Circumcentering reflection methods for nonconvex feasibility problems (2019). Preprint arXiv:1910.04384
11. J. Douglas, H. Rachford, On the numerical solution of heat conduction problems in two and three space variables. Trans. Amer. Math. Soc. **82**, 421–439 (1956)
12. D.J. Franklin, Projective algorithms for non-separable wavelets and clifford fourier analysis. Ph.D. Thesis, The University of Newcastle (Australia), 2018
13. D.J. Franklin, J.A. Hogan, M.K. Tam, Higher-dimensional wavelets and the Douglas-Rachford algorithm, in *13th International Conference on Sampling Theory and Applications (SampTA)* (IEEE, Piscataway, 2019), pp. 1–4
14. D. Franklin, J.A. Hogan, M.K. Tam, A Douglas–Rachford construction of non-separable continuous compactly supported multidimensional wavelets (2020). Preprint arXiv:2006.03302
15. S.B. Lindstrom, Computable centering methods for spiraling algorithms and their duals, with motivations from the theory of Lyapunov functions (2020). arXiv pp. arXiv–2001
16. S.B. Lindstrom, B. Sims, Survey: Sixty years of Douglas–Rachford. J. AustMS (to appear) (2018). Preprint arXiv:1809.07181
17. S. Mallat, Multiresolution approximations and wavelet orthonormal bases of $L_2(\mathbb{R})$. T. A. Math. Soc. **315**, 69–87 (1989)
18. Y. Meyer, *Wavelets and Operators.* (Cambridge University Press, Cambridge, 1993)
19. G. Pierra, Decomposition through formalization in a product space. Math. Program. **28**, 96–115 (1984)
20. X. Xia, Z. Zhang, On sampling theorem, wavelets, and wavelet transforms. IEEE Trans. Signal Proc. **41**, 3524–3535 (1993)

Some Topics on the Gabor Wavelet Transformation

Keiko Fujita

Abstract We studied the Gabor wavelet transform of analytic functionals on the sphere in general dimension. Then we studied the Gabor wavelet transformation on the two-dimensional sphere and its inverse transformation. In this note, we review our previous results and we consider the relationship among the Fourier transformation, the windowed Fourier transformation whose windows function is given by a Gaussian function, and the Gabor wavelet transformation on the two-dimensional unit sphere.

Keywords Gabor wavelet transformation

Mathematics Subject Classification (2010) Primary 44A15; Secondary 43A32

1 Introduction

In [4], we call the windowed Fourier transformation whose windows function is the Gaussian function by the "Gabor transformation". We studied the Gabor transform and the Gabor wavelet transform of analytic functional on the sphere in general dimension. Then, in [5], we constructed the inverse Gabor wavelet transformation concretely in the two dimensional sphere by using our results in [1]. Further in [6] we treated some examples of the Gabor wavelet transform.

In this note, to understand the Gabor wavelet transformation on the sphere, we compare the images of a harmonic polynomial under the Fourier transformation, under the windowed Fourier transformation and under the Gabor wavelet transformation on the two-dimensional unit sphere. Note that we call the Gabor wavelet transformation the Gabor transformation in [2, 4] and [5].

K. Fujita (✉)
University of Toyama, Toyama, Japan
e-mail: keiko@sci.u-toyama.ac.jp

P. Cerejeiras et al. (eds.), *Current Trends in Analysis, its Applications and Computation*, Research Perspectives,
https://doi.org/10.1007/978-3-030-87502-2_67

2 Gabor Wavelet Transformation

2.1 Fourier Transformation on the Sphere

Let S^2 be the unit sphere in $\mathbf{R}^3$, that is,

$$S^2 = \{(x_1, x_2, x_3) \in \mathbf{R}^3; x_1^2 + x_2^2 + x_3^2 = 1\}.$$

For $z = (z_1, z_2, z_3) \in \mathbf{C}^3$ and $w = (w_1, w_2, w_3) \in \mathbf{C}^3$, we set

$$z \cdot w = z_1 w_1 + z_2 w_2 + z_3 w_3, \quad z^2 = z \cdot z.$$

For an integrable function f on S^2, we define the Fourier transform of f by

$$(\mathcal{F} f)(w) = \int_{S^2} e^{-ix \cdot w} \overline{f(x)} d\Omega_x, \tag{1}$$

where $d\Omega$ is the normalized invariant measure on S^2. Note that $\mathrm{vol}(S^2) = 4\pi$.

For the square integrable functions f and g on S^2, we define a sesquilinear form $(f, g)_{S^2}$ by

$$(f, g)_{S^2} \equiv \int_{S^2} f(x) \overline{g(x)} d\Omega_x.$$

Then $(\cdot, \cdot)_{S^2}$ gives an inner product and we denote by $L^2(S^2)$ the space of square integrable functions on S^2 with the inner product $(\cdot, \cdot)_{S^2}$, and the norm $\|\cdot\|_{S^2}$ of f is given by $\|f\|_{S^2} = \sqrt{(f, f)_{S^2}}$.

2.2 The Gaussian Function on the Sphere

We consider the Gaussian function on the sphere. For $b > 0$, put

$$f_y(x) = \frac{1}{b} \exp(-(x - y)^2/(2b^2)), \quad y \in \mathbf{R}^3.$$

For $x, y \in S^2$,

$$f_y(x) = \frac{1}{b} \exp(-(1 - x \cdot y)/b^2).$$

For fixed $y \in S^2$, let θ be the angle between x and y. Then

$$f_y(x) = f_y(\theta) = \frac{1}{b}\exp(-(1-\cos\theta)/b^2).$$

For a sufficiently small $b > 0$ and fixed $y \in S^2$ the Gaussian function $f_y(x)$ looks like the Delta function with peak at y.

2.3 Windowed Fourier Transformation on the Sphere

Let $b \in \mathbf{R}\backslash\{0\}$ and put $w_b(x) = \frac{1}{b}\exp(-x^2/(2b^2))$. For $f \in L^2(S^2)$ and $\omega, \tau \in \mathbf{C}^3$, we define the windowed Fourier transformation $\mathcal{W}_b\mathcal{F}$ with the window function $w_b(x)$ by

$$\begin{aligned}\mathcal{W}_b\mathcal{F} : f \mapsto (\mathcal{W}_b\mathcal{F}f)(\tau,\omega) &= \int_{S^2} e^{-ix\cdot\omega} w_b(x-\tau)\overline{f(x)}d\Omega_x \\ &= \frac{1}{b}e^{\frac{-1-\tau^2}{2b^2}} \int_{S^2} e^{-ix\cdot(\omega+i\tau/b^2)}\overline{f(x)}d\Omega_x \\ &= \frac{1}{b}e^{\frac{-1-\tau^2}{2b^2}} \int_{S^2} e^{-ix\cdot\omega}e^{x\cdot\tau/b^2}\overline{f(x)}d\Omega_x \\ &= \frac{1}{b}e^{-\frac{1+\tau^2}{2b^2}}(\mathcal{F}f)(\omega+i\tau/b^2) = \frac{1}{b}e^{-\frac{1+\tau^2}{2b^2}}(\mathcal{F}g)(\omega), \end{aligned} \quad (2)$$

where $g(x) = \overline{e^{x\cdot\tau/b^2}}f(x)$. Note that we take $b = 1$ in our previous papers.

2.4 Gabor Wavelet Transformation on the Sphere

Let $\omega_0 \in \mathbf{R}^3 \setminus \{0\}$ be fixed. Put

$$G_{\omega_0}(x) = e^{-x^2/2}e^{-ix\cdot\omega_0}.$$

For $f \in L^2(S^2)$ and $a \in \mathbf{R}_+ = \{x : x > 0\}$, we define the Gabor wavelet transformation $\mathcal{G}_{\omega_0}$ by

$$\begin{aligned}\mathcal{G}_{\omega_0} : f \mapsto (\mathcal{G}_{\omega_0}f)(\tau,a) &= \frac{1}{a}\int_{S^2} G_{\omega_0}\left(\frac{x-\tau}{a}\right)\overline{f(x)}d\Omega_x \quad (3)\\ &= \frac{1}{a}\int_{S^2} e^{-i\frac{x-\tau}{a}\cdot\omega_0}e^{-\frac{1}{2}\left(\frac{x-\tau}{a}\right)^2}\overline{f(x)}d\Omega_x\end{aligned}$$

$$= \frac{1}{a} e^{-\frac{1+\tau^2}{2a^2}} e^{i\tau\cdot\frac{\omega_0}{a}} \int_{S^2} e^{-ix\cdot\frac{\omega_0}{a}} e^{\frac{x\cdot\tau}{a^2}} \overline{f(x)} d\Omega_x$$

$$= \frac{1}{a} e^{-\frac{1+\tau^2}{2a^2}} e^{i\tau\cdot\frac{\omega_0}{a}} \int_{S^2} e^{-ix\cdot(\frac{\omega_0}{a}+i\frac{\tau}{a^2})} \overline{f(x)} d\Omega_x$$

$$= \frac{e^{-\frac{1+\tau^2}{2a^2}} e^{i\tau\cdot\frac{\omega_0}{a}}}{a} (\mathcal{F}f)\left(\frac{\omega_0}{a} + i\frac{\tau}{a^2}\right) = \frac{e^{-\frac{1+\tau^2}{2a^2}} e^{i\tau\cdot\frac{\omega_0}{a}}}{a} (\mathcal{F}g)\left(\frac{\omega_0}{a}\right) \quad (4)$$

where $g(x) = \overline{\exp(x\cdot\tau/a^2) f(x)}$.

By (1), (2) and (4), if we can construct the inverse mapping of the Fourier transformation, we can find the inverse mappings of the windowed Fourier transformation, and of the Gabor wavelet transformation. We will treat the inverse mapping in Sect. 5.

3 Spherical Harmonics Expansion

To consider the images of (1), (2) and (4), we recall some notations.

Let $P_{k,2}(t)$ be the Legendre polynomial of degree k and of dimension 3:

$$P_{k,2}(t) = \left(\frac{-1}{2}\right)^2 \frac{1}{k!} \frac{d^k}{dt^k}(1-t^2)^k = \sum_{l=0}^{[k/2]} (-1)^l \frac{\Gamma(k-l+1/2)}{l!(k-2l)!\sqrt{\pi}} (2t)^{k-2l}.$$

We define the extended Legendre polynomial by

$$P_{k,2}(z,w) = (\sqrt{z^2})^k (\sqrt{w^2})^k P_{k,2}\left(\frac{z}{\sqrt{z^2}} \cdot \frac{w}{\sqrt{w^2}}\right), \quad z, w \in \mathbf{C}^3.$$

Note that $P_{k,2}(z,w) = P_{k,2}(w,z)$ and
$\Delta_z P_{k,2}(z,w) \equiv (\frac{\partial^2}{\partial z_1^2} + \frac{\partial^2}{\partial z_2^2} + \frac{\partial^2}{\partial z_3^2}) P_{k,2}(z,w) = 0$. Let

$$J_\nu(t) = \left(\frac{t}{2}\right)^\nu \sum_{l=0}^\infty \frac{1}{l!\Gamma(\nu+l+1)} \left(\frac{it}{2}\right)^{2l}, \quad \nu \neq -1, -2, \cdots,$$

be the Bessel function of order ν. We put

$$\tilde{J}_\nu(t) = \Gamma(\nu+1)\left(\frac{t}{2}\right)^{-\nu} J_\nu(t) = \sum_{l=0}^\infty \frac{\Gamma(\nu+1)}{l!\Gamma(\nu+l+1)} \left(\frac{it}{2}\right)^{2l}, \tilde{j}_k(t) = \tilde{J}_{k+1/2}(t). \quad (5)$$

By (5), we have $\tilde{j}_k(0) = 0$ and $\tilde{j}_k(-t) = \tilde{j}_k(t)$. When $\nu > 0$, $|\tilde{J}_\nu(t)| \le e^{|t|}$ for $t \in \mathbf{C}$ and $0 < \cos t < |\tilde{J}_\nu(t)|$ for $-\frac{\pi}{2} < t < \frac{\pi}{2}$. Further we know that $\lim_{\nu\to\infty} \tilde{J}_\nu(t) = 1$ for $t \in \mathbf{C}$ and $\nu > 0$. (See Lemma 5.13 in [7]).

Then by using the extended Legendre polynomials and the modified Bessel functions, the exponential function is represented as follows;

$$\exp(z \cdot w) = \sum_{k=0}^{\infty} \frac{\sqrt{\pi} N(k,2)}{2^{k+1}\Gamma(k+\frac{3}{2})} \tilde{j}_k(i\sqrt{z^2}\sqrt{w^2}) P_{k,2}(z,w), \tag{6}$$

where $N(k,2) = 2k+1$. Note that $\tilde{j}_k(-t) = \tilde{j}_k(t)$. For $\eta, \zeta \in \mathbf{C}^3$, we know

$$\begin{aligned}\int_{S^2} \exp(ix\cdot\eta)\exp(x\cdot\zeta)d\Omega_x &= \sum_{k=0}^{\infty} \frac{\pi N(k,2)}{2^{2k+2}\Gamma(k+\frac{3}{2})^2} \tilde{j}_k(\sqrt{\eta^2})\tilde{j}_k(i\sqrt{\zeta^2}) P_{k,2}(\eta,\zeta) \\ &= \tilde{j}_0\left(i\sqrt{(\zeta+i\eta)^2}\right).\end{aligned}$$

For this calculation see [5] and [7], for example. By (5), we note that

$$\tilde{j}_0(t) = \frac{\sqrt{\pi}}{2} \sum_{l=0}^{\infty} \frac{1}{l!\Gamma(3/2+l)} \left(\frac{it}{2}\right)^{2l}.$$

For $t \in \mathbf{R}$, the graph of $\tilde{j}_0(t)$ is as follows:

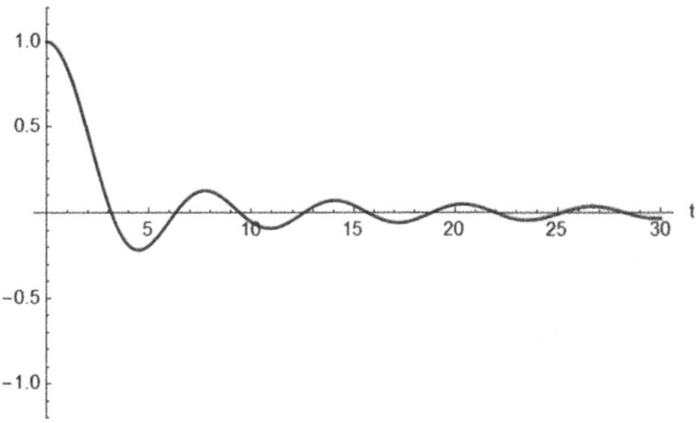

4 Expansion Formula

For $f \in L^2(S^2)$, define

$$f_k(x) = N(k, 2) \int_{S^2} f(\omega) P_{k,2}(x, \omega) d\Omega_\omega,$$

then we have

$$f(x) = \sum_{k=0}^{\infty} f_k(x),$$

in the sense of $L^2(S^2)$. Since $\Delta_x P_{k,2}(\omega, x) = 0$, $\Delta_x f_k(x) = 0$. That is, $f \in L^2(S^2)$ can be represented by the infinite sum of harmonic functions.

Therefore, we consider the images of $f_z(x) = P_{k,2}(x, z)$, $z \in \mathbf{R}^3$ under the Fourier transformation, the windowed Fourier transformation and the Gabor wavelet transformation. We recall $P_{k,2}(z, w) = P_{k,2}(w, z)$. By (6), we have

$$\begin{aligned}(\mathcal{F} f_z)(\omega) &= \int_{S^2} \sum_{l=0}^{\infty} \frac{\sqrt{\pi} N(l, 2)}{2^{l+1}\Gamma(l + \frac{3}{2})} \tilde{j}_l(i\sqrt{\omega^2}) P_{l,2}(x, \omega), P_{k,2}(x, z) d\Omega_x \\ &= C_k \tilde{j}_k(i\sqrt{\omega^2}) P_{k,2}(\omega, z),\end{aligned}$$

where we put

$$C_k = \frac{\sqrt{\pi}}{2^{k+1}\Gamma(k + \frac{3}{2})}.$$

By (2),

$$\begin{aligned}(\mathcal{W}_b \mathcal{F} f_z)(\tau, \omega) &= \frac{1}{b} e^{-\frac{1+\tau^2}{2b^2}} (\mathcal{F} f_z)(\omega + i\tau/b^2) \\ &= \frac{1}{b} e^{-\frac{1+\tau^2}{2b^2}} C_k \tilde{j}_k \left(i\sqrt{(\omega + i\tau/b^2)^2} \right) P_{k,2}(\omega + i\tau/b^2, z),\end{aligned}$$

and by (4),

$$\begin{aligned}(\mathcal{G}_{\omega_0} f_z)(\tau, a) &= \frac{1}{a}\, e^{-\frac{1+\tau^2}{2a^2}} e^{i\tau \cdot \frac{\omega_0}{a}} (\mathcal{F} f_z) \left(\frac{\omega_0}{a} + i\frac{\tau}{a^2} \right) \\ &= \frac{1}{a}\, e^{-\frac{1+\tau^2}{2a^2}} e^{i\tau \cdot \frac{\omega_0}{a}} C_k \tilde{j}_k \left(i\sqrt{(\frac{\omega_0}{a} + i\frac{\tau}{a^2})^2} \right) P_{k,2} \left(\frac{\omega_0}{a} + i\frac{\tau}{a^2}, z \right).\end{aligned}$$

Thus for $f \in L^2(S^2)$, puting $\tilde{C}_k = C_k N(k,2)$, $c_{b,\tau,\omega} = \omega + i\frac{\tau}{b^2}$, $c_{\omega_0,\tau,a} = \frac{\omega_0}{a} + i\frac{\tau}{a^2}$, we have

$$(\mathcal{F}f)(\omega) = \sum_{k=0}^{\infty} \tilde{C}_k \tilde{j}_k(i\sqrt{\omega^2}) \int_{S^2} f(z) P_{k,2}(\omega, z) d\Omega_z, \tag{7}$$

$$(\mathcal{W}_b \mathcal{F} f)(\tau,\omega) = \frac{1}{b} e^{-\frac{1+\tau^2}{2b^2}} \sum_{k=0}^{\infty} \tilde{C}_k \tilde{j}_k \left(i\sqrt{c_{b,\tau,\omega}^2}\right) \int_{S^2} f(z) P_{k,2}\left(c_{b,\tau,\omega}, z\right) d\Omega_z, \tag{8}$$

$$(\mathcal{G}_{\omega_0} f)(\tau,a) = \frac{e^{-\frac{1+\tau^2}{2a^2}} e^{i\tau\cdot\frac{\omega_0}{a}}}{a} \sum_{k=0}^{\infty} \tilde{C}_k \tilde{j}_k \left(i\sqrt{c_{\omega_0,\tau,a}^2}\right) \int_{S^2} f(z) P_{k,2}\left(c_{\omega_0,\tau,a}, z\right) d\Omega_z. \tag{9}$$

By (7), (8), and (9), for $f \in L^2(S^2)$ we have

$$(\mathcal{F}f)(\omega) = \frac{1}{b} e^{-\frac{1+\tau^2}{2b^2}} (\mathcal{W}_b \mathcal{F} f)(0,\omega), \quad (\mathcal{W}_1 \mathcal{F} f)(\tau,\omega_0) = e^{i\tau\cdot\omega_0} (\mathcal{G}_{\omega_0} f)(\tau,1).$$

5 Inverse Gabor Wavelet Transformation

In [3], we treated the inverse Gabor wavelet transformation as follows. See [3] for the details.

For $z, w \in \mathbf{C}^3$, put

$$E(z,w) = \sum_{k=0}^{\infty} \frac{i^k}{2^k k! \tilde{j}_k(\sqrt{z^2}\sqrt{w^2})} P_{k,2}(z,w).$$

For $0 < s < \infty$, let

$$K_\nu(s) = K_{-\nu}(s) = \int_0^\infty \exp(-s\cosh t)\cosh \nu t dt,$$

be the modified Bessel function. Put

$$\rho(s) = a_0 s^{1/2} K_{-1/2}(s) + a_1 s^{3/2} K_{1/2}(s) = \left(a_0 s^{1/2} + a_1 s^{3/2}\right) K_{1/2}(s),$$

where the constants a_0, a_1 are defined by

$$\begin{cases} a_0 \int_0^{\infty} s^{3/2} K_{1/2}(s) ds + a_1 \int_0^{\infty} s^{5/2} K_{1/2}(s) ds = 1, \\ a_0 \int_0^{\infty} s^{7/2} K_{1/2}(s) ds + a_1 \int_0^{\infty} s^{9/2} K_{1/2}(s) ds = 18. \end{cases}$$

In [1] we define a measure $d\mu$ on $\mathbf{R}^3$ by

$$\int_{\mathbf{R}^3} f(x) d\mu(x) = \int_0^{\infty} \int_{S^2} f(s\omega) d\Omega(\omega) s\rho(s) ds.$$

For $z \in S^2$, the mapping

$$F \mapsto \int_{\mathbf{R}^3} E(z, x) \overline{F(x)} d\mu(x)$$

gives the inverse mapping of the Fourier transformation defined by (1). By (3), $\mathcal{G}_{\omega_0} f(\tau, a)$ represented by using the Fourier transform as allows:

$$\mathcal{G}_{\omega_0} f(\tau, a) = a^{-1} e^{i\tau\cdot\omega_0/a} e^{-\frac{1+\tau^2}{2a^2}} (\mathcal{F} f) \left(\frac{a\omega_0 + i\tau}{a^2} \right).$$

Since

$$a e^{-i\tau\cdot\omega_0/a} e^{\frac{1+\tau^2}{2a^2}} (\mathcal{G}_{\omega_0} f)(\tau, a) = (\mathcal{F} f) \left(\frac{a\omega_0 + i\tau}{a^2} \right)$$

is a function in $y = (a\omega_0 + i\tau)/a^2 \in \mathbf{C}^3$.

Let $F_f(\tau, a) \in \{F(\tau, a) = \mathcal{G}_{\omega_0} f(\tau, a);\ f \in L^2(S^2)\}$, and fix $a > 0$. Put $y = (a\omega_0 + i\tau)/a^2$, then the mapping

$$F(\tau, a) \mapsto \int_{\mathbf{R}^3} E(x, y) \overline{a e^{-i\tau\cdot\omega_0/a} e^{\frac{1+\tau^2}{2a^2}} F(\tau, a)} d\mu(y),$$

gives an inverse mapping of the Gabor transformation defined by (3).

References

1. K. Fujita, Hilbert spaces related to harmonic functions. Tôhoku Math. J. **48**, 149–163 (1996). https://doi.org/10.2748/tmj/1178225416
2. K. Fujita, Gabor transformation on the circle, in *Proceedings of the 2014 International Conference on Wavelet Analysis and Pattern Recognition*, (2014), pp. 122–126. https://doi.org/10.1109/ICWAPR.2014.6961302

3. K. Fujita, On an inverse transformation of the Gabor transformation on the sphere, in *Proceedings of the Seventh International Conference on Information, Information* (2015), pp. 45–48. ISBN 4-901329-08-1
4. K. Fujita, Gabor transform of analytic functional on the sphere, in *Proceedings of the 9th ISAAC Congress Current Trends in Analysis and its Applications, Kraków*, 2013 (Springer International Publishing, Berlin, 2015), pp. 451–458. https://doi.org/10.1007/978-3-319-12577-0_50
5. K. Fujita, Gabor transformation on the Sphere and its inverse transformation, in *Proceedings of the 10th ISAAC Congress New Trends in Analysis and Interdisciplinary Applications, Trends in Mathematics* (Springer International Publishing, Berlin, 2017), pp. 573–580. https://doi.org/10.1007/978-3-319-48812-7_72
6. K. Fujita, Gabor wavelet transformation on the Sphere and its related topic, in *Analysis, Probability, Applications, and Computation, Trends in Mathematics* (Springer Nature, Switzerland AG, 2019), pp. 533–541. https://doi.org/10.1007/978-3-030-04459-6_51
7. M. Morimoto, Analytic Functionals on the sphere, in *Translation of Mathematical Monographs*, vol. 178 (American Mathematical Society, Providence, 1998)

Frames and Multirate Perfect Reconstruction Filter Banks in Multiple Dimensions

Kensuke Fujinoki

Abstract We give a brief introduction to multidimensional theory for multirate signal processing and frames of Hilbert spaces. We highlight some important relations between filter banks in a multirate system and frames for a sequence space. Constructions of multidimensional wavelet frames with the lifting scheme are also discussed.

Keywords Frames · Wavelets · Filter banks · Lifting scheme

Mathematics Subject Classification (2010) Primary 42C15; Secondary 42C40

1 Introduction

The theories of frames and signal processing are closely related. A frame was first introduced as a mathematical tool for nonharmonic Fourier analysis by Duffin and Schaeffer in 1952 [12] and was later revived by Daubechies et al. in 1986 [10], where they connected the concept of frames with time-frequency analysis and signal processing. Frames having specific structures, such as a Gabor frame [21] and a wavelet frame, have become important tools in both theories and applications in various areas of applied mathematics and signal processing. For more details on links between frames and signal processing including time-frequency analysis, see [3, 4, 9, 16, 17].

In this brief note, we survey some important relations between frames and signal processing from the point of view of multirate filter banks in arbitrary dimensions. We begin with a short introduction of frames in a Hilbert space. Quick tutorials on frame theory are also presented in [5, 15]. Then, we present some significant relations between multirate filter banks and frames including wavelets. Finally,

K. Fujinoki (✉)
Department of Mathematical Sciences, Tokai University, Hiratsuka, Kanagawa, Japan
e-mail: fujinoki@tokai-u.jp

P. Cerejeiras et al. (eds.), *Current Trends in Analysis, its Applications and Computation*, Research Perspectives,
https://doi.org/10.1007/978-3-030-87502-2_68

in addition, relations with the lifting scheme [27, 28] in multiple dimensions are discussed. Our general references are [1, 6, 29] for frames and multirate filter banks in multiple dimensions.

2 Frames

Definition 1 (Frame) Let $\mathbb{J}$ be a countable index set. A sequence of elements $\{f_j\}_{j\in\mathbb{J}}$ in a separable Hilbert space $\mathcal{H}$ is called a frame for $\mathcal{H}$ if there exist constants $0 < A \leq B < \infty$ such that

$$A\|f\|^2 \leq \sum_{j\in\mathbb{J}} |\langle f, f_j\rangle|^2 \leq B\|f\|^2, \tag{2.1}$$

for all $f \in \mathcal{H}$.

Here, $\langle\cdot,\cdot\rangle$ and $\|\cdot\|$ denote the inner product and norm on $\mathcal{H}$, respectively. Constants A and B are referred to as *frame bounds*. When $A = B$, $\{f_j\}_{j\in\mathbb{J}}$ is called a *tight frame*, and when $A = B = 1$, $\{f_j\}_{j\in\mathbb{J}}$ is called a *Parseval frame*. The inequality (2.1) is referred to as the frame condition or the frame inequality.

In the case of a Parseval frame, the frame inequality becomes

$$\sum_{j\in\mathbb{J}} |\langle f, f_j\rangle|^2 = \|f\|^2,$$

and any $f \in \mathcal{H}$ can be represented as a linear combination of $\{f_j\}_{j\in\mathbb{J}}$:

$$f = \sum_{j\in\mathbb{J}} \langle f, f_j\rangle f_j, \tag{2.2}$$

which is called the frame decomposition or frame expansion. Unlike the case of a basis, the representation (2.2) is not unique; i.e., the frame coefficients $\{\langle f, f_j\rangle\}_{j\in\mathbb{J}}$ for the superposition in (2.2) are not uniquely determined. The representation and frame coefficients are unique for every $f \in \mathcal{H}$ if and only if $\{f_j\}_{j\in\mathbb{J}}$ is a Riesz basis, and every Riesz basis is a frame. A frame that is not a Riesz basis is said to be *overcomplete* or *redundant*.

If $\{f_j\}_{j\in\mathbb{J}}$ is a frame but not a Riesz basis, then there exist other frames $\{\tilde{f}_j\}_{j\in\mathbb{J}}$ such that

$$f = \sum_{j\in\mathbb{J}} \langle f, \tilde{f}_j\rangle f_j = \sum_{j\in\mathbb{J}} \langle f, f_j\rangle \tilde{f}_j,$$

for all $f \in \mathcal{H}$, where $\{\tilde{f}_j\}_{j\in\mathbb{J}}$ is called a *dual frame* of $\{f_j\}_{j\in\mathbb{J}}$, and $\{f_j, \tilde{f}_j\}_{j\in\mathbb{J}}$ is a dual pair of frames for $\mathcal{H}$.

3 Multirate Filter Banks and Frames

Let $M \in \mathbb{Z}^{d\times d}$ be a d-dimensional matrix. We assume that det $M \neq 0$, and that all of the eigenvalues λ_i of M satisfy $|\lambda_i| > 1$. A set $LAT(M) := \{Mn : n \in \mathbb{Z}^d\}$ is said to be a d-dimensional lattice generated by M. The matrix M is called a sampling matrix because a discrete signal $\{x_n\}_{n\in\mathbb{Z}^d}$ is obtained from a function $f(x)$ defined on $\mathbb{R}^d$ using the sampling matrix M as $x_n = f(Mn)$.

A filter bank is a collection of filters that consists of analysis filters for the decomposition of an input signal and synthesis filters for the reconstruction of an original signal. A filter bank is usually implemented by combinations of downsampling and upsampling for efficient computations.

In this note, we deal with N-channel critically sampled filter banks, where $N = |\det M|$, and a signal is downsampled by means of M. Suppose that two sets of sequences $\{h_{m,n}\}_{n\in\mathbb{Z}^d, 0\le m\le N-1}$ and $\{\tilde{h}_{m,n}\}_{n\in\mathbb{Z}^d, 0\le m\le N-1}$ are collections of analysis filters and synthesis filters, respectively. Each filter has a finite impulse response (FIR) and belongs to $\ell^1(\mathbb{Z}^d) := \left\{x = \{x_n\}_{n\in\mathbb{Z}^d} : x_n \in \mathbb{C}, \sum_{n\in\mathbb{Z}^d} |x_n| < \infty\right\}$. For a signal $x \in \ell^2(\mathbb{Z}^d)$, outputs of the analysis filters are represented by filtering, which is a discrete convolution with the input signal:

$$y_{m,n} = \sum_{k\in\mathbb{Z}^d} x_k\, \overline{h_{m,n-k}}, \quad 0 \le m \le N-1,$$

where $\overline{h}$ denotes the complex conjugate of h. The output signals $\{y_{m,n}\}_{n\in\mathbb{Z}^d, 0\le m\le N-1}$ are then downsampled by M. This operation is written as $\check{y}_{m,n} = (\downarrow M)\, y_{m,n}$, where $(\downarrow M)\, x_n := x_{Mn},\ n \in \mathbb{Z}^d$.

In order to recover the original signal from $\{\check{y}_{m,n}\}_{n\in\mathbb{Z}^d, 0\le m\le N-1}$, we need to upsample these signals by $y_{m,n} = (\uparrow M)\, \check{y}_{m,n}$, where

$$(\uparrow M)\, x_n := \begin{cases} x_{M^{-1}n}, & n \in LAT(M), \\ 0, & \text{otherwise.} \end{cases}$$

Then, we apply the synthesis filters to the upsampled signals as follows:

$$x_n = \sum_{m=0}^{N-1} \sum_{k\in\mathbb{Z}^d} y_{m,k}\, \tilde{h}_{m,n-k}.$$

Definition 3.1 (Perfect Reconstruction Filter) An N-channel filter bank is said to have the *perfect reconstruction property* if the output signal of the filter bank is exactly the same as the input signal. In this case, filters $\{h_{m,n}\}_{n\in\mathbb{Z}^d, 0\le m\le N-1}$ and $\{\tilde{h}_{m,n}\}_{n\in\mathbb{Z}^d, 0\le m\le N-1}$ in the filter bank are called *perfect reconstruction filters.*

Perfect reconstruction filters are also called *biorthogonal filters*, and, in particular, for the case in which $h_m = \tilde{h}_m$ for $0 \le m \le N-1$, these filters are called *orthogonal filters*.

We denote the Fourier transform for $f \in L^1(\mathbb{R}^d)$ by $\hat{f}(\xi) = \int_{\mathbb{R}^d} f(x)\, e^{-2\pi i \xi \cdot x}\, dx$. For $x \in \ell^1(\mathbb{Z}^d)$, we define the discrete-time Fourier transform as

$$X(\xi) = \sum_{n \in \mathbb{Z}^d} x_n\, e^{-2\pi i \xi \cdot n}, \quad \xi \in \mathbb{R}^d. \tag{3.1}$$

As in (3.1), we write a filter in lower case and its discrete-time Fourier transform in upper case. In terms of perfect reconstruction filters, the following two theorems are known.

Theorem 3.2 *Filters* $h_0, \ldots, h_{N-1} \in \ell^1(\mathbb{Z}^d)$ *and* $\tilde{h}_0, \ldots, \tilde{h}_{N-1} \in \ell^1(\mathbb{Z}^d)$ *are perfect reconstruction filters if and only if their discrete-time Fourier transforms satisfy*

$$\sum_{m=0}^{N-1} \widetilde{H}_m(\xi)\, \overline{H_m(\xi)} = |\det M|, \tag{3.2}$$

$$\sum_{m=0}^{N-1} \widetilde{H}_m(\xi)\, \overline{H_m(\xi + (M^T)^{-1}\nu)} = 0, \qquad \nu \in \Gamma(M^T) \backslash \{0\}, \tag{3.3}$$

where $\Gamma(M) := \left\{ Mx \,:\, x \in [0,1)^d \right\} \cap \mathbb{Z}^d$ *and* M^T *is the transposed matrix of* M.

Theorem 3.3 *Let* $h_0, \ldots, h_{N-1} \in \ell^1(\mathbb{Z}^d)$ *and* $\tilde{h}_0, \ldots, \tilde{h}_{N-1} \in \ell^1(\mathbb{Z}^d)$ *be perfect reconstruction FIR filters that satisfy* (3.2) *and* (3.3). *Then, two collections of filters*

$$\left\{ h_{m,n-Mk} \,:\, n \in \mathbb{Z}^d \right\}_{k \in \mathbb{Z}^d, 0 \le m \le N-1}$$

and

$$\left\{ \tilde{h}_{m,n-Mk} \,:\, n \in \mathbb{Z}^d \right\}_{k \in \mathbb{Z}^d, 0 \le m \le N-1}$$

form a dual pair of frames for $\ell^2(\mathbb{Z}^d)$.

We hereinafter assume that $\{h_n, \tilde{h}_n\}_{n \in \mathbb{Z}^d}$ denotes a pair of analysis and synthesis low-pass (LP) filters and that $\{g_{m,n}, \tilde{g}_{m,n}\}_{n \in \mathbb{Z}^d, 1 \le m \le N-1}$, where $N = |\det M|$, denotes a pair of analysis and synthesis high-pass (HP) filters, all of which are perfect reconstruction FIR filters. We present some examples related to Theorems 3.2 and 3.3.

Example Let $d = 1$ and $M = 2$. In this case, we have two analysis filters $\{h_n, g_n\}_{n \in \mathbb{Z}}$ and two synthesis filters $\{\tilde{h}_n, \tilde{g}_n\}_{n \in \mathbb{Z}}$. The perfect reconstruction property simply becomes $\widetilde{H}(\xi)\,\overline{H(\xi)} + \widetilde{G}(\xi)\,\overline{G(\xi)} = 2$ and $\widetilde{H}(\xi)\,\overline{H(\xi + 1/2)} + \widetilde{G}(\xi)\,\overline{G(\xi + 1/2)} = 0$. This two-channel critically sampled filter bank, which decomposes a signal into its LP and HP bands, is a typical example of a multirate filter bank, which is often used in practice, such as in audio coding applications.

If $h_n = \tilde{h}_n$ and $g_n = \tilde{g}_n$, then the perfect reconstruction filter bank is called an *orthogonal filter bank*. In particular, for the case in which a half-band condition

$$|H(\xi)|^2 + |H(\xi + 1/2)|^2 = 2$$

holds for the LP filter and the HP filter is defined by $G(\xi) = ae^{-2\pi i(2p+1)\xi}\,\overline{H(\xi + 1/2)}$, $|a| = 1, p \in \mathbb{Z}$, so that the alias cancellation condition (3.3) is satisfied, these filters are called *conjugate mirror filters* [20, 25].

Theorem 3.3 implies that the set of filters $\{h_{n-2k} : n \in \mathbb{Z}\}_{k\in\mathbb{Z}} \cup \{g_{n-2k} : n \in \mathbb{Z}\}_{k\in\mathbb{Z}}$ forms an orthogonal basis for $\ell^2(\mathbb{Z})$, which is a nonredundant tight frame with a frame bound $A = 2$. In addition, the conjugate mirror filter is closely related to the discrete wavelet transform, which cascades two-channel critically sampled filter banks with conjugate mirror filters to divide the frequency domain into octave bands.

Example Let $d = 2$ and a sampling matrix be $M = \begin{pmatrix} 2 & 0 \\ 0 & 2 \end{pmatrix} = 2I_d$. This natural extension of Example 1 is often used for image processing because the standard digital image is sampled on a square lattice. There exist four filters ($N = \det M = 4$), which satisfy the perfect reconstruction property:

$$\widetilde{H}(\xi)\overline{H(\xi)} + \sum_{m=1}^{3} \widetilde{G}_m(\xi)\overline{G_m(\xi)} = 4,$$

$$\widetilde{H}(\xi)\overline{H(\xi + \nu/2)} + \sum_{m=1}^{3} \widetilde{G}_m(\xi)\overline{G_m(\xi + \nu/2)} = 0, \qquad \nu \in \Gamma(M^T) \setminus \{\nu_0\},$$

where $\Gamma(M^T) = \{\nu_i \in \mathbb{Z}^2 : i = 0, 1, 2, 3\}$ with $\nu_0 = (0, 0)^T$, $\nu_1 = (0, 1)^T$, $\nu_2 = (1, 0)^T$, and $\nu_3 = (1, 1)^T$. This case, in which the two-dimensional square lattice is downsampled by $M = 2I_d$, is a special case of Durand [13] and Yin and Daubechies [33].

Assume that a set of perfect reconstruction FIR filters $\{h, g_m, \tilde{h}, \tilde{g}_m\}_{1\le m\le N-1} \subset \ell^1(\mathbb{Z}^d)$ satisfies

$$H(0) = \widetilde{H}(0) = G_m((M^T)^{-1}\nu_m) = \widetilde{G}_m((M^T)^{-1}\nu_m) = |\det M|^{1/2}$$

and

$$H((M^T)^{-1}\nu_m) = \widetilde{H}((M^T)^{-1}\nu_m) = G_m(0) = \widetilde{G}_m(0) = 0.$$

Such filters are called *biorthogonal wavelet filters* [8]. Then, two sets $\{h_{n-Mk}\}_{k\in\mathbb{Z}^d} \cup \{g_{m,n-Mk}\}_{k\in\mathbb{Z}^d, 1\le m\le N-1}$ and $\left\{\tilde{h}_{n-Mk}\right\}_{k\in\mathbb{Z}^d} \cup \{\tilde{g}_{m,n-Mk}\}_{k\in\mathbb{Z}^d, 1\le m\le N-1}$ form not only dual frames for $\ell^2(\mathbb{Z}^d)$ but also biorthogonal bases for $\ell^2(\mathbb{Z}^d)$.

Using these filters with certain conditions, functions $\psi_1, \dots, \psi_{N-1} \in L^2(\mathbb{R}^d)$ and $\widetilde{\psi}_1, \dots, \widetilde{\psi}_{N-1} \in L^2(\mathbb{R}^d)$, which are called biorthogonal wavelets, are generated. In this case, two systems $\{\psi_{j,k,m}\}_{j\in\mathbb{Z}, k\in\mathbb{Z}^d, 1\le m\le N-1}$ and $\{\widetilde{\psi}_{j,k,m}\}_{j\in\mathbb{Z}, k\in\mathbb{Z}^d, 1\le m\le N-1}$ defined by

$$\psi_{j,k,m}(x) = |\det M|^{j/2}\psi_m(M^j x - k), \quad \widetilde{\psi}_{j,k,m}(x) = |\det M|^{j/2}\widetilde{\psi}_m(M^j x - k)$$

form biorthogonal bases for $L^2(\mathbb{R}^d)$. With the biorthogonal wavelet bases, any $f \in L^2(\mathbb{R}^d)$ can be represented as the wavelet expansion:

$$f = \sum_{j\in\mathbb{Z}} \sum_{k\in\mathbb{Z}^d} \sum_{1\le m\le N-1} \langle f, \widetilde{\psi}_{j,k,m}\rangle\, \psi_{j,k,m} = \sum_{j\in\mathbb{Z}} \sum_{k\in\mathbb{Z}^d} \sum_{1\le m\le N-1} \langle f, \psi_{j,k,m}\rangle\, \widetilde{\psi}_{j,k,m}.$$

For more details on relations between wavelets and filter banks, including signal processing applications, see [19, 26, 30–32].

Note that if the systems $\{\psi_{j,k,m}\}_{j\in\mathbb{Z}, k\in\mathbb{Z}^d, 1\le m\le N-1}$ and $\{\widetilde{\psi}_{j,k,m}\}_{j\in\mathbb{Z}, k\in\mathbb{Z}^d, 1\le m\le N-1}$ are overcomplete, e.g., $N > |\det M|$, then they cannot form biorthogonal bases, but become wavelet frames if the frame equality holds. In the case of wavelet frames, frame coefficients $\{\langle f, \widetilde{\psi}_{j,k,m}\rangle\}_{j\in\mathbb{Z}, k\in\mathbb{Z}^d, 1\le m\le N-1}$ or $\{\langle f, \psi_{j,k,m}\rangle\}_{j\in\mathbb{Z}, k\in\mathbb{Z}^d, 1\le m\le N-1}$ are not uniquely determined. Thus, there exist several representations of f, as mentioned in Sect. 2.

In order to construct a wavelet frame, the following result, the *unitary extension principle*, by Ron and Shen [22, 23] is well known.

Theorem 3.4 (Unitary Extension Principle) *Let $\psi_0 \in L^2(\mathbb{R}^d)$. Assume that the following two conditions hold:*

1. *There exists a bounded measurable 1-periodic function $H_0 \in L^\infty(\mathbb{T}^d)$ such that*

$$\hat{\psi}_0(\xi) = H_0((M^T)^{-1}\xi)\, \hat{\psi}_0((M^T)^{-1}\xi).$$

2. $\lim_{\xi\to 0} \hat{\psi}_0(\xi) = 1.$

Let $H_1, \dots, H_{N-1} \in L^\infty(\mathbb{T}^d)$ and define $\psi_1, \dots, \psi_{N-1} \in L^2(\mathbb{R}^d)$ by

$$\hat{\psi}_m(\xi) = H_m((M^T)^{-1}\xi)\, \hat{\psi}_0((M^T)^{-1}\xi), \quad 1 \le m \le N-1.$$

If

$$\sum_{m=0}^{N-1} |H_m(\xi)|^2 = 1,$$

$$\sum_{m=0}^{N-1} H_m(\xi)\,\overline{H_m(\xi + (M^T)^{-1}\nu)} = 0, \qquad \nu \in \Gamma(M^T)\backslash\{0\}$$

hold for a.e. $\xi \in \mathbb{T}^d$, *then the system* $\{\psi_{j,k,m}\}_{j\in\mathbb{Z},k\in\mathbb{Z}^d,1\le m\le N-1}$ *forms a tight wavelet frame for* $L^2(\mathbb{R}^d)$ *with a frame bound* $A = 1$.

Remark 3.5 The unitary extension principle can also be extended in order to construct a dual pair of wavelet frames by introducing dual functions $\widetilde{\psi}_0, \ldots, \widetilde{\psi}_{N-1} \in L^2(\mathbb{R}^d)$ and $\widetilde{H}_0, \ldots, \widetilde{H}_{N-1} \in L^\infty(\mathbb{T}^d)$.

Several wavelet frames have been constructed using the unitary extension principle, such as spline-based wavelet frames [2, 18, 22, 24], which are used for signal processing applications. A more flexible concept than the unitary extension principle, the *oblique extension principle*, was also proposed by Chui et al. [7] and Daubechies et al. [11] independently.

4 Lifting Scheme

In this section, we show relations with the lifting scheme [27, 28] and multirate filter banks as well as frames, as described above. The original lifting scheme is introduced in one dimension and gives an elementary modification of perfect reconstruction filters without losing the perfect reconstruction property.

Definition 4.1 (Lifting Scheme) Let $d = 1$ and $M = 2$. Given filters $h, g, \tilde{h}, \tilde{g} \in \ell^1(\mathbb{Z})$, modifications of filters $\tilde{h}$ and g by a filter $l \in \ell^1(\mathbb{Z})$ defined as

$$\widetilde{H}^l(\xi) = \widetilde{H}(\xi) + \widetilde{G}(\xi)\,\overline{L(2\xi)}, \qquad G^l(\xi) = G(\xi) - H(\xi)\,L(2\xi) \tag{4.1}$$

are called *lifting*, and modifications of filters h and $\tilde{g}$ by a filter $\tilde{l} \in \ell^1(\mathbb{Z})$ defined as

$$H^{\tilde{l}}(\xi) = H(\xi) + G(\xi)\,\widetilde{L}(2\xi), \qquad \widetilde{G}^{\tilde{l}}(\xi) = \widetilde{G}(\xi) - \widetilde{H}(\xi)\,\overline{\widetilde{L}(2\xi)} \tag{4.2}$$

are called *dual lifting*.

If a set of filters $\{h_n, g_n, \tilde{h}_n, \tilde{g}_n\}_{n\in\mathbb{Z}}$ has the perfect reconstruction property, then a new set $\{h_n, g_n^l, \tilde{h}_n^l, \tilde{g}_n\}_{n\in\mathbb{Z}}$ generated by the lifting (4.1) also has the perfect reconstruction property. The same holds for the dual lifting (4.2) for $\{h_n^{\tilde{l}}, g_n, \tilde{h}_n, \tilde{g}_n^{\tilde{l}}\}_{n\in\mathbb{Z}}$. In particular, if the filter $L(\xi)$ or $\widetilde{L}(\xi)$ is a trigonometric polynomial, the resulting perfect reconstruction filters $\{h_n, g_n^l, \tilde{h}_n^l, \tilde{g}_n\}_{n\in\mathbb{Z}}$ or $\{h_n^{\tilde{l}}, g_n, \tilde{h}_n, \tilde{g}_n^{\tilde{l}}\}_{n\in\mathbb{Z}}$ are all FIR filters.

Next, we consider the lifting scheme in our multidimensional setting.

Theorem 4.2 *Let $\{h, g_m, \tilde{h}, \tilde{g}_m\}_{1\le m\le N-1} \subset \ell^1(\mathbb{Z}^d)$ be a set of perfect reconstruction FIR filters, and let $\{L_m\}_{1\le m\le N-1}$ be a collection of trigonometric polynomials.*

(i) A new set $\{h, g_m^l, \tilde{h}^l, \tilde{g}_m\}_{1\le m\le N-1}$ given by

$$\widetilde{H}^l(\xi) = \widetilde{H}(\xi) + \sum_{m=1}^{N-1} \widetilde{G}_m(\xi)\, \overline{L_m(M^T\xi)},$$

$$G_m^l(\xi) = G_m(\xi) - H(\xi)\, L_m(M^T\xi), \quad 1 \le m \le N-1$$

is a set of perfect reconstruction FIR filters.

(ii) The set

$$\left\{h_{n-Mk} : n \in \mathbb{Z}^d\right\}_{k\in\mathbb{Z}^d} \cup \left\{g_{m,n-Mk}^l : n \in \mathbb{Z}^d\right\}_{k\in\mathbb{Z}^d, 1\le m\le N-1}$$

is a frame for $\ell^2(\mathbb{Z}^d)$, and the set

$$\left\{\tilde{h}_{n-Mk}^l : n \in \mathbb{Z}^d\right\}_{k\in\mathbb{Z}^d} \cup \left\{\tilde{g}_{m,n-Mk} : n \in \mathbb{Z}^d\right\}_{k\in\mathbb{Z}^d, 1\le m\le N-1}$$

is its dual frame.

Proof The proof for the item (i) for $d = 2$ and $M = 2I_d$ has been shown by the author in [14], where a matrix representation of the lifting is considered, and we can use the same technique in this general case.

Once the prefect reconstruction property is guaranteed by the item (i), taking into account Theorem 3.3 proves the claim of the item (ii). □

Due to the properties of the lifting scheme, we have similar results, as follows.

Corollary 4.3 *Let $\{h, g_m, \tilde{h}, \tilde{g}_m\}_{1\le m\le N-1} \subset \ell^1(\mathbb{Z}^d)$ be a set of perfect reconstruction FIR filters, and let $\{\widetilde{L}_m\}_{1\le m\le N-1}$ be a collection of trigonometric polynomials. Then, a new set $\{h^{\tilde{l}}, g_m, \tilde{h}, \tilde{g}_m^{\tilde{l}}\}_{1\le m\le N-1}$ given by*

$$H^{\tilde{l}}(\xi) = H(\xi) + \sum_{m=1}^{N-1} G_m(\xi)\, \widetilde{L}_m(M^T\xi),$$

$$\widetilde{G}_m^{\tilde{l}}(\xi) = \widetilde{G}_m(\xi) - \widetilde{H}(\xi)\, \overline{\widetilde{L}_m(M^T\xi)}, \quad 1 \le m \le N-1$$

is a set of perfect reconstruction FIR filters, and the two sets $\left\{h^{\tilde{l}}_{n-Mk}\right\}_{k\in\mathbb{Z}^d} \cup \left\{g_{m,n-Mk}\right\}_{k\in\mathbb{Z}^d, 1\le m\le N-1}$ *and* $\left\{\tilde{h}_{n-Mk}\right\}_{k\in\mathbb{Z}^d} \cup \left\{\tilde{g}^{\tilde{l}}_{m,n-Mk}\right\}_{k\in\mathbb{Z}^d, 1\le m\le N-1}$ *form a dual pair of frames for* $\ell^2(\mathbb{Z}^d)$.

Acknowledgments The author would like to thank the referees for their useful comments and suggestions. This study was supported in part by JSPS KAKENHI Grant Number 17K12716.

References

1. H. Arai, *Wavelets* (in Japanese). (Kyoritsu Shuppan, Tokyo, 2010)
2. H. Arai, S. Arai, 2D tight framelets with orientation selectivity suggested by vision science. JSIAM Lett. **1**, 9–12 (2009)
3. J. Benedetto, S. Li, The theory of multiresolution analysis frames and applications to filter banks. Appl. Comput. Harmon. Anal. **5**, 389–427 (1998)
4. P.G. Casazza, G. Kutyniok, (eds.), *Finite Frames: Theory and Applications* (Birkhäuser, Boston, 2012)
5. O. Christensen, A short introduction to frames, Gabor systems, and wavelet systems. Azerbaijan J. Math. **4**, 25–39 (2014)
6. O. Christensen, An introduction to frames and riesz bases, in *Applied and Numerical Harmonic Analysis*, 2nd edn. (Birkhäuser Basel, 2016)
7. C. Chui, W. He, J. Stöckler, Compactly supported tight and sibling frames with maximum vanishing moments. Appl. Comput. Harmon. Anal. **13**, 224–262 (2002)
8. A. Cohen, I. Daubechies, J. Feauveau, Bi-orthogonal bases of compactly supported wavelets. Commun. Pure Appl. Math. **45**, 486–560 (1992)
9. I. Daubechies, *Ten Lectures on Wavelets* (SIAM, Philadelphia, 1992)
10. I. Daubechies, A. Grossmann, Y. Meyer, Painless nonorthogonal expansions. J. Math. Phys. **27**, 1271–1283 (1986)
11. I. Daubechies, B. Han, A. Ron, Z. Shen, Framelets: MRA-based constructions of wavelet frames. Appl. Comput. Harmon. Anal. **14**, 1–46 (2003)
12. R. Duffin, A. Schaeffer, A class of nonharmonic Fourier series. Trans. Am. Math. Soc. **72**, 341–366 (1952)
13. S. Durand, M-band filtering and nonredundant directional wavelets. Appl. Comput. Harmon. Anal. **22**, 124–139 (2007)
14. K. Fujinoki, Two-dimensional directional lifting schemes, in ed. by *Analysis, Probability, Applications, and Computation*, ed. by K. Lindahl, T. Lindström, L. Rodino, J. Toft, P. Wahlberg. Trends in Mathematics (Birkhäuser, Cham, 2019), pp. 525–531
15. C. Heil, WHAT IS... a frame? Not. Am. Math. Soc. **60**, 1–3 (2013)
16. J. Kovačević, A. Chebira, Life beyond bases: the advent of frames (part I). IEEE Signal Process. Mag. **24**, 86–104 (2007)
17. J. Kovačević, A. Chebira, Life beyond bases: the advent of frames (part II). IEEE Signal Process. Mag. **24**, 115–125 (2007)
18. M.J. Lai, J. Stöckler, Construction of multivariate compactly supported tight wavelet frames. Appl. Comput. Harmon. Anal. **21**, 324–348 (2006)
19. S. Mallat, *A Wavelet Tour of Signal Processing*, 3rd edn. (Academic Press, Cambridge, 2008)
20. F. Mintzer, Filters for distortion-free two-band multirate filter banks. IEEE Trans. Acoust. Speech Signal Process. **33**, 626–630 (1985)
21. K.A. Okoudjou, An invitation to Gabor analysis. Not. Am. Math. Soc. **66**, 1–12 (2019)

22. A. Ron, Z. Shen, Affine system in $L_2(\mathbb{R}^d)$: the analysis of the analysis operator. J. Funct. Anal. **148**, 408–447 (1997)
23. A. Ron, Z. Shen, Compactly supported tight affine spline frames in $L_2(\mathbb{R}^d)$. Math. Comput. **67**, 191–207 (1998)
24. Z. Shen, Z. Xu, On B-spline framelets derived from the unitary extension principle. SIAM J. Math. Anal. **45**, 127–151 (2013)
25. M.J. Smith, T.P. Barnwell, Exact reconstruction techniques for tree structured subband coders. IEEE Trans. Acoust. Speech Signal Process. **34**, 434–441 (1986)
26. G. Strang, T. Nguyen, *Wavelets and Filter Banks* (Wellesley-Cambridge Press, Wellesley, 1996)
27. W. Sweldens, The lifting scheme: a custom-design construction of biorthogonal wavelets. Appl. Comput. Harmon. Anal. **3**, 186–200 (1996)
28. W. Sweldens, The lifting scheme: a construction of second generation wavelets. SIAM J. Math. Anal. **29**, 511–546 (1997)
29. P. Vaidyanathan, *Multirate Systems and Filter Banks* (Prentice Hall, Hoboken, 1992)
30. M. Vetterli, Filter banks allowing perfect reconstruction. Signal Process. **10**, 219–244 (1986)
31. M. Vetterli, C. Herley, Wavelets and filter banks: theory and design. IEEE Trans. Acoust. Speech Signal Process. **40**, 2207–2232 (1992)
32. M. Vetterli, J. Kovačević, *Wavelets and Subband Coding* (Prentice-Hall, Hoboken, 1995)
33. R. Yin, I. Daubechies, Directional wavelet bases constructions with dyadic quincunx subsampling. J. Fourier Anal. Appl. **40**, 1–36 (2017)

Deficiencies of Holomorphic Curves for Linear Systems in Projective Manifolds

Yoshihiro Aihara

Abstract In this note we shall give theorems on deficiencies of holomorphic curves $f : X \to M$, where X is a finite sheeted analytic covering space over $\mathbf{C}$ and M is a projective manifold. We first give an inequality of second main theorem type and a defect relation for f that generalizes the results in Aihara (Tohoku Math J 58:287–315, 2012). By making use of this defect relation, we give theorems on the structure of the set of deficient divisors of f. We also discuss methods for constructing holomorphic curves with deficient divisors.

Mathematics Subject Classification (2010) Primary 32H30

Keywords Holomorphic curve · Linear system · Deficiency

1 Introduction

The aim of this note is twofold. The first is to give generalizaition of the structure theorem for the set of deficient divisors in [1]. Let M be a projective algebraic manifold and $L \to M$ an ample line bundle. We denote by $|L|$ the complete linear system of L and let $\Lambda \subseteq |L|$ be a linear system. In the previous paper [1], we studied properties of the deficiencies of a holomorphic curve $f : \mathbf{C} \to M$ as functions on linear systems and gave the structure theorem for the set

$$\mathcal{D}_f = \{D \in \Lambda\,;\ \delta_f(D) > \delta_f(B_\Lambda)\}$$

of deficient divisors. For definitions, see Sect. 2. In the proof of the structure theorem for $\mathcal{D}_f$, we used an inequality of the second main theorem type and a defect relation for f and Λ. In this note, we will generalize these to the case where holomorphic

Y. Aihara (✉)
Fukushima University, Fukushima, Japan
e-mail: aihara@educ.fukushima-u.ac.jp

P. Cerejeiras et al. (eds.), *Current Trends in Analysis, its Applications and Computation*, Research Perspectives,
https://doi.org/10.1007/978-3-030-87502-2_69

curves defined on finite sheeted analytic covering spaces over $\mathbf{C}$. The second is to give methods for constructing holomorphic curves with deficient divisors. Details will be published elsewhere.

2 Preliminaries

We recall some known facts on Nevanlinna theory for holomorphic curves. For details, see [5] and [6].

Let $\varpi : X \to \mathbf{C}$ be a finite analytic (ramified) covering space over $\mathbf{C}$ and let s_0 be its sheet number, that is, X is a one dimensional complex analytic space and $\varpi : X \to \mathbf{C}$ is a proper surjective holomorphic mapping with discrete fibers. Let z be the natural coordinate in $\mathbf{C}$, and set

$$X(r) = \varpi^{-1}(\{z \in \mathbf{C};\ |z| < r\}) \quad \text{and} \quad C(r) = \varpi^{-1}(\{z \in \mathbf{C};\ |z| = r\}).$$

For a (1,1)-current φ of order zero on X we set

$$N(r, \varphi) = \frac{1}{s_0}\int_1^r \langle \varphi, \chi_{X(t)} \rangle \frac{dt}{t},$$

where $\chi_{X(r)}$ denotes the characteristic function of $X(r)$.

Let M be a compact complex manifold and let $L \to M$ be a line bundle over M. We denote by $\Gamma(M, L)$ the space of all holomorphic sections of $L \to M$ and by $|L| = \mathbf{P}(\Gamma(M, L))$ the complete linear system of L. Denote by $\|\cdot\|$ a hermitian fiber metric in L and by ω its Chern form. Let $f : X \to M$ be a holomorphic curve. We set

$$T_f(r, L) = N(r, f^*\omega)$$

and call it the characteristic function of f with respect to L. If

$$\liminf_{r \to +\infty} \frac{T_f(r, L)}{\log r} = +\infty,$$

then f is said to be *transcendental*. We define the order ρ_f of $f : X \to M$ by

$$\rho_f = \limsup_{r \to +\infty} \frac{\log T_f(r, L)}{\log r}.$$

We notice that the definition of ρ_f is independent of a choice of positive line bundles $L \to M$. Let $D = (\sigma) \in |L|$ with $||\sigma|| \leq 1$ on M. Assume that $f(X)$ is not contained in Supp D. We define the proximity function of D by

$$m_f(r, D) = \frac{1}{s_0} \int_{C(r)} \log \left(\frac{1}{||\sigma(f(z))||} \right) \frac{d\theta}{2\pi}.$$

Then we have the following first main theorem for holomorphic curves $X \to M$.

Theorem 2.1 (First Main Theorem) *Let $L \to M$ be a line bundle over M and $f : X \to M$ a non-constant holomorphic curve. Then*

$$T_f(r, L) = N(r, f^*D) + m_f(r, D) + O(1)$$

for $D \in |L|$ with $f(X) \not\subseteq \mathrm{Supp}\, D$, where $O(1)$ stands for a bounded term as $r \to +\infty$.

Let f and D be as above. We define Nevanlinna's deficiency $\delta_f(D)$ by

$$\delta_f(D) = \liminf_{r \to +\infty} \frac{m_f(r, D)}{T_f(r, L)}.$$

It is clear that $0 \leq \delta_f(D) \leq 1$. Then we have a defect function δ_f defined on $|L|$. If $\delta_f(D) > 0$, then D is called *a deficient divisor in the sense of Nevanlinna*.

Next, we recall some basic facts in value distribution theory for coherent ideal sheaves (cf. [6, Chapter 2]). Let $f : X \to M$ be a holomorphic curve and $\mathcal{I}$ a coherent ideal sheaf of the structure sheaf $\mathcal{O}_M$ of M. Let $\mathcal{U} = \{U_j\}$ be a finite open covering of M with a partition of unity $\{\eta_j\}$ subordinate to $\mathcal{U}$. We can assume that there exist finitely many sections $\sigma_{jk} \in \Gamma(U_j, \mathcal{I})$ such that every stalk $\mathcal{I}_p$ over $p \in U_j$ is generated by germs $(\sigma_{j1})_p, \ldots, (\sigma_{jl_j})_p$. Set

$$d_{\mathcal{I}}(p) = \left(\sum_j \eta_j(p) \sum_{k=1}^{l_j} |\sigma_{jk}(p)|^2 \right)^{1/2}.$$

We may assume that $d_{\mathcal{I}}(p) \leq 1$ for all $p \in M$. Set

$$\phi_{\mathcal{I}}(p) = -\log d_{\mathcal{I}}(p)$$

and call it the proximity potential for $\mathcal{I}$. It is easy to verify that $\phi_{\mathcal{I}}$ is well-defined up to addition by a bounded continuous function on M. We now define the proximity function $m_f(r, \mathcal{I})$ of f for $\mathcal{I}$, or equivalently, for the complex analytic subspace (may be non-reduced)

$$Y = (\mathrm{Supp}\,(\mathcal{O}_M/\mathcal{I}), \mathcal{O}_M/\mathcal{I})$$

by

$$m_f(r, \mathcal{I}) = \frac{1}{s_0}\int_{C(r)} \phi_{\mathcal{I}}(f(z))\frac{d\theta}{2\pi},$$

provided that $f(X)$ is not contained in $\operatorname{Supp} Y$. For $z_0 \in f^{-1}(\operatorname{Supp} Y)$, we can choose an open neighborhood U of z_0 and a positive integer ν such that

$$f^*\mathcal{I} = ((z - z_0)^\nu) \quad \text{on } U.$$

Then we see

$$\log d_{\mathcal{I}}(f(z)) = \nu \log|z - z_0| + h_U(z) \quad \text{for } z \in U,$$

where h_U is a C^∞-function on U. Thus we have the counting function $N(r, f^*\mathcal{I})$ as above. Moreover, we set

$$\omega_{\mathcal{I},f} = -dd^c h_U \quad \text{on } U,$$

where $d^c = (\sqrt{-1}/4\pi)(\bar{\partial} - \partial)$. We obtain a well-defined smooth (1, 1)-form $\omega_{\mathcal{I},f}$ on X. Define the characteristic function $T_f(r, \mathcal{I})$ of f for $\mathcal{I}$ by

$$T_f(r, \mathcal{I}) = \frac{1}{s_0}\int_1^r \frac{dt}{t}\int_{X(t)} \omega_{\mathcal{I},f}.$$

We have the first main theorem in value distribution theory for coherent ideal sheaves:

Theorem 2.2 (First Main Theorem) *Let $f : X \to M$ and $\mathcal{I}$ be as above. Then*

$$T_f(r, \mathcal{I}) = N(r, f^*\mathcal{I}) + m_f(r, \mathcal{I}) + O(1).$$

Let $L \to M$ be an ample line bundle and $W \subseteq \Gamma(M, L)$ a subspace with $\dim W \geq 2$. Set $\Lambda = \mathbf{P}(W)$. The base locus $\operatorname{Bs}\Lambda$ of Λ is defined by

$$\operatorname{Bs}\Lambda = \bigcap_{D \in \Lambda} \operatorname{Supp} D.$$

We define a coherent ideal sheaf $\mathcal{I}_0$ in the following way. For each $p \in M$, the stalk $\mathcal{I}_{0,p}$ is generated by all germs $(\sigma)_p$ for $\sigma \in W$. Then $\mathcal{I}_0$ defines the base locus of Λ as a complex analytic subspace B_Λ, that is,

$$B_\Lambda = (\operatorname{Supp}(\mathcal{O}_M/\mathcal{I}_0), \mathcal{O}_M/\mathcal{I}_0).$$

Hence $\mathrm{Bs}\,\Lambda = \mathrm{Supp}\,(\mathcal{O}_M/\mathcal{I}_0)$. We define the deficiency of B_Λ for f by

$$\delta_f(B_\Lambda) = \liminf_{r\to+\infty} \frac{m_f(r,\ \mathcal{I}_0)}{T_f(r,\ L)}.$$

Set

$$\mathcal{D}_f = \{D \in \Lambda;\ \delta_f(D) > \delta_f(B_\Lambda)\}.$$

We call $\mathcal{D}_f$ *the set of deficient divisors* in Λ.

By making use of the generalized Crofton's formula due to R. Kobayashi ([6, Theorem 2.4.12]), we have the following proposition ([1, Proposition 4.1]).

Proposition 2.3 *The set* $\mathcal{D}_f$ *is a null set in the sense of the Lebesgue measure on* Λ. *In particular* $\delta_f(D) = \delta_f(B_\Lambda)$ *for almost all* $D \in \Lambda$.

This proposition plays an important role in what follows.

3 Inequality of the Second Main Theorem Type

We will give an inequality of the second main theorem type for a holomorphic curve $f : X \to M$ that generalizes Theorem 3.1 in [1]. For simplicity, we assume that f is of finite type. Let $W \subseteq \Gamma(M,\ L)$ be a linear subspace with $\dim W = l_0 + 1 \geq 2$ and set $\Lambda = \mathbf{P}(W)$. We call Λ a linear system included in $|L|$. Let $D_1, \ldots, D_q$ be divisors in Λ such that $D_j = (\sigma_j)$ for $\sigma_j \in W$. We first give a definition of *subgeneral position*. Set $Q = \{1, \ldots, q\}$ and take a basis $\{\psi_0, \ldots, \psi_{l_0}\}$ of W. We write

$$\sigma_j = \sum_{k=0}^{l_0} c_{jk}\psi_k \quad (c_{jk} \in \mathbf{C})$$

for each $j \in Q$. For a subset $R \subseteq Q$, we define a matrix A_R by $A_R = (c_{jk})_{j\in R, 0\leq k\leq l_0}$.

Definition 3.1 Let $N \geq l_0$ and $q \geq N + 1$. We say that $D_1, \ldots, D_q$ are in N-subgeneral position in Λ if

$$\mathrm{rank}\, A_R = l_0 + 1 \quad \text{for every subset } R \subseteq Q \text{ with } \sharp R = N + 1.$$

If they are in l_0-subgeneral position, we simply say that they are in general position.

Note that the above definition is different than the usual one (cf. [6, p. 114])

Let $\Phi_\Lambda : M \to \mathbf{P}(W^*)$ be a natural meromorphic mapping, where W^* is the dual of W. Then we have the linearly non-degenerate holomorphic curve

$$F_\Lambda = \Phi_\Lambda \circ f : X \to \mathbf{P}(W^*).$$

We let $W(F_\Lambda)$ denote the Wronskian of F_Λ.

Definition 3.2 If $\rho_f < +\infty$, then f is said to be of finite type.

Set

$$\kappa(X, \Lambda; N) = 2N - l_0 + 1 + (s_0 - 1)l_0(2N - l_0 + 1).$$

By making use of the methods in [1] and [4], we have an inequality of the second main theorem type as follows.

Theorem 3.3 *Let $f : X \to M$ be a transcendental holomorphic curve that is non-degenerate with respect to Λ. Let $D_1, \ldots, D_q \in \Lambda$ be divisors in N-subgeneral position with $q > \kappa(X, \Lambda; N)$. Assume that f is of finite type. Then*

$$(q - \kappa(X, \Lambda; N))\,(T_f(r, L) - m_f(r, \mathcal{I}_0)) \leq \sum_{j=1}^{q} N(r, f^*D_j) + E_f(r)$$

as $r \to +\infty$, where

$$E_f(r) = -\kappa(X, \Lambda; N)N(r, f^*\mathcal{I}_0) - \binom{N}{l_0} N(r, W(F_\Lambda)_0) + o(T_f(r, L)).$$

In order to get a defect relation from Theorem 3.3, we define a constant $\eta_f(B_\Lambda)$ by

$$\eta_f(B_\Lambda) = \liminf_{r \to +\infty} \frac{E_f(r)}{T_f(r, L)}.$$

It is clear that $\eta_f(B_\Lambda) \leq 0$. Now, by Theorem 3.3, we have a defect relation.

Theorem 3.4 *Let Λ, f and $D_1, \ldots, D_q$ be as in Theorem 3. Then*

$$\sum_{j=1}^{q} (\delta_f(D_j) - \delta_f(B_\Lambda)) \leq (1 - \delta_f(B_\Lambda))\kappa(X, \Lambda) + \eta_f(B_\Lambda).$$

4 Structure Theorems for the Set of Deficient Divisors

In this section we give theorems on the structure of the set of deficient divisors. Let $L \to M$ be an ample line bundle and $f : \mathbf{C} \to M$ a transcendental holomorphic curve of finite type. Let $\Lambda \subseteq |L|$ be a linear system. Let

$$\mathcal{D}_f = \{D \in \Lambda\,;\ \delta_f(D) > \delta_f(B_\Lambda)\}.$$

By making use of the above defect relation, we have the structure theorem for the set $\mathcal{D}_f$ (see [1, §5]).

Theorem 4.1 *The set $\mathcal{D}_f$ of deficient divisors is a union of at most countably many linear systems included in Λ. The set of values of deficiency of f is at most a countable subset $\{e_i\}$ of $[0, 1]$. For each e_i, there exist linear systems $\Lambda_1(e_i), \ldots, \Lambda_s(e_i)$ included in Λ such that $e_i = \delta_f(B_{\Lambda_j(e_i)})$ for $j = 1, \ldots, s$.*

By Theorem 5, there exists a family $\{\Lambda_j\}$ of at most countably many linear systems in Λ such that $\mathcal{D}_f = \bigcup_j \Lambda_j$. Define $\mathcal{L}_f = \{\Lambda_j\} \cup \{\Lambda\}$. We call $\mathcal{L}_f$ *the fundamental family of linear systems for f*. Then we have the following.

Proposition 4.2 *If $\delta_f(D) > \delta_f(B_\Lambda)$ for a divisor D in Λ, then there exists a linear system $\Lambda(D) \in \mathcal{L}_f$ such that*

$$\delta_f(D) = \delta_f(B_{\Lambda(D)}).$$

5 Methods for Constructing Holomorphic Curves with Deficiencies

In this section we consider the case where $M = \mathbf{P}^n(\mathbf{C})$ and $L = \mathcal{O}_{\mathbf{P}^n}(d)$. The existence of $f : X \to \mathbf{P}^n(\mathbf{C})$ with $\mathcal{D} \neq \emptyset$ is a delicate matter. In fact, S. Mori [3] showed that a family

$$\{f \in \mathrm{Hol}(\mathbf{C}, \mathbf{P}^n(\mathbf{C}); \delta_f(H) = 0 \text{ for all } H \in |\mathcal{O}_{\mathbf{P}^n}(1)|\}$$

of holomprphic curves is dense in $\mathrm{Hol}(\mathbf{C}, \mathbf{P}^n(\mathbf{C}))$ with respect to a certain kind of topology. However, for any $\Lambda \subseteq |\mathcal{O}_{\mathbf{P}^n}(d)|$, there exists an algebraically non-degenerate holomorphic curve $f : \mathbf{C} \to \mathbf{P}^n(\mathbf{C})$ with $\mathcal{D}_f \neq \emptyset$. In fact, we have the following theorem [2, Theorem 3.2].

Theorem 5.1 *Let $D \in |\mathcal{O}_{\mathbf{P}^n}(d)|$. There exists a constant $\lambda(D)$ with $0 < \lambda(D) \leq d$ depending only on D that satisfies the following property: Let α be a positive real constant such that*

$$0 < \alpha \leq \frac{\lambda(D)}{d}.$$

Then there exists an algebraically non-degenerate holomorphic curve $f : \mathbf{C} \to \mathbf{P}^n(\mathbf{C})$ *such that*

$$\delta_f(D) = \alpha.$$

We will generalize the above theorem for holomorphic curves defined on X.

Theorem 5.2 *Let* $D \in \Lambda \subseteq |\mathcal{O}_{\mathbf{P}^n}(d)|$. *Then there exists a finite sheeted analytic covering space* $\varpi : X \to \mathbf{C}$ *and an algebraically non-degenerate transcendental holomorphic curve* $f : X \to \mathbf{C}$ *with* $\rho_f = 0$ *such that* $\mathcal{D}_f \neq \emptyset$. *Furthermore, there exists a family* $\{\Lambda_j\}$ *of finitely many linear systems such that*

$$\mathcal{D}_f = \bigcup_j \Lambda_j.$$

The set of values of δ_f *is a finite set* $\{e_j\}$ *with*

$$\delta_f(B_{\Lambda_j}) \leq e_j \leq \frac{\mu(\Lambda_j)}{d}.$$

Here $\mu(\Lambda_j)$ *are constants depending only on* Λ_j *with* $0 < \mu(\Lambda_j) \leq d$.

Remark 5.3 In the case where X is an affine algebraic variety, there always exists an algebraically non-degenerate transcendental holomorphic curve that satisfies the above propeties.

The proofs of the above theorems are based on Valiron's theorem on algebroid functions of order zero (see [7]). Hence the resulting holomorphic curves are of order zero.

In the case where $d = 1$ and $X = \mathbf{C}$, we can construct holomorphic curves with $\mathcal{D}_f \neq \emptyset$ by another way (cf. [1, §6]). By using exponential curves

$$f(z) = (\exp a_0 z, \ldots, \exp a_n z) \quad (a_0, \ldots, a_n \in \mathbf{C}),$$

we can construct holomorphic curves $\mathbf{C} \to \mathbf{P}^n(\mathbf{C})$ with $\mathcal{D}_f \neq \emptyset$. We denote by $\mathcal{C}_f$ the circumference of the convex polygon spanned by the set $\{a_0, \ldots, a_n\}$. If the convex polygon reduces to the segment with the end points with a_j and a_k, then we see $\mathcal{C}_f = 2|a_j - a_k|$. Let H be a hyperplane in $\mathbf{P}^n(\mathbf{C})$ defined by

$$H : L(z) = \sum_{j=0}^{n} \alpha_j \zeta_j = 0 \quad (\alpha_0, \ldots, \alpha_n \in \mathbf{C}),$$

where $\zeta = (\zeta_0 : \ldots : \zeta_n)$ is a homogeneous coordinate system in $\mathbf{P}^n(\mathbf{C})$. We define the set J_H of index by $J_H = \{j\,;\ \alpha_j \neq 0\}$. Let $\mathcal{C}_f(H)$ be the circumference of the convex polygon spanned by the set $\{a_j\,;\ j \in J_H\}$. Then we have the following lemma.

Lemma 5.4 *Let* f *and* H *be as above. Then*

$$T_f(r, \mathcal{O}_{\mathbf{P}^n}(1)) = \frac{\mathcal{C}_f}{2\pi} r + O(1).$$

and the deficiency of f *for* H *is given by*

$$\delta_f(H) = 1 - \frac{\mathcal{C}_f(H)}{\mathcal{C}_f}.$$

Furthermore, the constant $\mathcal{C}_f(H)$ *depends only on* f *and* J_H.

By making use of this lemma, we have the following theorem.

Theorem 5.5 *Let* $\Lambda \subseteq |\mathcal{O}_{\mathbf{P}^n}(1)|$. *Then there is a transcendental holomorphic curve* $f : \mathbf{C} \to \mathbf{P}^n(\mathbf{C})$ *non-degenerate with respect to* Λ *such that the set of values of* δ_f *is a finite set* $\{e_j\}$ *with* $0 < e_j < 1$. *Furthermore, there are finitely many linear systems* $\{\Lambda_j\}$ *included in* Λ *such that*

$$\delta_f(H) = e_j \quad \textit{for all} \quad H \in \Lambda_j \setminus (\bigcup_k \Lambda_{jk}),$$

where $\{\Lambda_{jk}\}$ *are linear systems included in* Λ_j.

References

1. Y. Aihara, Deficiencies of holomorphic curves in algebraic varieties. Tohoku Math. J. **58**, 287–315 (2012)
2. Y. Aihara, S. Mori, Deficiencies of meromorphic mappings for hypersurfaces. J. Math. Soc. Japan **97**, 233–258 (2005)
3. S. Mori, Defects of holomorphic curves into $\mathbf{P}^n(\mathbf{C})$ for rational moving targets and a space of meromorphic mappings. Complex Var. **43**, 363–379 (2000)
4. J. Noguchi, A note on entire pseudo-holomorphic curves and the proof of Cartan-Nochka's theorem. Kodai Math. J. **28**, 336–346 (2005)
5. J. Noguchi, T. Ochiai, *Geometric Function Theory in Several Complex Variables*. Translations of Mathematical Monographs, vol. 80 (American Mathematical Society, Providence, 1990)
6. J. Noguchi, J. Winkelmann, *Nevanlinna Theory in Several Complex Variables and Diophantine Approximation* (Springer, Tokyo, 2014)
7. G. Valiron, Sur valeures déficientes des fonctions algebroïdes méromorphes d'ordre null. J. d'Analyse Math. **1**, 28–42 (1951)

Correction to: Heat and Energy Consumption Management of a Public Object

S. Sh. Ixanov, Zh. Kh. Zhunussova, V. V. Nikulin, and L. Kh. Zhunussova

Correction to:
Chapter 31 in: Cerejeiras, Paula et al. (eds.), *Current Trends in Analysis, its Applications and Computation*, Research Perspectives, https://doi.org/10.1007/978-3-030-87502-2_31

The book was inadvertently published with an incorrect initial for one of the authors in this chapter as K. Kh. Zhunussova where it should have been L. Kh. Zhunussova.

This has now been corrected in the book.

The updated original version for this chapter can be found at
https://doi.org/10.1007/978-3-030-87502-2_31

P. Cerejeiras et al. (eds.), *Current Trends in Analysis, its Applications and Computation*, Research Perspectives,
https://doi.org/10.1007/978-3-030-87502-2_70

The manufacturer's authorised representative in the EU is Springer Nature Customer Service Centre GmbH, Europaplatz 3, 69115 Heidelberg, Germany. If you have any concerns regarding our products, please contact ProductSafety@springernature.com

Printed and bound by CPI Group (UK) Ltd, Croydon, CR0 4YY
28/09/2026
02224057-0010